Biochemistry of microbial degradation

Edited by

COLIN RATLEDGE
Department of Applied Biology, University of Hull, Hull, U.K.

Kluwer Academic Publishers

Dordrecht / Boston / London

Library of Congress Cataloging-in-Publication Data

Biochemistry of microbial degradation / edited by C. Ratledge.
    p.   cm.
   Includes bibliographical references and index.
   ISBN 0-7923-2273-8 (acid-free)
   1. Microbial metabolism. 2. Microbial surfactant. 3. Alkanes-
 -Metabolism.   I. Ratledge, Colin.
 QR88.B557 1993
 576'.1133--dc20   93-10260

ISBN 0-7923-2273-8

Published by Kluwer Academic Publishers,
P.O. Box 17, 3300 AA Dordrecht, The Netherlands.

Kluwer Academic Publishers incorporates
the publishing programmes of Martinus Nijhoff,
Dr W. Junk, D. Reidel, and MTP Press.

Sold and distributed in the U.S.A. and Canada
by Kluwer Academic Publishers,
101 Philip Drive, Norwell, MA 02061, U.S.A.

In all other countries, sold and distributed
by Kluwer Academic Publishers Group,
P.O. Box 322, 3300 AH Dordrecht, The Netherlands.

printed on acid-free paper

Printed in the Netherlands

Contents

Preface

Life on the planet depends on microbial activity. The recycling of carbon, nitrogen, sulphur, oxygen, phosphate and all the other elements that constitute living matter are continuously in flux: microorganisms participate in key steps in these processes and without them life would cease within a few short years. The comparatively recent advent of man-made chemicals has now challenged the environment: where degradation does not occur, accumulation must perforce take place. Surprisingly though, even the most recalcitrant of molecules are gradually broken down and very few materials are truly impervious to microbial attack.

Microorganisms, by their rapid growth rates, have the most rapid turn-over of their DNA of all living cells. Consequently they can evolve altered genes and therefore produce novel enzymes for handling "foreign" compounds – the xenobiotics – in a manner not seen with such effect in other organisms. Evolution, with the production of micro-organisms able to degrade molecules hitherto intractable to breakdown, is therefore a continuing event. Now, through the agency of genetic manipulation, it is possible to accelerate this process of natural evolution in a very directed manner. The time-scale before a new microorganism emerges that can utilize a recalcitrant molecule has now been considerably shortened by the application of well-understood genetic principles into microbiology. However, before these principles can be successfully used, it is essential that we understand the mechanism by which molecules are degraded, otherwise we shall not know where best to direct these efforts.

The biochemical understanding of degradation is therefore of paramount importance if we are to use microorganisms successfully. This then implies that microorganisms can be successfully used in a controlled manner to achieve degradation of unwanted materials and, of course, this is now increasingly proving to be the case. Indeed, new words such as "bio-remediation" and "land-farming" are now appearing to describe what are intrinsically old processes.

Not all compounds in the environment are, of course, undesirable but are nevertheless still broken down as part of the natural recycling processes. The

xvii

C. Ratledge (ed.), Biochemistry of Microbial Degradation, xvii–xix, 1994.

understanding of how these processes occur also has considerable economic importance. The advantages in being able to accelerate or decelerate these activities is fundamental to our control of the environment and also in the direct use of microorganisms in many biotechnological processes.

The biotechnological application of microorganisms is one of the major developments of the twentieth century. The controlled large-scale cultivation of microorganisms begins with the growth of selected microorganism on a chosen substrate. Routes of degradation of the major carbohydrates have, of course, been well-established for many years but numerous biotechnological processes use complex substrates whose routes of breakdown are not as well established. From such processes not only are many products produced, but also the enzymes that carried out the initial breakdown of the substrate. These enzymes are often extracellular and so can be recovered and then used elsewhere. Lipases, carbohydrases, proteases all fit this category.

An entire spectrum of microbial degradations can therefore be seen: from the breakdown of natural materials through the destruction of recalcitrant xenobiotics to the use of enzymes for accelerating key steps in other biotechnological-linked processes. Biodegradations involving inorganic nitrogen compounds, attack on metals to accelerate corrosion are also of increasing economic importance. Also increasing in awareness is the ability of anaerobic microorganisms to carry out degradation of molecules not previously considered as susceptible to anaerobic attack. Anaerobic degradations may prove to be much more wide-spread than we currently suspect.

All these aspects of biodegradation are covered in this volume. It has, however, been the overall aim of this book to move away from the more descriptive aspects of the suject and to focus on the fundamental underlying science: the biochemistry of the processes themselves. In these chapters the reader will not find an amassed mountain of information on bioremediation – what to do with oil spills, how to deal with soil and water contaminations, etc. – but rather the nature of the processes that occur within the microorganisms themselves will be covered in some detail. How microorganisms attack water-insoluble substrates when they live in aqueous surroundings; how do microorganisms derive both carbon and energy from molecules on which they can grow. How do new enzymes evolve: how are recalcitrant molecules broken down and how are new compounds not seen before in the environment then handled by the most ubiquitous of all organisms: the microbe.

It is this information which is central to the understanding of how microorganisms function. It is only from a basis of such knowledge can then advancement be made to improve the capabilities of microorganisms in a manner that brings advantage to us all.

It will be clear to any interested reader that it would be impossible to cover the microbial degradations of all compounds, natural as well as man-made, in even a substantial volume such as this. As editor I have therefore had to select the major topics of current and hopefully future interest to give a breadth of coverage rather than to concentrate on one single area of degradation. I am very conscious that there are omissions from this book but to have included everything would have doubled the size of the book without necessarily increasing our fundamental understanding of the biochemistry of microorganisms. Readers finding fault with the coverage can therefore rightly castigate the editor: for those, hopefully, greater number of readers finding within these covers the answers to their questions, they can commend the authors for their perspicacity and wisdom in catering for the readers' needs.

It is a pleasure for me to thank all the contributors for producing not only first-class manuscripts that have been focused exactly where they should be, but also for the promptness by which they have produced such worthy chapters. It has been a pleasure to have worked with them.

Colin Ratledge
University of Hull in the County of Yorkshire
Easter 1993

1. Biodegradation of components of petroleum

PHILIP MORGAN[1] and ROBERT J. WATKINSON[2]
[1]*Shell Internationale Petroleum Maatschappij B.V., P.O. Box 162, 2501 AN The Hague, The Netherlands;* [2]*Shell Research Ltd., Sittingbourne Research Centre, Sittingbourne, Kent, ME9 8AG, U.K.*

I. Introduction

Petroleum can be defined as a naturally-derived organic material that is composed primarily of hydrocarbon compounds and held in geological traps. The liquid fraction of this is crude oil. Crude oils are essentially produced by high pressure and temperature action on biological material over geological time-scales. The variability of all of these factors means that there is great variability in the chemical composition and properties of different crude oils. Furthermore, refined products from different crudes will have different chemical composition even though the products meet the relevant physical and performance standards. The aim of this paper is to review the biochemical and physiological mechanisms involved in the biodegradation of components of crude oil and petroleum products that are not gaseous under physiological conditions.

A. Composition of petroleum

Although the main components present in petroleum are hydrocarbons, a significant proportion of compounds of other chemical structure may be present in the mixture. These include nitrogen-, sulphur- and oxygen-containing (NSO) compounds derived from hydrocarbon skeletons. Some metals may also be present, often held in porphyrin-derived chemical structures. The components of crude oil can be split into a number of fundamental chemical groups on the basis of their structures. These groups are the aliphatic hydrocarbons, the cycloaliphatic hydrocarbons, the aromatic hydrocarbons and the NSO compounds. The relative proportions of these within oils from different sources varies greatly and there also exist numerous compounds within oil that contain structures from more than one of these groups (Banks and King 1984).

The aliphatic hydrocarbon fraction is primarily represented by alkanes. Alkanes in petroleum deposits range from methane to compounds with carbon chain lengths of 40 or more whose relative proportions differ widely in different

1

C. Ratledge (ed.), Biochemistry of Microbial Degradation, 1–31.

oils. Both n-alkanes and branched alkanes are present and the branched compounds are often recognizably derived from biological chemicals (Fig. 1). Unsaturated aliphatics are rare in crude oils but may be more common in some refined products. The cycloaliphatic compounds may be common and they are often alkyl-substituted. Of the monocyclic compounds, cyclopentane and cyclohexane compounds are the most prevalent. Multi-ring compounds are also common and their biological origins are often evident (Fig. 1). The complex variation in chemical structures of these and the numerous stereoisomers that can be produced makes the chemistry of this group very complex (Mackenzie 1984). The biochemistry of their degradation is similarly complex, as is evidenced by the review by Trudgill (Chapter 2) on the biochemistry of biodegradation of the terpene compounds. Aromatics are very important components and may range from benzene to multi-ring polycyclic aromatic hydrocarbons (PAHs). These compounds will not be considered further in this review since they form the subject of the chapter (11) by Smith in this book.

The non-hydrocarbon components are a similarly diverse group and the NSO compounds in petroleum may originate in a number of ways. Nitrogen and oxygen compounds apparently arise directly from biological precursors. In some crude oils they may be very common and include both aliphatic and aromatic carboxylic acids and complex structures derived from sterols. Since sulphur is depleted in biological material relative to its concentration in some crude oils, it is thought that geochemical reactions with inorganic sulphur compounds are largely responsible for the organic sulphur compounds present. Chemical structures present include thiols, thioesters and thiophenes. There also exists a fraction termed "resins" or "asphaltenes" which is composed of high molecular weight compounds consisting primarily of cross-linked NSO units. Figure 2 illustrates some common NSO compounds found in petroleum.

B. Biodegradation of crude oil and oil products

From the above discussion, it is clear that the biochemistry of petroleum biodegradation is very complex. However, the capacity of biological systems to degrade oil components is very widespread. This is illustrated in the environment by the fact that oil from natural seeps does not accumulate, by the observed elimination of anthropogenic oil inputs and by the widespread existence of oil and tar sands, which are composed of the less degradable fractions of oil remaining after extended periods of biodegradation of natural reserves close to the surface. Microbiologically, oil degradation is also widespread. Hydrocarbons *per se* can be metabolized by many genera of bacteria, filamentous fungi, yeasts and some unicellular algae (Watkinson and Morgan 1990).

The biodegradation of petroleum poses a variety of challenges to microorganisms. Firstly, crude oil and petroleum products are complex mixtures of compounds. Secondly, the lower molecular weight hydrocarbons

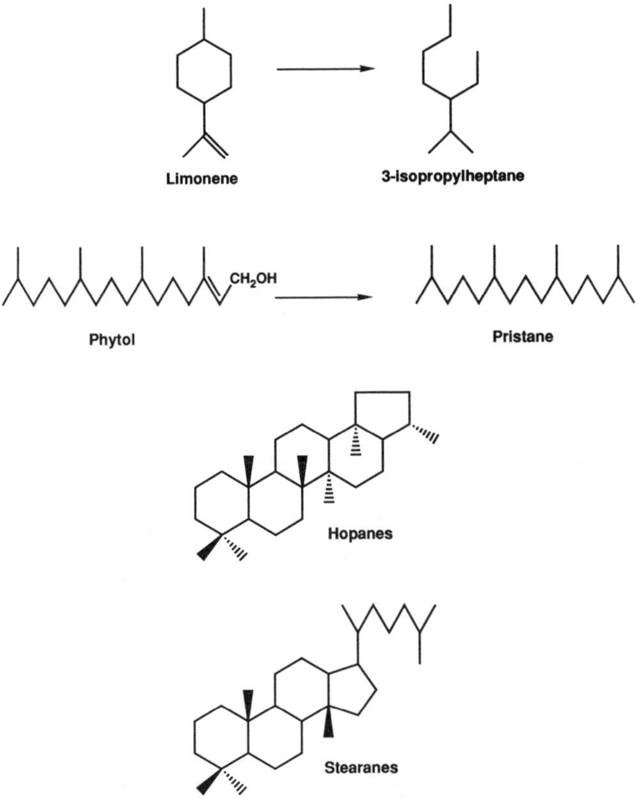

Fig. 1. Some branched alkane and complex cycloaliphatic components of crude oil illustrating their proposed origins in biological compounds. Both 3-isopropylheptane and the common branched alkane pristane have clear origins in limonene and phytol, respectively. The hopane and stearane structures illustrated each represent one form within major structural groups of cycloaliphatics which have a clear biological origin. Information from Mackenzie (1984).

may exhibit toxicity owing to their solvent effects on cell membranes. Thirdly, and most significantly, there is the issue of the physical properties of oil components, particularly hydrocarbons. The majority of hydrocarbons in a crude oil are liquid or solid at physiological temperatures and they exhibit extremely limited solubility in water. As can be seen from Table 1 the solubility decreases rapidly with increasing molecular weight. From a physiological viewpoint, the solubility of hydrocarbons can be considered insignificant except for a limited range of compounds of low molecular weight. As a consequence, microorganisms have had to develop a number of adaptations to be able to utilize hydrocarbons as substrates. The widespread biodegradation of petroleum pays tribute to the effectiveness of these mechanisms and the subsequent metabolic pathways.

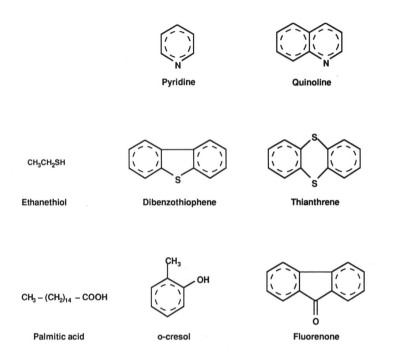

Fig. 2. Some NSO compounds found in crude oils.

Table 1. Physical properties of selected aliphatic and cycloaliphatic hydrocarbons. Data from Eastcott et al. (1988).

Compound	C Atoms	Mol. wt.	m.p. (°C)	Solubility (mg l^{-1})
n-hexane	6	86.2	−94.3	12.3
n-decane	10	128.3	−31.0	0.05
n-hexadecane	16	226.4	19.0	5.2×10^{-5}
n-eicosane	20	282.6	36.7	3.1×10^{-7}
n-hexacosane	26	366.7	56.4	1.3×10^{-10}
2-methylpentane	6	86.2	−154.0	13.8
2,2,4-trimethylpentane	8	114.2	−107.2	2.4
4-methyloctane	9	128.3	−	0.12
cyclohexane	6	84.2	6.3	57.5
methylcyclohexane	7	98.2	−	16.0
propylcyclopentane	8	112.2	−	2.0
pentylcyclopentane	10	140.3	−	0.12

II. Uptake of hydrocarbons by microorganisms

Microbial uptake of hydrocarbons can potentially occur by three mechanisms: incorporation of microdroplets much smaller than the cell itself; transport of macrodroplets; uptake of soluble-phase compounds. The uptake of hydrocarbons in aqueous solution is only feasible for compounds of low molecular weight since higher molecular weight compounds exhibit both negligible solubility (Table 1) and slow dissolution. Miller and Bartha (1989) elegantly demonstrated that the rate of metabolism of n-hexatriacontane (C_{36}) was limited by the rate of substrate uptake. They showed that degradation rates were very much slower when hydrocarbon supply was dependent upon solubilization than when the substrate was artificially microencapsulated. Cameotra and Singh (1990) studied the transport of short-chain n-alkanes by *Pseudomonas* sp. PG-I. They found that C_5 to C_8 compounds were utilized either from the vapour or soluble phases whereas n-nonane and n-decane could only be transported following solubilization of the liquid hydrocarbons. An extracellular product was secreted to enhance the solubilization process and when production of this was inhibited then soluble-phase uptake did not occur. Uptake of vapour-phase substrates could continue unhindered, however.

The uptake of hydrocarbons as microdroplets is the most common mechanism employed by unicellular microorganisms. Very often the degradative microorganisms produce biosurfactant compounds in order to emulsify the free-phase hydrocarbon and thereby enhance substrate availability. The production and properties of biosurfactants are reviewed elsewhere in this book (Hommel, Chapter 3). They are clearly very important. For example, Koch et al. (1991) obtained a Tn5-mutant of a hydrocarbon-degrading strain of *Pseudomonas aeruginosa* that could not grow on n-hexadecane. It was found that the mutation had resulted in a loss of biosurfactant production and that this resulted in complete cessation of hydrocarbon uptake. Goswami and Singh (1991) found that a "pseudosolubilisation" process (microemulsion formation) was involved in the uptake of n-hexadecane. They studied two *Pseudomonas* strains capable of growing on this hydrocarbon. One strain grew much faster than the other and only this organism produced a lipoprotein emulsifier and a glycoprotein "pseudosolubilising" factor. These worked synergistically and fully accounted for the greater growth rate of the organism that produced them.

Many hydrocarbon-degrading microorganisms have hydrophobic cell surfaces and may therefore associate with hydrocarbon droplets or even move into the organic phase during culture (Finnerty and Singer 1985). It has been widely demonstrated that extensive changes in cell membrane lipid composition can occur during growth on alkanes (Singer and Finnerty 1984a; Ratledge 1978) and this may in some cases represent a means of increasing cellular association with the hydrocarbon phase.

The mechanisms involved in transport of soluble and free-phase hydrocarbons across the cell envelope are poorly understood although free hydrocarbon droplets are frequently observed within the cell (Scott and Finnerty 1976a, b). There is conflicting evidence concerning the involvement of active transport systems. For a long period, transport was widely believed to be a passive process (Ratledge 1978). An analogous passive transport situation was studied by Parsons et al. (1987). The transport of n-alkyl esters of p-aminobenzoic acid was found to be limited by the rate of membrane diffusion when the n-alkyl chain was short and this limited biodegradation rate. When the alkyl chain was longer, membrane permeation was more rapid and the rate of degradation was limited by the metabolic enzymes themselves rather than the rate of substrate transport. However, there is evidence for an energy requirement for uptake of hydrocarbons at least in some yeasts (Bassel and Mortimer 1985). Less substantial evidence has been obtained using bacteria. Kappeli and Finnerty (1979) described an interesting active transport system used by *Acinetobacter* HO1-N growing on n-hexadecane: hydrocarbon droplets were encapsulated in membrane microvesicles which were then taken into the cell by an energy-requiring process (Singer and Finnerty 1984a). Structural adaptations that may be involved in hydrocarbon uptake have also been observed in other organisms. Scott and Finnerty (1976a) reported the presence of pores in the cell walls of some hydrocarbon-degrading yeasts and these were proposed to permit diffusion of free-phase hydrocarbons to the surface of the cell membrane. Intracellular structures have also been observed. For example, Watkinson (1980) described an extensive array of intracellular vesicles and tubules in a *Nocardia* sp. and it was hypothesized that these played a role in substrate uptake. However, such structures may have other roles in hydrocarbon metabolism rather than being adaptations for uptake. For example, microbodies rich in oxidative enzymes are found in hydrocarbon-grown yeasts (Fukui and Tanaka 1979; Hommel and Ratledge 1990) and in some filamentous fungi (Carson and Cooney 1988, 1989).

The uptake of hydrocarbons by filamentous fungi is very poorly understood. Kirk and Gordon (1988) observed that some strains of alkane-degrading marine fungi produced biosurfactants to emulsify their substrates and that the resultant microdroplets were then surrounded and penetrated by hyphae. However, the mode of hydrocarbon transport into the cells was not determined. Cooney et al. (1980) and Smucker and Cooney (1981) reported the accumulation of hydrocarbon droplets in hyphae of *Cladosporium resinae* during growth on n-alkanes but the mechanisms involved in uptake were not clear. Lindley and Heydeman (1986, 1988) investigated *C. resinae* in more detail and found that uptake of n-alkanes was a two-stage process. Firstly, the hydrocarbon was passively adsorbed onto the cell surface and then the adsorbed hydrocarbon was transported into the cell by a process which obeyed Michaelis-Menten kinetics. During growth on mixtures of n-alkanes, uptake proceeded sequentially in order of increasing molecular weight. This was found to be due to lower affinities for substrates of longer chain length: the K_m approximately doubled for each additional C_2H_4 unit on the chain. Uptake was limited by the rate of initial adsorption under poorly agitated conditions.

III. Degradation of aliphatic hydrocarbons

A. Aerobic degradation

1. n-Alkanes

n-Alkanes have generally been found to be readily degraded in both laboratory culture and the natural environment (Wakeham et al. 1986; Kennicutt 1988; Oudot et al. 1989). The fundamental details of the metabolic pathways involved are well documented (Britton 1984; Singer and Finnerty 1984a). In most cases the initial metabolic attack on n-alkanes is by a hydroxylase (monooxygenase) enzyme to produce the corresponding alkan-1-ol:

$$R\text{-}CH_3 + O_2 + NAD(P)H + H^+ \rightarrow R\text{-}CH_2OH + NAD(P)^+ + H_2O$$

Such hydroxylation reactions may be linked to a number of electron carrier systems – those involving rubredoxin (e.g. *Pseudomonas putida*) and cytochrome P-450 (e.g. *Candida* spp.) have been the most thoroughly investigated. Attack by dioxygenase enzymes has also been reported but is less common. In such cases the n-alkanes are converted to give the corresponding hydroperoxides which are reduced to yield an alkan-1-ol:

$$R\text{-}CH_3 + O_2 \rightarrow R\text{-}CH_2OOH + NAD(P)H + H^+ \rightarrow R\text{-}CH_2OH + \\ NAD(P)^+ + H_2O$$

Rarer still, but nevertheless well-documented, is sub-terminal oxidation to yield secondary alcohols. In a survey of the initial metabolic processes involved in degradation of C_8 to C_{18} n-alkanes by a variety of bacteria and fungi, Rehm and Reiff (1982) reported that certain *Aspergillus*, *Fusarium* and *Bacillus* strains could bring about sub-terminal oxidation. In these organisms, oxidation of the n-alkanes most commonly produced alcohols with the hydroxyl group in the 4-, 5- or 6-positions. Lesser quantities of 2- and 3-substituted compounds were produced.

Subsequent metabolism may follow a number of pathways as illustrated in Fig. 3. Following terminal oxidation, the alcohol is normally oxidized to the corresponding aldehyde and fatty acid by means of pyridine nucleotide-linked dehydrogenases in bacteria although alcohol oxidases have been reported in yeasts and moulds, e.g. in some *Candida* spp. (Blasig et al. 1988; Kemp et al. 1988; Hommel and Ratledge 1990). Less commonly (Rehm and Reiff 1982) ω-oxidation may occur resulting in the production of either or both α, ω-dioic acids and ω-hydroxy fatty acids. Mono- and di-terminal oxidation systems may occur in the same organism (Rehm et al. 1983; Blasig et al. 1988). The fatty acids produced by all of the pathways are then further metabolized by β-oxidation. Secondary alcohols produced by sub-terminal oxidation are further oxidized to the corresponding ester and hydrolytically cleaved to produce an acid and an alcohol, the alcohol then being oxidized and the ensuing fatty acid also subjected to β-oxidation. In many cases the fatty acids may be incorporated directly in to cell membranes (e.g. Blasig et al. 1989). There is also some weak evidence for n-alkane metabolism occurring *via* alkenes produced by the action of a NAD(P)-linked dehydrogenase. The alkenes are hydrolysed across the double bond and then further metabolized as described above. The existence of such a pathway is dubious and, at best, it is slow (Ratledge 1978; Singer and Finnerty 1984a).

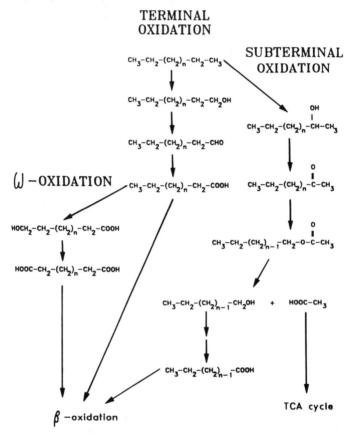

Fig. 3. Basic metabolic pathways involved in the metabolism of n-alkanes. The three metabolic routes illustrated have been unequivocally demonstrated to occur in microorganisms with terminal oxidation being the most common.

The enzymes involved in n-alkane biodegradation have been extensively characterized. Particular emphasis has been placed on the alcohol and aldehyde dehydrogenases involved in hydrocarbon degradation although it is often difficult to determine which of the groups of these enzymes present in a microbial cell are actually involved in hydrocarbon degradation. The available data have been excellently reviewed by Singer and Finnerty (1984a) and will not be repeated here. However, it is important to note that the enzymes in bacteria may be soluble or membrane-bound, may require NAD^+ or $NADP^+$

as co-factors, may be inducible or constitutive and can exhibit widely differing substrate ranges. In yeasts and filamentous fungi the dehydrogenases are all NAD-dependent but otherwise exhibit the same variability.

Highly reduced substrates such as n-alkanes have high heats of combustion (typically 13 to 15 kcal g^{-1}). However, contrary to expectation, evidence from growth yield studies indicates that yields on long chain length compounds may not be significantly greater than yields on shorter chains with lesser heats of combustion (Linton and Stephenson 1978; Anthony 1980). For example, growth on n-hexane yielded 1.21–1.59 g dry weight biomass per g substrate carbon yet on n-hexadecane the yield was only 1.35 to 1.59 g dry weight biomass per g substrate. It is likely that during growth on these compounds there is excess ATP and reducing power generated during the initial oxidation of the hydrocarbons relative to the carbon supply available and this limits the biomass yield.

2. Branched alkanes

In general terms, branched chain alkanes are more slowly degraded than their straight-chain counterparts and in hydrocarbon mixtures it is commonly observed that the degradation of branched-chain compounds is repressed by the presence of n-alkanes. As more intensive studies of the degradability of branched alkanes have been made, it has become apparent that many of these compounds are more degradable than had previously been thought. However, it is possible to make some statements about the relationship between structure and biodegradability for branched alkanes. It is generally true that highly branched compounds are more recalcitrant than simpler compounds and particularly recalcitrant are β-branched (anteiso-) and quaternary compounds owing to steric hindrance of oxidative enzymes (Schaeffer et al. 1979; Britton 1984). However, there are exceptions even to these simple statements, for example degradation of the quaternary compound 2,2-dimethylheptane has been reported (Singer and Finnerty 1984a).

Degradation of the isoprenoid alkane pristane (2,6,10,14-tetramethylpentadecane) has been particularly studied. This compound has long been used as a marker compound in studies of oil biodegradation since it was perceived to be highly persistent. Such an assumption has turned out to be ill-founded. Griffin and Cooney (1979) found that 14 of 35 mixed cultures, 5 of 21 bacterial isolates and 11 of 14 fungal isolates obtained from fresh water environments could degrade pristane. The metabolic pathways involved have been widely studied, for example in *Brevibacterium* sp. (Pirnik et al. 1974), *Corynebacterium* sp. (McKenna and Kallio 1971) and *Rhodococcus* sp. (Nakajima and Sato 1983). The metabolic pathways responsible for pristane degradation may involve β- or ω-oxidation (Fig. 4; Pirnik 1977).

Other complex branched chain alkanes have also been shown to be metabolized. Nakajima et al. (1985) isolated a *Rhodococcus* sp. capable of degrading phytane (2,6,10,14-tetramethylhexadecane), norpristane (2,6,10-trimethylpentadecane) and farnesane (2,6,10-trimethyldodecane) as sole

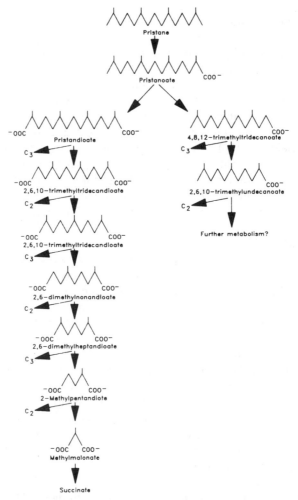

Fig. 4. Metabolic pathways involved in the degradation of the isoprenoid alkane pristane (2,6,10,14-tetramethylpentadecane).

sources of carbon and energy. Metabolism proceeded by oxidation of isopropyl units to yield terminal alcohols and thence fatty acids. Cox et al. (1976) described a *Mycobacterium* sp. that could degrade phytane, norpristane, 2,6,10-trimethyltetradecane and 2,6,10,14-tetramethylheptadecane. In these cases the initial metabolic attack did not only occur at the isopropyl units on the molecules but the products of initial oxidation were always terminal alcohols. Rontani and Giusti (1986) have reported the utilization of 2,2,4,4,6,8,8-heptamethylnonane as a sole carbon and energy source by a mixed marine microbial population. Degradation was relatively rapid and the only metabolic intermediates detected were straight-chain fatty acids. The metabolic pathway involved oxidation of the hydrocarbon at the β-position to yield the corresponding ketone. This was oxidized to an ester which in turn was hydrolytically cleaved to yield a fatty acid and alcohol that were subjected to β-oxidation. The only branched hydrocarbons that were resistant to attack by this population were those which were blocked at the β-position at both ends of the molecule.

Since they are key intermediates in the degradation of branched alkanes, information on the metabolism of branched-chain fatty acids also sheds light on the relationship between carbon skeleton structure and biodegradation. In compounds substituted with quaternary units the location of these plays a key role. In general, compounds with quaternary structures on carbon atoms 1 or 2 are more readily degradable than those with quaternary structures on carbon 3 or equivalent positions further along the carbon chain (Mohanrao and McKinney 1962; Dias and Alexander 1971; Hammond and Alexander 1972). Although α-oxidation of such fatty acids can occur in a few cases, this relative recalcitrance can be attributed to the mechanistics of β-oxidation. β-Oxidation involves dehydrogenation between carbon atoms 2 and 3 and the presence of a quaternary structure on the third atom sterically inhibits enzyme action. This will also hold for fatty acids produced during the metabolism of branched alkanes.

3. Alkenes

Metabolic attack on unsaturated aliphatic hydrocarbons may be initiated either *via* attack on the double bond or by the same mechanisms employed in n-alkane metabolism. Four main patterns of initial attack can be recognized (Fig. 5): oxygenase attack upon a terminal methyl group to produce the corresponding alken-1-ol; subterminal oxygenase attack to produce an alkenol with the hydroxyl group at a non-terminal carbon; oxidation across the double bond to give an epoxide; oxidation across the double bond to produce a diol. Many microorganisms display more than one of these pathways. There is evidence to suggest that mixed populations are particularly important in alkene degradation since it is often difficult to isolate microorganisms capable of utilizing alkenes as sole carbon and energy sources from alkene-degrading mixed cultures (Griffin and Cooney 1979).

Recent studies have concentrated primarily on the metabolism of short-chain (C_6 and below) alkenes. This work has been thoroughly reviewed elsewhere (Hartmans et al. 1989; see also Watkinson and Morgan 1990) and will not be discussed further in this paper. However, it should be noted that some organisms capable of growth on short-chain alkenes cannot grow on the corresponding alkanes since they can only initiate metabolic attack at the double bond.

Alkynes can also be biodegraded aerobically but have been little studied. It is thought that hydratases are responsible for the initial metabolic attack (Hartmans et al. 1989).

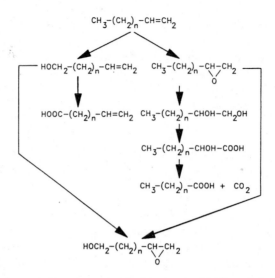

Fig. 5. Basic metabolic pathways involved in the microbial degradation of alkenes.

B. Anaerobic degradation

1. Alkanes

The potential for anaerobic biodegradation of hydrocarbons has been a contentious issue for many years. Convincing evidence for the anaerobic degradation of some simple aromatic petroleum hydrocarbons has been obtained (see Chapter 16 by Fuchs) but the potential degradability of aliphatic compounds under anoxic conditions has remained a matter for debate. Papers showing apparent anaerobic degradation of oil without analyzing individual components (e.g. Shelton and Hunter 1975) may have merely measured the degradation of non-hydrocarbon components of the oil. It is studies using defined substrates or those that analysed the removal of individual components within oil mixtures that are most revealing. However, one of the problems with much of the work on anaerobic hydrocarbon degradation has been that the experimental methods have not sufficiently demonstrated strict anaerobiosis or excluded the possibility that abiotic processes led to the observed results.

Ward and Brock (1978) reported the mineralization of radiolabelled n-hexadecane in lake sediments incubated under anaerobic conditions. The authors of this paper wisely cautioned against drawing firm conclusions from their results since only 5% of the added radiolabel was recovered as CO_2 in their experiments. This may have been due to the breakdown of contaminants in the substrate rather than the substrate itself or may have been the result of the utilization of trace concentrations of oxygen within the system. Delaune et al. (1980) and Hambrick et al. (1980) studied the effects of redox potential on the mineralization of radiolabelled octadecane in estuarine sediment slurries.

They reported that degradation occurred at redox potentials as low as −250 mV but the papers do not give sufficient detail on the experimental methods performed to be sure that anaerobiosis had been achieved and preclude abiotic interference. Thus the findings must remain open to question. Indeed, in a similar study, Mille et al. (1988) reported no degradation of n-nonadecane, n-eiscosane (C_{20}) or pristane in sediment samples incubated at redox potentials of −180 to −200 mV. Schink (1985) could detect no degradation of a range of straight-chain and branched alkanes under methanogenic conditions despite very long incubation periods. In contrast, Parekh et al. (1977) reported the apparent anaerobic denitrifying growth of a *Pseudomonas* isolate on n-alkanes. However, it was subsequently shown that the observed growth could be entirely attributed to organic growth supplements added to the medium and that there was no attack upon the hydrocarbons (Griffin and Traxler 1981). Taken as a whole these observations do not conclusively support the possibility of anaerobic degradation of aliphatic hydrocarbons and the issue remains open. Only one potential pathway had been described that could represent an initial metabolic step for anaerobic alkane degradation: Azoulay et al. (1963) reported that *Pseudomonas aeruginosa* Sol20 was capable of converting n-heptane to n-heptene under anaerobic conditions. Although this reaction could potentially represent a method of initiating metabolic attack on alkanes under anaerobic conditions, there was no further metabolism of n-heptene unless molecular oxygen was present. No further work on this system has been reported.

Recently, excellent evidence for anaerobic degradation of n-alkanes has at last been reported (Aeckersberg et al. 1991). Using impeccable anaerobic techniques, sulphate-reducing communities were enriched on C_{12} to C_{20} n-alkanes and populations developed which were capable of degrading these substrates. A pure culture of a bacterium, designated Hxd3, was obtained from one of these and a stoichiometric relationship existed between n-hexadecane degradation and sulphate-reduction:

$$C_{16}H_{34} + 12.25\,SO_4^{2-} + 8.5\,H^+ \rightarrow 16\,HCO_3^- + 12.25\,H_2S + H_2O$$

No intermediates were detected and the metabolic pathway involved remains unknown. However, dehydrogenation to produce an alkene was not thought to be likely as a first metabolic step since this is thermodynamically unfavourable. Thus we apparently have the first reliable report of anaerobic alkane degradation and further developments, particularly the determination of the metabolic pathway, are awaited with interest. However, it should be noted that degradation under anaerobic conditions is always likely to be an extremely slow process.

2. Alkenes

There has been one report of the anaerobic degradation of alkenes (Schink 1985). He performed long-term methanogenic enrichments on a range of alkenes and monitored degradation by measuring methane production. Only two compounds were degraded: 1-hexadecene and squalene. Hexadecene was degraded well with up to 91% of the expected methane being produced according to the stoichiometry:

$$C_{16}H_{32} + 12\ H_2O \rightarrow 12\ CH_4 + 4HCO_3^- + 4H^+$$

The only intermediate detected was a trace quantity of acetate. It was hypothesized that hexadecane was converted to hexadecan-1-ol and thence to palmitate since both of these compounds could also be converted to methane. With squalene as a substrate growth was slower and weaker with only 50% of the theoretical methane being produced:

$$4\ C_{30}H_{50} + 105\ H_2O \rightarrow 85\ CH_4 + 35\ HCO_3^- + 35\ H^+$$

Once again the only intermediate detected was acetate. Three potential reasons for the failure of the cultures to degrade the other alkenes tested were given: inability of cells to transport substrates; substrate toxicity; microbial conversion to products that could not be converted to methane.

IV. Degradation of cycloaliphatic compounds

Cycloaliphatic compounds may make up a large proportion of certain boiling point fractions of some oils and are important constituents of certain petroleum products. In contrast to the situation with aliphatic hydrocarbons, there have been relatively few reports of pure cultures capable of utilizing cycloaliphatics as pure carbon and energy sources. Rather, the biodegradation of cycloaliphatics generally involves either or both mixed populations and co-metabolism (Beam and Perry 1974a; Trudgill 1978, 1984a; Perry 1984). The majority of work that has been performed on cycloaliphatic compounds in oils has investigated simple compounds, particularly cyclohexane and its alkyl-substituted derivatives. The metabolic pathway present in pure cultures of cyclohexane-utilising bacteria was elucidated by Stirling et al. (1977) for *Nocardia* sp. and by Anderson et al. (1980) for *Pseudomonas* sp. and is illustrated in Fig. 6. Magor et al. (1986) studied the properties of the enzymes involved in the initial stages of cyclohexane degradation by *Xanthobacter* spp. The properties of these enzymes are summarized in Table 2.

Co-metabolism of cycloaliphatics has been widely detected using whole cells and cell extracts of alkane and other hydrocarbon degraders. The initiation of co-metabolic attack involves the conversion of the cycloaliphatics to alcohols or ketones by low specificity monooxygenase enzymes. These metabolites are then metabolized, often by other specialist organisms in the microbial community, *via* well-defined pathways which have been excellently reviewed elsewhere (Perry 1984; Trudgill 1984a). It has been suggested (Perry 1984) that because of this vulnerability of cycloaliphatic hydrocarbons to co-metabolism by microorganisms degrading other hydrocarbons there has been little requirement for specific cycloaliphatic-degrading enzymes or organisms to

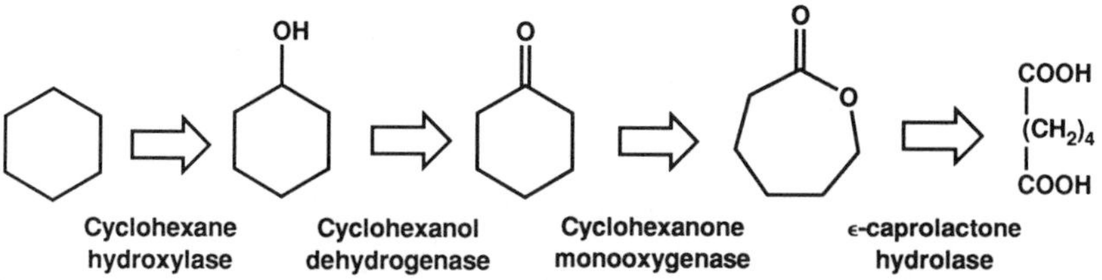

Cyclohexane   Cyclohexanol   Cyclohexanone   ε-caprolactone
hydroxylase   dehydrogenase   monooxygenase   hydrolase

Fig. 6. Metabolic pathway involved in the metabolism of cyclohexane as a sole carbon and energy source by *Nocardia* and *Pseudomonas* (Stirling et al. 1977; Anderson et al. 1980). The hydroxylase enzyme responsible for the initial metabolic step was only detected in *Pseudomonas*.

Table 2. Properties of enzymes involved in the initial stages of cyclohexane metabolism in two *Xanthobacter* spp. isolated from soil (Magor et al. 1986). The *X. autotrophicus* could not degrade cyclohexane but could metabolize cyclohexanol and cyclohexanone.

Enzyme	Organism	
	X. autotrophicus	*Xanthobacter* sp.
Cyclohexane hydroxylase	Not present	pH optimum 6.2–6.7 NADPH cofactor
Cyclohexanol dehydrogenase	pH optimum 10.1–10.5 NAD main cofactor NADP weak cofactor mol. wt. 20 kDa	pH optimum 9.0–9.4 NAD cofactor mol. wt. 43 kDa
Cyclohexanone monooxygenase	pH optimum 8.9–9.0 NADPH cofactor	pH optimum 7.2–7.7 NADH cofactor

evolve. The work of Lloyd-Jones and Trudgill (1989) has clearly illustrated the importance of consortia and the complexities of unravelling the roles of individual organisms within them. A stable consortium was obtained which could degrade cyclohexane and alkyl-substituted cyclohexanes of a chain length up to C_{12}. The consortium consisted of *Rhodococcus* sp. (80 to 90%), *Flavobacterium* sp. (5 to 10%) and *Pseudomonas cepacia* (5 to 10%) but despite extensive effort it was not possible to ascertain which organisms played which role in the metabolic pathway which was elucidated (Fig. 7).

The degradation of alkyl-substituted aliphatics has been studied in some detail. Methyl- and ethyl-substituted cyclohexanes may be used as carbon and energy sources by pure cultures (Tonge and Higgins 1974; Arai and Yamada 1969). With longer alkyl chains metabolic attack may occur either on the ring or on the alkyl chain depending upon the organisms and the substrate (Trudgill 1984a; Perry 1984). Beam and Perry (1974b) studied the degradation of n-alkylcyclohexanes with a chain-length of up to C_{12} by *Mycobacterium* and *Nocardia* strains in pure culture. These organisms could not degrade methyl- or ethyl-cyclohexane and metabolized the side chains of the longer chain length

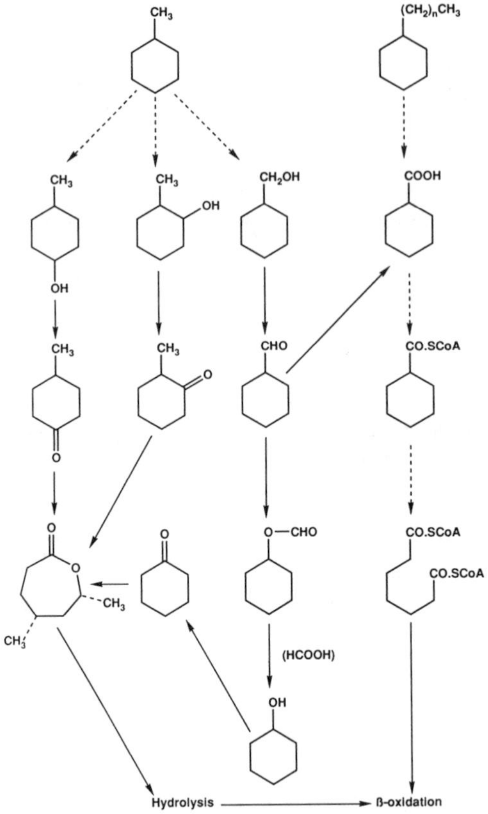

Fig. 7. Degradation of cyclohexane and n-alkylcyclohexanes in a defined mixed population (Lloyd-Jones and Trudgill 1989). Solid arrows indicate metabolic steps for which enzymes were detected and broken arrows indicate proposed reactions.

compounds by β-oxidation to yield cycloalkanoic fatty acids that accumulated in the cell membrane. β-Oxidation of the alkyl chain can also result in other end-products. With odd chain-length compounds β-oxidation results in the production of cyclohexane carbonylCoA the ring of which can be opened and subjected to further β-oxidation. In contrast, compounds of even chain length are frequently degraded only as far as cyclohexylacetic acid which accumulates in the medium. Feinberg et al. (1980) described a co-culture of *Mycobacterium rhodochrous* and an *Arthrobacter* sp. which could fully mineralize dodecylcyclohexane. Only *M. rhodochrous* was capable of metabolizing the parent compound and did this by β-oxidation to yield cyclohexylacetic acid which it secreted into the medium and this was then mineralized by the second member of the consortium.

Degradation of multi-ring cycloaliphatic compounds and cycloalkenes has been little studied but evidence has been obtained that indicates that they can be co-metabolized.

V. Degradation of nitrogen-, sulphur- and oxygen-containing components (NSO) of petroleum

The NSO fraction of petroleum may represent a large proportion of the available carbon sources in some oils. The metabolism of nitrogen-containing heterocyclic compounds is the subject of a review by Lingens (Chapter 14) in this book and therefore the metabolism of the nitrogen-containing components of petroleum will not be considered further in this paper. The metabolism of sulphur- and oxygen-containing petroleum components will be reviewed here but we will not discuss those compounds which are widely found in non-petroleum sources (e.g. fatty acids, thiols, phenolics).

A. Aerobic degradation

1. Sulphur-containing compounds

The degradation of thiophene, benzothiophene and dibenzothiophene has been widely reported in studies employing both pure substrates and oil mixtures. Aerobic degradation of these compounds by populations in environmental samples has been reported (e.g. Fedorak and Westlake 1983, 1984; Arvin et al. 1988; Mueller et al. 1991). Arvin et al. (1989) noted that the rates of degradation of such compounds in ground water samples were comparable to those measured for aromatic hydrocarbons in the same systems. Individual strains capable of degrading these compounds have been isolated and characterized (e.g. *Pseudomonas* sp. by Foght and Westlake 1988; *Pseudomonas* spp. by Bayona et al. 1986 and *Pseudomonas putida* by Mormile and Atlas 1988). However, there have been few reports of organisms capable of using dibenzothiophene as a sole carbon, sulphur and energy source but these do exist (e.g. *Brevibacterium* described by van Afferden et al. 1990). Metabolism of such compounds is nevertheless widely distributed: Fredrickson et al. (1991) were able to isolate a dibenzothiophene-degrading strain from a sample of subsurface sediment taken from a depth of 410 m.

The metabolic pathway involved in dibenzothiophene breakdown has been well-characterized. Two basic routes have been identified: degradation *via* a diol or *via* dibenzothiophene-5-oxide (Fig. 8; Ensley 1984; van Afferden et al. 1990). The diol route has been claimed to be present in dibenzothiophene-mineralizing bacteria but more usually there is accumulation of 3-hydroxy-2-formylbenzothiophene as a dead-end metabolite. Conversely, the oxidation of the sulphur group permits complete mineralization to occur *via* benzoic acid and has been well documented in a dibenzothiophene-mineralizing *Brevibacterium* sp. (van Afferden et al. 1990).

Degradation of other relevant sulphur-containing compounds has been less studied. Fedorak and Westlake (1983, 1984) studied the degradation of n-alkyl substituted benzothiophenes by mixed populations and pure cultures. Increasing alkyl chain length decreased the rate of degradation. Fedorak et al. (1988) investigated the degradation of alkyl-substituted tetrahydrothiophenes

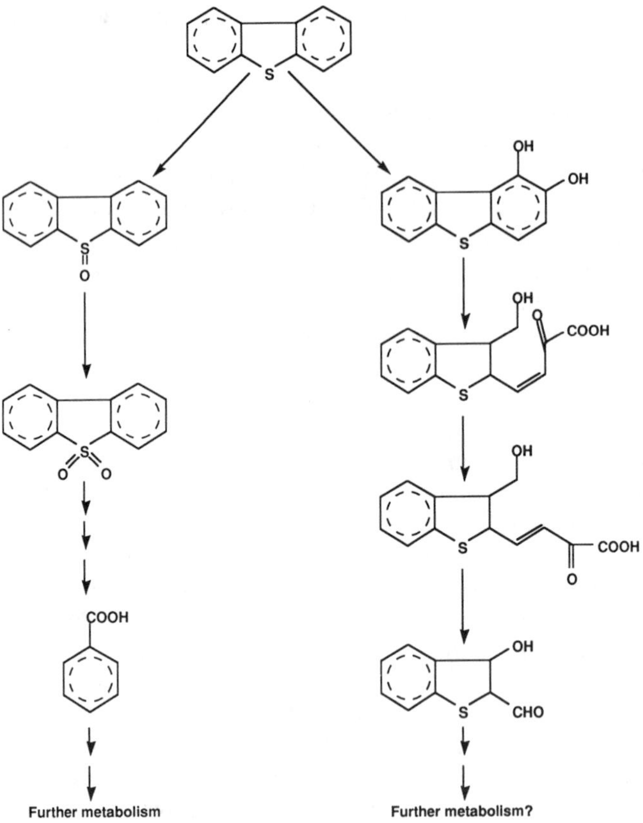

Fig. 8. Metabolic pathways for the aerobic degradation of dibenzothiophene. After Emsley (1984) and van Afferden et al. (1990).

and found that compounds substituted with alkyl chains of up to at least 30 carbon atoms could be degraded. The metabolism of undecyl- and dodecyl-tetrahydrothiophenes was studied in detail using both bacterial and fungal isolates. The initial attack on the molecules involved repeated oxidation of the alkyl chain to yield thiophene-carboxylates which were then fully mineralized. There have also been reports of mineralization of 2-acetylthiophene by micromycete fungi (Seigle-Murandi et al. 1991) and of thiophene carboxylates by other bacteria (e.g. *Vibrio* sp. by Evans and Venables 1990 and *Rhodococcus* sp. by Kanagawa and Kelly 1987).

Co-metabolism of sulphur-containing compounds in petroleum has also been reported. In some cases, the carbon and energy source may be substituted sulphur compounds (especially the thiophene carboxylates, Evans and Venables 1990) or may be other hydrocarbons. The latter is illustrated by the work of Fedorak and Grbić-Galić (1991) who reported that mixed populations and an isolate of *Pseudomonas* grown on 1-methylnaphthalene could co-metabolize benzothiophene and 3-methylbenzothiophene to their diones or sulphones. However, these products were not metabolized further.

2. *Oxygen-containing compounds*

With the exception of phenolic compounds, aliphatic fatty acids and naphthenic acids (cycloaliphatic fatty acids), the degradation of oxygen-containing organic components of petroleum has been relatively little studied. Specific compounds which have received some attention include fluorenone, furans and dibenzofuran. These compounds are widely metabolized under aerobic conditions in environmental samples at significant rates (Arvin et al. 1988; Jensen et al. 1988; Mueller et al. 1991). Pure cultures capable of mineralizing these compounds have also been reported. Furans are widely degraded owing to their production by plants and their metabolic pathways have been well characterized (Trudgill 1984b). More complex compounds have been less studied. Foght and Westlake (1988) isolated a *Pseudomonas* sp. capable of degrading a variety of oxygen heterocycles, including dibenzofuran. Strubel et al. (1989) isolated a bacterium (designated DBO230) that mineralized dibenzofuran *via* salicylate and catechol and could co-metabolize a number of other heterocyclic compounds including oxofluorene, xanthone, dibenzo-*p*-dioxin and dibenzothiophene. Fortnagel et al. (1990) obtained a mixed culture and a *Pseudomonas* sp. capable of utilizing dibenzofuran as a sole source of carbon and energy. An entirely new metabolic pathway was reported (Fig. 9) involving conversion of the dibenzofuran to 2,2′,3-trihydroxybiphenyl and thence *via* salicylic acid which branched into both gentisic acid and catechol pathways. Both salicylic acid and gentisic acid accumulated in the culture medium but were slowly metabolized once growth had ceased.

B. *Anaerobic degradation*

Anaerobic biodegradation of both sulphur- and oxygen-heterocyclic compounds has been reported. Kuhn and Suflita (1989) studied the degradation of petroleum-derived NSO compounds under methanogenic and sulphate-reducing conditions in ground water samples. They found that furan, 2-methylfuroic acid and 2-methylfuran were degraded under sulphate-reducing conditions but that only 2-furoic acid was degraded under methanogenic conditions. Brune et al. (1983) reported a *Desulfovibrio* sp. capable of transforming 2-furaldehyde to acetate.

More detailed studies have been performed with sulphur-containing compounds. Kuhn and Suflita (1989) reported that thiophene, methylthiophenes and thiophene-2-carboxylate were resistant to attack under methanogenic and sulphate-reducing conditions in aquifer samples except for some poor methanogenic attack upon the carboxylate. In contrast, Kurita et al. (1971) reported the isolation of an anaerobic thiophene-degrader from sludge collected near to an oil well but no physiological studies were performed.

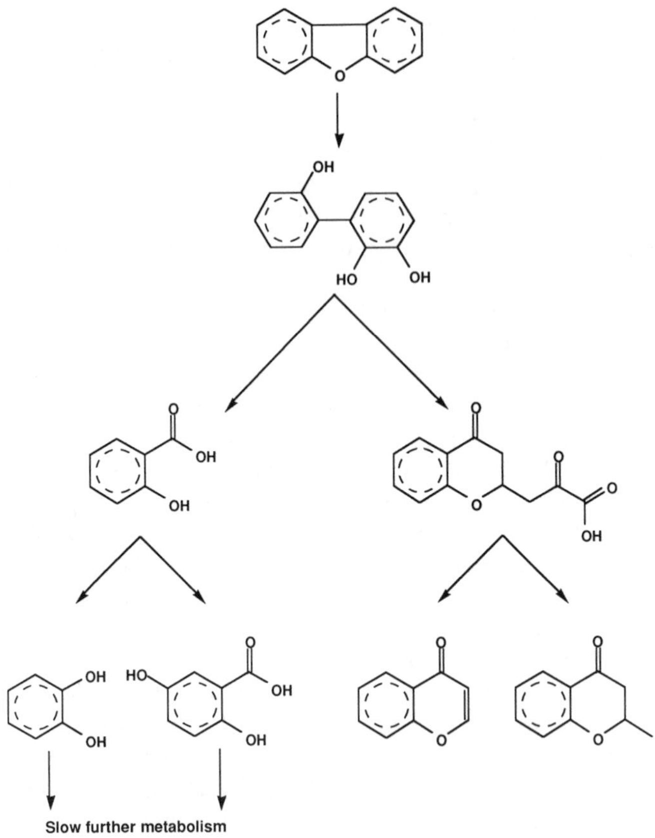

Fig. 9. Aerobic metabolic pathway for utilization of dibenzofuran as sole carbon and energy source in *Pseudomonas* sp. HH69 (Fortnagel et al. 1990).

Grbić-Galić (1990) reported the methanogenic degradation of benzothiophene and dibenzothiophene in soils, ground water and sewage sludge and described the metabolic pathways involved in benzothiophene degradation in one system (Fig. 10). The benzothiophene was initially converted to 2- and 7-oxobenzothiophenes which were subjected to ring cleavage. With 2-oxobenzothiophene the heterocyclic ring was cleaved to yield a variety of monoaromatic products including styrene, phenylacetic acid, benzoic acid, benzyl alcohol, phenol and p-hydroxybenzenesulphonic acid. With 7-oxobenzothiophene ring cleavage appeared to take place on the original aromatic ring to yield thiophene-2-ol. In both cases complete conversion to methane and CO_2 occurred.

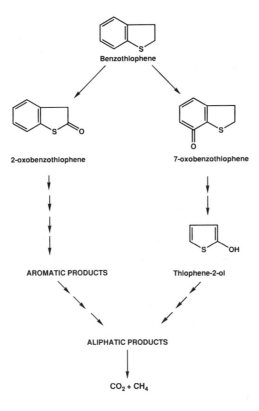

Fig. 10. Metabolic pathways involved in the anaerobic degradation of benzothiophene. After Grbić-Galić (1990).

VI. Genetics of biodegradation

A. Aliphatic and cycloaliphatic hydrocarbons

The genetics of aliphatic hydrocarbon-degrading organisms has only been studied in a relatively small number of organisms. In bacteria the only well-characterized system is the *OCT* plasmid which codes for a number of proteins involved in growth on n-alkanes ranging from C_6 to C_{10}. Initial studies performed with *Pseudomonas putida* PpG6 elucidated the fundamental pathways and genetic properties of the system (Chakrabarty et al. 1973; Nieder and Shapiro 1975; Grund et al. 1975; Benson et al. 1977). The metabolic pathway was found to involve an alkane hydroxylase (monooxygenase) and the alcohol so produced was further oxidized by an alcohol dehydrogenase and an aldehyde dehydrogenase to yield a fatty acid. The fatty acids were further oxidized by means of β-oxidation. Fennewald and colleagues (Fennewald and Shapiro 1977, 1979; Fennewald et al. 1979) then performed intensive studies on a strain of *Pseudomonas aeruginosa* containing the *OCT* plasmid and identified a number of loci on the plasmid and the chromosome that were involved in coding the pathway. On the chromosome the following genes were identified: *alc A*, soluble alcohol dehydrogenase with a substrate range of C_7 and above; *alc B*, soluble alcohol dehydrogenase with a substrate range of C_3 to C_6; *ald A* and *ald B*, the corresponding aldehyde dehydrogenases; *oic*, locus involved in β-oxidation. Among the loci present on the plasmid were a promoter sequence, regulatory genes and genes coding for sub-units of alkane hydroxylase and alcohol dehydrogenase.

Detailed further work has enabled the operation of the *OCT* system to be deduced (Fig. 11). Eggink et al. (1987a) cloned a region of the plasmid containing the promoter, the *alk BAC* operon and the regulatory region into non-degradative strains. Gene expression resulted in the production of functional alkane hydroxylase and alcohol dehydrogenase. The *alk BAC* region was further characterized by Eggink et al. (1987b). They found that six proteins could be obtained on translation of a 16.9 kilobase-pair fragment in a minicell preparation. Two of these (20 and 59 kDa) could not be assigned a definite function. The other proteins had the following functions: membrane-bound alkane hydroxylase (41 kDa, *alk B*); sub-units of soluble alkane hydroxylase (15 and 49 kDa, *alk A*); membrane-bound alcohol dehydrogenase (58 kDa, *alk C*). Kok et al. (1989) further studied the *alk B* region and found that the alkane hydroxylase was a 401 amino-acid polypeptide with 8 hydrophobic regions of suitable size for spanning the cell membranes. Characterization of the regulatory region (Owen 1986; Eggink et al. 1988) indicated that there were two gene products coded. The *alk S* gene product was found to be a 99 kDa polypeptide which directly promoted transcription of the *alk BAC* operon. The *alk T* product was a 48 kDa protein necessary for activity of alkane hydroxylase. This protein did not act as a regulator of gene expression and was concluded to be part of the alkane hydroxyase complex.

Singer and Finnerty (1984b) reviewed studies on the genetics of alkane degradation in *Acinetobacter* HO1-N. No plasmids coding for any part of the degradative pathway have been detected and, unlike the *OCT* system, there are only one set of enzymes involved in degradation of compounds of all chain lengths. Analysis of non-degradative mutants has indicated the region of the genome coding for the alkane hydroxylase is completely independent of that region coding for the aldehyde and alcohol dehydrogenases.

The genetics of hydrocarbon-metabolizing yeasts has been less studied. Certainly, yeast systems can be very complex: Bassel and Mortimer (1985) demonstrated that there were at least 16 different loci responsible for the uptake of alkanes in *Yarrowia lipolytica*. One particular area that has attracted intensive research is the alkane-inducible cytochrome P-450 involved in alkane hydroxylation in some *Candida* spp. In *C. tropicalis* sequence analysis has shown that this inducible cytochrome represents a member of a family distinct from that of other P-450 cytochromes previously reported (Sanglard et al. 1987; Sanglard and Loper 1989). The cytochrome family has been designated LII and

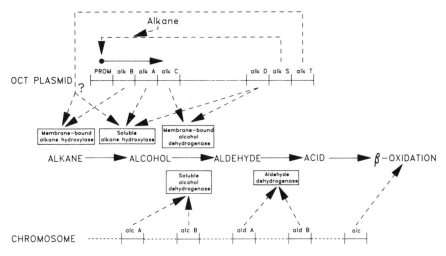

Fig. 11. Structure and function of the *OCT* plasmid and associated chromosomal genes in *Pseudomonas aeruginosa*. The *alk S* gene is induced by the presence of alkanes and codes for a regulatory protein that stimulates transcription of the *alk BAC* operon. The role of the *alk T* gene product is unclear but it appears to function in the alkane hydroxylase enzyme complexes. Chromosomal genes encode for the soluble alcohol dehydrogenase, aldehyde dehydrogenase and enzymes of β-oxidation. After Singer and Finnerty (1984b) and Eggink et al. (1987a, 1988).

the two alkane-inducible cytochromes detected in *C. tropicalis* have been designated *alk 1* and *alk 2* (Sanglard and Fiechter 1989). Sanglard et al. (1990) have cloned the genes for the *C. tropicalis* alkane-inducible cytochrome P-450 and the NADPH-cytochrome P-450 oxidoreductase into *Saccharomyces cerevisiae* and obtained expression. A similar, alkane-inducible cytochrome P-450 has been isolated from *C. maltosa* and characterized (Sunairi et al. 1988; Takagi et al. 1989; Schunck et al. 1989). This is a protein of 523 amino-acids and contains highly hydrophobic regions towards its *N*-terminus that presumably serve to anchor it in the membrane of the endoplasmic reticulum.

The genetics of cycloaliphatic hydrocarbon degradation have been little studied. Lloyd-Jones and Trudgill (1989) observed that two plasmids were present in a stable alkylcyclohexane-degrading consortium and that plasmid curing resulted in a loss of cycloaliphatic-degrading capability. It was not possible to ascertain the roles of these plasmids.

B. *Sulphur- and oxygen-containing compounds*

The genetics of biodegradation of sulphur- and oxygen-containing petroleum components has been little studied. Evans (1989) reported that a *Vibrio* capable of degrading certain substituted thiophenes and furans contained a plasmid of approximately 80 kilobase pairs but she could not prove that this coded for enzymes in the metabolic pathway. Monticello et al. (1985) isolated three dibenzothiophene-degrading *Pseudomonas* strains from contaminated soil and demonstrated that all three contained plasmids. In *P. alcaligenes* DBT2 and *P. putida* DBT4 the plasmid had a molecular weight of 5.5×10^7. The size of the plasmid in the other strain isolated (*P. stutzeri* DBT3) was not determined. Plasmid curing resulted in the loss of degradative capability by these strains. Foght and Westlake (1990) isolated a plasmid-carrying *Pseudomonas* strain (HL76) capable of co-metabolising dibenzothiophene. By hybridization studies with the plasmid of Monticello et al. (1985) they were able to demonstrate that HL76 contained no genes coding for dibenzothiophene degradation and that the co-metabolism was due to a low specificity naphthalene degradation pathway.

VII. Relevance of petroleum component biodegradation

A. Fate of oil in the environment

Obviously, the microbial degradation of petroleum oil components has a significant impact on the fate of oil and oil products entering the environment. Much of this is beyond the scope of this review and has been excellently covered elsewhere (e.g. Atlas 1981, 1989; Leahy and Colwell 1990). However, the relationship between structure and degradability has wide significance and merits brief discussion here. As has been shown in this review, many components of oil are known to be biodegradable but the rates and extents of degradation differ. Killops and Al-Juboori (1990) characterized the residual hydrocarbons remaining after extensive biodegradation of a crude oil and found that they could be classified into a number of general groups. About 10% of the residual carbon was aromatic with alkylated aromatics (chain length up to C_{19}) being particularly common. Cycloaliphatic compounds made up a large proportion of the balance and the majority of these were single ring compounds. Aliphatics made up a small proportion of the total and were represented by isoprenoid branched alkanes and some residual α, ω-dioic acids. It is probable that all of these compounds are fundamentally biodegradable but the rate of degradation will be slow. This has a number of important implications, in particular when the application of microbiological clean-up of an oil spill (bioremediation) is being considered (Morgan and Watkinson 1989; Pritchard and Costa 1991). Here the residual compounds and their slow degradation may impact upon the oil concentrations that can be reached within a suitable treatment period.

B. Biotechnological applications

The microbial degradation of hydrocarbons has a number of potential applications in biotechnology. However, the economics of using oil-derived substrates as feedstocks for biotechnological processes will always require careful investigation: extensive work on single cell protein production was rendered redundant by the oil price rises of the 1970s. It has been suggested that alkene-degrading mutants could be used for the production of high-value epoxides for chemical synthesis (Hartmans et al. 1989). This process may be particularly cost-effective when the substrate is asymmetric since stereospecific isomers may be produced by some microbial enzymes. There has also been an interest in the biotechnological production of fatty acids from alkane or cycloalkane substrates (Ratledge 1978; Finnerty 1984) but this is unlikely to be economic if the compounds can be obtained from other sources. The same is true of proposals to produce other widely available products (e.g. citrate, amino-acids) by microbial growth on hydrocarbon substrates (Miall 1980; Shennan 1984). More specialist metabolites may be of greater interest, e.g. biosurfactants (Hommel, Chapter 3), wax esters (Ratledge 1984) and intracellular storage polymers for use in biodegradable plastics (Lageveen et al. 1988).

Removal of sulphur-containing compounds from petroleum and coal has attracted attention as a potential means of desulphurising high sulphur fuels and thereby increasing their value and reducing sulphur dioxide emissions during their combustion. For example, Kargi (1987) reported that the thermophilic organism *Sulfolobus acidocaldarius* could oxidize thianthrene, thioxanthine and dibenzothiophene at 70°C and release sulphide ions. However, the large-scale application of this technology has not been adopted.

VIII. Conclusions

Despite the wealth of information available, there is still much to learn about the physiology and biochemistry of oil-component degradation. The only pathways which are understood in detail in a variety of organisms are those involved in the aerobic degradation of n-alkanes. For other compounds we have only a small part of the overall picture available to us. There remains much challenging and exciting research to be done on the subject of the biodegradation of petroleum components. A few key areas can be highlighted. Recent developments have demonstrated more extensive degradation of branched chain alkanes than was previously believed possible and there is merit in further study in this area. Very little is understood about the role of communities in hydrocarbon biodegradation yet they play a key role in biodegradation in nature, as illustrated by their central role in cycloaliphatic degradation. The control of degradation at the genetic and physiological level is still poorly understood with the exception of the *OCT* plasmid and this subject are represents a great challenge. There is also the question of those metabolic pathways which have been studied little or not at all. How do microbes degrade sulphur- and oxygen-containing compounds? How does the anaerobic degradation of alkanes proceed and how widespread is the phenomenon? To answer such questions sucessfully the microbial biochemist, physiologist and ecologist must all play a role.

References

Aeckersberg F, Bak F and Widdel F (1991) Anaerobic oxidation of saturated hydrocarbons to CO_2 by a new type of sulfate-reducing bacterium. Arch. Microbiol. 156: 5–14

Anderson MS, Hall RA and Griffin M (1980) Microbial metabolism of alicyclic hydrocarbons: cyclohexane catabolism by a pure strain of Pseudomonas sp. J. Gen. Microbiol. 120: 89–94

Anthony C (1980) Methanol as substrate: theoretical aspects. In: DEF Harrison, IJ Higgins and RJ Watkinson (eds) Hydrocarbons in Biotechnology (pp 35–57). Heyden, London

Arai Y and Yamada K (1969) Studies on the utilization of hydrocarbons by microorganisms. XII. Screening of alicyclic hydrocarbon-assimilating microorganisms and trans-4-ethylcyclohexanol formation from ethylcyclohexane. Agric. Biol. Chem. 33: 63–68

Arvin E, Jensen B, Aamand J and Jorgensen C (1988) The potential of free-living ground water bacteria to degrade aromatic hydrocarbons and heterocyclic compounds. Water Sci. Technol. 20: 109–118

Arvin E, Jensen B, Godsy EM and Grbić-Galić D (1989) Microbial degradation of oil and creosote related aromatic compounds under aerobic and anaerobic conditions. In: YC Wu (ed) International Conference on Physiochemical and Biological Detoxification of Hazardous Wastes (pp 828–847). Technomic, Lancaster, Pennsylvania

Atlas RM (1981) Microbial degradation of petroleum hydrocarbons: an environmental perspective. Microbiol. Rev. 45: 180–209

Atlas RM (1989) Biodegradation of hydrocarbons in the environment. In: GS Omenn (ed) Environmental Biotechnology. Reducing Risks from Environmental Chemicals Through Biotechnology (pp 211–222). Plenum Press, New York

Azoulay E, Choteau J and Davidovics G (1963) Isolement et characterization des enzymes responsables de l'oxidation des hydrocarbures. Biochim. Biophys. Acta 77: 554–567

Banks RE and King PJ (1984) Chemistry and physics of petroleum. In: GD Hobson (ed) Modern Petroleum Technology (pp 279–327). Wiley, Chichester

Bassel JB and Mortimer RK (1985) Identification of mutations preventing n-hexadecane uptake among 26 n-alkane non-utilizing mutants of Yarrowia (Saccharomycopsis) lipolytica. Curr. Genet. 9: 579–586

Bayona JM, Albaiges J, Solanas AM, Pares R, Garrigues P and Ewald M (1986) Selective aerobic degradation of methyl-substituted polycyclic hydrocarbons in petroleum by pure microbial cultures. Int. J. Environ. Anal. Chem. 23: 289–303

Beam HW and Perry JJ (1974a) Microbial degradation of cycloparaffinic hydrocarbons via co-metabolism and commensalism. J. Gen. Microbiol. 82: 163–166

Beam HW and Perry JJ (1974b) Microbial degradation and assimilation of n-alkyl-substituted cycloparaffins. J. Bacteriol. 118: 394–399

Benson S, Fennewald M, Shapiro J and Huettner C (1977) Fractionation of inducible alkane hydroxylase activity in Pseudomonas putida and characterization of hydroxylase-negative plasmid mutations. J. Bacteriol. 132: 614–621

Blasig R, Mauersberger S, Riege P, Schunk WH, Jockish W, Franke P and Mueller HG (1988) Degradation of long-chain n-alkanes by the yeast Candida maltosa. II. Oxidation of n-alkanes and intermediates using microsomal membrane fractions. Appl. Microbiol. Biotechnol. 28: 589–597

Blasig R, Huth J, Franke P, Borneleit P, Schunk WH and Mueller HG (1989) Degradation of long-chain n-alkanes by the yeast Candida maltosa. III. Effect of solid n-alkanes on cellular fatty acid composition. Appl. Microbiol. Biotechnol. 31: 571–576

Britton LN (1984) Microbial degradation of aliphatic hydrocarbons. In: DT Gibson (ed) Microbial Degradation of Organic Compounds (pp 89–129). Marcel Dekker, New York

Brune G, Schoberth SM and Sahm H (1983) Growth of a strictly anaerobic bacterium on furfural (2-furaldehyde). Appl. Environ. Microbiol. 46: 1187–1192

Cameotra SS and Singh HD (1990) Uptake of volatile n-alkanes by Pseudomonas PG-1. J. Biosci. 15: 313–322

Carson DB and Cooney JJ (1988) Spheroplast formation and partial purification of microbodies from hydrocarbon-grown cells of *Cladosporium resinae*. J. Ind. Microbiol. 3: 111–117

Carson DB and Cooney JJ (1989) Characterization of partially purified microbodies from hydrocarbon-grown cells of *Cladosporium resinae*. Can. J. Microbiol. 35: 565–572

Chakrabarty AM, Chou G and Gunsalus IC (1973) Genetic regulation of octane dissimilation plasmid in *Pseudomonas*. Proc. Natl. Acad. Sci. U.S.A. 70: 1137–1140

Cooney JJ, Siporin C and Smucker RA (1980) Physiological and cytological responses to hydrocarbons by the hydrocarbon-using fungus *Cladosporium resinae*. Bot. Mar. 23: 227–232

Cox RE, Maxwell JR and Myers RN (1976) Monocarboxylic acids from oxidation of acyclic isoprenoid alkanes by *Mycobacterium fortuitum*. Lipids 11: 72–76

Delaune RD, Hambrick GA and Patrick WH (1980) Degradation of hydrocarbons in oxidized and reduced sediments. Mar. Pollut. Bull. 11: 103–106

Dias FF and Alexander M (1971) Effect of chemical structure on the biodegradability of aliphatic acids and alcohols. Appl. Microbiol. 22: 1114–1118

Eastcott L, Shiu WY and Mackay D (1988) Environmentally relevant physical-chemical properties of hydrocarbons: a review of data and development of simple correlations. Oil Chem. Pollut. 4: 191–216

Eggink G, Lageveen RG, Altenburg B and Witholt B (1987a) Controlled and functional expression of the *Pseudomonas oleovorans* alkane utilizing system in *Pseudomonas putida* and *Escherichia coli*. J. Biol. Chem. 262: 17712–17718

Eggink G, van Lelyveld PH, Arnberg A, Arfman N and Witholt B (1987b) Structure of the *Pseudomonas putida alkBAC* operon. Identification of transcription and translation products. J. Biol. Chem. 262: 6400–6406

Eggink G, Engel H, Meijer WG, Otten J, Kingma J and Witholt B (1988) Alkane utilization in *Pseudomonas oleovorans*. Structure and function of the regulatory locus *alkR*. J. Biol. Chem. 263: 13400–13405

Ensley BD (1984) Microbial metabolism of condensed thiophenes. In: DT Gibson (ed) Microbial Degradation of Organic Compounds (pp 309–317). Marcel Dekker, New York

Evans JS (1989) Genetic and biochemical studies of microbial degradation of thiophenes. Ph.D. thesis, University of Wales College of Cardiff

Evans JS and Venables WA (1990) Degradation of thiophene-2-carboxylate, furan-2-carboxylate, pyrrole-2-carboxylate and other thiophene derivatives by the bacterium *Vibrio* YC1. Appl. Microbiol. Biotechnol. 32: 715–720

Fedorak PM and Grbić-Galić D (1991) Aerobic microbial cometabolism of benzothiophene and 3-methylbenzothiophene. Appl. Environ. Microbiol. 57: 932–940

Fedorak PM and Westlake DWS (1983) Microbial degradation of organic sulfur compounds in Prudhoe Bay crude oil. Can. J. Microbiol. 29: 291–296

Fedorak PM and Westlake DWS (1984) Degradation of sulfur heterocycles in Prudhoe Bay crude oil by soil enrichments. Water Air Soil Pollut. 21: 225–230

Fedorak PM, Payzant JD, Montgomery DS and Westlake DWS (1988) Microbial degradation of n-alkyl tetrahydrothiophenes found in petroleum. Appl. Environ. Microbiol. 54: 1243–1248

Feinberg EL, Ramage PIN and Trudgill PW (1980) The degradation of n-alkylcycloalkanes by a mixed bacterial culture. J. Gen. Microbiol. 121: 507–511

Fennewald M and Shapiro J (1977) Regulatory mutations of the *Pseudomonas* plasmid *alk* regulon. J. Bacteriol. 132: 622–627

Fennewald M and Shapiro J (1979) Transposition of Tn7 in *Pseudomonas aeruginosa* and isolation of *alk::Tn7* mutations. J. Bacteriol. 139: 264–269

Fennewald M, Benson S, Oppici M and Shapiro J (1979) Insertion element analysis and mapping of the *Pseudomonas* plasmid *alk* region. J. Bacteriol. 139: 940–952

Finnerty WR (1984) The application of hydrocarbon-utilizing microorganisms for lipid production. AOCS Monogr. 11: 199–215

Finnerty WR and Singer ME (1985) Membranes of hydrocarbon-utilizing microorganisms. In: BK Ghosh (ed) Organization of Prokaryotic Cell Membranes, Volume III (pp 1–44). CRC Press, Boca Raton, Florida

Foght JM and Westlake DWS (1988) Degradation of polycyclic aromatic hydrocarbons and aromatic heterocycles by a *Pseudomonas* species. Can. J. Microbiol. 34: 1135–1141

Foght JM and Westlake DWS (1990) Expression of dibenzothiophene-degradative genes in two *Pseudomonas* species. Can. J. Microbiol. 36: 718–724

Fortnagel P, Harms H, Wittich R-M, Krohn S, Meyer H, Sinnwell V, Wilkes H and Francke W (1990) Metabolism of dibenzofuran by *Pseudomonas* sp. strain HH69 and the mixed culture HH27. Appl. Environ. Microbiol. 56: 1148–1156

Fredrickson JK, Brockman FJ, Workman DJ, Li SW and Stevens TO (1991) Isolation and characterization of a subsurface bacterium capable of growth on toluene, naphthalene, and other aromatic compounds. Appl. Environ. Microbiol. 57: 796–803

Fukui S and Tanaka A (1979) Peroxisomes of alkane- and peroxisome-grown yeasts: metabolic functions and practical applications. J. Appl. Biochem. 1: 171–201

Goswami P and Singh HD (1991) Different modes of hydrocarbon uptake by two *Pseudomonas* species. Biotechnol. Bioeng. 37: 1–11

Grbić-Galić D (1990) Anaerobic microbial transformation of nonoxygenated aromatic and alicyclic compounds in soil, subsurface, and freshwater sediments. In: J-M Bollag and G Stotzky (eds) Soil Biochemistry, Volume 6 (pp 117–189). Marcel Dekker, New York

Griffin WM and Cooney JJ (1979) Degradation of model recalcitrant hydrocarbons by microorganisms from freshwater ecosystems. Dev. Ind. Microbiol. 20: 479–488

Griffin WM and Traxler RW (1981) Some aspects of hydrocarbon metabolism by Pseudomonas. Dev. Ind. Microbiol. 22: 425–435

Grund A, Shapiro J, Fennewald M, Bacha P, Leahy J, Markbreiter K, Nieder M and Toepfer M (1975) Regulation of alkane oxidation in *Pseudomonas putida*. J. Bacteriol. 123: 546–556

Hambrick GA, Delaune RD and Patrick WH (1980) Effect of estuarine sediment pH and oxidation-reduction potential on microbial hydrocarbon degradation. Appl. Environ. Microbiol. 40: 365–369

Hammond MW and Alexander M (1972) Effect of chemical structure on microbial degradation of methyl-substituted aliphatic acids. Environ. Sci. Technol. 6: 732–735

Hartmans S, de Bont JAM and Harder W (1989) Microbial metabolism of short-chain unsaturated hydrocarbons. FEMS Microbiol. Rev. 63: 235–264

Hommel R and Ratledge C (1990) Evidence for two fatty alcohol oxidases in the biosurfactant-producing yeast *Candida* (*Torulopsis*) *bombicola*. FEMS Microbiol. Lett. 70: 183–186

Jensen BK, Arvin E and Gundersen AT (1988) Biodegradation of nitrogen- and oxygen-containing aromatic compounds in groundwater from an oil-contaminated aquifer. J. Contam. Hydrol. 3: 65–75

Kanagawa T and Kelly DP (1987) Degradation of thiophenes by bacteria isolated from activated sludge. Microb. Ecol. 13: 47–57

Kappeli O and Finnerty WR (1979) Partition of alkane by an extracellular vesicle derived from hexadecane-grown *Candida tropicalis*. J. Bacteriol. 140: 707–712

Kargi F (1987) Biological oxidation of thianthrene, thioxanthine and dibenzothiophene by the thermophilic organism *Sulfolobus acidocaldarius*. Biotechnol. Lett. 9: 478–482

Kemp GD, Dickenson FM and Ratledge C (1988) Inducible long chain alcohol oxidase from alkane-grown *Candida tropicalis*. Appl. Microbiol. Biotechnol. 29: 370–374

Kennicutt MC (1988) The effect of biodegradation on crude oil bulk and molecular composition. Oil Chem. Pollut. 4: 89–112

Killops SD and Al-Juboori MAHA (1990) Characterization of the unresolved complex mixture (UCM) in the gas chromatograms of biodegraded petroleums. Org. Geochem. 15: 147–160

Kirk PW and Gordon AS (1988) Hydrocarbon degradation by filamentous marine higher fungi. Mycologia 80: 776–782

Koch AK, Kapelli O, Fiechter A and Keiser J (1991) Hydrocarbon assimilation and biosurfactant production in *Pseudomonas aeruginosa* mutants. J. Bacteriol. 173: 4212–4219

Kok M, Oldenhuis R, van der Linden MPG, Raatjes P, Kingma J, van Lelyveld PH and Witholt B (1989) The *Pseudomonas oleovorans* alkane hydroxylase gene. Sequence and expression. J. Biol. Chem. 264: 5435–5441

Kuhn EP and Suflita JM (1989) Anaerobic biodegradation of nitrogen-substituted and sulfontated benzene aquifer contaminants. Hazard. Waste Hazard. Mater. 6: 121–133

Kurita S, Endo T, Nakamura H, Yagi T and Tamiya N (1971) Decomposition of some organic sulfur compounds in petroleum by anaerobic bacteria. Appl. Microbiol. 17: 185–198

Lageveen RG, Huisman GW, Preusting H, Ketelaar P, Eggink G and Witholt B (1988) Formation of polyesters by *Pseudomonas oleovorans*: effect of substrates on formation and composition of poly-(R)-3-hydroxyalkanoates and poly-(R)-3-hydroxyalkenoates. Appl. Environ. Microbiol. 54: 2924–2932

Leahy JG and Colwell RR (1990) Microbial degradation of hydrocarbons in the environment. Microbiol. Rev. 54: 305–315

Lindley ND and Heydeman MT (1986) Mechanism of dodecane uptake by whole cells of *Cladosporium resinae*. J. Gen. Microbiol. 132: 751–756

Lindley ND and Heydeman MT (1988) The uptake of n-alkanes from alkane mixtures during growth of the hydrocarbon-utilizing fungus *Cladosporium resinae*. Appl. Microbiol. Biotechnol. 23: 384–388

Linton JD and Stephenson RJ (1978) A preliminary study on growth yields in relation to the carbon and energy content of various organic growth substrates. FEMS Microbiol. Lett. 3: 95–98

Lloyd-Jones G and Trudgill PW (1989) The degradation of alicyclic hydrocarbons by a microbial consortium. Int. Biodet. 25: 197–206

Mackenzie AS (1984) Applications of biological markers in petroleum geochemistry. In: J Brooks and D Welte (eds) Advances in Petroleum Geochemistry, Volume 1 (pp 115–214). Academic Press, London

Magor AM, Warburton J, Trower MK and Griffin M (1986) Comparative study of the ability of three *Xanthobacter* species to metabolise cycloalkanes. Appl. Environ. Microbiol. 52: 665–671

McKenna EJ and Kallio RE (1971) Microbial metabolism of the isoprenoid alkane pristane. Proc. Natl. Acad. Sci. U.S.A. 68: 1552–1554

Miall LM (1980) Organic acid production from hydrocarbons. In: DEF Harrison, IJ Higgins and RJ Watkinson (eds) Hydrocarbons in Biotechnology (pp 25–34). Heyden, London

Mille G, Mulyono M, El Jammel T and Bertrand J-C (1988) Effects of oxygen on hydrocarbon degradation studies in vitro in surficial sediments. Estuarine Coast. Shelf Sci. 27: 283–295

Miller RM and Bartha R (1989) Evidence from liposome encapsulation for transport-limited microbial metabolism of solid alkanes. Appl. Environ. Microbiol. 55: 268–274

Mohanrao GJ and McKinney RE (1962) A study of the biochemical characteristics of quaternary carbon compounds. Int. J. Air. Water Pollut. 6: 153–168

Monticello DJ, Bakker D and Finnerty WR (1985) Plasmid-mediated degradation of dibenzothiophene by *Pseudomonas* species. Appl. Environ. Microbiol. 49: 756–760

Morgan P and Watkinson RJ (1989) Hydrocarbon degradation in soils and methods for soil biotreatment. Crit. Rev. Biotechnol. 8: 305–333

Mormile MR and Atlas RM (1988) Mineralization of the dibenzothiophene biodegradation products 3-hydroxy-2-formyl benzothiophene and dibenzothiophene sulfone. Appl. Environ. Microbiol. 54: 3183–3184

Mueller JG, Lantz SE, Blattmann BO and Chapman PJ (1991) Bench-scale evaluation of alternative biological treatment processes for the remediation of pentachlorophenol- and creosote-contaminated materials: slurry-phase bioremediation. Environ. Sci. Technol. 25: 1055–1061

Nakajima K and Sato A (1983) Microbial oxidation of isoprenoid hydrocarbons. Part IV. Microbial metabolism of isoprenoid alkane pristane. Nippon Nogei Kagaku Kaishi 57: 299–305

Nakajima K, Sato A, Takahara Y and Iida T (1985) Microbial oxidation of isoprenoid hydrocarbons. Part V. Microbial oxidation of isoprenoid alkanes, phytane, norpristane and farnesane. Agric. Biol. Chem. 49: 1993–2002

Nieder M and Shapiro J (1975) Physiological function of the *Pseudomonas putida* PpG6 (*Pseudomonas oleovorans*) alkane hydroxylase: monoterminal oxidation of alkanes and fatty acids. J. Bacteriol. 122: 93–98

Oudot J, Ambles A, Bourgeois S, Gatellier C and Sebyera N (1989) Hydrocarbon infiltration and biodegradation in a land-farming experiment. Environ. Pollut. 59: 17–40

Owen DJ (1986) Molecular cloning and characterization of sequences from the regulatory cluster of the *Pseudomonas* plasmid *alk* system. Mol. Gen. Genet. 203: 64–72

Parekh VR, Traxler RW and Sobek JM (1977) n-Alkane oxidation enzymes of a pseudomonad. Appl. Environ. Microbiol. 33: 881–884

Parsons JR, Opperhuizen A and Hutzinger O (1987) Influence of membrane permeation on biodegradation kinetics of hydrophobic compounds. Chemosphere 16: 1361–1370

Perry JJ (1984) Microbial metabolism of cyclic alkanes. In: RM Atlas (ed) Petroleum Microbiology (pp 61–97). Macmillan, New York

Pirnik MP (1977) Microbial oxidation of methyl branched alkanes. CRC Crit. Rev. Biotechnol. 5: 413–422

Pirnik MP, Atlas RM and Bartha R (1974) Hydrocarbon metabolism by *Brevibacterium erythrogenes*: normal and branched alkanes. J. Bacteriol. 119: 868–878

Pritchard PH and Costa CF (1991) EPA's Alaska oil spill bioremediation project. Environ. Sci. Technol. 25: 372–379

Ratledge C (1978) Degradation of aliphatic hydrocarbons. In: RJ Watkinson (ed) Developments in Biodegradation of Hydrocarbons – 1 (pp 1–46). Applied Science, London

Ratledge C (1984) Microbial conversions of alkanes and fatty acids. J. Am. Oil Chem. Soc. 6: 447–453

Rehm HJ and Reiff I (1982) Regulation der mikrobiellen alkanoxidation mit hinblick auf die produktbildung. Acta Biotechnol. 2: 127–138

Rehm HJ, Hortmann L and Reiff I (1983) Regulation der Fettsaurebildung bei der mikrobiellen alkanoxidation. Acta Biotechnol. 3: 279–288

Rontani JF and Giusti G (1986) Study of the biodegradation of poly-branched alkanes by a marine bacterial community. Mar. Chem. 20: 197–205

Sanglard D and Fiechter A (1989) Heterogeneity within the alkane-inducible cytochrome P450 gene family of the yeast *Candida tropicalis*. FEBS Lett. 256: 128–134

Sanglard D and Loper JC (1989) Characterization of the alkane-inducible cytochrome P450 (P450alk) gene from the yeast *Candida tropicalis*: identification of a new P450 gene family. Gene 76: 121–136

Sanglard D, Chen C and Loper JC (1987) Isolation of the alkane inducible cytochrome P450 (P450alk) gene from the yeast *Candida tropicalis*. Biochem. Biophys. Res. Comm. 144: 251–257

Sanglard D, Beretta I, Wagner M, Kaeppeli O and Fiechter A (1990) Functional expression of the alkane-inducible monooxygenase system of the yeast *Candida tropicalis* in *Saccharomyces cerevisiae*. Biocatalysis 4: 19–28

Schaeffer TL, Cantwell SG, Brown JL, Watt DS and Fall RR (1979) Microbial growth on hydrocarbons: terminal branching inhibits biodegradation. Appl. Environ. Microbiol. 38: 742–746

Schink B (1985) Degradation of unsaturated hydrocarbons by methanogenic enrichment cultures. FEMS Microbiol. Ecol. 31: 69–77

Schunck WH, Kaergel E, Gross B, Wiedmann B, Mauersberger S, Koepke K, Kiessling U, Strauss M and Gaestel M (1989) Molecular cloning and characterization of the primary structure of the alkane hydroxylating cytochrome P-450 from the yeast *Candida maltosa*. Biochem. Biophys. Res. Comm. 161: 843–850

Scott CCL and Finnerty WR (1976a) A comparative analysis of the ultrastructure of hydrocarbon-oxidizing microorganisms. J. Gen. Microbiol. 94: 342–350

Scott CCL and Finnerty WR (1976b) Characterisation of intracytoplasmic hydrocarbon inclusions from the hydrocarbon-oxidizing *Acinetobacter* sp. HO1-N. J. Bacteriol. 127: 481–489

Seigle-Murandi F, Krivobok S, Steiman R, Thiault GA and Benoit-Guyod JL (1991) Biotransformation of 2-acetylthiophene by micromycetes. Appl. Microbiol. Biotechnol. 34: 436–440

Shelton TB and Hunter JV (1975) Anaerobic decomposition of oil in bottom sediments. J. Water Poll. Control Fed. 47: 2256–2270

Shennan JL (1984) Hydrocarbons as substrates in industrial fermentations. In: RM Atlas (ed) Petroleum Microbiology (pp 643–683). Macmillan, New York

Singer ME and Finnerty WR (1984a) Microbial metabolism of straight-chain and branched alkanes. In: RM Atlas (ed) Petroleum Microbiology (pp 1–59). Macmillan, New York

Singer ME and Finnerty WR (1984b) Genetics of hydrocarbon-utilizing microorganisms. In: RM Atlas (ed) Petroleum Microbiology (pp 299–354). Macmillan, New York

Smucker RA and Cooney JJ (1981) Cytological responses of *Cladosporium resinae* when shifted from glucose to hydrocarbon medium. Can. J. Microbiol. 27: 1209–1218

Stirling LA, Watkinson RJ and Higgins IJ (1977) Microbial metabolism of alicyclic hydrocarbons: isolation and properties of a cyclohexane-degrading bacterium. J. Gen. Microbiol. 99: 119–125

Strubel V, Rast HG, Fietz W, Knackmuss H-J and Engesser KH (1989) Enrichment of dibenzofuran utilizing bacteria with high co-metabolic potential towards dibenzodioxin and other anellated aromatics. FEMS Microbiol. Lett. 58: 233–238

Sunairi M, Suzuki R, Takagi M and Yano K (1988) Self-cloning of genes for n-alkane assimilation from *Candida maltosa*. Agric. Biol. Chem. 52: 577–579

Takagi M, Ohkuma M, Kobayashi N, Watanabe M and Yano K (1989) Purification of cytochrome P-450alk from n-alkane-grown cells of *Candida maltosa*, and cloning and nucleotide sequencing of the encoded gene. Agric. Biol. Chem. 53: 2217–2226

Tonge GM and Higgins IJ (1974) Microbial metabolism of alicyclic hydrocarbons: growth of *Nocardia petroleophila* (NCIB 9438) on methylcyclohexane. J. Gen. Microbiol. 81: 521–524

Trudgill PW (1978) Microbial degradation of alicyclic hydrocarbons. In: RJ Watkinson (ed) Developments in Biodegradation of Hydrocarbons – 1 (pp 47–84). Applied Science, London

Trudgill PW (1984a) Microbial degradation of the alicyclic ring. Structural relationships and metabolic pathways. In: DT Gibson (ed) Microbial Degradation of Organic Compounds (pp 131–180). Marcel Dekker, New York

Trudgill PW (1984b) The microbial metabolism of furans. In: DT Gibson (ed) Microbial Degradation of Organic Compounds (pp 295–308). Marcel Dekker, New York

van Afferden M, Schacht S, Klein J and Truper HG (1990) Degradation of dibenzothiophene by *Brevibacterium* sp. DO. Arch. Microbiol. 153: 324–328

Wakeham SG, Canuel EA and Doering PH (1986) Behavior of aliphatic hydrocarbons in coastal seawater: mesocosm experiments with [^{14}C]octadecane and [^{14}C]decane. Environ. Sci. Technol. 20: 574–580

Ward DM and Brock TD (1978) Anaerobic metabolism of hexadecane in sediments. Geomicrobiol. J. 1: 1–9

Watkinson RJ (1980) Interaction of microorganisms with hydrocarbons. In: DEF Harrison, IJ Higgins and RJ Watkinson (eds) Hydrocarbons in Biotechnology (pp 11–24). Heyden, London

Watkinson RJ and Morgan P (1990) Physiology of aliphatic hydrocarbon-degrading microorganisms. Biodegradation 1: 79–92

2. Microbial metabolism and transformation of selected monoterpenes

Department of Biochemistry, University College of Wales, Aberystwyth, Dyfed SY23 3DD, U.K.

I. Introduction

Monoterpenoids (C_{10}) are major components of plant oils and are synthesized from two isoprene units. Parent structures are acyclic, monocyclic or bicyclic and the latter consist of fused C_6-C_5, C_6-C_4 and C_6-C_3 ring systems. The bicyclic monoterpenes are of particular interest from a metabolic standpoint because of the specific problems of ring cleavage that have to be overcome before microorganisms can utilize them as carbon sources for growth. In addition, because of their structural complexity, gratuitous biotransformation by organisms growing on other carbon sources leads to a considerable range of products.

Metabolic studies have a particular relevance because of the importance of some natural terpenes as renewable resources for the perfume, flavour and pharmaceutical industries. Some representative monoterpene structures are shown in Fig. 1. In this chapter a brief review of earlier important work on cyclic monoterpene metabolism is followed by a consideration of selected areas of current interest.

Some of the earliest studies of the metabolism of cyclic monoterpenes were initiated by Gunsalus and his co-workers at the University of Illinois and by Bhattacharyya and his group at the National Chemical Laboratory, Poona, India. Studies of the catabolism of the C_6-C_5 bicyclic monoterpene (+) camphor by strains of *Pseudomonas putida* (Bradshaw et al. 1959; Conrad et al. 1961; Conrad et al. 1965a,b; Trudgill et al. 1966a,b) and *Mycobacterium rhodochrous* (Chapman et al. 1966) provided early examples of subcellular enzymology and, although only limited success was achieved in understanding the catabolic pathways, the location of catabolic genes on a transmissible plasmid (Rheinwald et al. 1973) and key roles for methylene group hydroxylation and biological Baeyer-Villiger monooxygenases in ring cleavage strategies were all established (Figs. 2 and 3). No further experimental studies with *M. rhodochrous* T1 have been published but the (+) camphor 5-hydoxylase from *P. putida* ATCC 17453 has been the subject of extensive and detailed research (see Gunsalus and Marshall 1971; Gunsalus and Lipscomb

33

C. Ratledge (ed.), Biochemistry of Microbial Degradation, 33–61.

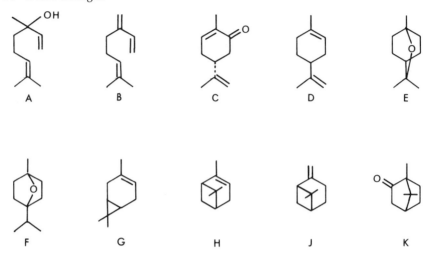

Fig. 1. Some naturally occurring monoterpenes. (A) linalool; (B) β-myrcene; (C) (−)-carvone; (D) limonene; (E) 1,8-cineole; (F) 1,4-cineole; (G) car-3-ene; (H) α-pinene; (J) β-pinene; (K) camphor.

1973; and Gunsalus et al. 1974 for reviews). More recently, the oxygenating component of 2,5-diketocamphane 1,2-monooxygenase has been purified to homogeneity from *P. putida* (Taylor and Trudgill 1986) and the enzymology of the second ring cleavage step, which involves CoA ester formation prior to biological Baeyer-Villiger oxygenation, has been elucidated (Ougham et al. 1983; Trudgill 1986). These particular areas of research probably represent the most detailed metabolic and enzymological studies to date of monoterpene catabolism by organisms capable of growth with them as sole carbon sources.

Bhattacharyya and his research group have taken a somewhat different approach to the problem in their investigations of the transformation of a number of monterpenes including geraniol, linalool, limonene, α-pinene and β-pinene, primarily by *Pseudomonas* strains. Very thorough and detailed analyses of metabolites accumulated in culture media have been made. However, primarily because of the large number of metabolites accumulated, it was often difficult to identify a clear cut route leading to central metabolites. Exceptions to this were the catabolism of limonene through perillic acid and a β-oxidative ring cleavage cycle (Dhavalikar and Bhattacharyya 1966; Dhavalikar et al. 1966) and evidence for the degradation of α-pinene through limonene or a closely related compound (Shukla and Bhattacharyya 1968; Shukla et al. 1968).

II. Current studies

Current work is concentrated in four main areas:
- a continued interest in cytochrome P450 systems involved in the degradation of monoterpenes and related compounds (Unger et al. 1986),

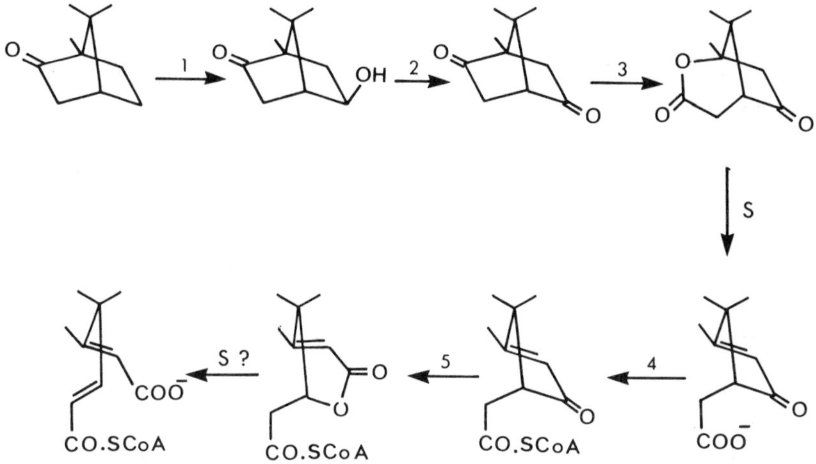

Fig. 2. Established steps of (+)-camphor degradation by *Pseudomonas putida* ATCC 17453. Enzymes; (1) camphor 5-hydroxylase; (2) 5-*exo*hydroxycamphor dehydrogenase; (3) 2,5-diketocamphane 1,2-monooxygenase; (4) 2-oxoΔ^3-4,5,5-trimethylcyclopentenylacetyl-CoA synthetase (5) 2-oxoΔ^3-4,5,5-trimethylcyclopentenylacetyl-CoA monooxygenase (a biological Baeyer-Villiger rection. S = spontaneous reaction.

– expanding our understanding of catabolic pathways that are academically fascinating and environmentally important for carbon cycling in the biosphere,
– understanding the enzymology of metabolic transformations and exploiting

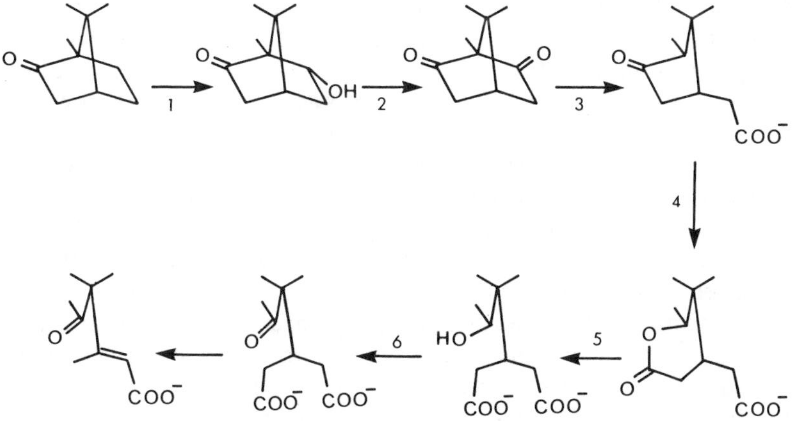

Fig. 3. Proposed steps of (+)-camphor degradation by *Mycobacterium rhodochrous* T1. Enzymes: (1) (+)-camphor 6-*endo*-hydroxylase; (2) 6-*endo*-hydroxycamphor dehydrogenase; (3) 2,6-diketocamphane lyase; (4) 3-oxo-4,5,5-trimethylcyclopentanylacetic acid monooxygenase; (5) a lactone hydrolase; (6) a dehydrogenase. Based on results of Chapman et al. (1966).

this commercially for the production of novel synthons from bulk renewable terpene resources such as car-3-ene, α- and β-pinenes and limonene isomers, – miscellaneous biotechnological, environmental and other novel aspects.

The biosynthesis of monoterpenes does not fall within the remit of this chapter but the interested reader is directed to the work of Croteau and his group (Croteau et al. 1989) as a lead-in reference to active research in this area.

III. Cytochrome P450 oxygenases and monoterpene metabolism

Although mammalian liver microsomal preparations and a purified reconstituted P450 hydroxylating system were shown to be capable of the hydroxylation of C10 methyl group of geraniol and nerol (Light and Coscia 1978) the involvement of bacterial cytochrome P450 oxygenases in the dissimilation of monoterpenes is well documented only for a small number of compounds. The described microbial systems consist of three protein components; a flavoprotein NAD(P)H dehydrogenase which passes single electrons sequentially, via an iron sulphur centre, to a P450 cytochrome which acts an hydroxylase. The best characterized example is the camphor 5-exohydroxylase of *P. putida* ATCC 17453. It will not be considered in detail in this chapter as it has been the subject of a number of reviews to which the reader is referred (White and Coon 1980; Sligar and Murrey 1986; Unger et al. 1986). All protein components are available in the pure state and have been used by investigators to obtain detailed information on the mechanism and other aspects of cytochrome P450 catalysis. The sequence of substrate additions to the enzyme and electron transfer during reaction is shown in Fig. 4. A second "P450 hydroxylase" which is structurally close to the $P450_{cam}$ has been described (Ullah et al. 1990) and catalyses 8-methyl hydroxylation as the first step in linalool oxidation by derivative Ppg777 of *P. putida* var. incognita. In addition cytochrome P450 hydroxylases have been authenticated from *P. putida* var. incognita (Suhara et al. 1985) and other organisms grown with p-cymene, from organisms grown on n-alkanes and, most recently, from a cyclohexane-grown *Xanthobacterium* sp. (Warburton et al. 1990; Trickett et al. 1991). In contrast, although the involvement of hydroxylases, possibly cytochrome P450 systems, in the microbial degradation of a number of monoterpenes including car-3-ene, 1,8-cineole and limonene, has been predicted or inferred from substrate oxidation, accumulated intermediates and additional enzymology, their more general involvement in monoterpene metabolism has still to be substantiated. A recent mini-review by Guengerich (1991) provides a convenient starting point for a more general consideration of the metabolic roles of cytochrome P450 enzymes.

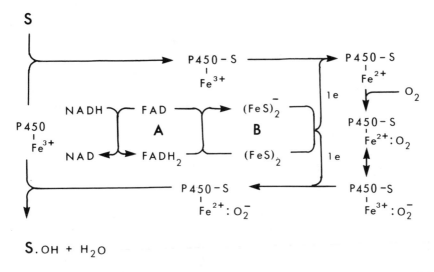

Fig. 4. Cytochrome P-450$_{cam}$ oxygenation-reduction cycle. (A) putidaredoxin reductase; (B) putidaredoxin; (S) substrate (2,5- diketocamphane).

IV. 1,8-Cineole and 1,4-cineole

Recent metabolic studies with bacteria have been concentrated on a restricted number of cyclic terpenes where knowledge is limited and microorganisms are presented with particular structural complications in the growth substrate.

1,8-Cineole and 1,4-cineole have structurally identical monocyclic monoterpene carbon skeletons but each has the additional structural complication of an ether linkage superimposed thereon (Fig. 1, E and F). 1,8-Cineole [1,3,3-trimethyl-2-oxabicyclo(2,2,2)octane] is isomeric with α-terpineol but contains neither a hydroxyl group nor a double bond. The trivial name eucalyptol is therefore rather misleading.

1,8-Cineole is quite widely distributed, occurring in nearly 300 higher plants. However *Eucalyptus* spp. are the major sources of the compound (Nishimura et al. 1980) The realization that different species of Eucalyptus accumulate different patterns of monoterpenes is said to have evolved from a chance observation by Penfold while waiting for a car tyre to be changed. He crumbled the leaves of several adjacent *Eucalyptus* trees and noticed that the odour of one was quite different from that of the others. In a study that followed from this observation, Penfold and Morrison (1927) investigated 280 species of *Eucalyptus* and classified them into four basic types producing 1,8-cineole, piperitone, phellandrene or acyclic terpenes such as geraniol as major monoterpene components. Some species were only distinguishable by the patterns of terpene accumulation. 1,8-Cineole is found in considerable quantities in eucalyptus oil from the leaves of *Eucalyptus radiata* var.

Australiana (Nishimura et al. 1982).

The likely metabolic strategies that may be adopted by microorganisms capable of growth with the compound as sole source of carbon were not apparent from a consideration of pathways and reactions elucidated for other cyclic terpenes such as (+)-camphor, limonene and α-pinene, primarily because of uncertainty relating to the mode of cleavage of the ether linkage. Observations made by the two groups of workers, who studied metabolite accumulation, were only of limited use in predicting the initial steps in a pathway to central metabolism.

A. Metabolism of 1,8-cineole by Pseudomonas flava

1. Metabolite accumulation

Until recently, reports of microbial 1,8-cineole metabolism were confined to the isolation of metabolites from culture media and, following investigations of the dynamics of their production and disappearance, to suggestions of a metabolic progression (MacRae et al. 1979; Carman et al. 1986). *P. flava*, isolated by enrichment culture on 1,8-cineole, accumulated the metabolites shown in Fig. 5. It is reasonable to suggest that initial attack occurs by hydroxylation at carbon 2, followed by dehydrogenation to form 2-oxocineole. Further metabolic steps are obscure, although a biological Baeyer-Villiger oxygenation can be envisaged and, since virtually all biological Beayer-Villiger reactions follow the rule of the chemical reaction in which when there are two different alkyl groups either side of the carbonyl the more highly substituted carbon becomes bonded to the new oxygen atom, the point of attack can be predicted. The source of the lactone (R)-5,5-dimethyl-4-(3'-oxobutyl)-4,5-dihydrofuran-2(3H)-one, the only ring cleavage product accumulated, was not immediately evident.

B. Metabolism of 1,8-cineole by Rhodococcus C1

1. Metabolite accumulation

Rhodococcus strain C1 was isolated by enrichment culture on 1,8-cineole (Williams et al. 1989) and grew in shake flasks with a doubling time of about 8 h. Gas liquid chromatography (GLC) analysis of diethyl ether extracts of neutral culture media showed the short-term accumulation of two compounds during the 15 to 18 h region of a 24 h growth regime. No acidic metabolites were detected. Sufficient quantities of the compounds for unequivocal identification and for metabolic studies were obtained from a 40 litre culture. One of these compounds gave a mass fragmentation pattern almost identical with that reported for the two isomers of 2-hydroxycineole and an ^{1}H-NMR spectrum identical with that reported for *2-endo*-hydrocycineole (MacRae et al. 1979).

---

*Strict application of chemical nomenclature does not allow use of *endo* and *exo* in the (2,2,2)bicyclo system in relation to 1,8-cineole derivatives. However, to simplify presentation of

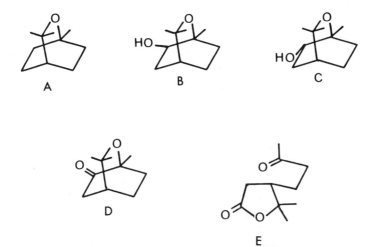

Fig. 5. Metabolites accumulated by *Pseudomonas flava* during growth on 1,8-cineole. Compounds are : (A) 1,8-cineole (growth substrate); (B) 2-*endo*-hydroxycineole; (C) 2-*exo*-hydroxycineole; (D) 2-oxocineole; (E) (*R*)-5,5-dimethyl-4-(3'-oxobutyl)-4,5-dihydrofuran 2(3H)-one. Data from MacRae et al. (1979).

These authors have reported that 2-*endo*-hydroxycineole is laevorotatory, $[\alpha]_D$ − 26° (c = 0.2 in ethanol). The alcohol accumulated by *Rhodococcus* C1 was dextrorotatory $[\alpha]_D$ + 27.5° in the same solvent. These results are only compatible with the compound being the optical isomer 6-*endo*-hydroxycineole. The second metabolite reacted with acidic 2,4-dinitrophenylhydrazine (Friedemann and Haugen 1943) to give an orange insoluble 2,4-dinitrophenylhydrazone and a mass spectrum almost identical with that reported for 2-oxocineole (MacRae et al. 1979). The ¹H-NMR spectrum displayed all the diagnostic features attributable to 2-oxocineole. However, while 2-oxocineole is dextrorotatory the ketone accumulated by *Rhodococcus* C1 was laevorotatory, compatible with the compound being 6-oxocineole.

Rhodococcus C1 was also capable of growth with 6-*endo*-hydroxycineole and 6-oxocineole as sole carbon sources and it is logical to conclude that, by attacking the compound at carbon 6, it degrades 1,8-cineole in a manner that is analogous but enantiomeric to that utilized by *P. flava*. Further investigations of the route of 1,8-cineole degradation were made with subcellular systems.

---

results we have a) designated as *endo* those alcohols in which the hydroxyl substitution is on the opposite side of the reference plane to the lowest priority bridge [see Fig. 5] and b) made use of the numbering system consistent with the trivial name 1,8-cineole.

2. 1,8-Cineole hydroxylation

Although it is logical to assume that initial attack upon the molecule is by a monooxygenase that inserts an oxygen atom at the 6-endo position to form 6-*endo*-hydroxycineole we failed to identify any such hydroxylating system in subcellular preparations from induced cells. The use of a variety of different cell disruption procedures, inclusion of potential stabilizing agents and a range of of electron donors and acceptors (in case the oxygen atom was derived from water) were all unsuccessful.

3. 6-Endo-*hydroxycineole dehydrogenase*

In contrast, an induced NAD-linked 6-*endo*-hydroxycineole dehydrogenase was readily detected (specific activity 0.62 μmol/min.mg protein, pH optimum 10.5) and GLC analysis showed 6-oxocineole to be the only product. The reverse reaction (specific activity 1.1 μmol/min.mg protein, pH optimum 7.5) yielded only 6-*endo*-hydroxycineole.

4. Further metabolism of 6-oxocineole

The catabolic transformation of cyclic hydrocarbons and alcohols into cyclic ketones is usually a preparation for ring oxygen insertion by a biological Baeyer-Villiger reaction (see Trudgill 1984 and 1986 for reviews). These enzymes are flavoproteins and display typical monooxygenase characteristics, requiring an electron donor and incorporating one atom of oxygen into the organic product. When a crude extract of 1,8-cineole-grown *Rhodococcus* C1 was incubated with 6-oxocineole and NADPH an oxygen-dependent oxidation of the NADPH was observed. Polarographic assays in the presence of limited amounts of either NADPH or 6-oxocineole allowed an approximate (1:1:1) stoichiometry of the reaction to be established. GLC analysis of diethyl ether extracts of acidified reaction mixtures showed that the 6-oxocineole had been converted into a single more polar metabolite. GLC-mass spectral analysis of the product gave a spectrum identical with that reported for the lactone (*R*)-5,5-dimethyl-4-(3'-oxobutyl)-4,5-dihydrofuran-2(3H)-one isolated by MacRae et al. (1979) (Fig. 5). Although the initial products of biological Baeyer-Villiger oxygenation of cyclic ketones are lactones there is no way in which this particular lactone can be formed directly from 6-oxocineole and the identity of the initial product of oxygenation was therefore uncertain.

5. Reaction with partially purified 6-oxocineole oxygenase

Purification of the NADPH-linked 6-oxocineole oxygenase from *Rhodococcus* C1 was a valid approach to determining the identity of the immediate oxygenation product. Unfortunately, the enzyme was not very stable and, although addition of ethanol (5% v/v) did have a stabilizing influence, we were unable, using either conventional procedures or by taking advantage of the short processing times of fast protein liquid chromatography, to purify the enzyme more than 5 to 7-fold without rapid loss of activity.

When 15 mg of partially purified enzyme were incubated with 60 μmol of

6-oxocineole and excess NADPH the consumption of 50 μmoles of oxygen was observed. Direct ferric hydroxamate assay for lactones (Cain 1961) showed them to be absent. When the aqueous reaction mixture was acidified (pH ~ 1) and extracted with diethyl ether, the ether phase contained "ferric hydroxamate positive" material which co-chromatographed with the 5,5-dimethyl-4-(3'-oxobutyl)-4,5-dihydrofuran-2(3H)-one obtained with crude cell extracts. This reinforced the view that this lactone is an artifact of isolation and not the immediate product of oxygenation.

5,5-Dimethyl-4-(3'-oxobutyl)-4,5-dihydrofuran-2(3H)-one is the lactone of 3-(1-hydroxy-1-methylethyl)-6-heptanoic acid. 4-Hydroxy acids readily lactonize in acidic solution and organic solvents. Wallach (1895) made use of precisely this spontaneous ring closure of 3-(1-hydroxy-1-methylethyl)-6-heptanoic acid in his investigations into the structure of α-terpineol. The working hypothesis that this acid is a catabolic intermediate from which the lactone is formed during extraction procedures was supported by the observation that 3-(1-hydroxy-1-methylethyl)-6-heptanoic acid prepared by alkali hydrolysis of an aqueous solution of the racemic lactone [synthesized chemically from α-terpineol essentially as described by MacRae et al. (1979)] would support growth of *Rhodococcus* C1 and was oxidized rapidly by 1,8-cineole-grown cells.

6. Chemical Baeyer-Villiger oxygenation of 6-oxocineole

Magnesium monoperoxyphthalate is an efficient chemical Baeyer-Villiger reagent that gives clean reaction products in good yield. When the 6-oxocineole, isolated from culture medium, was incubated with this reagent in dimethylformamide, followed by addition of water, acidification and extraction into diethyl ether, the reaction product had the same GLC retention time as 5,5-dimethyl-4-(3'-oxobutyl)-4,5-dihydrofuran-2(3H)-one.

These observations clearly suggested that the initial step in 6-oxocineole degradation is indeed a biological Baeyer-Villiger oxygenation. As already stated, chemical Baeyer-Villiger reagents insert the ring oxygen in such a manner that, when there are two different alkyl groups either side of the carbonyl, the more highly substituted carbon atom becomes bonded to the new oxygen. From this it is possible to predict that the ring oxygenation product is 1,6,6-trimethyl-2,7-dioxabicyclo(3,2,2)nonan-3-one. However, the steps between this lactone and 3-(1-hydroxy-1-methyethyl)-6-oxoheptanoic acid and any enzymes involved were not immediately clear.

7. 2,5-Diketocamphane 1,2-monooxygenase and 6-oxocineole

2,5-Diketocamphane 1,2-monooxygenase of *P. putida* ATCC 17453 is a well described NADH-linked biological Baeyer-Villiger monooxygenase that is known to insert an oxygen atom into camphor between the carbonyl group and the bridgehead. The enzyme consists of NADH oxidase and oxygenating flavoprotein components that reversibly dissociate and which have been purified to homogeneity from (+)-camphor-grown cells (Taylor and Trudgill

1986). Although the enzyme is known to have a broad ketone substrate specificity, it might be expected that 2-oxocineole would be preferred to 6-oxocineole. Preliminary experiments established that 6-oxocineole is a good substrate. This is presumably a result of the greater spatial symmetry of the (2,2,2)-bicyclo skeleton when compared with 2,5-diketocamphane. Reactions in which 0.65 unit of enzyme was incubated with 10 μmol 6-oxocineole and excess NADPH were used for product accumulation and analysis. Direct assay for lactones at the end of the reaction showed none to be present and, as occurred with the oxygenase preparations from *Rhodococcus* C1, acidification and diethyl ether extraction yielded, once again, the lactone 5,5-dimethyl-4-(3′oxobutyl)-4,5-dihydrofuran-2(3H)-one. This observation is of particular interest since camphor-grown *P. putida* ATCC 17453 is not known to have a lactone hydrolase. The clear implication is that the initial product of Baeyer-Villiger oxygenation is unstable. It was possible to confirm this fact in further investigations with a pH-stat. When the 2,5-diketocamphane 1,2-monooxygenase complex is incubated with ($^+$)-camphor oxygenation takes place, a proton is consumed and the lactone product accumulates. In contrast the lactone formed from 2,5-diketocamphane is unstable and spontaneously ring opens to give an unsaturated acid so that overall proton balance is maintained (Fig. 6). The experimental consequences of this are shown in Fig. 7. When 6-oxocineole is used as substrate proton balance is maintained confirming that, subsequent to the initial oxygenation step, further non-enzymic reaction(s) occur in which a proton is regenerated.

8. Reaction sequence and ring cleavage

A reaction sequence for the cleavage of 6-oxocineole that is compatible with with the experimental observations is shown in Fig. 8. The presumed lactone, 1,6,6-trimethyl-2,7-dioxabicyclo(3,2,2)nonan-3-one, is a strained molecule. We suggested originally that spontaneous cleavage is hydrolytic (Williams et al. 1989). This would form a hemiacetal with consequent further spontaneous cleavage of the ether linkage and formation of the anticipated acyclic acid. A possible alternative involving internal electron shifts would result in vinyl ether formation. Vinyl ethers are relatively unstable, especially in dilute acidic solution, and spontaneous cleavage would again yield the anticipated acid. It is not known which of these alternative routes is followed.

The further metabolism of 3-(1-hydroxy-1-methylethyl)-6-oxoheptanoic acid has not been investigated but in this context it is of interest that 1,8-cineole-grown *Rhodococcus* C1 also oxidizes levulinic acid and acetone which can be derived theoretically from the open chain acid (Fig. 8) by fairly conventional metabolic steps.

MacRae et al. (1979) established the absolute configuration of the lactone (Fig. 5) accumulated in culture media by *P. flava*. Although we have not established the absolute configuration of of the lactone formed by *Rhodococcus* C1 and 3-(1-hydroxy-1-methylethyl)-6-oxoheptanoic acid from which it is formed logic would dictate that they will have the (*S*) configuration.

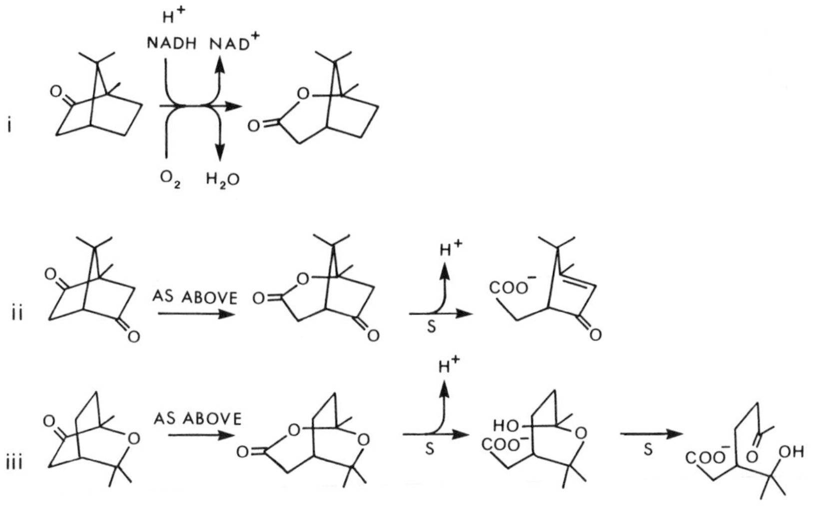

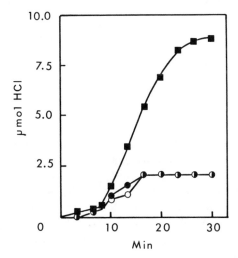

Fig. 6. Reactions catalysed by 2,5-diketocamphane 1,2-monooxygenase from *P. putida* ATCC 17453 with (i) (+)-camphor and (ii) 2,5-diketocamphane as substrates. The proposed oxygenation reaction with 6-oxocineole (iii) is followed by spontaneous cleavage to form the acyclic compound 3-(1-hydroxy-1-methylethyl)-6-oxoheptanoic acid. An alternative postulated spontaneous cleavage sequence leading to the same product is shown in Fig. 8. Adapted from Williams et al. (1989).

Fig. 7. A comparison of proton yields in the oxygenation reactions illustrated in Fig. 6. The reaction vessel of a pH-stat contained 0.5 unit of 2,5-diketocamphane 1,2-monooxygenase (coupled to NADH oxidase) from *P. putida* ATCC17453 and 10 μmol of NADH in 11 ml of distilled water adjusted to pH 6.7 with a minimum amount of phosphate buffer. After establishment of the endogenous rate 10 μmol of (+)-camphor ($\blacksquare$), 2,5-diketocamphane ($\bullet$) or 6-oxocineole ($\circ$) were added and proton release monitored by controlled addition of 2 mM-HCl to maintain a constant pH. Adapted from Williams et al. (1989).

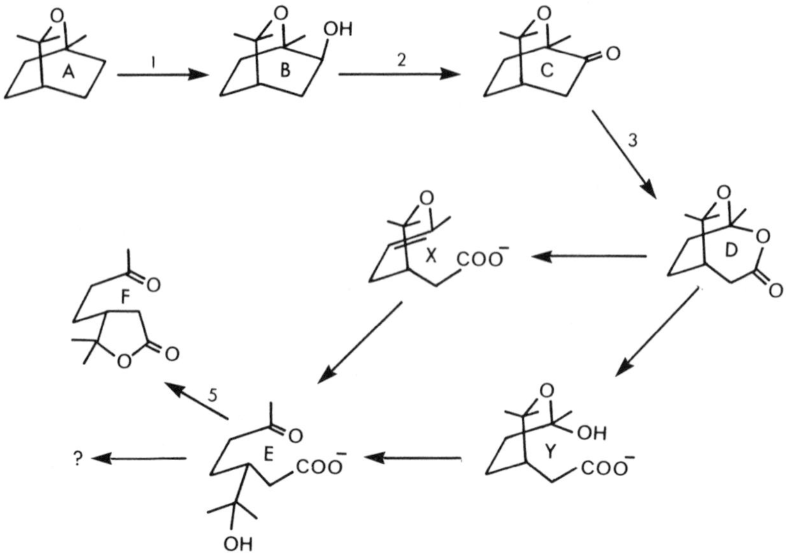

Fig. 8. Proposed ring-cleavage reactions in the metabolism of 1,8-cineole by *Rhodococcus* strain C1. Compounds are: (A) 1,8-cineole; (B) 6-*endo*-hydroxycineole; (C) 6-oxocineole; (D) 1,6,6-trimethyl-2,7-dioxabicyclo(3,2,2)nonan-3-one; (F) 5,5-dimethyl-4-(3′-oxobutyl)-4,5-dihydrofuran-2(3H)-one. Compounds (X) 2,6,6-trimethyl-5-acetyl-4,5-dihydropyran-2-ol and (Y) 2,6,6-trimethyl-5-acetyltetrahydropyran-2-ol are proposed alternative transient intermediates of spontaneous ring cleavage. Enzymes are: (1) putative 1,8-cineole 6-*endo*-hydroxylase; (2) NAD-linked dehydrogenase; (3) NADPH-linked monooxygenase (a biological Baeyer-Villiger monooxygenase). Reaction (5) is a spontaneous hydroxyacid ring closure (with loss of water) that occurs upon acidification of reactions and extraction with diethyl ether (Wallach 1895). Adapted from Williams et al. (1989).

C. Metabolism of 1,4-cineole

In principle the sequence of ring cleavage steps elucidated for 1,8-cineole is also theoretically applicable to the less-widely distributed isomer 1,4-cineole which is a flavour constituent of lime juice (Kovats 1963). A sequence of more highly strained intermediates would yield 3-hydroxy-3-(1′-methylethyl)-6-oxoheptanoic acid. In this context it is of interest that Rosazza et al. (1987) have reported the 2-*endo*-hydroxy and 2-oxo-derivatives as products of 1,4-cineole fermentation by a considerable variety of bacteria and fungi grown in rich media. There is no information available on catabolic routes leading from 1,4-cineole to central metabolites.

D. Biotechnological applications of microbial cineole metabolism

The herbicide cinmethylin is the racemic 2-*exo*-(o-methylbenzyl)ether of 1,4-cineole. 2-*Exo*-hydroxy-1,4-cineole is a key synthetic intermediate and it may be made by epoxidation of terpen-4-ol followed by acid-catalysed intramolecular esterification (Payne 1984). An alternative biotechnological approach to its production could involve the microbial formation of the 2-*exo*-hydroxy-1,4-cineole from the renewable parent terpene 1,4-cineole. As has already been stated, the established reactions for the cleavage of 1,8-cineole are, in principle, also applicable to 1,4-cineole: incubation of broth-grown cultures of *Aspergillus niger* with 1,8-cineole results in the accumulation of the 2-hydroxycineoles in addition to other products (Nishimura et al. 1982). It therefore follows that two alternative approaches to the target molecule, one dependent upon exploitation of a segment of a catabolic pathway the other upon fermentation, are potentially available on the basis of existing knowledge. Rosazza et al. (1987) examined a variety of microorganisms for their potential to hydroxylate 1,4-cineole. Hydroxylation at the 2-position was the most commonly observed microbial transformation although hydroxylations at carbons 3 and 8 were also observed in addition to the formation of 2-oxocineole. In addition, the production of specific 2-hydroxycineole isomers from the enantiomeric 2-oxo-1,4-cineoles, with the potential for synthesis of specific cinmethylin isomers, can be achieved by making use of strains of *Curvularia lunata* and *Penicillium frequetans* (Goswami et al. 1987).

V. α-Pinene and car-3-ene

A. Introduction

Naturally occurring bicyclic monoterpenes are homocyclic and are fused C_6-C_5 (borneol, camphor), C_6-C_4 (α-pinene, β-pinene) or C_6-C_3 (car-3-ene) carbon rings. As was reported in Section III the degradation of camphor has been studied in most detail and preparation for ring cleavage is dominated by hydroxylation and "Baeyer-Villiger" oxygenation reactions which form lactones although, in the case of *P. putida* ATCC 17453, lactone hydrolase activity has not been detected. (+)-Camphor- grown *P. putida* ATCC 17453 hydroxylates the compound at C-5 and oxidizes this to 2,5-diketocamphane. This appears to be a two-fold strategy leading to sequential biological Baeyer-Villiger oxygenations and spontaneous lactone cleavage steps that open both carbocyclic rings (Fig. 3). The keto group of camphor has a directing influence on catabolism and it transpires that camphor metabolism provides few clues to the pathways used to degrade the unsaturated hydrocarbons α-pinene and car-3-ene. However, in retrospect the reported cytochrome $P450_{cam}$catalysed epoxidation of 5,6-dehydrocamphor (Gelb et al. 1982) does provide a clue to the strategy used by one group of organisms to degrade α-pinene.

Although these compounds are renewable resources of considerable interest to the perfume and flavour industries, it is only recently that some understanding of the enzymology of α-pinene degradation has been acquired.

There is no significant published information on the bacterial degradation of car-3-ene although limited microbial transformation has been described (Prema and Bhattacharyya 1962a; Stumpf et al. 1990).

B. Early studies of α-pinene metabolism by Pseudomonas strains

Shukla and Bhattacharyya (1968) and Shukla et al. (1968) isolated a variety of metabolites from culture medium when *Pseudomonas* strain PL was grown with α-pinene as sole carbon source. Accumulated acidic metabolites, primarily associated with oxidation of the C10 methyl group, included perillic acid, which had previously been identified as a metabolite produced from limonene by this organism (Dhavalikar and Bhattacharyya 1966) and the acyclic acids 3-isopropylpimelic acid and 3-isopropenylpimelic acid. Although Shukla and Bhattacharyya (1968) presented, on the basis of accumulation experiments, a rather complex pattern of metabolic pathways leading from α-pinene to other compounds, their evidence, including limited subcellular studies, is consistent with a primary degradative route that involves initial cleavage of the cyclobutane ring to form limonene or 1-*p*-menthene; oxidation of the C7 methyl group (C10 of α-pinene) to carboxyl and β-oxidative ring cleavage (Fig. 9).

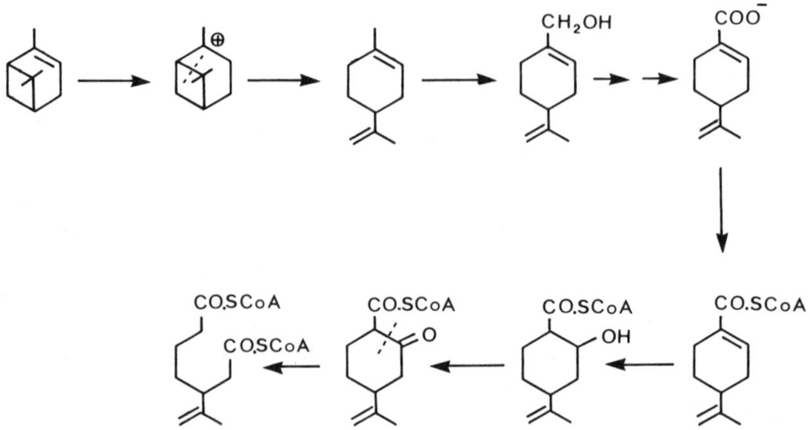

Fig. 9. Proposed pathway for cleavage of α-pinene by *Pseudomonas* PL. A prototrophic rearrangement of the α-pinene to form limonene has been suggested; followed by methyl group oxidation and ring cleavage mediated by a β-oxidation cycle. Based on the results of Shukla and Bhattacharyya (1968) and Shukla et al. (1968).

In contrast to these observations Gibbon and Pirt (1971), Gibbon et al. (1972) and Tudroszen et al. (1977), again on the basis of metabolites accumulated by parent *Pseudomonas* strains and mutants, proposed a quite different catabolic route in which cleavage of the C6 ring occurred between carbon atoms 3 and 4 leading to the formation of 2-methyl-5-isopropylhexa-2,5-dienoic acid (Fig. 10). Although a biological Baeyer-Villiger oxygenation step was proposed no supporting subcellular evidence was reported.

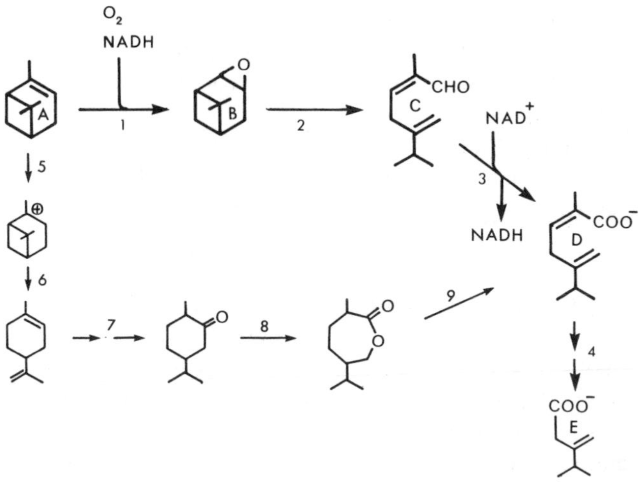

Fig. 10. Partial catabolism of α-pinene by *Nocardia* sp. strain P18.3, *P. fluorescens* NCIMB 11671 and *P. putida* NCIMB 10684. Compounds are: (A) α-pinene; (B) α-pinene epoxide; (C) *cis*-2-methyl-5-isopropylhexa-2,5-dienal; (D) *cis*-2-methyl-5-isopropylhexa-2,5-dienoic acid; (E) 3-isopropylbut-3-enoic acid. Reactions are: (1) NADH-dependent α-pinene monooxygenase (*P. putida* NCIMB 11671); (2) α-pinene epoxide lyase (all three organisms); (3) NAD-linked 2-methyl-5-isopropylhexa-2,5-dienal dehydrogenase (*Nocardia* P18.3); (4) proposed β-oxidation cycle to give the acid identified by Tudroszen et al. (1977). Reactions (5)–(9) were originally proposed for *P. putida* NCIMB 10684 by Gibbon and Pirt (1971).

C. α-*Pinene metabolism by* P. fluorescens *NCIMB 11671*

Recently Best et al. (1987) reported that extracts of α-pinene-grown *P. fluorescens* NCIMB 11671 catalysed an NADH-linked consumption of oxygen in the presence of α-pinene. Rates of NADH and oxygen consumption suggested a 1:1 stoichiometry and the activity was located in the soluble protein fraction. The neutral reaction product was extracted into diethyl ether and mass spectral, ^{1}H- and ^{13}C-NMR analyses identified the compound as 2-methyl-5-isopropylhexa-2,5-dien-1-al, although the isomeric configuration was not reported. It is of particular interest that, on oxidation, this aldehyde would yield 2-methyl-5-isopropylhexa-2,5-dienoic acid, both isomers of which were reported (Gibbon and Pirt 1971; Gibbon et al. 1972; Tudroszen et al. 1977) as being accumulated from α-pinene by *Pseudomonas* PX1 (NCIMB 10684), *P. putida* PIN11 and mutants thereof.

A particularly remarkable feature was the rapid cleavage of both carbocyclic rings of α-pinene that was catalysed by very small amounts of soluble protein from induced cells. It was provisionally assumed that an initial hydroxylation of

α-pinene had occurred. However, the inclusion of atebrin in assays, as a supposed inhibitor of enzymes with flavin prosthetic groups and thus of a flavoprotein hydroxylase, resulted in an inhibition of ring cleavage and the accumulation of α-pinene epoxide. Subsequent experiments showed that
- the initial oxygenation step formed α-pinene epoxide,
- α-pinene and (+)-limonene were also substrates for the enzyme,
- the enzyme was sensitive to inhibitors that react with thiol groups and that complex Fe^{2+}.

Further preliminary studies with α-pinene epoxide showed that the soluble protein extract of α-pinene-grown *P. fluorescens* NCIMB 11671 catalysed a very rapid cleavage of α-pinene epoxide with the formation of the acyclic aldehyde 2-methyl-5-isopropylhexa-2,5-dienal.

D. α-*Pinene metabolism by* Nocardia *sp. strain P18.3*

We have isolated a Gram-positive organism *Nocardia* sp. strain P18.3 by elective culture with α-pinene. Cells grown on the terpene oxidize the growth substrate and α-pinene epoxide at comparable rates. Although we have been unable to demonstrate the putative epoxide-forming monooxygenase in this organism, incubation of extracts of α-pinene-grown cells with α-pinene epoxide resulted in its very rapid conversion into the cis isomer of 2-methyl-5-isopropylhexa-2,5-dienal [specific activity 16 μmol/min.mg protein] (Griffiths et al. 1987a). Quite rapid transition to the trans isomer took place non-enzymically in the glycine/NaOH buffer (pH 9) which was used to suppress spontaneous breakdown of the α-pinene epoxide. An induced NAD^+-linked dehydrogenase oxidized both isomers of the aldehyde to 2-methyl-5-isopropylhexa-2,5-dienoic acid.

E. *Distribution of catabolic pathways for* α-*pinene*

The enzymic cleavage of α-pinene epoxide by extracts of α-pinene-grown bacteria would appear to be a useful diagnostic test. An examination (Table 1) of the distribution of the enzyme in available *Pseudomonas* strains provides additional support for there being two distinct degradative pathways. The reports by Gibbon and Pirt (1971) and Tudroszen et al. (1977) are substantially correct with regard to the identities of metabolites accumulated by *Pseudomonas* PX1 (NCIMB 10684), although it is likely that (+)-trans-carveol arises non-biologically from α-pinene epoxide, since carveol isomers are formed spontaneously, especially in slightly acidic solution (Griffiths et al. 1987a). They are, however, incorrect as regards the mechanism of ring cleavage and this illustrates the dangers inherent in formulating catabolic pathways on this type of evidence alone. The induction of the epoxide-cleaving enzyme firmly places *Pseudomonas* PX1 in the same group as *P. fluorescens* NCIMB 11671 and *Nocardia* P18.3 (Fig. 11). In contrast, *Pseudomonas* strains PL and PIN 18 (NCIMB 10687) degrade α-pinene by the alternative route

(Fig. 9); a dichotomy first appreciated by Gibbon and Pirt (1971) and Gibbon et al. (1972).

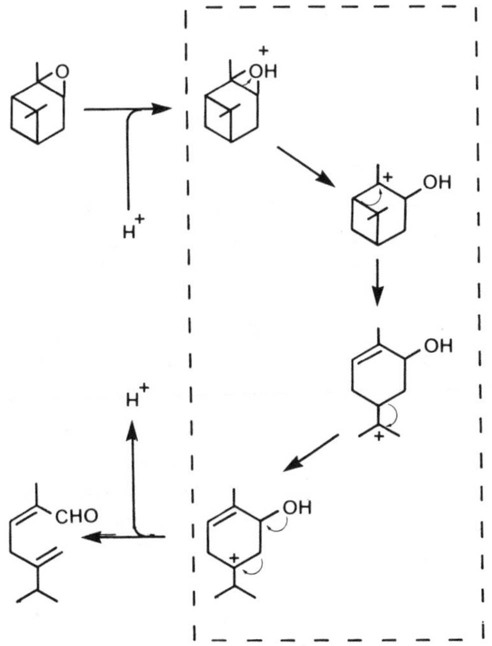

Fig. 11. Proposed reaction scheme for the decyclization of α-pinene epoxide by α-pinene epoxide lyase. Compounds within the box are suggested enzyme-bound carbo-cations. Adapted from Griffiths et al. (1987b).

Table 1. Distribution of α-pinene epoxide lyase in extracts of α-pinene and succinate-grown bacteria.

Strain	α-Pinene epoxide lyase (μmol cleaved/min. mg protein) from cells grown on:	
	α-Pinene	Succinate
Nocardia sp. strain P18.3	16	0.2
Pseudomonas sp. strains		
NCIMB 10684	5.3	0.3
NCIMB 11671	3.9	0.09
NCIMB 10687	0.01	0.001
PL	0.01	0.001

Adapted from Griffiths et al. (1987a).

F. α-Pinene epoxide lyase

This well-documented cleavage of α-pinene epoxide, in which a C-O bond and two C-C bonds are cleaved rapidly without any apparent requirement for additional co-factors was worthy of more detailed investigation.

1. Purification from Nocardia P18.3

The enzyme was purified from *Nocardia* P18.3 (Griffiths et al. 1987b) and shown to constitute about 6% of the soluble cell protein. Sephacryl S-200 chromatography gave an M_r of 40000 Da and ultracentrifugal determination (Yphantis 1964) a value of 50000 Da. SDS-polyacrylamide gel electrophoresis yielded two electrophoretically dissimilar bands (M_r values 17000 and 22000 Da).

The enzyme was devoid of detectable prosthetic groups, had no cofactor requirements and there was no significant variation in activity over a pH range from 7 to 10. The K_m for α-pinene epoxide was approx. 9 μM and the turnover number 15000. Production of cis-2-methyl-5-isopropylhexa-2,5-dienal in pyrophosphate buffer was stoichiometric. Structurally related terpenes such as β-pinene, pinan-3-ol, pinan-3-one, and carvone were inhibitors and, of a range of group-specific inhibitors, sensitivity was shown only towards sulphydryl active agents such as p-hydroxymercuribenzoate and 5,5'-dithiobis-2-nitrobenzoate and to atebrin.

A particularly interesting feature of the enzyme was that, although it was stable to dialysis and prolonged storage at 4 °C, it exhibited suicidal catalysis under a range of assay conditions and in different buffers. Product inhibition was not a factor and each molecule of enzyme turned over an average of 12000 molecules of substrate before being inactivated.

We have proposed a mechanism for the decyclization of α-pinene epoxide which is compatible with the observed properties of the enzyme (Griffiths et al. 1987b). Donation of a proton from a site in the catalytic centre initiates a series of concerted rearrangements, leading to formation of the cis isomer of the product (Fig. 11). If this mechanism is substantially correct then the enzyme should be classified as a lyase (EC 4.99.-.-). The proposed mechanism is compatible with the terpene inhibitor profile and the high turnover number. It may explain the origin of the small amounts of carvone accumulated by *P. fluorescens* NCIMB 11671 (Rhodes et al. 1983) since the elimination of a proton at any intermediate stage in catalysis would lead to the production of compounds such as carveol, α-terpineol or trans-sobrerol. It is also possible that inactivation of the enzyme during catalysis may be associated with a low frequency side reaction of one of the unstable carbo-cation intermediates with an essential reactive site in the catalytic centre.

2. Purification from P. putida PX1

We have also purified the α-pinene epoxide lyase from *P. putida* PX1 (NCIMB 10684). It also constitutes about 6% of the soluble cell protein and has an M_r of

42000 Da by ultracentrifugal analysis. It differs from the *Nocardia* enzyme in being formed from two electrophoretically identical subunits and in having a much higher K_m for α-pinene epoxide (210 μM). This, in conjunction with a less rapid inactivation during catalysis, made the enzyme a more suitable subject for kinetic and inhibitor studies.

3. Inhibition studies with atebrin and related compounds

As had previously been reported for the lyase from *P. fluorescens* NCIMB 11671 the enzymes from *Nocardia* P18.3 and *P. putida* NCIMB 10684 were very sensitive to inhibition by atebrin (quinacrine, 6-chloro-9-[(4-diethylamino)-1-methylbutyl]amino-2-methoxyacridine).

The α-pinene epoxide lyase form *P. putida* NCIMB 10684 was selected for detailed study. Kinetic analyses showed that the inhibition was non-competitive with a K_i of about 0.6 μM. A dissociation constant for atebrin was measured fluorimetrically (excitation 424 nm; fluorescence 497 nm in pyrophosphate buffer, pH 9.0) by following the increase in fluorescence that occurred upon adding enzyme to atebrin. Values of between 0.3 and 0.9 μM obtained correlate closely with the K_i obtained from kinetic studies. Inhibition by atebrin was reversible; dialysis of inhibited enzyme against buffer resulted in good recovery of activity.

In addition to atebrin, chlorpromazine (2-chloro-10-(3-dimethylaminopropyl)phenothiazine) and promethazine (10-(2-dimethyl-aminopropyl)phenothiazine) were also potent inhibitors (Fig. 12) while Methylene Blue, Pyronin Y and diphenyliodonium chloride all gave 50% inhibition at 10 μM or less (α-pinene epoxide at 2 mM, enzyme at 54 nM, pH 9).

Non-competitive inhibition by compounds that exhibit K_i values two to three orders of magnitude lower than the K_m for the substrate is typical of compounds that act as transition state inhibitors, taking advantage of additional binding interactions of a transition state by incorporating key structural elements of it in the stable structure of the inhibitor. One result is that transition state analogue inhibitors may have a much higher affinity for the enzyme active site than traditional ground state inhibitors. This is exemplified, for example, by the inhibition of 2,3-oxidosqualene cyclases by 2-aza-2,3-dihydrosqualene, 2-aza-2,3-dihydrosqualene-N-oxide and several derivatives thereof that show obvious structural analogy with carbocation intermediates of the cyclization process (see Benveniste 1986 for review).

In the case of α-pinene epoxide lyase the precise structural analogy between the potent inhibitors and proposed carbocation intermediates formed in catalysis (Fig. 12) is not obvious. Structural variations exist but the most potent compounds have both a ring N and a dimethyl or diethyl substituted amine group. The situation is further complicated by the fact that the two ionizable N atoms of atebrin have pK_a values of 7.5 (aromatic) and 10.1 (diethylamine) but the compound is a less effective inhibitor at pH 7 than at pH 9. In conclusion, although we are uncertain as to exactly how these inhibitors mimic transition state carbo-cations we are encouraged in our proposed catalytic mechanism

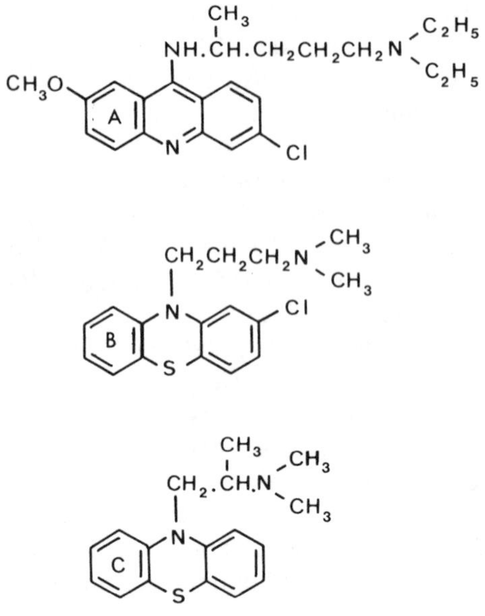

Fig. 12. Compounds that act as transition state inhibitors of α-pinene epoxide lyase. (A) atebrin (6-chloro-9-[(4-diethylamino)-1-methylbutyl]-amino-2-methyoxyacridine); (B) chlorpromazine (2-chloro-10-(3-dimethylaminopropyl)-phenothiazine; (C) promethazine (10-(2-dimethylamino-propyl)-phenothiazine).

(Fig. 12) by the proven requirement for both a positive charge and a cyclic component if the inhibitor is to be optimally effective (Trudgill 1990).

G. *Industrial application and biotechnological manipulation*

Turpentines, which constitute important renewable resources (>200,000 tonnes per annum) are the main sources of α-pinene, β-pinene and car-3-ene. The pinenes are used extensively as raw materials in the perfume and flavour industries (Bauer and Garbe 1985). Fragrances and flavours derived from them by chemical means are used in products as varied as soaps and ice cream.

Although the costs of α- and β-pinenes makes them attractive starting points for fragrance and flavour synthesis the chemical routes currently employed do not always give particularly high yields. Substitution of microbiologically-based procedures for bulk sythesis is beset with many difficulties that range from microbiological selection to chemical engineering problems. However, the level of structural complexity of the pinenes and the chirality of the molecules does allow, at a minimum, for an alternative microbiological role in the smaller scale production of novel molecules for trial screening and use as chiral synthons for developmental purposes. Successful use of microbial systems may well arise from the innovative interweaving of steps which enzymes are able to perform efficiently and specifically with those that are chemically facile and effective.

An early example of biotechnological interest in the pinenes is illustrated by a patent application (Rhodes et al. 1983) relating to the production of l-carvone from either α- or β-pinene by *P. fluorescens* NCIMB 11671 grown in liquid culture medium. l-Carvone is a major constituent of spearmint oil and the commercial market includes its use is in chewing gum and toothpaste. This particular patent application is interesting for two reasons:

- the concentrations of l-carvone accumulated in culture media were low and in the order of 10 mg per litre,
- l-carvone is now known not to be a metabolite in the degradation of α-pinene by *P. fluorescens* NCIMB 11671 (Best et al. 1987; Griffiths et al. 1987a).

It is possible that l-carvone may arise in small amounts as a side-product during cleavage of α-pinene epoxide by the lyase (EC 4.99.-.-) (Griffiths et al. 1987b). Spontaneous cleavage of α-pinene epoxide in acid solution leads to the formation of isomers of carveol (Griffiths et al. 1987a) and the proposed mechanism of enzymic ring cleavage (Fig. 11) could be regarded as a constrained variation of the spontaneous cleavage reactions. It follows from this that, while there is little prospect of producing significant amounts of l-carveol (or l-carvone) by pathway manipulation, modification of the lyase may be a potentially more useful approach to directing production of the monocyclic monoterpene of choice from the renewable pinenes. If there is a lesson to be learned form this it is that it is useful to know about pathways and key enzymology in some detail before embarking on investigations of biotechnological applicability.

In an alternative application of α-pinene catabolism crude extracts and the purified α-pinene oxide lyase from α-pinene-grown *P. fluorescens* NCIMB 11671 or *Nocardia* strain P18.3 can be used to catalyse the conversion of α-pinene oxide into *cis*-2-methyl-5-isopropylhexa-2,5-dienal. Spontaneous isomerization to the trans isomer can easily be achieved as can isomerization to the respective 2-methyl-5-isopropylhexa-2,4-dienals. In addition, suspensions of α-pinene grown cells can also be exploited in the conversion of α-pinene oxide into the 2,5-dienals in high yield (D.J. Best, personal communication). Interest shown in the isomers of 2-methyl-5-isopropylhexa-2,5-dienal as potential perfume materials is illustrated by a patent relating to the biological production of these compounds (Harries et al. 1988) and by subsequent interest shown in these compounds by the British Society of Perfumers.

VI. Metabolism of monoterpenes by fungi

A. Biotransformations by fungi cultured in rich media

Although little information is available on the metabolism of this class of compounds by fungi using them as sole carbon sources there are many reports of the transformation of monoterpenes by fungi grown in rich media. Early reports included those of Prema and Bhattacharyya (1962a,b) in which a strain of *Aspergillus niger*, grown on rich medium, converted α-pinene into (+)-verbanone, *cis*-verbanol and *trans*-sobererol. Car-3-ene, in what was until recently the only report of the its microbial metabolism, was converted into an hydroxyketone which defied precise structural identification but was either the 2- or 5-keto derivative with an hydroxyl group at an unestablished position. Chaminic acid, which is found in the heartwood of the tree *Chamaecyparis nootkatensis* (Carlsson et al. 1952; Erdtman et al. 1956), is said to possess insecticidal properties (Gensler and Solomon 1973). Recently the microbial conversion of car-3-ene into the (+) isomer of the acid and car-2-ene into (−)-isochaminic acid has been reported for a wide range of organisms including fungi (Stumpf et al. 1990) (Fig. 13). No attempts to exploit this transformation commercially have yet been made.

Noma and Nonomura (1974) showed the hydroxylation of (+)- and (−)-dihydrocarvone by *A. niger* and Madayastha and Murthy (1988) showed that this organism converted cintronellol, geraniol and linalool into their 8-hydroxy derivatives. The ability of *A. niger* to hydroxylate accessible methyl and methylene carbon atoms in monoterpenes was further exemplified by the work of Yamazaki et al. (1988) who reported multiple-site hydroxylations of β-myrcene by a strain of this organism (Fig. 13).

Draczynska et al. (1985) reported the transformation of isomers of α- and β-pinene by a culture of the honey fungus *Armillariella mellea*. The composition of products obtained indicated that allylic hydroxylation was the most characteristic reaction, yielding verbenol and verbenone from the unmodified pinene skeleton and, following cleavage of the four-membered ring to form the menthane skeleton, yielding *trans*-sobrerol and 7-hydroxy α-terpineol.

B. Transformations by fungi in the environment

1. Armillariella mellea, *other fungi and bark beetle pheromones*
A study of the relationship between plant pathogenic *Ar. mellea*, coniferous hosts and bark beetles reveals a major role for cyclic monoterpene metabolites in regulating attack by beetles such as *Ips typographus* (European spruce bark beetle) and *Dendroctonus ponderosae* (mountain pine beetle). The commercial implications are considerable since, for example, *Dendroctonus* spp. kill millions of pine trees annually in North America (Borden 1990). Madziara-Borusiewick and Strzelecka (1977) observed that *Picea excelsa* infected by *Ar. mellea* were more frequently attacked by *I. typographus* than sound trees. One of the monoterpene hydrocarbon-derived aggregation pheromone components of *I. typographus* is *cis*-verbenol (Bigersson et al. 1984) and, in contrast, *trans*-verbenol is an aggregation pheromone for *D. ponderosae*.

In the context of this chapter, the source of these pheromone components is of particular interest. The resin of *Picea* spp. contains α-pinene and the

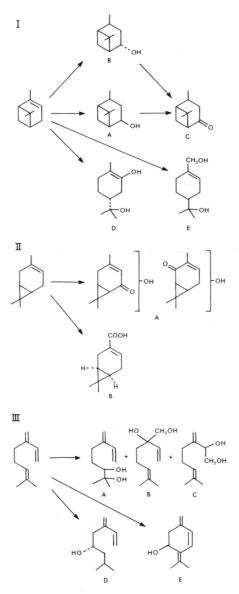

Fig. 13. Representative transformations of monoterpenes catalysed by fungi. I. Transformations of α-pinene: Products formed by *Aspergillus niger*, *Armillariella mellea*, other fungi and bark beetles. (A) *cis*-verbenol; (B) *trans*-verbenol; (C) verbenone; (D) *trans*-sobrerol; (E) 7-hydroxy-α-terpineol [from (−)-α-pinene by *A. mellea*] (Draczynska et al. 1985). II. Tranformations of car-3-ene. (A) either 2- or 5-oxocar-3-ene with an hydroxyl group of uncertain location produced by *A. niger*; (B) (+)-chaminic acid formed by bacteria and fungi (Stumpf et al. 1990). III. Transformation of β-myrcene into: (A) 2-methyl-6-methylene-7-octene-2,3-diol; (B) 6-methyl-2-ethenyl-5-heptene-1,2-diol; (C) 7-methyl-3-methylene-6-octene-1,2-diol by *A. niger* and into (D) ipsenol and (E) ipsdienol by bark beetles.

conversion of α- and β-pinenes into the verbenols and verbenone by *A. mallea* (Draczynska et al. 1985) may be an important signal that leads to beetle colonization of weakened and dying trees.

It is also known that bark beetles often have bacteria, yeasts and other fungi closely associated with them (Barras and Perry 1975; Leufvén et al. 1984) and that some associations involve the carrying of microorganisms from one plant host to another on specialized structures on the bodies of the insects, termed mycetangia (Franke-Grossman 1967). This ectosymbiosis, coupled with the knowledge that fungi are able to convert monoterpenes into bark beetle pheromone components [α-pinene into *cis*-verbenol, *trans*-verbenol and verbenone; β-myrcene into ipsenol and ipsdienol (Fig. 13)] has led to frequent suggestions that microrganisms have a role to play in pheromone production and secretion by the insects. The natural situation is, however, complicated. It is not disputed that the pheromones of bark beetles include several compounds derived through oxygenation of monoterpene hydrocarbons from host trees when an attack is initiated (Hendry et al. 1980). Indeed, it seems likely that pine bark beetles may have adapted to their hosts by the development of metabolic pathways, either directly or by symbiotic association, that detoxify monoterpenes and thus overcome the host's chemical defence system and only later have exploited the resultant metabolites as pheromones (Hughes 1973a,b). Of particular interest to us is the conversion of α-pinene into *cis*- and *trans*-verbenols. The finding of pheromone components in the hindgut of bark beetles (Byers 1983) and the isolation of yeasts from the gut of *I. typographus* that are capable of the conversion of (1*R*)- and (1*S*)-cis-verbenol to verbenone and of (1*R*)-cis-verbenol to *trans*-verbenol (Leufvén et al. 1984) suggests a possible microbial role in the production and destruction of bark beetle pheromones. This is reinforced by the knowledge that microorganisms associated with the beetles can produce some of the terpene pheromones from the hydrocarbons (Brand et al. 1975, 1976; Chararas 1980). This view received some support from experiments with antibiotic-fed *I. paraconfusus* Lanier (Western pine beetle) and other pine beetle species (Chararas 1980) in which beetle-to-beetle attraction was slightly reduced, and from experiments with *I. paraconfusus* fed with streptomycin and reported to be unable to convert β-myrcene into ipsenol and ipsdienol (Fig. 13) although the conversion of α-pinene into cis-verbenol was not inhibited. In contrast, in an earlier report, Hughes (1973a,b) had reported that ingestion of terpene hydrocarbons is not necessary for their transformation as exposure of *D. ponderosae* to vapour or topical application of tree resins or individual terpenes led to their conversion to pheromones. He concluded that the gut was not a obligatory site of transformation.

Recent experiments with axenically reared *I. paraconfusus* and *D. ponderosae* (Hunt and Borden 1989; Gries et al. 1990) showed that they produced most of their complement of terpene alcohols in the absence of readily culturable microorganisms. An exception to this was a possible involvement of microorganisms in the production of ipsenol and ipsdienol by male *I. paraconfusus*. Hunt and Borden (1989) suggested that the more aggressive behaviour of *Dendroctonus* sp., exemplified by their ability to attack healthy resinous trees, may be associated with an ability to produce aggregation pheromones from host monoterpenes rapidly without reliance on microbial intervention. In contrast verbenone, the *anti*aggregation pheromone of *D. ponderosae* was not produced in quantifiable amounts by axenically reared or streptomycin-fed *D. ponderosae* and Hunt and Borden (1989) presented evidence for symbiotic yeasts in the gut of the beetle being responsible for its formation.

2. Botrytis cinerea, *monoterpenes and wines*

The fungus *Botrytis cinerea* is of particular interest to wine producers (Dittrich 1977), wine drinkers and terpenologists alike. Infection of unripe grapes with the organism results in spoilage of the crop while growth of the organism on ripe grapes (the noble rot) delivers sweet wines and is responsible for the degradation of the monoterpenes geraniol, linalool, nerol and α-terpineol observed in wines made from "botryized" Muscat grapes (Boidron 1978; Shimizu et al. 1982). In a broad review Schreier (1984) reported that monoterpenes are only one class of a wide range of flavour components of wines. The biotransformation of geraniol, linalool and nerol by *Botrytis cinerea* has been studied by Bock et al. (1986), Rapp et al. (1986) and Brunerie et al. (1987). Terpene metabolism and the range of products formed were strain-dependent, predominant metabolites were *E*-3,7-dimethyl-2-octen-1,8-diol and 2*Z*,6*E*-3,7-dimethyl-2,6-octan-1,8-diol from geraniol and nerol, and 2,6-dimethyl-1,8-octandiol and (*E*)-2,6-dimethyl-2-octen-1,8- diol from cintronellol. Other minor metabolites included the products of hydroxylation at carbon atoms 2 and 3 of linalool, ring-closed compounds and citronellic acid. A futher complicating factor is the report by Williams et al. (1980) that heating or mild acidification of grape juice results in the formation of furans and other cyclic products from polyols initially formed from linalool by the fungus.

Acknowledgements

All workers who have contributed to our growing understanding of this complex metabolic field are gratefully acknowledged. It is their efforts that have made this chapter possible. The SERC and Unilever are gratefully thanked for funds provided to support our work on bicyclic monoterpene metabolism.

References

Barras SJ and Perry TJ (1975) Interrelationships among microorganisms, bark or ambrosia beetles and woody host tissue: an annotated bibliography 1965–1974. U.S. Dept. Agric. For. Serv.

Gen. Tech. Rep. SO-10

Bauer K and Garbe D (1985) Common Fragrance and Flavour Materials, Preparation, Properties and Uses. Verlag Chemie, Weinheim

Benveniste P (1986) Sterol biosynthesis. In: WR Briggs, RL Jones and V Walbot (eds) Ann. Rev. Plant Physiol. 37: 275–308. Annual Reviews Inc. Palo Alto, CA

Best DJ, Floyd NC, Magalhaes A, Burfield A and Rhodes PM (1987) Initial steps in the degradation of alpha-pinene by *Pseudomonas fluorescens* NCIMB 11671. Biocatalysis 1: 147–159

Bigersson G, Schlyter F, Lofqvist J and Bergstrom G (1984) Quantitative variation of pheromone components in the spruce bark beetle *Ips typographus* from different attack phases. J. Chem. Ecol. 10: 1029–1054

Bock G, Benda I and Schreier P (1986) Biotransformation of linalool by *Botrytis cinera*. J. Food Sci. 51: 659–662

Boidron JN (1978) Relation between terpinic substrates and grape quality (Role of *Botrytis cinerea*). Ann. Technol. Agric. 27: 141–145

Borden JH (1990) Practical applications of insect pheromones and other attractants. In: R Ridgeway, RM Silverstein and M Iscoe (eds) Behavior Modifying Chemicals for Insect Management (pp 281–315). Marcel Dekker, New York

Bradshaw WH, Conrad HE, Corey EJ, Gunsalus IC and Lednicer D (1959) Microbial degradation of (+)-camphor. J. Amer. Chem. Soc. 81: 5507

Brand JM, Bracke JW, Markovetz AJ, Wood DL and Browne LE (1975) Production of verbenol pheromone by a bacterium isolated from bark beetles. Nature 254: 136–137

Brand JM, Bracke JW, Britton LN, Markovetz AJG and Barras SJ (1976) Bark beetle pheromones: production of verbenone by a mycangial fungus of *Dendroctonus frontalis*. J. Chem. Ecol. 2: 195–199

Brunerie P, Benda J, Bock G and Schreier P (1987) Bioconversion of citronellol by *Botrytis cinerea*. Appl. Microbiol. Biotechnol. 27: 6–10

Byers JA (1983) Bark beetle conversion of a plant compound to a sex specific inhibitor of pheromone attraction. Science 220: 624–626

Cain RB (1961) The metabolism of protocatechuic acid by a *Vibrio*. Biochem. J. 79: 298–312.

Carlsson B, Erdtman H, Frank A and Harvey WE (1952) The chemistry of natural order *Cupressales* VIII. Heartwood constituents of *Chamaecyparis nootkatensis* – carveol, nootkatin and chamic acid. Acta Chem. Scand. 6: 690–696

Carman RM, MacRae IC and Perkins MV (1986) The oxidation of 1,8-cineole by *Pseudomonas flava*. Aust. J. Chem. 39: 1739–1746

Chapman PJ, Meerman G, Gunsalus IC, Srinivasan R and Rinehart KL (1966) A new acyclic metabolite in camphor oxidation. J. Amer. Chem. Soc. 88: 618–619

Chararas C (1980) Physiologie des invertebres – attraction primaire et secondaire chez trois espèces de *Scolytidae* (*Ips*) et méchanisme de colonisation. CR Acad. Sci. Paris Ser. D 290: 375–378

Conrad HE, Dubus R and Gunsalus IC (1961) An enzyme system for cyclic ketone lactonization. Biochem. Biophys. Res. Commun. 6: 295–297

Conrad HE, Dubus R, Namtvedt MJ and Gunsalus IC (1965a) Mixed function oxidation II. Separation and properties of the enzymes catalysing camphor lactonization. J. Biol. Chem. 240: 495–503

Conrad HE, Leib K, and Gunsalus IC (1965b) Mixed function oxidation III. An electron transport complex in camphor lactonization J. Biol. Chem. 240: 4029–4037

Croteau R, Satterwhite DM, Wheeler CJ and Felton NM (1989) Biosynthesis of monoterpenes: stereochemistry of the enzymatic cyclization of geranyl pyrophosphate to (+)-α-pinene and (−)-α-pinene. J.Biol. Chem. 264: 2075–2080

Dhavalikar RS and Bhattacharyya PK (1966) Microbialtransformations of terpenes: Part VII – fermentation of limonene by a soil pseudomonad. Indian J. Biochem. 3: 144–157

Dhavalikar RS, Rangachari PN and Bhattacharyya PK (1966) Microbial transformations of terpenes: Part IX – pathways of limonene degradation by a soil pseudomonad. Indian J.

Biochem. 3: 158–164

Dittrich HH (1977) Mikrobiologie des Weines. Ulmer, Stuttgart

Draczynska B, Cagara CZ, Siewinski A, Rymkiewicz A, Zabza A and Leufvén A (1985) Biotransformation of pinenes XVII. Transformation of α- and β-pinenes by means of *Armillariella mellae* (honey fungus), a parasite of woodlands. J. Basic Microbil. 8: 487–492

Erdtman H, Harvey WE and Topliss JG (1956) The chemistry of natural order cupressales XVI. Heartwood constituents of *Chamaecypris nootkatensis* (Lamb) spach. The structure of chamic and chaminic acids. Acta Chem. Scand. 10: 1381–1392

Franke-Grossman H (1967) Ectosymbiosis in wood-inhabiting insects. In: SM Henry (ed) Symbiosis, Vol II: Association of Invertebrates, Birds, Ruminants and other Biota (pp 141–163). Academic Press, London

Friedemann T and Haugen GE (1943) Pyruvic acid II. The determination of ketoacids in blood and urine. J. Biol. Chem. 147: 415–422

Gelb MH, Malkonen P and Sligar SG (1982) Cytochrome P450$_{cam}$ catalysed epoxidation of dehydrocamphor. Biochem. Biophys. Res. Commun. 104: 853

Gensler WJ and Solomon PH (1973) Synthesis of chaminic acids. J. Org. Chem. 38: 1726–1730

Gibbon GH and Pirt SJ (1971) Degradation of α-pinene by *Pseudomonas* PX1. FEBS Lett. 18: 103–105

Gibbon GH, Millis NF and Pirt SJ (1972) Degradation of α-pinene by bacteria. In: G Terui (ed) Fermentation Technology Today (pp 609–612). Proceedings of the 4th International Fermentation Symposium. Osaka, Japan

Goswami A, Steffek RP, Liu W-G and Rosazza JPN (1987) Microbial reductions of enantiomeric 2-oxo-1,4-cineoles Enz. Microb. Technol. 9: 521–525

Gries G, Leufvén A, Lafontaine JP, Pierce HD, Borden JH, Vanderwel D and Oehlschlager AC (1990) New metabolites of α-pinene produced by the mountain pine beetle, *Dendroctonus ponderosae* (*Coleptera*: *Scolytidae*). Insect Biochem. 20: 365–371

Griffiths ET, Bociek SM, Harries PC, Jeffcoat R, Sissons DJ and Trudgill PW (1987a). Bacterial metabolism of α-pinene: pathway from α-pinene oxide to acyclic metabolites in *Nocardia* sp. strain P18.3. J. Bacteriol. 169: 4972–4979

Griffiths ET, Harries PC, Jeffcoat R and Trudgill PW (1987b) Purification and properties of α-pinene oxide lyase from *Nocardia* sp. strain P18.3. J. Bacteriol. 169: 4980–4983

Guengerich FP (1991) Reactions and significance of cytochrome P-450 enzymes. J. Biol. Chem. 266: 10019–10022

Gunsalus IC and Lipscomb JD (1973) Structure and reactions of a microbal monooxygenase: the role of putidaredoxin. In: W Lovenberg (ed) Iron-Sulfur Proteins, Vol 1 (pp 151–171). Academic Press, New York

Gunsalus IC and Marshall VP (1971) Monoterpene dissimilation: chemical and genetic models. CRC Crit. Rev. Microbiol. 1: 291–310

Gunsalus IC, Meeks JR, Lipscomb JD, Debrunner P and Munck E (1974) Bacterial monooxygenases – the P450 cytochrome system. In: O Hayaishi (ed) Molecular Mechanisms of Oxygen Activation (pp 559–613). Academic Press, New York

Harries PC, Jeffcoat R, Griffiths ET and Trudgill PW (1988) Monoterpene aldehyde or alcohol derivatives and their uses as perfumes and flavouring agents. European Patent No 85202335.5

Hendry LB, Piatek B, Browne LE, Wood DL, Byers JA, Fish RH and Hicks RA (1980) In vivo conversion of a labelled 'host plant' chemical to pheromones of the bark beetle *Ips paraconfusus*. Nature 284: 485

Hughes PR (1973a) *Dendroctonus*: production of pheromones and related compounds in response to host monoterpenes. Z. Angew. Entomol. 73: 294–312

Hughes PR (1973b) Effect of α-pinene exposure on *trans*-verbenol synthesis in *Dendroctonus ponderosae* Hopk. Experientia 60: 261–262

Hunt DWA and Borden JH (1989) Terpene alcohol pheromone production by *Dendroctonus ponderosae* and *Ips paraconfusus* (*Coleoptera*: *Scoltidae*) in the absence of readily identifiable microorganisms. J. Chem. Ecol. 15: 1433–1465

Kovats ES (1963) Zur kenntnis des sog. (destillierten) limetten-ols (*Citrus medica*, L., var acida,

Brandis; *Citrus aurantifolia*, Swingle). Helv. Chim. Acta 46: 2705–2731

Leufvén A, Bergstrom G and Falsen E (1984) Interconversion of verbenols and verbenone by indentified yeasts isolated from the spruce bark beetle *Ips typographus*. J. Chem. Ecol. 10: 1349–1361

Light HJ and Cosica CJ (1978) Cytochrome P-450 LM2-mediated hydroxylation of monoterpene alcohols. Biochemistry 17: 5638–5646

MacRae IC, Alberts V, Carman RM and Shaw IM (1979) Products of 1,8-cineole oxidation by a pseudomonas. Aust. J. Chem. 32: 917–922.

Madayastha KM and Murthy NSRK (1988) Transformation of acetates of citronellol, geraniol and linalool by *Aspergillus niger*: regiospecific hydroxylation of citronellol by a cell-free system. Appl. Microbiol. Biotechnol. 28: 324–329

Madziara-Borusiewick K and Strzelecka H (1977) Conditions of spruce (*Picea excelsa* lk.) infestation by the engraver beetle (*Ips typographus*) in mountains of Poland. Z. Ang. Ent. 4: 409–415

Nishimura H, Paton DM and Calvin M (1980) *Eucalyptus radiata* oil as renewable biomass. Agric. Biol. Chem. 44: 2495–2496

Nishimura H, Noma Y and Mizutani J (1982) *Eucalyptus* as biomass. Novel compounds from microbial conversion of 1,8-cineole. Agric. Biol. Chem. 46: 2601–2604

Noma Y and Nomomura S (1974) Conversion of $(-)$-carvone and $(+)$-carvone by a strain of *Aspergillus niger*. Agric. Biol. Chem. 38: 741–744

Ougham HJ, Taylor DG and Trudgill PW (1983) Camphor revisited: involvement of a unique monooxygenase in the metabolism of 2-oxo-δ^3-4,5,5-trimethylcyclopentenylacetic acid by *Pseudomonas putida*. J. Bacteriol. 153: 140–152

Payne GB (1984) Preparation of 2-*exo*-hydroxy-7-oxabicyclo(2.2.1-) heptanes. U.S. Patent No 4,487,945

Penfold AR and Morrison FR (1927) Occurance of a number of varieties of *Eucalyptus dives* as determined by chemical analysis of their essential oils. J. Proc. Royal Soc. New South Wales 61: 54–67

Prema BR and Bhattacharyya PK (1962a) Microbiological transformation of terpenes II. Transformations of α-pinene. Appl. Microbiol. 10: 524–528

Prema BR and Bhattacharyya PK (1962b) Microbiological transformation of terpenes III. Transformation of some mono- and sesqui-terpenes. Appl. Microbiol. 10: 529–531

Rapp A, Mandery H and Niebergall H (1986) Neue Monoterpendiol in Traubenmost und Wein sowie in Kulturen von *Botrytis cinerea*. Vitis 25: 79–84

Rheinwald JG, Chakrabarty AM and Gunsalus IC (1973) A transmissable plasmid controlling camphor oxidation in *Pseudomonas putida*. Proc. Nat. Acad. Sci. U.S. 70: 885–889

Rhodes PM, Winskill N and Moore WJ (1983) Microbial process for the preparation of l-carvone. European Patent no. 82305540.5

Rosazza JPN, Steffens JJ, Sariaslani FS, Goswami A, Beale JM, Reeg S and Chapman R (1987) Microbial hydroxylation of 1,4-cineole. Appl. Environ. Microbiol. 53: 2482–2486

Schreier P (1984) Formation of wine aroma. In: L Nykanen and P Lehtonen (eds) Flavour Research of Alchoholic Beverages, Vol 3 (pp 9–37). Foundation for Biotechnological and Industrial Fermentation

Shimizu J, Uehara M and Watanabe M (1982) Transformation of terpenes in grape must by *Botrytis cinerea*. Agric. Biol. Chem. 46: 1339–1344

Shukla OP and Bhattacharyya PK (1968) Microbial transformation of terpenes: Part IX – pathways of degradation of α- and β-pinenes in a soil pseudomonad (PL-strain). Indian J. Biochem. 5: 92–101

Shukla OP, Moholay MN and Bhattacharyya PK (1968) Microbial transformations terpenes: Part X – fermentation of α- and β-pinenes by a soil pseudomonad (PL-strain). Indian J. Biochem. 5: 79–91

Sligar SG and Murrey RI (1986) Cytochrome P-450$_{cam}$ and other bacterial cytochrome P-450-enzymes. In: PR Oritz de Montellano (ed) Cytochrome P-450: Structure, Mechanism and Biochemistry (pp 429–503). Plenum, New York

Stumpf B, Wray V and Kieslich K (1990) Oxidation of carenes to chaminic acids by *Mycobacterium smegmatis* DSM 43061. Appl. Microbiol. Biotechnol. 33: 251–254

Suhara K, Gunsalus K and Gunsalus IC (1985) P-450$_{cam}$ induced by p-cymene. Proc. 58th Annu. Meeting Japan. Biochem. Soc.

Taylor DG and Trudgill PW (1986) Camphor revisited: studies of 2,5-diketocamphane 1,2-monooxygenase from *Pseudomonas putida* ATCC 17453. J. Bacteriol. 165: 489–497

Trickett JM, Hammonds EJ, Worrall TL, Trower MK and Griffin M (1991) Characterisation of cyclohexane hydroxylase; a three-component enzyme system from a cyclohexane-grown *Xanthobacter* sp. FEMS Microbiol. Lett. 82: 329–334

Trudgill PW (1984) Microbial degradation of the alicyclic ring. In: DT Gibson (ed) Microbial Degradation of Organic Compounds (pp 131–180). Marcel Dekker, New York

Trudgill PW (1986) Terpenoid metabolism by Pseudomonas. In: JR Sokatch (ed) The Bacteria, a Treatise on Structure and Function, Vol X (pp 483–525). Academic Press, New York

Trudgill PW (1990) Microbial metabolism of monoterpenes – recent developments. Biodegradation 1: 93–105

Trudgill PW, Dubus R and Gunsalus IC (1966a) Mixed function oxidation V. Flavin interaction with a reduced diphosphopyridine nucleotide dehydrogenase, one of the enzymes participating in camphor lactonization. J. Biol. Chem. 241: 1194–1205

Trudgill PW, Dubus R and Gunsalus IC (1966b) Mixed function oxidation VI. Purification of a tightly coupled electron transport complex in camphor lactonization. J. Biol. Chem. 241: 4288–4290

Tudrozen NJ, Kelly DP and Millis NF (1977) α-Pinene metabolism by *Pseudomonas putida*. Biochem. J. 168: 312–318.

Ullah AJH, Murray RI, Bhattacharyya PK, Wagner GC and Gunsalus IC (1990) Protein components of a cytochrome P-450 linalool 8-methyl hydroxylase. J. Biol. Chem. 265: 1345–1351

Unger BP, Sliger SG and Gunsalus IC (1986) Pseudomonas cytochrome P-450. In: JR Sokatch (ed) The Bacteria, a Treatise on Structure and Function, Vol X (pp 557–589). Academic Press, New York

Wallach O (1895) Zur constitutionsbestimmung des terpineols. Chem. Ber. 28: 1755–1777

Warburton EJ, Magor AM, Trower MK and Griffin M (1990) Characterisation of cyclohexane hydroxylase; involvement of a cytochrome P-450 system from a cyclohexane-grown *Xanthobacter* sp. FEMS Microbiol. Lett. 66: 5–10

White RE and Coon MJ (1980) Oxygen activation by cytochrome P-450. Ann. Rev. Biochem. 49: 315–356

Williams PJ, Strauss CR and Wilson B (1980) Hydroxylated linalool derivatives as precursors of volatile monoterpenes of muscat grapes. J. Agric. Food Chem. 28: 766–771

Williams DR, Trudgill PW and Taylor DG (1989) Metabolism of 1,8-cineole by a *Rhodococcus* species: ring cleavage reactions. J. Gen. Microbiol. 135: 1957–1967

Yamazaki Y, Hayashi Y, Hori N and Mikami Y (1988) Microbial conversion of β-myrcene by *Aspergillus niger*. Agric. Biol. Chem. 52: 2921–2922

Yphantis DA (1964) Equilibrium ultracentrifugation of dilute solutions. Biochemistry, 3: 297–317

3. Formation and function of biosurfactants for degradation of water-insoluble substrates[1]

ROLF K. HOMMEL

Institut für Biochemie, FB Biowissenschaften der Universität, Talstr. 33, 0-04103 Leipzig, Germany

[1] Dedicated to Professor Dr H.-P. Kleber on the occasion of his 55th birthday.

I. Introduction

The degradation of water-immiscible substrates has been intensively studied over the last two decades. Many reports have sought answers to questions of how microorganisms adapt to enable them to metabolize such strongly hydrophobic substrates as n-alkanes. Enzymes involved in the respective metabolisms, their encoding and aspects of their regulation, as well as morphological alterations caused by alkane substrate, are well documented and are discussed in Chapter 1. This interest on alkane metabolism originates partly from commercial applications (Bühler and Schindler 1984; Ratledge 1988). Additionally, increasing environmental pollution by petroleum, its composites or oily derivatives are a new challenge to maintain the environment by biological means. The slow degradation of petroleum composites originates in their chemically inert structure, hydrophobicity and partial toxicity and thereof, special complex mechanisms of cell adaptations have been demanded.

Whereas, degradative aspects are well understood the key mechanisms of alkane uptake are still unclear processes. Growth of various alkane-degrading microorganisms is accompanied by appearance of low-molecular, extracellular and cell-bound compounds of definite structure which possess a hydrophilic and a hydrophobic moiety. In supernatants of other cultivations, large complexes of carbohydrates, proteins and lipids, as bioemulsifiers have been reported. These substances have been partly purified, structurally elucidated and characterized in some aspects. Mainly based on their properties to reduce surface tension of the supernatant or to emulsify alkanes or to stabilize water-alkane emulsions, it was generally assumed that biosurfactants or bioemulsifiers will play an essential role in alkane degradation by facilitating the uptake or are constituents of this process. However, the physiological role of each compound itself has to be considered individually. Beside a brief survey on bioemulsifiers this paper will mainly focus on biosurfactants produced mainly by alkane-degrading microorganisms, their structure, biosynthesis, production kinetics and aspects of participation in alkane degradation.

63

C. Ratledge (ed.), Biochemistry of Microbial Degradation, 63–87.

II. Structures of extracellular amphiphiles

The presence of both a hydrophilic and a lipophilic moiety is recognized within the biosurfactant molecule of definite molecule structure which gives the typical properties of the surfactants (cf. Table 1). This results in interactions at liquid/liquid, liquid/solid and liquid/gaseous interfaces and interactions of the molecules with itself forming aggregates such as micelles or reversed micelles. This differentiates these molecules from other extracellular compounds like biopolymers which generally do not reduce interfacial tension but may prevent oil droplets from coalescing.

Table 1. Selected properties of some structurally elucidated biosurfactants. (Reprinted from Hommel, 1990, with permission of Kluwer Academic publishers.)

Type	Charge		Location σ_s [mNm^{-1}]	CMC* [mgl^{-1}]	σ_i [mNm^{-1}]	Reference
Trehalose mycolates						Kretschmer et al. 1982
mono-	non-ionic	c.b.[1]	32	2	16	
di-	non-ionic	c.b.	36	4	17	
Sugar esters of mycolates						Li et al. 1984
Mannose	non-ionic	c.b.	40	5	19	
Glucose	non-ionic	c.b.	40	10	9	
Maltose	non-ionic	c.b.	33	1	1	
Rhamnolipids						Syldatk et al. (1985a)
R I	anionic	c.f.[2]	31	20	<1	
R II	anionic	c.f.	25	200	<1	
R III	anionic	c.f.	31	20	3	
R IV	anionic	c.f.	30	200	<1	
Sophorose lipids						
Lactone	non-ionic	c.f.	35	60	9	Stüwer et al. 1987
Mixture	non-ionic	c.f.	25		<0.9	Hommel et al. 1987
Acidic	anionic	c.f.			3	Lang et al. 1984
Surfactin	non-ionic	c.f.	27	5	1	Cooper and Zajic 1980

[1] Cell bound.

[2] Cell-free (extracellular).

* CMC = critical micelle concentration; σ_i = interfacial tension; σ_s = surface tension.

A. Bioemulsifiers

A typical representative of the bioemulsifiers is emulsan, excreted by *Acinetobacter calcoaceticus* RAG-1 grown on hydrocarbons or ethanol (recently reviewed by Gutnick et al. 1991). The high molecular weight complex consists of a polysaccharide, fatty acids and associated protein. The extracellular polysaccharides of the *Acinetobacter* strains BD4 and BD413 seem to represent partially degraded capsular constituents. Other polymeric bacterial bioemulsifiers display similar composition, e.g. the emulsifier of *Pseudomonas* PC1 (Reddy et al. 1983). The particulate vesicle forming complexes of *A. calcoaceticus* strains seem to be originated in the outer membrane (Käppeli and Finnerty 1979; Claus et al. 1984; Borneleit et al. 1988). The high molecular weight bioemulsifier from *Candida lipolytica*, liposan, (Cirigliano and Carman 1985) is composed of hetero-polysaccharide, 17% (w/w) protein and no lipid. Other emulsifiers from alkane-degrading yeasts also contain comparable ratios of carbohydrate and protein (Iguchi et al. 1969; Roy et al. 1979). Their composition is similar to the cell wall mannoproteins from *Saccharomyces cerevisiae*, *C.* (*Yarrowia*) *lipolytica* and other yeasts (Cameron et al. 1988). The emulsifiers isolated from *Bacillus cereus* grown on carbohydrates originated from the capsular layer and exhibited surface activity only when complexed with an excreted monoacylglycerol (Cooper and Goldenberg 1987). It was suggested that reported effects of bioemulsifiers on surface tension may base on similar aggregates.

B. Structures of biosurfactants

Biosurfactants can be divided according to whether their charge is either neutral, negative (free carboxylic residues) or positive (amine functions). This division also recognizes the character of the hydrophilic moiety of the molecule that is linked to the hydrophobic portion, usually the hydrocarbon tail of one or more saturated, unsaturated, hydroxylated or branched fatty acids. Biosurfactants therefore belong to either glycolipids, acylpolyols, lipopeptides (acylpeptides), fatty acids, phospholipids or neutral lipids. The spectrum of known structures is very broad and has been reviewed by numerous workers, for example by Haferburg et al. (1986) and Lang and Wagner (1987). The last three groups listed above cover conventional structures of compounds that are cited in most textbooks and so need not to be repeated here. The following sections will therefore focus on the glycolipids, acylpolyols and lipopeptides biosurfactants.

1. Glycolipids
In glycolipids, like in the rhamnose lipids, cellobiose lipids and sophorose lipids (Figs. 1 and 2), the sugar moiety and the fatty acid are linked glycosidically i.e. at the C-1 atom via an ether linkage. In sophorose lipids, restricted in appearance to some strains of the genus *Candida*, sophorose is linked at C-1 to a hydroxyfatty acid and may be acetylated in the 6 and/or 6' positions. The fatty acid moiety of the homogenous anionic lipid of *C. bogoriensis* is 13-hydroxydocosanoic acid (Fig. 1a) (Esders and Light 1972b). The other sophorosides contain ω- and mainly [ω−1]-hydroxylated fatty acids, with varied the chain length, degree of unsaturation, branching (within broad limits) depending on composition of the hydrophobic carbon source used (Tulloch

1976; Spencer et al. 1979) and appear (after growth on vegetable oil, e.g.) as oily, heavier-than-water mixtures of several sophorolipids. These mixtures of anionic or neutral (lactonized) sophorosides (Fig. 1b,c) display further differences in saturation and unsaturation of the fatty acid, types of lactonization (1,6'; 1,6''; 1,4'') and acetylation (6', 6''; 6'; 6'') (Tulloch 1976; Asmer et al. 1988; Weber et al. 1990). The number of individual sophorolipids may be decreased by addition of non-metabolized organic acids and amino acids providing two main lactonic lipids covering up to 95% of the crude crystalline glycolipid produced (Stüwer et al. 1987).

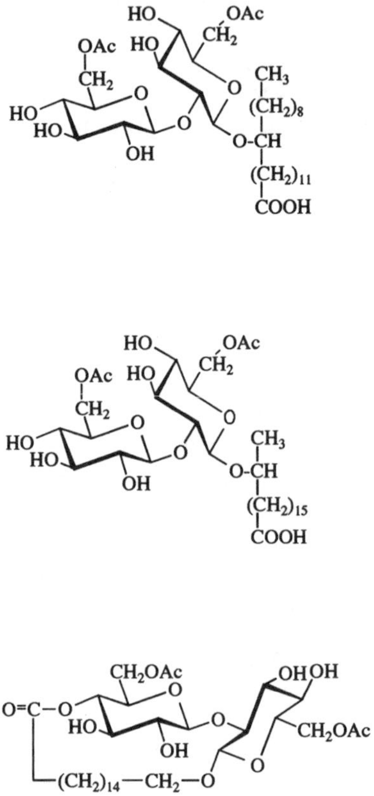

Fig. 1. Sophorose lipids from (a) *Candida bogoriensis*, (b) *Candida bombicola* (acidic from) and (c) *Candida apicola* (lactonized).

The anionic cellobiose lipids, in which β-D-cellobiose is esterified in 2' and 6 position with short-chain acids and at C-1 β-glycosidically linked to "ustilagic acids", 15,16-dihydroxy- or 2,15,16-trihydroxyhexadecanoic acid (Fig. 2b), were isolated from supernatants of the fungus *Ustilago maydis* PRL-617 after growth on glucose and vegetable oil (Bhattacharjee et al. 1970). Similar lipids

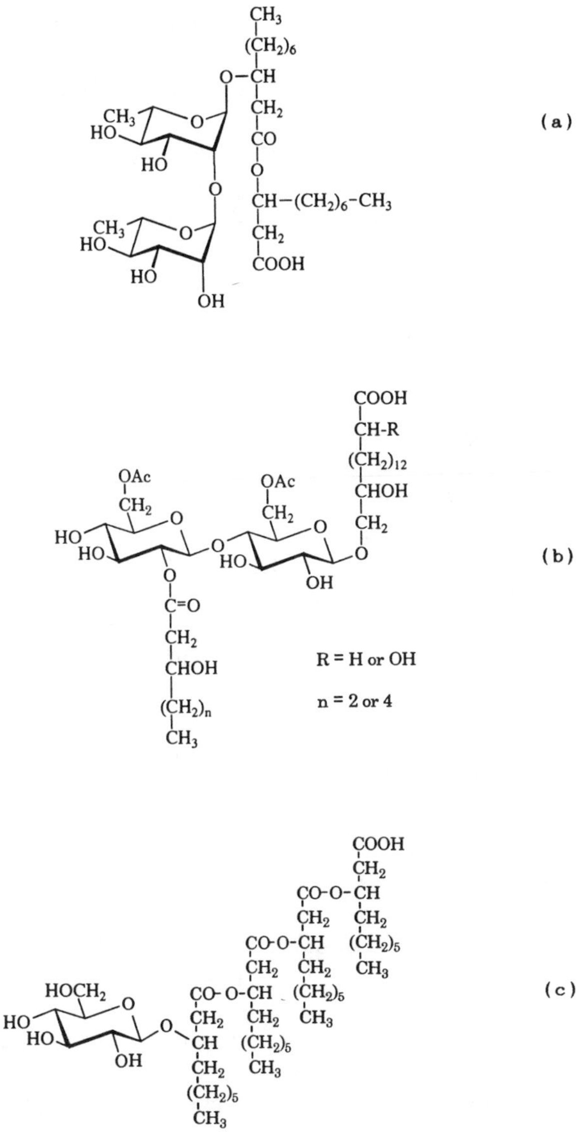

Fig. 2. Glycolipids from (a) *Pseudomonas aeruginosa*: rhamnolipid RH1, (b) *Ustilago maydis*: cellobiose lipid, and (c) *Alcaligenes* sp.: glucose-lipid.

have been reported by Frautz et al. (1986) of *U. maydis* ATCC 14826.

The sugar moiety of rhamnolipids produced by *Pseudomonas aeruginosa* strains consists of one or two rhamnose (6-deoxy-L-mannose) units glycosidically linked with one or two β-hydroxydecanoic acid units (Fig. 2a) (Syldatk et al. 1985a,b). With resting cells, altered combinations of the building blocks were described in formation of the more hydrophilic di-rhamnosyl mono-β-D-hydroxydecanoic acid and rhamnosyl mono-β-D-hydroxy-decanoic acid (cf. Fig. 6). Yamaguchi et al. (1976) reported a structurally different rhamnolipid in which an additional decenoyl moiety linked in 2' and 2'' position of the mono- and di-rhamnosyl unit, respectively.

The marine bacterium *Alcaligenes* sp. MM1 produced a glucose lipid consisting of a four fatty acid subunit of polyhydroxy β-hydroxydecanoate glycosidically linked with C-1 of glucose (Fig. 2c) (Schulz et al. 1991).

2. Acylpolyols

Acylpolyols predominantly appear as products of Actinomycetes. In the trehalose mycolates, trehalose is esterified in 6 and/or 6' position (Fig. 3a) with non-hydroxylated fatty acids and long-chain α-alkyl-β-hydroxylated fatty acids, termed mycolic acids. Additional esterification of trehalose hydroxy groups with short-chain fatty acids and/or mycolic acids are known: e.g. the trehalose 2,3,6'-trimycolate which only contains C_{62-74} polyunsaturated acids in *Rhodococcus aurantiacus* (Tomiyasu et al. 1986); α,α-trehalose-2,3,4,2'-tetraester (Fig. 3b) with two decanoyl, one octanoyl and one succinoyl residues in *Rhodococcus erythropolis* after growth on n-alkanes (Ristau and Wagner 1983), a lipid with the same residues, however, in 2,3,6,2' position in *Mycobacterium paraffinicum* grown on n-alkanes (Batrakov et al. 1981) or a succinoyl trehalose lipid produced by *R. erythropolis* (Uchida et al. 1989); and also a soil isolate, coryneform bacterium 51T7, produced a tetraester after being grown on hydrocarbons (Martin et al. 1991) with short chain fatty acid (C_{8-11}) substituents in 2,3,4,2' position. The tetraester of *Arthrobacter* sp. S11 has three fatty acids (C_8 to C_{14}) in 3,4,2' position and one succinoyl group in 2 position (Schulz et al. 1991; Passeri et al. 1991). The pentasaccharides of *Mycobacterium smegmatis* ATCC 356 contain two trehalose units one of which had two pyruvate ketal residues and fatty acids ranging from C_{14} to C_{22} (Saadat and Ballou 1983). A similar pentasaccharide linked to seven fatty acids and succinate was reported from *Nocardia corynebacteroides* SM1 (besides two trehalose mycolates) after growth on n-alkanes (Powalla et al. 1989). The α,α-trehalose unit of the pentasaccharide, suggested to be its biosynthetic precursor, is substituted in the same positions as the tetraester of *R. erythropolis*. Beside trehalose mycolates, a large number of other sugar mycolates were found both as cell wall components or extracellularly in Nocardiae and related microorganisms (Ioneda et al. 1981) and with *Arthrobacter* sp. DSM 2567 (Göbbert et al. 1988). *Arthrobacter paraffineus* KY 4303 and several, other related bacteria produced sucrose and fructose mono- and di-mycolates after growth on the respective sugars (Itoh and Suzuki 1974), whereas after growth on n-alkanes trehalose mono- and di-mycolates were obtained. Further acylpolyols are known: e.g. the diacyl mannosyl erythritol of *Schizonella melanogamma* (Deml et al. 1980) and those from different *Candida* sp. are shown in Fig. 3c (Kawashima et al. 1983).

Mixtures of different types of biosurfactants have been reported from

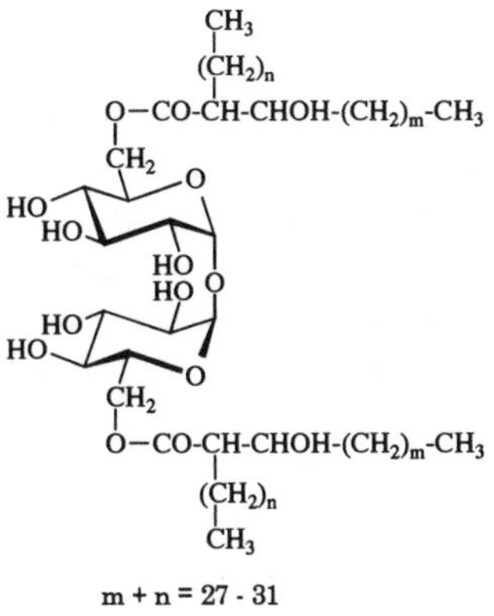

$m + n = 27 - 31$ (a)

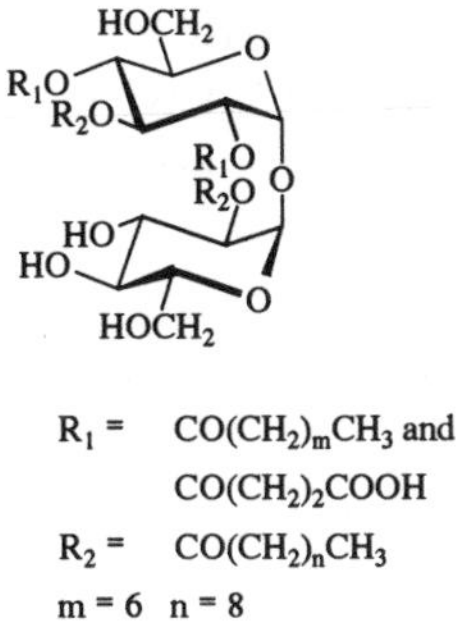

$R_1 = CO(CH_2)_mCH_3$ and
$CO(CH_2)_2COOH$
$R_2 = CO(CH_2)_nCH_3$
$m = 6 \quad n = 8$ (b)

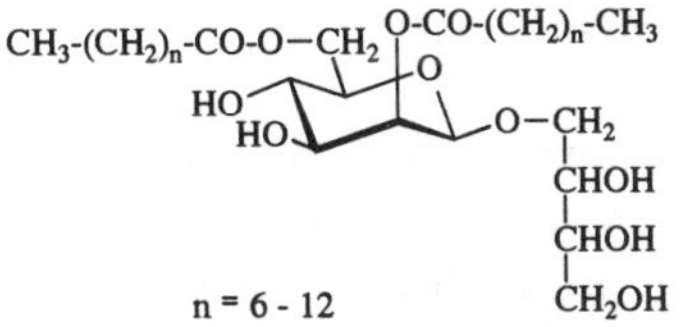

$CH_3-(CH_2)_n-CO-O-CH_2$ ... $O-CO-(CH_2)_n-CH_3$
$O-CH_2$
$CHOH$
$CHOH$
CH_2OH
$n = 6 - 12$ (c)

Fig. 3. Acylpolyols from *Rhodococcus erythropolis*: (a) trehalose dicorynomycolate, (b) trehalose tetraester, and from *Schizonella melanogamma*: diacetyl mannosyl erythritol.

Corynebacterium spp. and related bacteria mainly after growth on hydrocarbons and/or kerosene (MacDonald et al. 1981; Duvnjak and Kosaric 1985) which include free corynomycolic acids, lipopeptides, phospholipides, neutral lipids and fatty alcohols. Corynomycolic acids were characterized as containing a lipophilic backbone of some of these elucidated compounds including lipopeptides. A peptido-trehalose lipid excreted by the coryneform bacterium species H13A grown on alkanes has also been reported (Singer 1985).

3. Lipopeptides

Among biosurfactants, the acylpeptides or lipopeptides are one of the most interesting group. The main type of hydrophobic moiety is represented by different β-hydroxy- or amino-fatty acids of closely related isomers within a restricted number of carbon atoms (Vater 1986). Both the D- and the L-isomers of one amino acid may constitute the molecule. Although the amino acid sequence of the peptides reveals strongly conserved parts, the substitution of different amino acids is known without alteration of the conserved sequence of the configuration. Most of the known lipopeptides have been isolated from strains of the genus *Bacillus*, like surfactin (Fig. 4). The acylpeptides of *Bacillus* sp. may be divided into the iturin group (β-amino-fatty acid with an octa-amino acid ring having a sequence of configuration as L-Asx-D-Tyr-D-Asn-LLDL); the octapeptin and polymyxin group (with a ring linkage within the amino acid sequence – with mostly only 7 of the 8 to 10 amino acid residues of the peptide being involved – and with the occurrence of 2,4-diamino butyric acid) and the lactone group (in which there is an ester linkage between a free hydroxyl group of the fatty acid and the C-terminal amino acid, with 5 to 11 amino acids participating in lactone formation).

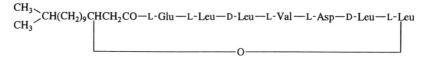

Fig. 4. Surfactin from *Bacillus subtilis*.

III. Biosynthetic routes to biosurfactants

Biosynthetic paths to biosurfactants include the formation of the hydrophilic and the lipophilic moiety of the molecule. These are obviously different and involve partly opposite metabolic routes mainly to carbohydrate containing biosurfactants. The division of metabolic routes suggested by Haferburg et al. (1986) and Syldatk and Wagner (1987) regarding the general biosynthetic principles reflects this behaviour:

a) both the hydrophilic and the lipid moiety are synthesized independently of the growth substrate (*de novo* synthesis);

b) the synthesis of the lipid moiety depends on the hydrophobic carbon source and is derived from it; the hydrophilic moiety is synthesized *de novo*;

c) the hydrophilic moiety reflects the carbon source used for growth or maintenance, the lipid moiety is synthesized *de novo*;

d) the synthesis of both residues depends on carbon substrates used.

Whereas, examples of the latter principle pathway are not known examples for the other three possibilities are given below. Detailed information of the enzymes involved in biosynthetic paths to biosurfactants, aspects of regulation and genetic determination are mainly restricted to some exceptional lipids.

A. Glycolipids

Growth experiments with alkanes, fatty acids and derivatives of different chain length have revealed direct incorporation of these compounds into the glycolipids of *Torulopsis* (*Candida*) strains (Tulloch 1976). The direct incorporation of alkanes makes two hydroxylation steps evident: the primary alkane hydroxylation and the hydroxylation at the opposite terminus in ω and/ or [ω−1]-position. Heinz et al. (1969, 1970) reported the hydroxylation of octadecanoate to 17-L-hydroxyoctadecanoic acid by following the incorporation of $^{18}O_2$ with *Torulopsis* sp. strain 319-67 and the NADPH-dependent conversion of oleic acid to 17-hydroxyoleic acid in a cell-free system of *T. bombicola* in presence of NADPH. Kleber et al. (1989) detected cytochrome P-450 in *T. apicola* IMET 43747 growing on glucose as well as, on hexadecane and on a mixture of both in the transition phase from the logarithmic to stationary phase of growth coinciding with the beginning of glycolipid production. Inhibition with cerulenin and NMR studies with ^{13}C-labelled hydrophobic substrates revealed a biphasic behaviour with regard to alkane hydroxylation in this strain (Hommel et al. 1990; Weber et al. 1991). It is suggested therefore that two different monooxygenase systems might be involved in the formation of hydroxy hexadecanoic acids. Selective ω-hydroxylation (initial alkane hydroxylation) is expressed more in the logarithmic phase than in stationary cells where (at least one) ω- and [ω−1]-hydroxylating monooxygenase systems must be induced. Both in *T. bombicola* ATCC 22214 and also in *T. apicola*, microsomal fatty alcohol oxidases that are not repressed in glucose grown cells have been reported, which are constitutive in the former and induced in the transition phase to the stationary growth phase in the latter strain, been reported (Hommel and Ratledge 1990; Hommel et al. 1993). These enzymes may participate in further oxidation of the fatty alcohol in conjunction with a microsomal fatty aldehyde dehydrogenase (Weber et al. 1991; Hommel et al. 1993) (Fig. 5). From these enzymatic and NMR studies, it is likely that the induction of the enzymes permitting the oxidation of the fatty alcohols and the diterminal hydroxylation will be the rate-limiting steps in the generation of hydroxyfatty acids in *T. apicola*. The comparison of glycolipid yields obtained on mixed substrate cultivations with alkanes or vegetable oils reveal significantly higher yields of sophorolipid production with the latter ones (Stüwer et al. 1987). This may be deduced to the higher energy demands of cells oxidizing alkanes rather than acylglycerols (cf. Fig. 5) which is in line with energetic considerations of sophorolipid production (Linton 1991). Similar kinetics of sophorose lipid production by strains of *T. bombicola* (e.g., Asmer et al. 1988; Cooper and Paddock 1984; Ito and Inoue 1982) suggest comparable pathways and regulations as in *T. apicola*. Both *T. apicola* and *T. bombicola* also produce sophorolipids when grown on water-soluble carbon sources (e.g., Göbbert et al. 1984; Hommel et al. 1987, 1990) via *de novo* synthesis of the fatty acid moiety and without replacing the sophorose unit. Under such conditions, the flux of glucose through glycolysis must allow both the *de novo* fatty acid

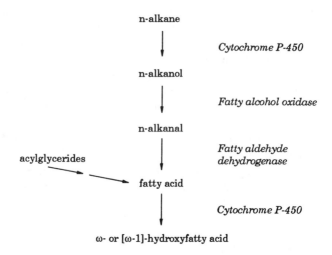

n-alkane

↓   *Cytochrome P-450*

n-alkanol

↓   *Fatty alcohol oxidase*

n-alkanal

acylglycerides   ↓   *Fatty aldehyde dehydrogenase*

fatty acid

↓   *Cytochrome P-450*

ω- or [ω-1]-hydroxyfatty acid

Fig. 5. Path of microsomal alkane oxidation to hydroxyfatty acids in *Candida apicola*. (Reprinted from Hommel and Ratledge 1992, by courtesy of Marcel Dekker Inc.)

synthesis and the synthesis of sophorose.

Two indistinguishable cytosolic glycosyltransferases responsible for establishment of glycosidic linkages in the sophorose lipid have been characterized in *C. bogoriensis* (Esders and Light 1972a,b; Breithaupt and Light 1982), which showed their maximal activities in the transition phase to stationary growth. Low glucose concentrations de-repressed both enzymes. The synthesis of the sophorolipid involves stepwise transfer of UDP-glucose, first to 13-hydroxydocosanoic acid to form 13-(β-D-glucopyranosyl)hydroxydocosanoic acid and finally to form 13-[2'-*O*-β-D-glucopyranosly-β-D-glucopyranosyl)oxy]docosanoate. In the same yeast, an acetyl-CoA-dependent acetyltransferase was reported (Bucholtz and Light 1976) catalysing the acetylation reaction in the 6' and 6'' position. For *T. bombicola* ATCC 22214 Asmer et al. (1988) suggested involvement of acetylation and lactonization steps in biosynthesis of the lactonic di-acetylated lipid.

Detailed studies concerning excretion and localization of the whole biosynthetic pathway and its regulation are not yet available. Recently, Rilke et al. (1992) documented the probable existence of periplasmic vesicles in *T. apicola* derived from the cytoplasmic membranes and which could be separated from the cytoplasmic membrane after cell protoplasting. In *Serratia marcescens*, biosurfactants have been reported to be main lipid components of extracellular vesicles (Matsuyama et al. 1986).

Glucose, fructose, glycerol were incorporated into both constituents of *P. aeruginosa* rhamnolipids but acetate was only incorporated into the fatty acid. TDP-L-Rhamnose, derived from TDP-D-glucose, acts as the precursor of rhamnolipids. The subsequent path of the *de novo* synthesis includes two different TDP-L-rhamnose specific rhamnosyl transferases, the TDP-L-rhamnose:β-hydroxdecanoyl-β-hydroxydecanoate rhamnosyl transferase and the TDP-L-rhamnose:L-rhamnose-β-hydroxdecanoyl-β-hydroxydecanoate rhamnosyl transferase, catalyzing the stepwise rhamnosyl transfer reaction (Fig. 6). The synthesis of the fatty acid proceeds via fatty acid synthetase complex and the acyl-CoA compounds were enzymatically linked (Burger et al. 1963). Recently, increased glutamine synthetase activity was reported to be directly correlated with enhanced rhamnolipid production (Mulligan and Gibbs 1989). Both phenomena were repressed by increased ammonium and glutamine concentrations indicating the control of rhamnolipid metabolism was linked to nitrogen metabolism. Mulligan et al. (1989) obtained a chloramphenicol tolerant mutant, *P. aeruginosa* ATCC 9027 var. RCII, which gave higher yields of rhamnolipid and exhibited biphasic growth. Whereas, the logarithmic phase was connected with amino acid catabolism, the stationary growth phase was linked with glucose metabolism. At the transition point of this reverse diauxie, both transhydrogenase and glucose-6-phosphate dehydrogenase activity were induced and rhamnolipid biosynthesis was initiated.

B. Acylpolyols

Trehalose 6-phosphate is synthesized by a synthetase reaction of α-D-glucopyranosyl units; in *R. erythropolis* UDP-glucose and glucose 6-phosphate were the precursors (Kretschmer and Wagner 1983). The tetraester of *R. erythropolis* (Ristau and Wagner 1983) was assumed to be the biosynthetic precursor of the pentasaccharide lipid comparable to those described in other bacteria (Saadat and Ballou 1983; Powalla et al. 1989). The biosynthetic pathway (Fig. 7) of the alkane-induced, trehalose mycolates of *R. erythropolis* (Kretschmer and Wagner 1983; Rapp et al. 1979) differs from the usual sugar-linked trehalose mycolate synthesis by the appearance of free intermediates. By comparison of individual *in vitro* enzyme activities, it was suggested that the condensation of the trehalose mycolate is the rate-limiting step. The addition of precursors, like glucose or free corynomycolates, improved biosynthetic productivity. In the tetraester of coryneform bacterium, 51T7, the short chain (C_{8-11}) fatty acids were derived by direct incorporation (C_{10}) or chain shortening of the hydrocarbon substrate (C_{11-16}) (Martin et al. 1991). The appearance of related sugar mono- and di-mycolates (Itoh and Suzuki 1974; Göbbert et al. 1988) after growth or incubating resting cells on the respective sugar was attributed to a membrane-bound transesterase catalysing the transfer of mono-, di- or tri-saccharides to mycolic acids in different

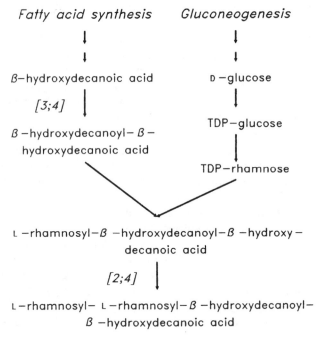

Fig. 6. Scheme of *de novo* synthesis of rhamnolipids by *Pseudomonas aeruginosa* towards RH1 (cf. Fig. 2a); inhibition of biosynthetic pathway providing other rhamnolipids are marked by numbers corresponding to the number of the appropriate rhamnolipid: 2: [RH2]-L-rhamnosyl-hydroxydecanoyl-hydroxydecanoic acid, 3: [RH3]-di-L-rhamnosyl-hydroxydecanoic acid, 4: [RH4]-L-rhamnosyl-hydroxydecanoic acid. (Reprinted from Hommel and Ratledge 1992, by courtesy of Marcel Dekker Inc.)

coryneform and related bacteria. With the alkane substrate, obviously the *de novo* synthesis of trehalose for mono- and di-mycolates is induced or dominating (Göbbert et al. 1988).

4-*O*-β-D-Mannosyl-D-erythritol was suggested as the direct precursor of the diacyl mannosylerythritol of *Schizonella melanogamma* and of different *Candida* sp. (Kobayashi et al. 1987) which then will become acylated when the cells grown on hydrophobic substrates.

C. Lipopeptides

The characteristic structural hydroxy- or amino-fatty acids of lipopeptides reflect the capability mainly of *Bacillus* sp. to synthesize branched-chain, hydroxylated fatty acids. The chain initiation for the synthesis of branched-chain fatty acids starts from acyl-CoA esters of endogenous precursors such as pyruvate or α-ketobutyrate and exogenous ones such as isoleucine, valine, isobutyrate, isovalerate which results in the respective iso- or the anteiso-derivatives of the appropriate chain length (Kaneda 1977). Recent reports on

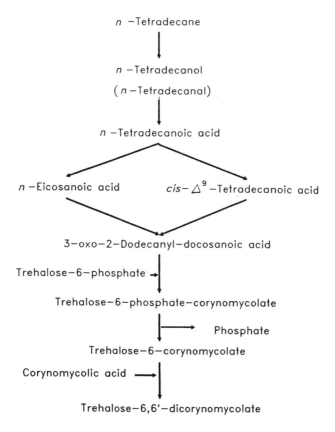

n −Tetradecane

↓

n −Tetradecanol

(*n* −Tetradecanal)

↓

n −Tetradecanoic acid

n −Eicosanoic acid   *cis*− △^{9} −Tetradecanoic acid

3−oxo−2−Dodecanyl−docosanoic acid

Trehalose−6−phosphate →|

Trehalose−6−phosphate−corynomycolate

|——→ Phosphate

Trehalose−6−corynomycolate

Corynomycolic acid ——→|

Trehalose−6,6'−dicorynomycolate

Fig. 7. Scheme of *de novo* synthesis of trehalose mono- and di-corynomycolates by *Rhodococcus erythropolis* on n-tetradecane. (Reprinted from Hommel and Ratledge 1992, by courtesy of Marcel Dekker Inc.)

biosynthesis of bacillomycin L and surfactin (Kluge et al. 1988) reveal a non-ribosomal peptide synthesis consisting with thio-template mechanism. Ullrich et al. (1991) partially purified a multienzyme complex from *B. subtilis* that catalysed ATP P_i-exchange reactions depending on the appropriate L- but not D-amino acids. The addition of L-Val or L-Ile to the culture media resulted in substituting the C-terminal amino acid Leu[7] by Val[7] in *B. subtilis* S499 (Peypoux et al. 1991; Peypoux and Michel 1992). The same strain also co-produced surfactin and iturin A (Sandrin et al. 1990). In iturin A, C_{14} or C_{15} β-amino acids were involved in the closing of the heptapeptide ring with an LDDLLDL-sequence of amino acids in contrast to surfactin with an LLDLLDL-sequence of different amino acids. Incorporation of D(−)β-hydroxyfatty acyl-CoA into surfactin, and forming the ester linkage of the lactone, is assumed to be formed in a similar way to peptide bond formation. In *B. subtilis* ATCC 21332 increased surfactin synthesis was accompanied with decreased isocitrate

ꞁtivity due to glucose repression in presence of citrate (de
Nakano et al. (1988) revealed the chromosomal location of
ꞁ. The transformation of the surfactin production locus, *sfp*, to
ꞁg strains established surfactin formation and *sfp*[0] strains were
ꞁ harbour surfactin biosynthesis genes. This phenotype was thought to
ꞁused by lack of complementary structural or appropriate regulator genes.

IV. Microbial growth and kinetics of biosurfactant formation

Uptake of, and growth of microorganisms on, alkanes have been strongly
connected with the ability to release surfactants in order to facilitate their
uptake (Boulton and Ratledge 1984; Cooper and Zajic 1980; Syldatk et al.
1984) as prerequisites of alkane metabolism. However, biosurfactants were not
exclusively produced when microorganisms grow on water-immiscible carbon
sources and likewise, not every water-insoluble carbon source induces
production of biosurfactants.

A. Models of alkane uptake

Basing on macroscopic observations three different models of alkane uptake
(borderline cases) have been proposed (for more details cf. Chapter 1) which
however, do not satisfactorily explain the passage over the cell wall and the
appropriate membranes into the cytoplasm. (i) Due to the low solubility of
alkanes, the uptake of monodispersed-dissolved alkanes may not explain the
observed specific growth rates on alkanes. This model may be valid for short-
chain hydrocarbons which however become increasingly toxic as the chain
length decreases below C_9. (ii) The increased hydrophobicity of the cell surface
induced by hydrocarbons is a prerequisite to establish direct contact of cells
with large oil drops. This has been proposed to occur with *Pseudomonas* M1
which has a strong capacity to adhere firmly to hydrocarbon phase (Goswami
and Singh 1991) and with *R. erythropolis* and *R. aurantiacus* growing in alkane
drops (Rapp et al. 1979; Ramsay et al. 1988) which then have a hydrophobic
cell wall by the synthesis of trehalose mycolates. Otherwise, alterations in
hydrophobicity of the cell envelope are not only connected with alkane
utilization. (iii) The interfacial properties of surface-active metabolites are the
basis of the model of contact with accommodated oil. Surfactants are able to
form micelles, microemulsions harbouring the pseudo-solubilzied alkane.
Emulsified alkanes are much faster assimilated by cells and surfactants
promote growth on lipophilic substrates varying from strain to strain and
depending on surfactants used. Comparable effects of biosurfactants will be
discussed in Section V. The model of direct interaction of alkane droplets with
the cell surface of *Pseudomonas oleovorans*, as proposed by Witholt et al.
(1990), is based on the physical extraction of lipopolysaccharides or other
related amphiphilic outer membrane components (glycoproteins or

lipoproteins, phospholipids, e.g.). This is then comparable to vesiculation by other bacteria (e.g. Borneleit et al. 1988) and to biosurfactant extraction from coryneform bacteria (e.g. Duvnjak and Kosaric 1985; Duvnjak et al. 1982). In each case, the cell surface molecules are used to make the alkane available for the cells (cf. also Section II.A).

B. Growth and biosurfactant production

Hisatsuka et al. (1971), Itoh et al. (1971) and Itoh and Suzuki (1972) documented the strong stimulatory effect of rhamnolipids on growth of *P. aeruginosa* on hydrocarbons. A rhamnolipid negative mutant only grew weakly on alkanes. Alkane assimilation could only be restored by adding rhamnolipid. A protein-like factor was shown to interact with rhamnolipid in emulsification of alkanes (Hisatsuka et al. 1977). In another *P. aeruginosa* strain, a rhamnose-containing peptidoglycolipid was strongly associated with the lag period and the exponential growth phase on hexadecane (Koronelli et al. 1983). Cameotra and Singh (1990) documented the necessity of an emulsifying factor which could be inhibited by EDTA for the growth of *P. aeruginosa* PG-1 on nonane and longer alkanes; shorter alkanes were assumed to be taken up from the gas phase. The pseudosolubilzation of alkane by extracellular macromolecular emulsifying factors was also shown for *Pseudomonas* N1 (Goswami and Singh 1991). Koch et al. (1991) isolated different transposon Tn5-GM-induced mutants of *P. aeruginosa* PG 201 that were unable to grow on hexadecane. The addition of small amounts of rhamnolipid restored growth and uptake of the alkane similar to the stimulatory effect reported by Zhang and Miller (1992). One of these mutants, 59C7, however, over-produced rhamnolipids in glucose-containing media. Whereas, in biosurfactant-producing strains, no cellular lipid accumulation (cf. Chapter 4 this book) was reported, so far the rhamnolipid-producing strain 44T1 reached up to 38% (w/w) lipids of its biomass both on glucose and on hydrophobic carbon sources (de Andres et al. 1991). The cellular fatty acid profiles revealed *de novo* lipid biosynthesis in cells grown on glucose but direct incorporation of the corresponding fatty acids when cells were grown on alkanes. Rhamnolipid production is not only linked to hydrophobic carbon sources (cf. e.g., Syldatk et al. 1985a,b; Haferburg et al. 1989). In batch cultures the production of rhamnolipids was not growth associated and was independent of type of carbon source being used (Syldatk et al. 1984). Limitations of one component of the growth medium were prerequisites for biosynthesis to commence. High C:N ratios and iron concentrations within a small concentration range promoted rhamnolipid production (Guerra-Santos et al. 1986; Haferburg et al. 1989; Syldatk et al. 1985a) allowing the production in both continuous cultures (Guerra-Santos et al. 1984; Reiling et al. 1986) and immobilized cells (Müller-Hurtig et al. 1987).

Only during growth on carbohydrates were the proteolipids of *Bacillus* spp. produced. Cooper (1984) reported the repression of surfactin synthesis by addition of hydrocarbons whereas growth was not affected. Excessive production of surfactin did not appear to be associated with actively growing cells; it started in the logarithmic growth phase and continued into the stationary phase (Cooper 1984; Vater 1986). The proteolipids of *Bacillus licheniformis* were co-produced in chemostat culture on glucose with small amounts of long-chain saturated fatty acids and small amounts of hydrocarbons with chain length of C_{20-22} being added (Jenny et al. 1991) without affecting surfactant production.

In cultivations of different corynebacteria and related bacteria on hydrocarbons or kerosene, the surface tension of the medium was lowered in the lag phase of growth (MacDonald et al. 1981). Growth was accompanied by distinct maxima in surface activity attributed to the sequential appearance of free corynomycolic acids, lipopeptides, phospholipides, neutral lipids and fatty alcohols. Hexadecane-grown cultures displayed the highest surface activities (Atkit et al. 1981) that were reduced by addition of glucose. *Corynebacterium lepus* and *Arthrobacter paraffineus* were also able to produce biosurfactants on glucose as carbon source. The appropriate products, however, remained cell-bound and could only be released by treatment the cells with hydrocarbons (Duvnjak and Kosaric 1985; Duvnjak et al. 1982). After growth on water-soluble carbon sources the peptido-trehalose lipid of the coryneform species, H13A, also remained cell-bound (Singer 1985). In alkane cultivations, it was though excreted in the stationary phase of growth.

R. erythropolis showed a biphasic growth pattern on alkanes (Rapp et al. 1979). The cells started to grow in the alkane droplets and cell-bound trehalose dimycolates were produced, from which only 10% were found in the culture medium. At a definite ratio of cells/glycolipid/alkane, the hydrophobic cells formed aggregates in the aqueous phase which coincided with growth limitation caused by diminished O_2 transfer. In contrast, biosurfactant production of *R. aurantiacus*, also growing in alkane droplets, was associated with initial growth at an exponential rate (Ramsay et al. 1988). A subsequent linear growth phase was attributed to limitations in hydrocarbon transport proceeding without alterations in O_2 uptake rate or in biosurfactant production. The synthesis of tetramycolates of *R. erythropolis* started, in contrast to that of the trehalose dimoycolates, after nitrogen depletion in the stationary growth phase (Ristau and Wagner 1983). Synthesis could also be initiated by limitations of metal ions or by lowering the growth rate. A biphasic behaviour of a trehalose lipid was also reported from a coryneform bacteria growing on hydrocarbons (Martin et al. 1991): 90% of the trehalose lipid were cell-associated during 80 h of cultivation, after that they were mainly liberated. The production of a pentasaccharide ester of *Nocardia corynebacteroides* was greater with alkanes as carbon source than with glucose and, moreover, continued into the stationary growth phase (Powalla et al. 1989).

Candida sp. B-7 produced an acylated mannosyl-erythritol lipid only when grown on alkanes or triacylglycerols (Kawashima et al. 1983). In both cases growth rates were strongly increased by addition of the lipid product. The non-acylated lipid that was formed by *Candida* sp. KSM-1529 grown on glucose was acylated in presence of hydrocarbons (Kobayashi et al. 1987). *C. antarctica* T-34 accumulated up to 12% of mannosyl-erythritol lipid in oil globules within the cell presumed as one of the storage material when grown on carbohydrates (Kitamoto et al. 1992). Recent isolates of *C. antarctica*, however, produced a structurally modified mannosyl-erythritol lipid only when grown on vegetable oils but not on hydrocarbons (Kitamoto et al. 1990).

The other *Candida* (*Torulopsis*) yeasts share a common feature of product formation: sophorolipid formation does not require the presence of a hydrophilic carbon source and is not growth-associated either on *n*-alkanes (Hommel et al. 1987), on carbohydrates (Göbbert et al. 1984; Hommel et al. 1987), or on mixtures of both (Cooper and Paddock 1984; Stüwer et al. 1987). The bulk biosurfactant production starts in the transition phase from exponential to stationary growth phase. With *C. bogoriensis*, lipid production was only reported with glucose as carbon source; high concentrations of glucose in the culture promoted glycolipid biosynthesis (Cutler and Light 1979). *T. bombicola* KSM-36 produced lactonic sophorolipids after growth on safflower oil and glucose but not on hydrocarbons as sole source of carbon (Ito and Inoue 1982). Highest yields of sophorose lipids have been obtained with *T. bombicola* ATCC 22214 and *T. apicola* IMET 43734 on mixtures of carbohydrates and hydrophobic carbon sources (vegetable oils, oleic acid, alkanes) (Cooper and Paddock 1984; Asmer et al. 1988; Stüwer et al. 1987). By adding sophorose lipid to cultivations of *T. bombicola* ATCC 2217 and strain KSM-36, both microorganisms could then utilize a number alkanes not normally assimilated. The specific growth rates and the biomass yields were enhanced which could not be achieved by adding synthetic surfactants (Ito and Inoue 1982). The stimulatory effect of sophorolipid addition to alkane utilization was generally restricted to the producing strain itself or related sophorolipid producers (Ito and Inoue 1982; Ito et al. 1980; Hommel et al. 1987; Hommel 1990). Ishigami et al. (1989) recorded increased membrane fluidity of hydrocarbon-assimilating *C. bombicola* ATCC 22214 after addition of corynomycolic acids suggesting mediated uptake of small alkane droplets. Other yeasts or bacteria tested, including the bioemulsifier-producing *C. lipolytica* and *Acinetobacter calcoaceticus*, were partly inhibited in presence of the lactonic sophoroside which did not emulsify hydrocarbons. Due to its inability to stabilize alkane-in-water emulsions *in vitro*, Cooper and Paddock (1984) excluded the lactonic sophorolipid as a stimulator of alkane uptake. Otherwise, the strong interfacial activities with *n*-alkanes (cf. Table 1) will enable this sophorolipid to form the observed fine, dispersed non-coalescing alkane droplets in cultivation broths of the producing strains which will facilitate alkane transport into the cell.

V. Biosurfactants in natural habitats

The marine organisms are of special interest regarding their potential oil pollutant degradation. Kirk and Gordon (1988) described the formation of alkane droplets by emulsifying agents produced by filamentous marine fungi which in turn surrounded the droplets and subsequently penetrated into by hyphae. The formation of an emulsifier by a marine denitrifying *Pseudomonas* sp. All in anaerobic and aerobic degradation of *n*-heptadecene has been reported (Gilewicz et al. 1991). The emulsifying activity increased throughout the exponential growth phase and reached a maximum at the end of this phase. Degradation of crude oil in sea water with a natural microbial population demonstrated the importance of emulsifying factors or surface active compounds (Mattei et al. 1986). Based on light microscopic observations of oil polluted sea water, Poremba et al. (1989) assumed the participation of both cell-bound and microemulsion-forming extracellular biosurfactants in the degradation of crude oil by marine bacteria. Recently, three marine isolates of hydrocarbon-degrading microorganisms were reported to produce a growth-associated cell-bound glucose lipid along with either a trehalose tetraester or a high molecular weight extracellular emulsifying agent (Schulz et al. 1991).

Soil bacteria can degrade hydrocarbons and produce biosurfactants (Aktit et al. 1981). In oil polluted soils, trehalose esters were produced by the natural soil population after water-soluble substrates had been exhausted resulting in degradation of alkanes (Oberbremer and Müller-Hurtig 1989; Oberbremer et al. 1990). The degradation of oil pollutants by soil bacteria was stimulated by the addition of biosurfactants; the adaptation phase was shortened and the extent of hydrocarbon degradation enhanced which coincided with *de novo* synthesis of surface-active glycolipids. The addition of the trehalose mycolate-producing *R. erythropolis* as a starter culture did not achieve the same effect as the purified trehalose dicorynomycolates (Goclik et al. 1990).

According to Syldatk et al. (1984), ionic biosurfactants should pseudo-solubilize the alkane by micelle formation by which the surface area will be increased; neutral biosurfactants (e.g. trehalose lipids) will be spread over the charged cell surface forming a hydrophobic film that will enable direct attachment to hydrocarbon droplets. Results of recent studies and the strong micelle-forming properties of many biosurfactants (Table 1) do not fully support this assumption.

Microbial alkane uptake and the role of biosurfactants (and also the extracellular high molecular weight emulsifiers) is a complex dynamic process. Ratledge (1988) proposed a hypothetic model of alkane uptake based on the interplay of biosurfactants, micellized or emulsified alkane and a specific amphiphatic receptor/channel. Both directly attached and solubilized alkanes may partition via a strong hydrophobic channel into cytoplasm. Action of biosurfactants may be retarded by hydrophobic interactions which should be specific for the individual microorganism explaining the mostly specific promoting effects of biosurfactants on growth and on mineralization of hydrophobic growth substrates of the producing microorganism itself. The extraction of small amounts of lipid molecules from the cell membranes by hydrocarbons (e.g., Cooper and Goldenberg 1987; Duvnjak et al. 1982; Duvnjak and Kosaric 1985; Witholt et al. 1990) may also allow the pseudo-solubilization of the carbon source due to very low critical micelle concentrations (cf. Table 1).

The question whether the production of biosurfactants is a prerequisite for, or result of, alkane uptake or both, or has other evolutionary roots, must be answered in the complexity of microbial life and for each individual microorganism itself. Any explanation must also consider the observed antimicrobial properties of many biosurfactants (Haferburg et al. 1986; Lang et al. 1989) as well as detailed knowledge of biosynthetic paths and their regulatory scheme which for some strains resemble secondary metabolites. The mostly non-growth associated biosurfactant formation and dependence on nutrient limitation indicates that the enzymes involved in biosurfactant biosynthesis are normally repressed in actively growing cells.

In summary, a general definition of the real physiological role of biosurfactants cannot yet be given. The empirically collected data, however, support the assumption that biosurfactants play an important role in alkane utilization not only of the producing strain itself but also of other microorganisms in natural habitats wherein biosurfactants may be an additional biological tool to defend man-made oil pollutions in soil and aqueous media.

Acknowledgements

The author wishes to thank the diploma students, Silke Stegner, Olaf Rilke, Dirk Lassner, and the PhD students, Olaf Stüwer, Annette Baum, Simone König, who spent their time for studying the complex behaviour of biosurfactants as well as Dr L. Weber, Dr A. Weiss and Dr K. Huse for successful collaboration. Additional thanks is contributed to Prof. Dr H.-P. Kleber, for supporting and guiding research work, Prof. Dr C. Ratledge (Hull) for many fruitful ideas as well as Prof. Dr F. Wagner and Dr S. Lang (both Braunschweig) for joint discussions.

References

Asmer HJ, Lang S, Wagner F and Wray V (1988) Microbial production, structure elucidation and bioconversion of sophorose lipids. J. Am. Oil. Chem. Soc. 65: 1460–1466

Atkit J, Cooper DG, Manninen KI and Zajic JE (1981) Investigation of potential biosurfactant production among phytopathogenic corynebacteria and related soil microbes. Current Microbiol. 6: 145–150

Batrakov SG, Rozynov BV, Koronelli TV and Bergelson DL (1981) Two novel types of trehalose lipids. Chem. Phys. Lipids. 29: 241–266

Bhattacharjee SS, Haskins RH and Gorin PAJ (1970) Location of acyl groups on two partly acylated glycolipids from strains of *Ustilago* (smut fungi). Carbohydrate Res. 13: 235–246

Borneleit P, Hermsdorf T, Claus R, Walther P and Kleber H-P (1988) Effect of hexadecane-induced vesiculation on the outer membrane of *Acinetobacter calcoaceticus*. J. Gen. Microbiol. 134: 1983–1992

Boulton CA and Ratledge C (1984) The physiology of hydrocarbon-utilizing microorganisms. In: A Wiseman (ed) Topics Enz. Ferment. Biotechnol., Vol 9 (pp 11–77). John Wiley and Sons, New York, Chichester, Brisbane, Toronto

Breithaupt TB and Light RJ (1982) Affinity chromatography and further characterization of the glycosyltransferases involved in hydroxydocosanoic acid sophoroside production in *Candida bogoriensis*. J. Biol. Chem. 257: 9622–9628

Bucholtz ML and Light RJ (1976) Acetylation of 13-sophorosyloxydocosanoic acid by an acetyltransferase purified from *Candida bogoriensis* J. Biol. Chem. 252: 424–430

Bühler M and Schindler J (1984) Aliphatic hydrocarbons. In: H-J Rehm and G Reed (eds) Biotechnology, Vol 6a (pp 329–385). Verlag Chemie, Weinheim

Burger MM, Glaser L and Burton RM (1963) The enzymatic synthesis of a rhamnose-containing glycolipid by extracts of *Pseudomonas aeruginosa*. J. Biol. Chem. 238: 2595–2602

Cameotra SS and Singh HD (1990) Uptake of volatile *n*-alkanes by *Pseudomonas* PG-1. J. Biosci. 15: 313–322

Cameron DR, Cooper DG and Neufeld RJ (1988) The mannoprotein of *Saccharomyces cerevisiae* is an effective bioemulsifier. Appl. Environ. Microbiol. 54: 1420–1425

Cirigliano MC and Carman GM (1985) Purification and characterization of liposan, a bioemulsifier from *Candida lipolytica*. Appl. Environ. Microbiol. 50: 846–850

Claus R, Käppeli O and Fiechter A (1984) Possible role of extra-cellular membrane particles in hydrocarbon utilization by *Acinetobacter calcoaceticus* 69-V. J. Gen. Microbiol. 130: 1035–1039

Cooper DG (1984) Unusual aspects of biosurfactant production. In: C Ratledge, P Dawson and J Rattray (eds) Biotechnology for the Oils and Fats Industry. Am. Oil. Chem. Soc. Monograph no 11 (pp 281–287). Am. Oil. Chem. Soc., Champaign, Illinois

Cooper DG and Goldenberg BG (1987) Surface-active agents from two *Bacillus* species. Appl. Environ. Microbiol. 53: 224–229

Cooper DA and Paddock DA (1984) Production of a biosurfactant from *Torulopsis bombicola*. Appl. Environ. Microbiol. 47: 173–176

Cooper DA and Zajic JE (1980) Surface active compounds from microorganisms. Adv. Appl. Microbiol. 26: 229–253

Cutler AJ and Light RJ (1979) Regulation of hydroxydocosanoic acid sophoroside production in *Candida bogoriensis* by the levels of glucose and yeast extract in the growth medium. J. Biol. Chem. 254: 1944–1950

de Andres C, Espuny MJ, Robert M, Mercade ME, Manresa A and Guinea J (1991) Cellular lipid accumulation by *Pseudomonas aeruginosa* 44T1. Appl. Microbiol. Biotechnol. 35: 813–816

Deml G, Anke T, Oberwinkler F, Gianetti BM and Steglich W (1980) Schizonellin A and B, new glycolipids from Schizonella melanogramma. Phytochemistry 19: 83–87

de Roubin MR, Mulligan CN and Gibbs BF (1989) Correlation of enhanced surfactin production with decreased isocitrate dehydrogenase activity. Can. J. Microbiol. 35: 854–859

Duvnjak Z and Kosaric N (1985) Production and release of surfactant by *Corynebacterium lepus* in hydrocarbon and glucose media. Biotechnol. Lett. 7: 793–796

Duvnjak Z, Cooper DG and Kosaric N (1982) Production of surfactant by *Arthrobacter paraffineus* ATCC 19558. Biotechnol. Bioeng. 24: 165–175

Esders TW and Light RJ (1972a) Glycosyl- and acety-transferases involved in the biosynthesis of glycolipids from *Candida bogoriensis*. J. Biol. Chem. 247: 1375–1386

Esders TW and Light RJ (1972b) Characterization and in vivo production of three glycolipids from *Candida bogoriensis*. 13-Glucopyranosylglucopyranosyloxydocosanoic acid and its mono- and di-acetylated derivatives. J. Lipid Res. 13: 663–671

Frautz B, Lang S and Wagner F (1986) Formation of cellobiose lipids by growing and resting cells of *Ustilago maydis*. Biotechnol. Lett. 11: 757–762

Gilewicz M, Monpert G, Acquaviva M, Mille G and Bertrand J-C (1991) Anaerobic oxidation of 1-*n*-heptedecene by a marine denitrifying bacterium. Appl. Microbiol. Biotechnol. 36: 252–256

Göbbert U, Lang S and Wagner F (1984) Sophorose lipid formation by resting cells of *Torulopsis bombicola*. Biotechnol. Letters 6: 225–230

Göbbert U, Schmeichel A, Lang S and Wagner F (1988) Microbial transesterification of sugar corynomycolates. J. Am. Oil. Chem. Soc. 65: 1519–1525

Goclik E, Müller-Hurtig R and Wagner F (1990) Influence of the glycolipid-producing bacterium *Rhodococcus erythropolis* on the degradation of a hydrocarbon mixture by an original soil population. Appl. Microbiol. Biotechnol. 34: 122–125

Goswami P and Singh HD (1991) Different modes of hydrocarbon uptake by two *Pseudomonas* species. Biotechnol. Bioeng. 37: 1–11

Guerra-Santos L, Käppeli O and Fiechter A (1984) *Pseudomonas aeruginosa* biosurfactant production in contiuous culture with glucose as carbon source. Appl. Environ. Microbiol. 48: 301–305

Guerra-Santos LH, Käppeli O and Fiechter A (1986) Dependence of *Pseudomonas aeruginosa* continuous culture biosurfactant production on nutritional and environmental factors. Appl. Microbiol. Biotechnol. 24: 443–448

Gutnick DL, Allon R, Levy C, Petter R and Minas W (1991) Applications of *Acinetobacter* as an industrial microorganism. In: J Towner, E Bergogne-Berezin and CA Fewson (eds) The Biology of *Acinetobacter* (pp 411–441). Plenum Press, New York, London

Haferburg D, Hommel R, Claus R and Kleber H-P (1986) Extracellular microbial lipids as biosurfactants. Adv. Biochem. Engin./Biotechnol. 33: 53–93

Haferburg D, Hommel R and Kleber H-P (1989) Biotechnologie extrazellulärer microbieller Glycolipide. Wiss. Z. Univ. Leipzig, Math.-natwiss. Reihe 38: 303–311

Heinz E, Tulloch AP and Spencer JFT (1969) Stereospecific hydroxylation of long chain compounds by a species of *Torulopsis*. J. Biol. Chem. 244: 882–888

Heinz E, Tulloch AP and Spencer JFT (1970) Hydroxylation of oleic acid by cell-free extracts of a species of *Torulopsis*. Biochim. Biophys. Acta. 202: 49–55

Hisatsuka K, Nakahara T, Minoda Y and Yamada K (1977) Formation of protein-like activator for *n*-alkane oxidation and its properties. Agric. Biol. Chem. 41: 445–450

Hisatsuka K, Nakahara T, Sano N and Yamada K (1971) Formation of rhamnolipid by *Pseudomonas aeruginosa* and its function in hydrocarbon fermentation. Agric. Biol. Chem. 35: 686–692

Hommel RK (1990) Formation and physiological role of biosurfactants produced by hydrocarbon-utilizing microorganisms. Biosurfactants in hydrocarbon utilization. Biodegradation 1: 107–119

Hommel R and Ratledge C (1990) Evidence for two fatty alcohol oxidases in the biosurfactant-producing yeast *Candida* (*Torulopsis*) *bombicola*. FEMS Microbiol. Letters. 70: 183–186

Hommel R and Ratledge C (1992) Biosynthetic Mechanisms to Low Molecular Weight Surfactants and their Precursor Molecules. In: N Kosaric (ed) Biosurfactants: Production-Properties-Application (pp 3–63). Marcel Dekker, New York

Hommel R, Stüwer O, Stuber W, Haferburg D and Kleber H-P (1987) Production of water-soluble surface-active exolipids by *Torulopsis apicola*. Appl. Microbiol. Biotechnol. 26: 199–205

Hommel R, Stegner S, Ziebolz C, Weber L and Kleber H-P (1990) Effect of cerulenin on growth and glycolipid production of *Candida apicola*. Microbios. Letters 45: 41–47

Hommel R, Lassner D, Weiss J and Kleber H-P (1992) The microsomal fatty alcohol oxidase of the sophorose lipid producing strain *Candida* (*Torulopsis*) *apicola*. (submitted to Appl. Environ. Microbiol.)

Iguchi J, Takeda I and Okasawa H (1969) Emulsifying factor of hydrocarbon assimilating yeast. Agric. Biol. Chem. 33: 1657–1658

Ioneda T, Silva CL and Gesztesi J-L (1981) Mycolic acid containing glycolipids of Nocardiae and related organisms. Zbl. Bakt. Suppl. 11: 401–406

Ishigami Y, Kamada T, Gama Y, Kaise M, Iwahasi H and Someya J (1989) Correlation of synthetic corynomycolic acids as biosurfactant between their surface-active properties and its function of biomembranes. J. Jap. Oil Chem. Soc. 38: 1001–1006

Ito S and Inoue S (1982) Sophorolipids from *Torulopsis bombicola*: possible relation to alkane uptake. Appl. Environ. Microbiol. 43: 1278–1283

Ito S, Kinta M and Inoue S (1980) Growth of yeasts on n-alkanes: inhibition by a lactonic sophorolipid produced by *Torulopsis bombicola*. Agric. Biol. Chem. 44: 2221–2223

Itoh S and Suzuki T (1972) Effects of rhamnolipids on growth of *Pseudomonas aeruginosa* mutant deficient in n-paraffin-utilizing ability. Agric. Biol. Chem. 36: 2233–2235

Itoh S and Suzuki T (1974) Fructose-lipids of *Arthrobacter*, *Corynebacteria*, *Nocardia* and *Mycobacteria* grown on fructose. Agric. Biol. Chem. 38: 1443–1449

Itoh S, Honda H, Tomita F, and Suzuki T (1971) Rhamnolipid produced by *Pseudomonas aeruginosa* grown on n-paraffin. J Antibiot 24: 855–859

Jenny K, Käppeli O and Fiechter A (1991) Biosurfactants from *Bacillus licheniformis*: structural analysis and characterization. Appl. Microbiol. Biotechnol. 36: 5–13

Käppeli O and Finnerty WR (1979) Partition of alkane by an extracellular vesicle derived from hexadecane-grown *Acinetobacter* sp. J. Bacteriol. 140: 707–712

Kaneda T (1977) Fatty acids of the genus *Bacillus*: Example of branched chain preference. Bacteriol. Rev. 41: 391–418

Kawashima H, Nakahara T, Oogaki M and Tabuchi T (1983) Extra-cellular production of a mannosyl-erythritol lipid of a mutant of *Candida* sp. from n-alkanes and triacylglycerols. J. Ferment. Technol. 61: 143–148

Kirk PW and Gordon AS (1988) Hydrocarbon degradation by filamentous marine higher fungi. Mycologia 80: 776–782

Kitamoto D, Akiba S, Hioki C and Tabuchi T (1990) Extracellular accumulation of mannosylerythritol lipids by a strain of *Candida antarctica*. Agric. Biol. Chem. 54: 31–36

Kitamoto D, Nakane T, Najao N, Nakahara T and Tabuchi T (1992) Intracellular accumulation of mannosylerythritol lipids as storage materials by *Candida antarctica*. Appl. Microbiol. Biotechnol. 36: 768–772

Kleber H-P, Asperger O, Stüwer O, Stüwer B and Hommel R (1989) Occurrence and regulation of cytochrome P-450 in *Torulopsis apicola*. In: I Schuster (ed) Cytochrome P-450: Biochemistry and Biophysics, (pp 169–172). Taylor and Francis, London, New York, Philadelphia

Kluge B, Vater J, Salnikow J and Eckart K (1988) Studies on the biosynthesis of surfactin, a lipopeptide antibiotic from *Bacillus subtilis* ATCC 21332. FEBS Lett 231: 107–110

Kobayashi T, Ito S and Okamoto K (1987) Production of mannosylerytritol by *Candida* sp. KSM-1529. Agric. Biol. Chem. 51: 1715–1716

Koch AK, Käppeli O, Fiechter A and Reiser J (1991) Hydrocarbon assimilation and biosurfactant production in *Pseudomonas aeruginosa* mutants. J. Bacteriol. 173: 4212–4219

Koronelli TV, Komarova TI and Debisov YV (1983) The chemical composition of *Pseudomonas aeruginosa* proteoglycolipid and its role in the process of hydrocarbon assimilation. Mikrobiologiya 53: 767–770

Kretschmer A and Wagner F (1983) Characterization of biosynthetic intermediates of trehalose dicorynomycolates from *Rhodococcus erythropolis* grown on *n*-alkanes. Biochim. Biophys. Acta. 753: 306–313

Kretschmer A, Bock H and Wagner F (1982) Chemical and physical characterization of interfacial-active lipids from *Rhodococcus erythropolis* grown on n-alkanes. Appl. Environ. Microbiol. 44: 864–870

Lang S and Wagner F (1987) Structure and properties of biosurfactants. In: N Kosaric, WL Cairns and WL Gray (eds) Surfactant Science Series. Biosurfactants and Biotechnology, Vol 25 (pp 21–45). Marcel Dekker, New York, Basel

Lang S, Gilbon A, Syldatk C and Wagner F (1984) Comparison of interfacial active properties of glycolipids from microorganisms. In: KL Mittal and B Lindman (eds) Surfactants in Solution, Vol 2 (pp 1365–1376). Plenum Press, New York

Lang S, Katsiwela E and Wagner F (1989) Antimicrobial effects of biosurfactants. Fat. Sci. Technol. 9: 363–366

Li Z-Y, Lang S, Wagner F, Witte L and Wray V (1984) Formation and identification of interfacial-active glycolipids from resting microbial cells. Appl. Environ. Microbiol. 48: 610–617

Linton JD (1991) Metabolite production and growth efficiency. Antonie van Leewenhoek 60: 293–311

MacDonald CR, Cooper DG and Zajic JE (1981) Surface-active lipids from *Nocardia erythropolis* grown on hydrocarbons. Appl. Environ. Microbiol. 41: 117–123

Martin M, Bosch P and Parra JL (1991) Structure and bioconversion of trehalose lipids. Carbohydrate Res. 220: 93–100

Mattei G, Rembeloarisoa E, Giusti G, Ronatani JF and Bertrand J-C (1986) Fermentation procedure of a crude oil in continuous culture on sea water. Appl. Microbiol. Biotechnol. 23: 302–304

Matsuyama T, Murakami T, Fujita M, Fujita S and Yano I (1986) Extracellular vesicle formation and biosurfactant production by *Serratia marcescens*. J. Gen. Microbiol. 132: 865–875

Mulligan CN and Gibbs BF (1989) Correlation of nitrogen metabolism with biosurfactant production by *Pseudomonas aeruginosa*. Appl. Environ. Microbiol. 55: 3016–3019

Mulligan CN, Mahmourides G and Gibbs BF (1989) Biosurfactant production by a chloramphenicol-tolerant strain of *Pseudomonas aeruginosa*. J. Biotechnol. 12: 37–44

Müller-Hurtig R, Matulovic U, Feige I and Wagner F (1987) Comparison of the formation of rhamnolipis with free and immobilized cells of *Pseudomonas* spec. DSM 2874 with glycerol as C-substrate. Proc. 4th European Congress on Biotechnology, Vol 2 (pp 257–260). Elsevier Sci. Publ., Amsterdam

Nakano MM, Marahiel MA and Zuber P (1988) Identification of a genetic locus required for biosynthesis of the lipopeptide antibiotic surfactin in *Bacillus subtilis*. J. Bacteriol. 170: 5662–5668

Oberbremer A and Müller-Hurtig R (1989) Aerobic stepwise hydrocarbon degradation and formation of biosurfactants by an original soil population in a stirred reactor. Appl. Microbiol. Biotechnol. 31: 582–586

Oberbremer A, Müller-Hurtig R and Wagner F (1990) Effect of the addition of microbial surfactants on hydrocarbon degradation in a soil population in a stirred reactor. Appl. Microbiol. Biotechnol. 32: 485–489

Passeri A, Lang S, Wagner F and Wray V (1991) Marine Biosurfactants, II. Production and characterization of an anionic trehalose tetraester from the marine bacterium *Arthrobacter* sp. EK1. Z. Naturforsch. 46c: 204–209

Peypoux F and Michel G (1992) Controlled biosynthesis of Val^7- and Leu^7-surfactins. Appl. Microbiol. Biotechnol. 36: 515–517

Peypoux F, Bonmatin J-M, Labbe H, Das BC, Ptak M and Michel G (1991) Isolation and characterization of a new variant of surfactin, the $[Val^7]$surfactin. Eur. J. Biochem. 202: 101–106

Poremba K, Gunkel W, Lang S and Wagner F (1989) Mikrobieller Ölabbau im Meer. Biologie in unserer Zeit 19: 145–148

Powalla M, Lang S and Wray V (1989) Penta- and disaccharide lipid formation by *Nocardia corynebacteroides* grown on n-alkanes. Appl. Microbiol. Biotechnol. 31: 473–479

Ramsay B, McCarthy J, Guerra-Santos L, Käppeli O and Fiechter A (1988) Biosurfactant production and diauxic growth of *Rhodococcus aurantiacus* when using n-alkanes as the carbon source. Can. J. Microbiol. 34: 1209–1212

Rapp P, Bock H, Wray V and Wagner F (1979) Formation, isolation and characterization of trehalose dimycolates from *Rhodococcus erythropolis* grown on n-alkanes. J. Gen. Microbiol. 115: 491–503

Ratledge C (1988) Hydrocarbons. Products of hydrocarbon-microorganism interaction. In: DR Houghton, RN Smith and HOW Eggins (eds) Biodeterioration 7 (pp 219–236). Elsevier Applied Science, London, New York

Reddy PG, Singh HD, Roy PK and Baruah JN (1983) Isolation and functional characterization of hydrocarbon emulsifying and solubilizing factors produced by a *Pseudomonas* species. Biotech. Bioeng. 25: 387–401

Reiling HE, Thanei-Wyss U, Guerra-Santos LH, Hirt R, Käppeli O and Fiechter A (1986) Pilot plant production of rhamnolipid biosurfactant by *Pseudomonas aeruginosa*. Appl. Environ. Microbiol. 51: 985–989

Rilke O, Baum A, Weiss J, Hommel R and Kleber H-P (1992) Kinetics of enzymatic lysis, formation, and regeneration of protoplasts of *Candida (Torulopsis) apicola*. World J. Micobiol. Biotechnol. 8: 14–20

Ristau E and Wagner F (1983) Formation of novel anionic trehalose-tetraesters from *Rhodococcus erythropolis* under growth limiting conditions. Biotechnol. Letters 5: 95–100

Roy PK, Singh HD, Bhagat SD and Baruah JN (1979) Characterization of hydrocarbon emulsification and solubilization occurring during the growth of *Endomycopsis lipolytica* on hydrocarbons. Biotechnol. Bioeng. 21: 955–974

Saadat S and Ballou CE (1983) Pyruvylated glycolipids from *Mycobacterium smegmatis* J. Biol. Chem. 258: 1813–1818

Sandrin C, Peypoux F and Michel G (1990) Production of surfactin and iturin A, lipopeptides with surfactant and antifungal properties, by *Bacillus subtilis*. Biotechnol. Appl. Biochem. 12: 370–375

Schulz D, Passeri A, Schmidt M, Lang S, Wagner F, Wray V and Gunkel W (1991). Marine biosurfactants. I. Screening for biosurfactants among crude oil degradation marine microorganisms from the north sea. Z. Naturforsch. 46c: 197–230

Singer ME (1985) Microbial biosurfactants. In: JE Zajic and EC Donaldson (eds) Microbes and Oil Recovery. Int. Bioresources J. 1: 19–38

Spencer JFT, Spencer DM and Tulloch AP (1979) Extracellular glycolipids of yeasts. In: AH Rose (ed) Economic Microbiology. Secondary Products of Metabolism, Vol 3 (pp 522–540). Academic Press, London, New York, San Francisco

Stüwer O, Hommel R, Haferburg D and Kleber H-P (1987) Production of crystalline surface-active glycolipids by a strain of *Torulopsis apicola*. J. Biotechnol. 6: 259–269

Syldatk C and Wagner F (1987) Production of biosurfactants. In: N Kosaric, WL Cairns and WL Gray (eds) Surfactant Sience Series. Biosurfactants and Biotechnology, Vol 25 (pp 21–45). Marcel Dekker, New York, Basel

Syldatk C, Matulovic U and Wagner F (1984) Biotenside – Neue Verfahren zur mikobiellen Herstellung grenzflächenaktiver, anionischer Glycolipide. Biotech. Forum 1: 58–66

Syldatk C, Lang S, Matulovic U and Wagner F (1985a) Production of four interfacial active rhamnolipids from n-alkanes or glycerol by resting cells of *Pseudomonas* species DSM 2874. Z. Naturforschung 40c: 61–67

Syldatk C, Lang S, Wagner F, Wray V and Witte L (1985b) Chemical and physical characterization of four interfacial-active rhamnolipids from *Pseudomonas* spec. DSM 2874 grown on n-alkanes. Z. Naturforschung 40c: 51–60

Tomiyasu I, Yoshinaga J, Kurano F, Kato Y, Kaneda K, Imaizumi S and Yano I (1986) Occurrence of a novel glycolipid, trehalose 2,3,6'-trimycolate in a psychrophilic, acid-fast bacterium, *Rhodococcus aurantianus (Gordona aurantiaca)*. FEBS Lett. 203: 239–242

Tulloch AP (1976) Structures of extracellular glycolipids produced by yeasts. In: LA Wittling (ed) Glycolipid Methodology (pp 329–344). Am. Oil Chem. Soc., Champaign, IL

Uchida Y, Tsuchiya R, Chino M, Hirano J and Tabuchi T (1989) Extracellular accumulation of mono- and di-succinoyl trehalose lipids by a strain of *Rhodococcus erythropolis* grown on n-alkanes. Agric. Biol. Chem. 53: 757–761

Ullrich C, Kluge B, Palacz Z and Vater J (1991) Cell-free biosynthesis of surfactin, a cyclic lipopeptide produced by *Bacillus subtilis*. Biochemistry 30: 6503–6508

Vater J (1986) Lipopeptides, an attractive class of microbial surfactants. Progr. Colloid. and Polymer Sci. 72: 12–18

Weber L, Stach J, Haufe G, Hommel R and Kleber H-P (1990) Elucidation of the structure of an unusual cyclic glycolipid from *Torulopsis apicola*. Carbohydrate Res 206: 13–19

Weber L, Döge C, Haufe G, Hommel R and Kleber H-P (1991) Oxygenation of hexadecane in the biosynthesis of cyclic glycolipids in *Torulopsis apicola*. Biocatalysis 5: 267–272

Witholt B, de Smert M-J, Kingma J, van Beilen JB, Kok M, Lageveen RG and Eggink G (1990) Bioconversions of aliphatic compounds by *Pseudomonas oleovorans* in multiphase bioreactors: background and economic potential. Trends Biotechnol. 8: 46–52

Yamaguchi M, Sato M and Yamada K (1976) Microbial production of sugar lipids. Chemical Ind. 17: 741–742

Zang Y and Miller RM (1992) Enhanced octadecane dispersion and biodegradation by a *Pseudomonas* rhamnolipid surfactant (biosurfactant). Appl. Environ. Microbiol. 58: 3276–3282

4. Biodegradation of oils, fats and fatty acids

COLIN RATLEDGE

Department of Applied Biology, University of Hull, Hull HU6 7RX, U.K.

I. Introduction

Oils and fats, being naturally produced materials found in all living cells, are readily broken down both *in situ* as well as when in the environment. World production of oils and fats is currently over 70 millions tonnes/year (Gunstone 1989) and is predicted to rise at about 2.5 to 3.0 million tonnes per year throughout the 1990s (Mielke 1992) mainly due to increases in world population but also because of the continual industrialization of developing nations. About 75% of the total fats and oils are derived from plants, of which soybean, palm, sunflower and rape oils account for over 70%, and the remainder are from animals which encompass lard, tallow and marine oils.

The majority of the oils and fats are used in the food industry for manufacture of a wide variety of products ranging from margarines to chocolate and from inclusion in most confectionery products to direct use as salad and cooking oils. However, a significant proportion ($\sim 20\%$) is used in the oleochemicals industry for the manufacture of an equally wide range of products: soaps, detergents, toiletries, plastics, paints, inks etc. For these purposes the oils and fats may be converted from their natural triacylglycerol (triglyceride) structure into their component fatty acids which, in turn, may then be fractionated and used as individual fatty acids for the manufacture of alcohols, amides, dimer acids, esters, etc. In food uses, however, the triacylglycerol structure of the original oil or fat is preserved though there can be fractionation of an oil to give individual triacylglycerol fractions suitable for a particular function. Fatty acids as such are not used in the food industry as these impart undesirable "soapy" characteristics to a food and, if present in the original oil, must be removed during the initial refinement procedures. (A useful account of the various stages of oilseed processing has recently been prepared by Carlson and Scott 1991.)

Oils and fats are defined as triacylglycerol materials and are differentiated as either oils or fats solely by virtue of whether they are either liquid or solid at ambient temperature. The main non-triacylglycerol components of the oils and fats are usually phospholipids (~ 2 to 3%), carotenoids and steroids (the latter

C. Ratledge (ed.), Biochemistry of Microbial Degradation, 89–141.

two are termed "non-saponifiable" material). Steroids will include sterols, such as stigmasterol, cholesterol, etc., as well as sterol esters and hopanoids. Carotenoids are responsible for most of the yellowness of oils. Some chlorophyll (also classed as a non-saponifiable) may also be co-extracted and give some oils such as olive oil and rapeseed oil, a greenish appearance. Most oils and fats when first extracted will contain some free fatty acids. These arise either by action of the natural lipases within the plant or animal tissues or by chemical hydrolytic reactions stimulated by the high temperatures and pressures used in the extracting process.

All non-triacylglycerol components are removed from edible oils during the subsequent refining process which is often accompanied by a de-odourization step involving passage through activated charcoal. Some of these materials are recovered and are sold for particular purposes: for example phospholipids (usually termed "lecithin") are recovered, further refined and used as emulsifiers and for other purposes in various branches of the food industry (Szuhaj 1989). The sterols, principally stigmasterol and cholesterol, are recovered for use in the pharmaceutical industries. β-Carotene is also now recovered from palm oil on a commercial scale (Iwasaki and Murakoshi 1992). Very little material could be considered as genuinely waste.

The principal sources of oleo-materials in the environment arises from the non-edible uses of oils and fats. Fatty acids, as the major components of most detergents and soaps, are invariably released after use into the waste-waters from both domestic and industrial sources. Domestic sewage and waste-water typically contain about 40 to 100 mg fatty acid/l but it is industrial effluents that are of greater concern. Most industrial effluents containing fats, oils and greases receive some form of preliminary chemical treatment before passing into a biodegradation stage (Forster 1992). These pre-treatments can vary from a simple fat-trap or oil interceptor to being a floatation system using dissolved air to achieve fat/water separation. If these systems are able to operate without sudden overload, as much as 90% of the initial Biological Oxygen Demand (BOD) level can be removed before the waste-water reaches the sewage plant. The specifications of municipal sewage treatment plants that have to deal with these materials have been set out by Gode (1991) as they relate to European standards. It is calculated that about 50% of the pollutants in domestic washing waste-waters arise from the detergents and washing aids but constitute only about 30% for institutional laundering (Gode 1991).

Full descriptions of the biodegradation of anion detergents and complexes such as nitriloacetic acid are given elsewhere in this book by White and Russell (Chapter 5) and by Egli (Chapter 6). In this chapter I intend to cover the main routes by which oils and fats are broken down to their component fatty acids and then to describe the biochemical routes by which microorganisms are then able to utilize these materials. As there is considerable interest in the potential use of microorganisms for upgrading oils and fats into products of increased monetary values, these biotechnological aspects of oil and fat biotransformation will also be covered. However, the main emphasis will be on how microorganisms are able to effect the complete dissimilation of a major group of biological molecules.

For the applications of microorganisms in sewage treatment plants, which has been alluded to above, the reader is referred to the brief review by Forster (1992) and to the more extensive monograph by Barnes et al. (1984). It should though be appreciated that degradation of fatty acids in sewage systems takes place as a mixed substrate system wherein the effects of fatty acids on cell metabolism may well be minimal and not elicit many of the responses indicated in the final parts of this chapter. Very little biochemical work has been carried out following the fate of fatty acids in mixed substrate fermentation systems.

II. Microbial utilization of oils and fats

Although it has been tacitally assumed that microorganisms are able to breakdown oils and fats, the long term stability of these materials during storage without adventitious biodegradation led to the general view that oils and fats were difficult substrates for microorganisms to attack. The use of oils in antibiotic fermentations, initially as anti-foaming agents, but then as feedstocks in their own right gradually led to the view that some organisms could indeed degrade these materials (Perlman and Wagman 1952; Pan et al. 1959). However like most substrates, if the concentration can be kept below a critical level and the culture conditions optimized, then it would appear that a large number of microorganisms can indeed grow on, and therefore degrade, oils and fats. The realization that lipid degradation is a more generalized phenomenon than at first appreciated has arisen mainly from the continuing and developing interest in microbial lipases (see Section III).

A. Utilization of triacylglycerols

The majority of work on the utilization of lipid materials has concentrated on growth of microorganisms on various oils, fats and greases. Many of these materials, unless carefully refined, will contain some trace fatty acids but, as will be pointed out below (II.B), these materials should not adversely affect growth of a microorganism.

As stated in the Introduction to this section, the realization that many microorganism could utilize oils and fats was very slow to be appreciated. Perlman and Wagman (1952) stated that " . . . lipid metabolism has not received much attention (in comparison with other areas of metabolism)" but noted that several groups had observed the addition of vegetable and animal fats, or fatty acids, to media had increased the yield of penicillin in submerged culture. It was though uncertain whether the fats were acting as a source of energy. These workers and also Pan et al. (1959) were possibly the first to show that vegetable oils and animal fats could be utilized as sole sources of carbon and energy. However their studies were confined to the antibiotic-producing strains of *Streptomyces griseus* and *Penicillium* and the general applicability of their findings was not appreciated. Indeed, it is surprising to realize that work carried out on the formation of microbial lipases in the late 1940s, 1950s and into the 1960s, did not consider adding even small amounts of oils to culture media to promote lipase production (Peters and Nelson 1948a,b; Ramakrishnan 1957; Alford and Pierce 1963; Alford et al. 1964) presumably in the mistaken view that such materials would not be utilized by the various microorganisms under examination.

Although lipolytic bacteria and yeasts have long been known (Huss 1908 and Harrison 1927; respectively), the first reports of microorganisms being cultivated on oils and fats, apart from the above mentioned work on antibiotic-production media, were those in which yeasts were grown on whole fatty fish or fish-processing wastes (Burkholder et al. 1968; Li et al. 1970) and subsequently on isolated fish oils themselves (Hottinger et al. 1974a,b; Green et al. 1976). However although the organisms – principally *Candida lipolytica* and *Geotrichum candidum* – were undoubtedly utilizing the fish oils, the media used was complex and poorly defined. As Tan and Gill (1984) subsequently commented, "Uncertainties as to the physical and chemical conditions in such systems, which may affect the nutrient status of the growing organisms, limits the evaluation and general application of these findings."

The first report of yeasts being grown on oils and fats in a defined medium appears to be in an otherwise unpublished Ph.D. thesis from the Netherlands (van der Veen 1974). Yeasts that successfully grew on both fatty acids and animal lard were, in diminishing order of yield: *Endomycopsis lipolytica*, *Saccharomycopsis lipolytica* (both these species would now be classed as *Yarrowia lipolytica*), *Candida maltosa*, *C. tropicalis*, *Pichia vini*, *C. cacoi*, *Torulopsis haemulonii*, *C. cloacae*, *C. parapsilosis* and finally *C. edax* which only grew on the lard substrate.

These studies and others had been stimulated by the considerable activity in the 1960s and 70s in the production of Single Cell Protein (SCP) by yeasts being grown on n-alkanes (see Chapter 1). The realization that many microorganisms could grow as well, if not better, on alkanes than on glucose suggested that microorganisms should also be able to grow on fatty acids and fats particularly as fatty acids are intermediates in the oxidation of alkanes (cf. Chapter 1). However it was probably always evident that although alkanes as a feedstock for SCP production were marginally cost-effective, it was unlikely that the higher-valued oils could be used in their place and still produce an economic product.

However in spite of these earlier studies, the majority of work on microbial utilization of oils and fats has only been conducted since 1982. The objectives have been both as a means of producing a cheap animal feedstuff (Kajs and Vanderzant 1980; Koh and Minoda 1984; Koh et al. 1983, 1985; Laborbe et al. 1989; Lee et al. 1992; Martinet et al. 1982b; Montet et al. 1983; Nakahara et al. 1982; Tan and Gill 1984, 1985a,b) and as a means of treating and dealing with lipid waste materials from oil- or fat-producing factories (Fernandes et al. 1988;

Rydin et al. 1990).

The key feature which Tan and Gill (1985b) observed was that for successful growth of a microorganism to occur on a lipid, the culture medium had to be maintained close to the optimum pH of the lipase that was needed to hydrolyse the oil. The high dependancy of microbial growth on fats upon the culture pH is shown in Fig. 1. This key fact can then explain the failure of some workers to find many oleo-degrading microorganisms when carrying out surveys: for example, Koh et al. (1983) isolated over 200 yeasts from almost 250 samples but found only a few were effective in assimilating palm oil. Unfortunately these authors carried out their work using media at pH 5.5 and thus, in view of the pH optimum of most yeast lipases lying between 6.5 and 7.0, their poor isolation rate is perhaps not too surprising.

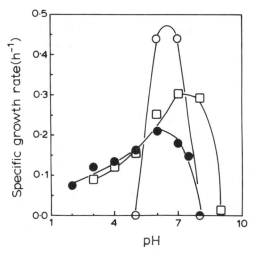

Fig. 1. Effect of culture pH on the growth on fats of: *P. fluorescens* (○), *A. niger* (●) and *Yarrowia lipolytica* (□). (K.H. Tan, personal communication.)

A further factor that may have to be taken into account for successful microbial growth is the dispersal of the oil or fat within the cultivation vessel. With *Pseudomonas fluorescens* and *Yarrowia* (*Sacch.*) *lipolytica* growing on olive oil, Tan and Gill (1984; 1985b) found that varying the agitation rate from 200 to 1000 rev/min made little difference to cell growth but with *Aspergillus niger* (K.H. Tan, personal communication) growing on the same substrate there was considerable dependence (see Fig. 2). In most cases though sufficient dispersal of the oil is achieved in shake-flask cultures provided there is a low ratio of medium-to-vessel volumes. For solid fats, pre-dispersal using vigorous shaking or ultrasound may be required but to prevent such microemulsion coalescing, a suitable surfactant may have to be added (Koh et al. 1983; and see also Ratledge and Tan 1990). The use of emulsificants is thought unnecessary in stirred systems. Many microorganisms are capable of producing their own surfactant which may be sufficient to aid micro-droplet formation (see Chapter 3).

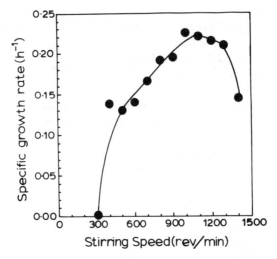

Fig. 2. Effect of culture agitation rate on the growth of *A. niger* on olive oil. (K.H. Tan, personal communication.)

Whilst the presence of a lipase is essential for a microorganism to grow on an oil or fat, it may not be always necessary for the inoculum culture to be pre-grown on an oil or fat to induce the enzyme. Some organisms, e.g. *Yarrowia lipolytica*, appear to initiate growth on an oil within an hour or less even when inoculated from a glucose-growing culture (Tan and Gill 1984). Here the lipase must be constitutive but in other cases, as with *Pseudomonas fluorercens* (Tan and Gill 1985b), a period of adaptation is needed before growth commences so that for the most rapid growth rate to be initiated the inoculum needs to be grown on an oil.

Once growth has been established on an oil, all other nutrients must continue to be present otherwise only partial utilization will ensue. In some of the earlier work (Li et al. 1970; Burkholder et al. 1968) it is apparent that growth of the inoculated yeasts on the whole fish material was limited by the supply of assimilatable nitrogen to the yeast: the fact that organic nitrogen compounds were present was misleading as evidently these compounds were not effectively utilized as N sources.

B. Utilization of fatty acids

It is important to distinguish between microbial utilization of triacylglycerols (i.e. oils and fats) and of fatty acids. The inhibitory effect of fatty acids and of fatty alcohols on the growth of bacteria and yeasts is well-established as these are the main components of soaps which themselves are known for their microbiocidal actions. Growth of microorganisms on soapstocks (i.e. salts of various long chain fatty acids produced by alkali-treatment of any oil or fat) relies on the concentration of the fatty acids in the aqueous growth medium being below a critical level. The early work of Rahn (1945) and Prince (1959) being below a critical level. The early work of Rahn (1945) and Prince (1959) amongst others established that the inhibitory properties of fatty acids towards microorganisms was pH dependent and that it was the concentration of the undissociated fatty acid that was of key significance. Bell and Trancart (1973) confirmed these findings observing less inhibition by C_4 and C_6 alkanoic acids at pH 5.7 than pH 4.0 with *Candida tropicalis*. Earlier, Bell (1971), working with the complete range of fatty acids from C_2 to C_{12} at pH 4.0, had observed that the critical inhibitory concentration of a fatty acids towards *C. tropicalis* was not exceedable above C_{11} as fatty acids of longer chain length were too insoluble to reach their critical concentration in aqueous growth medium (see Fig. 3). Below C_{11}, the critical inhibitory concentration increased logarithmically as the chain length decreased. The aqueous-phase critical concentration of, for example, decanoic acid was 0.1 mM.

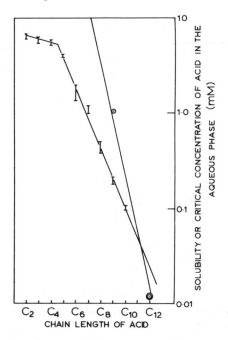

Fig. 3. Influence of chain length on the critical concentration (in the aqueous phase only) of straight-chain fatty acids, and on their solubility in the aqueous phase at pH 4.0. I = critical concentration; O solubility. (From Bell 1971.)

Failure of yeasts, and other microorganisms too, to grow on fatty acids above C_{11} is not attributed to their toxicity but to their insolubility that makes initial attack difficult (Bell 1973). If the long-chain fatty acids are, however, dissolved in a non-metabolizable, water-insoluble hydrocarbon, growth can take place provided the cell has the appropriate envelope structure to permit adherence to the hydrocarbon (see Chp 2). Thus, *C. tropicalis* could successfully utilize such fatty acids but *Saccharomyces cerevisiae* could not (Bell, 1973).

The findings of Bell (1971; 1973), to explain the lack of toxicity of fatty acids above C_{11}, can therefore provide a basis for explaining growth of most microorganisms on fatty acids found in most oils and fats *providing* that the pH of cultivation is maintained at 5.5 or higher. Clearly though some organisms will remain sensitive to even low concentrations of fatty acids as the critical inhibitory concentration of a fatty acid will vary from organism to organism. Thus Atlas and Bartha (1973) observed inhibition of a *Brevibacterium* sp., and to a lesser extent of a *Flavobacterium* sp., by fatty acids up to C_{14}, but not of C_{16} or C_{18} chain length. Hunkova and Fencl (1977) considered that the diminished cell yields of yeasts utilizing shorter chain (C_9) alkanes, rather than a longer-chain (C_{16}) substrate, could be attributed to the accumulation of fatty acids at above their critical inhibition concentrations. Respiration rates of the yeasts with nonane were twice those obtained with hexadecane indicating that rapidity of oxidation was not a key factor in explaining the diminished growth yields with the shorter chain alkanes. These authors then offered an appraisal of fatty acid toxicity that took into account not only the species of microorganism but also the composition of the growth medium, its pH, and the chain length of the fatty acid. They suggested that each acid has four ranges of concentration that had microbiological relevance:
1) *Sub-threshold concentrations.* These concentrations do not affect microbial growth nor affect oxidation of another substrate.
2) *Threshold concentrations.* These cause a decrease in growth rate, inhibit cell division and such activities as phosphate uptake. Oxidation of a carbon substrate though is not inhibited.
3) *Above threshold concentrations.* At this level, oxidation of a carbon substrate is now partially inhibited but some growth may still occur.
4) *Microbiocidal concentrations.* At this concentration cells cease to grow and cell lysis then begins.

The effects of groups 2 and 3 will be characterized by their reversibility; that of no. 4 produces irreversible changes.

These studies of the 1970s provided the basis by which other studies on the microbial utilization of fatty acids could be based. For example, Vuillemin et al. (1981) found that out of 857 microorganisms utilizing C_{10} to C_{18} fatty acids, only *Proteus morganii* and *Torulopsis glabrata* were unable to effect degradations. Thus it would appear that if care is taken to maintain the pH of growth media about 6.0 or even to pH 7.0, then the majority of microorganisms will grow on the longer chain ($C_{12} - C_{18}$) fatty acids. Various attempts have been made to take advantage of this and, in particular, to attempt the salvaging of waste fatty acids from vegetable oil-production units and refineries. Most of these studies have used yeasts as the preferred organisms as the object has usually been to achieve the production of a saleable single cell protein biomass. Soapstocks from groundnut oil (Ba et al. 1981), palm oil (Martinet et al. 1982a) and of mixed origins (Montet et al. 1985; Laborbe et al. 1987) have all been examined using a wide variety of yeasts and some moulds. The only yeasts failing to grow out of 23 examined were *Kluyveromyces fragilis* and *K. lactis*

(Ba et al. 1981). The preferred organisms for SCP production appears to be *Yarrowia* (*Candida*) *lipolytica* (Ba et al. 1983; Lepidi et al. 1975; Anelli et al. 1975) *Geotrichum candidum* (Laborbe et al. 1987) and *Torulopsis holmii* (Fiorentini et al. 1976).

The early work of the Italian group (Anelli et al. 1975; Lepidi et al. 1975; Fiorentini et al. 1976) found that best yields of biomass were by using a co-culture of *Y. lipolytica* and *T. holmii* which gave a combined yield of 1 g dry cells per g of an oleine feedstock. The protein content of the yeast was about 60% (w/w) and the lipid content about 13% (w/w). These yeasts were originally isolated from the waste water or surface mud of a drain canal of a vegetable oil refinery. With the work of the French group, *Geotrichum candidum* produced a yield of about 0.9 g per g soapstock with a protein content of 40% (w/w) (Montet et al. 1985).

In summary, it is now quite clear that numerous microorganisms can utilize long-chain fatty acids as sole sources of carbon and energy. Problems arise with fatty acids C_{10} and shorter but these can be minimized by maintaining a high culture pH so that the concentration of the undissociated acid is minimal. However if the fatty acids are present along with a water-insoluble substrate such as a vegetable oil (earlier authors have only usually considered alkanes) then the fatty acids will clearly partition between the aqueous and oil phases thereby diminishing any toxic effects still further. Fatty acids that are solid at microbial growth temperatures may still be used effectively either by microemulsification (see Chapter 3) or by partitioning into a second, water-insoluble substrate. In this way the concentration of the acid in the aqueous phase is kept below criticality.

C. Assimilation of fatty acids from triacylglycerols

The hydrolysis of oils and fats proceeds via the action of extracellular lipases or by lipases that are attached to the cell surface and thus appear to be cell-associated. The oil is not assimilated as such. Tan and Gill (1985a) showed quite clearly that *Yarrowia lipolytica* when using lard, beef or mutton tallows assimilates the component fatty acids at different rates. When unsaturated fatty acids (oleic and linoleic acids) were present on the outer (*sn*-1 and *sn*-3) positions of the glycerol moiety of the lard, and as these are the positions preferentially hydrolysed by the lipase of this yeast, it was these acids that became abundantly available at a relatively early stage of growth. With the tallows, the oleoyl residues are located at the central (*sn*-2) position and thus were released at a slower rate than the saturated acyl groups on the *sn*-1 and *sn*-3 positions. Consequently with tallows as substrates, the saturated acyl groups (palmitate and stearate) were found to a higher proportion in the yeast lipids than the unsaturated fatty acids. With olive oil as substrate, no distinction could be made of differential uptake because of the high proportion of oleoyl residues on all three positions of the glycerol (Tan and Gill 1984, 1985b).

Because of the regio-specificity of the lipases from many microorganisms (see Fig. 7), it has been possible to show that this is reflected in the isomeric forms of the diacyl- and monoacyl-glycerols that are formed in the culture media – even when olive oil is used as substrate. Again referring to the careful analytical work of Tan and Gill (1984; 1985a,b), they showed that *sn*-1,2-diacylglycerols and *sn*-2-monoacylglycerols were the predominant partial acylglycerols that were produced (see Figs. 4 and 5) during the growth of *Y. lipolytica* on both olive oil and animal fats. Accumulation of the monoacylglycerols occurred only when the diacylglycerols were being consumed. Free fatty acids could rise to a relatively high concentration before disappearing. Tan has pointed out (see Ratledge and Tan 1990) that the extensive hydrolysis of monoacylglycerols, in the presence of large concentrations of diacylglycerols, is at variance with the reported specificity of the purified lipase from *Y. lipolytica*. With the isolated lipase, tri- and di-acylglycerols are hydrolysed at equal rates but monoacylglycerols hardly at all. This is explained by realizing that in the enzyme assay systems *sn*-1-monoacylglycerols are used rather than the naturally produced *sn*-2-isomers. Similar observations have been made for the differential utilization of fatty acids from the three positions of animals fats by *Pseudomonas fluorescens* and again, though this is at variance with the known specificity of the isolated lipase, this is due to the initial specificity of the first two hydrolytic reactions yielding the *sn*-2-monoacylglycerol not the *sn*-1-isomer.

Related results have also been obtained by Bati et al. (1984) who grew *Yarrowia (Candida) lipolytica* on corn oil and found little difference initially between the gross fatty acyl composition of the yeast and that of the initial corn

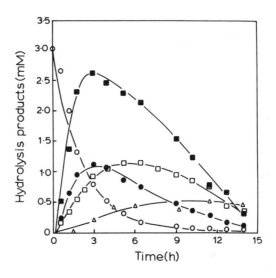

Fig. 4. Changes in supernatant lipid composition during growth of *Yarrowia lipolytica* on olive oil. Triacylglycerol (○), diacylglycerol (●), monoacylglycerol (□), free fatty acids (■) and glycerol (Δ). (From Tan and Gill 1984.)

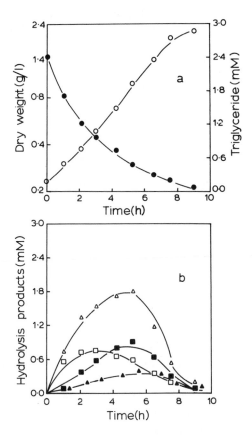

Fig. 5. Concentrations of products of triacylglycerol hydrolysis during growth of *Yarrowia lipolytica* or beef tallow. (a): cell dry weight (○); triacylglycerols (●); (b): diacylglycerols (□), monoacylglycerols (■), free fatty acids (Δ) and glycerol (▲). (From Tan and Gill 1985a.)

oil. However closer inspection of the yeast oil showed that it had a quite different distribution of fatty acyl groups. For example, palmitic acid (16 : 0) was distributed between the *sn*-1, -2 and -3 positions in corn oil in the ratio of 50: 45: 5 but in the yeast oil it was now at 35 : 12 : 53; more strikingly, oleic acid was originally distributed as 33 : 33 : 33 but in the recovered yeast oil was now at 36 : 50 : 14. Similar results have been obtained with another yeast, *Candida* (*Apiotrichum*) *curvata* grown on both lard and corn oil (Lee et al. 1992). The results of the triacylglycerol analysis are shown in Table 1. These results and others (see Ratledge 1989) would suggest that this pattern of acyl group redistribution probably occurs in all cells that grow on triacylglycerols.

   Although there are few exceptions, most microorganisms appear to use the component fatty acyl groups of the lipid with only the minimum amount of modification of the acyl groups themselves. That is, there is very little desaturation or elongation of the incoming fatty acids. Clearly, the incoming

Table 1. Stereospecific analysis of triacylglycerol oil from Candida (Apiotrichum) curvata compared with its growth substrate of either corn oil or lard (from Lee et al. 1992).

On corn oil	Relative % (w/w) of principal fatty acyl residues				
	16:0	18:0	18:1	18:2	18:3
Substrate	11.4	2.1	26.7	58.8	1.1
sn-1	21.3	2.9	23.6	50.8	1.5
sn-2	1.3	0.3	26.7	70.6	1.0
sn-3	11.5	3.1	29.8	55.1	0.7
Yeast oil	8.7	2.8	27.0	59.5	1.0
sn-1	17.1	2.6	18.6	61.0	0.7
sn-2	0.4	0.2	41.3	57.8	0.4
sn-3	8.6	5.6	21.1	59.7	1.9
On lard*					
	16:0	18:0	18:1	18:2	18:3
Substrate	24.5	14.7	42.6	11.1	0.6
sn-1	12.8	24.0	46.1	10.5	0.9
sn-2	66.1	3.3	14.8	5.6	–
sn-3	–	16.8	66.9	17.2	0.9
Yeast oil	18.0	6.9	52.1	17.5	0.5
sn-1	32.5	8.2	41.6	13.3	–
sn-2	3.2	0.5	80.8	14.0	0.3
sn-3	18.3	12.0	34.0	25.2	1.2

* Contained small amount ($< 3\%$) of 16:1 which also appeared in the yeast oil.

fatty acids must repress the synthesis of the desaturase enzymes in the same way that they repress *de novo* fatty acid biosynthesis (see Gill and Ratledge 1973). These observations with fatty acids parallel the earlier observations that the fatty acyl composition of microorganisms after growth on n-alkanes was directly relatable to the chain-length of the substrate (Ratledge 1980; Thorpe and Ratledge 1972). If the chain length of the substrate is too short to give fatty acids of sufficient chain length to be satisfactorily incorporated into the microbial lipid then some modification is essential. Thus Lee et al. (1992) observed that following growth of *Apiotrichum curvatum* on tricaprin (i.e. tri-decylglycerol), only 15% of the resulting fatty acids in the yeast oil were of C_{10} chain length. The remainder of the fatty acids were C_{14} (in small part) with C_{16} and C_{18} fatty acids predominating. Clearly with fatty acids that are too short, chain elongation is essential to produce the fatty acids required for proper cell function. When myristic acid (C_{14}) was used as such, the yeast oil now contained 90% of its acyl groups as this fatty acid. With palmitic acid (16 : 0) as substrate, elongation to stearic acid (18 : 0) occurred as well as desaturation of this to both oleic (18 : 1) and linoleic (18 : 2) acids. These workers then went on to examine growth of this yeast in a wide range of different fatty acids. Even unusual fatty acids not found in microorganisms, such as petroselenic acid (18 : 1 *c* 6) or eleostearic acid (18 : 3 *t* 5, *t* 8, *c* 11), were significantly incorporated into the yeast lipids indicating that there is little problem in accommodating such fatty acids, at least in the storage triacylglycerol fraction of the yeast. No examination of the phospholipid components of the yeast was made to see if the acyl groups here were equally diverse but there is no real reason to consider that at least some of these acids could not have been incorporated into metabolically-active phospholipid membranes. Such is certainly the case with *Saccharomyces cerevisiae* (Pilkington and Rose 1991).

In summary, most microorganisms now appear able to utilize either fatty acids *per se* or triacylglycerols. The latter requires the presence of lipases that must at least be located at the cell surface as the triacylglycerol oil is unable to penetrate into the cell interior. Fatty acids from either source can be assimilated as such into the lipids of the cell. Fatty acids may though be modified by elongation and/or desaturation to ensure that the functional lipids of membranes etc are maintained. Fatty acids that have the necessary chain length and degree of flexibility by virtue of their double bonds can be accommodated even though they may not be synthesizable by that particular microorganism. When a cell grows on a fatty acid or oil, some of the substrate is also metabolized to yield energy and carbon intermediates for cell synthesis. Thus there is a partition between accommodation of the substrate directly into the cell lipids and its degradation. How this is accomplished is detailed in Section IV.

III. Lipases and phospholipases

From sections II.A, and II.C, it will be apparent that the first step in the microbial degradation of an oil or fat is its hydrolysis to release the component fatty acyl groups. Triacylglycerols and their partially hydrolysed products, di- and mono-acylglycerols, are not assimilated as such (see II.C).

Although it is obviously necessary for a microorganism to be able to elaborate a lipase for it to be able to grow on an oil or fat, it would appear that most microorganisms are able to do so. Also most microorganisms must be able to degrade their own lipid material if this should be required during a period of starvation or if changes in the cell constitution are required because of changes in the prevailing environmental conditions. Few if any cell components are not subjected to turn-over at some stage. Thus it should be the exception, rather than the rule, to find a microorganism devoid of any lipase or phospholipase activity.

A. *Lipases*

Lipases (triacylglycerol acylhydrolase, EC 3.1.1.3) are distributed throughout all living cells – bacteria, eukaryotic microorganisms, lower and higher plants and animals. The succession of hydrolytic reactions are each reversible (Fig. 6) but under normal aqueous conditions, where water is in abundance, the synthetic reactions (see Figs. 6 and 7) cannot occur against the concentration of free water (55 mol/litre). However under low water conditions (with water at or less than 1% of the reactive volume), ester synthesis can occur. Such reactions are now of considerable biotechnological importance as a wide range of esters can be produced using "natural" conditions which then has the advantage of producing desirable products that are "nature-identical".

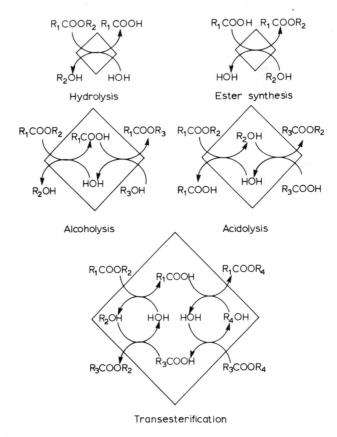

Fig. 6. Schematic representation of the reactions catalysed by lipases. R_1, R_2 etc. are alkyl chains. The diamonds enclose the processes that take place the active site at the enzyme but do not contribute to the overall stoichiometry. The compounds located outside the diamonds denote reactants or products for the reaction in question. (From Malcata et al. 1992.)

1. Reactions of lipases

Lipases will also catalyse the exchange-reactions of interesterification (or acidolysis) and transesterification (see Figs. 6 and 8) in which replacement reactions occur that are stereospecific and lead to the formation of new triacylglycerols. If the fatty acid (or fatty acyl group) to be transesterified (or interesterified) is chosen correctly then, under low or zero water conditions, high yields of the desired new triacylglycerol can be produced. In this way, for example, the commercially desirable cocoa butter substitute can be produced

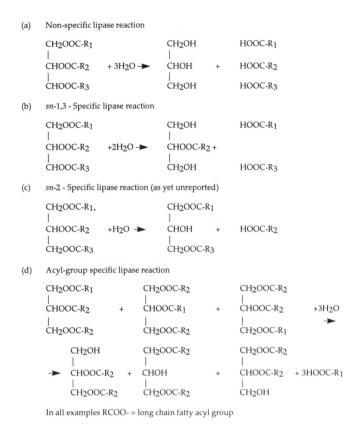

Fig. 7. Hydrolytic reactions of lipases.

from cheaper palm oil fractions being reacted with stearic acid or tristearoylglycerol. A discussion of such biosynthetic reactions is though beyond the scope of this book and the reader is referred to the following reviews and papers for further information on lipase-catalysed reactions in low water systems: Baratti et al. (1986), Dordick (1989), Halling (1990), Klibanov (1990), Laane et al. (1987) Lazar et al. (1986), Macrae (1983), Macrae and Hammond (1985), Miller et al. (1991), Ratledge (1989). Recent developments not only include reactions carried out in organic media but also in solvents such as supercritical CO_2 (Vermuë et al. 1992). A most useful description of the kinetics and reaction mechanisms has been recently provided by Malcata et al. (1992).

Figure 6 presents a synopsis of the reactions catalysed by lipases. These are then amplified, in part, in Figs. 7 and 8, where they relate to the degradation of lipids. Part of the reason for lipases functioning so well in organic media is their natural tendency to assosiate at any oil-water interface (see Fig. 10). This property has been reviewed (Brockman 1984) and forms the basis for the

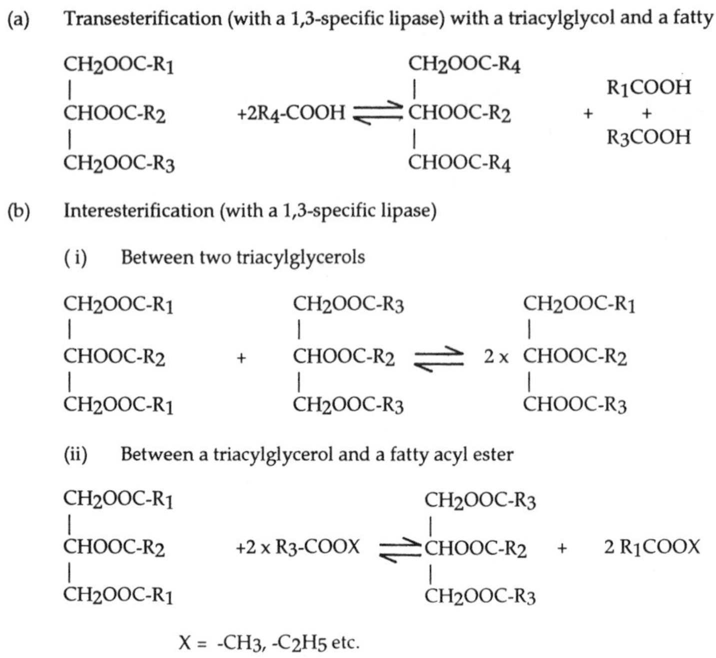

(a)   Transesterification (with a 1,3-specific lipase) with a triacylglycol and a fatty acid.

$$
\begin{array}{ll}
CH_2OOC\text{-}R_1 & CH_2OOC\text{-}R_4 \\
| & | \qquad\qquad R_1COOH \\
CHOOC\text{-}R_2 \quad +2R_4\text{-}COOH \rightleftharpoons CHOOC\text{-}R_2 \qquad + \quad + \\
| & | \qquad\qquad R_3COOH \\
CH_2OOC\text{-}R_3 & CHOOC\text{-}R_4
\end{array}
$$

(b)   Interesterification (with a 1,3-specific lipase)

(i)   Between two triacylglycerols

$$
\begin{array}{lll}
CH_2OOC\text{-}R_1 & CH_2OOC\text{-}R_3 & CH_2OOC\text{-}R_1 \\
| & | & | \\
CHOOC\text{-}R_2 \quad + & CHOOC\text{-}R_2 \rightleftharpoons & 2 \times \; CHOOC\text{-}R_2 \\
| & | & | \\
CH_2OOC\text{-}R_1 & CH_2OOC\text{-}R_3 & CHOOC\text{-}R_3
\end{array}
$$

(ii)   Between a triacylglycerol and a fatty acyl ester

$$
\begin{array}{ll}
CH_2OOC\text{-}R_1 & CH_2OOC\text{-}R_3 \\
| & | \\
CHOOC\text{-}R_2 \quad +2 \times R_3\text{-}COOX \rightleftharpoons CHOOC\text{-}R_2 \quad + \quad 2\, R_1COOX \\
| & | \\
CH_2OOC\text{-}R_1 & CH_2OOC\text{-}R_3
\end{array}
$$

X = -CH_3, -C_2H_5 etc.

Fig. 8. Trans- and inter-estification reactions of lipases.

enzyme being able to handle both water-soluble and water-insoluble substrates with equal alacrity (Malcata et al. 1992). Lipases will also function perfectly well in reverse-phased micelles whereby the lipase can be effectively trapped within the water pool of the micelles and are then protected from the non-polar environment by a layer of water. The theory behind this concept has been simply but elegantly described by Dickinson and Fletcher (1989) and a useful example for the hydrolysis of olive oil described by Han and Rhee (1986).

The present considerable interests in lipases have arisen from principally from their applications in industry (see Harwood 1989; Macrae 1986). In particular, these requirements are for lipases for use in fat-splitting reactions [see for example, Taylor and Craig (1991) where unstable fatty acids need to be produced and which would therefore involve some loss using conventional saponification – alkali-treatment – reactions], for use in low-temperature washing powders (which simply involves fat splitting *in situ* to remove greases and other oily residues on fabrics) and for the modification of oils and fats by the above mentioned reactions of interesterification, transesterification and ester synthesis (Macrae and Hammond 1985).

A great deal is now known about the widespread distribution of lipases, of how they function, of their specification and now more recently of their three-dimensional structure and genetics of synthesis. We have now reached the stage where direct modification of lipase properties by alteration of the amino acids of the primary sequence is now possible by site-directed mutagenesis, otherwise known as "Protein engineering". The following sections are intended to highlight some of these developments.

The distribution of lipases in microorganisms has been the subject of several recent reviews and research papers (Brune and Gotz 1992; Godtfredsen 1989; Iwai and Tsujisaka 1984; Koritala et al. 1987; Lazar and Schröder 1992 Stead 1986; Sztajer et al. 1988). It would now seem that it would be more unusual to find a microorganism *without* a lipase than to report yet another example of a lipase-producing microorganism. Nevertheless, the quest for unusual lipases, such as alkalophilic or thermostable ones, still continues (see for examples Savitha and Ratledge 1992; Ibrahim et al. 1987) and even for lipases with novel stereo- or regio-specificities (see below). Numerous methods have been reported for the assay of lipases (e.g. Kouker and Jaeger 1987; Lima et al. 1991; Miles et al. 1992; Shelley et al. 1987) but because of the diversity of procedures it is often difficult to make accurate comparisons between reports from different laboratories. These problems have been recently addressed by Vorderwülbecke et al. (1992) and some suggestions made for rationalization of lipase methodology. Mechanistically, as all lipases function at an oil-water interface (see Fig. 10), the contact of the enzyme with the water-insoluble substrate will always present a difficulty for *in vitro* assays. Various procedures for overcoming this problem have varied from using triacetyl- and tributyl-glycerols as substrate, which are water-soluble and thus overcome the obvious difficulty of immiscibility, to forming emulsions with any of a number of surfactant materials. All these changes can lead to a variation in reported activities for lipases that make subsequent comparisons impossible.

Lipases have been described which show either a non-specific hydrolysis of triacylglycerols (see Figs. 6 and 7a) or specificity for the acyl subsituent (Fig. 7d) or a specificity for the position at which hydrolysis occurs (Fig. 7b, c). Lipases in the first class – showing no regio-specificity – appear to be the most commonly found (Macrae and Hammond 1985) occurring both in bacteria (Brune and Gotz 1992) and fungi (Lazar and Schröder 1992). Lipases showing substrate specificity are unusual though triacylglycerols having very long chain polyunsaturated fatty acids, as found in oils from fish and marine organisms, may be slowly released by a number of lipases (Macrae and Hammond 1985) but the best known example is the lipase from *Geotrichum candidum*. This lipase only catalyses the hydrolysis of fatty acyl substituents with a double bond at the 9, 10-position and then only provided that there is no double bond between the $\Delta 9$ bond and the carboxylic ester group. Fatty acyl groups with *cis*- (or *Z*-) double bonds are hydrolysed more quickly than those with *trans*- (or *E*-) double bonds. Thus triacylglycerols containing (as examples) oleoyl- (18:1 $c9$), linoleoyl- (18:2 $c9$, $c12$) or, α-linolenoyl- (18:3, $c9$, $c12$, $c15$) groups are readily hydrolysed whereas those with substituents such as palmitoyl (16:0), stearoyl (18: 0), petroselenoyl (18:1, $c6$), γ-linolenoyl- (18:3, $c6$, $c9$, $c12$), or arachidonoyl (20:4, $c5$, $c8$, $c11$, $c14$) are not. Some suggestions have been made that other fungal enzymes may show similar specificities (see Ratledge 1989;

Lazar and Schröder 1992). This property may be particularly useful for the selective release and separation of long-chain polyunsaturated fatty acids from marine sources and for which there is a growing industrial interest (Kendrick and Ratledge 1990; 1992; and see also Kyle and Ratledge 1992). Some recent work on the lipase of this organism (Sidebottom et al. 1991) has indicated that not all strains of *G. candidum* produce an enzyme of such specificity.

The final group of lipases are those showing regio-specificity (see Fig. 7b and c). Only lipases showing 1, 3-regio-specificity are known; 2-specific lipases remain to be found. These enzymes are particularly useful for catalysing trans- and inter-esterification reactions (Fig. 8) and thus are used for the creation of improved triacyglycerols. A number of lipases show this type of selectivity. (Interesterification is also referred to as acidolysis – see Fig. 6.)

All the lipases mentioned so far will hydrolyse tri-, di- and mono-acylglycerols at approximately equal rates. It was therefore somewhat surprising that a novel lipase has been recently described that is strictly specific for di- and mono-acylglycerols and which would not attack any triacylglycerol (Yamaguchi and Mase 1991). This lipase was isolated from *Penicillium camembertii* (being the organism associated with the cheese of the same name). Earlier, a similar enzyme had been reported from *Pen. cyclopium* but this showed some limited hydrolytic activity towards triacylglycerols (Okumura et al. 1980). The enzyme from *P. camembertii* occurred as two isoenzymes (form A: $M_r = 37$ kDa; form B: $M_r = 39$ kDa). The major component, the B-form, was purified and shown to hydrolyse both *sn*-1 and *sn*-2 monoacylglycerols (the latter at about 70 to 60% of the rate of the former) and *sn* 1, 3- and *sn* 1, 2-diacylglycerols at varying rates depending on the acyl group constituents. No reaction occurred with any triacyglycerol from triacetyl- to tristearoyl- and trioleoyl-glycerols. Further work (Isobe et al. 1992) has resolved the original two lipases into four components which were each then crystallizabled. One of these four components corresponded to the original form A but the other three all arose from form B indicating a multiplicity of forms. Given that the enzymes are now available as crystals, further work should be able to establish the three dimensional structure of the enzyme. (The three-dimensional structures of other lipases are discussed below.)

Multiple forms of lipases from the same organism, as reported above for *Pen. camembertii*, have been reported earlier from *Rhizopus miehei* (Boel et al. 1988) and *Geotrichum candidum* (Baillargeon et al. 1989; Jacobsen et al. 1989a,b; Sidebottom et al. 1991).

2. Genetic studies

With the high level of commercial interest in lipases, there has been a large number of recent papers which have described the cloning and expression of lipase genes from a number of bacteria and fungi (Aoyama et al. 1988; Chung et al. 1991a,b; Feller et al. 1990; 1991; Gotz et al. 1985; Haas et al. 1991; Ihara et al. 1991; Jorgensen et al. 1991; Kawaguchi et al. 1989; Kugimiya et al. 1986; Lee and Iandolo 1986; Shimada et al. 1990; Tan and Miller 1992). An extensive monograph on this subject has also recently appeared (Alberghina et al. 1991).

Similarity of the substrate-binding domains in both bacterial and fungal lipases has been suggested by Tan and Miller (1992) who compared the published DNA sequences, and thus derived putative amino acid sequences, of 13 organisms: 8 bacteria and 5 fungi (see Fig. 9). Although this suggested similarity of the five or six amino acids involved with substrate binding between the lipases of the disparate organisms was very striking, the remainder of the lipase molecule from *Pseudomonas fluorescens*, examined by Tan and Miller (1992), showed little similarity to the lipases synthesized by *Pseudomonas fragi, P. cepacia* or *Pseudomonas* nov. sp. 109 (Ihara et al. 1991; Jorgensen et al. 1991). Curiously though, there was very high homology of the lipase amino acid sequence with that reported for another strain of *P. fluorescens* (Chung et al. 1991a).

Ref	Organism	Substrate Binding Region										
1.	*Moraxella Ta*144	Leu	Gly	Ala	Ile	Gly	Trp	Ser	Met	Gly	Gly	Gly
2.	*Staphylacoccus hyicus*	Val	His	Phe	Ile	Gly	His	Ser	Met	Gly	Gly	Gln
3.	*Staphylacoccus aureus*	Val	His	Leu	Val	Gly	His	Ser	Met	Gly	Gly	Gln
4.	*Pseudomonas* nov. sp. 109	Val	Asn	Leu	Ile	Gly	His	Ser	His	Gly	Gly	Pro
5.	*Pseudomonas fragi* 3458	Val	Asn	Leu	Ile	Gly	His	Ser	Gln	Gly	Ala	Leu
6.	*Pseudomonas fragi* 12049	Val	Asn	Leu	Ile	Gly	His	Ser	Gln	Gly	Ala	Leu
7.	*Pseudomonas cepacia*	Val	Asn	Leu	Val	Gly	His	Ser	Gln	Gly	Gly	Leu
8.	*Pseudomonas fluorescens* B52	Val	Val	Val	Ser	Gly	His	Ser	Leu	Gly	Gly	Leu
9.	*Rhizopus delemar*	Val	Ile	Val	Thr	Gly	His	Ser	Leu	Gly	Gly	Ala
10.	*Rhizomucor miehei*	Val	Ala	Val	Thr	Gly	His	Ser	Leu	Gly	Gly	Ala
11.	*Geotrichum candidum* I	Val	Met	Ile	Phe	Gly	Glu	Ser	Ala	Gly	Ala	Met
11.	*Geotrichum candidum* II	Val	Met	Ile	Phe	Gly	Glu	Ser	Ala	Gly	Ala	Met
12.	*Candida cylindracea*	Val	Thr	Ile	Phe	Gly	Glu	Ser	Ala	Gly	Ser	Met

Fig. 9. Similarity (or Homology) amongst the substrate-binding domains of bacterial and fungal lipases as presented by Tan and Miller (1992) by comparison of data published by: (1) Feller et al. (1990); (2) Gotz et al. (1985); (3) Lee and Iandolo (1986); (4) Ihara et al. (1991); (5) Kugimiya et al. (1986); (6) Aoyama et al. (1988); (7) Jorgensen et al. (1991); (8) Tan and Miller (1992); (9) Haas et al. (1991); (10) Boel et al. (1988); (11) Shimada et al. (1990); (12) Kawaguchi et al. (1989).

Gilbert et al. (1991b) considered that the lipase they isolated from *Pseudomonas aeruginosa*, but not included in the comparison of Tan and Miller (1992) given in Fig. 9, was similar in properties including its *N*-terminal amino acid sequence to several other *Pseudomonas* lipases. These included other strains of *Ps. aeruginosa* (originally called *Ps. fluorescens* and studied by Nishioka et al. 1991; *Ps. pseudoalcigenes* (Andreoli et al. 1989; *Ps. cepacia* spp. (Nakanishi et al. 1989; Jorgensen et al. 1991) and an unnamed *Pseudomonas* sp. ATCC 21808 (Kordel and Schmid 1991). It was, however, regarded as distinct from those of *Ps. fragi* (Aoyama et al. 1988; Kugimiya et al. 1986), *Ps. fluorescens* (Bozoglu et al. 1984) and the lipases from several other bacteria. The lipase from *Ps. aeruginosa* (Gilbert et al. 1991a,b) possibly does not have the serine residue at the active site (see Fig. 9) and thus, together with the other evidence, suggests that as with other hydrolytic enzymes, such as the proteases, there may be families of lipases showing disparate active sites but nevertheless all catalysing similar reactions.

The unique mono- and di-acylglycerol lipase from *Pen. camembertii* mentioned in III.A.(1) has also been examined at the gene level (Yamaguchi et al. 1992) and, although this enzyme is characterized by its inability to hydrolyse triacylglycerols, it continues to show similarity of its putative primary structure to those lipases derived from other flamentous fungi such as *Mucor miehei*, and *Humicola lanuginosa* (Yamaguchi et al. 1992).

3. Molecular structures
The three-dimensional structures of several lipases have now been published: from *Mucor (Rhizomucor) miehei* (Brady et al. 1990; Brzozowski et al. 1991; Derewenda et al. 1992), *Geotrichum candidum* (Schrag et al. 1991) and also from human pancreas (Winkler et al. 1990; van Tilbeurgh et al. 1992). All the structures can now account for the well-established association of a lipase at an oil-water interface (Brockman 1984) as each shows a lipophilic sequence of amino acids that allows the enzyme to become attached to an oil droplet (see Fig. 10). This loop acts as a "lid" and, as it is pulled away from the remainder of the enzyme that continues to reside in the aqueous phase, it then opens up the catalytic site of the enzyme to receive the substrate. This phenomenon is described as interfacial activation. With pancreatic lipase, because of its requirement to function in the presence of bile salts, there is a small protein known as colipase that serves to activate the lipase still further by enhancing its association with the lipid interface (van Tilbeurgh et al. 1992). Such a colipase is not seen with the microbial lipases though the role of the surfactants produced by microorganisms when grown not only on hydrocarbons but any water-insoluble material including a triacylglycerol oil (see Chapter 3) could well have a significant role to play in enhancement of lipase activity. It is perhaps surprising that no one has yet examined the role of some of these glycolipids on lipase functionality as it has been known for some time that lipase production can be stimulated when cells are grown on alkanes (Breuil et al. 1978).

Although all the structures of the lipases so far elaborated show a serine protease-like triad of amino acids – serine, histidine and aspartic acid or glutamic acid – at the catalytic site, there are obviously modifications of this if

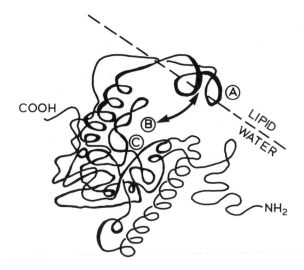

Fig. 10. Diagrammatic representation of the three-dimensional structure of a lipase (adapted from van Tilbeurgh et al. 1992; see also *Nature* 1992, Vol. 359, p. 107.) Domain A is the "lid" which is a short amphipathic helix which opens from region B to expose the active site, C, of the enzyme to its lipid substrate. This is known as interfacial activation.

the predictions shown in Fig. 9 are verified. The data in Fig. 9 indicates that there is a serine-histidine pair of amino acids but the third key amino acid is evidently glycine not Asp or Glu. Again, if the deductions of Gilbert et al. (1991b) are correct, not all lipases may use serine as the key amino acid at the active site (see above). It is not though essential to determine the complete three-dimensional structure of a lipase to determine if serine is involved at the catalylic site. This may be done simply but indirectly using a serine-reactive reagent such as 3,4-dichloroisocoumarin (Harper et al. 1985; Gilbert et al. 1991b) – though a negative result may not reveal very much. The alternative is to use site-directed mutagenesis which is obviously much more elaborate than using an inhibitor but nevertheless is simpler than carrying out a complete structural determination. Using this latter method, Yamaguchi et al. (1991) replaced two key serine residues (Ser[83] and Ser[145]) along with histidine at 259 with glycine in the lipase from *Penicillium camembertii* and rendered the lipase inactive. Also implicated to be involved at the active site was Asp[199]. These residues would then correspond to the amino acid-catalytic triad found in the three-dimensional structure of *M. miehei* lipase, that is Ser[144], Asp[203] and His[257] (Brady et al. 1990).

In summary, it is now clear that the initial difficulties in obtaining crystals of lipase have been overcome and substantial progress can be expected with this group of enzymes. It will be of considerable interest to learn if all lipases have a similar and conserved catalytic site involving a serine residue or whether, as some workers have suggested, there may be other types awaiting discovery. It is already evident that there may be little similarity of amino acid sequences away from the catalytic site. This perhaps is not significant or even surprising: but some consistency in functionality particularly of the "lid" region (see Fig. 10) might be expected. These fundamental studies should then lead to a more complete understanding of how lipases operate in their environment and help to explain differences in substrate- or regio-specificity. The prospects of being able to re-design the catalytic properties of a lipase – perhaps to introduce novel properties – are now realistic goals. Enhanced stabilities or activities under different operational conditions using techniques, such as site-directed mutagenesis, are now intriguing possibilities.

B. Phospholipases

In contrast to the huge volume of work that has been carried out in the last decade on lipases, there has been relatively little carried out with the phospholipases. This is principally because there is much less industrial interest in these enzymes than in lipases. Also phospholipases are intracellular enzymes and therefore are more difficult to isolate and purify.

Phospholipids are ubiquitous in all living cells forming the components of all membrane systems. They also possess a number of surfactant properties and therefore have numerous applications in the food industry (Szuhaj 1989). They also have important applications in the pharmaceutical industries and can be used for a number functions (Hanin and Ansell 1987), including a role as surfactant in the lungs of neonatal children (see for example, Bangham 1992).

Because of the inhertent turnover of all biological molecules, enzymes for the degradation of phospholipids may be expected to have an equally ubiquitous distribution as lipases and indeed phospholipases have been found in all forms of life (Dennis 1983; Waite 1987). They are frequently found to be membrane-associated enzymes (van den Bosch 1980). The activity of phospholipases must be under stringent control to prevent undesirable attack upon membrane structures during normal cell growth. The ability of phospholipases to disrupt membrane structures is attributed to their production of the partial (monoacyl-) phospholipids or lyso-phospholipids which are excellent surfactants that readily lyse biological membranes. Not surprisingly, phospholipases are the active components of snake venoms, some bacterial toxins and arthropod poisons (Dennis 1983). There has also been some suggestion that fungal pathogenicity, including that of *Candida albicans*, may be due at least in part to the presence of these enzymes (Barrett-Bee et al. 1985). Phospholipases may also function intracellularly as part of the cell separation process that occurs after yeast bud formation (Price and Cawson 1977).

Four types of phospholipase, attacking the diacylglycerolphospholipid molecule at distinct sites, are known (Fig. 11). The phospholipases A_1 and A_2 hydrolyse, respectively, the fatty acyl groups at the *sn*-1 and *sn*-2 positions;

phospholipases C and D hydrolyse the phosphoester bond on the glycerol sides or the polar head-group side, respectively. There is no phospholipase B. Following the action of phospholipase A_1 or A_2, a lysophospholipid is produced with the remaining acyl group at either the *sn*-1 or *sn*-2 position. Removal of the second acyl group is then considered to be carried out by a lysophospholipase of which two types are named: L_1 and L_2 depending on the position from which the last acyl group is removed. Enzymes loosely described as lecithinases (Zvyagintseva and Pitryuk 1975) are probably mixtures of phospholipase A_1 and A_2. Such enzymes have been described as generally distributed in yeasts but methods for their detection have relied upon zone-clearing by the yeasts on lecithin/agar plates. Clearly this technique is only of limited usefulness as it can detect extracellular activities whereas phospholipases are probably normally intracellular enzymes.

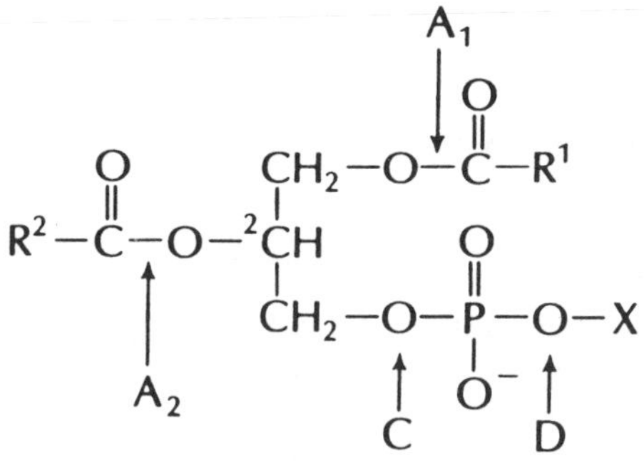

X = polar head group
R^1CO- and R^2CO- are fatty acyl groups

Fig. 11. Sites of cleavage of a phospholipid by phospholipases A, A_2, C and D. (It is uncertain which bond, C-O or P-O, is cleaved by the C and D enzymes – see Dennis 1983.)

Phospholipids may also be hydrolysed by lipases if the phospholipid and lipase can be appropriately emulsified together (Yamaoka et al. 1992). These authors by using reverse micelles, with the creation of a low-water environment, were able to show that the lipases from *Rhizopus delemar*, *Candida cylindracea* and *Pseudomonas* sp. could each catalyse the acylation of lysophosphactidylcholine to the same extent as the hydrolytic reaction.

Phospholipase A_2 is commercially produced from porcine pancreas and is used industrially to hydrolyse soybean lecithin to give lysolecithins for use as food emulisificants (Macrae 1986). The gene coding for this enzyme has now been transferred and expressed in *Escherichia coli* (Bhat et al. 1991) with a view to being able to modulate its activity by site-directed mutagenesis (Goodenough et al. 1991; Pickersgill et al. 1991).

Phospholipase C is the only genuine extracellularly occurring phospholipase and has been found in a member of spore-forming bacteria: *Clostridium* and *Bacillus* (Ikezawa and Tuguchi 1981; Little 1981; Low 1981; Takahashi et al. 1981). Some selectively for the polar head group may occur so that in some *Bacillus* spp. phosphatidylinositol is preferentially attacked (Ikezawa and Tuguchi 1981).

Phospholipase D is also produced commercially using cabbage as the starting material (see Ratledge 1989). It has been detected intracellularly in a number of moulds and yeasts including *Saccharomyces cerevisiae* (Chakravarti et al. 1987). There has been a report of it occurring extracellularly in the bacterium, *Actinomadura* (Kokusho et al. 1987).

The main interest in phospholipases A_1 and A_2, apart from generating lysophospholipids, is in their ability to catalyse acyl transfers. With phospholipase D, the main interest is in the transfer and exchange of the polar head group (see Ratledge 1989). Like the lipases, phospholipases can also function in organic solvents (Juneja et al. 1987).

IV. Uptake and oxidation of fatty acids

A. Transport processes

The preceding sections have indicated how microorganisms are able to grow on triacylglycerols and also unesterified fatty acids. With triacylglycerols and other acylated lipids, release of the fatty acids is a prerequisite for degradation as they are too bulky for direct uptake into the cell. Thus, degradation begins by the physical transport of long-chain fatty acids into the microbial cell.

In *Esch. coli*, transport of long-chain fatty acids across the cytoplasmic membrane requires the presence of specific transport proteins and is an energy-mediated process (Nunn 1986). Two functional genes, *fad*L and *fad*O, are required to achieve this uptake (Nunn et al. 1986). The former gene codes for a membrane-associated protein with a molecular size of 43 kDa which functions probably as a receptor and transporter of long-chain fatty acids. There is a correlation between the presence of this protein and fatty acid binding activity. This was the first membrane protein to have been shown, both genetically and biochemically, to be involved in the direct uptake of fatty acids (Maloy et al. 1981). Once the long-chain fatty acids (C_9 to C_{18}) have adsorbed on the receptor protein, they are transferred across the cytoplasmic membrane into the periplasmic space between the outer membrane and the inner membrane (see Fig. 12). Here they encounter the protein produced by the second gene, *fad*D, involved in the uptake process. This protein is an acyl-CoA synthetase which then activates the fatty acids into their CoA esters and these are then released into the cystoplasm.

If medium-chain fatty acids (C_7 to C_{11}) are being utilized by *E. coli*, these may be taken up either by the *fad*L-mediated process or, if this gene is

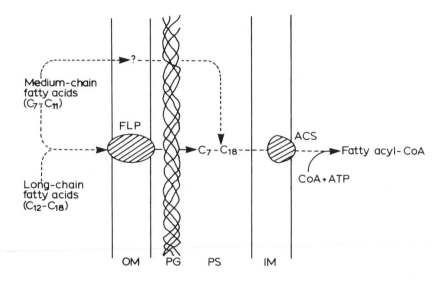

Fig. 12. Proposed model of fatty acid transport into *Esch. coli*. All fatty acids traverse the outer membrane via a 43 kDa membrane protein (FLP) which is the probable receptor. Medium-chain fatty acids (C_7 to C_{11}) may also enter the cell by a separate diffusional process. To traverse the inner membrane, fatty acids must be activated by the acyl-CoA synthetase (ACS) protein. OM = outer membrane (composed of a lipopolysaccharide/phospholipid bilayer); PG =peptidoglycan; PS = periplasmic space; IM = inner membrane (phospholipid/phospholipid bilayer). (Adapted from Nunn 1986.)

inactivated, by a simple diffusion mechanism (Nunn 1986). The fatty acyl-CoA synthetase, however, is still necessary for the entry of these shorter-chain fatty acids into the cytosol and thus this enzyme shows broad substrate specificity (Kameda and Nunn 1981).

In yeast cells, as exemplified by *Saccharomyces cerevisiae* (= *uvarum*) and *Yarrowia* (*Saccharomycoposis*) *lipolytica*, there are possibly two uptake processes (Kohlwein and Paltauf 1983). One involves a saturable, but energy-independent, process suggestive of a carrier-operated system, perhaps similar to the *fadL* gene product in *E. coli*, but it is not directly linked to the fatty acyl-CoA synthetase system. This lack of coupling can then lead to accumulation of small amounts of free fatty acid within the cell under certain circumstances. The second system in the yeasts for fatty acid uptake is by direct diffusion not involving any carrier proteins. This process only operates when high concentrations of free fatty acids are being used as substrates. At low concentrations of fatty acids, only the carrier-mediated process is considered to operate. Although both processes lead to the direct entry of fatty acids into the cell, the toxicity of fatty acids to cytoplasmic enzymes and organelles is prevented by the presence of apparently specific intracellular fatty acid-binding proteins that are associated with the cytoplasmic membrane (Kohlwein and Paltauf 1983).

B. Activation of fatty acids

It is important in all cells, whether microbial or otherwise, that free fatty acids within the cytoplasm are effectively eliminated. In *E. coli*, and presumably other Gram-negative bacteria, this is arranged by ensuring translocation across the inner membrane is mediated by the fatty acyl-CoA synthetase (Fig. 12). In yeasts, the putative presence of fatty acid-binding proteins can ensure a temporary holding of the fatty acids before they are transferred to the fatty acyl-CoA synthetase.

The fatty acyl-CoA synthetase therefore fulfills a dual role: it activates the fatty acid, prior either to its transfer into cell lipids or to its oxidative degradation, and also effectively detoxifies the fatty acid as the CoA esters are not inhibitory. Fatty acyl-CoA esters can though be powerful intracellular modulators and have been shown to be strong inhibitors of a number of key processes including the biosynthesis of acetyl-CoA for fatty acid formation in oleaginous yeasts (Evans and Ratledge 1985; Ratledge and Evans 1989).

The reaction catalysed by the fatty acyl-CoA synthetase is:

$$RCOOH + HS\text{-}CoA + ATP \rightarrow RCO\text{-}S\text{-}CoA + AMP + PP_i + H_2O \qquad (1)$$

Although the reaction is theoretically revisible, this is prevented in practice by the reaction generating AMP and pyrophosphate rather than ADP and ortho-phosphate; the pyrophosphate is readily hydrolysed irreversibly by a pyrophosphatase which then ensure irreversibility of the first reaction.

The enzyme activity, together with the subsequent β-oxidation system (see below), is induced by growing microorganisms on oils, fatty acids or alkanes (Nunn 1986; Ratledge and Evans 1989). There are though some exceptions: for example in *Caulobacter crescentus* there is a high constitutive activity of the enzyme in succinate-grown cells and, although some increase in activity occurs when a fatty acid is used as a substrate, there is only moderate catabolite repression (2- to 3-fold) when glucose is used and even this effort is abolished if cAMP is added (O'Connell et al. 1986).

Two separate long-chain fatty acyl-CoA synthetases – I and II – occur in *Yarrowia lipolytica* (Mishina et al. 1978a,b; Numa 1981; Tanaka and Fukui 1989). A similar arrangement may also occur in other yeasts (Holdsworth and Ratledge 1991). Synthetase I occurs in both glucose- and oleate-grown cells and is considered constitutive being implicated in the formation of fatty acyl-CoA esters destined for incorporation into acylglycerols, including phosphoacylglycerols. Synthetase II only occurs when cells are grown on oleate, or alkanes. Unlike synthetase I, synthetase II will also activate long-chain dicarbocylic acids and ω-hydroxy fatty acids albeit at a slower rate than for saturated or unsaturated fatty acods (Mishina et al. 1978a). The two enzymes can also be distinguished immunologically (Hosaka et al. 1979). The functional differences of the two enzymes though is to be seen in their different cellular locations: synthetase I is located in the endoplasmic reticulum (or microsomal) fraction and synthetase II is within the peroxisome (or microbody) organelle. These locations are shown in Fig. 15 (see later). Synthetase II is therefore associated with the β-oxidation process for the degradation of fatty acids; synthetase I serves to provide acyl-CoA esters for lipid synthesis and accordingly must function whatever substrate is being used as lipids must always be formed *in situ*. Details of how triacylglycerols and phospholipids are synthesized from fatty acyl-CoA esters have been reviewed by Pieringer (1989).

C. β-Oxidation of saturated fatty acids

The principal route by which fatty acids are oxidized in cells is by the sequence known as the β-oxidation cycle (Fig. 13). This cycle involves four separate reactions.

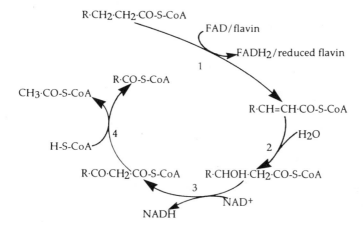

Fig. 13. β-Oxidation cycle. Individual enzyme reactions: 1. Fatty acyl-CoA dehydrogenase (bacteria and mitochondrial, FAD-linked) or fatty acyl-CoA oxidase (peroxisomal, flavo-enzyme); 2. 2,3-enoyl-CoA hydratase (also known as crotonase); 3. 3-hydroxyacyl-CoA dehydrogenase (NAD$^+$-dependent); 4. 3-oxoacyl-CoA thiolase.

1. Fatty acyl-CoA dehydrogenase/oxidase
The reaction catalysed in microorganisms is ostensibly the same:

$$R\text{-}CH_2\text{·}CH_w\text{·}CO\text{-}S\text{-}CoA + FAD/flavin \rightarrow R\text{-}CH = CH\text{·}CO\text{-}S\text{-}CoA + FADH_2/\text{reduced flavin} \tag{2}$$

The product is the *trans* (or *E*) isomer.

The difference between bacteria and yeasts is that the former use a dehydrogenase in which the FADH$_2$ is reoxidized through the process of oxidative phosphosylation with the production of ATP. In yeasts (but perhaps not in moulds), the enzyme is a flavo-complex in which the flavin is only re-oxidized by molecular oxygen:

$$\text{flavin (red.)} + O_2 \rightarrow \text{flavin (ox.)} + H_2O_2 \tag{3}$$

The hydrogen peroxide is then cleaved by catalase:

$$H_2O_2 \rightarrow H_2O + \tfrac{1}{2}O_2 \tag{4}$$

This latter system therefore does not generate usable metabolic energy (as ATP) but instead liberates it from reaction (4) in the form of heat. Although this may appear to be energetically wasteful, fatty acids (and alkanes) are considered to be "energy-rich, carbon-poor" substrates (Linton and Stephenson 1978). In other words, the energy content of a fatty acid is more than the cell needs for anabolic reactions in proportion to the carbon units that will be generated from its degradation (see Ratledge 1987; 1991 and Boulton and Ratledge 1984 where this point is discussed in greater detail).

In the few moulds that have been studied for β-oxidation (*Aspergillus tamarii, Neurospora crassa*), reaction (2) is catalysed by a dehydrogenase and not an oxidase (Kunau et al. 1988). The activity though is still an induced one and is associated with the microbodies (Kionka and Kunau 1985) but, unlike the organelles from the yeasts, plants or animals, these microbodies do not contain catalase.

The fatty acyl-CoA oxidase was first reported in *Candida tropicalis* by Kawamoto et al. (1978) and subsequently in *Yarrowia* (*Candida*) *lipolytica* by Mishina et al. (1978b). Dommes et al. (1983) reported that in *C. tropicalis*, activities of both the oxidase and the corresponding dehydrogenase could be detected. The latter enzyme was predominant in glucose-grown cells and was located in the mitochondrion rather than peroxisomes, which are not in evidence in glucose-grown cells (Veenhuis and Harder 1991). When the yeast was grown on oleate the dehydrogenase, still within the mitochondrion, increased four-fold in activity but the oxidase went from an insignificant activity to being 13 times more active than the dehydrogenase. Thus, a low activity β-oxidation cycle occurs in the mitochondrion of this yeast, and it could be suggested that this may be to deal with small amounts of fatty acids that may be turned over *in situ*. However when yeasts are induced to consume their own storage triacylglycerols, peroxisomes are seen to be induced (Holdsworth and Ratledge 1988) and with them peroxisome-associated enzymes (Holdsworth et al. 1988). The occurrence of the oxidase has also been recognized in *S. cerevisiae* (Skoneczny et al. 1988; Kunau et al. 1988).

The nucleotide sequences of the DNA coding for fatty acyl-CoA oxidase have been reported for both *C. tropicalis* (Rachubinski et al. 1985; Murray and Rachubinski 1987) and *Candida maltosa* (Hill et al. 1988). Somewhat confusingly, Okazaki et al. (1986) reported that there were two genes encoding for fatty acyl-CoA oxidases in *C. tropicalis*. One gene product was considered to be the true oxidase and the other product to be probably part of another peroxisomal enzyme. Such a proposal would be in keeping with current concepts about multiplicity of peroxisomal enzymes and multifunctional proteins in the peroxisomes (Kunau et al. 1988; Veenhuis and Harder 1991; and see also section IV.D).

In contrast to the work carried out on the yeast enzyme, very little has been reported on the corresponding bacterial acyl-CoA dehydrogenase enzyme though mutants of *E. coli* lacking this activity were reported some time ago (Klein et al. 1971). Interestingly, however, this enzyme does not form part of the multi-enzyme complex that carries out the remaining activities of the β-oxidation cycle (Nunn 1986; Yang et al. 1986). (This aspect is discussed later in Section IV.D.)

2. 2,3-Enoyl-CoA hydratase (crotonase)
The reaction catalysed is the same in all cells:

$$\text{R-CH=CH·CO-S-CoA} + H_2O \rightarrow \text{R-CHOH·CH}_2\text{·CO-S-CoA} \quad (5)$$

The product is the L-isomer (in contrast to the D-isomer that is formed during the biosynthesis of fatty acids – see Schweizer 1989).

This enzyme has been purified from a number of yeasts and moulds where it has been shown to be part of a single, multi-functional protein, see Section IV.D. The organization of the hydratase in *E. coli* is also discussed in Section IV.D in relation to the structural organization of the β-oxidation cycle enzymes.

3. 3-Hydroxyacyl-CoA dehydrogenase
This reaction is again the same in all cells.

$$\text{R-CHOH·CH}_2\text{·CO-S-CoA} + NAD^+ \rightarrow \text{R-CO.CH}_2\text{·CO-S-CoA} + \atop \text{NADH} \quad (6)$$

The enzyme is entirely specific for the L-isomer. In bacteria, the NADH generated is linked directly via oxidative phosphorylation to the generation of ATP. In yeasts, as the reaction takes place in the peroxisome (see Fig. 15), direct coupling to oxidative phosphorylation is not possible as this is, of course, an activity associated with the mitochondrion. Accordingly, the NADH is re-oxidized by linkage to glycerol-3- phosphate dehydrogenase (see Fig. 15):

$$\text{NADH} + \text{dihydroxyacetone phosphate} \rightarrow NAD^+ + \text{glycerol 3-phosphate} \quad (7)$$

4. 3-Oxoacyl-CoA thiolase (thiolase)
The final reaction of the cycle involves the cleavage of the acyl chain to split off acetyl-CoA leaving an acyl-CoA ester that is now two C atoms shorter than the initial acyl chain.

$$\text{R-CO·CH}_2\text{·CO-S-CoA} + \text{CoA} \rightarrow \text{R-CO-S-CoA} + CH_3\,\text{CO-S-CoA} \quad (8)$$

The shortened fatty acyl-CoA ester then repeats the sequence (see Fig. 13) until the acyl chain reaches the C_4 level. The final thiolase reaction is therefore the cleavage of acetoacetyl-CoA ($CH_3\text{·CO·CH}_2\text{·CO-S-CoA}$) with the formation of 2 acetyl-CoAs.

The overall sequence of reactions for the degradation of, say, palmitic acid (C16) would therefore be:

$$\text{palmitate} + 8\,\text{CoA} + \text{ATP} + 7\,\text{FAD} + 7\,\text{NAD}^+\,7\,\text{H}_2\text{O} \rightarrow 8\,\text{acetyl-CoA} \\ + \text{AMP} + \text{PP}_i + 7\,\text{FADH}_2 + 7\,\text{NADH} \tag{9}$$

The acetyl-CoA will of course be used, in part, for provision of cell intermediates (see below and Fig. 15) but also part will degraded via the tricarboxylic acid cycle and glyoxylate by-pass to generate both CO_2, water and ATP. As each $CH_3\,COOH$, when completely oxidized, yields $2CO_2 + 2H_2O$, the net yield of water in the degradation of palmitic acid would be 9 mol H_2O. Thus the degradation of fatty acids not only yields utilizable metabolic energy, but also serves as a provider of water which may be of key significance for some organisms growing under low-water conditions which would, of course, include degradation of oils and fats.

D. β-Oxidation of unsaturated fatty acids

For unsaturated acids, such as oleic acid, 18:1 (c9), the β-oxidation cycle can proceed only for three complete sequences before a metabolic blockage occurs. At this point, the product from oleate is 3-cis-dodecanoyl-CoA (see Fig. 14A) which has the double bond in the wrong protein (it should be at the 2-position) and in the wrong configuration (it should be trans not cis). Accordingly, an isomerase (cis-Δ3, trans-Δ2-enoyl-CoA isomerase) then converts this to the corresponding 2-trans isomer. This then continues in the β-oxidation sequence as before: hydratase, dehydrogenase and thiolase to give decanoyl-CoA which is then handled without further deviations.

For an acid such as linoleic acid, (18:2 c9, c12), a further enzyme to the 2,3-enoyl-CoA isomerase is needed which is a 2,4-dienoyl reductase (see Fig. 14B). This converts the intermediate 2 (trans), 4 (cis)-deca-diene-oyl-CoA ester to the 3 (trans)-decenoyl-CoA. The 2, 3-enoyl-CoA isomerase now acts again and converts this to the 2-enoyl isomer and β-oxidation can then proceed as before.

These two enzymes, 2, 3-enoyl-CoA isomerase and the 2, 4- dienoyl-CoA reductase are induced simultaneously with the four "standard" β-oxidation enzymes in Candida tropicalis grown on oleate (even though the latter activity is only required if linoleic acid is being degraded) (Dommes et al. 1983).

The pathway proposed by Dommes et al. (1983), and reviewed by Schulz and Kunau (1987) for the degradation of linoleic acid, differs from a previous proposal by Stoffel and Caesar (1965) which is known as the epimerase pathway. Here the key differences lies in the fate of cis-4-decenoyl-CoA (see Fig. 14B). In the epimerase pathway, this intermediate was considered to undergo a further sequences of the β-oxidation cycle producing cis-2-octenoyl-CoA. This was hydrated to yield the 3-D-hydroxyoctanyl-CoA which had then

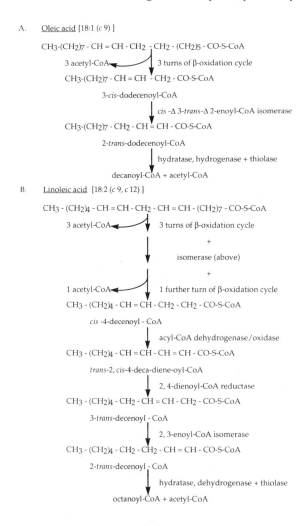

A. Oleic acid [18:1 (*c* 9)]

CH_3-$(CH_2)_7$ - CH = CH - CH_2 - CH_2 - $(CH_2)_5$ - CO-S-CoA

3 acetyl-CoA ⟵ 3 turns of β-oxidation cycle

CH_3-$(CH_2)_7$ - CH = CH - CH_2 - CO-S-CoA

3-*cis*-dodecenoyl-CoA

cis -Δ 3-*trans*-Δ 2-enoyl-CoA isomerase

CH_3-$(CH_2)_7$ - CH_2 - CH = CH - CO-S-CoA

2-*trans*-dodecenoyl-CoA

hydratase, hydrogenase + thiolase

decanoyl-CoA + acetyl-CoA

B. Linoleic acid [18:2 (*c* 9, *c* 12)]

CH_3 - $(CH_2)_4$ - CH = CH - CH_2 - CH = CH - $(CH_2)_7$ - CO-S-CoA

3 acetyl-CoA ⟵ 3 turns of β-oxidation cycle

+

isomerase (above)

+

1 acetyl-CoA ⟵ 1 further turn of β-oxidation cycle

CH_3 - $(CH_2)_4$ - CH = CH - CH_2 - CH_2 - CO-S-CoA

cis -4-decenoyl - CoA

acyl-CoA dehydrogenase/oxidase

CH_3 - $(CH_2)_4$ - CH = CH - CH = CH - CO-S-CoA

trans-2, *cis*-4-deca-diene-oyl-CoA

2, 4-dienoyl-CoA reductase

CH_3 - $(CH_2)_4$ - CH_2 - CH = CH - CH_2 - CO-S-CoA

3-*trans*-decenoyl - CoA

2, 3-enoyl-CoA isomerase

CH_3 - $(CH_2)_4$ - CH_2 - CH_2 - CH = CH - CO-S-CoA

2-*trans*-decenoyl - CoA

hydratase, dehydrogenase + thiolase

octanoyl-CoA + acetyl-CoA

Fig. 14. β-Oxidation of sequences of unsaturated fatty acids. A. Oleic acid. B. Linoleic acid.

to be epimerized to the 3-L-hydroxy ester to enable the cycle to continue. The envisaged sequence was thus:

10:1 (*c*4) → 8:1 (*c*2) → 3-D HO-8:0 → 3-L HO-8:0 → 3-keto-8:0 → 6:0   (10)

Although this sequence was originally proposed for mammalian liver tissue, and indeed this epimerase is reported to occur in bacteria degrading fatty acids (Nunn 1986), Dommes et al. (1983) could find no evidence for its occurrence in yeast. Related work by the same group though would even cast serious doubts upon whether this epimerase route does occur in mammalian tissues (Dommes et al. 1981; Schulz and Kunau 1987; Hiltunen et al. 1989).

E. Non-mitochondrial organization of β-oxidation cycle enzymes

There is a remarkable degree of structural organization in the enzymes of the β-oxidation system (Kunau et al. 1988). (In mitochondria, all enzymes appear as separable, mono-functional proteins.)

The system in bacteria (see Nunn 1986) consists of a multi-functional complex of two protein sub-units assembled in an $\alpha_2\beta_2$ structure. The α-subunit has a molecular size of 78 kDa and the β-unit is 42 kDa; the total complex is therefore 240 kDa. One sub-unit (the β) carries out only the thiolase activity (rn. 8) and the other carries out the crotonase and 3-hydroxyacyl-CoA dehydrogenase reactions (rns. 5 and 6) as well as the 2, 3-enoyl-CoA isomerase and 3-hydroxyacyl-CoA epimerase reactions for handling unsaturated fatty acids (Yang and Schultz 1983). The location of the acyl-CoA dehydrogenase is not yet established. Such an assemblage of activities is highly efficient in catalysing the β-oxidation sequences of reactions (Yang et al. 1986). This system is considered to channel the product of one reaction directly into the following enzyme. Free intermediates are not thought to be produced so tight is the coupling. However, whether this view can be upheld by closer scrutiny is uncertain in view of results with a different bacterium, *Corynebacterium* sp. 7E1C: Broadway et al. (1992) succeeded in detecting all the saturated acyl-CoA from C_{16} to C_8 during the oxidation of palmitic acid (16:0) by cell-free extracts. Furthermore, if this organism was grown on succinate, trace amounts of the unsaturated, enoylacyl-CoA and 3-hydroxy-acyl-CoA esters were also detected using HPLC from the degradation of palmitic acid. These results then challenge the previous concept that intermediates are channelled through the cycle without ever appearing as free entities. All that was required in this work was careful analysis and an understanding of the appropriate enzyme kinetics. Similar work, however, would have to be carried out using *E. coli*, and perhaps with yeast systems, to verify that these results were of general applicability in microorganisms.

In yeasts, the peroxisomal β-oxidation enzymes have been studied in some detail where they have long been known to be associated with the peroxisomal (or microbody) organelle (see Tanaka and Fukui 1989; Veenhuis and Harder 1991 and Kunau and Hartig 1992 for recent reviews). The peroxisomal location for the β-oxidation cycle appears to be widespread in yeasts (Veenhuis et al. 1987; McCammon et al. 1990).

In the organization of the β-oxidation cycle enzymes themselves, it was initially thought (Ueda et al. 1987) that there was a bi-functional protein ($M_r = 105$ kDa) that catalysed both the crotonase (rn. 5) and the 3-hydroxylacyl dehydrogenase (rn. 6) activities. Other and more recent evidence though suggests that this is a trifunctional protein of the same size that catalyses, in addition to the above reactions, 3-hydroxyacyl-CoA epimerase (Moreno de la Garza et al. 1985; Hiltunen et al. 1992). The other enzyme activities are though still considered to be associated with separable mono-functional proteins (Kunau et al. 1988). In all five yeasts and fungi that have been examined (*C. tropicalis*, *Y. lipolytica*, *S. cerevisiae*, *N. crassa* and *Asp. tamarii*) there are the same trifunctional proteins and apparently similar arrangements.

The spatial organization of the β-oxidation enzymes in yeasts has also been studied in some detail (Veenhuis et al. 1987; Tanaka and Fukui 1989; Veenhuis and Harder 1991). In contrast, in bacteria very little is known if there is any specific intracellular location for these enzymes. It is usually supposed with pathways which are in opposition, as with β-oxidation and fatty acid synthesis, that they do not operate simultaneously and, moreover, do not interfere with each other. Control of synthesis versus degradation is achieved by stringent control mechanisms being in operation to prevent futile cycling between the opposed pathways. In yeasts, fatty acid biosynthesis is prevented during fatty acid degradation by the fatty acyl-CoA esters inhibiting the early steps of fatty acid biosynthesis (Gill and Ratledge 1973; Boulton and Ratledge 1984; Ratledge and Evans 1989) and, moreover and more importantly, by the latter sequence occurring within the peroxisomes whereas the former is a cytosolic activity.

The cellular organization in yeasts of the enzymes relating to fatty acid and alkane breakdown is shown in Fig. 15. The first enzymes for alkane oxidation (alkane hydroxylase, fatty alcohol oxidase and fatty aldehyde dehydrogenase) are generally considered to be located within the membrane system of the endoplasmic reticulum which is described as the microsomal fraction upon cell disruption and isolation of the component parts (Mauersberger et al. 1987; Blasig et al. 1988; Kemp et al. 1988). The fatty alcohol oxidase, like the fatty acyl-CoA oxidase reaction (rn. 2), uses a flavin co-factor (Dickinson and Wadforth 1992) and therefore needs to be coupled to catalase to regenerate the oxidized flavin (rns. 3 and 4). However, it is usually thought that catalase is a peroxisomal enzyme and so the exact position of the fatty alcohol oxidase between the microsomal fraction and peroxisomal is as yet unclear (Mauersberger et al. 1987; Kemp 1988). It is possible that the fatty alcohol is oxidized on or within the peroxisomal membrane; the fatty aldehyde may also be oxidized there and the resultant fatty acid could be proportioned between the microsomal membranes and the peroxisome interior by virtue of the two fatty acyl-CoA synthetases (see IV.B).

The manner whereby the products of β-oxidation are re-cycled and linked to those within the mitochondrion are also shown in Fig. 15. The final "end-product" that leaves the perioxome to contribute to cell biosynthesis is uncertain. Figure 15 indicates this to be malate but Tanaka and Fukui (1989) considered α-ketoglutarate could be the end-metabolite. In practice, however, the cell will probably draw its intermediates as required from either the peroxisome or mitochondrion and a simplistic view (as in Fig. 15) is possibly unwarranted. It should though be pointed out that for any cell to be able to grow on fatty acids, the glyoxylate cycle must be induced so that C_4 units can be effectively synthesized from the acetyl-CoA. The glyoxylate cycle (or by-pass) requires the induction of two enzymes: isocitrate lyase and malate synthase

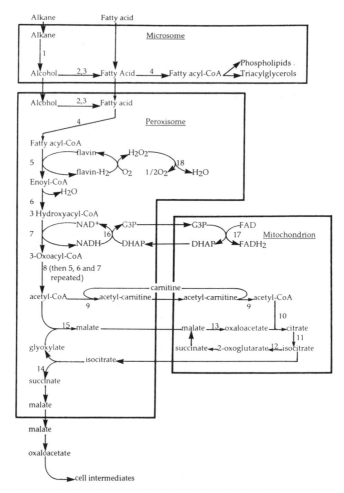

Fig. 15. Spatial organization of enzymes involved in the oxidation of alkanes, fatty alcohols and fatty acids in eukaryotic organisms. Enzymes: 1. alkane hydroxylase; 2. fatty alcohol oxidase; 3. fatty aldehyde dehydrogenase; 4. fatty acyl-CoA synthetase; 5. fatty acyl-CoA oxidase; 6. enoyl-CoA dehydrogenase; 7. 3-hydroxyacyl-CoA dehydrogenase; 8. 3-oxoacyl-CoA thiolase; 9. carnitine acetyl-transferase; 10. citrate synthase; 11. aconitase; 12. isocitrate dehydrogenase; 13. malate dehydrogenase; 14. isocitrate lyase; 15. malate synthase; 16. glyceraldehyde-3-phosphate dehydrogenase (NAD⁺-linked); 17. glyceraldehyde-3-phosphate dehydrogenase (FAD-linked); 18. catalase.

(reactions 14 and 15 in Fig. 15). These enzymes are peroxisomally-located in yeasts (Kawamoto et al. 1978; Tanaka and Fukui 1989) but it is unknown if they have any particular location or association in bacteria. The glyoxylate cycle enzymes occur along with the tricarboxylic acid cycle enzymes and not in place of them.

F. Other routes of fatty acid oxidation

1. Formation of methyl ketones in fungi

The partial oxidation of medium chain-length fatty acids (C_4 to C_{12}), or triacylglycerols that contain them, to methyl ketones has been known for many years (Stärkle 1924; Stokoe 1928). These conversions are commercially important as they form the basis of flavour development in blue cheeses (see Kinsella and Hwang 1976) and ketonic randicity in butter fat (Stärkle 1924), coconut oil (Stokoe 1928) and, more recently, dessicated coconut (Kellard et al. 1985) and coconut itself (Kinderlerer 1986; 1987).

The methyl ketone arises from fatty acids of chain length one carbon atom longer than the product, thus:

$$CH_3 (CH_2)_n CH_2 CH_2 COOH \rightarrow CH_3 (CH_2)_n CO CH_3 + CO_2$$
$$\text{where } n = 0 \text{ to } 8 \tag{11}$$

These biotransformations have been studied by Kinderlerer and her group using a variety of moulds (Kellard et al. 1985; Kinderlerer and Kellard 1984; Hatton and Kinderlerer 1991; Kinderlerer and Hatton 1991; Kinderlerer 1986; 1987; 1992; 1993). The view has been advanced that this route of oxidation represents a rapid detoxification mechanism as, ordinarily, the medium chain-length fatty acids are much more toxic than the longer chain acids (see Section II.B). However, no shorter chain-length products are found in these oxidations and, where a range of methyl ketones have been produced (Table 2 and see Kinderlerer 1987; 1993), these may have arisen either from the initial mixed fatty acids in the original coconut oil or by a conventional β-oxidation cycle operating to produce a homologous series of fatty acids which then act as the substrate for the production of the methyl ketones.

The production of the methyl ketones by the fungi is, however, not obligatory. Fungi that produce them can grow on the medium-chain length saturated fatty acids (Hatton & Kinderlerer, 1991) and the extent of growth

Table 2. Aliphatic methyl ketones (μmol) produced by fermentation of coconut oil (1.5 g) for 3 days at 25° by different isolates of *Aspergillus ruber* and *A. repens* from International Mycological Institute (from Kinderler 1987).

Strain no.	A. repens			A. ruber	
	298304	298305	298308	298306	298309
2-Pentanone	3	2	2	ND	8
2-Hexanone	0.1	0.1	0.2	ND	0.1
2-Heptanone	32	17	45	0.3	80
2-Octanone	0.6	0.5	0.3	0.1	0.4
2-Nonanone	23	14	44	0.2	33
2-Undecanone	53	33	95	2	82
2-Tridecanone	ND	0.1	ND	ND	1

ND = not detected.

and ketone production can be varied. Ketone production is favoured if growth is constrained in some way by, for example, the prevailing O_2 or CO_2 concentration, the pH, the nature of the fatty acids or their concentration (Lawrence 1966; Lawrence and Hawke 1968).

The mechanism of production is uncertain: the considered view (Kinderlerer 1993) is that the initial oxidation of the saturated fatty acid may involve a β-oxidation system (IV.C) but which is incomplete. The first cycle is broken when the 3-oxoacyl-CoA ester is deacylated and the free oxo-acid is then decarboxylated to the methyl ketone:

$$R\text{-}CH_2 \cdot CH_2 \cdot CO \cdot S \cdot CoA \rightarrow \rightarrow R\text{-}CO \cdot CH_2 \cdot CO\text{-}S\text{-}CoA \rightarrow R\text{-}CO \cdot CH_2 \cdot COOH$$
$$+ \text{HS-CoA} \quad (12)$$
$$R\text{-}CO \cdot CH_2 \cdot COOH \rightarrow R\text{-}CO \cdot CH_3 + CO_2$$

Occasionally, the corresponding secondary alcohol ($R\text{-}CHOH \cdot CH_3$) may be also released. This is probably formed by reduction of ketone (Engel et al. 1989).

The oxidations to methyl ketone can also be carried out by fungal spores as well as mycelium (Franke and Heinen 1958; Creuly et al. 1992). The activity appears to be fairly common amongst moulds and is not confined to those found associated with blue cheeses or with deteriorated oil-containing foods. The phenomenon is, however, apparently confined to fatty acids with a chain-length no longer than C_{12} (i.e. lauric acid).

The subsequent degradation of the methyl ketones is also uncertain though Yagi et al. (1991) have isolated a number of hydroxy compounds (7- and 8-hydroxynonan-2-ones) from 2-nonanone as well as dihydroxy derivatives. These may simply represent attack by residual hydroxylases at the opposite end of the molecule [see ω- and (ω−1)-oxidations below).

2. α-Oxidation

The occurrence of an α-oxidation pathway for fatty acids degradation in microorganisms is uncertain though evidence has been presented that this route occurs in plants (Kindl 1984). The essential reactions of this cycle are:

$$R\text{-}CH_2 \cdot COO^- \rightarrow R\text{-}CHOH \cdot COO^- \rightarrow R \cdot CHO + CO_2 \rightarrow R \cdot COO^- \quad (13)$$

Free fatty acids, and not their CoA esters, are apparently involved. The mechanism of the first reaction is uncertain. O_2 or H_2O_2 may be the oxidizing substrate with the formation of hydroperoxide ($R\text{-}CHOOH \cdot COO^-$) as the intermediate.

The presence of small amounts of 2-hydroxy-12:0 and 2-hydroxy-14:0 in a number of Gram-negative bacteria, including *Salmonella, Klebsiella, Serratia, Pseudomonas, Acinetobacter* and *Alcaligenes*, have been reported as part of the lipopolysaccharides of the cell envelope (Wilkinson 1988). The occurrence of such acids may suggest, but not prove, the presence of an α-oxidation route of fatty acids in bacteria.

In yeasts, very long chain α-hydroxy fatty acids have been reported as components of sphingolipids (see Ratledge and Evans 1989). These include 2-hydroxy-26:0 but their biosynthetic origin is also unclear.

3. ω-Oxidation

Fatty acids can be oxidized at their terminal methyl group by both bacteria and yeasts using an ω-hydroxylase system which is similar in mechanism to that involved in the initial oxidation of alkanes (see Chapter 1). However there is evidence that in yeasts (*Candida maltosa* and *C. tropicalis*) there may be a second cytochrome P-450 hydroxylase for the terminal oxidation of a fatty acid (Schunck et al. 1989; Sanglard and Loper 1989). There has been considerable commercial interest in the production of these ω-hydroxy acids as precursors of macrocyclic lactones (for example see Antczak et al. 1991; Ercoli et al. 1992, and references therein) and in their further oxidation to long-chain dicarboxylic acids (Casey and Macrae 1992; Casey et al. 1990; Furukawa et al. 1986; Hill et al. 1986). Numerous dicarboxylic acid-producing microbes have been patented (see Buhler and Schindler 1984). Such organisms have increased yields of the dicarboxylic acids by having their β-oxidation cycle blocked by mutagenesis.

The route of oxidation is thus:

$$CH_3\text{-}(CH_2)_n \cdot COOH \rightarrow HOCH_2 \cdot (CH_2)_n \cdot COOH \rightarrow$$
$$OCH \cdot (CH_2)_n \cdot COOH \rightarrow COOH \cdot (CH_2)_n \cdot COOH \tag{14}$$

Some suggestion has been made (Yi and Rehm 1982a,b,c) that the α, ω-diol could be an minor intermediate when alkanes were being used as feedstock. Such an intermediate would not though occur starting from a fatty acid.

Yi and Rehm (1988a,b) have demonstrated the conversion of unsaturated fatty acids, oleic acid (18:1, *c*9) and elaidic acid (18:1, *t*9), to the corresponding unsaturated dicarboxylic acids: Δ 9-*cis*-1,18-octadecenedioic acid and Δ 9-*trans*-1, 18-octadecenedioic acid, respectively, by *Candida tropicalis* (see Fig. 16). It would not be unreasonable to expect other mono-, di- or even poly-unsaturated fatty acids might not be oxidized similarly.

Despite the enzymology of monoterminal alkane oxidation (i.e. alkane to fatty acid) having been investigated in considerable detail (cf. Chapter 1), there has been very little published on the enzymology of long-chain dicarboxylic acid metabolism. Modrzakowski and Finnerty (1989) reported the induction of a dicarboxyl-CoA synthetase in the bacterium, *Acinetobacter* sp. HO1-N, grown on one of four alkyl ethers that were degraded via the corresponding dicarboxylic acid. Activation of a range of dicarboxylic acids from malonate to sebacate was examined and, as these activities varied according to the chain length of the diethyl growth substrate, it could be that more than one acyl-CoA synthetase was being expressed in the organism. Growth of the bacterium on hexadecane was less effective in inducing activity of the diacyl-CoA synthethase than growing it on the ethers.

Yi and Rehm (1988a), in their study of the metabolism of Δ 9-*cis*-18:1-dioic acid (see above) by whole cells of *C. tropicalis*, observed the formation of a

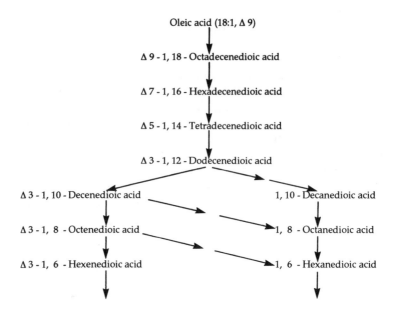

Fig. 16. Oxidation of Δ9-1,18-octadecenedioic acid formed from oleic acid by *Candida tropicalis* mutant S₇₆ (from Yi and Rehm 1988a). (The positions of the double bonds of the metabolites are tentative.)

series of dicarboxylic acids all with even numbers of carbon atoms down to hexenedioic acid (see Fig. 16). These acids are presumed to arise by β-oxidation following the same route as indicated in Section IV.C. As with monocarboxylic acids, the dicarboxylic acids must also be converted to their CoA esters. Only a mono-CoA ester is, however, formed. The activating enzyme, acyl-CoA synthetase, may exhibit a broad substrate specificity (Zhihua and Xiuzhen 1986) and it is possibly the same enzyme that functions with monocarboxylic acids.

Broadway et al. (1993) investigated the enzymology of dicarboxylic acid (DC) formation in *Corynebacterium* sp. 7EIC grown on alkanes (C_{10} to C_{16}) and fatty acid methyl esters (C_{10} to C_{16}) observing formation of the corresponding DC acids was maximal with dodecane. When DC acids (from C_{16} to C_8) were presented to washed-whole cells, the intermediate DCs (cf. Fig. 16) could always be detected. Rates of β-oxidation of 10-, 12-, 14- and 16-DC acids with cell-free extracts were about 10% of the rates observed with 12:0, 14:0 and 16:0 monocarboxylic acids, a pattern that was repeated when the CoA esters were used instead. (Interestingly, di-CoA esters were not oxidized at all.) When the activity of the acyl-CoA synthetase was examined, it was only 30% less active with DC-12:0 than with 12:0 indicating that the differences in β-oxidation rates with the monocarboxylic fatty acids and the DCs must be in the specificity of one or more of the β-oxidation enzymes. As suggested the work of Zhihua and Xiuzhen (1986) using *C. tropicalis*, the acyl-CoA synthetase of *Corynebacterium* sp. 7EIC exhibited a very broad substrate specificity and functioned with mono- and di-carboxylic acids from C_4 to C_{16} and with the ω-hydroxy fatty acids from C_{10} to C_{16} (the hydroxy acids from C_4 to C_8 were not tested).

Thus there is good evidence from both yeasts and bacteria that ω-oxidation of fatty acids to the corresponding dicarboxylic acids is then followed by degradation via an unmodified β-oxidation cycle culminating in the formation of succinyl-CoA ($COOH \cdot CH_2 \, CH_2 \cdot CO$-S-CoA) which would then be oxidized by the tricarboxylic acid (Krebs') cycle.

4. (ω−1)- and (ω−2)-Oxidation
The occurrence of (ω−1)-hydroxy fatty acids in the lipids of some bacteria, including methane-oxidizing bacteria (Skerratt et al. 1992), and yeasts suggests that there may be specific oxidation of fatty acids at their penultimate terminal carbon atom. The formation of (ω−1)-hydroxy fatty acids has been briefly studied in the yeast, *Candida bombicola* which produces a number of acylated derivatives of sophorose (see Chapter 3). On the basis of finding that there were different ratios of the ω- and (ω−1)-hydroxyfatty acids according to the growth conditions, Hommel et al. (1992) suggested that there could be a separate cytochrome P-450 hydroxylase for oxidation at the ω−1 C atom. Related evidence has also been reported for the sub-terminal oxidation of fatty acids, alcohols and amides by *Bacillus megaterium*(Sligar and Murray 1986).

The strongest evidence of there being a sub-terminal hydroxylase comes from work with *Bacillus megaterium* though this does not necessarily imply that the same would be true for eukacyotic microorganims or indeed other bacteria. Fulco and his group (Buchanan and Fulco 1978; Miura and Fulco 1974; 1975; Nahri and Fulco 1982; 1986; 1987; Ruettinger and Fulco 1981) have described a cytochrome P-450 complex in *B. megaterium* that hydroxylates saturated and monosaturated C_{12} to C_{18} fatty acids to an isomeric mixture of (ω−1)-, (ω−2)- and (ω−3)-monohydroxyfatty acids. No ω-hydroxyfatty acids could be detected on any occasion thereby indicating that the hydroxylase was distinct from ω-hydroxylase recognised in other bacteria although, like both the alkane and fatty acid α- and ω-hydroxylases, three components of the system were recognized: an NADPH-dependent-oxidizing protein, a ferredoxin and the cytochrome P-450 protein itself. In the isolated enzyme system, the ω-2 carbon atom was the most abundantly oxidized; hydroxylation at the (ω−1)- and (ω−3)-positions was therefore due "wobble" in the specificity of the enzymes. The structure of the enzyme has been investigated in some detail (Nahri and Fulco 1986; 1987) and the gene has been cloned and sequenced (Boddupalli et al. 1992): the enzyme has an M_r of 119 kDa and has two distinct domains which have been separated and studied individually. One domain contains the haem prosthetic group and the other the flavoprotein. Progress towards elucidating the three-dimensional structure is underway (see Boddupalli et al. 1992).

Lanser et al. (1992) have investigated the hydroxylation of oleic acid (18:1) in whole cells of two strains of *Bacillus pumilus*. Like Miura and Fulco (1974; 1975), they could not detect formation of any ω-hydroxyoleic acid and also found hydroxylation had occurred at the three adjoining C atoms, ω−1, ω−2 and ω−3. However, the most abundant hydroxylated product with both strains was 17-hydroxyoctadecenoic acid.

The reason for the differences in enzyme specificity between *B. megaterium* and *B. pumilus* is unknown but clearly resolution of this could reveal which component is responsible and from this the opportunity to carry out site-directed syntheses to give single products would be a useful attribute.

The further oxidation of a (ω−1)-hydroxyfatty acid is uncertain. It could be converted to the corresponding keto-acid by a dehydrogenase or oxidase. Indeed the fatty alcohol oxidase of yeasts is active with alkan-2-ols (Dickinson and Wadforth 1992) and would be expected to operate with the equivalent (ω−1)-hydroxy fatty acid substrate (Hommel and Ratledge 1990). The recent report of the occurrence of a 27-keto-28:0 fatty acid in *Legionella pneumophila* (Moll et al. 1992) might indicate that this is feasible. The fate of the keto acid though is obscure. However, with *L. pneumophila*, Moll et al. (1992) also identified the C_{27}-dicarboxylic acid. This might have arisen by a variation of the α-oxidation system shown in reaction no. 13 but this would then require oxidation of the terminal methyl group before it could have been decarboxylated.

It is possible therefore that (ω−1) hydroxy fatty acids are not oxidized further at this end of the molecule but are degraded by the β-oxidation cycle operating conventionally at the other end.

5. Mid-chain oxidations

Hammond (1988) has comprehensively reviewed the variety of enzymatic modifications that may occur at the mid-chain of fatty acids. This review included modification by both plant and animal systems as well as microorganisms. Also included were mechanisms of fatty acid desaturation (e.g. stearate to oleate), of the formation of hydroperoxy acids with lipoxygenases and of the formation of cyclopropane and cyclopropene fatty acids. These reactions and related ones are strictly outside the scope of this review and the review of Hammond is therefore commended to the reader.

The work of Fulco and colleagues (see immediately previous section) with the solubilized cytochrome P-450 from *Bacillus megaterium* showed that besides carrying out oxidation at the (ω−1)-, (ω−2)- and (ω−3)- position, epoxidation of unsaturated fatty acids could also occur (Ruettinger and Fulco 1981; Michaels et al. 1980). Thus from oleic acid, 9, 10-epoxystearic acid was produced. The epoxidation also occurred with *trans* (*E*) – double bonds. The bacterium was further able to hydroxylate the product to yield 9, 10-dihydroxystearic acid:

$$-CH{:}CH- \xrightarrow{+\frac{1}{2}O_2} \underset{O}{-CH-CH-} \xrightarrow{+H_2O} -CH(OH){\cdot}CH(OH)- \tag{15}$$

Subsequent work by Soda (1987) indicated that he was successful in converting oleic acid to 12-hydroxy-oleic acid (i.e. ricinoleic acid – the major fatty acid of castor oil) using a strain of a soil bacterium. However in an attempt to repeat these findings, Lanser et al. (1992) were completely unsuccessful and only isolated the 15-, 16- and 17- hydroxy-oleic acids (see IV.F.4). So far as the present reviewer is aware, no one has yet verified the potentially very interesting result of Soda though several groups have tried. Thus, Hou et al. (1991), using an unspecified bacillus, have isolated 7,10-dihydroxy-8(E)-octadecenoic acid from oleic acid and Koritala et al. (1989) using *Nocardia cholesterolicum* isolated 10-hydroxystearic acid from the same substrate. The former biotransformation is clearly the more complex involving double bond migration and possible hydroxylation at two positions, (reaction 16). The latter reaction is clearly a simple hydration reaction across the double bond (reaction 17).

$$-{}^{10}CH={}^{9}CH \cdot {}^{8}CH_2 \cdot {}^{7}CH_2 - \rightarrow - CH(OH) \cdot CH = CH \cdot CH(OH) - \qquad (16)$$

$$-{}^{10}CH={}^{9}CH - + H_2O \rightarrow - CH(OH) \cdot CH_2 - \qquad (17)$$

Reaction no. 17 is catalysed by the enzyme, oleate hydratase, and is widely distributed amongst bacteria (aerobes and anaerobes) and a number of soil microorganisms (see Hammond 1988; Koritala et al. 1989). Koritala and Bagley (1992) have recently extended their work with *N. cholesterolicum* using both linoleic (18:2, Δ 9,12), and linolenic (18:3, Δ 9, 12, 15) acids. In both cases, hydration occurred across the Δ 9-double bond yielding the 10-hydroxy 18:1 (Δ12) and 10-hydroxy 18:2 (Δ12, 15) fatty acids, a finding that was originally reported by Schroepfer and colleagues (Schroepfer et al. 1970; Niehaus et al. 1970a,b) studying the purified oleate hydratase from *Pseudomonas*.

In the ergot fungus, *Claviceps purpurea* and other species, ricinoleic acid (12 hydroxyoleic acid) occurs naturally in the sclerotial stage of cell development (see Ratledge 1989). This too is formed by a hydratase (Morris 1970) but, unlike the oleate hydratase – rn 17, this one is specific for the Δ12 double bond in linoleic acid which is therefore the precursor of ricinoleic acid in the fungus. (In the castor plant, ricinoleic acid is formed by a specific hydroxylase reaction with oleic acid: i.e. oleate Δ12 hydroxylase – see Hammond 1988, for references.) Similar hydratase reactions may account for the occurrence of mid-chain hydroxy groups in fatty acids of yeasts, such as a *Candida* sp. producing 15-hydroxylinoleic acid (probably from linolenic acid) (Fuji and Tonomura 1971) though the origin of 7-hydroxy derivatives of 10:0, 12:0 and 14:0 fatty acids in several species of *Mucor* (Tahara et al. 1980) is less clear and would, if a hydratase is involved, require the prior formation of the Δ6 or Δ7 unsaturated fatty acid.

The fate of the mid-chain hydroxyfatty acids is uncertain. In many cases, for example see Koritala et al. (1989), very high (>90%) conversion of substrate to product have been recorded indicating that there can be minimal degradation by this route. It is unlikely that the fatty acids would be cleaved at the hydroxy-carrying C atom and it is possible that these acids, like most other ones, are subsequently degraded by the β-oxidation cycle with appropriate modifications when double bonds or hydroxyl groups are encountered (see Section IV.D). Their occurrence therefore are of more interest to those concerned with biotransformation reactions of fatty acids rather than their degradation.

References

Alberghina L, Verger R and Schmid RD (eds) (1991) Lipases: Structure, Mechanism and Genetic Engineering, (GBF Monograph no. 16). VCH, Weinheim

Alford JA and Pierce DA (1963) Production of lipase by *Pseudomonas fragi* in a synthetic medium. J. Bacteriol. 86: 24–29

Alford JA, Pierce DA and Suggs FG (1964) Activity of microbial lipases on natural fats and synthetic triglycerides. J. Lipid Res. 5: 390–394

Andreoli PM, Cox MMJ, Farin F and Wohlfarth S (1989) Molecular cloning and expression of genes encoding lipolytic enzymes. European Patent 0334462

Anelli G, Lepidi AA and Galoppini C (1975) Biomasses from fat by-products. II: Some factors of yield. Riv. ital. Sost. grasse 52: 117–118

Antczak U, Gora J, Antczak T and Galas E (1991) Enzymatic lactonization of 15-hydroxypentadecanoic acid and 16-hydroxyhexadecanoic acids to macrocylic lactones. Enzyme Microb. Technol. 13: 589–593

Aoyama S, Yoshida and Inouye S (1988) Cloning, sequencing and expression of the lipase gene from *Pseudomonas fragi* IFO-12049 in *E. coli*. FEBS Lett. 242: 36–40

Atlas RM and Bartha R (1973) Inhibition by fatty acids of the degradation of petroleum. Antonie van Leeuwenhoek 39: 257–271

Ba A, Ratomahenina R, Graille J and Galzy P (1981) Etude de la croissance de quelques souches de levures sur les sous-products du raffinage de l'huile d'arachide. Oleagineux 36: 439–445

Ba A, Ratomahenina R, Galzy P, Graille J and Pina M (1983) Consideration on fatty acids metabolism of by yeast. Riv. ital. Sost. grasse 60: 673–675

Baillargeon MW, Bistline RG and Sonnet PE (1989) Evaluation of strains of *Geotrichum candidum* for lipase production and fatty acid specificity. Appl. Microbiol. Biotechnol. 30: 92–96

Bangham AD (1992) "Surface tension" in the lungs. Nature 359: p 110

Baratti J, Buono G, Deleuze H, Langrand G, Secchi M and Triantaphylides C (1986) Enantioselective synthesis of fatty acid esters by lipases. In: AR Baldwin (ed) World Conference on Emerging Technologies in the Fats and Oils Industry (pp 355–358). Amercian Oil Chemists' Society, Champaign, Illinois, U.S.A.

Barnes D, Forster CF and Hrudey SE (eds) (1984) Surveys in Industrial Wastewater Treatment, Vol 1. Pitman Publ. Co. Ltd., London

Barrett-Bee, K, Hayes Y, Wilson RG and Ryley JF (1985) A comparison of phospholipase activity, cellular adherence and pathogencity of yeasts. J. Gen. Microbiol. 131: 1217–1221

Bati N, Hammond EG and Glatz BA (1984) Biomodification of fats and oils: trials with *Candida lipolytica*. J. Am. Oil Chem. Soc. 61: 1743–1747

Bell GH (1971) The action of monocarboxylic acids on *Candida tropicalis* growing on hydrocarbon substrates. Antonie van Leeuwenhoek 37: 385–400

Bell GH (1973) The metabolism of long-chain fatty acids and alcohols by *Candida tropicalis* and *Saccharomyces cerevisiae*. Antonie van Leeuwenhoek 39: 385–400

Bell GH and Trancart P (1973) The action of n-butyric, n-hexanoic and iso-butyric acids on the respiration of *Candida tropicalis*. Antonie van Leeuwenhoek 39: 129–136

Bhat KM, Sumner IG, Perry BN, Collins ME, Pickersgill RW and Goodenough PW (1991) A novel method for the production of porcine phospholipase A_2 expressed in *E. coli*. Biochem. Biophys. Res. Comm. 176: 371–377

Blasig R, Mauersberger S, Riege P, Schunk WH, Jockisch W, Franke P and Müller HG (1988) Degradation of long-chain *n*-alkanes by the yeast *Candida maltosa*. II. Oxidation of *n*-alkanes and intermediates using microsomal membrane fractions. Appl. Microbiol. Biotechnol. 28: 589–597

Boddupalli SS, Oster T, Estabrook RW and Peterson JA (1992) Reconstruction of the fatty acid hydroxylation faction of cytochrome P-450 $_{BM-3}$ utilizing its individual recombinant hemo- and flavoprotein domains. J. Biol. Chem. 267: 10375–10380

Boel E, Huge-Jensen B, Christensen M, Thim L, and Fiil NP (1988) *Rhizomucor miehei* triglyceride lipase is synthesized as a precursor. Lipids 23: 701–706

Boulton CA and Ratledge C (1984) The physiology of hydrocarbon – utilizing microorganisms. Topics Enz. Ferment. Technol. 9: 11–77

Bozoglu F, Swaisgood HE and Adams DM (1984) Isolation and characterisation of an extracellular heat-stable lipase produced by *Pseudomonas fluorescens* MC 50. J. Agric. Fd. Chem. 32: 2–6

Brady L, Brzozowski AM, Derewenda ZS, Dodson E, Dodson G, Tolley JP, Turkenburg L, Christiansen L, Huge-Jensen B, Norskov L, Thim L and Menge U (1990) A serine protease triad forms the catalytic centre of a triacylglycerol lipase. Nature 343: 767–770

Breuil C, Schindler DB, Sijher JS and Kushner DJ (1978) Stimulation of lipase production during bacterial growth on alkanes. J. Bacteriol. 133: 601–606

Broadway NM, Dickinson FM and Ratledge C (1992) Long-chain acyl-CoA ester intermediates of β-oxidation of mono- and di-carboxylic fatty acids by extracts of *Corynebacterium* sp. strain 7EIC. Biochem. J. 285: 117–122

Broadway NM, Dickinson FM and Ratledge C (1993) The enzymology of dicarboxylic acid formation by *Corynebacterium* sp. strain 7EIC grown on alkanes. J. Gen. Microbiol. 139: 1337–1344

Brockman HL (1984) General features of lipolysis: reaction scheme, interfacial structure and experimental approaches. In: B Borgström and HL Brockman (eds) Lipases (pp 3–46). Elsevier Science Publishers, Amsterdam

Brune KA and Gotz F (1992) Degradation of lipids by bacterial lipases. In: G Winkelmann (ed) Microbial Degradation of Natural Products (pp 242–266). VCH, Weinheim

Brzozowski AM, Derewenda U, Derewenda ZS, Dodson GG, Lawson DM, Turkenberg JP, Bjorkling F, Huge-Jensen B, Patker SA and Thim L (1991) A model for interfacial activation in lipases from the structure of a fungal lipase-inhibitor complex. Nature 351: 491–494

Buchanan JF and Fulco AJ (1978) Formation of 9, 10-epoxypalmitate acid by a soluble system from *Bacillus megaterium*. Biochem. Biophys. Res. Commun. 85: 1254–1260

Buhler M and Schindler J (1984) Aliphatic hydrocarbons. In: HJ Rehm and G Reed (eds) Biotechnology, Vol 6a (pp 329–385). Verlag Chemie, Weinheim

Burkholder L, Burkholder PR, Chu A, Kostyk N and Roels OA (1968) Fish fermentation. Food Technol. 22: 1278–1284

Carlson KF and Scott JD (1991) Recent developments and trends: processing of oilseeds, fats and oils. INFORM 2, 1034–1060

Casey J and Macrae A (1992) Biotechnology and the olechemical industry. INFORM 3, 203–207

Casey J, Dobb R and Mycock G (1990) An effective technique for enrichment and isolation of *Candida cloacae* mutants defective in alkane metabolism. J. Gen. Microbiol. 136: 1197–1202

Chakravarti DN, Chakravarti B and Chakrabarti P (1987) Studies on phospholipase activities in *Saccharomyces cerevisiae*. Biochem. Arch. 3: 61–67

Chung GH, Lee YP, Jeohn GH, Yoo OJ and Rhee JS (1991a) Cloning and nucleotide sequence of thermostable lipase gene from *Pseudomonas fluorescens* SIK W1. Agric. Biol. Chem. 55: 2359–2365

Chung GH, Lee YP, Yoo OJ and Rhee JS (1991b) Overexpression of a thermostable lipase gene from *Pseudomonas fluorescens* in *Escherichia coli*. Appl. Microbiol. Biotechnol. 35: 237–241

Creuly C, Larroche C and Gros JB (1992) Bio-conversion of fatty acids into methyl kestones by spores of *Penicillium roquefortii* in a water-organic solvent, two-phase systems. Enzyme Microb. Technol. 14: 669–678

Dennis EA (1983) Phospholipases. In: PD Boyer (ed) The Enzymes 3rd edition, Vol 16 (pp 307–353). Academic Press, New York

Derewenda U, Brzozowski AM, Lawson DM and Derewenda ZS (1992) Catalysis at the interface: the anatomy of a conformational change in a triglyceride lipase. Biochemistry 31: 1532–1541

Dickinson FM and Fletcher PDI (1989) Enzymes in organic solvents. Enzyme Microb. Technol. 11: 55–56

Dickinson FM and Wadforth C (1992) Purification and some properties of alcohol oxidase from alkane-grown *Candida tropicalis*. Biochem. J. 282: 325–331

Dommes V, Baumgart C and Kunau WH (1981) Degradation of unsaturated fatty acids in peroxisomes: existence of a 2,4-dienoyl-CoA reductase pathway. J. Biol. Chem. 256: 8259–8262

Dommes P, Dommes V and Kunau WH (1983) β-Oxidation in *Candida tropicalis*. Partial purification and biological function of an inducible 2, 4-dienoyl coenzyme A reductase. J. Biol. Chem. 258: 10846–10852

Dordick JS (1989) Enzymatic catalysis in monophasic organic solvents. Enzyme Microb. Technol. 11: 194–211

Engel EH, Heidlas J, Albrecht W, Tressl R (1989) Biosynthesis of chiral flavour and aroma compounds in plants and microorganisms. In: R Teranishi, RG Butlery, F Shahidi (eds) Flavor Chemistry Trends and Developments. Amer. Chem. Soc. Symp. Series 388: 8–22

Ercoli B, Fuganti C, Grasselli P, Servi S, Allegrone G, Barbeni M and Pisciotta A (1992) Sterochemistry of the biogeneration of C-10 and C-12 gamma lactones in *Yarrowia lipolytica* and *Pichia ohmeri*. Biotechnol. Lett. 14: 665–668

Evans CT and Ratledge C (1985) The physiological significance of citric acid in the control of metabolism in lipid-accumulating yeasts. Biotechnol. Gen. Enz. Rev. 3: 349–375

Feller G, Thiry M and Gerday C (1990) Sequence of a lipase gene from the Antarctic psychrotroph *Moraxella* TA144. Nucleic Acids Res. 18: 6431

Feller G, Thiry M, Arpigny JL and Gerday C (1991) Cloning and expression in *Escherichia coli* of three lipase-encoding genes from the psychrotrophic Antarctic strain *Moraxella* TA144, Gene 102: 111–115

Fernandes F, Viel M, Sayag D and Andre L (1988) Microbial breakdown of fats through in-vessel co-composting of agricultural and urban wastes. Biological Wastes 26: 33–48

Fiorentini R, Anelli G, Lepidi AA, and Galoppini C (1976) Biomasses from fat by-products. III: Lipid composition of yeast cells. Rev. ital. Sost. grasse 53: 102–104

Forster CF (1992) Oils, fats and greases in wastewater treatment. J. Chem. Tech. Biotech. 55: 402–404

Franke W and Heinen W (1958) Zur Kenntnis des Fettsäureabbaus durch Schimmelpilze 1. Mittleilung: Uber die Methylketonbildung durch Schimmelpilze. Arch. Mikrobiol. 31: 50–59

Fuji T and Tonomura K (1971) Fatty acids extracellularly produced by *Candida* sp. growing on ethanol. Agric. Biol. Chem. 35: 1188–1193

Furukawa T, Matsuyoshi T and Kise S (1986) Selection of high brassylic acid producing strains of *Torulopsis candida* by single-cell cloning and by mutants. J. Ferment. Technol. 64: 97–101

Gilbert EJ, Drozd JW and Jones CW (1991a) Physiological regulation and optimization of lipase activity in *Pseudomonas aeruginosa* EF2. J. Gen. Microbiol. 137: 2215–2221

Gilbert EJ, Cornish A and Jones CW (1991b) Purification and properties of extracellular lipase from *Pseudomonas aeruginosa* EF2. J. Gen. Microbiol. 137: 2223–2229

Gill CO and Ratledge C (1973) Inhibition of glucose assimilation and transport by *n*-decane and other alkanes in *Candida* 107. J. Gen. Microbiol. 75: 11–22

Gode P (1991) Wastewaters from hospital laundries. The legal framework and its proper application in practice. Henkel-Referate 27: 109–114 (Henkel KGaA, Dussledorf, Germany)

Godtfredsen SE (1989) Microbial lipases. In: WM Fogarty and CI Kelly (eds) Microbial Enzymes and Biotechnology, 2nd edition (pp 255–274). Applied Science Publishers, London

Goodenough PW, Bhat KM, Collins ME, Perry BN, Pickersgill RW, Sumner IG, Warwicker J, De Haas GH and Verheij HM (1991) Changes in activity of porcine phospholipase A_2 brought about by charge engineering of a major structural element to alter stability. Protein Eng. 4: 929–934

Gotz F, Popp F and Schleifer KH (1985) Complete nucleotide sequence of the lipase gene from

Staphylococcus hyicus cloned in *Staphylococcus carnosus*. Nucleic Acids Res. 13: 5895–5906

Green JH, Paskell SL and Goodmintz D (1976) Lipolytic fermentation of stickwater by *Geotrichum candidum* and *Candida lipolytica*. Appl. Environ. Microbiol. 31: 569–575

Gunstone FD (1989) Oils and fats – past, present and future. In: RC Cambie (ed) Fats for the Future (pp 1–16). Ellis Horwood Ltd, Chichester

Haas MJ, Allen J and Berka TR (1991) Cloning, expression, and characterization of a cDNA encoding a lipase from Rhizopus delemar. Gene 109: 107–114

Halling PJ (1990) Lipase-catalysed modifications of oils and fats in organic two-phase systems. Fat Sci. Technol. 92: 74–79

Hammond RC (1988) Enzymic modification at the mid-chain of fatty acids. Fat Sci. Technol. 90: 18–27

Han D and Rhee JS (1986) Characteristics of lipase-catalyzed hydrolysis of olive oil in AOT-isooctane reversed miscelles. Biotech. Bioeng. 28: 1250–1255

Hanin I and Ansell GB (eds) (1987) Lecithin: technological, biological and therapeutic aspects. Plenum Press, New York

Harper JW, Hemmi K and Powers JC (1985) Reaction of serine proteases with substituted isocoumarins: discovery of 3, 4-dichloroisocoumarin, a new general mechanism based serine protease inhibitor. Biochemistry 24, 1831–1841

Harrison FC (1927) A systematic study of some torulae. Trans. Roy. Soc. Can., ser 3. V, 21: 341–379

Harwood J (1989) The versatility of lipases for industrial uses. Trends Biochem. Sci. 14: 125–126

Hatton PV and Kinderlerer JL (1991) Toxicity of medium chain fatty acids to *Penicillium crustosum* Thom and their detoxification to methyl ketones. J. Appl. Bacteriol. 70: 401–407

Hill DE, Boulay R and Rogers D (1988) Complete nucleotide sequence of the peroxisomal acyl-CoA oxidase from the alkane-utilizing yeast *Candida maltosa*. Nucleic Acid Res. 16: 365–366

Hill FF, Venn I and Lukas KL (1986) Studies on the formation of long-chain dicarboxylic acids from pure n-alkanes by a mutant of *Candida tropicalis*. Appl. Microbiol. Biotechnol. 24: 168–174

Hiltunen JK, Palosaari PM and Kunau WH (1989) Epimerization of 3-hydroacyl-CoA esters in rat liver. Involvement of two enoyl-CoA hydratase. J. Biol. Chem. 264: 13536–13540

Hiltunen JK, Wenzel B, Beyer A, Erdmann R, Fossa A and Kunau WH (1992) Peroxisomal multifunctional β-oxidation protein of *Saccharomyces cerevisiae*. J. Biol. Chem. 267: 6646–6653

Holdsworth JE and Ratledge C (1988) Lipid turnover in oleaginous yeasts. J. Gen. Microbiol. 134: 339–346

Holdsworth JE and Ratledge C (1991) Triacylglycerol synthesis in the oleaginous yeast *Candida curvata* D. Lipids 26: 111–118

Holdsworth JE, Veenhuis M and Ratledge C (1988) Enzyme activities in oleaginous yeasts accumulating and utilizing exogenous or endogenous lipids. J. Gen. Microbiol. 134: 2907–2915

Hommel R and Ratledge C (1990) Evidence for two fatty alcohol oxidases in the biosurfactant-producing yeast *Candida* (*Torulopsis*) *bombicola*. FEMS Microbiol Lett. 70: 183–186

Hommel R, Kirste S, Weber L and Kleber HP (1992) German Patent DD 298273

Hosaka K, Mishina M, Tanaka T, Kamiryo T and Numa S (1979) Acyl-coenzyme A synthetase I from *Candida lipolytica*. Eur. J. Biochem. 93: 197–203

Hottinger HH, Richardson T, Amundson CH and Stuiber DA (1974a) Utilization of fish oil by *Candida lipolytica* and *Geotrichum candidum*. I. Basal conditions. J. Milk Food Technol. 37: 463–468

Hottinger HH, Richardson T, Amundson CH and Stuiber DA (1974b) Utilization of fish oil by *Candida lipolytica* and *Geotrichum candidum*. II. Optimization of conditions. J. Milk Food Technol. 37: 522–528

Hou CT, Bagby MO, Plattner RD and Koritala S. (1991) A novel compound, 7, 10-dihydroxy-8(E)-octadeconoic acid from oleic acid by bioconversion. J. Am. Oil. Chem. Soc. 68, 99–104

Hunkova Z and Fencl Z (1977) Toxic effects of fatty acids on yeast cells: dependence of inhibitory effects on fatty acid concentration. Biotech. Bioeng. 19: 1623–1641

Huss H (1908) Eine fettspaltende Bakterie (*Bactridium lipolyticum* n. sp.) Zentr. Bakt. Parasitenk II, 20: 474–484

Ibrahim CO, Hayashi M and Nagai S (1987) Purification and some properties of a thermostable lipase from *Humicola lanuginosa* No. 3. Agric. Biol. Chem. 51: 37–45

Ihara F, Kageyama Y, Hirata M, Nihira T and Yamada Y (1991) Purification, characterization and molecular cloning of lactonizing lipase from *Pseudomonas* species. J. Biol. Chem. 266: 18135–18140

Ikezawa H and Tuguchi R (1981) Phosphatidylinositol – specific phospholipase C from *Bacillus cereus* and *Bacillus thuringiensis*. Methods in Enzymology 71: 731–741

Isobe K, Nokihara K, Yamaguchi S, Mase T and Schmid RD (1992) Crystallization and characterization of monoacylglycerol and diacylglycerol lipase from *Pencillium camembertii*. Eur. J. Biochem. 203: 233–237

Iwai M and Tsujisaka Y (1984) Fungal lipases. In: B Borgström and HL Brockman (eds) Lipases (pp 443–469). Elsevier Science Publishers, Amsterdam

Iwasaki R and Murakoshi M (1992) Palm oil yields carotene for world markets. INFORM 3: 210–217

Jacobsen T, Olsen J, Allermann K, Poulsen OM and Han J (1989a) Production partial purification, and limmunochemical characterization of multiple forms of lipase from *Geotrichum candidum*. Enzyme Microb. Technol. 11: 90–95

Jacobsen T, Jensen B, Olsen J and Allerman K (1989b) Extracellular and cell-bound lipase activity in relation to growth of *Geotrichum candidum*. Appl. Microbiol. Biotechnol. 32: 256–261

Jorgensen S, Skov K, and Diderichsen B (1991) Cloning, sequence, and expression of a lipase gene from *Pseudomonas cepacia*: lipase production in heterologous hosts requires two *Pseudomonas* genes. J. Bacteriol. 173: 559–567

Juneja LR, Hibi N, Inagaki N, Yamane T and Shimizu S (1987) Comparative study on conversion of phosphatidylcholine to phosphatidylglycerol by cabbage phospholipase in micelle and emulsion systems. Enzyme Microb. Technol. 9: 350–354

Kajs TM and Vanderzant C (1980) Batch-scale utilization studies of tallow by food yeasts. Dev. Ind. Microbiol. 21: 481–488

Kameda K and Nunn WD (1981) Purification and characterization of acyl coenzyme A synthetase from *Escherichia coli*. J. Biol. Chem. 256: 5702–5707

Kawaguchi Y, Honda H, Taniguchi-Morimura J and Iwasaki S (1989) The codon CUG is read as serine in an asporogenic yeast *Candida cylindracea*. Nature 341: 164–166

Kawamoto S, Nozaki C, Tanaka A and Fukiu S (1978) Fatty acid β-oxidation system in microbodies of *n*-alkane-grown *Candida tropicalis*. Eur. J. Biochem. 83: 609–613

Kellard B, Busfield DM and Kinderlerer JL (1985) Volatile off-flavour compounds in desiccated coconut J. Fd . Sci. Agric. 36: 415–420

Kemp GD, Dickinson FM and Ratledge C. (1988) Inducible long chain alcohol oxidase from alkane-grown *Candida tropicalis*. Appl. Microbiol. Biotechnol. 29: 370–37

Kendrick A and Ratledge C (1990) Microbial lipid technology: microbial formation of polyunsaturated fatty acids. Lipid Technology 2: 62–66

Kendrick A and Ratledge C (1992) Microbial polyunsaturated fatty acids of potential commercial interest. SIM (Soc. Ind. Microbiol.) Indust. Microbiol. News 42: 59–65

Kinderlerer JL (1986) Bacterial and fungal spoilage of coconut. In: B Flanagan (ed) Spoilage and Mycotoxins of Cereals and Other Stored Products. International Biodeterioration 22: 41–47

Kinderlerer JL (1987) Conversion of coconut oil to methyl ketones two *Aspergillus* species. Phytochem. 26: 1417–1420

Kinderlerer JL (1992) Biodeterioration of the lauric acid oils by fungi. J. Chem. Tech. Biotech. 55: 400–402

Kinderlerer JL (1993) Fungal strategies for toxification of medium chain fatty acids. Intern. Biodeterioration. (in press)

Kinderlerer JL and Hatton PV (1991) The effect of temperature, water activity and sorbic acid on ketone rancidity produced by *Penicillium crustosum* Thom in coconut and palm kernel oils. J. Appl. Bacteriol. 70: 502–506

Kinderlerer JL and Kellard B (1984) Ketonic rancidity in coconut due to xerophilic fungi. Phytochem. 23: 2847–2849

Kindl H (1984) Lipid degradation in higher plants. In: S Numa (ed) Fatty Acid Metabolism and its Regulation (pp 181–204). Elsevier Science Publishers, Amsterdam

Kinsella JE and Hwang DH (1976) Enzymes of *Penicillium roquefortii* involved in the biosynthesis of cheese flavour. CRC Crit. Rev. Fd. Sci. and Nutr. 8: 191–228

Kionka C and Kunau WH (1985) Inducible β-oxidation pathway in *Neurorpora crassa*. J. Bacteriol. 161: 153–157

Klein K, Steinberg R, Fiethen B and Overath P (1971) Fatty acid degradation in *Escherichia coli*. An inducible system for the uptake of fatty acids and further characterization of *old* mutants. Eur. J. Biochem. 19: 442–450

Klibanov AM (1990) Asymmetric transformations by enzymes in organic solvents. Accounts. Chem. Res. 23: 114–120

Koh JS and Minoda Y (1984) Microbial utilization of natural oil and fats. Oil Chemistry 33 672–675

Koh JS, Kodama T and Minoda Y (1983) Screening of yeasts and cultural conditions for cell production from palm oil. Agric. Biol. Chem. 47: 1207–1212

Koh JS, Yamakawa T, Kodama T and Minoda Y (1985) Rapid and dense culture of *Acinetobacter calcoaceticus* on palm oil. Agric. Biol. Chem. 49: 1411–1416

Kohlwein SP and Paltauf F (1983) Uptake of fatty acids by the yeasts, *Saccharomyces uvarum* and *Saccharomycopsis lipolytica*. Biochim. Biophys. Acta. 792: 310–317

Kokusho Y, Kato S, Machida H and Iwasaki S (1987) Purification and properties of phospholipase D from *Actinomadura* sp. strain no. 362. Agric. Biol. Chem. 51: 2515–2523

Kordel M and Schmid RD (1991) Inhibition of the lipase from *Pseudomonas* sp. ATCC 21808 by diethyl *p*-nitrophenyl phosphate; hints for one buried active site for lipolytic and esterolytic activity. In: L Alberghina, R Verger and RD Schmid (eds) Lipases: Structure, Mechanism and Genetic Engineering, GBF Monographs no 16 (pp 385–388). VCH, Weinheim: VCH

Koritala S and Bagby MO (1992) Microbial conversion of linoleic acid and linolenic acids to unsaturated fatty acids. J. Am. Oil Chem. Soc. 69: 575–578

Koritala S, Hesseltine CW, Pryde EH and Mounts TL (1987) Biochemical modifications of fats by microorganisms: a preliminary survey. J. Am. Oil. Chem. Soc. 64: 509–513

Koritala S, Hosie L, Hou CT, Hesseltine CW and Bagby MO (1989) Microbial conversion of oleic acid to 10-hydroxystearic acid. Appl. Microbiol. Biotechnol. 32: 299–304

Kouker G, and Jaeger KE (1987) Specific and sensitive plate assay for bacterial lipases. Appl. Environ. Microbiol. 53: 211–213

Kugimiya W, Otani Y, Hashimoto Y and Takagi Y (1986) Molecular cloning and nucleotide sequence of the lipase gene from *Pseudomonas fragi*. Biochem. Biophys. Res. Commun. 141: 185–190

Kunau WH and Hartig A (1992) Peroxisome biogenesis in *Saccharomyces cerevisiae*. Antonie van Leeuwenhoek 62: 63–68

Kunau WH, Bühne S, de la Garza M, Kionka C, Mateblowski M, Schultz-Borchard U and Thieringer R (1988) Comparative enzymology of β-oxidation. Biochem. Soc. Trans. 16: 418–420

Kyle D and Ratledge C (eds) (1992) Industrial Applications of Single Cell Oils. American Oil Chemists' Society, Champaign, Illinois, U.S.A.

Laane C, Tramper J and Lilly MD (eds) (1987) Biocatalysis in Organic Media. Elsevier, Amsterdam

Laborbe JM, Rieu Y, Ratomahenina R, Galzy P, Montet D, Pina M and Graille J (1987) Essai de multiplication de cellules de *Geotrichum candidum* Link CBS 178-53 sur pâte de neutralisation. Oleagineux 42: 83–86

Laborbe JM, Dwek C, Ratomahenina R, Pina M, Graille J and Galzy P (1989) Production of a single cell protein from palm oil using *Candida rugosa*. MIRCEN. J. Microbiol. Biotech. 5: 517–523

Lanser AC, Plattner RD and Bagby MO (1992) Production of 15-, 16- and 17- hydroxy-9-octadecenoic acids by bioconversion of oleic acid with *Bacillus pumilus*. J. Am. Oil. Chem. Soc. 69: 363–366

Lawrence RC (1966) The oxidation of fatty acids by spores of *Penicillium roquforti* J. Gen. Microbiol. 44: 393–405

Lawrence RC and Hawke JC (1968) The oxidation of fatty acids by mycelium of *Penicillium roqueforti* J. Gen. Microbiol. 51: 289–302

Lazar G. and Schröder FR (1992) Degradation of lipids by fungi. In: G Winkelmann (ed) Microbial Degradation of Natural Products (pp 267–291). VCH, Weinheim

Lazar G, Weiss A and Schmid RD (1986) Synthesis of esters by lipases. In: AR Baldwin (ed) World Conference on Emerging Technologies in the Fats and Oils Industry (pp 346–354). Amercian Oil Chemists' Society, Champaign, Illinois, U.S.A.

Lee CY and Iandolo JL (1986) Lysogenic conversion of staphylococcal lipase is caused by insertion of the bacteriophage L54a genome into the lipase structural gene. J. Bacteriol. 166: 385–391

Lee I, Hammond EG, Glatz BA (1992) Triacylglycerol assembly from lipid substrates by *Apiotrichum curvatum*. In: DJ Kyle and C Ratledge (eds) Industrial Applications of Single Cell Oils (pp 139–155). American Oil Chemists' Society, Champaign, Illinois, U.S.A.

Lepidi AA, Anelli G and Galoppini C (1975) Biomasses from fat by-products. I: Preliminary selection of yeasts. Riv. ital. Sost. grasse 52: 95–97

Li CF, Stuiber D, Richardson T and Amundson CH (1970) Fermentation of fish lipids. Food Prod. Develop. 4: 28–32

Lima N, Teixeira JA and Mota M (1991) Deep agar-diffusion test for preliminary screening of lipolytic activity of fungi. J. Microbiol. Meth. 14: 193–200

Linton JD and Stephenson RJ (1978) A preliminary study on growth yields in relation to the carbon and energy contents of various organic growth substrates. FEMS Microbiol. Lett. 3: 95–98

Little C (1981) Phospholipase C from *Bacillus cereus*. Methods in Enzymology. 71: 725–730

Low MG (1981) Phosphatidylinositol-specific phospholipase C from *Staphylococcus aureus*. Methods in Enzymology, 71 741–746

Macrae AC (1986) Lipase-catalyzed interesterification of oils and fats. In: AR Baldwin (ed) World Conference on Emerging Technologies in the Fat and Oils Industry (pp 189–198). American Oil Chemists' Society, Illinois, U.S.A.

Macrae AR (1983) Extracellular microbial lipases. In: WM Fogarty (ed) Microbial Enzymes and Biotechnology (pp 225–250). Applied Science Publishers, London

Macrae AR and Hammond RC (1985) Present and future applications of lipases. Biotechnol. Gen. Eng. Rev. 3: 193–217

Malcata FX, Reyes HR, Garcia HS, Hill CG and Amundson CH (1992) Kinetics and mechanisms of reactions catalysed by immobilized lipases. Enzyme Microb. Technol. 14: 426–446

Maloy SR, Ginsburgh CL, Simons RW and Nunn WD (1981) Transport of long and medium chain fatty acids by *Escherichia coli*. J. Biol. Chem. 256: 3735–3742

Martinet F, Ba A, Ratomahenina R, Graille J. and Galzy P (1982a) Utilisation de savons pour la production de levures aliment. Oleagineux 37: 193–198

Martinet F, Ratomahenina R, Graille J and Galzy P (1982b) Production of food yeasts from the solid fraction of palm oil. Biotechnol. Lett. 4: 9–12

Mauersberger S, Kärgel E, Matyashova RN and Müller HG (1987) Sub-cellular organization of alkane oxidation in the yeast *Candida maltosa*. J. Basic Microbiol. 27: 565–582

McCammon MT, Veenhuis M, Trapp SB and Goodman JM (1990) Association of glyoxylate and β-oxidation enzymes with peroxisomes of *Saccharomyces cerevisiae*. J Bacteriol. 172: 5816–5827

Michaels BC, Ruettinger RT and Fulco AJ (1980) Hydration of 9, 10-epoxypalmitic acid by a soluble enzyme from *Bacillus megaterium*. Biochem. Biophys. Res. Commun 92: 1189–1195

Mielke S (1992) Trends in supply, consumption and prices. In: VKS Shukla and Gunstone FD (eds) Oils and Fats in the Nineties (pp 10–22). International Food Science Centre, Lystrup, Denmark, International Food Science Centre, pp 10–22

Miller DA, Prausnitz JM and Blanch HW (1991) Kinetics of lipase-catalysed interesterification of triglycerides in cyclohexane. Enzyme Microb. Technol. 13: 98–103

Miles RJ, Siu ELT, Carrington C, Richardson AC, Smith BV and Price RG (1992) The detection of lipase activity in bacteria using novel chromogenic substrates. FEMS Microbiol. Lett. 90: 283–288

Mishina M, Kamiryo T, Tashiro S and Numa S (1978a) Separation and characterization of two long-chain acyl-CoA synthetases from *Candida lipolytica* Eur. J. Biochem. 82: 347–354

Mishina M, Kamiryo T, Tashiro S, Hagihara T, Tanaka A, Fukui S, Osumi M and Numa S (1978b) Subcellular localization of two long-chain acyl-CoA synthetases in *Candida lipolytica*. Eur. J. Biochem 89: 321–328

Miura Y and Fulco AJ (1974) ($\omega-2$) Hydroxylation of fatty acids by a soluble system from *Bacillus megaterium* J. Biol. Chem. 249: 1880–1887

Miura Y and Fulco AJ (1975) $\omega-1$, $\omega-2$ and $\omega-3$ Hydroxylation of long chain fatty acids, amides and alcohols by a soluble enzyme system from *Bacillus megaterium*. Biochim. Biophys. Acta. 388: 305–317

Modrzakowski MC and Finnerty WR (1989) Intermediary metabolism of *Acinetobacter* grown on dialkyl ethers. Canad. Microbiol. 35: 1031–1036

Moll H, Sonesson A, Jantzen E, Marre R and Zähringer U (1992) Identification of 27-oxo-octacosanoic acid and heptacosane-1, 27-dioic acid in *Legionella pneumophila*. FEMS Microbiol. Lett. 97: 1–6

Montet D, Ratomahenina R, Ba A, Pina M, Graille J and Galzy P (1983) Production of single cell protein from vegetable oils – rape seed oil and palm oil culture medium for 9 lipolytic yeast strains including *Candida rugosa*. J. Ferment. Technol. 61: 417–420

Montet D, Ratomahenina R, Laborbe JM, Pina M, Graille J and Galzy P (1985) Production de proteines d'organismes unicellulaires à partir de pâtes de neutralisation d'origine industrielle. Oleagineux 40: 505–509

Moreno de la Garza M, Schultz-Borchard U, Crabb JW and Kunau WH (1985) Peroxisomal β-oxidation system of *Candida tropicalis*. Purification of a multi-functional protein possessing enoyl-CoA hydratase, 3-hydroxyacyl -CoA dehydrogenase and 3-hydroxyacyl-CoA epimerase activities. Eur. J. Biochem. 148: 285–291

Morris LJ (1970) Mechanisms and stereochemistry in fatty acid metabolism. Biochem. J. 118: 681–693

Murray WW and Rachubinski RA (1987) The primary structure of a peroxisomal fatty acyl-CoA oxidase from the yeast *Candida tropicalis* pK233. Gene 51: 119–128

Nahri LO and Fulco AJ (1982) Phenobarbital induction of a soluble cytochrome P-450-dependent fatty acid monooxygenase in *Bacillus megaterium*. J. Biol. Chem. 257: 2147–2150

Nahri LO and Fulco AJ (1986) Characterization of a catalytically self-sufficient 119000-Dalton cytochrome P-450 monooxygenase induced by barbiturates in *Bacillus megaterium*. J. Biol. Chem. 261: 7160–7169

Nahri LO and Fulco AJ (1987) Identification and characterization of two functional domains in cytochrome P-450 $_{BM-3}$, a catalytically self-sufficient monooxygenase induced by barbiturates in *Bacillus megaterium*. J. Biol. Chem. 262: 6683–6690

Nakahara T. Sasaki K and Tabuchi T (1982) Production of yeast cells from palm oil at the high temperature of 40 °C. J. Ferment, Technol. 60: 89–91

Nakanishi J, Kurono Y, Kolde Y and Beppu T (1989) Recombinant DNA, bacterium of the genus *Pseudomonas* containing it, and process for preparing lipase using it. European Patent 0331376

Niehaus WG, Kisic A, Torkelson A, Bednarczyki DJ and Schroepfer GJ (1970a) Stereospecific hydration of *cis*– and trans-9, 10-epoxyoctadecanoic acid. J. Biol. Chem. 245: 3802–3809

Niehaus WG, Kisic A, Torkelson A, Bednarczki DJ and Schroepfer GJ (1970b) Stereospecific hydration of the Δ9 double bond of oleic acid. J. Biol. Chem. 245: 3790–3797

Nishioka T, Chihara-Shiomi M, Yoshikawa K, Inagaki M, Yamamoto Y, Hiratake J, Baba N and Oda J (1991) Lipase from *Pseudomonas* sp.: reactions, cloning and amino acid sequence analysis. In: L Alberghina, R Verger and RD Schmid (eds) Lipases: Structure, Mechanism and Genetic Engineering, GBF Monographs no 16 (pp 253–262). VCH, Weinheim

Numa S (1981) Two long-chain acyl coenzyme A snythetases: their different roles in fatty acid metabolism Trend. Biochem. Sci. 6: 113–115

Nunn WD (1986) A molecular view of fatty acid catabolism in *Escherichia coli*. Microbiol. Rev. 50: 179–192

Nunn WD, Colburn RW and Black PN (1986) Transport of long-chain fatty acids in *Escherichia coli*. J. Biol. Chem. 261, 167–171

O'Connell M, Henry S and Shapiro L (1986) Fatty acid degradation in *Caulobacter crescentus*. J. Bacteriol. 168, 48–54.

Okazaki K, Takechi T, Kambara N, Fukui S, Kubota I and Kamiryo T (1986) Two acyl-coenzyme A oxidases in peroxisomes of the yeast *Candida tropicalis*: primary structures deduced from genomic DNA sequence. Proc. Natl. Acad. Sci. U.S.A. 83: 1232–1236

Okumura S, Iwai M and Tsujisaka Y (1980) Purification and properties of partial glyceride hydrolase of *Pencillium cyclopium* M1. J. Biochem. (Tokyo) 87: 205–211

Pan SC, Bonanno S and Wagman GH (1959) Efficient utilization of fatty oils as energy source in penicillin fermentation. Appl. Microbiol 7: 176–180

Perlman D and Wagman GH (1952) Studies on the utilization lipids by *Streptomyces griseus*. J. Bacteriol. 63: 253–262

Peters II and Nelson FE (1948a) Factors influencing the production of lipase by *Mycotorula lipolytica*. J. Bacteriol. 55: 581–591

Peters II and Nelson FE (1948b) Preliminary characterization of the lipase of *Mycotorula lipolytica* J. Bacteriol, 55: 592–600

Pickersgill RW, Sumner IG, Collins ME, Warwicker J, Perry B, Bhat KM and Goodenough PW (1991) Modification of the stability of phospholipase A_2 by charge engineering, FEBS Lett. 281: 219–222

Pieringer RA (1989) Biosynthesis of non-terpenoid lipids. In: C Ratledge and SG Wilkinson (eds) *Microbial Lipids*, Vol 2 (pp 51–114). Academic Press, London

Pilkington BJ and Rose AH (1991) Incorporation of unsaturated fatty acids by *Saccharomyces cerevisiae*: conservation of fatty-acyl saturation in phosphatidylinositol. Yeast 7: 489–494

Price MF and Cawson RA (1977) Phospholipase activity in *Candida albicans*. Sabouraudia 15: 179–186

Prince HN (1959) Effect of pH on the antifungal activity of undecylenic acid and its calcium salt. J Bacteriol. 78: 788–791

Rachubinski RA, Fujiki Y and Lazarow PB (1985) Cloning of cDNA coding for peroxisomal acyl-CoA oxidase from the yeast *Candida tropicalis* pK223. Proc. Natl. Acad. Sci. U.S.A. 82: 3973–3977

Rahn O (1945) Injury and Death by Chemical Agents. Biodynamica, Normandy, MO, U.S.A.

Ramakrishnan CV (1957) *In vitro* lipase synthesis in *Aspergillus niger*. Naturwissenschaften 44: 401–402

Ratledge C (1980) Microbial lipids derived from hydrocarbons. In: DEF Harrison, IJ Higgins and R Watkinson (eds) Hydrocarbons in Biotechnology (pp 133–153). Heyden and Son Ltd, London

Ratledge C (1987) Biochemistry of growth and metabolism. In: JD Bu'Lock and B Kristiansen (eds) Basic Biotechnology (pp 11–55). Academic Press, London

Ratledge, C. (1989) Biotechnology of oils and fats. In: C Ratledge and SG Wilkinson (eds) Microbial Lipids, Vol 2 (pp 567–668). Academic Press, London

Ratledge C (1991) Yeast physiology: a micro-synopsis. Bioproc. Eng. 6: 195–203

Ratledge C and Evans CT (1989) Lipids and their metabolism. In: AH Rose and J Harrison (eds) The Yeasts, 2nd edition, Vol 3 (pp 367–455). Academic Press, London

Ratledge C and Tan KH (1990) Oils and fats: production, degradation and utilization by yeasts. In: H Verachtert and R De Mot (eds) Yeast Biotechnology and Biocatalysis (pp 223–253). Marcel Dekker Inc., New York

Ruettinger RT and Fulco AJ (1981) Epoxidation of unsaturated fatty acids by a soluble cytochrome P-450 dependent systems from *Bacillus megaterium*. J. Biol. Chem. 256: 5728–5738

Rydin S, Molin G and Nilsson I (1990) Conversion of fat into yeast biomass in protein-containing waste-water. Appl. Microbiol. Biotech. 33: 473–476

Sanglard D and Loper JC (1989) Characterization of the alkane-inducible cytochrome – P450 (P450 alk) gene from the yeast *Candida tropicalis*: identification of a new P-450-gene family. Gene 76: 121–136

Savitha J and Ratledge C (1992) An inducible, intracellular, alkalophilic lipase in *Aspergillus flavipes* grown on triacylglycerols. World J. Microbiol. Biotechnol. 8: 129–131

Schrag JD, Li Y, Wu S and Cygler M (1991) Ser-His-Glu triad forms the catalytic site of the lipase from *Geotrichum candidum*. Nature 351: 761–764

Schroepfer GJ, Niehaus WG and McCloskey JA (1970) Enzymatic conversion of linoleic acid to 10 D-hydroxy-Δ12-*cis*-octadecenoic acid. J. Biol. Chem. 245: 3798–3801

Schulz H and Kunau WH (1987) Beta-oxidation of unsaturated fatty acids: a revised pathway. Trends Biochem. Sci. 12: 403–407

Schunck WH, Kargel E, Gross B, Wiedmann S, Mauersberger S, Kopke K, Kiessling U, Strauss M, Gasestel M and Müller HG (1989) Molecular cloning and characterization of the primary structure of the alkane hydroxylating cytochrome P-450 from the yeast *Candida maltosa*. Biochem. Biophys. Res. Comm. 161: 843–850

Schweizer E (1989) Biosynthesis of fatty acids and related compounds. In: C Ratledge and SG Wilkinson (eds) Microbial Lipids, Vol 2 (pp 3–50). Academic Press, London

Shelley AW, Deeth HC and Mac Rae IC (1987) Review of methods of enumeration, detection and isolation of lipolytic microorganisms with special reference to dairy applications. J. Microbiol. Meth. 6: 123–137

Shimada Y, Sugihara A, Iizumi T and Tominaga Y (1990) cDNA cloning and characterization of *Geotrichum candidum* lipase II. J. Biochem. 107: 703–707

Sidebottom CM, Charton E, Dunn PPJ, Mycock G, Davies C, Sutton JL, Macrae AR and Slabas AR (1991) *Geotrichum candidum* produces several lipases with markedly different substrate specificities. Eur. J. Biochem. 202: 485–491

Skerratt JH, Nichols PD, Bowman JP and Sly LI (1992) Occurrence and significance of long-chain (ω−1)-hydroxy fatty acids in methane-utilizing bacteria. Org. Geochem. 18: 189–194

Skoneczny M, Chelstowska A and Rytka J (1988) Study of the coinduction by fatty acids of catalase A and acyl-CoA oxidase in standard and mutant *Saccharomyces cerevisiae* strains. Eur. J. Biochem. 174: 297–302

Sligar SG and Murray RI (1986) Cytochrome P-450$_{cam}$and other bacterial P-450 enzymes. In: PR Ortiz de Montellano (ed) Cytochrome P-450 (pp 429–503). Plenum Press, New York

Soda K (1987) Biotransformation of oleic acid to ricinoleic acid. J. Am. Oil Chem. Soc. 64: 1254

Stärkle M (1924) Die Methylketone im oxydativen Abbau einiger Triglyceride (bzw Fettsäuren) durch Schimmelpilze unter Berücksichtigung der besonderen Ranzidität des Kokosfettes. I. Die Bedeutung der Methylketone im Biochemismus der Butterranzidität II. Uber die Entstehung und Bedeutung der Methylketone als Aromastoffe im Roqueforkäse. Biochem. Z. 151: 371–415

Stead D (1986) Microbial lipases: their characteristics, role in food spoilage and industrial uses. J. Dairy Res. 53: 481–505

Stoffel W and Caesar H (1965) Der Stoffwechsel der ungesättigten Fettsäuren. V. Zur beta-oxidation der mono- und polyenfettsäuren. Der mechanism der enzymatischen Reaktionen an delta-2-cis-Enoyl-CoA-Verbindungen. Hoppe -Seyler's Z. Physiol. Chem. 342: 76–83

Stokoe WN (1928) The rancidity of coconut oil produced by mould action. Biochem. J. 22: 80–93

Sztajer H, Maliszewska I and Wieczorek J (1988) Production of exogenous lipases by bacteria, fungi and actinomycetes. Enzyme Microb. Technol. 10: 492–497

Szuhaj BF (ed) (1989) Lecithins: sources, manufacture and uses. American Oil Chemists' Society, Champaign, Illinois, U.S.A.

Tahara S, Hosokawa K and Mizutani J (1980) Occurrence of 7-hydroxyalkanoic acids in *Mucor* species. Agric. Biol. Chem. 44: 193–197

Takahashi Y, Sugahara T and Ohsaka A (1981) Phospholipase C from *Clostridium perfringens*. Methods in Enzymology 71: 710–725

Tan KH and Gill CO (1984) Effect of culture conditions on batch growth of *Saccharomycopsis lipolytica* on olive oil. Appl. Microbiol. Biotechnol. 20: 201–206

Tan KH and Gill CO (1985a) Batch growth of *Saccharomycopsis lipolytica* on animal fats. Appl. Microbiol. Biotechnol. 21: 292–298

Tan KH and Gill CO (1985b) Effect of culture conditions on batch growth of *Pseudomonas fluorescens* on olive oil. Appl. Microbiol. Biotechnol. 23: 27–32

Tan Y and Miller KJ (1992) Cloning, expression, and nucleotide sequence of a lipase gene from *Pseudomonas fluorescens* B52. Appl. Environ. Microbiol. 58: 1402–1407

Tanaka A and Fukui S (1989) Metabolism of alkanes. In: AH Rose and J Harrison (eds) The

The Yeasts, 2nd edition, Vol 3 (pp 261–287). Academic Press, London

Taylor F and Craig JC (1991) Scale-up of flat-plate reactors for enzymatic hydrolysis of fats and oils. Enzyme Microb. Technol. 13: 956–959

Thorpe RF and Ratledge C (1972) Fatty acid distribution in triglycerides of yeasts grown on glucose or n-alkanes. J. Gen. Microbiol 72: 151–163

Ueda M, Morikawa T, Okada H and Tanaka A (1987) Relationship between enoyl-CoA hydratase and a peroxisomal bifunctional enzyme, enoyl-CoA-hydratase/3-hydroxyacyl-CoA dehydrogenase, from an n-alkane-utilizing yeast, Candida tropicalis. Agric. Biol. Chem. 51: 2197–2205

van den Bosch H (1980) Intracellular phospholipases A. Biochim. Biophys. Acta. 604: 191–246

van Tilbeurgh H, Sarda L, Verger R and Cambillan C (1992) Structure of the pancreatic-procolipase complex. Nature 359: 159–162

van der Veen J (1974) Production of single cell protein by cultivation of yeasts on fats and fat products. Chemisch Doctorandus, Rotterdam, Netherlands

Veenhuis M and Harder W (1991) Microbodies. In: AH Rose and J Harrison (eds) The Yeasts, 2nd edition, Vol 4 (pp 601–653). Academic Press, London

Veenhuis M, Mateblowski M, Kunau WH and Harder W (1987) Proliferation of microbodies in Saccharomyces cerevisiae. Yeast 3: 77–84

Vermuë MH, Tramper J, de Jong JPJ and Oostrom WHM (1992) Enzymic transesterification in near-critical carbon dioxide: effect of pressure, Hildebrand solubility parameter and water content. Enzyme Microb. Technol. 14: 649–655

Vorderwülbecke T, Kieslich K and Erdmann H (1992) Comparsion of lipases by different assays. Enzyme Microb. Technol. 14: 631–639

Vuillemin N, Dupreyron C, Leluan C and Bory J (1981) Degradation of fatty acids by microorganisms. Ann. Pharm. Françaises 39: 155–159

Waite M (1987) The Phospholipases (Handbook of Lipid Research, Vol 5). Plenum Press, New York

Wilkinson SG (1988) Gram-negative bacteria. In: C Ratledge and SG Wilkinson (eds) Microbial Lipids, Vol 1 (pp 299–488). Academic Press, London

Winkler FK, D'Arcy A and Hunziker W (1990) Structure of human pancreatic lipase. Nature 343: 771–774

Yagi T, Hatona A, Hatano T, Fukui F and Fukui S (1991) Formation of mono hydroxy-n-nonane-2-ones from tricaprin by Fusarium arenaceum f. sp. fabae IFO-7158. J. Ferment. Bioeng. 71: 176–180

Yamaguchi S and Mase T (1991) Purification and characterization of mono- and diacylglycerol lipase isolated from Penicillium camembertii. Appl. Microbiol. Biotechnol. 34: 720–725

Yamaguchi S, Mase T and Takeuchi K (1992) Cloning and structure of the monoacylglycerol and diacylglycerol lipase-encoding gene from Penicillium camembertii U-150. Gene 103: 61–67

Yamaoka M, Fortkamp J, Morr M and Schmid RD (1992) The effect of 2, 4, 6, 8-tetramethyldecanoic acid on the hydrolysis and acylation of phospholipid by lipase in isooctane. Lipids 67: 826–828

Yang SY and Schultz H (1983) The large subunit of the fatty acid oxidation complex from Escherichia coli is a multifunctional polypeptide. J. Biol. Chem. 258: 9780–9785

Yang SY Cuebas D and Schulz H (1986) Channeling of 3-hydroxy-4-trans-decenoyl coenzyme A on the bifunctional β-oxidation enzyme from rat liver peroxisomes and on the large subunit of the fatty acid oxidation complex from Escherichia coli. J. Biol. Chem. 261: 15390–15395

Yi ZH and Rehm HJ (1982a) Metabolic formation of dodecanedioic acid from n-dodecane by a mutant of Candida tropicalis. Eur. J. Appl. Microbiol. Biotechnol. 14: 254–258

Yi ZH and Rehm HJ (1982b) Degradation pathways from n-tridecane to α, ω-dodecanedioic acid in a mutant of Candida tropicalis. Eur. J. Appl. Microbiol. Biotechnol. 15: 144–146

Yi ZH and Rehm HJ (1982c) A new metabolic pathway from n-dodecane to α, ω-dodecanedioic acid in a mutant of Candida tropicalis. Eur. J. Appl. Microbiol. Biotechnol. 15: 175–179

Yi ZH and Rehm HJ (1988a) Formation and degradation of Δ^9-1,18-octadecanedioic acid from oleic acid by Candida tropicalis. Appl. Microbiol. Biotechnol. 28: 520–526

Yi, ZH and Rehm HJ (1988b) Bioconversion of elaidic acid to Δ^9-trans-1,18-octadecanedioic acid from oleic acid by *Candida tropicalis*. Appl. Microbiol. Biotechnol. 29: 305–309

Zhihua, Y and Xiuzhen H (1986) Formation and characterization of chain (*sic.*) diacyl-CoA synthetase from *Candida tropicalis*. Acta Microbiol. Sinica 26: 333–340

Zvyagintseva IS and Pitryuk IA (1975) Lecithinase activity of several yeast species. Mikrobiologiya (USSR) Intern. ed. 44: 1119–1121

5. Biodegradation of anionic surfactants and related molecules

G.F. WHITE and N.J. RUSSELL
Department of Biochemistry, University of Wales, Cardiff, P.O. Box 903, Cardiff CF1 1ST, U.K.

Abbreviations: CoA – coenzyme A; MBAS – methylene-blue active substance; LAS – linear alkylbenzene sulphonate; SDS – sodium dodecyl sulphate; SDTES – sodium dodecyltriethoxy sulphate

I. Introduction

The term *surfactant* is used to describe a large group of structurally-diverse molecules, whose common feature is that they possess (as their name indicates) *surf*ace-*acti*ve properties, i.e. they tend to concentrate at the surfaces and interfaces between aqueous solutions and gases or solids or non-aqueous liquid phases. This property results from the fact that they are amphiphiles, containing polar and non-polar moieties; there is a certain balance of the polar (hydrophilic) and non-polar (hydrophobic) parts of the molecule, which together with their specific chemical identity, gives different surfactants particular properties. As such, surfactants have a broad range of uses as cleaning, wetting and emulsifying agents in a large variety of domestic and industrial situations.

A distinction should be drawn between the term "surfactant" and the more generally used term "detergent", which refers to a commercial formulation or product that is designed with particular cleansing properties. Most modern detergent formulations contain, besides up to one-third surfactant, larger amounts of a "builder" (to act as a chelating agent) plus smaller amounts of perfumes, colouring agents, whiteners, enzymes and other components.

The required amphiphilic properties of surfactants can be achieved by a seemingly endless variety of chemical structures. However, most of this diversity lies with the hydrophilic moieties and their means of linkage to the hydrophobe on which basis surfactants are classified (Fig. 1), and in practice a relatively small number of surfactants account for the majority of usage worldwide. Surfactants are classified broadly according to the chemical nature of the polar group as being anionic, non-ionic, cationic or zwitterionic; within these large groups there is further sub-division on the basis of the chemistry of

C. Ratledge (ed.), Biochemistry of Microbial Degradation, 143–177.
© 1994 *Kluwer Academic Publishers. Printed in the Netherlands.*

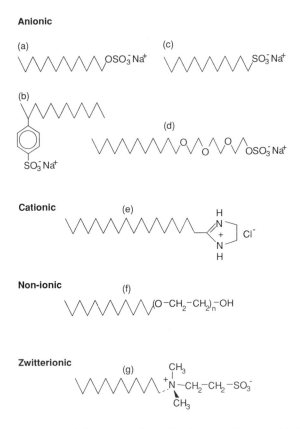

Anionic

(a)

(b)

(c)

(d)

Cationic   (e)

Non-ionic   (f)

Zwitterionic   (g)

Fig. 1. Illustrative examples of the structural classification of surfactants. (a) Sodium dodecyl sulphate (SDS); (b) 2-dodecylbenzene sulphonate; (c) sodium dodecane sulphonate; (d) sodium dodecyl triethoxy sulphate; (e) octadecylimidazolinium chloride; (f) dodecyl ethoxylate; (g) dodecylsulphobetaine.

the link between the hydrophile and hydrophobe (Fig. 1). In contrast, the hydrophobic moiety invariably contains medium-to-long alkyl chains.

A measure of the scale on which the developed world uses synthetic surfactants is amply illustrated by the fact that the combined annual production figure for Europe, Japan and U.S.A. is 3 to 4 million metric tonnes (Werdelmann 1984). Irrespective of the particular use made of a surfactant, its chemical structure remains unchanged when performing its cleansing or other task. Consequently most surfactants find their way either directly into the natural environment or into some form of treatment plant where they should be degraded; however, there may not be total breakdown and a proportion of intact parent surfactant (and/or biodegradation intermediates) may enter natural water courses where biodegradation can also occur. The wide range of applications of surfactants, together with large-scale usage of some types, makes them one of the most common components in domestic and industrial waste streams. To protect the environment from potential pollution, legislation and voluntary agreements in the European Community (EC, formerly EEC) and U.S.A. (EEC 1973; Soap and Detergents Association 1965) require that surfactants meet certain criteria of biodegradability. In man-made disposal plants, e.g. activated sludge and filter-bed systems, and natural water systems, e.g. rivers and lakes, bacteria are considered to be largely responsible for the biodegradation of surfactants although other microorganisms may also contribute (Biedlingmaier and Schmidt 1983). This chapter reviews what is known about the biochemical mechanisms and pathways present in competent biodegradative bacteria for the breakdown of anionic surfactants. These surfactants represent the majority of those in common use, both domestically and industrially. Most studies have been conducted using pure isolates in the laboratory in order to identify the biodegradative pathways and in some instances to isolate the key enzymes responsible. Therefore, this will be the emphasis of this review.

II. An overview of surfactant biodegradation

The anthropocentric view of biodegradation of such man-made compounds as synthetic surfactants is that it is a beneficial activity because it prevents undesirable pollution of the natural environment, while still allowing the compounds to be used. From the microorganisms' viewpoint, surfactants represent a potential source of energy and reduced carbon for growth. Bacteria utilize essentially two strategies to access the carbon in a surfactant (Fig. 2), the bulk of which (at least in ionic surfactants) is generally present in the hydrophobic moiety. The first strategy involves initial separation of the hydrophile from the hydrophobe which is then attacked oxidatively. In the second mechanism, the alkyl chain is oxidized directly whilst it is still attached to the hydrophile. Both strategies lead to an immediate loss of amphiphilicity in the molecule which therefore no longer behaves as a surfactant. For some surfactants, only one or the other route has so far been observed, whereas for others both of these general mechanisms have been found to operate. Although where dual pathways exist, they often occur (as might be expected) in separate organisms, in some cases multiple pathways are present in a single organism (e.g. see Section III.D). On the other hand, for the more complex surfactants which might demand considerable metabolic diversity to achieve overall biodegradation, different organisms with complementary metabolic capabilities may be needed in a consortium for complete breakdown of the molecule, albeit via a single strategic route.

Essentially two types of mechanism can be used to achieve the separation of hydrophile from hydrophobe, and the key factor determining which one is employed is the overall chemical stability of the link between the two halves of the molecule. Surfactants such as alkyl sulphates and sulphosuccinates, which

Hydrophile/hydrophobe
separation

Direct oxidation

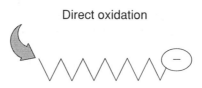

Fig. 2. Strategies for bacterial attack on anionic surfactants.

contain ester bonds between hydrophobic and hydrophilic moieties, are readily cleaved at the ester link via simple hydrolysis to give long-chain alcohols and anionic products from the acidic hydrophilic groups. These reactions are energetically favourable and do not require cofactors. In contrast, the alkane sulphonate surfactants contain a C-S link, which is much more resistant to hydrolysis (Wagner and Reid 1931) compared with the alcohol-*O*-acid links in sulphate and carboxylate esters. Consequently, simple hydrolysis does not occur and instead bacteria utilize a more complex oxidative mechanism, catalysed by monooxygenase enzymes and requiring O_2 and a reduced co-factor (see Section III.C). Similar considerations apply to ether-linked surfactants (Section III.D).

Whether hydrophile/hydrophobe separation is by simple hydrolytic cleavage or involves monooxygenation, the products derived from the hydrophobic part of the molecule are fatty acids, alcohols or aldehydes. Since these are normal bacterial metabolites, they can be catabolized readily via the ubiquitous pathway of fatty acid β-oxidation to give acetyl-CoA (see Chapter 4) which can be channelled into either energy production or the biosynthesis of cellular components.

The second general strategy of attack on surfactants also employs β-oxidation but instead seeks to metabolize the surfactant alkyl chain whilst it is still attached to the hydrophilic moiety. However, before β-oxidation can occur the end of the alkyl chain distal to the hydrophilic group must first be oxidized by ω-oxidation to give a carboxyl group which can be activated with the coenzyme A that is needed for the β-oxidation process. The ω-oxidation step is achieved in an energy-dependent monooxygenation reaction that uses NADH and O_2, which is the same pathway as that used in the microbial breakdown of alkanes and other hydrocarbons which lack a hydrophile (see Chapter 1). The β-oxidation of linear alkyl chains may then proceed either to completion or at least until the approaching hydrophilic group is close enough to the site of oxidation to interfere. β-Oxidation can cope with limited amounts of branching in the alkyl chain, e.g. methyl groups at α-positions, but not with geminal dimethyl substituted carbon atoms, i.e. quaternary carbons, in the alkyl chain. This is the structural aspect of tetrapropylene benzene sulphonates which makes them very resistant to biodegradation by hydrophobe oxidation.

Generally, because it requires the investment of energy, the hydrophobe oxidation mode of attack is most often seen for surfactants in which the linkage between the hydrophilic and hydrophobic parts is relatively difficult to break. Thus, for the alkylbenzene sulphonates where hydrophile removal would require a difficult disruption of the resonance-stabilized benzene ring, it is the only biodegradative pathway so far observed. In the alkane sulphonates and alkyl ethoxysulphates it occurs as an *alternative* to hydrophile separation (by oxidative desulphonation and ether-cleavage respectively). At the other end of the "biodegradability spectrum", alkyl sulphates apparently are not broken down by ω-/β-oxidation, at least in those competent bacterial isolates which have been studied in laboratories. It is energetically less demanding to release sulphate by sulphatase action (simple hydrolysis) rather than to expend energy on ω-oxidation in order to make the alkyl chain available for breakdown by β-oxidation. Thus alkylsulphatase-positive bacteria have an advantage over potential ω-oxidizers, at least in enrichment cultures.

The initial biodegradative step(s) whereby the surface-active property is removed constitutes *primary biodegradation*. This may well leave residues which, from the bacterial perspective, still contain much useful carbon and, from an environmental viewpoint, could be toxic (possibly more so than the parent molecule). The subsequent complete breakdown of residues formed by primary degradation into CO_2, H_2O, mineral salts and other "normal" cellular constituents, which may be low-molecular weight intermediates of metabolism or macromolecular components of cellular structure and function, is termed *ultimate (complete) biodegradation*. The process of conversion of organic materials (including surfactants) to CO_2 and H_2O is also known as *mineralisation*, but this term should not be used synonymously with ultimate biodegradation which in the normal lifetime of a biodegrading microorganism probably never results in complete conversion of surfactant to CO_2 and H_2O (Swisher 1987).

Where primary attack is by hydrophile/hydrophobe separation, ultimate biodegradation usually follows easily because it occurs via the common oxidative pathways used to break down fatty acids intracellularly. For example, removal of SO_4^{2-} from primary alkyl sulphates yields an aliphatic alcohol which is readily oxidized by dehydrogenase enzymes found in bacteria to give first the corresponding aldehyde and then the acid which can subsequently be activated and β-oxidized (*vide supra*). Therefore, in this instance the measurement of primary biodegradation by sulphate release or disappearance of surfactant [e.g. with the methylene-blue active substance (MBAS) test] can be used as an indicator of the complete (ultimate) biodegradability of the compound. However, separate measurements of ultimate biodegradation are generally also made to confirm ultimate biodegradability, for instance by CO_2 release.

In other situations arising from either hydrophile/hydrophobe separation or from direct oxidation of the hydrophobe, residual hydrophilic moieties may still contain carbon. Complete mineralization of a surfactant means that these residues must be metabolized. This process may also mobilize elements such as sulphur for metabolism into other cellular components. For some anionic surfactants primary biodegradation leads directly to full mineralization of the hydrophile; for example, sulphate and sulphite are released from alkyl sulphate and alkane sulphonate hydrophiles, respectively. With some other surfactants, organic residues do remain. For instance, the primary etherase cleavage of the alkylpolyethoxy sulphate surfactant sodium dodecyltriethoxy sulphate (see Section III.D) produces mono-, di- and tri-ethylene glycol monosulphates plus their oxidized derivatives acetic acid 2-sulphate, acetic acid 2-(ethoxysulphate) and acetic acid 2-(diethoxy sulphate). Although these intermediates persist in pure cultures (Hales et al. 1982, 1986), their accumulation in mixed environmental cultures is only transient (Griffiths et al. 1986), indicating that they do indeed undergo further metabolism.

If the mode of primary attack is by ω-/β-oxidation, then this will normally proceed unhindered until the alkyl chain becomes too short for further rounds of oxidation. In practice, this means that primary biodegradation produces organic compounds comprised of the hydrophilic group attached to a small residue of the alkyl chain such as happens during linear alkylbenzene sulphonate (LAS) biodegradation. In the particular case of LAS, details of the biodegradation of the hydrophilic residues are now emerging (see Section III.B), but for other surfactants the ultimate biodegradation of the organic intermediates of primary biodegradation has been poorly studied in comparison to work on the parent surfactants.

III. Biodegradation pathways, enzymes and mechanisms

A. Alkyl sulphates

1. Pathways of primary and ultimate biodegradation

Alkyl sulphates such as sodium dodecyl (lauryl) sulphate (SDS or SLS) are readily broken down by bacteria and constitute the most readily biodegradable surfactants in common use. Early studies by Payne (Payne 1963; Payne and Feisal 1963; Payne et al. 1965, 1967) and Hsu (1963, 1965) established the role of bacteria in the biodegradation process and indicated that the metabolic pathway was initiated by sulphatase-mediated hydrolysis of ester to alcohol plus inorganic sulphate. More recently, Thomas and White (1989) used sodium [1-^{14}C]dodecyl sulphate to show that the major radiolabelled metabolite of biodegradation by *Pseudomonas* sp. C12B was dodecanol with traces of dodecanal; these were oxidized to the fatty acid dodecanoic, which was elongated to tetradecanoic acid. These fatty acids were incorporated as the acyl chains in membrane phospholipids (Fig. 3). In addition, water-soluble products were formed, believed to be intermediates of the tricarboxylic acid cycle. Thus the surfactant was utilized both for energy production (70 to 80% based on $^{14}CO_2$ release) and for incorporation into cellular material (20 to 30%). Such extensive conversion to CO_2 is considered (Gilbert and Watson 1977; Payne and Wiebe 1978) to reflect extensive or even complete biodegradation. Thus, the primary biodegradative step of sulphate removal essentially signals the complete biodegradation of the molecule and, for practical purposes, desulphation can be used as a measure of ultimate biodegradation.

The biodegradation of secondary alkyl sulphates follows broadly the same pattern. Desulphation is again the first step, followed by oxidation of the liberated secondary alcohol which in this case leads not to an aldehyde

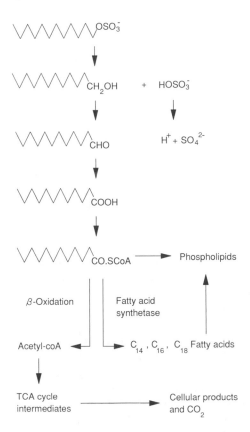

Fig. 3. Pathways of incorporation of SDS into bacterial cellular products. Summarised from the data on [1-^{14}C] SDS incorporation into *Pseudomonas* C12B (Thomas and White 1989).

(alkanal) but to the corresponding ketone. At this point the pathways for primary and secondary compounds differ (Fig. 4). The ketones undergo O_2-dependent hydroxylation at the carbon adjacent to the carbonyl group, then oxidation to the vicinal dione which in turn undergoes C-C bond cleavage to yield an aldehyde and a carboxylic acid; both of these metabolites are able to rejoin the fatty acid β-oxidation route into central metabolic pathways (Lijmbach and Brinkhuis 1973).

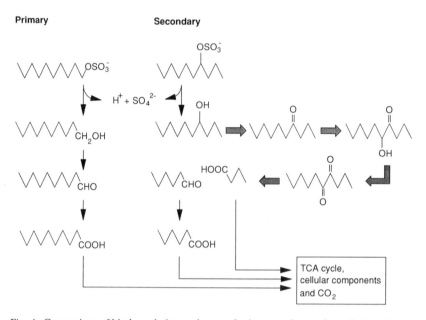

Fig. 4. Comparison of biodegradation pathways of primary and secondary alkyl sulphates.

From the foregoing discussion, it is clear that removal of the sulphate ion is the key initial step in the biodegradation of both primary and secondary alkyl sulphate surfactants. An obvious corollary is that it is the properties of the alkylsulphatase enzymes which will determine the biodegradability of the surfactant, and it is on these enzymes that attention has mainly focussed.

2. Enzymology of alkyl sulphate biodegradation

It is well known that primary and secondary alkyl sulphates are readily biodegradable (Swisher 1987), and that bacteria capable of their degradation are widely distributed (Rotmistrov et al. 1978; White et al. 1985, 1989; Anderson et al. 1988, 1990). Paradoxically, however, relatively few bacterial isolates have been examined to establish the relevant enzymology. The alkylsulphatases which initiate primary biodegradation catalyse the release of sulphate through cleavage of the C-O or O-S bond of the sulphate ester

R-O-SO$_3^-$ linkage by means of a hydrolytic reaction mechanism to give a primary alcohol as the co-product (see Section III.A.5). The alkylsulphatases most extensively studied are those in *Pseudomonas* C12B and *Comamonas terrigena* which have been subjected to a long-standing series of investigations conducted cooperatively by K.S. Dodgson, G.F. White and co-workers in Cardiff and W.J. Payne's group in Georgia, U.S.A., who originally isolated the organisms from river-bank soil beneath a sewage-treatment-plant outfall.

Under appropriate growth conditions, *Pseudomonas* C12B can produce up to five alkylsulphatases, two primary (P1 and P2) and three secondary (S1, S2 and S3), whereas *C. terrigena* produces no primary but two secondary alkylsulphatases (CS1 and CS2). These enzymes can be separated and visualized by gel zymography (see Fig. 5, Section III.A.4). With the exception of S3, the secondary enzymes remove sulphate from the penultimate carbon of the alkyl chain; they all exhibit stereo-specificity for enantiomeric forms of the ester substrates and liberate enantiomerically-pure secondary alcohols. The S3 enzyme acts on symmetrical or near-symmetrical esters and its stereo-specificity for S(−)isomers becomes more pronounced as the asymmetry of the substrate increases. Even primary alkylsulphatases from the same organism operate by different molecular mechanisms of ester-bond cleavage. Thus the few alkylsulphatases so far characterized reveal a rich diversity of specificity; preliminary indications are that this diversity also extends to the long-chain alcohol dehydrogenases which oxidize the resulting alcohols.

The majority of the alkylsulphatase enzymes listed above have been purified to homogeneity and characterized in considerable detail in terms of molecular properties (molecular weight, sub-unit structure), physiological aspects (cellular localisation, regulation) and catalytic properties (kinetics, mechanisms of substrate binding and catalytic steps). This information has already been collated in earlier reviews (Dodgson et al. 1982; Dodgson and White 1983). Consequently, the following treatment is generally brief and focusses on information relevant to the enzymes' role in surfactant biodegradation. One of the enzymes (P1), however, has been purified and characterized more recently: so for this enzyme more detail is given here, both for completeness and to illustrate the level of detailed knowledge which is available for these enzymes.

3. Primary alkylsulphatases
The constitutive P1 alkylsulphatase of *Pseudomonas* C12B has been purified 1500-fold to homogeneity by a combination of ammonium sulphate precipitation and column chromatography on DEAE-cellulose (DE52), Sephacryl S-300 and butyl-agarose (Bateman et al. 1986). The enzyme is a tetrameric protein with a subunit M$_r$ of 88 kDa and a pH optimum of 6.1. It is active against alkyl sulphates with chain lengths from C$_6$ to C$_{14}$ and displays decreasing K$_m$ values from 4.7 mM for hexan-1-sulphate to 0.07 mM for dodecan-1-sulphate. The V$_{max}$ is highest for octan-1-sulphate with a value of 131.9 μmol SO$_4^{2-}$ released (mg protein)$^{-1}$.min^{-1}, decreasing to 43.9 and 27.4

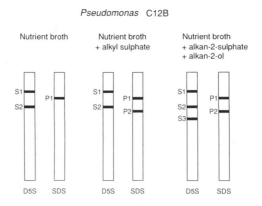

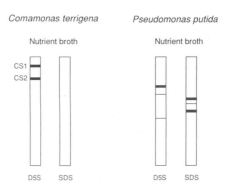

Fig. 5. Illustration of the use of gel zymography to separate and visualize alkylsulphatases. Extracts of cells grown in the indicated media were subjected to non-denaturing electrophoresis on 6.8% w/v polyacrylamide gels and the gel subsequently developed by incubating in SDS or a mixture of isomeric decyl sulphates (D5S) to locate primary (designated "P") and secondary ("S") alkylsulphatases respectively, as white bands of insoluble alcohol droplets. "C" distinguishes the enzymes of *Comamonas terrigena* from those of *Pseudomonas* C12B.

for the C_6 and C_{10} homologues respectively. The values of ΔG° for binding of substrate decrease from -13.5 kJ mol^{-1} for hexan-1-sulphate to -24.3 kJ mol^{-1} for dodecan-1-sulphate, indicating that there is a strong hydrophobic interaction between the substrate and enzyme active site. The P1 enzyme is competitively inhibited by secondary alkyl sulphates and primary alkane sulphonates, and is also weakly active towards aryl sulphates. On the basis of ^{18}O incorporation from $H_2^{18}O$ into the co-product of the reaction SO_4^{2-}, it was demonstrated that the P1 enzyme cleaves the O-S bond of the C-O-S ester linkage (see Section III.A.5). This mode of bond cleavage is compatible with the noted activity towards aryl sulphates because enzymic hydrolysis of the latter (and of almost all other alcohol-*O*-acid esters) invariable proceeds via *O*-acid cleavage (Bartholomew et al. 1977).

The properties of the P2 enzyme have been reviewed thoroughly elsewhere (Dodgson et al. 1982; Dodgson and White 1983), so a summary only for comparison with the P1 enzyme is given here. The purified P2 enzyme is a dimer with a subunit M_r value of 79 kDa. It has an optimum pH of 8.3 and is active towards primary alkyl sulphates with chain lengths from C_4 to C_{14} with maximum activity towards nonan-1-sulphate, and K_m values which decrease with increasing alkyl chain length. The P2 enzyme is competitively inhibited by secondary alkyl sulphates and primary alkane sulphonates. Implications of the similarities and differences between P1 and P2 are discussed later (Section III.A.5).

Information about other long-chain primary alkylsulphatases is scarce. *Pseudomonas aeruginosa* produces a single primary enzyme, and some aspects of its regulation have been studied in Fitzgerald's laboratory (see Section III.A.6). However this enzyme has yet to be purified and characterized. A strain of *Pseudomonas putida*, isolated in Cardiff for its ability to utilize 2,4-dichlorophenoxyethyl sulphate ("Crag" herbicide) as sole source of sulphur, liberates sulphate from this alkyl sulphate by means of a sulphatase enzyme. Use of the gel zymographic technique showed that crude extracts of broth-grown cells contained three primary enzymes active on SDS together with three secondary alkylsulphatases which were revealed by staining gels with a mixture of secondary dodecyl sulphate isomers. One of the primary alkylsulphatases was also active towards Crag herbicide and this enzyme was purified to homogeneity from crude extracts using a combination of heat treatment, chromatography on DEAE-cellulose, Sephacryl gel-exclusion media and butyl agarose (Lillis et al. 1983). The native enzyme is tetrameric with a M_r of 295 kDa, its pH optimum is 6.0; it exhibits saturation kinetics with Crag herbicide and with the C_6 to C_9 primary alkyl sulphates so far tested (Bateman et al. 1986). 2-Butoxyethyl sulphate is a relatively poor substrate for the enzyme but the surfactant sodium dodecyltriethoxy sulphate (see Section III.D) and aryl sulphates are not substrates. Incorporation of ^{18}O from $H_2^{18}O$ into inorganic sulphate, but not the alcohol co-product, shows that O-S bond scission occurs for all esters tested (Crag herbicide, octan-1-sulphate, 2-butoxyethyl sulphate), a surprising result in view of the lack of activity with aryl sulphates (cf. the P1 enzyme). Although the enzyme is active towards C_6 to C_9 primary alkyl sulphates, it differs from the P1 and P2 enzymes in that K_m decreases from C_6 to C_7 but remains constant for C_7 to C_9. Thus although this enzyme shows some similarities with both P1 and P2, it also differs significantly from both (Bateman et al. 1986).

4. Secondary alkylsulphatases

In many avenues of science, progress owes much to serendipity, and the discovery of stereospecific alkylsulphatases is a good example. In the early 1970s, Dodgson and Payne developed a zymographic technique for detecting alkylsulphatase activity by incubating electrophoresis gels in solutions of long-chain alkyl sulphates (Fig. 5). The alcohol (e.g. dodecanol) liberated by localized alkylsulphatase activity in the gel forms an insoluble white band of alcohol droplets (Payne et al. 1974). In attempts to detect activity towards near-symmetrical secondary alkyl sulphates in extracts of *Pseudomonas* C12B, electrophoresed gels were incubated in dodecan-5-sulphate prepared by sulphation of dodecan-5-ol using sulphuric acid. Fortuitously, but unknown at the time, this sulphation method produces a mixture of isomers including dodecan-2-sulphate and dodecan-5-sulphate. As a result, Payne et al. (1974) were surprised to see not one but three white bands appearing on the gels. Later work revealed the mixed nature of the ester and eventually the complementary substrate specificities of these enzymes emerged (*vide infra*). Had the ester used to stain gels been pure dodecan-5-sulphate, two of the three novel secondary enzymes would have remained hidden.

The breakthrough in understanding the multiplicity of secondary alkylsulphatases in *Pseudomonas* C12B and *C. terrigena* (see Section III.A.2) came with the purification and characterization of the CS2 enzyme in the latter organism. This constitutive alkylsulphatase was shown to be specific not only for secondary alkyl sulphates with the sulphate at C-2, but also for the R(+)enantiomers (Barrett et al. 1980; Matcham et al. 1977). The CS1 enzyme, which is also constitutive, has been purified only on a small scale, but this was enough to demonstrate that the enzyme is specific for S(−)isomers.

Subsequent studies with *Pseudomonas* C12B showed that, in terms of their specificities for R(+) and S(−)alkan-2-sulphates, the properties of its S1 and S2 secondary enzymes paralleled those of the CS2 and CS1 enzymes in *C. terrigena*. Moreover, the similarity between S1 and CS2 extends to include the stereo-specificity of the ester-bond-cleavage step, because in both cases the alcohol product was liberated from its ester by inversion of configuration (Walden inversion, Fig. 6) at C-2 (Bartholomew et al. 1977).

The remaining secondary enzyme from *Pseudomonas* C12B, namely S3, has also been purified and shown to act preferentially on symmetrical secondary alkyl sulphates, e.g. undecan-6-sulphate. The enzyme will also hydrolyse near-symmetrical esters, e.g. decan-5-sulphate, and more asymmetrical esters, e.g. octan-2-sulphate, for which the enzyme exhibits a preference for S(−)isomers. This enzyme thus completes the alkyl sulphate-degrading armoury of *Pseudomonas* C12B which is able to degrade mixtures of primary and isomeric secondary alkyl sulphates with chain lengths ranging from C_6 to at least C_{14}.

5. Alkylsulphatase mechanism

There are two features of the alkylsulphatases particularly germane to their action on surfactants: *viz.* alkyl chain-length specificity and mode of ester-bond cleavage. Despite their diverse specificities for stereo- and positional isomers, the primary and secondary alkylsulphatases have in common a minimum chain-length requirement around C_5 to C_6. The upper chain-length limit is less well-defined because of the difficulties of synthesis and the low solubility of the higher chain-length homologues. However, most of the enzymes so far examined will operate on C_{12} substrates and some on C_{14}. For the enzymes so

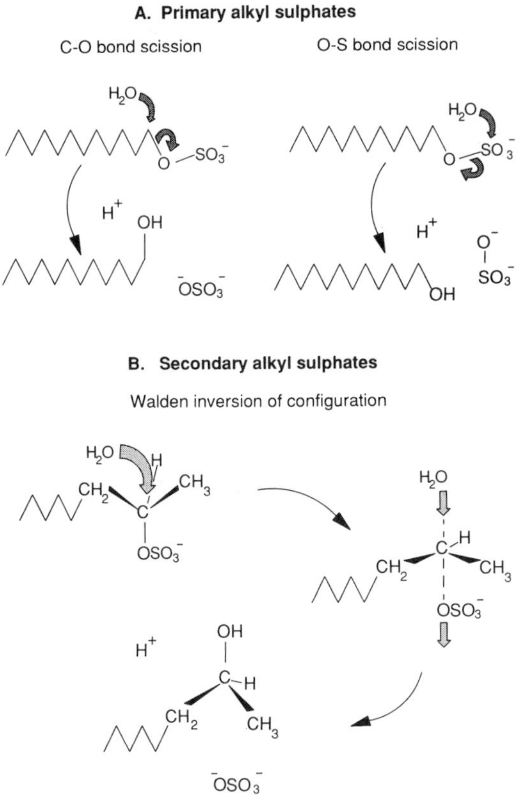

Fig. 6. Reaction mechanisms for the hydrolysis of alkyl sulphates by bacterial alkylsulphatases.

far examined (P1 and P2 enzymes in *Pseudomonas* C12B, and CS2 in *C. terrigena*), the substrate K_m values decrease progressively as the chain length increases. Assuming rapid pre-equilibrium (i.e. Michaelis-Menten kinetics), it has been possible to estimate $\Delta G°$ values for binding incremental methylene (-CH_2-) units to the active site (Table 1). Corresponding values for competitive inhibition of the primary enzymes P1 and P2 by alkan-2-sulphates and primary alkane sulphonates, and of the S1 and CS2 enzymes by primary alkyl sulphates and primary alkane sulphonates, are also included. The values cluster around $-2kJ$ mol^{-1} per methylene unit, which indicates a significant reliance on hydrophobic interactions to bind the ligand to the enzyme.

For the secondary enzymes, CS2 and S3, further analysis of substrate and inhibitor specificity has enabled active-site models to be postulated. The methylene units in the alkyl chains of R- and S-alkan-2-sulphates, primary alkyl sulphates and alkane sulphonates all bind with the same affinity to the enzyme (Table 1) but the plots of $\Delta G°$ versus alkyl chain length are displaced from each other (Barrett et al. 1980; Dodgson and White 1983). This is interpreted in

Table 1. Free energy increments for binding methylene (-CH$_2$-) units in the alkyl chains of substrates and inhibitors, to primary and secondary alkylsulphatases. Free energy increments are calculated from the slope of plots of free energy of ligand binding *versus* chain length for each group of compounds. Free energies of ligand binding were calculated from the equation $\Delta G^{o\prime} = -RT\ln K$, where K is the Michaelis-Menten constant K$_m$ for enzyme-substrate interactions or K$_i$ for competitive inhibition (data in italics).

Enzyme	Reference	$\Delta G^{o\prime}$ per methylene group (kJ mol^{-1})		
		Alkan-1-sulphates	Alkan-2-sulphates	Alkane sulphonates
P1	(Bateman et al. 1986)	−1.7	*−1.95*	*−1.3*
P2	(Cloves et al. 1980a)	−2.5	−2.4	*−1.2*
S1	(Bartholomew et al. 1978)	−2.2	–	*−3.3*
CS2	(Barrett et al. 1980)	−2.4	−2.4(R)	−2.4
			−2.4(S)	
Crag	(Bateman et al. 1986)	−2.6[a]	–	–

[a]Calculated from the K$_m$ values for C$_6$ and C$_7$; no change in K$_m$ for C$_7$ to C$_9$.

terms of a three-point interaction between enzyme and substrate to account for the observed stereo-specificity: (i) a relatively unrestricted hydrophobic binding site for the longer alkyl chains attached at C-2 of the substrates; (ii) a small hydrophobic pocket just large enough to accommodate the C-1 methyl group; (iii) a sulphate-binding site. These three interactions allow only R(+)alkan-2-sulphates to bind with alkyl, methyl and sulphate in their respective sites, thus constraining the groups around C-2 which are to undergo inversion of configuration during hydrolysis of the C-O ester bond (Walden inversion, *vide infra*). Other esters bind, but with at least one site occupied incorrectly, so that the esterified C-2 is not fully restrained and therefore not susceptible to hydrolysis.

The S3 enzyme activity is highest with symmetrical alkyl sulphates but it retains activity as the sulphate group is moved from the centre of the chain towards one end, even to C-2. However, the more asymmetric the ester, the greater is the enzyme's preference for S(−) enantiomers: not surprisingly then, a similar three-point attachment has been postulated for this enzyme as for the CS2, except that the smaller hydrophobic pocket, while still restricted, will in this case accommodate up to four methylene groups (Shaw et al. 1980; Dodgson and White 1983).

Turning to the catalytic step in alkyl sulphate hydrolysis, the C-O-S ester linkage presents two possible bonds for hydrolytic attack: either C-O or O-S (Fig. 6). The occurrence of Walden inversion for at least three of the secondary enzymes (S1, CS2 and S3) strongly implies C-O bond cleavage (Fig. 6). This has been unequivocally confirmed by incorporation of ^{18}O from H$_2$^{18}O into the alcohol, but not into the sulphate co-product. The Walden inversion at C-2 is also consistent with the three-point interaction model, discussed above, between enzyme and groups at C-2 of the substrate which accounts for substrate stereo-specificity.

For primary alkyl sulphates, inversion of configuration is not observable because substrates and products are not chiral. However, ^{18}O-labelling studies with the P2 enzyme established that in fact it does operate by C-O cleavage (Cloves et al. 1977). Thus a picture was beginning to emerge which suggested that all the long-chain alkylsulphatases in *Pseudomonas* C12B and *C. terrigena*, and possibly those in other organisms too, might operate in this way (Dodgson et al. 1982). This sweeping generalisation was soon to be disproved because the next two long-chain alkylsulphatases examined, P1 in *Pseudomonas* C12B and Crag herbicide sulphatase in *Pseudomonas putida* (Lillis et al. 1983; Bateman et al. 1986), were both found to operate by O-S cleavage! Nevertheless, the discovery of alcohol-*O* ester bond cleavage by some enzymes was an important discovery because all other ester hydrolases (carboxylesterases, fatty acyl esterases, phosphatases, bisphosphatases, arylsulphatases, choline sulphatase, lactate-2-sulphatase, butylsulphatase) so far examined had been shown to operate by cleavage of the *O*-acid bond (Bartholomew et al. 1977; Crescenzi et al. 1984; White and Matts 1992).

Despite their starkly different modes of ester-bond hydrolysis, the dependence of V_{max} on alkyl chain length for P1 and P2 are remarkably similar to each other (Fig. 7) and to that of CS2, the only secondary alkylsulphatase for which data is available. All give maximum activity with the C_9 homologue and plots of activity *versus* alkyl chain-length are very similar. The data for the Crag alkylsulphatase are also consistent with this pattern in the C_6 to C_9 chain-length

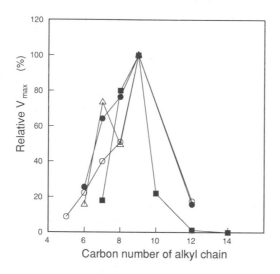

Fig. 7. Dependence of V_{max} of alkylsulphatases on the length of the alkyl chain of its substrate.
   The relative values of V_{max} (%) are given for the following alkylsulphatases: ●, P1 primary enzyme from *Pseudomonas* C12B; ○, P2 primary enzyme from *Pseudomonas* C12B; ■, CS2 secondary enzyme from *Comamonas terrigena*; △, Crag (2,4-dichlorophenoxyethyl sulphate) sulphatase from *Pseudomonas putida* FLA.

range, although it should be noted that no measurements have been made with longer esters (Bateman et al. 1986).

From molecular weights and V_{max} values for purified enzymes, estimates of k_{cat} (i.e. V_{max} / E_o, the first order rate constant for turnover of the enzyme-substrate complex) have been made (Table 2). Where possible, values are given for both C_8 (the usual assay substrate) and C_9 (optimum substrate). For the primary alkylsulphatases and the S3 secondary enzymes the values of k_{cat} are 10^2 to $10^3 s^{-1}$, whereas for the remaining enzymes they are about an order of magnitude higher. These values are typical of many other enzyme-catalysed reactions (Cornish-Bowden and Wharton 1988).

6. Alkylsulphatase regulation

Besides having different mechanisms of ester-bond cleavage, the P1 and P2 primary alkylsulphatases also differ in their mode of regulation. In batch culture, synthesis of the P2 enzyme occurs transiently during the exponential growth phase (Fitzgerald and Payne 1972; Fitzgerald 1974; Thomas and White 1990) in response to the presence of its alkyl sulphate substrates (C_5 to at least C_{14}). It is also induced gratuitously by primary alkane sulphonates (C_7 to C_{12}) which are surfactants (see Section III.C) but not substrates of this enzyme. When the latter were used to assess concentration effects, a triphasic change in induced enzyme activity (rapid increase at low concentrations, followed by a decrease, then a slower increase) was observed (Cloves et al. 1980b). The triphasic plots were interpreted tentatively as reflecting separate uptake systems with different affinities for inducers. Closer study of the first phase indicated that there was a hyperbolic dependence of induction on inducer concentration; corresponding half-saturation constants decreased with increasing alkane sulphonate chain length. The plot of free-energy of ligand binding *versus* substrate chain length was linear with a slope corresponding to $-1.4 kJ mol^{-1}$ per methylene unit, remarkably close to the value for methylene group-binding to the P2 enzyme itself (Table 1) and indicative of a hydrophobic contribution to binding. No further progress has been made on the regulation of P2 synthesis which now clearly warrants further investigation at the genetic level (see Section III.A.7).

The original designation of the P1 enzyme as constitutive (Dodgson et al. 1982) has been abandoned in the light of more recent work (Bateman 1985; Thomas and White 1990) which showed that the enzyme was not produced until well into the stationary phase of batch culture; activity of a constitutive enzyme in a batch culture would be expected to follow the growth curve throughout exponential phase. Moreover, the enzyme was not produced on some simple growth media and therefore cannot be considered as being constitutive. The increase in activities produced by growing cells on lower concentrations of carbon source (e.g. pyruvate, succinate and especially nutrient broth) suggested that there was the involvement of carbon catabolite repression superimposed on specific repression by an as yet unidentified component(s) in the broth. The nature of the sulphur source in the growth medium had no effect

Table 2. Estimates of k_{cat} for some purified alkylsulphatase enzymes acting on octan- and nonan-1-sulphates.

Enzyme	Type	Source	k_{cat} (s^{-1})		Reference
			Octan-1-sulphate	Nonan-1-sulphate	
P1	Primary	*Pseudomonas* C12B	411	536	(Bateman et al. 1986)
P2	Primary	*Pseudomonas* C12B	56	109	(Cloves et al. 1980a)
Crag	Primary	*Pseudomonas putida*	402	772	(Bateman et al. 1986; Lillis et al. 1983)
S1	Secondary	*Pseudomonas* C12B	2358	–	(Bartholomew et al. 1978)
CS2	Secondary	*Comamonas terrigena*	3200	4000	(Matcham et al. 1977)
S3	Secondary	*Pseudomonas* C12B	–	121	(Shaw et al. 1980)

on the amount of activity induced.

The indications that the P1 enzyme serves in a carbon-scavenging role is compatible with its periplasmic location, a property shared with the P2 enzyme (Thomas et al. 1988). Such a location would be equally appropriate if the enzymes' function was to protect the cytoplasmic membrane and/or intracellular constituents from the disruptive effects of sulphated surfactants.

A strain of *Pseudomonas aeruginosa*, isolated from soil, was found capable of synthesising a single, substrate-inducible, primary alkylsulphatase (Fitzgerald and Kight 1977) although the enzyme has not been purified. Like the P2 enzyme from *Pseudomonas* C12B, the alkylsulphatase from *P. aeruginosa* is inducible by primary alkyl sulphate homologues (C_5 to C_{16}) and alkane sulphonates (C_7 to C_9 tested) but, in contrast, the induction is markedly inhibited by several sulphur sources. In addition, the induction is apparently subject to carbon catabolite repression and to inhibition by tricarboxylic acid cycle intermediates (Fitzgerald et al. 1978; Kight-Olliff and Fitzgerald 1978). Thus this regulation manifests elements of both the P1 and P2 regulator systems but any significance of this is unknown.

Knowledge about the regulation of synthesis of the secondary enzymes is even more scant than for the primary enzymes. The CS1 and CS2 enzymes of *C. terrigena* and S1 and S2 from *Pseudomonas* C12B are all classed as constitutive, being based on invariate production during growth on a variety of carbon and sulphur sources (Dodgson et al. 1982). The S3 enzyme on the other hand is inducible and induction requires the concerted presence of *two* inducers (Dodgson et al. 1974; Humphreys et al. 1986), namely a secondary alcohol (e.g. tetradecan-2-ol) and a long-chain alkan-2-sulphate (e.g. tetradecan-2-sulphate). Induction does not occur when one or other or both components are omitted. Tetradecan-2-ol can be replaced with any one of most secondary alcohols but primary alcohols are ineffective. Keeping the tetradecan-2-ol component constant, the C_{14} ester can be replaced by C_{12} to C_{16} alkan-2-sulphates but shorter homologues and symmetrical, near-symmetrical and primary alkyl sulphates are ineffective. This "concerted induction" is not unique (Blackwell and Turner 1978; Mandrand-Berthelot et al. 1977) but so far no mechanistic explanation has emerged.

7. Conclusions

This brief review of properties of alkylsulphatases illustrates both the remarkable similarities and the stark contrasts which are scattered among the catalytic and physiological properties of the enzymes described. For example, the various enzymes show exclusive specificity for either primary or secondary esters, although there is a universal dependence on hydrophobic binding of substrates and similar dependence of V_{max} on chain length; C-O bond cleavage is the norm for all the secondary enzymes, but C-O and O-S splitting primary alkylsulphatases co-exist in the same organism; these same two primary enzymes, P1 and P2, have similar substrate specificities, yet one is inducible and the other derepressible; all but one of the secondary enzymes are apparently constitutive but the exception (S3) is regulated by concerted induction. These comparisons raise questions about the relationships among these various enzymes, both at the protein and at the genetic level. Apart from gross properties such as molecular weight and sub-unit structure, little is known about the properties of the respective proteins. Still less is known about genetic inter-relationships, and experiments are currently underway in our laboratory to rectify this deficiency. Mapping the respective structural and regulatory genes, followed by cloning, sequencing, site-directed mutagenesis and over-expression to produced large amounts of protein for structural analysis are all now realistic possibilities which would enhance dramatically our understanding of these biodegradative systems.

B. *Linear alkylbenzene sulphonates*

1. *Overall biodegradation pathway*
Linear alkylbenzene sulphonates (LAS) contain alkyl chains that are largely devoid of the methyl and other branches which retard or block biodegradation. The distinctive chemical natures of the three parts of the LAS molecule require correspondingly different biodegradative modes of attack: *viz.* the alkyl chain by ω-/β-oxidation, the benzene ring by ring cleavage, and the sulphonate residue by desulphonation (Fig. 8). The sequence in which these events occur has not been unequivocally determined and appears to depend on, for example, the length of the alkyl chain. For the most commonly-used LAS, i.e. those mixtures containing chain lengths centred around C_{12}, it appears that initial biodegradative attack is by ω-oxidation with subsequent β-oxidation of the alkyl chain, followed by opening of the benzene ring and desulphonation; the order of the latter two stages is not clear (*vide infra*).

In terms of ultimate biodegradation, one must also consider the fate of the intermediates generated by these three types of reactions. The products of ω-/β-oxidation are short-chain sulphophenyl alkanoates or sulphobenzoate. Some bacterial species are capable only of carrying out biodegradation of LAS as far as this stage (Cain 1987), whereas others can apparently complete the ring opening and desulphonation stages as well (Willetts and Cain 1972). In bacteria capable of complete biodegradation of LAS, ω-/β-oxidation is accomplished first and is accompanied by induction of enzymes of the glyoxylate by-pass, which facilitates the biosynthetic use of the acetyl-CoA generated; once the later stages of ring opening and desulphonation begin, the glyoxylate by-pass enzymes are repressed as the benzene ring carbons become available for assimilation into cellular material via intermediates which are not exclusively acetyl-CoA.

2. *Alkyl chain biodegradation*
The β-oxidation enzymes (see Chapter 4) appear to be those of the normal fatty acid oxidation pathway, which are able to cope with methyl branches on the α-carbons when the product is propionyl-CoA instead of the usual acetyl-CoA.

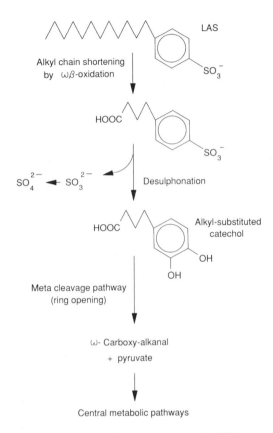

Fig. 8. Overall pathway of linear alkylbenzene sulphonate (LAS) biodegradation in bacteria.

The rate of ω-/β-oxidation increases with the length of the alkyl chain, known as the "distance principle" (Huddleston and Allred 1963; Setzkorn et al. 1964), and also depends on the position of the benzene ring. The alkylbenzene sulphonates originally sythesized and marketed in the 1950s were derived from tetrapropylene, giving them a quaternary-substituted alkyl chain with a *gem*-dimethyl branching pattern. These alkyl chains were particularly resistant to β-oxidation which accounted for their environmental persistence and eventual withdrawal from the market. They are broken down slowly by an alternative metabolic pathway, which has not been identified despite the fact that analogous methyl branching occurs in a number of naturally-occurring compounds such as pantothenic acid.

The β-oxidation pathway is also blocked if there is branching on the β-carbon. However, alkylbenzene sulphonates containing such β-substituted alkyl chains are broken down, albeit more slowly than linear (unsubstituted) LAS, so an alternative pathway must exist. This biodegradative pathway has not been identified: a simple solution to the problem would be to remove one carbon atom from the chain by α-oxidation, but this type of oxidative one-carbon removal is uncommon in bacteria (in comparison with plants and animals). It appears that once activated by ω-oxidation the terminal end of the alkyl chain is invariably attacked by β-oxidation. Little is known about the detailed mechanism of ω-oxidation of surfactants as the inability of researchers to prepare a cell-free system has precluded further characterization and purification of this enzyme.

3. Desulphonation and ring cleavage

Once the alkyl-chain length has been reduced to less than five carbon atoms, the next stages can occur. Although there is not complete agreement (e.g. see Feigel and Knackmuss 1988, 1990), the balance of opinion and evidence is that desulphonation precedes and probably facilitates opening of the benzene ring in a concerted action (*vide infra*). This conclusion is based on studies, not with surfactants *per se* but with some analogous sulphonated compounds. There is justification for this approach, since it appears that once the alkyl side-chain in an alkylbenzene sulphonate has been shortened to about five carbon atoms bacteria deal with the molecule as an arylsulphonate, not an alkylbenzene sulphonate (Cain 1987). Evidence to support this assumption comes from a recent study of alkylbenzene biodegradation by *Pseudomonas* sp. NCIB 10643 (Smith and Ratledge 1989). The initial step for a range of *n*-alkylbenzenes (C_2 to C_7) was found to occur by dioxygenase attack on the aromatic ring, rather than via side-chain oxidation, to give 3-substituted-catechol products which were further metabolized by *meta* ring-cleavage (*vide infra*).

The use of such analogous compounds as toluenesulphonic acid (Locher et al. 1989) and sulphobenzoate (Locher et al. 1991) has proved useful in probing the mechanism of desulphonation, which is performed by a complex enzyme system containing a dioxygenase – i.e. 2 atoms of oxygen are introduced into the substrate molecule in order to activate the otherwise stable aromatic ring, so that it can be cleaved for further metabolism. These studies with aromatic sulphonate analogues confirmed the original proposal by Cain and Farr (1968) for dioxygenolytic cleavage of arylsulphonate surfactants, which was made on the basis of the appropriate incorporation of $^{18}O_2$ into catechol product. Subsequently, Brilon et al. (1981a,b) isolated and identified the dioxygenolytic products from naphthalenesulphonic acid substrates, which are also analogues of aromatic sulphonated surfactants. The dioxygenase enzyme has proved difficult to study in cell-free extracts of many bacterial species because it may not be present in all growth phases during batch culture and also because of its instability, even in the cold. Thurnheer et al. (1990), who obtained an active enzyme preparation *in vitro* using a cell-free preparation from an *Alcaligenes* sp., were able to measure the reaction stoichiometry as being 1 mol of O_2 per mol of substrate.

Recently, success has been achieved in purifying a two-component desulphonating enzyme system that is active towards 4-sulphobenzoate from a different species, namely *Comamonas testosteroni* strain T-2 (Locher et al.

1991). This bacterium degrades toluene 4-sulphonate by dioxygenation of the methyl group via the corresponding alcohol to give 4-sulphobenzoate; this compound is then desulphonated to give protocatechuate (3,4-dihydroxybenzoate) which is broken down by *meta* ring cleavage (*vide infra*). Thus, desulphonation of 4-sulphobenzoate is a key reaction in the ultimate breakdown pathway.

Using FPLC anion-exchange chromatography, Locher et al. (1991) have identified the two components of the 4-sulphobenzoate dioxygenase system, a red one termed A and a yellow one termed B. These fractions were purified by hydrophobic interaction chromatography and gel filtration. Fraction A exhibits dioxygenase activity and is a dimer with a subunit M_r value of 50 kDa; each subunit contains one iron-sulphur centre with the structure (2Fe-2S). Fraction B is a reductase and is a monomer with a subunit M_r value of 36 kDa; associated with each molecule of B protein is one molecule of FMN and two atoms each of Fe and S. Both the reductase and oxygenase proteins contain a high proportion (almost a half) of apolar amino acids, which accounts for their hydrophobic behaviour during purification. Partial N-terminal sequences have been determined but, as yet, no three-dimensional data is available. The organization of the component proteins appears to be the simplest of those (relatively few) multicomponent dioxygenase systems so far described.

Unlike some other dioxygenases which metabolize apolar substrates (Zamanian and Mason 1987), the enzyme system from *C. testosteroni* has a particularly narrow substrate specificity with a high affinity for 4-sulphobenzoate (Locher et al. 1991). The desulphonation activity appears to be a function entirely of the dioxygenase enzyme system and, in comparison with analogous reactions with benzene or benzoic acid, does not require an accompanying dehydrogenase to re-aromatize the dihydrodiol intermediate. The explanation for this difference is that the postulated sulphono-dihydrodiol product of dioxygenolytic cleavage of the C-S bond in an aromatic sulphonate is the bisulphite addition complex of a cyclic ketone. This is so unstable that it breaks down spontaneously by release of sulphite, a good leaving group. Electronic rearrangement then follows to reform the aromatic status and give protocatechuate (Fig. 9). In comparison, the dihydrodiol from benzoate is stable and requires an NAD^+-linked dehydrogenase enzyme to remove the hydride ion (Fig. 9). The desulphonation mechanism, proposed by Locher et al. (1991) for 4-sulphobenzoate is similar to that proposed by Brilon et al. (1981b) for the catabolism of naphthalene sulphonic acids by some *Pseudomonas* spp., and by Markus et al. (1984, 1986) for the dechlorination of 4-chlorophenylacetate by another dioxygenase. An analogy can also be drawn between the desulphonation mechanism proposed for 4-sulphobenzoate and that proposed by Thysse and Wanders (1974) for linear alkane sulphonates (see Section III.C).

In relation to the sequence of desulphonation and ring cleavage, a claim for initial desulphonation has been made by Feigel and Knackmuss (1988, 1990) on the basis of studies of the breakdown of sulphanilic acid

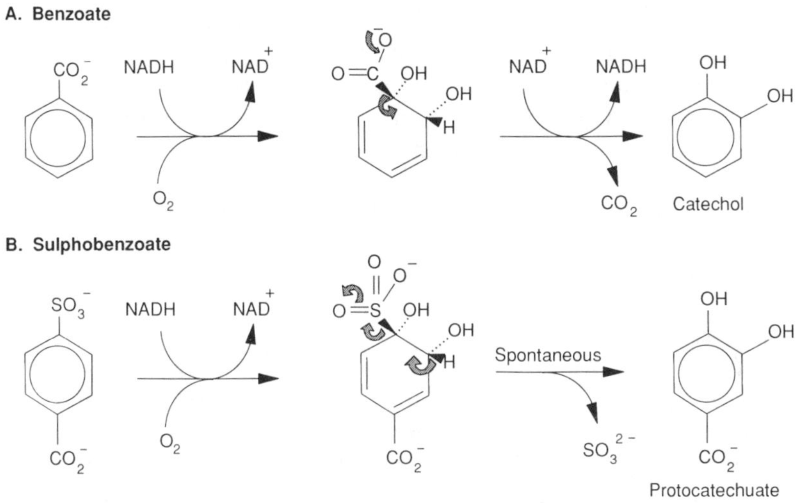

A. Benzoate

B. Sulphobenzoate

Fig. 9. Desulphonation mechanism proposed for benzoate and 4-sulphobenzoate biodegradation in bacteria. Adapted from Locher et al. (1991).

(4-aminobenzenesulphonic acid) by an unidentified bacterial species. Catechol 4-sulphonic acid was identified as a stable product, showing that removal of the *amino group* without ring cleavage is the first step in the biodegradation pathway. However, the authors go on to state that this also indicates that there was no release of the sulphonate group before ring cleavage (Feigel and Knackmuss 1988), although the ring system is still intact in catechol 4-sulphate and no subsequent breakdown products were identified.

Whenever it occurs, the immediate product of desulphonation is generally believed to be sulphite, which usually oxidizes rapidly to sulphate, particularly under acidic conditions and in the presence of some metal ions such as Fe^{3+}. In fact, this oxidation is so fast that it is often difficult to detect sulphite unless special steps are taken to do so, and some early studies reported sulphate as being the direct product of desulphonation (e.g. Focht and Williams 1970; Ripin et al. 1971; Willetts 1973). Another explanation for the difficulty of detecting sulphite could be the presence of a particularly active sulphite oxidase enzyme.

Removal of the sulphonate residue leaves phenolic hydroxyl groups on the aromatic ring system. This can then be opened by a number of well-characterized metabolic pathways that biodegrade catechols, protocatechuate and related compounds. These pathways include the *meta* (extradiol) and *ortho* (intradiol) cleavage pathways (see Chapter 11); whilst not present universally in bacteria, they are common in a wide range of environmentally-relevant species. The *ortho*-cleavage pathway gives β-ketoadipate as an intermediate, which is broken down via acetoacetate to acetyl-CoA, while the *meta*-cleavage pathway gives acetyl-CoA (via pyruvate) as a product together with an aliphatic aldehyde. It should be noted that the aldehyde products will be simple alkyl aldehydes only in the case of the biodegradation of short-chain alkylbenzene sulphonates. For longer-chain LAS surfactants, the alkyl residue after ω-/β-oxidation will terminate in carboxylic acid groups so that the aldehydic products will be ω-carboxy-alkanals such as adipic acid γ-semialdehyde. These, and the aliphatic aldehydes, depending on their exact structure, are metabolized to intermediates such as pyruvate or succinate via common metabolic pathways inside bacteria (Schlegel 1986), and thus can be used for both energy production and biosynthetic reactions.

4. Consortia

Not all of the necessary three stages in LAS breakdown may be present in a single bacterial species, and in the natural environment it is likely that several species cooperate in the ultimate removal of this surfactant. Jimenez et al. (1991) obtained a four-membered bacterial consortium, consisting of three *Pseudomonas* spp. and an *Aeromonas* sp., that was capable of mineralizing LAS; pure cultures of each single species or mixtures of any three did not mineralize LAS, although each of three isolates could carry out primary biodegradation. This study confirmed earlier work which indicated that consortia of bacteria were involved in LAS biodegradation in natural environments (Hrsak et al. 1982; Sigoillot and Nguyen 1990).

5. Genetics

The *meta*-cleavage pathway, in common with those for a number of aromatic and other xenobiotics, has long been known to be encoded by genes which are plasmid-encoded (Burlage et al. 1989; Assinder and Williams 1990). This fact raised the possibility that other surfactant-biodegradation genes such as that encoding the "desulphonase" enzyme might also be plasmid-encoded. Cain's group have presented data to support the involvement of plasmids in arylsulphonate biodegradation in *Pseudomonas* spp. (summarized in Cain 1987), but the selection techniques which they used were for overall utilization of the surfactant rather than for the "desulphonase" *per se*. The ability of bacteria to biodegrade arylsulphonates could be lost ("cured") by exposure to supra-maximal temperatures or mitomycin C or other treatments known to enhance plasmid loss; the putative plasmid DNA was isolated and there appeared to be a correlation between the acquisition of desulphonating activity and transfer of plasmid in conjugation experiments between competent and non-competent strains. However, subsequent studies on a number of LAS-degrading strains failed to demonstrate a clear relationship between the ability to biodegrade the surfactant and the presence of plasmid DNA. Thus, it remains an open question as to the extent of the involvement of plasmid-encoded genes in the biodegradation of LAS.

C. Aliphatic sulphonates

Aliphatic sulphonates fall into two groups, based on whether the sulphonate residue is attached to a primary or secondary carbon atom; the secondary alkane sulphonates include the sulphosuccinates and the α-sulpho-fatty acid methyl esters (Fig. 10). A number of naturally-occurring primary (but not secondary) aliphatic sulphonates are known: these include taurine, an end-product of L-cysteine metabolism excreted by animals, and sulphoquinovose, which is present in the sugar-sulpholipid of plants and some algae, and thus is widely distributed throughout the environment. However, despite the ubiquitous presence of such naturally-occurring, primary, aliphatic sulphonates in the environment, which would presumably have encouraged the evolution of bacteria and other microorganisms to break them down, much less is known about the biodegradation of this group of surfactants compared with alkyl sulphates or LAS. Within the group more is known about the biodegradation of primary aliphatic sulphonates and, moreover, the mechanism involves a monooxygenation step in order to de-stabilize the C-S bond. This mechanism is in direct contrast to the desulphonation of taurine which is catalysed by a pyridoxal-phosphate-dependent sulpho-lyase (Shimamoto and Berk 1980), indicating that the catabolic pathways for natural sulphonates may not be typical of those for biodegradation of synthetic aliphatic sulphonate surfactants.

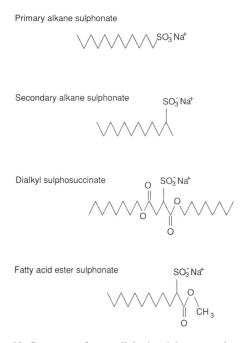

Fig. 10. Structures of some aliphatic sulphonate surfactants.

The earliest report of the mechanism of aliphatic sulphonate biodegradation was that by Cardini et al. (1966), who isolated several bacterial species capable of growth on dodecane 1-sulphonate as the sole carbon/energy source. In contrast with LAS, the primary biodegradative event for aliphatic sulphonates was found to be desulphonation, rather than ω-/β-oxidation at the other end of the molecule. This conclusion was based on the detection of sulphite and the corresponding aldehyde as products, and the fact that the bacteria were able to oxidize the corresponding long-chain aldehyde and acid but not alcohol. It was presumed that the bacteria were able to oxidize the long-chain aldehyde via oxidation to the fatty acid which was activated and entered the β-oxidation pathway.

Support for this scheme was provided by the subsequent studies of Thysse and Wanders (1972, 1974) who isolated two species of bacteria, one by enrichment on short-chain (C_4 to C_7) primary alkane sulphonates and the second on long-chain (C_8 to C_{12}) primary alkane sulphonates. Using cell-free extracts, a requirement for NAD(P)H and O_2 could be demonstrated, indicating that a monooxygenase was involved in the desulphonation reaction. It was postulated that an aldehyde-sulphite adduct was formed and that this spontaneously lost sulphite (cf. mechanism of 4-sulphobenzoate desulphonation, *vide supra*) to give the second product, an aldehyde, which could be oxidized to the corresponding acid and activated with coenzyme A ready to enter the β-oxidation cycle.

Compared with the preparations of a number of other workers, the extracts prepared by Thysse and Wanders were unusually stable in that they could be stored for at least 24 hours at 0°C without loss of activity. The monooxygenase was also unusual for such an enzyme in having a wide substrate specificity which extended to some secondary alkane sulphonates; however, the extracts could have contained more than one enzyme.

There is less information about the biodegradation of the secondary alkane sulphonate surfactants, α-sulpho-fatty acid methyl esters and sulphosuccinate. A number of studies have shown that these compounds undergo both primary and complete biodegradation in mixed environmental cultures (Hammerton 1956; Cordon et al. 1970; Maurer et al. 1977; Lotzsch et al. 1979; Schoberl and Bock 1980), but little is known about the enzymic mechanisms involved.

Steber and Weirich (1989) investigated the biodegradation of α-sulpho-fatty acid methyl esters in a model sewage-treatment plant. They detected short-chain sulphonated intermediates that retained the ester methyl group, and which presumably were formed by ω-/β-oxidation, but they were unable to demonstrate the subsequent desulphonation step. Hence, for α-sulpho-fatty acid methyl esters, separation of the hydrophilic sulphonate group occurs at a late stage after the hydrophobic group has been extensively degraded (cf. Fig. 2).

In contrast, the biodegradation of dialkyl sulphosuccinate surfactants is initiated by hydrolytic cleavage of the hydrophile from the hydrophobe (Fig. 2). This may reflect their structural analogy with the alkyl sulphates, in that they contain an aliphatic alcohol esterified to succinic instead of sulphuric acid. Hydrolysis of the ester bond by esterases would release the hydrophobe as an alcohol to leave the sulphosuccinate moiety. This, of course, is a secondary aliphatic sulphonate, and recently in our laboratories we have demonstrated using specifically radiolabelled sulphosuccinates that this compound undergoes biodegradation by initial desulphonation with the release of sulphite before subsequent metabolism of the remainder of the molecule to CO_2 (Quick et al. 1993). Unfortunately, just as for the α-sulpho-fatty acid methyl esters discussed above, we have been unable to purify the desulphonating enzyme because of the difficulty of obtaining stable activity in cell-free extracts.

D. Alkyl ethoxy sulphates

Inspection of the structure of alkyl ethoxy sulphates (Fig. 1) might lead one to predict that, like alkyl sulphates, primary biodegradation would be achieved by sulphatase action. However, this turns out to be a minor contribution to primary biodegradation; instead it is scission of the alkyl-glycol and glycol-glycol ether bonds which predominates both in pure cultures and in mixed environmental samples (White and Russell 1988). Commercial preparations of alkyl ethoxy sulphates contain typically four, but sometimes up to ten, ether bonds, so the patterns of ether-cleavage intermediates are likely to be complex and this, indeed, is the case.

The most extensively studied bacterial isolate, *Pseudomonas* DES1, degrades sodium dodecyl triethoxy sulphate (SDTES) which, if labelled in the hydrophilic moiety with ^{35}S, releases into the culture fluid small amounts of $^{35}SO_4^{2-}$ as well as five organic ^{35}S-labelled metabolites (Hales et al. 1982). These have been identified as mono-, di-, and tri-ethylene glycol sulphates, together with the oxidation products (carboxylic acids) of the last two (Fig. 11). A single etherase is believed to be responsible for the ether-cleavage of SDTES and a single, substrate-inducible alkylsulphatase releases inorganic sulphate but only from the parent surfactant and not from the glycol sulphate metabolites (Hales 1981; Hales et al. 1982).

Parallel experiments with the same organism and the same substrate, but labelled in the hydrophobic moiety with ^{14}C at C-1 of the alkyl chain, revealed the production of a correspondingly complex set of hydrophobic fragments arising from sulphatase- or etherase-scission (Griffiths et al. 1987). In fact, up to 20 radiolabelled, ether-extractable metabolites were found, of which eight predominated. However, unlike the hydrophilic metabolites, none of the hydrophobic metabolites accumulated and collectively they accounted for <20% of the total radiolabel throughout the biodegradation period with the bulk of the label appearing as CO_2 or cellular components. The main metabolites were identified as dodecanol, mono-, di- and tri-ethylene glycol dodecyl ether, together with the four corresponding oxidation products in which the primary alcoholic -CH_2OH groups were oxidized to carboxylic acids (Fig. 11). Other metabolites containing alkyl-ether bonds and aldehydic

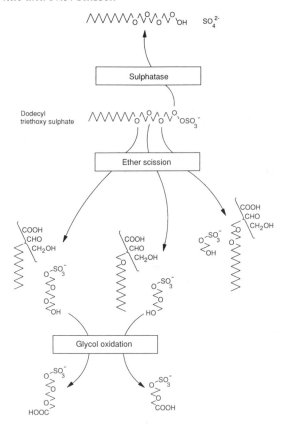

Fig. 11. Biodegradation pathways of dodecyl triethoxy sulphate in *Pseudomonas* DES1. Adapted from White and Russell (1988).

but no alcoholic groups were also detected. Although there was strong evidence for interconversion of alcohol ↔ aldehyde ↔ carboxylic acid, there were no obvious precursor-product relationships, so no clear indication emerged as to the initial product(s) of etherase cleavage.

The pattern of ^{35}S-labelled metabolites produced in *Pseudomonas* DES1 was compared with that in three other surfactant-degrading *Pseudomonas* spp. designated C12B, SC25A, and TES5. All four isolates used ether-scission as the predominant primary biodegradation pathway (Hales et al. 1986). In DES1 and C12B, the etherase system (accounting for 60 to 70% of primary biodegradation) liberated mono-, di- and tri-ethylene glycol sulphates in substantial proportions, the last two compounds undergoing further metabolism to the corresponding carboxylic acids. Primary biodegradation via sulphate hydrolysis (30 to 40%) was achieved in each case by single sulphatase enzymes (P2 in the case of *Pseudomonas* C12B; see Section III.A.3). For the other isolates, etherase was even more dominant in primary biodegradation (>80%) and was confined almost completely to the alkyl-ether bond. Small amounts (~10%) of surfactant were degraded by sulphate hydrolysis, and *Pseudomonas* TES5 also produced compounds tentatively identified as ω-/β-oxidation intermediates.

All of the glycol sulphates, the oxidized glycol sulphates and ω-/β-oxidation intermediates detected in the pure-culture studies, were also observed during biodegradation by mixed cultures derived from OECD or river water die-away experiments (Griffiths et al. 1986). Primary biodegradation of the related non-ionic octadecyl ethoxylate has also been shown to occur via both ether-scission and ω-/β-oxidation in similar mixed cultures (Steber and Wierich 1985). Very little sulphate was liberated from SDTES in mixed cultures but, in contrast with pure cultures, large amounts of SO_4^{2-} accompanied the eventual disappearance of the sulphated glycol intermediates. No organic metabolites persisted. Thus, the glycol sulphates, which accumulated in the pure culture studies, were degraded by other organisms in mixed cultures. No serious attempt has yet been made to isolate and investigate the latter, although bacteria capable of degrading related short-chain alkyl sulphates (e.g. methyl sulphate, propyl-2-sulphate, butyl sulphate) have been isolated (White et al. 1987) and the corresponding biodegradation pathways established (Crescenzi et al. 1985; Davies et al. 1990; White and Matts 1992).

Biodegradation of alkyl ethoxy sulphates is thus a good example of the versatility which bacteria bring to the biodegradation of xenobiotics. Although these compounds are degraded by the alkylsulphatases that dominate alkyl sulphate biodegradation, they are inferior substrates for those enzymes. Slower degradation by this route presumably allows the alternative, metabolically more demanding pathways to become manifest, in this particular case the ether-scission and ω-/β-oxidation routes. Considerable effort has been devoted to the mechanism of polyethylene glycol metabolism and, while extracellular hydrolytic cleavage does occur (Haines and Alexander 1975), it is very slow and intracellular oxidative cleavage is more commonly reported (Kawai et al. 1978, 1980; Thelu et al. 1980; Steber and Wierich 1985; Kawai and Yamanak 1986; Obradors and Aguilar 1991). Thus, both these alternative pathways are oxidative, and therefore metabolically more demanding than simple sulphate hydrolysis, but they contribute because they compete effectively with lowered sulphatase activity towards ethoxy sulphate esters. This multiple pathway metabolism is seen in both pure cultures and in mixed environmental samples.

IV. Postscript

It should be clear from the foregoing that our understanding of the biodegradative pathways for the different classes of surfactant is extremely uneven. It is only for LAS and alkyl sulphates that a substantial body of information including enzymological data, is currently available. Notwithstanding the fact that these two classes represent the bulk of anionic (and total) surfactant usage, there is an urgent need to understand better how other surfactants are biodegraded. Perhaps we shall see more progress in this field, with the altered climate of opinion amongst the general public concerning the desirability for biodegradable products, backed by legislative requirements placed on the surfactant industry for completely biodegradable compounds (rather than simple loss of surfactancy). It is only through a proper molecular understanding of the metabolic pathways and the enzymes involved that rational decisions can be made about the design of novel biodegradable surfactants. The kind of approach now used widely in the pharmaceutical industry involving the computerised modelling of drug (surfactant) interactions with target proteins (biodegradative enzymes) remains a distant dream.

Undoubtedly the better progress in purification and characterization of alkylsulphatase enzymes compared with "desulphonases" reflects, not a deficiency of intent or effort, rather the fact that alkylsulphatases are soluble and amenable to conventional purification techniques. In contrast, it is usually difficult even to retain desulphonating activity when bacteria are disrupted. Perhaps molecular genetics can help to overcome this problem as it has done so for several other types of membrane-bound proteins that cannot be purified using conventional methods. Sadly, to date progress in this aspect of surfactant-degrading enzymes has been extremely slow and only a single gene has been cloned, that for an alkylsulphatase from *Pseudomonas* ATCC 19151 (Hsu 1965) by Davison's group (Davison et al. 1990, 1992). The Cardiff group are currently attempting to clone further alkylsulphatase genes from *Pseudomonas* C12B. Clearly the similarities and contrasts among the multiplicity of enzymes in this isolate (see Section III.A.7) not only raises many intriguing questions about the genetic relationships among the enzymes, but also renders the task much more difficult. For LAS, it is still unclear whether the desulphonating activity *per se* or just the later ring-opening enzymes are encoded by plasmid genes. A molecular genetical approach would have the benefit of providing a means of exploring the regulation of not just expression but also the regulation of biodegradative enzymes. Combined with suitable laboratory models this would advance our understanding of the fate of surfactants in the environment so that untoward pollution can be avoided.

References

Anderson DJ, Day MJ, Russell NJ and White GF (1988) Temporal and geographical distributions of epilithic sodium dodecyl sulfate SDS-degrading bacteria in a polluted South Wales river. Appl. Environ. Microbiol. 54: 555–560

Anderson DJ, Day MJ, Russell NJ and White GF (1990) Die-away kinetic analysis of the capacity of epilithic and planktonic bacteria from clean and polluted river water to biodegrade sodium dodecyl sulfate. Appl. Environ. Microbiol. 56: 758–763

Assinder SJ and Williams PA (1990) The TOL plasmids: determinants of the catabolism of toluene and the xylenes. Adv. Microbial Physiol. 31: 1–69

Barrett CH, Dodgson KS and White GF (1980) Further studies on the substrate specificity and inhibition of the stereospecific CS2 secondary alkylsulphohydrolase of *Comamonas terrigena*. Biochem. J. 191: 467–473

Bartholomew B, Dodgson KS, Matcham GWJ, Shaw DJ and White GF (1977) A novel mechanism of enzymic ester hydrolysis. Inversion of configuration and carbon-oxygen bond cleavage by secondary alkylsulphohydrolases from detergent-degrading micro-organisms. Biochem. J. 167: 575–580

Bartholomew B, Dodgson KS and Gorham SG (1978) Purification and properties of the S1 secondary alkylsulphohydrolase of the detergent-degrading microorganism *Pseudomonas* C12B. Biochem. J. 169: 659–667

Bateman TJ (1985) Primary alkylsulphatase activity in the detergent-degrading bacterium *Pseudomonas* C12B. PhD thesis, University of Wales

Bateman TJ, Dodgson KS and White GF (1986) Primary alkylsulphatase activities of the detergent-degrading bacterium *Pseudomonas* C12B. Purification and properties of the P1 enzyme. Biochem. J. 236: 401–408

Biedlingmaier S and Schmidt A (1983) Alkylsulfonic acids and some S-containing detergents as sulfur sources for growth of *Chlorella fusca*. Arch. Microbiol. 136: 124–130

Blackwell CML and Turner JM (1978) Microbial metabolism of amino alcohols: formation of coenzyme B12-dependent ethanolamine-ammonia lyase and its concerted induction in *Escherichia coli*. Biochem. J. 76: 751–757

Brilon C, Beckmann W, Hellwig M and Knackmuss H-J (1981a) Enrichment and isolation of naphthalenesulfonic acid-utilizing Pseudomonads. Appl. Environ. Microbiol. 42: 39–43

Brilon C, Beckmann W and Knackmuss H-J (1981b) Catabolism of naphthalene sulfonic acids by *Pseudomonas* sp. A3 and *Pseudomonas* sp. C22. Appl. Environ. Microbiol. 42: 44–55

Burlage RS, Hooper SW and Sayler GS (1989) The TOL (pWWO) catabolic plasmid. Appl. Environ. Microbiol. 55: 1323–1328

Cain RB (1987) Biodegradation of anionic surfactants. Biochem. Soc. (UK) Trans. 15: 7S–22S

Cain RB and Farr DR (1968) Metabolism of arylsulphonates by micro-organisms. Biochem. J. 106: 859–877

Cardini G, Catelani D, Sorlini C and Trecanni V (1966) La degradazione microbica dei detergenti di sintesi. Ann. Microbiol. Enzimol. 16: 217–223

Cloves JM, Dodgson KS, Games DE, Shaw DJ and White GF (1977) The mechanism of action of primary alkylsulphohydrolase and arylsulphohydrolase from a detergent degrading micro-organism. Biochem. J. 167: 843–846

Cloves JM, Dodgson KS, White GF and Fitzgerald JW (1980a) Purification and properties of the P2 primary alkylsulphohydrolase of the detergent-degrading bacterium *Pseudomonas* C12B. Biochem. J. 185: 23–31

Cloves JM, Dodgson KS, White GF and Fitzgerald JW (1980b) Specificity of P2 primary alkylsulphohydrolase induction in the detergent-degrading bacterium *Pseudomonas* C12B. Biochem. J. 185: 13–21

Cordon TC, Maurer EW and Stirton AJ (1970) Course of biodegradation of anionic detergents by analyses for carbon, methylene blue active substance and sulfate ion. J. Am. Oil Chem. Soc. 47: 203–206

Cornish-Bowden A and Wharton CW (1988) Enzyme Kinetics. IRL Press, Oxford

Crescenzi AMV, Dodgson KS and White GF (1984) Purification and some properties of the D-lactate-2-sulphatase of *Pseudomonas syringae* GG. Biochem. J. 223: 487–494

Crescenzi AMV, Dodgson KS, White GF and Payne WJ (1985) Initial oxidation and subsequent desulphation of propan-2-yl sulphate by *Pseudomonas syringae* GG. J. Gen. Microbiol. 131: 469–477

Davies I, White GF and Payne WJ (1990) Oxygen-dependent desulphation of monomethyl sulphate by *Agrobacterium* sp. M3C. Biodegradation 1: 229–241

Davison J, Brunel F and Phanopoulos A (1990) The genetics of vanillate and sodium dodecyl sulphate degradation in *Pseudomonas*. In: S Silver and AM Chakrabarty (eds) *Pseudomonas*: Biotransformations, Pathogenesis, and Evolving Biotechnology (pp 159–164). American Society for Microbiology, Washington, DC

Davison J, Brunel F, Phanopoulos A, Prozzi D and Terpstra P (1992) Cloning and sequencing of *Pseudomonas* genes determining sodium dodecyl sulfate biodegradation. Gene 114: 19–24

Dodgson KS and White GF (1983) Some microbial enzymes involved in the biodegradation of sulphated surfactants. Top. Enz. Ferment. Biotechnol. 7: 90–155

Dodgson KS, Fitzgerald JW and Payne WJ (1974) Chemically defined inducers of alkylsulphatases present in *Pseudomonas* C12B. Biochem. J 138: 53–62

Dodgson KS, White GF and Fitzgerald JW (1982) Sulfatases of Microbial Origin, Vols 1 and 2. CRC Press, Boca Raton

EEC (1973) Directive on the biodegradability of anionic surfactants. Directive number 73/405/EEC, EEC, Brussels

Feigel BJ and Knackmuss H-J (1988) Bacterial catabolism of sulfanilic acid via catechol-4-sulfonic acid. FEMS Microbiol. Lett. 55: 113–118

Feigel BJ and Knackmuss H-J (1990) Catabolic pathway of sulfanilic acid. Forum Mikrobiol. 13: 75

Fitzgerald JW (1974) Hydrolysis of hexan-1-yl sulphate by the P2 electrophoretic form of primary alkylsulphatase. Microbios 11: 153–158

Fitzgerald JW and Kight LC (1977) Physiological control of alkylsulfatase synthesis in *Pseudomonas aeruginosa*: effects of glucose, glucose analogs, and sulfur. Can. J. Microbiol. 23: 1456–1464

Fitzgerald JW and Payne WJ (1972) Induction in a *Pseudomonas* species of a sulphatase active on short-chain alkyl sulphates. Microbios 5: 87–100

Fitzgerald JW, Kight-Olliff LC, Stewart GJ and Beauchamp NF (1978) Reversal of succinate-mediated catabolite repression of alkylsulfatase in *Pseudomonas aeruginosa* by 2,4-dinitrophenol and by sodium malonate. Can J. Microbiol. 24: 1567–1573

Focht DD and Williams FD (1970) The degradation of p-toluenesulphonate by a *Pseudomonas*. Can. J. Microbiol. 16: 309–316

Gilbert PA and Watson GK (1977) Biodegradability testing and its relevance to environmental acceptability. Tenside Deterg. 14: 171–177

Griffiths ET, Hales SG, Russell NJ, Watson GK and White GF (1986) Metabolite production during the biodegradation of the surfactant sodium dodecyltriethoxy sulphate under mixed-culture die-away conditions. J. Gen. Microbiol. 132: 963–972

Griffiths ET, Hales SG, Russell NJ and White GF (1987) Identification of hydrophobic metabolites formed during biodegradation of alkyl ethoxylate and alkyl ethoxy sulphate surfactants by *Pseudomonas* sp. DES1. Biotechnol. Appl. Biochem. 9: 217–229

Haines JR and Alexander M (1975) Microbial degradation of polyethylene glycols. Appl. Microbiol. 29: 621–625

Hales SG (1981) Microbial degradation of linear ethoxylate sulphates. PhD thesis, University of Wales

Hales SG, Dodgson KS, White GF, Jones N and Watson GK (1982) Initial stages in the biodegradation of the surfactant sodium dodecyltriethoxy sulfate by *Pseudomonas* sp. strain DES1. Appl. Environ. Microbiol. 44: 790–800

Hales SG, White GF, Dodgson KS and Watson GK (1986) A comparative study of the biodegradation of the surfactant sodium dodecyltriethoxy sulphate by four detergent-degrading bacteria. J. Gen. Microbiol. 132: 953–961

Hammerton C (1956) Synthetic detergents and water supplies. II. Chemical constitution of anionic surface active compounds and their susceptibility to biochemical oxidation. Proc. Soc. Water Treatment Exam. 5: 160–174

Hrsak D, Bosnjak M and Johanides V (1982) Enrichment of linear alkylbenzenesulphonate (LAS) degrading bacteria in continuous culture. J. Appl. Bacteriol. 53: 413–422

Hsu Y-C (1963) Detergent (sodium lauryl sulphate)-splitting enzyme from bacteria. Nature 200: 1091–1092

Hsu Y-C (1965) Detergent-splitting enzyme from *Pseudomonas*. Nature 207: 385–388

Huddleston RL and Allred RC (1963) Microbial oxidation of sulfonated alkylbenzenes. Dev. Ind. Microbiol. 4: 24–38

Humphreys PGM, Shaw DJ, Dodgson KS and White GF (1986) Concerted induction of the S3 alkylsulphatase of *Pseudomonas* C12B by combinations of alkyl sulphates and alcohols. J. Gen. Microbiol. 132: 727–736

Jimenez L, Breen A, Thomas N, Federle TW and Sayler G (1991) Mineralization of linear alkylbenzene sulfonate by a four-member aerobic bacterial consortium. Appl. Environ. Microbiol. 57: 1566–1569

Kawai F and Yamanak H (1986) Biodegradation of polyethylene glycol by symbiotic mixed culture (obligate mutualism). Arch. Microbiol. 146: 125–129

Kawai F, Kimura T, Fukaya T, Tani Y, Ogata K, Ueno T and Fukami H (1978) Bacterial oxidation of polyethylene glycol. Appl. Environ. Microbiol. 35: 679–684

Kawai F, Kimura T, Tani Y, Yamada H and Kurachi M (1980) Purification and characterisation of polyethylene glycol dehydrogenase involved in the bacterial metabolism of polyethylene glycol. Appl. Environ. Microbiol. 40: 701–705

Kight-Olliff LC and Fitzgerald JW (1978) Inhibition of enzyme induction in *Pseudomonas aeruginosa* by exogenous nucleotides. Can. J. Microbiol. 24: 811–817

Lijmbach GWM and Brinkhuis E (1973) Microbial degradation of secondary *n*-alkyl sulfates and secondary alcohols. Antonie van Leeuwenhoek 39: 415–423

Lillis V, Dodgson KS and White GF (1983) Initiation of activation of a pre-emergent herbicide by a novel alkylsulfatase of *Pseudomonas putida* FLA. Appl. Environ. Microbiol. 46: 988–994

Locher HH, Leisinger T and Cook AM (1989) Degradation of *p*-toluenesulphonic acid via sidechain oxidation, desulphonation and meta ring cleavage in *Pseudomonas* (*Comamonas*) *testosteroni* T-2. J. Gen. Microbiol. 135: 1969–1978

Locher HH, Leisinger T and Cook AM (1991) 4-Sulphobenzoate 3,4-dioxygenase. Biochem. J. 274: 833–842

Lotzsch K, Neufahrt A and Taeuber G (1979) Comparative study of the biodegradation of secondary alkanesulfonates using ^{14}C-labelled preparations. Tenside Deterg. 16: 150–155

Mandrand-Berthelot M-A, Novel G and Novel M (1977) L'induction gratuite de la β-glucuronidase d'Escherichia coli K12 et son double mecanisme de repression. Biochemie 59: 163–170

Markus A, Klages U, Krauss S and Lingens F (1984) Oxidation and dehalogenation of 4-chlorophenylacetate by a two-component enzyme system from *Pseudomonas* sp. CBS3. J. Bacteriol. 160: 618–621

Markus A, Krekel D and Lingens F (1986) Purification and some properties of component A of the 4-chlorophenylacetate 3,4-dioxygenase from *Pseudomonas* species strain CBS. J. Biol. Chem. 261: 12883–12888

Matcham GJW, Bartholomew B, Dodgson KS, Payne WJ and Fitzgerald JW (1977) Stereospecificity and complexity of microbial sulphohydrolases involved in the biodegradation of secondary alkyl sulphate detergents. FEMS Microbiol. Lett. 1: 197–200

Maurer EW, Weil JK and Linfield WM (1977) The biodegradation of esters of α-sulfo fatty acids. J. Am. Oil Chem. Soc. 54: 582–584

Obradors N and Aguilar J (1991) Efficient biodegradation of high-molecular-weight polyethylene glycols by pure cultures of *Pseudomonas stutzeri*. Appl. Environ. Microbiol. 57: 2383–2388

Payne WJ (1963) Pure culture studies of the degradation of detergent compounds. Biotechnol. Bioeng. 5: 355–365

Payne WJ and Feisal VE (1963) Bacterial utilization of dodecyl sulfate and dodecylbenzene sulfonate. Appl. Microbiol. 11: 339–344

Payne WJ and Wiebe WJ (1978) Growth yield and efficiency in chemosynthetic microrganisms. Annu. Rev. Microbiol. 32: 155–183

Payne WJ, Williams JP and Mayberry WR (1965) Primary alcohol sulfatase in a *Pseudomonas* species. Appl. Microbiol. 13: 698–701

Payne WJ, Williams JP and Mayberry WR (1967) Hydrolysis of secondary alcohol sulphate by a bacterial enzyme. Nature 214: 623–624

Payne WJ, Fitzgerald JW and Dodgson KS (1974) Methods for visualization of enzymes in polyacrylamide gels. Appl. Microbiol. 27: 154–158

Quick A, Hales S, Russell NJ and White GF (1993) Desulphonation of a secondary alkane sulphonate: bacterial biodegradation of sulphosuccinate. In preparation

Ripin MJ, Noon KF and Cook TM (1971) Bacterial metabolism of aryl sulfonates. 1. Benzene sulfonate as growth substrate for *Pseudomonas testosteronii* H-8. Appl. Microbiol. 21: 495–499

Rotmistrov MN, Stavskaya SS, Krivetz IA and Samoylenko LS (1978) Microorganisms which decompose alkyl sulphates. Mikrobiologiya 47: 338–341

Schlegel, HG (1986) General Microbiology. University Press, Cambridge

Schoberl P and Bock KJ (1980) Surfactant degradation and its metabolites. Tenside Deterg. 17: 262–266

Setzkorn EA, Huddleston RL and Allred RC (1964) An evaluation of the river die-away technique for studying detergent biodegradability. J. Am. Oil Chem. Soc. 41: 826–830

Shaw DJ, Dodgson KS and White GF (1980) Substrate specificity and other properties of the inducible S3 secondary alkylsulphohydrolase from the detergent degrading bacterium *Pseudomonas* C12B. Biochem. J. 187: 181–190

Shimamoto G and Berk RS (1980) Taurine catabolism. II. Biochemical and genetic evidence for sulfoacetaldehyde sulfolyase involvement. Biochim. Biophys. Acta 632: 121–130

Sigoillot JC and Nguyen MH (1990) Isolation and characterisation of surfactant-degrading bacteria in a marine environment. FEMS Microbiol. Ecol. 73: 59–68

Smith MR and Ratledge C (1989) Catabolism of alkylbenzenes by *Pseudomonas* sp. NCIB 10643. Appl. Microbiol. Biotechnol. 32: 68–75

Soap and Detergents Association U.S. (1965) A procedure and standards for the determination of the biodegradability of alkylbenzene sulfonate and linear alkylate sulfonate. J. Am. Oil Chem. Soc. 42: 986–993

Steber J and Wierich P (1985) Metabolites and biodegradation pathways of fatty alcohol ethoxylates in microbial biocenoses of sewage treatment plants. Appl. Environ. Microbiol. 49: 530–537

Steber J and Wierich P (1989) The environmental fate of fatty acid α-sulfomethyl esters. Tenside Deterg. 26: 406–411

Swisher RD (1987) Surfactant Biodegradation. Surfactant Science Series, Vol 18, 2nd edition. Marcel Dekker, New York

Thelu J, Medina L and Pelmont J (1980) Oxidation of poly(oxethylene) oligomers by an inducible enzyme from *Pseudomonas* P400. FEMS Microbiol. Lett. 8: 187–190

Thomas ORT and White GF (1989) Metabolic pathway for the biodegradation of sodium dodecyl sulfate by *Pseudomonas* sp. C12B. Biotechnol. Appl. Biochem. 11: 318–327

Thomas ORT and White GF (1990) Immobilization of the surfactant-degrading bacterium *Pseudomonas* C12B in polacrylamide gel beads: II. Optimizing SDS-degrading activity and stability. Enzyme Microb. Technol. 12: 969–975

Thomas ORT, Matts PJ and White GF (1988) Localisation of alkylsulphatases in bacteria by electron microscopy. J. Gen. Microbiol. 134: 1229–1236

Thurnheer T, Zurrer D, Hoglinger O, Leisinger T and Cook AM (1990) Initial steps in the degradation of benzene sulfonic acid, 4-toluene sulfonic acid, and orthanilic acid in *Alcaligenes* sp. strain O-1. Biodegradation 1: 55–64

Thysse GJE and Wanders TH (1972) Degradation of n-alkane-1-sulfonates by *Pseudomonas*. Antonie van Leeuwenhoek 38: 53–63

Thysse GJE and Wanders TH (1974) Initial steps in the degradation of n-alkane-1-sulfonates by *Pseudomonas*. Antonie van Leeuwenhoek 40: 25–37

Wagner FC and Reid EE (1931) The stability of the carbon-sulphur bond in some aliphatic sulphonic acids. J. Am. Chem. Soc. 53: 3407–3413

Werdelmann BW (1984) Tenside in unserer Welte – heute und morgen. Second World Surfactants Congress, Vol 1 (pp 3–21). Syndicat National des Fabricants d'Agents de Surface et de Produits Auxiliares Industriels, Paris

White GF and Matts PJ (1992) Biodegradation of short-chain alkyl sulphates by a coryneform species. Biodegradation 3: 83–91

White GF and Russell NJ (1988) Mechanisms of bacterial biodegradation of alkyl sulphate and alkylpolyethoxy sulphate surfactants. In: DR Houghton, RN Smith and HOW Eggins (eds)

Biodeterioration 7 (pp 325–332). Elsevier Applied Science, London

White GF, Russell NJ and Day MJ (1985) A survey of sodium dodecyl sulphate SDS-resistance alkylsulphatase production in bacteria from clean and polluted river sites. Environ. Pollut. A37: 1–11

White GF, Dodgson KS, Davies I, Matts PJ, Shapleigh JP and Payne WJ (1987) Bacterial utilization of short-chain primary alkyl sulphate esters. FEMS Microbiol. Lett. 40: 173–177

White GF, Anderson DJ, Day MJ and Russell NJ (1989) Distribution of planktonic bacteria capable of degrading sodium dodecyl sulphate SDS in a polluted South Wales river. Environ. Pollut. A57: 103–115

Willetts AJ (1973) Microbial metabolism of alkylbenzene sulphonates. Fungal metabolism of 1-phenylundecane-p-sulphonate and 1-phenyldodecane-p-sulphonate. Antonie van Leeuwenhoek 39: 585–597

Willetts AJ and Cain RB (1972) Microbial metabolism of alkylbenzene sulfonates. Bacterial metabolism of undecylbenzene-p-sulphonate and dodecylbenzene-p-sulphonate. Biochem. J. 129: 389–402

Zamanian M and Mason JR (1987) Benzene dioxygenase in *Pseudomonas putida*. Sub-unit composition and immuno-cross-reactivity with other aromatic dioxygenases. Biochem. J. 244: 611–616

6. Biochemistry and physiology of the degradation of nitrilotriacetic acid and other metal complexing agents

THOMAS EGLI

Swiss Federal Institute for Environmental Science and Technology (EAWAG),
CH-8600 Dübendorf, Switzerland

Abbreviations: ATMP – Aminotrimethylphosphonate [$N(CH_2PO_3H_2)_3$]; DTPA – Diethylenetriaminepentaacetate [$(CH_2COOH)_2NC_2H_4N(CH_2COOH)C_2H_4N(CH_2COOH)_2$]; DTPMP – Diethylenetriaminepentamethylphosphonate [$(CH_2PO_3H_2)_2NC_2H_4N(CH_2PO_3H_2)C_2H_4N(CH_2PO_3H_2)_2$]; EDTA – Ethylenediaminetetraacetate [$(CH_2COOH)_2NC_2H_4N(CH_2COOH)_2$]; EDTMP – Ethylenediaminetetramethylphosphonate [$(CH_2PO_3H_2)_2NC_2H_4N(CH_2PO_3H_2)_2$]; FAD – Flavin adenine dinucleotide; FMN – Flavin mononucleotide; HEDP – Hydroxyethylidenediphosphonate [$HOC(PO_3H_2)_2CH_3$]; HEDTA – Hydroxyethylethylenediaminetetraacetate [$(CH_2COOH)_2NC_2H_4N(CH_2COOH)(CH_2CH_2OH)$]; IDA – iminodiacetate [$HN(CH_2COOH)_2$]; IDA-DH – Iminodiacetate dehydrogenase; NTA – Nitrilotriacetate [$N(CH_2COOH)_3$], NTA-DH – NTA dehydrogenase; NTA-MO – NTA monooxygenase; NtrR – Nitrate reductase; PMS – phenazine methosulfate; SDS-PAGE – sodium dodecylsulfate polyacrylamide gel electrophoresis; STP pentasodiumtriphosphate; Q_n – Ubiquinone (with n isoprene units in the side chain); μ_{max} – maximum specific growth rate (h^{-1})

I. Introduction

Synthetic organic, metal-sequestering compounds, such as the aminopolycarboxylic acids (mainly EDTA and NTA) or the organophosphonic acids HEDP or ATMP, are included in a wide range of different consumer products. Approximately 20% of the total amount produced is used in household detergents to prevent precipitation of bivalent ions from washing suds thereby avoiding deposition of scale on both textile fibers and washing machine parts and to support performance of surfactants (also known as tensides). The rest is used in various other applications such as water treatment, descaling of boilers, in the photographic industry, in agricultural fertilizers, in the dying of textiles, during pulp and paper production, for metal finishing and rubber processing, or in food, pharmaceuticals and cosmetics (McCrary and Howard 1979; Chemical Economics Handbook Program 1983; Egli 1988).

179

C. Ratledge (ed.), Biochemistry of Microbial Degradation, 179–195.

All these compounds are used in aqueous systems and therefore their susceptibility to biodegradation during wastewater treatment and in the aquatic environment is an important criterion for assessing their environmental impact and toxicology (Maki et al. 1980). The various environmental aspects of the use of natural and synthetic chelating agents has been the subject of many reviews (e.g., Tiedje 1980; Anderson et al. 1985; Bernhardt 1990; Egli et al. 1990; Egli 1992).

Of all the synthetic chelating agents, NTA has clearly received most attention. This is firstly, because one of its most controversial applications is that as a substitute for sodium triphosphate in laundry detergents (Mottola 1974; Martell 1975; Tiedje 1980; Bernhardt et al. 1984) and, secondly, because microorganisms capable of degrading NTA can easily be isolated, whereas isolation of microbes capable of degrading other synthetic chelating agents such as EDTA, ATMP, HEDP or EDTMP has been reported only recently (reviewed in Egli et al. 1990).

The water-insoluble, ion exchanging sodium aluminium silicates (Zeolite type A) and polymeric carboxylic acids, which are frequently used in combination or together with a complexing agent as builder systems, are, strictly speaking, not complexing agents (Opgenorth 1987; Diessel et al. 1988). Recent studies indicated that neither is particularly susceptible to biodegradation (Opgenorth 1987; Egli et al. 1990) and, therefore, they will not be considered here.

II. Nitrilotriacetic acid

A. NTA-degrading microorganisms

Earlier work suggested that biodegradation of NTA was primarily a trait inherent to specialist *Pseudomonas* strains (see Egli et al. 1990). However, the information presently available clearly shows that the ability to grow at the expense of NTA is especially associated with three distinctly different groups of Gram-negative bacteria. Extensive characterization (Egli et al. 1988; Wehrli and Egli 1988; Wanner et al. 1990; Wilberg et al. 1993) revealed that all three groups are clearly different from the "true" (fluorescent) pseudomonads, as defined by De Vos and De Ley (1983). Two out of the three groups consist of obligately aerobic members of the α-subclass of *Proteobacteria*. Recently, it has been proposed that they should be assigned to two new genera (species), *Chelatobacter* (*heintzii*) and *Chelatococcus* (*asaccharovorans*), respectively (Auling et al. 1993). The third "group" consists of the facultatively aerobic, denitrifying strain TE 11. It is a member of the γ-subgroup of *Proteobacteria* and is closely related to, but clearly different from, members of the genus *Xanthomonas* (Wanner et al. 1990). It is possible that this strain will also form the nucleus of a new genus in this branch.

In addition, isolation of *Bacillus*, *Listeria*, *Rhodococcus* and yeast species has been reported in the literature, indicating that the ability to utilize NTA is not restricted to Gram-negative bacteria (reviewed in Egli et al. 1990). However, all these strains either grow badly or are poorly characterized and detailed biochemical and physiological studies are restricted to *Chelatobacter* and *Chelatococcus* species.

B. Biochemistry of NTA degradation

First indications of the possible intermediates of intracellular NTA metabolism were obtained by Focht and Joseph (1971) who found in respiratory studies that NTA-grown cells of *Cb. heintzii* strain ATCC 27109 were able to utilize both IDA and glycine without a lag. Their work, and similar research by Tiedje et al. (1973), indicated that IDA, glyoxylate and glycine were possible intermediates but that sarcosine (*N*-methylglycine), *N*-methyliminodiacetate or acetate were not.

Further evidence for the pathway of NTA metabolism outlined in Fig. 1 was presented by Cripps and Noble (1972, 1973). In cell-free extracts of strain T23 (most probably a *Chelatobacter* sp.), they were able to confirm that glyoxylate, glycine, tartronic semialdehyde and glycerate accumulated following incubation with NTA. Both the exclusive release of $^{14}C\text{-}CO_2$ from [1-^{14}C]nitrilotriacetate and the requirement for thiamine pyrophosphate, Mg^{2+} and NADH supported the involvement of glyoxylate carboxyligase and tartronic semialdehyde dehydrogenase in the pathway of glyoxylate to glycerate. Comparison of enzyme induction patterns in glucose- and NTA-grown cells showed enhanced activities in NTA-grown cells of all enzymes postulated to be involved in the proposed pathway (Cripps and Noble 1973). The only metabolic step proposed in Fig. 1 that these researchers were unable to detect in cell-free extracts was the conversion of IDA to glycine and glyoxylate.

1. Transport of NTA

Whereas considerable information is available on the intracellular metabolism of NTA, its transport into the cells has received very little attention. Wong et al. (1973) found that the transport of ^{14}C-labelled NTA was inhibited by addition of either cyanide or azide and concluded that active transport was responsible for the translocation of NTA through the membrane. Firestone and Tiedje (1975) investigated the uptake of different NTA-metal complexes by whole cells of *Cb. heintzii* strain ATCC 29600. Because they found that incubation with different metal-NTA complexes resulted in no measurable metal accumulation inside the cells, they hypothesized that the metal-NTA complex was destabilized at the cytoplasmic membrane and free NTA was the species transported into the cell. However, such suggestions are rather speculative and no convincing evidence has been presented supporting any particular type of transport for NTA.

Although no details are presently available on the transport of NTA, it is

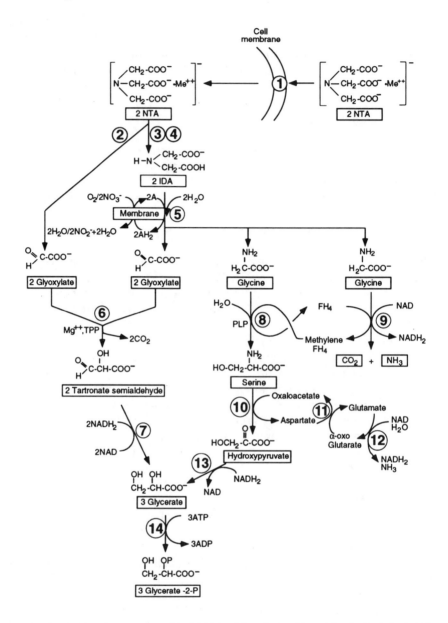

Fig. 1. Metabolic pathway proposed for NTA in obligately aerobic and facultatively denitrifying Gram-negative bacteria. (1) Transport enzyme; (2) NTA monooxygenase; (3) NTA dehydrogenase; (4) NTA dehydrogenase/nitrate reductase complex; (5) IDA dehydrogenase; (6) glyoxylate carboligase; (7) tartronate semialdehyde reductase; (8) serine hydroxymethyl transferse; (9) glycine synthase (decarboxylase); (10) serine: oxaloacetate aminotransferase; (11) transaminase; (12) glutamate dehydrogenase; (13) hydroxypyruvate reductase; (14) glycerate kinase; (methylene)FH$_4$, (N^5,N^{10}-methylene) tetrahydrofolic acid; PLP, pyridoxal phosphate; TPP, thiamine pyrophosphate.

interesting to compare it to the transport of the structurally similar tricarboxylic acid, citrate, a compound which also exhibits chelating properties. Transport of citrate has been studied in both Gram-negative and Gram-positive bacteria and four different mechanisms are presently known (Kay et al. 1987; Furlong 1987; Bergsma and Konings 1983): i) transport together with a bivalent metal ion; ii) transport depending on a binding protein with no specific requirement of a cation; iii) transport depending on on Na^+ only; and iv) transport depending on both K^+ and Na^+. Such studies show that transport of NTA is not necessarily dependent on the co-transport of a divalent metal ion and its subsequent excretion.

2. Initial oxidation of NTA

Presently, two different enzymes, a monooxygenase (NTA-MO) and a dehydrogenase (NTA-DH), have been identified which are able to split NTA oxidatively. In the strains investigated so far, activity of NTA-MO was detected in all Gram-negative obligate aerobic bacteria, whereas NTA-DH was present in the denitrifying strain TE 11 (Table 1). The biochemical evidence presently available shows that the initial step in both aerobic and anoxic metabolism of NTA yields the same products, i.e., iminodiacetate and glyoxylate (Fig. 2).

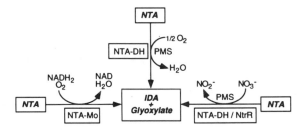

Fig. 2. Enzyme reactions presently known to catalyse the oxidative cleavage of NTA.

3. NTA monooxygenase

In all the obligately aerobic *Chelatobacter* and *Chelatococcus* strains studied so far the activity of a soluble, NADH-dependent monooxygenase was reported (Table 1). First attempts to purify NTA-MO resulted in the isolation of a protein that exhibited NTA-stimulated NADH oxidation but had completely lost the ability to oxidatively cleave NTA to glyoxylate and IDA (Firestone et al. 1978). This result indicated that NTA-MO is a labile, multi-component enzyme. Using an enzyme assay based on the consumption of NTA, the isolation of a functional, NTA-hydroxylating enzyme complex exhibiting classical monooxygenase activity was reported only recently (Uetz et al. 1992). The enzyme consisted of two components, cA (dimer, native molecular mass 99 kDa) and cB (dimer, native molecular mass 88 kDa), which were both

Table 1. Distribution of enzymes in different Gram-negative NTA-degrading bacteria grown under aerobic or denitrifiying conditions. Data from Uetz and Egli (1993), Jenal-Wanner (1991), J. Kemmler and T. Egli (unpublished work) and Uetz (1992).

| Bacterium | Growth substrate[a] | NTA-MO | | | NTA-DH[d] | IDA-DH[e] | Succ-DH |
		sp. A.[b]	cA[c]	cB(FMN)[c]			
Chelatococcus asaccharovorans[f]	NTA/O_2	+	+	−	−	+	+
Chelatobacter heintzii[g]	NTA/O_2	+	+	+	−	+	+
	IDA/O_2	+	+	+	−	+	+
	Succ/O_2	−	−	−	−	−	+
Unknown genus strain TE 11	NTA/O_2	−	−	−	+	+	+
	NTA/NO_3	−	−	−	+	+	+

[a] O_2, aerobic growth; NO_3^-, denitrifying growth conditions.

[b] +, sp. A. > 20 nmol NTA min^{-1} (mg protein)$^{-1}$; −, sp. A. < 1 nmol NTA min$^-$ (mg protein)$^-$.

[c] Cross-reaction with antibodies raised against cA and cB of NTA-MO purified from *Chelatobacter heintzii* ATCC 29600.

[d] Specific activity and cross-reaction with antibodies raised against NTA-DH purified from strain TE 11.

[e] +, sp. A. > 5 nmol NTA min^{-1} (mg protein)$^{-1}$; −, sp. A. < 1 nmol NTA min^{-1} (mg protein)$^{-1}$.

[f] Strain TE 2.

[g] Strains ATCC 29600, ATCC 27109, TE 4–8 and TE 10.

purified to more than 95% homogeneity and it was possible to reconstitute the functional, NTA-hydroxylating enzyme complex from pure cA and cB. The latter contained FMN and exhibited NTA-stimulated NADH-oxidation but was unable to hydroxylate NTA, whereas no catalytic activity has yet been shown for cA on its own. Activity of NTA-MO was dependent on the presence of Mg^{2+} ions (which could be replaced by Co^{2+} or Mn^{2+}, but not by Ca^{2+} or other metal ions) and results from inhibition studies suggested that metal-complexed NTA was the substrate for the enzyme. Under optimum conditions, the K_m for Mg^{2+}-complexed NTA was 0.5 mM. Out of the 26 substrates tested, including IDA and many other compounds structurally related to NTA, NTA was the only substrate for NTA-MO. This indicates that NTA-MO is not one of the broad-spectrum monooxygenases with fortuitous substrate specificity for NTA.

The two components of NTA-MO are only weakly associated and under unfavourable conditions (high salt concentrations, the presence of FAD, low protein concentration or ratio of the two components) the oxidation of NADH is partly or completely uncoupled from the hydroxylation of NTA resulting in the formation of H_2O_2 (Fig. 3). Under *in vitro* conditions molar ratios of cA:cB > 4 were required to obtain tight coupling. At present, the catalytic role of cA is unknown except that it is absolutely required for NTA-hydroxylating activity. Since it does not contain any flavin, it is presently thought that it acts as a protein which can modify the active site at cB in such a way that NTA is accepted as a substrate and is hydroxylated. Immuno-quantification of the two components in NTA-grown cells suggests that the two components are present at approximately a 1:1 (mol:mol) ratio and that NTA-MO may account for some 7% of the total cellular protein. The observations with reconstituted NTA-MO (equimolar ratio of cA:cB), that the degree of coupling increases *in vitro* with increasing protein concentration and that *in vivo* concentrations of NTA-MO are probably 100 times higher than those used experimentally, suggest that *in vivo* uncoupling is probably not relevant.

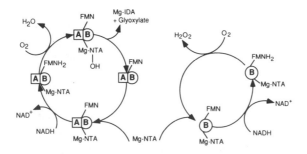

Fig. 3. Proposed scheme for the reactions catalyzed by NTA-MO. On the right hand the reactions mediated by pure component B (cB) are shown, whereas the left-hand circle illustrates the reactions catalyzed by the functional NTA-MO consisting on the two components cA and cB. From Uetz et al. (1992).

The immunological cross-reaction pattern with antibodies raised against the purified components of NTA-MO from *Cb. heintzii* ATCC 29600 suggested that in all members of *Cb. heintzii* NTA-MO consisted of two components identical with, or closely related to, cA and cB in strain ATCC 29600 (Table 1). However, in the phylogenetically but only distantly related *Chc. asaccharovorans* strain TE 2, only cA was detected whereas the catalytically active flavin component B was missing. Isolation of the cA-cross-reacting protein from *Chc. asaccharovorans* strain TE 2 and its subsequent biochemical characterization strongly indicated that it was most probably homologous to cA from *Cb. heintzii* ATCC 29600. It was possible to form a functional hybrid monooxygenase consisting of cA from *Chc. asaccharovorans* strain TE 2 plus cB from *Cb. heintzii* ATCC 29600 which exhibited similar properties to the original enzyme isolated from *Cb. heintzii* ATCC 29600 (Uetz 1992). From this it was concluded that this protein fulfilled a similar function and also formed part of a two-component NTA-monooxygenase present in *Chc. asaccharovorans*. To date, it has not been possible to purify a protein from *Chc.asaccharovorans* with a function similar to cB. But this might be due to its unstability.

4. NTA dehydrogenase

It is obvious that, in contrast to the majority of obligately aerobic NTA-utilizing bacteria exhibiting NTA-MO activity, the initial step in the breakdown of NTA in the denitrifying NTA-degrading bacteria has to proceed via an O_2-independent enzyme. A first comparison of cells of isolate TE 11 grown under aerobic and denitrifying conditions suggested that two different enzyme systems, namely a soluble "NTA-DH" that passed the electrons derived from NTA on to nitrate and a soluble "NTA-oxidase", might be responsible for the initial step in the metabolism of NTA in this strain (Kemmler and Egli 1990; Kemmler et al. 1990, 1991). This conclusion was reached because cell-free extracts containing the two enzymes exhibited completely different kinetic characteristics, such as temperature and pH optimum and with respect to metal and PMS requirement. However, subsequent purification of both enzymes revealed the presence of a quite unusual enzyme system for the oxidation of NTA in this bacterium.

From membrane-free extracts of NTA-grown, denitrifying cells an enzyme complex consisting of a NTA-dehydrogenase and a nitrate reductase was purified (Wanner et al. 1989; Jenal-Wanner 1991). The active complex was able to catalyse a PMS-dependent transfer of electrons from NTA to nitrate. The two proteins were tightly associated and activity of NTA-DH could only be measured in the presence of NtrR and *vice versa*. Only recently, NTA-DH was purified from aerobically-grown cells of strain TE 11 (Kemmler 1992). Comparison of the two NTA-DH proteins established that they were identical based on their molecular weight, flavin component, immunological cross reaction and the sequence of the last 14 amino acids at the *N*-terminus. Although NTA-DH from both aerobically grown cells and from the NTA-DH/NtrR complex from denitrifying cells catalysed the same overall reaction, i.e., the PMS-dependent oxidation of NTA to glyoxylate and IDA, association of NTA-DH with NtrR results in a marked difference with respect to the observed kinetic properties (Table 2). The substrate for NTA-DH seems to be uncomplexed NTA since, in contrast to NTA-MO, neither the requirement of bivalent metal ions for NTA-DH activity nor its inhibition by EDTA was reported. Even though a range of compounds, many of them structurally similar to NTA, have been tested, NTA is still the only known substrate for both NTA-DH and the NTA-DH/NtrR complex. By analogy to NTA-MO, neither NTA-DH on its own nor the NTA-DH/NtrR complex was able to use IDA as a substrate.

The enzyme system detected in strain TE 11 is the first report of a dehydrogenase that can function either on its own or in combination with a nitrate reductase. The NTA-DH/NtrR complex is seemingly comparable to the redox cascade systems known from oxygenases but several interesting questions still remain open. For example, it is presently unknown how the electrons derived from the oxidation of NTA are passed on to NtrR. The

Table 2. Some properties of key enzymes involved in the degradation of NTA in obligately aerobic and facultatively denitrifying Gram-negative bacteria. Data were taken from Uetz et al. (1992), Uetz and Egli (1993), Jenal-Wanner (1991), and J. Kemmler and T. Egli (unpublished work). NTA-MO, NTA monooxygenase; cA, component A; cB, component B; NTA-DH, NTA dehydrogenase; NtrR, Nitrate reductase; IDA-DH, IDA dehydrogenase.

Enzyme properties	NTA-MO cA / cB	NTA-DH/NtrR	NTA-DH	IDA-DH
Localisation	soluble	soluble[a]	soluble[a]	particulate
Molecular mass				
SDS (kDa)	47 / 36	80/63	78	–
Native (kDa)	99 / 88	170 / 105	162	–
Substrate	NTA	NTA	NTA	IDA
Products	IDA + Glyoxylate	IDA + Glyoxylate	IDA + Glyoxylate	Glycine + Glyoxylate
Cosubstrate	NADH	–	–	–
e^--Acceptor	O_2	NO_3^-	O_2	Respiration chain (Q_{10}?)
e^--Mediators	–	PMS[b]	PMS[b]	–
Redox component	– / FMN	FAD[c]/(cyt b)[d]	FAD[c]	(cyt b)[d]
Metal requirement	Mg^{2+} (Co^{2+}, Mn^{2+})	none	none	$(Ca^{2+}, Ba^{2+}, Mg^{2+})$[e]
K_m(NTA)	0.5 mM[f]	0.095 mM[g]	0.19 mM[g]	–
K_m(IDA)	–	–	–	8 mM[h]

[a] Might be membrane-associated.
[b] *In vivo* e^--mediator unknown.
[c] Covalently bound FAD.
[d] Indicated by spectral properties.
[e] Stimulation observed, but not strictly necessary for activity.
[f] Measured for Mg^{2+}-complexed NTA.
[g] Measured for uncomplexed NTA.
[h] Measured for Ca^{2+}-complexed IDA.

observation that even in the presence of membranes the NTA-DH/NtrR complex was not active in the absence of PMS (Jenal-Wanner 1991) supports the speculation that electrons from NTA-DH might be passed on to NtrR directly and not via some presently unknown redox components in the membrane. The "soluble" nature of both NTA-DH and the NTA-DH/NtrR complex within the cell still requires further investigation. It is possible that in aerobically grown cells NTA-DH is loosely associated with the cytoplasmic membrane feeding the electrons directly to one of the electron transport components, whereas in cells grown under denitrifying conditions it is closely associated with a loosely membrane-bound dissimilatory NtrR. First indications that the NTA-DH/NtrR complex might indeed be composed of NTA-DH and the dissimilatory NtrR were obtained recently in our laboratory

(J. Kemmler and T. Egli, unpublished work): addition of purified NTA-DH from aerobically grown cells to cell-free extracts obtained from cells grown under denitrifying conditions with glucose/ammonium (and therefore containing neither NTA-DH nor assimilatory NtrR) allowed reconstitution of the typical NTA-DH/NtrR activity.

5. *Iminodiacetate dehydrogenase*

The failure of both purified NTA-MO and NTA-DH (on its own or in combination with NtrR) to oxidatively split IDA into glycine and glyoxylate indicates that the subsequent conversion of IDA is catalysed by an additional enzyme in the pathway, rather than by the NTA-oxidizing enzyme as suggested by Firestone et al. (1978). Recently, we were able to demonstrate activity of a membrane-bound IDA-DH in all members of the presently known NTA-degrading genera (Uetz and Egli 1993). Detailed investigation in *Cb. heintzii* ATCC 29600 confirmed that the enzyme was different for other membrane-bound dehydrogenases, e.g., succinate dehydrogenase, and catalysed the stoichiometric conversion of IDA to glycine and glyoxylate (Table 2). Although activity was stimulated by the addition of Ca^{2+}, Ba^{2+} or Mg^{2+}, IDA oxidation was not significantly decreased by EDTA indicating that the involvement of metal ions is not strictly necessary. Nevertheless, the K_m for IDA was considerably lower in the presence of Ca^{2+} compared to in its absence (8 mM and 50 mM, respectively). At present, the substrate specificity of IDA-DH appears to be very limited. From a range of compounds tested only IDA was accepted as a substrate.

Activity of IDA-DH was strictly linked to the electron transport chain. It was very sensitive to KCN but under these conditions INT could act as the terminal electron acceptor. The transfer of electrons from IDA to both O_2 and to INT was inhibited by dibromomethylisopropyl-benzoquinone, a ubiquinone analogue which is known to inhibit reoxidation of ubiquinones. It was possible to extract IDA-DH from membranes and to incorporate the enzyme into soybean phospholipid vesicles. Under such conditions IDA-DH activity was successfully reconstituted using either ubiquinone Q_1 or ubiquinone Q_{10} (the major quinone in this genus; Egli et al. 1988), as the intermediate electron carrier and INT as the terminal electron acceptor. All this suggests that IDA-DH is an integral membrane protein which *in vivo* feeds electrons from the oxidation of IDA into the ubiquinone poole (Uetz 1992). Unfortunately, in contrast to other membrane-bound dehydrogenases, IDA-DH is not able to reduce an artificial electron acceptor such as PMS after solubilization with detergents, which implies that purification of this protein will be fraught with difficulties.

It is striking that all the bacterial strains investigated up to now, belonging to three different bacterial genera, contained a membrane-associated IDA-DH (Uetz and Egli 1993). This stresses the important role of this enzyme in the biodegradation of NTA. Nevertheless, presence of IDA-DH activity does not necessarily imply that in all the bacterial strains investigated this activity is brought about by the same protein. The low affinity towards IDA suggests that the enzyme has evolved from an already existing protein or that IDA is not the natural substrate for this enzyme.

IDA-DH might not only be of importance for the degradation of NTA but it might also be involved in the metabolism of other xenobiotic compounds. For example, IDA was reported to be an intermediate in the the biodegradation pathway of EDTA. Furthermore, an enzyme able to oxidize the herbicide glyphosate, $HOOC-CH_2-NH-CH_2-PO_3H_2$, exhibiting a higher affinity to IDA than to glyphosate, was recently cloned from a *Pseudomonas* strain (Kishore and Barry 1992). The IDA-DH from *Cb. heintzii* ATCC 29600 did not accept the herbicide as a substrate (T. Uetz and T. Egli, unpublished work).

C. Growth physiology of NTA-degrading bacteria

To understand the behaviour and the performance of NTA-degrading bacteria in sewage treatment plants and in nature, the influence of environmental parameters on growth and on the regulation of enzymes involved in the metabolism of NTA in these organisms must first be understood. In this respect, it is not only parameters such as temperature and pH that have to be considered but also the availability of alternative nutrients, particularly carbon and/or nitrogen sources. Unfortunately, there is little information available in the literature and the few studies that have been done concentrated on obligately aerobic NTA-degraders.

With respect to growth, most studies have focussed on the effects of temperature, pH or NTA concentration on the specific growth rate exhibited by batch cultures (summarized in Egli 1992). The data suggest that all the NTA-degrading strains isolated so far are relatively slow-growing bacteria. The maximum specific growth rates observed in batch culture for the growth of obligately aerobic strains with NTA as the sole source of carbon, nitrogen and energy were in the range of 0.05 to 0.18 h^{-1}. The μ_{max} values reported for strains able to grow under denitrifying conditions were considerably lower, 0.03 to 0.05 h^{-1}. In general, growth was not only slow during cultivation with NTA but also with other carbon sources, including complex medium.

A general phenomenon observed in batch as well as in carbon-limited chemostat culture with all strains investigated in our laboratory was their ability to utilize NTA simultaneously with a suitable carbon source. As a rule, during batch cultivation simultaneous growth with NTA plus a metabolizable carbon source resulted in an increase in the μ_{max} of the culture (Egli et al. 1988). Evidence that synthesis of enzymes involved in the metabolism of NTA can be induced in cells growing exponentially in batch culture has been demonstrated first for *Chc. asaccharovorans* TE 1 (Hamer et al. 1985). This is shown in Fig. 4 where cells growing exponentially with acetate and ammonia were pulsed with NTA. For a further 6 to 7 hours, cells continued to grow exponentially at the original rate but after some 4 to 5 hours the cells started to utilize NTA and this resulted in an enhanced exponential growth rate. Similar results have been obtained for a pulse of NTA to a culture of *Cb. heintzii* ATCC 29600 growing exponentially with succinate and ammonia and it was confirmed that both NTA-MO and IDA-DH became induced within one hour after addition of NTA and that after induction of the two enzymes NTA was consumed simultaneously with NTA (Uetz and Egli 1993).

In *Cb. heintzii* ATCC 29600 expression of NTA-specific enzymes was also investigated in carbon-limited chemostat culture as a function of culture conditions. During NTA-limited growth specific activity of NTA-MO increased with decreasing dilution rates (Egli et al. 1990) which is a pattern often observed for enzymes at the beginning of a catabolic pathway. In view of the fact that in treatment plants and in nature the cells have to grow in the presence of many alternative carbon substrates and cannot be expected to grow with NTA only, the cells were cultivated under carbon limitation (ammonia was present in excess) at a constant dilution rate with mixtures of NTA/glucose of different composition (Bally et al. 1992). At all NTA/glucose ratios both substrates were utilized simultaneously. Significant induction of specific activity of NTA-MO was found only when the fraction of NTA-carbon relative to the total carbon of the mixed substrate exceeded approximately 3% (Fig. 5) and this observation was confirmed by immunological quantification of the two NTA-MO components cA and cB. Similar results were obtained for IDA-DH activity. This result raises the question as to what level the cells are induced under ecologically relevant growth conditions. Nevertheless, it has recently been demonstrated that uninduced cells of *Cb. heintzii* become induced when exposed in diffusion chambers to sewage treatment plant conditions (McFeters et al. 1990).

To date, very little information on the regulation of expression of NTA-DH

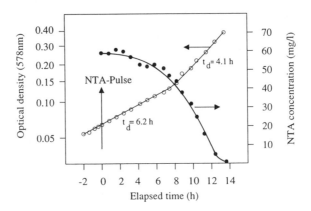

Fig. 4. Effect of a pulse of NTA on a culture of *Chelatococcus asaccharovorans* TE 1 growing exponentially in batch culture with acetate as the carbon/energy source and ammonia as the nitrogen source. ○, growth (OD_{578}); ●, NTA concentration (mg l^{-1}). Adapted from Hamer et al. (1985).

and NtrR in facultatively denitrifying NTA-degrading bacterium TE 11 is available. However, the fact that NTA-DH (assayed both via enzyme activity and immunological cross-reaction) was absent during growth of this strain with a wide range of different substrates, many of them structurally similar to NTA, indicates that the synthesis of this enzyme is tightly regulated (J. Kemmler and T. Egli, unpublished work).

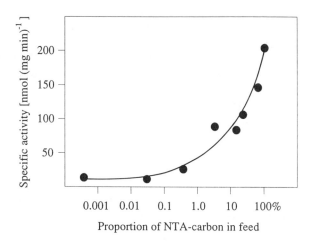

Fig. 5. Specific activity of NTA-MO during growth of *Chelatobacter heintzii* ATCC 29600 in carbon-limited chemostat culture at a constant dilution rate of $D = 0.06 \text{ h}^{-1}$ with mixtures of glucose and NTA as the sole sources of carbon and energy. Ammonium was included in the medium as a source of nitrogen. The composition of the glucose/NTA mixture in the inflowing medium is is given as the proportion of NTA-carbon in % with respect to the total carbon from both glucose and NTA. Data from Bally et al. (1992).

III. Other chelating agents

A. Aminopolycarboxylic acids

Although the biological transformation of aminopolycarboxylic acids other than NTA, such as EDTA, DTPA, or HEDTA, was described in the literature more than a decade ago (summarized in Egli et al. 1990) it was only recently that the successful isolation of EDTA-degrading bacteria was reported (Lauff et al. 1990; Nörtemann 1992). The two cultures described exhibited distinctly different properties with respect to their EDTA degradation characteristics. The pure culture isolated by Lauff et al. (1990) could degrade EDTA exclusively when complexed with iron and was able to utilize EDTA at concentrations as high as 100 mM. In contrast, the EDTA-degrading bacterium isolated by Nörtemann (1992) performed best in a mixed culture (together with an unidentified Gram-positive bacterium) at EDTA concentrations below 2 mM and speciation did not significantly affect degradation (in fact ferric EDTA was only incompletely degraded). Interestingly, both EDTA-degraders were members of the α-subgroup of *Proteobacteria* and Lauff et al. (1990) identified their isolate as a strain of *Agrobacterium radiobacter*. No studies on the biochemistry of EDTA degradation have been reported yet.

B. *Phosphonates*

Several bacterial strains are able to utilize the metal-complexing phosphonic acids ATMS, HEDP, EDTMP and DTPMP, and have been described in the literature (summarized in Egli et al. 1990). Investigations by Schowanek and Verstraete (1990a) suggest that microbes with the capability to degrade these complexing agents are widely distributed in polluted natural waters. The results presently available suggest that all the bacterial strains isolated up to now are using the phosphonate group from these compounds as a source of phosphorus, whereas the fate of the carbon and/or the nitrogen remains uncertain. However, the author is not aware of any biochemical investigation where the enzymic nature of the cleavage reaction of the C-P bond in complexing phosphonic acids has been identified. In analogy to other phosphonic acids one can speculate that either a C-P lyase (Wackett et al. 1987) or a phosphonatase (La Nauze et al. 1970) type of enzymic reaction might be involved. Whereas it was found for several alkylphosphonate-degrading bacteria that both transport and subsequent cleavage of the carbon-phosphorus bond is inhibited/repressed in the presence of *ortho*-phosphate (Lerbs et al. 1990; Daughton et al. 1979), the report by Schowanek and Verstraete (1990b) indicates that some bacteria are also able to degrade phosphonates simultaneously with, and in the presence of high concentrations of *ortho*-phosphate.

Acknowledgements

Over the years, research on NTA in our laboratory has been supported by the Swiss National Science Foundation (Grants No 3.204-0.85 and 31-26326.89), the Research Commission of ETH Zürich (Grant No 0.330.089.86), as well as Lever AG Switzerland and Unilever, Port Sunlight, United Kingdom. Their financial support is gratefully acknowledged. The author is also indebted to C.A. Mason for his linguistic help.

References

Anderson RL, Bishop WE and Campbell RL (1985) A review of the environmental and mammalian toxicology of nitrilotriacetic acid. CRC Critical Reviews in Toxicology 15: 1–102

Auling G, Busse H-J, Egli T, El-Banna T and Stackebrandt E (1993) Description of the Gram-negative, obligately aerobic, nitritotriacetate (NTA)-utilizing bacteria as *Chelatobacter heintzii*, gen. nov., sp. nov., and *Chelatococcus asaccharovorans*, gen. nov., sp. nov. System. Appl. Microbiol. 16: 104–112

Bally M, Uetz T and Egli T (1992) Physiology of biodegradation of nitrilotriacetate (NTA) by *Chelatobacter heintzii*. Abstract. 6th International Symposium on Microbial Ecology, Barcelona, Spain

Bergsma J and Konings W (1983) The properties of citrate transport in membrane vesicles. Eur. J. Biochem. 191: 151–156

Bernhardt H (1990) Bewertung von organischen Phosphatersatzstoffen aus ökologischer Sicht. Vom Wasser 74: 159–176

Bernhardt H, Berth W, Förster U, Hamm A, Janicke W, Kandler J, Kanowski S, Kleiser HH, Koppe P, Opgenorth HJ, Reichert JK and Stehfest H (1984) NTA: Studie über die aquatische Umweltverträglichkeit von Nitrilotriacetat (NTA). Verlag Hans Richarz, Sankt Augustin, Germany

Chemical Economics Handbook Program: Chelating Agents Product Review (1983) Stanford Research Institute International

Cripps RE and Noble AS (1972) The microbial metabolism of nitrilotriacetate. Biochem. J. 130: 31P–32P

Cripps RE and Noble AS (1973) The metabolism of nitrilotriacetate by a Pseudomonad. Biochem. J. 136: 1059–1068

Daughton CG, Cook AM and Alexander M (1979) Phosphate and soil binding: factors limiting bacterial degradation of ionic phosphorus-containig pesticides. Appl. Environ. Microbiol. 37: 605–609

De Vos P and De Ley J (1983) Intra- and intergeneric similarities of *Pseudomonas* and *Xanthomonas* ribosomal ribonucleic acid cistrons. Int. J. Syst. Bacteriol. 33: 487–509

Diessel P, Stabenow J and Trieselt W (1988) Wirkungsweise von Copolycarboxylaten in Waschmitteln. Tens. Surfact. Deterg. 25: 268–274

Egli T (1992) Biodegradation of nitrilotriacetic acid. Habilitationsschrift, ETH-Zürich, Switzerland

Egli T (1988) (An)aerobic breakdown of chelating agents used in household detergents. Microbiol. Sci. 5: 36–41

Egli T, Weilenmann H-U, El-Banna T and Auling G (1988) Gram-negative, aerobic, nitrilotriacetate-utilizing bacteria from wastewater and soil. Syst. Appl. Microbiol. 10: 297–305

Egli T, Bally M and Uetz T (1990) Microbial degradation of chelating agents used in detergents with special reference to nitrilotriacetic acid (NTA). Biodegradation 1: 121–132

Firestone MK and Tiedje JM (1975) Biodegradation of metal-nitrilotriacetate complexes by a *Pseudomonas* species: mechanism of reaction. Appl. Microbiol. 29: 758–764

Firestone MK, Aust SD and Tiedje JM (1978) A nitrilotriacetic acid monooxygenase with conditional NADH-oxidase activity. Arch. Biochem. Biophys. 190: 617–623

Focht DD and Joseph HA (1971) Bacterial degradation of nitrilotriactic acid. Can. J. Microbiol. 17: 1553–1556

Furlong CE (1987) Osmotic shock-sensitive transport systems. In: FC Neidhardt, JL Ingraham, KB Low, B Magasanik and HE Umbarger (eds) *Escherichia coli* and *Salmonella typhimurium*, Vol I (pp 768–796). American Society for Microbiology, Washington, DC

Hamer G, Egli T and Mechsner K (1985) Biological treatment of industrial wastewater: A microbiological basis for process performance. J. Appl. Bacteriol. Symp. Suppl. 127S-140S

Jenal-Wanner U (1991) Anaerobic degradation of nitrilotriacetate in a denitrifying bacterium: purification and characterization of the nitrilotriacetate dehydrogenase / nitrate reductase enzyme complex. PhD thesis, ETH-Nr. 9531. Swiss Federal Institute of Technology, Zürich, Switzerland

Kay WW, Sweet GD, Widenhorn K and Somers JM (1987) Transport of organic acids in prokaryotes. In: BP Rosen and S Silver (eds) Ion Transport in Prokaryotes (pp 269–301). Academic Press, San Diego

Kemmler J (1992) Biochemistry of nitrilotriacetate degradation in the facultatively denitrifying bacterium TE 11. PhD thesis No 9983, Swiss Federal Institute of Technology, Zürich, Switzerland

Kemmler J and Egli T (1990) Nitrilotriacetate-abbauende Mikroorganismen. gwf Wasser Abwasser 131: 251–255

Kemmler J, Wanner U, Egli T, Snozzi M and Hamer G (1990) Degradation of nitrilotriacetic acid by two different enzymes in a denitrifying bacterium (strain TE 11). Abstract. Symposium on Environmental Biotechnology, Braunschweig, Germany

Kemmler J, Wanner U, Egli T, Snozzi M and Hamer G (1991) Abbau von Nitrilotriacetat durch zwei verschiedenen Enzyme in einem denitrifizierenden Bakterium. gwf Wasser Abwasser 132: 345–355

Kishore GM and Barry GF (1992) Glyphosate tolerant plants. Patent PCT/US91/04514

La Nauze JM, Rosenberg H and Shaw DC (1970) The enzymic cleavage of the carbon phosphorus bond: purification and properties of phosphonatase. Biochim. Biophys. Acta 212: 332–350

Lauff JJ, Steele DB, Coogan LA and Breitfeller JM (1990) Degradation of ferric chelate of EDTA by a pure culture of an *Agrobacterium* sp. Appl. Environ. Microbiol. 56: 3346–3353

Lerbs W, Stock M and Parthier B (1990) Physiological aspects of glyphosate degradation in *Alcaligenes* sp. strain GL. Arch. Microbiol. 153: 146–150

Maki AW, Dickson KL and Cairns J Jr. (eds) (1980) Biotransformation and fate of chemicals in the aquatic environment. American Society for Microbiology, Washington, DC

Martell AE (1975) The influence of natural and synthetic ligands on the transport and function of metal ions in the environment. Pure and Applied Chemistry 44: 81–113

McCrary AL and Howard WL (1979) Chelating agents. In: M Grayson and D Eckroth (eds) Kirk-Othmer Encyclopedea of Chemical Technology, Vol 5, 3rd edition (pp 339–368). Wiley, New York

McFeters GA, Egli T, Wilberg E, Alder A, Schneider RP, Snozzi M and Giger W (1990) Activity and adaptation of nitrilotriacetate (NTA)-degrading bacteria: field and laboratory studies. Water Res. 24: 875–881

Mottola HA (1974) Nitrilotriacetic acid as a chelating agent: applications, toxicology and bioenvironmental impact. Toxicol. Environ. Chem. Rev. 2: 99–161

Nörtemann B (1992) Total degradation of EDTA by mixed cultures and a bacterial isolate. Appl. Environ. Microbiol. 58: 671–676

Opgenorth HJ (1987) Umweltverträglichkeit von Polycarboxylaten. Tens. Surfact. Deterg. 24: 366–369

Schowanek D and Verstraete W (1990a) Phosphonate utilization by bacterial cultures and enrichments from environmental samples. Appl. Environ. Microbiol. 56: 895–903

Schowanek D and Verstraete W (1990b) Phosphonate utilization by bacteria in the presence of alternative phosphorus sources. Biodegradation 1: 43–53

Tiedje JM (1980) Nitrilotriacetate: hindsights and gunsights. In: AW Maki, KL Dickson and J Cairns Jr. (eds) Biotransformation and Fate of Chemicals in the Aquatic Environment (pp 114–119). American Society for Microbiology, Washington, DC

Tiedje JM, Mason BB, Warren CB and Malec EJ (1973) Metabolism of nitrilotriacetate by cells of *Pseudomonas* species. Appl. Microbiol. 25: 811–818

Uetz T (1992) Biochemistry of nitrilotriacetate degradation in obligately aerobic, Gram-negative bacteria. PhD thesis No. 9722, Swiss Federal Institute of Technology, Zürich, Switzerland

Uetz T and Egli T (1993) Characterization of an inducible, membrane-bound iminodiacetate dehydrogenase from *Chelatobacter heintzii* ATCC 29600. Biodegradation 3: 423–434

Uetz T, Schneider R, Snozzi M and Egli T (1992) Purification and characterization of a two component monooxygenase that hydroxylates nitrilotriacetate from "*Chelatobacter*" strain ATCC 29600. J. Bacteriol. 174: 1179–1188

Wackett LP, Shames SL, Venditti CP and Walsh CT (1987) Bacterial carbon-phosphorus lyase: products, rates and regulation of phosphonic and phoshinic acid metabolism. J. Bacteriol. 169: 1753–1756

Wanner U, Egli T and Snozzi M (1989) A dehydrogenase as the first step in the anaerobic pathway for nitrilotriacetate (NTA) degradation. In: G Hamer, T Egli and M Snozzi (eds) Mixed and Multiple Substrates and Feedstocks (pp 165–167). Hartung-Gorre, Constance, Germany

Wanner U, Kemmler J, Weilenmann H-U, Egli T, El-Banna T and Auling G (1990) Isolation and growth of a bacterium able to degrade nitrilotriacetic acid under denitrifying conditions. Biodegradation 1: 31–42

Wehrli E and Egli T (1988) Morphology of nitrilotriacetate-utilizing bacteria. Syst. Appl. Microbiol. 10: 306–312

Wilberg E, El-Banna T, Auling G and Egli T (1993) Serological studies on nitrilotriacetic acid (NTA)-utilizing bacteria: distribution of *Chelatobacter heintzii* and *Chelatococcus asaccharovorans* in sewage treatment plants and aquatic ecosystems. System. Appl. Microbiol. 16: 147–152

Wong PTS, Liu E and McGirr DJ (1973) Mechanism of NTA degradation by a bacterial mutant. Water Res. 7: 1367–1374

7. Enzymes and mechanisms involved in microbial cellulolysis

THOMAS M. WOOD and VICENTA GARCIA-CAMPAYO
Rowett Research Institute, Bucksburn, Aberdeen AB2 9SB, U.K.

I. Introduction

Cellulose-degrading enzymes have been studied with a new intensity in the last 10 years. To some extent this interest is connected with potential oil shortages in the next century and the realisation that forest, agricultural and municipal wastes can be converted into ethanol from glucose generated by cellulose-degrading enzyme systems. However, there is now an appreciation that a more immediate use for cellulose-degrading enzymes may be found in industries that are involved in processing paper and pulp, feed and food. In these industries the emphasis is on the controlled degradation of the cellulose in the plant cell wall and this has placed a new importance on the need for a better understanding on the multiple enzymes that comprise the cellulose-degrading system of some bacteria and fungi. A sound understanding of the enzymes and their modes of action is an essential prerequisite for using the cellulases with maximum effectiveness in these industrial processes. However, the same understanding is needed for meaningful improvements in other areas, such as the digestibility of feedstuffs in the animal and in soil fertility.

Microbial cellulolysis is a very complex process indeed, particularly that part of the mechanism that relates to the degradation of hydrogen bond-ordered (pseudo-crystalline) cellulose. It is now known that a number of interacting enzyme species are required for the degradation of the recalcitrant hydrogen bond-ordered areas of the cellulose. However, the nature of these enzymes appears to vary according to the source. Thus, cellulolysis by soft rot and white rot aerobic fungi (reviewed in Eriksson and Wood 1985) and some aerobic bacteria (Creuzet et al. 1983; Yablonsky et al. 1988) involves the interaction of enzymes loosely defined as exoglucanases (normally cellobiohydrolases, i.e. 1,4-β-D-glucan cellobiohydrolase), endoglucanases (endo-1,4-β-D-glucan 4-glucanohydrolase) and β-glucosidases (reviewed in Eriksson and Wood 1985). Brown rot fungi, on the other hand, which produce endoglucanases but no exoglucanases may have a different mechanism, perhaps involving H_2O_2 (Koenigs 1975). Some anaerobic bacteria (Lamed and Bayer 1988), and possibly anaerobic fungi (Wood et al. 1988; Wilson and Wood 1992b), use a

197

C. Ratledge (ed.), Biochemistry of Microbial Degradation, 197–231.
© 1994 *Kluwer Academic Publishers. Printed in the Netherlands.*

multicomponent enzyme complex (so-called, cellulosome), the exact composition of which remains to be described in each case.

Much progress has been made recently in understanding the cellulose-degrading enzymes acting in fungi and bacteria but there are still many unresolved problems. In particular, the mechanism by which the various enzymes interact on the face of the cellulose crystallite continues to be a contentious issue. In this review the most significant new insights obtained into the enzymes and their modes of action have been highlighted. The reader is directed to a variety of other reviews for other interpretations of the published data (Béguin 1990; Béguin et al. 1988, 1990, 1992; Coughlan 1985; Coughlan and Ljungdahl 1988; Eriksson and Wood 1985; Enari and Niku Paavola 1987; Eveleigh 1987; Gilkes et al. 1991a, 1991b; Goyal et al. 1991; Knowles et al. 1987; Lamed and Bayer 1987; Walker and Wilson 1991; Wood 1989, 1991, 1992; Wood and Garcia-Campayo 1990; Wood et al. 1988; Woodward 1991).

II. Fungal cellulases

A. Cellulases of aerobic fungi

1. Composition of the cellulase system
The distinguishing feature of the aerobic fungal cellulase that can solubilize crystalline cellulose is that it contains a cellobiohydrolase. This is in addition to the randomly-acting endoglucanases and β-glucosidases/cellobiases found in all culture filtrates of all aerobic fungi. Only a few fungi synthesise and release into the culture medium appreciable amounts of the cellobiohydrolase enzyme. A fact that appears to be forgotten is that some of the most potent cellulolytic fungi, such as *Myrothecium verrucaria*, release into solution in an active form only the endoglucanase and β-glucosidase components. Of those fungi that release the complete cellulase system into the culture solution the most notable are: *Trichoderma reesei, Trichoderma viride, Fusarium solani, Penicillium funiculosum/pinophilum, Phanerochaete chrysosporium/Sporotrichum pulverulentum* (reviewed in Eriksson and Wood 1985) and *Sclerotium rolfsii* (Lachke and Deshpande 1988).

All fungal cellulase systems studied so far have been shown to contain each of the main types of enzyme in multiple forms (reviewed by Wood 1991; Coughlan 1985; Enari and Niku Paavola 1987). The actual number of components depends on the source of the fungus and the manner in which it has been cultured. *Trichoderma viride* and *Trichoderma reesei* cellulases have been most extensively studied (reviewed by Eriksson and Wood 1985; Wood 1991; Coughlan 1985; Coughlan and Ljungdahl 1988; Goyal et al. 1991; Wood and Garcia-Campayo 1990). They have been shown to contain 4 to 8 endoglucanases, 2 cellobiohydrolases and 1 to 2 β-glucosidases. *Penicillium funiculosum/pinophilum* cellulase contains 2 cellobiohydrolases (Wood et al. 1980; Wood and McCrae 1986a), 5 to 8 endoglucanases (Bhat et al. 1989) and

2 β-glucosidases (Wood et al. 1980). Other cellulases, are equally heterogeneous (Streamer et al. 1975; Wood 1991). It seems that only some of these components are genetically determined (Teeri et al. 1990; Enari and Niku Paavola 1987); others may be artefacts resulting from differential glycosylation of a common polypeptide chain (Wood and McCrae 1972; Gum and Brown 1977), from partial proteolysis Eriksson and Pettersson 1982), from aggregation of the enzymes with each other or with part of the fungal cell wall (Sprey and Lambert 1983), or from manipulation of the enzymes during purification (Enari and Niku Paavola 1987). These artefacts make elucidation of the mechanism of action extremely difficult; consequently there is considerable discussion on the substrate specificity of the enzymes, on the mode of action of the individual enzymes, particularly the cellobiohydrolases, and on the nature of the co-operation between the various enzymes.

2. Classification of the enzymes

Classification of the enzymes found in aerobic fungal cellulase systems is based on their activity towards a wide range of substrates. β-Glucosidases are easy to classify: they hydrolyse cellobiose and some soluble cello-oligosaccharides to glucose; both nitrophenylglucoside or 4-methylumbelliferylglucoside are substrates. However, classification of the cellobiohydrolases and the endoglucanases is more difficult as the enzymes have overlapping specificities on substrates which are themselves very poorly defined. It is possible, however, to make some generalisations.

In the main, cellobiohydrolases are considered to degrade cellulose by splitting off cellobiose from the non-reducing end of the cellulose chain (Wood and McCrae 1972). Substituted celluloses (carboxymethylcellulose – CM-cellulose) are not attacked to a significant extent, but "swollen" partially degraded, amorphous cellulose and soluble cello-oligosaccharides are readily hydrolysed. Cotton fibre, which is highly hydrogen bond-ordered and has a degree of polymerisation in the order of 10,000, is hydrolysed slowly, presumably because it has relatively few chain-ends available for attack. Avicel, however, which is also highly hydrogen bond ordered, but has a degree of polymerisation of only 200 and therefore has more chain-ends available for attack, is more rapidly degraded. Some cellobiohydrolases hydrolyse synthetic substrates such as *p*-nitrophenyl-β-D-cellobioside and *p*-nitrophenyl-β-D-lactoside (van Tilbeurgh et al. 1982).

Endoglucanases, often called CM-cellulose, have little apparent capacity to hydrolyse hydrogen bond-ordered cellulose but attack CM-cellulose or acid-swollen cellulose, barley β-glucan or soluble cello-oligosaccharides at various points in the chain, resulting in a rapid fall in the degree of polymerisation. A test which is often used to measure and characterise a particular endoglucanase is to relate the increase in the specific fluidity of a solution of CM-cellulose to the unit increase in reducing power. The principal end-products of endoglucanase action are glucose and cellobiose, but cello-oligosaccharides of degree of polymerisation from 3 to 6 are generated as intermediates or, in some cases, as end-products.

Such information on the enzymes is useful for a working background, but exceptions to these generalisations are frequent and it is becoming increasingly clear that it is difficult to classify the various enzymes strictly as cellobiohydrolases and endoglucanases. For example, some purified cellobiohydrolases are reported to attack barley β-glucans and even CM-cellulose (Henrissat et al. 1985), which have long been held to be degradable only by enzymes classified as randomly-acting endoglucanases. On the other hand, some enzymes classified as endoglucanases are reported to be able to hydrolyse crystalline cellulose (Beldman et al. 1985; Enari and Niku-Paavola 1987) which, of course, is presumed to be property of the cellobiohydrolase. Yet another endoglucanase is reported to have no action on amorphous cellulose prepared by milling cellulose powder in ethanol (Niku-Paavola et al. 1985). There are other apparent anomalies too frequent to report here.

Clearly, the substrate specificities of the various enzymes continues to be a contentious issue. The possibility that the problem arises from the enzymes having overlapping substrate specificities is real, but to some extent the problem may be connected with the fact that enzymes which have been reported to be pure may be associated with traces of other components of the cellulase system that are difficult to remove by conventional chromatography. The purity of the cellobiohydrolases in particular has been questioned and, as a consequence, the mode of action of these enzymes has been the subject of great debate.

3. Distinguishing features of cellobiohydrolases type I and II

The cellobiohydrolases (CBH) have been the most extensively discussed cellulase component in the literature and opinions on the properties are particularly diverse and controversial (Wood and Garcia-Campayo 1990). Studies on the cellulases of *Trichoderma reesei* (Fägerstam and Petterson 1980) and *Penicillium pinophilum* (Wood et al. 1989; Wood and McCrae 1986a) have shown that cellobiohydrolase enzymes exist in two immunologically unrelated forms (so-called, CBH I and CBH II). CBH I is the principal enzyme component in both cellulases, constituting approximately 60% of the extracellular protein: CBH II constitutes approximately 20% (Wood 1991). However, CBH II is the principal conidial-bound cellulase in *T. reesei* (Messner et al. 1991). CBH I and CBH II produced by these fungi and by *Phanerochaete chrysosporium* (Uzcategui et al. 1991) exist in several iso-forms, but the principal CBH I and CBH II components from all three sources differ in isoelectric point, CBH I by convention being isoelectric at the more acidic pH. CBH I (active) can also be distinguished from CBH II (inactive) using 4-methylumbelliferyl-β-D-lactoside (van Tilbeurgh and Claeyssens 1985; Claeyssens 1988), and CBH I releases 4-methylumbelliferone from 4-methylumbelliferyl-β-D-cellobioside while CBH II does not (Fig. 1). With 4-methylumbelliferyl-β-D-cello-oligosaccharides of a longer degree of polymerisation there are other important differences in substrate specificity of

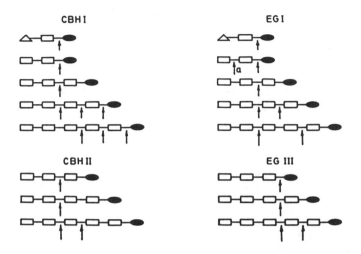

Fig. 1. Degradation pattern of 4-methylumbelliferyl-β-D-glycosides by CBH I, CBH II, EG I and EG III from *T. reesei.* △, β-galacctopyranosyl; ▢, β-glucopyranosyl; ⬬, 4-methylumbelliferyl group. Arrows show bonds hydrolysed. EG I possessed transferase activity: a product of this activity was 4-methylumbelliferylglucoside. From Claeyssens and Tomme (1989).

the cellobiohydrolases of *T. reesei* (Fig. 1). From these comparisons it would appear that CBH II shows a more strict substrate specificity, three or four β-1,4-linked glycosyl residues being required for hydrolysis: the same type of study carried out on *P. pinophilum* CBH I and CBH II has shown them to possess similar substrate specificities and modes of action (Claeyssens et al. 1989). Studies on radioactively-labelled cello-oligosaccharides have provided similar conclusions (van Tilbeurgh et al. 1982, 1985; Claeyssens et al. 1989). Equally suitable for differentiating CBH I and CBH II, but more difficult to measure, is the stereochemical course of hydrolysis: using model compounds, CBH I has been found to release β-glucose while CBH II generates α-glucose (see below for details).

The molecular masses of CBH I and CBH type II enzymes are in the range 45 to 65 kDa (Wood 1990; Uzcategui et al. 1991). Both cellobiohydrolases are glycoproteins with carbohydrate content varying between 9% and 19% (Wood 1990; Uzcategui et al. 1991). The enzymes have been shown to be both *N*-and *O*-glycosylated in *T. reesei* CBH I (Salovouri et al. 1987); the carbohydrate chains consisting mainly of mannose. CBH I of *T. reesi* consists of 496 amino acids: it has 12 disulphide bridges and no free cysteine residues (Fägerstam et al. 1984). The amino acid sequence determined by Edman degradation is in close agreement with the structure derived by cloning and gene sequencing (Shoemaker et al. 1983; Teeri et al. 1983). CBH I from *P. pinophilum* has been found to have at least one sequence of 17 amino acids identical to one in CBH I of *T. reesei* (J. van Beeuman and T.M. Wood, unpublished data). A peptide from CBH II of *P. pinophilum* has 10 out of 12 residues identical to the region

390–401 in *T. reesei* CBH II (J. van Beeuman and T.M. Wood, unpublished data).

4. Mode of action of the cellobiohydrolases

It has been held for many years that cellobiohydrolases are exoglucanases that remove cellobiose consecutively from the non-reducing end of the cellulose chain. Typically, crystalline and amorphous celluloses have been reported to be degraded to cellobiose, the rate depending on the degree of polymerisation and the crystallinity of the cellulose. Recently, however, there have been several reports which have indicated that the cellobiohydrolases may not attack exclusively the penultimate glycosidic link at the non-reducing end of the polymer chain. Unfortunately, the case is weakened by the apparent lack of agreement as to whether it is CBH I or CBH II or both that possess this property.

Thus, a CBH I from *T. reesei* attacked barley β-glucan in a random manner typical of an endoglucanase (Henrissat et al. 1985), but CBH I from *P. pinophilum* effected only a slow change in the degree of polymerisation of the β-glucan (Wood et al. 1989), as would be expected from an exo-acting enzyme. On the other hand, neither CBH I nor CBH II from *T. reesei* (Fig. 1) or *P. pinophilum* acted exclusively on the penultimate glycosidic bond at the non-reducing end of the chain of a series of 4-methylumbelliferyl cello-oligosaccharides (Claeyssens et al. 1989). Clearly, the site of action of these CBH enzymes may differ on soluble and insoluble substrates; and this has been noted on several occasions.

Support that the CBH I enzyme may not act from the end of the chain has been obtained by electron microscopy. Thus, Chanzy and Henrissat (1985) noted that CBH I from *T. reesei*, labelled with colloidal gold, was found to be attached to microcrystals of the alga, *Valonia microphysa*, along the length of the microfibril. The significance of using *Valonia* microcrystals is that the cellulose microfibril is composed of only one crystal and has only one reducing and one non-reducing end: other crystalline substrates, which contain a variety of chain lengths of the fibre, would be less suitable for such an analysis. White and Brown (1981), in a similar study, used cellulose from the bacterium *Acetobacter xylinum*: they reached similar conclusions to those reported by Chanzy and Henrissat. However, there were important differences between the two studies: whereas Chanzy et al. (1983) reported that their CBH I preparation caused some fibrillation of the cellulose microfibrils, White and Brown (1981) observed that disaggregation of the microfibril was a property of an endoglucanase that they isolated from *T. reesei* cellulase, but it was not a property of their CBH I preparation. Neither CBH I nor CBH II of *P. pinophilum* dissociated a cotton fibre substrate into short fibres (Wood and McCrae 1972, 1986a).

Conclusions as to the mode of action of CBH II are equally diverse. Thus, electron microscopic evidence showing that CBH II from *T. reesei* attacked *Valonia* cellulose microcrystals only from the non- reducing end supports the claim that it is a true exoglucanase (Chanzy and Henrissat 1985). Using biochemical studies, a similar mode of action was deduced for a CBH II from *P. pinophilum* purified by affinity chromatography (Wood et al. 1989). By contrast, Enari and Niku Paavola (1987) and Kyriacou et al. (1987), also using biochemical studies, conclude that CBH II from *T. reesei* are endo-acting, albeit "less randomly-acting" than a typical endoglucanase.

How can these conflicting results be rationalised? Clearly, enzymes from different sources, or from the same source cultured using a variety of carbon sources, may indeed have different substrate specificities and some of the variations at least may therefore be quite easily explained. However, when the same enzyme from the same source appears to have completely different properties another reason must be sought. One possibility is that the apparent differences may be a consequence of the existence of aggregates or enzyme-enzyme complexes between CBH and endoglucanases which are extremely difficult to break up into their constituent parts. Enzyme-enzyme complexes have been shown to exist in a cellulase from *T. reesei*. In this case, electrophoretically homogeneous complexes between endoglucanase, xylanase and β-glucosidase were found to be heterogeneous after treatment with a urea/octyl glucoside dissociation reagent (Sprey and Lambert 1983). Similar complexes have been found to exist between cellobiohydrolases and endoglucanases in electrophoretically homogeneous enzyme preparations isolated from cultures of *P. pinophilum* and *T. reesei* (Wood et al. 1989). These complexes, which appeared to contain only one enzyme species after rigorous use of ion exchange, isoelectric focusing, gel filtration and affinity chromatography on a column of cellulose, were found to be heterogeneous after affinity chromatography on a column that had been prepared by coupling *p*-aminobenzyl 1-thio-β-D-cellobioside to Affigel 10 (van Tilbeurgh et al. 1984). A CBH II from *P. pinophilum* purified in this way had the properties of a typical exoglucanase: the enzyme could effect only a slow decrease in the viscosity (a parameter related to chain length) of a solution of barley β-glucan in contrast to the rapid decrease shown by purified endoglucanases from the same fungus: CBH I and II from *T. reesei* prepared in the same way were similar in this respect (Wood et al. 1988).

Whether or not enzyme-enzyme complexes can explain the differing opinions on the substrate specificity, it is clear that at least some of the confusion regarding the properties of the CBH enzymes is caused by the difficulty in obtaining single enzyme species. Expression of the cellulase genes in a heterologous host makes it possible to produce each enzyme free of contaminating glycosidases. It may therefore be significant that a CBH I gene from *T. reesei* expressed in yeast showed no capacity to hydrolyse barley β-glucan (Knowles et al. 1988b), which is readily degraded by endoglucanases, while CBH II from a recombinant clone did. The implication is that CBH I from *T. reesei* may indeed be a exocellobiohydrolase and the CBH II may be have some endo-type action. The fact that the recombinant CBH II had no apparent activity to CM-cellulose, which is typical of an endoglucanase, confuses the issue however. Perhaps this observation will force a more profound evaluation of the use of non-cellulosic substrates (barley β-glucan) and model cellulosic substrates (CM-cellulose) in studies of the mode of action.

5. Properties of endoglucanases

The properties of the endoglucanases have caused less controversy than the cellobiohydrolases. Unlike the cellobiohydrolases there are only a few exceptions to the general conception of the substrate specificity and mode of action discussed above. A significant anomaly was the observation that a *T. viride* endoglucanase could degrade crystalline and amorphous cellulose but was inactive on CM-cellulose (Beldman et al. 1985). A very unusual view of the endoglucanases is that they have no action on insoluble amorphous celluloses but are produced by the fungus to hydrolyse soluble substrates only (Enari and Niku Paavola 1987). Some endoglucanases synthesise longer chain oligomers from degradation products (Claeyssens and Tomme 1989).

As already stated, most cellulase preparations contain a multiplicity of endoglucanases. Typical are the five endoglucanases (EG I to EG V) isolated from *P. pinophilum* cellulase (Bhat et al. 1989) and from the cellulase (T_1, T_{2a}, T_{2b}, T_{3a}, T_{3b}) produced by *Sporotrichum pulverulentum* (Eriksson and Petterson 1975): *Trichoderma reesei* produces three (EG I, EG II, EG III) (reviewed in Goyal et al. 1991). The five endoglucanases produced by *P. pinophilum* differed in isoelectric pH (3.7 to 7.4), molecular mass (25 to 62 kDa) and their carbohydrate content (7 to 18%); as did E I (54 kDa, pI 4.7; 4% carbohydrate), E II (called E III in some publications) (43 kDa; pI 5.5; 15% carbohydrate) and E III (20 to 24 kDa; pI 7.7) of *T. reesei*. Antiserum prepared with purified EG I of *P. pinophilum* reacted only with EG I, and EG II antiserum reacted only with EG II (Bhat et al. 1989). The implication of these results was that the enzymes were quite distinct. Enari and Niku-Paavola (1987), however, observed that the endoglucanases they found in a culture filtrate of *T. reesei* mutant strain VTT-D-80133 existed in a series of immunologically related components and they concluded that the enzymes originated in a common ancestor (Niku-Paavola et al. 1985). Hakansson et al. (1978), in contrast, working with a cellulase preparation from strain QM 9414 of the same fungus, reported the presence of two immunologically unrelateed endoglucanases. Gene cloning has indicated the presence of at least two distinct genes coding for endoglucanase (EG I and EG III) activities in *T. reesei* (Teeri et al. 1990). EG I is similar to CBH I in its action on 4-methylumbelliferylcello-oligosaccharides (Fig. 1). EG I is, however, quite different from EG III in its mode of action on the 4-methylumbelliferyl cello-oligosaccharides: EG III can be differentiated from EG I and other endoglucanases by the fact that it splits 4-methylumbelliferyl cellotrioside at the heterosidic bond (van Tilbeurgh et al. 1988).

The isolation of two different genes resolves the problem of the origin of the multiplicity in endoglucanase activity in *T. reesei* cellulase and shows conclusively that some of the multiplicity of components is genetically determined. The results obtained with the cellulase of *P. pinophilum* are not as definitive, but it has been shown that all five major endoglucanases appear in the culture in the early logarithmic growth phase and then increase in concentration with only minor alterations in the relative proportions (Bhat and Wood 1989). Differences in functional properties of the endoglucanases of *P. pinophilum* suggest that the enzymes have different roles to play in cellulolysis (Bhat et al. 1989; Wood et al. 1988). This was noted using acid-swollen cellulose and soluble cello-oligosaccharides as substrates. Thus, attack on the former by E I and E V was accompanied by a rapid fall in the degree of polymerisation, which is compatible with attack at points remote from the chain end: E III and E IV, in contrast, appear to attack near the chain ends. There was a marked difference in the length of chain needed for the various enzymes for hydrolysis. Thus, E II required at least four residues (cellotetraose) in the chain while E V and E I needed five and six respectively (Table 1). Hurst et al. (1977) have reported an endoglucanase from *Aspergillus niger* that required a chain of a minimum of five glucose residues for its substrate, but no endoglucanase requiring more than five residues has been reported hitherto.

Table 1. Hydrolysis of cello-oligosaccharides by the endoglucanases I-V of *Penicillium pinophilum.*

Substrate	I	II	III	IV	V
	Reducing sugar released (μg)				
Cellobiose	0	0	0	0	0
Cellotriose	0	0	345	257	0
Cellotetraose	0	209	313	201	0
Cellopentaose	0	241	747	884	345
Cellohexaose	178	n.d.	n.d.	n.d.	270

From Bhat et al. (1989).
n.d. = not determined.

Clearly, information on the mode of action on soluble substrates is useful for classifying and characterising endoglucanases and in providing some insight into the various active sites. However, as already stated in the discussion of the cellobiohydrolases, it is not at all certain that there will be a parallel between action on soluble substrates and on insoluble celluloses. The difficulty in deducing a mode of action from one substrate and assuming it applies on another has been shown using $[1-^3H]$-labelled, reduced and 4-methylumbelliferyl cello-oligosaccharides (Bhat et al. 1990). The preferred bond of cleavage of these substrates by the major endoglucanases of *P. pinophilum* varied according to the substrate used (Bhat et al. 1990).

6. Oxidative enzymes

Some fungi synthesize cellobiose oxidase and/or cellobiose dehydrogenase in addition to exoglucanases, endoglucanases and β-glucosidases for the hydrolysis of cellulose (Ayers et al. 1978; Eriksson et al. 1974; Eriksson 1981). As yet only the white rot fungus *Sporotrichum pulverulentum/Phanerochaete chrysosporium* has been reported to synthesize cellobiose oxidase (Ayers et al. 1978; Eriksson 1981) although there is some indirect evidence that *T. koningii* (Wood and McCrae 1978) and *T. viride* (Eriksson et al. 1974) may also produce it. Cellobiose dehydrogenase has been reported to be produced by several types of fungus, namely the white rot fungus *S. pulverulentum* (Westermark and Eriksson 1974; Eriksson 1981), the thermophilic fungus *S. thermophile*, phylogenetically belonging to the Ascomycetes (Coudray et al. 1982), a species of the imperfect fungus, *Monilia* (Dekker 1980) and the phytopathogenic fungus *Sclerotium rolfsii* (Lachke and Deshpande 1988; Sadana and Patil 1985).

The cellobiose oxidase, which is a haem protein and contains an FAD group, oxidizes cellobiose and higher cello-oligosaccharides to their corresponding onic acids using molecular oxygen. It is not known with any certainty if this enzyme also oxidizes reducing end groups in the cellulose chain. However, Morpeth (1985) was able to detect changes in the enzyme which were consistent with some enzyme action. Superoxide anion is a product of the reaction (Eriksson 1981; Morpeth 1985). It was suggested that superoxide anion might be involved in the initial attack on cellulose (Eriksson 1981). However, it is possible that the H_2O_2, which is produced by the dismutation of superoxide anion by superoxide anion dismutase, may also be effective in initiating attack on the cellulose chain (Eriksson 1981). Koenigs (1974, 1975) and Highley (1980) have suggested that initial attack on hydrogen bond-ordered cellulose by brown rot fungi involves a H_2O_2/Fe^{2+} system. Additional evidence for the involvement of oxidases in cellulolysis has been reported by Vaheri (1982a,b), who observed that culture solutions of *T. reesei* grown on cellulose contained gluconic acid and cellobionic acids.

The other oxidative enzyme acts in a different manner. The enzyme produced by *S. pulverulentum* reduces quinones and phenoxy radicals in the presence of cellobiose and is therefore of importance in the degradation of both lignin and cellulose (Westermark and Eriksson 1974): quinones are not electron acceptors for the cellobiose dehydrogenase of *Monilia* sp. or *S. rolfsi* (Dekker 1988; Sadana and Patil 1988). *S. pulverulentum* cellobiose quinone oxidoreductase has flavin as sole prosthetic group, is relatively specific for cellobiose, but is relatively unspecific for the quinone structure in that it is able to reduce both *ortho*- and *para*-quinones: the end-product of enzyme action is cellobionolactone.

Not surprisingly, some of the fungi which produce oxidative enzymes also produce lactonases. As lactones are powerful inhibitors of some of the enzymes of the cellulose-degrading enzyme systems produced by some fungi the presence of lactonases will ensure that the inhibitory effects of the lactones are removed. There is some evidence that cellobionolactone is involved in the induction of cellulases (Iyayi et al. 1989). Wood and Wood (1992) have shown evidence that cellobiose quinone oxidoreductase from *P. chrysosporium* is a breakdown product of cellobiose oxidase.

7. Synergism between enzyme components

When the observed action of two or more enzymes acting in solution together is greater than the sum of the individual actions it is concluded that the enzymes act synergistically. There are three types of synergistic action involved in the process by which crystalline cellulose is rendered soluble; that between endoglucanase and cellobiohydrolase (so-called, endo-exo synergism) (Wood and McCrae 1972, 1979), that between two cellobiohydrolases (so-called, exo-exo synergism) (Fägerstam and Pettersson 1980; Wood and McCrae 1986b; Wood et al. 1989), and that between β-glucosidase and the other two types of enzyme. Synergism between the first two types results in the solubilization of hydrogen bond-ordered cellulose; synergism of the third type is concerned with the hydrolysis (by β-glucosidase) of cellobiose which is a powerful inhibitor of CBH action. Only synergism of the third type is well understood. Unfortunately, despite intense research activity the molecular basis for synergistic action that results in the solubilization of crystalline cellulose is still not known. It is possible that the lack of agreement is a direct consequence of the wide diversity of opinion regarding the individual roles of the "purified" enzymes. However, the choice of substrate as an example of crystalline cellulose has also been responsible for some confusion as it has been shown (Henrissat et al. 1985) that the degree of synergistic activity observed between cellobiohydrolase and endoglucanase varies with substrate used. However, it appears, in general, that synergism is most marked when crystalline cellulose is the substrate, it is low or non-existent with amorphous-highly hydrated cellulose, and it is absent with soluble cellulose derivatives (Wood and McCrae 1979).

The original model for synergistic activity between enzyme components envisaged an enzyme (so-called C_1) whose sole function was to cause some relaxation in the intramolecular hydrogen bonding as a preliminary to action by the hydrolytic enzymes (Reese et al. 1950). Very few now believe that such an enzyme exists but there is no doubt that the disaggregation and subsequent hydration of the closely packed cellulose chains in the cellulose crystallite is an essential prerequisite of cleavage of the glycosidic bond by cellulase enzymes. The discovery that CBH I and CBH II consist of two domains (see below), one binding and one hydrolytic, has been interpreted to indicate that the "swelling" and hydrolytic function may reside in one enzyme. Knowles et al. (1988a,b) envisaged the binding domain causes some dissociation of the individual chains as a preliminary to hydrolysis of the cellulose by the hydrolytic domain. However, Stahlberg et al. (1991) could find no evidence that pretreatment of cellulose with the isolated binding domain resulted in improved activity of the hydrolytic domain: instead they concluded that CBH I is in fact bound simultaneously through both its domains.

If the hypothesis of Knowles is tenable this would suggest that some degree of synergism is manifested between different components of the same enzyme. Attempts to explain the synergism between two different types of enzyme, namely endoglucanase and cellobiohydrolase, have envisaged the mechanism to involve sequential action where a randomly acting endoglucanase initiates the attack in the amorphous areas of the cellulose to create non-reducing ends for the endwise-acting cellobiohydrolase (Wood and McCrae 1972). However, this model, while in essence is still true, is granted only qualified acceptance and it is generally regarded as an over-simplification. It does not account, for example, either for the fact that there is little or no synergism observed between some endoglucanases and cellobiohydrolases (Wood 1975), that synergism exists between two cellobiohydrolases (Fagerstam and Pettersson 1980), and that only one face of the cellulose I unit cell, defined as the 101 bar face by Miller indices, is preferentially attacked by the cellulases (Sagar 1985).

The discussion of the synergism between endoglucanase and cellobiohydrolase has been particularly extensive. There is no doubt that adsorption of the enzyme on the cellulose is essential for solubilisation (Coughlan 1985; Klyosov 1988, 1990). Klyosov and his colleagues (Klyosov 1988) concluded that only those endoglucanases that have a strong affinity for crystalline cellulose can act synergistically with the cellobiohydrolase: the results of Bhat et al. (1989) obtained with the endoglucanases and cellobiohydrolases of *P. pinophilum* support this hypothesis. Ryu et al. (1984) are of the opinion that endo-exo synergism can be described in terms of competitive adsorption of the two types of enzyme, optimum co-operation being apparent when the enzymes were present in the ratio in which they were present in the culture filtrate. Henrissat et al. (1985), Wood and McCrae (1986b) and Wood et al. (1989) show that the ratios of the different enzymes are a major consideration for the manifestation of maximum cooperation. Woodward et al. (1988a,b), concluded that the concentration of the mixture of CBH I and EG II (EG III) from *T. reesei* was more important than the ratio of the enzymes (Fig. 2). Optimum concentrations of CBH I and EG II (later called EG III) were needed for maximum degree of synergism.

The synergism between randomly-acting endoglucanases and endwise-acting cellobiohydrolases is logical; however, the reported synergism between two cellobiohydrolases is difficult to understand. Fägerstam and Pettersson (1980), working with CBH I and CBH II from *T. reesei*, were the first to demonstrate this type of synergism using a crystalline cellulose substrate. This unexpected finding was confirmed by Henrissat et al. (1985) and Kyriacou et al. (1987), working on the same cellulase, and by Wood and McCrae (1986b) using CBH I and II from *P. pinophilum*. Henrissat et al. (1985) envisaged that competitive adsorption, or the formation of a binary complex between CBH I and CBH II, might increase the effectiveness of the enzymes. There is now some evidence that such complexes do in fact exist in solution (Tomme et al. 1990), but no one has yet succeeded in identifying enzyme- enzyme complexes

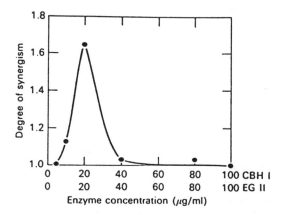

Fig. 2. Hydrolysis of Avicel by different concentrations of EG II (now called EG III) and CBH I from *T. reesei*. At each concentration the enzymes were in the ratio of 1:1. From Woodward et al. (1988b).

on the cellulose crystallite.

Synergism between two endwise-acting cellobiohydrolases is indeed difficult to explain. Wood and McCrae (1986b) have postulated that the mechanism can be discussed in terms of the stereochemistry of the cellulose chains, based on the fact that there are likely to be two different naturally-occuring configurations of non-reducing end group in the cellulose crystallite. In essence, they envisage that CBH I and CBH II to differ in their substrate stereospecificities and that the apparent cooperation can be discussed in terms of CBH I attacking only one of the two stereospecifically-different non-reducing end groups while CBH II attacks the other. Thus, synergistic action would be observed if the sequential removal of cellobiose from one type of non-reducing end by CBH I exposed, on a neighbouring chain, a non-reducing chain of different configuration which would be a substrate for CBH II, and *vice versa*. An extension of this hypothesis has been used to describe the synergistic effect between endoglucanases and CBH I and CBH II (Wood et al. 1988). Thus it was suggested that the endoglucanases, in initiating the attack on the cellulose chain, would generate different configurations of non-reducing end-group which would be substrates for the two stereospecific cellobiohydrolases (Wood et al. 1988). Clearly, if this hypothesis is acceptable, two stereospecific endoglucanases and two stereospecific CBH would be required for maximum efficiency in degrading hydrogen bond ordered cellulose.

Clearly there is a great deal of debate and uncertainty regarding the mechanism of synergistic action between the various enzymes and rationalisation of the plethora of observations is difficult. However, it is abundantly clear that there can be no agreement on the matter until there is a consensus of opinion regarding the substrate specificity and mode of action of the cellobiohydrolases. As purification techniques improve so views on the properties of the individual enzymes and on their synergistic interation may be modified. Of particular interest in this regard is the elegant technique developed by van Tilbeurgh et al. (1984) involving affinity chromatography on a column of p-aminobenzyl-1-thio-β-cellobioside. Thus, Wood et al. (1989) have shown that preparations of *P. pinophilum* CBH II isolated by conventional separation techniques, including affinity chromatography on a column of cellulose, and shown to be electrophoretically homogeneous, were in fact contaminated by trace amounts of endoglucanase. When the contaminating endoglucanase was removed by further affinity chromatography only low degrees of synergistic activity were observed between CBH I and CBH II, or between the individual cellobiohydrolases and any of the endoglucanases, using crystalline cellulose in the form of the cotton fibre as substrate. Synergistic action was only apparent, under the conditions tested, when CBH I and II and a specific endoglucanase were present in admixture. The optimum ratio of the cellobiohydrolase components was 1:1 and, significantly, the addition of a trace of endoglucanase was needed for extensive degradation of the substrate. Thus, it appears that three enzymes are required for a reasonable rate of hydrolysis of crystalline cellulose in the form of cotton fibre (Table 2). An electrophoretically homogeneous preparation of *T. reesei* CBH II could be further purified in the same way (Wood et al. 1989; Wood 1989). Cellobiohydrolase preparations from *T. reesei* purified in different ways have been shown to be able to degrade cotton fibre without the need for the addition of endoglucanase (Enari and Niku Paavola 1987): perhaps in this case one or other of the enzymes were complexed with small amounts of contaminating endoglucanase.

Table 2. Synergism between cellobiohydrolases and endoglucanases of *Penicillium pinophilum* in solubilizing cellulose (cotton fibre).

	Solubilization (%) with and without endoglucanases (E)					
CBH added	None	+EI	+EII	+EIII	+EIV	+EV
None	–	< 1	< 1	< 1	< 1	< 1
I	4	6	5	5	3	5
II	6	4	1	5	3	5
I + II	13	17	13	35	16	38
Increase in activity (%) with CBH I + II + one E	–	4	0	22	3	22

From Wood et al. (1989).

The implication of these results is that many of the contradictory statements in the literature for the synergistic activity of the components of the cellulase systems of some fungi may be the result of incomplete resolution of enzyme complexes. Clarification of this issue may result from studies involving recombinant enzymes. Unfortunately, no reports have appeared as yet to indicate that synergistic activity occurs between recombinant enzymes in solubilizing hydrogen bond-ordered cellulose. No synergism was observed between CBH I isolated from a cellulase produced by the wild-type strain and any of the four *Clostridium thermocellum* endoglucanases (A, B, C and D) purified from *E. coli* clones (Schwartz et al. 1986; Béguin et al. 1983, 1987; Pétré et al. 1986; Joliff et al. 1986) expressing the corresponding genes (P. Béguin and T.M. Wood, unpublished data). However, synergism was observed between CBH I of *T. koningii* and the crude cellulase of *C. thermocellum* (Gow and Wood 1988).

B. Cellulases of anaerobic fungi

The gut of some herbivores is a good source of cellulose-degrading fungi (Mountford 1987; Bauchop 1989). These fungi are anaerobic but the extracellular enzymes that they produce work efficiently in the presence of O_2. Relatively few anaerobic fungi have been isolated and characterized as yet: the species *Neocallimastix frontalis*, *Neocallimastix patriciarum*, *Piromonas communis* and *Sphaeromonas communis* have attracted much attention. In the main, studies on the extracellular enzymes have been confined to crude enzyme preparations (Barichievich and Calza 1990; Lowe et al. 1987; Kopecny and Williams 1988; Hebraud and Fevre 1988; Wilson and Wood 1992a; Williams and Orpin 1987), there having been only one report in which single enzyme species (β-glucosidase) has been isolated and characterized (Li and Calza 1991). This β-glucosidase had a molecular mass of 125 kDa and a pI of 7.1.

Most cellulase preparations from rumen fungi studied so far have, in reality, been shown to have significant activity only on hydrated or partially-hydrated substrates such as carboxymethylcellulose or amorphous "swollen" cellulose (Barichievich and Calza 1990; Lowe et al. 1987; Kopecny and Williams 1988; Hebraud and Fevre 1988; Pearce and Bauchop 1985). A strain of *N. frontalis* (strain RK21), isolated from the rumen of a sheep, has been shown to release an enzyme system which, in some comparisons (Wood et al. 1986; Wood et al. 1988), was more effective in hydrolysing hydrogen bond-ordered cellulose in the form of cotton fibre than either the extracellular cellulase of *T. reesei* or the cellulosome of *C. thermocellum* (Wilson and Wood 1992b) (Fig. 3). However, the reason for the efficiency displayed by the cellulase of the rumen fungus in converting cellulose to glucose is as yet poorly understood. It appears that the bulk of the activity resided in a high molecular mass cellulosome-type (Lamed et al. 1983a,b) enzyme complex (approximately 700 kDa) comprising subunits ranging from 68 to 135 kDa (Wilson and Wood 1992b). The high molecular mass complex could be separated from the low molecular mass cellulase components by affinity chromatography on the microcrystalline cellulose, Avicel (Wilson and Wood 1992a,b). The complex, which represented only a small proportion of the extracellular protein ($< 4\%$), contained endoglucanase, β-glucosidase and another enzyme which has yet to be

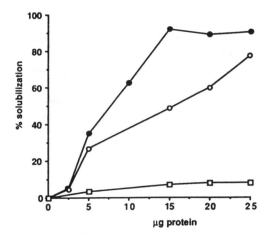

Fig. 3. Comparison of the kinetics of hydrolysis of crystalline cellulose in the form of cotton fibre by a cellulase component from *N. frontalis*, *C. thermocellum*, and *T. reesei*. ●, crystalline cellulose solubilizing component from *N. frontalis*; ○, *the cellulosome of C. thermocellum*; □, the cellulase of *T. reesei*. Incubation was carried out at optimal conditions for each cellulase: 40 °C, pH 6.0 for *N. frontalis*; 60 °C, pH 6.0 for *C. thermocellum*; 50 °C, pH 5.0 for *T. reesei*. From Wilson and Wood (1992b).

identified. The presence of the "other" enzyme was demonstrated by differential thermal inactivation of the enzyme activities and by the fact that some enzyme preparations, although rich in endoglucanase and β-glucosidase activity, were virtually devoid of activities to hydrogen bond-ordered cellulose. Some evidence was obtained that the integrity of the multicomponent enzyme complex was maintained by components of the fungal cell wall in that treatment with chitinase inactivated the activity to crystalline cellulose (Wilson and Wood, 1992a). Reymond et al. (1991), working with *N. frontalis*, found that a cDNA clone hybridized to a DNA probe encoding part of the CBH I gene of *T. reesei*, and it was implied that *N. frontalis* also produced a CBH-type enzyme.

C. Cellulases of bacteria

In comparison with the cellulases of the fungi, very little is known about the mechanisms by which bacteria degrade cellulose. To a large extent this is a consequence of the fact that many bacteria, unlike fungi, degrade the cellulosic fibre by erosion of the surface and use cell-bound enzymes. It appears possible however, that prokaryotes such as the *Actinomycetes* and the *Corynebacteria* (*Cellulomonas*) may degrade cellulose using a mechanism of action involving cell-free enzymes similar to those operating in the fungi (Béguin 1990). There may be certain situations where cell-bound enzymes will be more efficient (Yablonsky et al. 1988). These include situations where the microorganism will be exposed to predatory microorganisms, or where it is operating in an aquatic environment or in ecosystems such as that in the rumen. Cell-free enzymes, on the other hand, would be more effective in aiding the spread of mycelia through the plant cell wall by predigestion by the extracellular enzymes.

Cell-bound enzymes are more difficult to study. Some of the most efficient cellulolytic bacteria release practically no extracellular enzyme. Most bacteria that do secrete cellulases in fact appear to release only a variety of endoglucanases which show little activity to crystalline cellulose. Cell-free enzymes with activity to crystalline cellulose are found in cultures of the anaerobes, *Clostridium thermocellum* (Lamed et al. 1985; Johnson et al. 1982), *Acetivibrio cellulolyticus* (MacKenzie et al. 1987), *Clostridium cellulovorans* (Shoseyov and Doi 1990), *Bacteroides cellulosolvens* (Lamed et al. 1991) and *Clostridium stercorarium* (Creuzet et al. 1983); and the aerobes *Thermomonospora fusca* (Wilson 1988) and *Microbispora bispora* (Yablonski et al. 1988). The cellulases systems of *M. bispora, C. stercorarium* and *T. fusca* resemble fungal systems in that the various enzymes exist in solution as separate enzyme species. In contrast, the cellulases of *C. thermocellum, A. cellulolyticus, B. cellulosolvens* and *C. cellulovorans* exist in each case in multicomponenet enzyme complexes. The activity to crystalline cellulose by the cellulases of *C. thermocellum* (Johnson et al. 1982) and *A. cellulolyticus* (MacKenzie et al. 1987) appears to be dependent on the presence of Ca^{2+} ions and thiol reducing agents (Johnson et al. 1982). In the case of *C. thermocellum*, the multicomponent complex (so-called, cellulosome; Lamed et al. 1983a,b) contains endoglucanases and, according to a recent publication, it also contains a cellobiohydrolase-type enzyme (Morag et al. 1991; Mel'nik et al. 1991). The CBH was reported to have a molecular mass of 68 kDa. Ca^{2+} stabilized the activity of the cellobiohydrolase at high temperatures, but its presence was not essential for activity towards crystalline cellulose (Morag et al. 1991).

Cellobiohydrolases (or at least exoglucanases) have also been identified in a culture filtrates of *Cellulomonas fimi* (Miller et al. 1988), *Clostridium stercorarium* (Creuzet et al. 1983), *Ruminococcus flavefaciens* (Gardner et al. 1987), *Microbispora bispora* (Yablonsky et al. 1988), *Streptomyces flavogriseus* (MacKenzie et al. 1984) on the basis of their capacity to hydrolyse nitrophenyl- or 4-methylumbelliferyl cellobioside, the corresponding lactosides, amorphous cellulose or to release very small amounts of reducing sugar from Avicel. However, only two of these (*C. stercorarium; M. bispora*) have been characterized as "fungal-like" cellobiohydrolases which release cellobiose virtually exclusively from crystalline cellulose, and which act synergistically with the endoglucanases in effecting the extensive solubilization crystalline cellulose. The cellobiohydrolase of *C. thermocellum* (Morag et al. 1991) had no activity towards *p*-nitrophenyl-β-D-cellobioside or 4-methylumbelliferyl cellobioside, relatively low activity towards carboxymethylcellulose and hydrogen bond-ordered cellulose in the form of Avicel, but it was highly active on amorphous cellulose and an arabinoxylan polysaccharide: the capacity of the cellobiohydrolase to act synergistically with endoglucanase-type enzymes was not tested. The activity to the arabinoxylan is an unusual activity for an endwise-acting cellobiohydrolase. Xylans certainly have been reported to be substrates for endoglucanase-type enzymes, but arabinosyl substituents on the backbone of the polysaccharide would be expected to stop enzyme action by an endwise-acting cellobiohydrolase in much the same way that the carboxymethyl substituents on cellulose do. This enzyme clearly requires further investigation.

Of the bacterial extracellular cellulases studied, *C. thermocellum* cellulase has received most attention (Lamed and Bayer 1987; Béguin et al. 1990). The multicomponent cellulase complex (cellulosome) of *C. thermocellum* is a very stable structure comprising 14 to 18 polypeptides and has a molecular mass of 2–4 MDa (Lamed et al. 1983a,b). The cellulosomes cover the surface of protruberances (Fig. 4) which are associated with the bacterial cell at periodic intervals (Lamed and Bayer 1987, 1988; Bayer et al. 1985; Bayer and Lamed 1986; Lamed et al. 1987). The cellulosomes dissociate from the cell after a time and are to be found in clusters of different sizes (polycellulosomes – up to 100 MDa) covering the whole of the residual cellulose (Mayer 1988; Mayer et al. 1987; Coughlan et al. 1985; Bayer et al. 1985). A number of other aerobic, anaerobic, mesophilic and thermophilic cellulolytic bacteria (*Clostridium cellulovarens*, *Clostridium cellobioparum*, *Actinobacter cellulolyticus*, *Bacteroides cellulosolvens*, *Ruminococcus albus*, *Cellulomonas* sp.) have been shown to have these cellulosomes (Coughlan and Ljungdahl 1988; Béguin 1990): however, no polycellulosomal protruberances are found on non-cellulolytic bacteria.

The cellulosome of *C. thermocellum* can be fragmented into a number of subunits using SDS under different conditions of temperature and

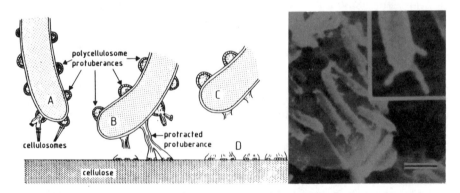

Fig. 4. Interaction of *C. thermocellum* cells with cellulose. The diagram on the left shows (A) the cell immediately prior to making contact with the cellulose; (B) the cell making contact through an extended protruberance; (C) the cell following detachment from the celluose leaving clusters of the cellulosome on the surface of the cellulose. This information is based on studies in the electron microscope. The diagram on the right shows a scanning electron micrograph of cationized ferritin-stained cells of *C. thermocellum* attached to microcrystalline cellulose. Bar 1.0 μm. From Lamed and Bayer (1988).

concentration (Lamed and Bayer 1988). One subunit (S1) of molecular mass 210 kDa is of special interest (Lamed et al. 1983a,b). It is highly antigenic, it contains about 40% carbohydrate, it is non-cellulolytic and it is relatively easily removed from the rest of the cellulosome by low concentrations of SDS (Lamed and Bayer 1988). A cellulose-binding role has been considered for S1. With some concentrations of SDS the intact cellulosome dissociates readily into a number of subunits at 25 °C, some of which show endoglucanase activity after renaturation, but activity to crystalline cellulose is lost completely even when the dissociating reagents are removed and a multicomponent cellulase complex is reformed (Lamed and Bayer 1988). Surprisingly, much more drastic conditions (a mixture of SDS, EDTA and DTT at 60 °C) were used by Wu et al. (1988) to effect the dissociation into a number of components, two of which had M_r values 82 kDa (designated S_S) and 250 kDa (S_L), respectively. Only S_S showed CM-cellulase activity, but neither could degrade crystalline cellulose in the form of Avicel when acting alone. However, when S_S and S_L were recombined some of the activity was recovered: the specific activity of the combination of S_S and S_L was less than 1% of that of the crude enzyme.

Bhat and Wood (1992), using a mixture of SDS, EDTA and DTT under carefully controlled conditions of pH and at low temperature (25 °C), succeeded in effecting the dissociation of the cellulosome into peptides of molecular masses ranging from 45 to 120 kDa: dialysis under carefully controlled conditions resulted in the reassociation of the subunits into a complex which could effectively solubilize crystalline cellulose in the form of cotton fibre. However, surprisingly, when Ca^{2+} ions (which have been shown to be necessary for activity of the undissociated cellulosome to crystalline cellulose) were incorporated into the dialysis buffer the reassociated complex showed poor activity to crystalline cellulose (K.M. Bhat and T.M. Wood, unpublished data). Presumably, Ca^{2+} bind to certain of the dissociated subunits, altering their conformation and preventing their reassociation into a complex with the correct tertiary structure for attacking crystalline cellulose. Thus the orientation and sequence of the subunits are important for activity to the hydrogen bond-ordered cellulose.

Clearly, SDS is useful for the dissociation of the cellulosome. Other dissociating agents have been tested with varying degrees of success (Lamed and Bayer 1987). The cellobiohydrolase of *C. thermocellum* was isolated from a proteinase K digest of the cellulosome (Morag et al. 1991).

Another component appears to have a role to play in binding the cellulosome and cellulose (Ljungdahl et al. 1988). The authors speculate that a yellow affinity substance (YAS) may be a factor used by the bacterial cell to recognize cellulose. The structure of the YAS is not known precisely, but there is some evidence that it may be a carotinoid (Ljungdahl 1989).

It is obvious that the mechanism by which the cellulosome hydrolyses cellulose is not well understood, but it may be that the same types of enzyme are involved in bacterial and fungal cellulolysis. On the basis of evidence obtained in the electron microscope and other evidence, Lamed and Bayer (1987) and

Mayer et al. (1987) have proposed a model in which the subunits of the cellulosome attack the cellulose chains simultaneously at regular intervals along the cellulose chain to release cello-oligosaccharides which are several cellobiose units in length. Subsequently the cello-oligosaccharides are degraded to cellobiose. An endwise-acting cellobiohydrolase could fit into this model as it acts processively in degrading cellulose: this was concluded when it was found that hydrolysis by *C. thermocellum* cellulase, unlike aerobic fungal cellulases, does not result in a significant reduction in average molecular weight during hydrolysis (Puls and Wood 1988). Thus one cellulose chain at a time would appear to be processed. An extended model of the original envisages that the catalytic subunits are held in position by a 'scaffolding' protein (Fig. 5), possibly the S_L component, which would be involved in binding the protein to the cellulose (Béguin 1990).

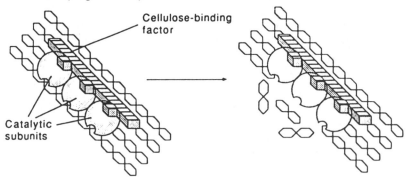

Fig. 5. Hypothetical model for the hydrolysis of cellulose by the cellulosome of *C. thermocellum*. The model is based on the assumption that attachment of the various catalytic subunits to cellulose is mediated by a cellulose binding factor. Hydrolysis of the cellulose chain at regular intervals results in the release of cellobiose and short chain cello-oligosaccharides. From Béguin et al. (1990).

III. Structure/functional relationships in fungal and bacterial cellulases

All cellulolytic microorganisms of fungal or bacterial origin which have been studied so far contain multiple genes for cellulose degradation. A remarkable fifteen different endoglucanase genes, two xylanase genes and two β-glucosidase genes of *C. thermocellum* have been cloned in *E. coli* (Hazlewood et al. 1988; Gräbnitz and Staudenbauer 1988) and several have been sequenced [celA, celB, celC, celD, celE, celF, celH, xynZ and blgB– (Béguin 1990). The products of these have been purified and one (endoglucanase D) has been crystallised (Joliff et al. 1986). Other bacteria contain a similar multiplicity in the genes. Thus, up to 10 have been cloned from *Ruminococcus albus* (Howard and White 1988) and 6 in *Bacteroides (Fibrobacter) succinogenes* (Crosby et al. 1984). Other examples of multiple genes in bacteria have been found in *M. bispora* (Yablonsky et al. 1988, 1989), *Erwinia chrysanthemi* (Chippaux 1988),

Pseudomonas fluorescens var. *cellulosa* (Gilbert et al. 1987), *T. fusca* (Hu and Wilson 1988) and *C. fimi* (Miller et al. 1988). Of the cellulolytic fungi, the molecular genetics of *T. reesei* have been studied in most detail and the genes encoding two cellobiohydrolases [CBH I (Shoemaker et al. 1983; Teeri et al. 1983), CBH II (Teeri et al. 1987b; Chen et al. 1987)] two endoglucanases [EG I (Penttilä et al. 1986; van Arsdell et al. 1987) and EG III (Saloheima et al. 1988)] and the corresponding cDNAs have now been cloned, sequenced and the primary structure of the enzymes deduced. Sequence analaysis of EG I and EG III indicated that there was no homology between the enzymes, but EG III shows remarkable homology with an endoglucanase from the fungus *Schizophylum commune* (Saloheimo et al. 1988). Interestingly, EG I is about 45% homologous with CBH I.

A comparison of primary protein sequences derived from the nucleotide sequences has shown that cellulase components from different sources have a common design in both fungi and bacteria. Thus, all cellulase components from these widely different sources appear to be composed of two separate functional domains which are conserved to different degrees and integrated into the proteins in different orders. Each enzyme appears to consist of a non-conserved catalytic core protein which is linked by a flexible hinge region, usually rich in proline, threonine and, in some cases serine, to a highly conserved tail region which is situated at either the C- or N-terminal end of the molecule (Fig. 6).

The domain structure has now been characterized to different degrees in the cellobiohydrolases of *T. reesei* (Knowles et al. 1988b; Teeri et al. 1987b) and *Sporotrichum pulverulentum* (Johansson et al. 1989), and there is some evidence for it in *P. pinophilum* CBH I (Claeyssens and Tomme 1989). Among the bacteria, it has been demonstrated in the enzymes from *C. fimi* (Warren et al. 1986; Gilkes et al. 1988, 1991a,b), *M. bispora* (Yablonski et al. 1988), *Bacteroides succinogenes* (McGavin and Forsberg 1989), *C. thermocellum* (Hall et al. 1988; Durrant et al. 1991), *Clostridium cellulolyticum* (Faure et al. 1989), *Thermomonospora fusca* (Changas and Wilson 1988), *Clostridium cellulovorans* (Foong et al. 1991) and *Pseudomonas fluorescens* (Gilbert et al. 1990).

Studies of the structures of *T. reesei* cellulase components derived from gene sequencing have been particularly thorough (Shoemaker et al. 1983; Teeri et al. 1983; Pentillä et al. 1986; Chen et al. 1987; van Arsdell et al. 1987; Teeri et al. 1987a,b; Saloheimo et al. 1988). Thus, it has been shown that CBH I, CBH II, EG I and EG III contain highly conserved regions (designated A, B in Fig. 6) which are either at the C-terminal (CBH I, EG I) or the N-terminal (CBH II, EG III) end of the enzymes. Block A, which comprises approximately 30 amino acid residues is the better conserved in all four components (70% homology). This block is rich in glycine and cysteine and is stabilized by 2 to 3 disulphide bridges (Bhikhabai and Pettersson 1984): it exists as a small domain which is separate from the catalytic domain. The domain structure was later supported by structural studies involving small angle X-ray scattering: CBH I

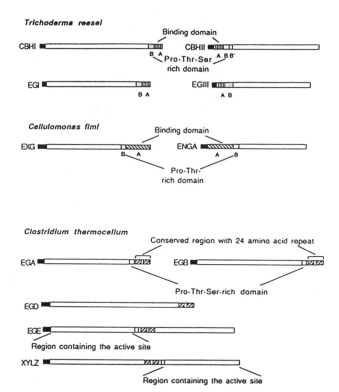

Fig. 6. Diagram showing the position of conserved domains at various positions in the sequences of several cellulases.

Abbreviations: CBH I, CBH II, EG I, EG III – cellobiohydrolases I and II and endoglucanases I and III, respectively, of *T. reesei*; EXG (also termed Cex in the text), ENGA (also termed CenA in the text) – exoglucanase and endoglucanase A, respectively of *C. fimi*; EGA, EGB, EGD, EGE – *C. thermocellum* endoglucanases A, B, D, and E, respectively; XYLZ – *C. thermocellum* xylanase Z. The black zones indicate the signal peptides. From Béguin et al. (1988).

(Schmuck et al. 1986; Abuja et al. 1988a) and CBH II (Abuja et al. 1988b) were shown to consist of a large ellipsoid head and a long tail reminiscent of a tadpole (Fig. 7). The small domain (block A) is joined to the catalytic "core" by region B. This region, which is rich in proline, serine and threonine, is heavily *O*-glycosylated (Fägerstam et al. 1984; Bhikhabhai and Pettersson 1984; van Tilbeurgh et al. 1986; Tomme et al. 1988). Glycosylation may protect against proteolytic attack (Claeyssens and Tomme 1989). Homology in region B is 50 to 60%. This region is assumed to function as a flexible hinge linking the small terminal domain A to the larger "core" domain. CBH II has a double B region (Abuja et al. 1988b; Esterbauer et al. 1991). An endoglucanase (CenA) from *C. fimi*, and expressed in *E. coli*, has a gross structural and functional organization similar to that of CBH I and CBH II of *T. reesei* but the model of the enzyme structure incorporates a constrained angle of 135° between the long

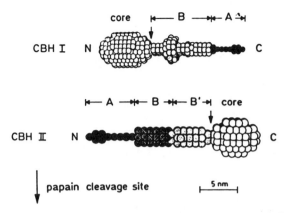

Fig. 7. Domain structures of the CBH I and CBH II from *T. reesei* as deduced from studies involving small-angle-X-ray scattering. From Abuja et al. (1988a,b).

axes of the core and tail regions (Pilz et al. 1990). The binding domain (so-called PT box-cellulose binding domain: Miller et al. 1988) of CenA (approximately 100 residues in length) could be excised precisely by an extracellular *C. fimi* protease (Gilkes et al. 1988). The binding domain is situated at the N-terminus of CenA but a *C. fimi* exoglucanase expressed in *E. coli* (Gilkes et al. 1991a,b) has the binding domain at the C-terminus. Regions involved in cellulose binding, but which are not concerned with catalysis, have been shown to exist at the *C*-terminal region of endoglucanases E and F of *C. thermocellum* (Durrant et al. 1991) and endoglucanase Z of *E. chrysanthemi* (Navarro et al. 1991).

Studies involving partial proteolysis have provided evidence of the possible functions of the domains of *T. reesei* CBH I and CBH II. Thus, the conserved region appears to be involved in substrate binding and the "core" in catalysis. This was readily demonstrated when it was found that removal of blocks A and B from both enzymes by limited proteolysis had a dramatic effect on the capacity of the "core" proteins of both CBH I and II to degrade microcrystalline cellulose (Tomme et al. 1988) (Table 3). Indeed, the specific activity of the CBH I "core" was only 10% of that shown by the intact enzyme: CBH II "core" retained 40% of the original activity. However, the "cores" retained much of the capacity (93% in the case of CBH I "core") of the intact enzyme to bind to amorphous cellulose: all the activity to small soluble cello-oligosaccharides was retained. From these data, it was concluded that the tails are involved in binding the intact enzymes to the substrate. Chemical modification of the binding domain showed that tyrosine is directly or indirectly involved in binding CBH I to cellulose (Claeyssens and Tomme 1990). The binding domain of CBH I contained two disulphide bridges: that of CBH II contained three (Johansson et al. 1989).

The core protein released from *T. reesei* CBH I was 56 kDa, the smaller domain 10 kDa (van Tilbeurgh et al. 1986): the large and small domains in CBH

Table 3. Residual activities and adsorption capacities of catalytic domains of the cellobiohydrolases of *Trichoderma reesei*.

	Soluble substrate	Insoluble substrate	
	cello-oligosaccharides	amorphous	microcrystalline
CBH I core	100	98(93)	10(34)
CBH II core	100	70(50)	40(40)

Values of adsorption (in parenthesis) and specific activity are quoted as a percentage of those obtained for the whole intact enzyme (i.e. catalytic "core" and the binding domain: see Fig. 7). The amorphous cellulose used was that "swollen" in phosphoric acid: the microcrystalline cellulose was Avicel. From Claeyssens and Tomme (1988).

II were 58 kDa and 13 kDa, respectively. Monoclonal antibodies raised against the respective cores of CBH I and II were highly specific (Mischak et al. 1989); no cross reactivities were observed.

Further studies carried out on the binding domain of *T. reesei* CBH I have established that the three-dimensional structure of the binding peptide must be retained intact for biological function (Johansson et al. 1989). Using NMR spectroscopy, it has been shown that the binding domain has the shape of a wedge with overall dimensions $3 \times 1.8 \times 1$ nm (Teeri et al. 1990). One surface is flat and hydrophobic (Kraulis et al. 1989), the other is flat and hydrophilic. It has been postulated that this structure could interact with crystalline cellulose to effectively hydrate the cellulose chain as a preliminary to hydrolysis by the core enzyme (Teeri et al. 1990).

Attention is now being directed at the "core" protein. Crystals of both CBH I and II "cores" have been obtained (Bergfors et al. 1989), but only the crystal structure of CBH II "core" protein has been elucidated (Rouvinen et al. 1989, 1990). The "core" is approximately 5 nm in diameter and consists of a seven-stranded, singly wound α/β-barrel. It seems that substrate binding occurs in a large channel which is formed by extended loops from the barrel. The active site of the enzyme has been located at the COOH end of the β-barrel in an enclosed tunnel using an inhibitor diffused in the crystal (Teeri et al. 1990; Rouvinen et al. 1990). It has been suggested that two aspartic acid residues which are located at the centre of the tunnel may be involved in catalysis (Rouvinen et al. 1990). However, using chemical modifications, Tomme and Claeyssens (1989) have identified Glu-126 to be a catalytically important carboxyl residue in CBH I of *T. reesei*: however, site-specific mutagenesis has shown that modification of Glu-126 did not completely inactivate the enzyme (Mituishi et al. 1990). The Glu-126 was reported to be located in a hydrophobic region between two large domains and to be equivalent to Glu-35 in egg white lysozyme (Tomme and Claeyssens 1989). Glu-127 was proposed as a potential active site in EG I (Claeyssens and Tomme 1989). A suggestion, based on limited homologies of the primary structures of the cellulases to the active sites of different lyzozymes (Knowles et al. 1987), that Glu-65 and Asp-74 were involved in the active site was not supported by chemical modification studies.

Recent work on endoglucanase EGD, prepared from *E. coli* expressing recombinant DNA of *C. thermocellum*, is as definitive as the work just described on *T. reesei* cellobiohydrolase enzymes. The three-dimensional structure of the crystallized EGD comprises three domains, (1) an N-terminal domain of 100 residues consisting of β-pleated sheet which is not involved in catalysis, (2) a domain of about 450 residues which consists of 12 α-helices, and (3) a region of about 65 residues, consisting of two homologous segments of 23 residues each separated by 9 to 15 amino acids, which is involved in anchoring the various catalytic components to the high molecular mass multi-enzyme complex. Deletion experiments were used to show that the conserved duplicated segments are not involved in catalysis or substrate binding (Chauvaux et al. 1990). Evidence for the conserved duplicated segments has been found in xylanase Z and endoglucanases EGA, EGB, EGE, EGH as well as EGD (Béguin et al. 1990; Navarro et al. 1991). It is considered significant that this particular "anchoring" segment is not present in endoglucanase Z of *C. stercorarium* which produces a cellulase that is not organized into a high molecular mass complex (Navarro et al. 1991).

The catalytic site of EGD of *C. thermocellum* appears to lie in a cleft formed on the surface of molecule by three of the loops which connect the α-helices (Béguin 1990). A combination of chemical modification and/or site specific mutagenesis on EGD has identified His-516 and Glu-555 as being concerned with catalysis (Tomme et al. 1991; Chauvaux et al. 1992).

Thus, some of the amino acids important in catalysis in bacterial and fungal cellulolysis have clearly been identified by chemical modification and by site-specific mutagenesis. Henrissat et al. (1989) have used another method, hydrophobic cluster analysis, to try to predict active site residues in a variety of cellulases. By this means the catalytic "core" domains of many cellulases have been classified into six families, designated A to F (Gaboriand et al. 1987; Henrissat et al. 1989; Béguin 1990, Henrissat and Mornon 1990). Each family has been divided into various sub-families. Family A is the largest group and contains cellulases from Gram-positive and Gram-negative aerobic and anaerobic microorganisms. *T. reesei* EG III is a member of this family, as are *C. thermocellum* endoglucanases C and E. *T. reesei*, CBH I from *Phanerochaete chrysosporium* and EG I from *T. reesei*. Apart from their importance in establishing structure/activity relationships among the cellulases, these classifications are interesting in that they suggest that there has been some exchange of DNA sequences during evolution and that this exchange has been very wide indeed (Béguin 1990). A comparison of the homologies of the catalytic and binding domains of cellulases from different microorganisms show that the catalytic "core" domain and the binding domain originated in different ancestors (Henrissat and Mornon, 1990).

IV. Stereochemical course of hydrolysis by cellulases

According to Reese et al. (1967) exoglucanases act by inversion of configuration. The question as to whether cellulases act by retention or inversion of anomeric configuration has been approached from time to time as this has an important bearing on our understanding and interpretation of the mode of action of the cellobiohydrolases in particular. It now seems that this problem has been resolved. NMR spectroscopy of the hydrolysis of β-D-cellobiosylfluroide (Knowles et al. 1988a) suggests that the two cellobiohydrolases have different mechanisms: CBH I acts by retention of configuration while CBH II inverts. Similar conclusions have been reported by Claeyssens et al. (1990): CBH I of *T. reesei* liberated β-cellobiose from β-methyl cellobiose while CBH II, in contrast, liberated α-cellobiose from cello-oligosaccharides. EG I from *T. reesei*, which is structurally similar to CBH I (see above), also acts by retention of configuration. Thus, while CBH I and EG I have a hydrolytic double inversion mechanism similar to that operating in lysozyme, CBH II may utilize a β-amylase type mechanism involving single displacement (Claeyssens and Tomme 1989). Exoglucanase Cex from *C. fimi* acted by retention of anomeric configuration while endoglucanase CenA from the same bacterium inverted the configuration (Wilters et al. 1986; Meinke et al. 1991).

V. Conclusions

Clearly, remarkable advances have been made in the last few years in the understanding of the cellulases and the study is in the middle of a very exciting phase. Further rapid advances in our understanding of the individual enzymes are in prospect now that crystallisation of several enzymes has been possible. The availability of cellulase components from recombinant clones promises to provide valuable information on the properties of the individual enzymes and may open the way for more definitive work on the mechanism of synergistic action between the components that results in the solubilization of hydrogen bond-ordered cellulose.

Acknowledgements

The authors thank the Scottish Office Agriculture and Fisheries Department and the Commission of the European Communities (contracts JOUB. 0042-UK; MA1D-0014-UK; MA2B-CT91-0023) for funding.

References

Abuja PM, Pilz I, Claeyssens M and Tomme P (1988a) Domain structure of cellobiohydrolase II as studied by small X-ray scattering: close resemblance to cellobiohydrolase I. Biochem. Biophys. Res. Commun. 156: 180–185

Abuja PM, Schmuck M, Pilz I, Tomme P, Claeyssens M et al. (1988b) Structural and functional domains of cellobiohydrolase I from *Trichoderma reesei*. A small angle X-ray scattering study of the intact enzyme and its core. Eur. Biophys. J. 15: 339–342

Ayers AR, Ayers SB and Eriksson KE (1978) Cellobiose oxidase: purification and partial characterization of a haemoprotein from *Sporotrichum pulverulentum*. Eur. J. Biochem. 90: 171–181

Barichievich EM and Calza RE (1990) Supernatant protein and cellulase activities of the anaerobic ruminal fungus *Neocallimastix frontalis* EB 188. Appl. Environ. Microbiol. 56: 43–48

Bauchop T (1989) Biology of gut anaerobic fungi. BioSystems. 23: 53–64

Bayer EA and Lamed R (1986) Ultrastructure of the cell surface cellulosome of *Clostridium thermocellum* and its interaction with cellulose. J. Bacteriol. 167: 828–836

Bayer EA, Setter E and Lamed R (1985) Organization and distribution of the cellulosome in *Clostridium thermocellum*. J. Bacteriol. 163: 552–559

Béguin P (1990) Molecular biology of cellulose degradation. Ann. Rev. Microbiol. 44: 219–248

Béguin P, Cornet P and Millet J (1983) Identification of the endoglucanase encoded by the *celB* gene of *Clostridium thermocellum*. Biochimie 65: 495–500

Béguin P, Millet J and Aubert JP (1987) The cloned *cel* (cellulose degradation) genes of *Clostridium thermocellum* and their products. Microbiol. Sciences. 4: 277–280

Béguin P, Grepinet O, Millet J and Aubert JP (1988) Recent aspects in the biochemistry and genetics of cellulose degradation. In: G Durand, G Bobichon and L Florent (eds) Proc. 8th International Biotechnology Symposium held in Paris, Vol 2 (pp 1015–1029)

Béguin P, Millet J, Chauvaux S, Mishra S, Tokatlidis K and Aubert JP (1990) *Clostridium thermocellum* as a model system for anaerobic cellulose degradation. In: H Heslot, J Davies, J Florent, L Bobichon, G Durand and L Penasse (eds) Proc. 6th International Symposium on Genetics of Industrial Microorganisms held in Strasbourg. Vol 2 (pp 947–958)

Béguin P, Millet J, Chauvaux S, Salamitou S, Tokatlidis K, Navas J, Fujino T, Lemaire M, Raynaud O, Daniel MK and Aubert JP (1992) Bacterial cellulases. Biochem. Soc. Trans. 20: 42–46

Beldman G, Searle-Van Leewen MF, Rombouts FR and Voragen FGJ (1985) The cellulase of *Trichoderma viride*. Purification, characterization and comparison of all detectable endoglucanases, exoglucanases and β-glucosidases. Eur. J. Biochem. 146: 301–308

Bergfors T, Rouvinen J, Lethovaara P, Caldentey X, Tomme P, Claeyssens M, Pettersson G, Teeri T, Knowles J and Jones TA (1989) Crystallization of the core protein of cellobiohydrolase II from *Trichoderma reesei*. J. Molec. Biol. 209: 167–169

Bhat KM and Wood TM (1989) Multiple forms of endo-1,4-β-D-glucanase in the extracellular cellulase of *Penicillium pinophilum*. Biotechnol. Bioeng. 33: 1242–1248

Bhat KM and Wood TM (1992) The cellulase of the anaerobic bacterium *Clostridium thermocellum*: isolation, dissociation and reassociation of the cellulosome. Carbohydr. Res. 227: 293–300

Bhat KM, McCrae SI and Wood TM (1989) The endo-1,4-β-D-glucanase system of *Penicillium pinophilum* cellulase: isolation, purification, and characterization of five major endoglucanase components. Carbohyd. Res. 190: 279–297

Bhat KM, Hay AJ, Claeyssens M and Wood TM (1990) Study of the mode of action and site-specificity of the endo-(1→4)-β-D-glucanases of the fungus *Penicillium pinophilum* with normal, 1-^{3}H-labelled, reduced and chromogenic cello-oligosaccharides. Biochem. J. 266: 371–378

Bhikhabhai R and Pettersson G (1984) The disulphide bridges in a cellobiohydrolase and an endoglucanase from *Trichoderma reesei*. Biochem. J. 222: 729–736

Changas GS and Wilson DB (1988) Cloning of the *Thermomonospora fusca* endoglucanase E2 gene in *Streptomyces lividans*: affinity purification and functional domains of the cloned gene product. Appl. Environ. Microbiol. 54: 2521–2526

Chanzy H and Henrissat B (1985) Unidirectional degradation of *Valonia* cellulose microcrystals subjected to cellulase action. FEBS Lett. 184: 285–288

Chanzy H, Henrissat B, Vuong R and Schülein M (1983) The action of 1,4-β-D-glucan cellobiohydrolase on *Valonia* cellulose microcrystals. An electron microscope study. FEBS Lett. 153: 113–118

Chauvaux S, Béguin P, Aubert JP, Bhat KM, Gow LA, Wood TM and Bairoch A (1990) Calcium-binding affinity and calcium-enhanced activity of *Clostridium thermocellum* endoglucanase D. Biochem. J. 265: 261–265

Chauvaux S, Béguin P and Aubert JP (1992) Site-directed mutagenesis of essential carboxylic residues in *Clostridium thermocellum* endoglucanase CelD. J. Biol. Chem. 267: 4472–4478

Chen CM, Gritzali M and Stafford DW (1987) Nucleotide sequence and deduced primary structure of cellobiohydrolase II from *Trichoderma reesei*. Bio/Technol. 5: 274–278

Chippaux M (1988) Genetics of cellulase in *Erwinia chrysanthemi*. In: JP Aubert, P Béguin and J Millet (eds) FEMS Symposium No. 43, Biochemistry and Genetics of Cellulose Degradation (pp 219–234). Academic Press, London

Claeyssens M (1988) The use of chromophoric substrates and specific assays in the study of structure-activity relationships of cellulolytic enzymes. In: JP Aubert, P Béguin and J Millet (eds) FEMS Symposium No. 43, Biochemistry and Genetics of Cellulose Degradation (pp 393–398). Academic Press, London

Claeyssens M and Tomme P (1989) Structure-activity relationships in cellulolytic enzymes. In: MP Coughlan (ed) Enzyme Systems for Lignocellulose Degradation (pp 37–49). Elsevier Applied Science, London

Claeyssens M and Tomme P (1990) Structure-function relationships of cellulolytic proteins from *Trichoderma reesei*. In: CP Kubicek, DE Eveleigh, H Esterbauer, W Steiner and EM Kubicek-Pranz (eds) *Trichoderma reesei* Cellulases: Biochemistry, Genetics, Physiology and Applications (pp 1–11). Royal Chemical Society, London

Claeyssens M, Van Tilbeurgh H, Tomme P, Wood TM and McCrae SI (1989) Comparison of the specificities of the cellobiohydrolases isolated from *Penicillium pinophilum* and *Trichoderma reesei*. Biochem. J. 261: 819–825

Claeyssens M, Tomme P, Boewer CF and Hehre EJ (1990) Stereochemical course of hydrolysis and hydration reactions catalysed by cellobiohydrolases I and II from *Trichoderma reesei*. FEBS Lett. 263: 89–92

Coudray MR, Canevascini G and Meier H (1982) Characterization of a cellobiose dehydrogenase in the cellulolytic fungus *Sporotrichum (Chrysosporium) thermophile*. Biochem. J. 203: 277–284

Coughlan MP (1985) The properties of fungal and bacterial cellulases with comment on their production and application. Biotechnol. Genet. Eng. Revs. 3: 39–109

Coughlan MP and Ljungdahl LG (1988) Comparative biochemistry of fungal and bacterial cellulolytic enzyme systems. In: JP Aubert, P Béguin and J Millet (eds) FEMS Symposium No. 43, Biochemistry and Genetics of Cellulose Degradation (pp 11–30). Academic Press, London

Coughlan MP, Hon-Nami K, Hon-Nami H, Ljungdahl LG, Paulin JJ and Rigsby WE (1985) The cellulase complex of *Clostridium thermocellum* is a very large. Biochem. Biophys. Res. Comm. 130: 904–909

Creuzet N, Berenger JF and Frixon C (1983) Characterization of exoglucanase and synergistic hydrolysis of cellulose in *Clostridium stercorarium*. FEMS Microbiol. Lett. 20: 347–350

Crosby B, Collier B, Thomas DY, Teather RM and Erfle JD (1984) Cloning and expression in *Escherichia coli* of cellulase genes from *Bacteroides succinogenes*. In: S Hasnain (ed) Proc. 5th Canadian Bioenergy R & D Seminar (pp 573–576). Elsevier Applied Science Publishers Ltd, Barking, England

Dekker RFH (1980) Induction and characterization of a cellobiose dehydrogenase produced by a species of *Monilia*. J. Gen. Microbiol. 120: 309–316

Dekker RFH (1988) Cellobiose dehydrogenase produced by *Monilia* sp. In: WA Wood and

ST Kellog (eds) Methods in Enzymology, Vol 160 (pp 454–463). Academic Press, New York

Durrant AJ, Hall J, Hazlewood GP and Gilbert HJ (1991) The non-catalytic C-terminal domain of endoglucanase E from *Clostridium thermocellum* contains a cellulose-binding domain. Biochem. J. 273: 289–293

Enari TM and Niku-Paavola ML (1987) Enzymatic hydrolysis of cellulose: is the current theory of the mechanisms of hydrolysis valid? CRC Crit. Rev. Biotechnol. 5: 67–87

Eriksson KE (1981) Cellulases of fungi. In: A Hollaender (ed) Trends in the Biology of Fermentations (pp 19–31). Plenum Press, New York

Eriksson KE and Pettersson (1975) Extracellular enzyme system utilized by the fungus *Sporotrichum pulverentum (Chrysosporium lignorum)* for the breakdown of cellulose. Separation, purification and physico-chemical characterization of five endo-1,4-β-glucanases. Eur. J. Biochem. 51: 193–206

Eriksson KE and Pettersson B (1982) Purification and partial characterization of two acidic proteases from the white rot fungus *Sporotrichum pulverulentum*. Eur. J. Biochem. 124: 635–642

Eriksson KE and Wood TM (1985) Biodegradation of cellulose. In: T Higuchi (ed) Biosynthesis and Biodegradation of Wood Components (pp 469–504). Academic Press, London

Eriksson KE, Pettersson B and Westermark U (1974) Oxidation: an important enzyme reaction in fungal degradation of cellulose. FEBS Lett. 49: 282–284

Esterbauer H, Hayn M, Abuja PM and Claeyssens M (1991) In: GF Leatham and ME Himmel (eds) Enzymes in Biomass Conversion (pp 301–312). Amer. Chem. Soc., Washington, DC

Eveleigh DE (1987) Cellulase: a perspective. In: BS Hartley, PMA Broda and PJ Senior (eds) Technology in the 1990s: Utilization of Lignocellulosic Wastes. Phil. Trans. R. Soc. Lon. 321: 435–447

Fägerstam LG and Pettersson LG (1980) The 1,4-β-glucan cellobiohydrolases of *Trichoderma reesei* QM9414. FEBS Lett. 119: 97–101

Fägerstam LG, Pettersson LG and Engström JA (1984) The primary structure of a 1,4-β-glucan cellobiohydrolase from the fungus *Trichoderma reesei* QM9414. FEBS Lett. 167: 309–315

Faure E, Belaich A, Bagnara C, Gaudin C and Belaich JP (1989) Sequence analysis of the *Clostridium cellulolyticum* celCCA endoglucanase gene. Gene 65: 51–58

Foong F, Hamamoto T, Shoseyov O and Doi RH (1991) Nucleotide sequence and characteristics of endoglucanase gene *engB* from *Clostridium cellulovorans*. J. Gen. Microbiol. 137: 1729–1736

Gardner RM, Doewer KC and White BA (1987) Purification and characterization of an exo-β-1,4-glucanase from *Ruminococcus flavefaciens* FD-1. J. Bacteriol. 169: 4581–4588

Gaboriaud C, Bissery V, Benchetrit T and Mornon JP (1987) Hydrophobic cluster analysis: an efficient new way to compare and analyse amino acid sequences. FEBS Lett. 224: 149–154

Gilbert HJ, Jenkins G, Sullivan DA and Hall J (1987) Evidence for multiple carboxymethylcellulase genes in *Pseudomonas fluorescens* subsp. *cellulosa*. Molec. Gen. Genet. 210: 551–556

Gilbert HJ, Hall J, Hazlewood GP and Ferreira LMA (1990) The N-terminal region of an endoglucanase from *Pseudomonas fluorescens* subspecies *cellulosa* constitutes a cellulose-binding domain that is distinct from the catalytic centre. Molec. Microbiol. 4: 759–767

Gilkes NR, Henrissat B, Kilburn DG, Miller RC and Warren RAJ (1991a) Domains in microbial β-1,4-glycanases: sequence conservation, function and enzyme families. Microbiol. Revs. 55: 303–315

Gilkes NR, Kilburn DG, Miller RC and Warren RAJ (1991b) Bacterial cellulases. Bioresource Technol. 36: 21–35

Gilkes NR, Warren RAJ, Miller RC Jr and Kilburn DG (1988) Precise excision of the cellulose binding domains from two *Cellulomonas fimi* cellulases by a homologous protease and the effect on catalysis. J. Biol. Chem. 263: 10401–10407

Goyal A, Ghosh B and Eveleigh D (1991) Characterization of fungal cellulases. Bioresource Technol. 36: 37–50

Gow LA and Wood TM (1988) Breakdown of crystalline cellulose by synergistic action between cellulase components from *Clostridium thermocellum* and *Trichoderma koningii*. FEMS

Microbiol. Lett. 50: 247–252

Gräbnitz F and Staudenbauer WL (1988) Characterization of two β-glucosidase genes from *Clostridium thermocellum*. Biotechnol. Lett. 10: 73–78

Gum EK and Brown RD (1977) Comparison of four purified extracellular 1,4-β-D-glucan cellobiohydrolase enzymes from *Trichoderma viride*. Biochim. Biophys. Acta 492: 225–231

Hakansson U, Fägerstam L, Pettersson G and Andersson L (1978) Purification and characterization of a low molecular weight 1,4-β-glucanohydrolase from cellulolytic fungus *Trichoderma viride* QM 9414. Biochim. Biophys. Acta. 524: 385–392

Hall J, Hazlewood GP, Barker PJ and Gilbert HJ (1988) Conserved reiterated domains in *Clostridium thermocellum* endoglucanases are not essential for activity. Gene 69: 29–38

Hazlewood GP, Romaniec MP, Davidson K, Grépinet O and Béguin P (1988) A catalogue of *Clostridium thermocellum* endoglucanase, β-glucosidase and xylanase genes cloned in *Escherichia coli*. FEMS Microbiol. Lett. 51: 231–236

Hébraud M and Fevre M (1988) Characterization of glycoside and polysaccharide hydrolases secreted by the rumen anaerobic fungi *Neocallimastix frontalis, Sphaeromonas communis* and *Piromonas communis*. J. Gen. Microbiol. 134: 1123–1129

Henrissat B and Mornon JP (1990) Comparison of *Trichoderma* cellulases with other β-glucanases. In: CP Kubicek, DE Eveleigh, H Esterbauer, W Steiner and EM Kubicek-Pranz (eds) *Trichoderma reesei* Cellulases: Biochemistry, Genetics, Physiology, and Applications (pp 12–29). Royal Society of Chemistry, London

Henrissat B, Driguez H, Viet C and Schülein M (1985) Synergism of cellulases from *Trichoderma reesei* in the degradation of cellulose. Bio/Technol. 3: 722–726

Henrissat B, Claeyssens M, Tomme P, Lemesle L and Mornon JP (1989) Cellulase families revealed by hydrophobic cluster analysis. Gene 81: 83–95

Highly TL (1980) Degradation of cellulose by *Poria placenta* in the presence of compounds that affect hydrogen peroxide. Materia u. Organismen. 15: 81–90

Howard GT and White B (1988) Molecular cloning and expression of cellulase genes from *Ruminococcus albus* 8 in *Escherichia coli* bacteriophage. Appl. Environ. Microbiol. 54: 1752–1755

Hu YJ and Wilson DG (1988) Cloning of *Thermomonospora fusca* genes coding for beta 1-4 endoglucanases E_1, E_2 and E_5 Gene. 71: 331–337

Hurst PI, Sullivan PA and Shepherd MG (1977) Chemical modification of a cellulase from *Aspergillus niger*. Biochem. J. 167: 549–556

Iyayi CB, Bruchmann EE and Kubicek CP (1989) Induction of cellulase formation in *Trichoderma reesei* by cellobiono-1,5-lactone. Arch. Microbiol. 151: 326–330

Johansson G, Ståhlberg J, Lindeberg G, Engström Å and Pettersson G (1989) Isolated fungal cellulase terminal domains and a synthetic minimum analogue bind to cellulose. FEBS Lett. 243: 389–393

Johnson EA, Sakajah M, Halliwell G, Madia A and Demain AL (1982) Saccharification of complex cellulosic substrates by the cellulase system from *Clostridium thermocellum*. Appl. Environ. Microbiol. 43: 1125–1132

Joliff G. Béguin P, Juy M, Millet J, Ryter A, Poljak R and Aubert JP (1986) Isolation, crystallization and properties of a new cellulase of *Clostridium thermocellum* overproduced in *Escherichia coli*. Bio/Technol. 4: 896–900

Klyosov AA (1988) Cellulases of the third generation. In: JP Aubert, P Béguin and J Millet (eds) FEMS Symposium No. 43, Biochemistry and Genetics of Cellulose Degradation (pp 87–99). Academic Press, London

Klyosov AA (1990) Trends in biochemistry and enzymology of cellulose degradation. Biochemistry. 29: 10577-10585

Knowles JKC, Lehtovaara P and Teeri TT (1987) Cellulase families and their genes. Trends Biotechnol. 5: 255–261

Knowles JKC, Lehtovaara P, Murray M and Sinnott M (1988a) Stereochemical course of action of the cellobioside hydrolases I and II of *Trichoderma reesei*. J. Chem. Soc., Chem. Commun. 1988: 1401–1402

Knowles JKC, Teeri TT, Lehtovaara P, Penttilä M and Saloheimo M (1988b) The use of gene technology to investigate fungal cellulolytic enzymes. In: JP Aubert, P Béguin and J Millet (eds) FEMS Symposium No. 43, Biochemistry and Genetics of Cellulose Degradation (pp 153–169). Academic Press, London

Koenigs JW (1974) Production of hydrogen peroxide by wood-decaying fungi in wood and its correlation with weight loss, depolymerisation and pH changes. Arch. Microbiol. 99: 129–145

Koenigs JW (1975) Hydrogen peroxide and iron: a microbial cellulolytic system. Biotechnol. Bioeng. Symp. 5: 151–159

Kopečný J and Williams AG (1988) Synergism of rumen microbial hydrolases during degradation of plant polymers. Folia Microbiol. 33: 208–212

Kraulis PJ, Clore GM, Nilges M, Jones TA, Pettersson G, Knowles JKC and Gronenborn AM (1989) Determination of the three dimensional structure of the C-terminal domain of cellobiohydrolase I from *Trichoderma reesei*. A study using nuclear magnetic resonance and hybrid distance geometry-dynamical simulated annealing. Biochemistry 28: 7241–7257

Kyriacou AK, MacKenzie CR and Neufield RJ (1987) Detection and characterization of specific and non-specific endoglucanases of *Trichoderma reesei*. Evidence demonstrating endoglucanase activity by cellobiohydrolase II. Enzyme Microb. Technol. 9: 25–32

Lachke AH and Deshpande MV (1988) *Sclerotium rolfsii*: status in cellulase research. FEMS Microbiol. Revs. 54: 177–194

Lamed R and Bayer EA (1987) The cellulosome of *Clostridium thermocellum*. Adv. Appl. Microbiol., 33: 1–46

Lamed R and Bayer EA (1988) The cellulosome concept: exocellular/extracellular enzyme reactor centers for efficient binding and cellulolysis. In: JP Aubert, P Béguin and J Millet (eds) FEMS Symposium No. 43, Biochemistry and Genetics of Cellulose Degradation (pp 101–116). Academic Press, London

Lamed R, Setter E and Bayer EA (1983a) Characterization of a cellulose-binding, cellulase-containing complex in *Clostridium thermocellum*. J. Bacteriol. 156: 828–836

Lamed R, Setter E, Kenig R and Bayer EA (1983b) The cellulosome: a discrete cell surface organelle of *Clostridium thermocellum* which exhibits separate antigenic, cellulose-binding and various cellulolytic activities. Biotechnol. Bioeng. Symp. 13: 163–181

Lamed R, Kenig R and Setter E (1985) Major characteristics of the cellulolytic system of *Clostridium thermocellum* coincide with those of the purified cellulosome. Enzyme Microb. Technol. 7: 32–41

Lamed R, Naimark J, Morgenstern E and Bayer E (1987) Specialized cell surface structures in cellulolytic bacteria. J. Bacteriol. 169: 3792–3800

Lamed R, Morag F, Mor-Yosef O and Bayer EA (1991) Cellulosome-like entities in *Bacteroides cellulosolvens*. Curr. Microbiol. 22: 27–33

Li X and Calza RE (1991) Purification and characterization of an extracellular β-glucosidase from the rumen fungus *Neocallimastix frontalis* EB188. Enzyme Microb. Technol. 13: 622–628

Ljungdahl LG (1989) Mechanisms of cellulose hydrolysis by enzymes from anaerobic and aerobic bacteria. In: MP Coughlan (ed) Enzyme Systems for Lignocellulose Degradation (pp 5–16). Elsevier Applied Science, London

Ljungdahl LG, Coughlan MP, Mayer F, Mori Y and Hon-nami K (1988) Macrocellulase complexes and yellow affinity substance from *Clostridium thermocellum*. In: WA Wood and ST Kellog (eds) Methods in Enzymology, Vol 160 (pp 483–500). Academic Press, New York

Lowe SE, Theodorou MK and Trinci APJ (1987) Cellulases and xylanase of an anaerobic rumen fungus grown on wheat straw, wheat straw holocellulose, cellulose and xylan. Appl. Environ. Microbiol. 53: 1216–1233

MacKenzie CR, Bilous D and Johnson KG (1984) Purification and characterization of an exoglucanase from *Streptomyces flavogriseus*. Can. J. Microbiol. 30: 1171–1178

MacKenzie CR, Patel GB and Bilous D (1987) Factors involved in hydrolysis of microcrystalline cellulose by *Acetivibrio cellulolyticus*. Appl. Environ. Microbiol. 53: 304–308

Mayer F (1988) Cellulolysis: ultrastructural aspects of bacterial systems. Electron Microsc. Rev. 1: 69–85

Mayer F, Coughlan MP, Mori Y and Ljungdahl LG (1987) Macromolecular organization of the cellulolytic enzyme complex of *Clostridium thermocellum* as revealed by electron microscopy. Appl. Environ. Microbiol. 53: 2785–2792

McGavin M and Forsberg CW (1989) Catalytic and substrate-binding domains of endoglucanase 2 from *Bacteroides succinogenes*. J. Bacteriol. 121: 3310–3315

Meinke A, Braun C, Gilkes NR, Kilburn DG, Miller RC and Warren RAJ (1991) Unusual sequence organization in CenB, an inverting endoglucanase from *Cellulomonas fimi*. J. Bacteriol. 171: 308–314

Mel'nik MS, Rabinovich ML and Voznyi YV (1991) Cellobiohydrolase of *Clostridium thermocellum* produced by the recombinant strain of *Escherichia coli*. Biokhimiya (Moscow). 56: 1787–1797

Messner R, Kubicek-Pranz EM, Gsur A and Kubicek CP (1991) Cellobiohydrolase II is the main conidial-bound cellulase in *Trichoderma reesei* and other *Trichoderma* strains. Arch. Microbiol. 155: 601–606

Miller RC Jr, Gilkes NR, Greenberg NM, Kilburn DG, Langsford ML and Warren RAJ (1988) *Cellulomonas fimi* cellulases and their genes. In: JP Aubert, P Béguin and J Millet (eds) FEMS Symposium No. 43, Biochemistry and Genetics of Cellulose Degradation (pp 235–248). Academic Press, London

Mischak H, Hofer F, Messner R, Weissinger E and Hayn M (1989) Monoclonal antibodies against different domains of cellobiohydrolase I and II from *Trichoderma reesei*. Biochim. Biophys. Acta 990: 1–7

Mitsuishi Y, Nitisinprasert S, Saloheimo M, Biese I, Reinikainen T, Claeyssens M, Keränen S, Knowles JKC and Teeri TT (1990) Site-directed mutagenesis of the putative catalytic residues of *Trichoderma reesei* cellobiohydrolase I and endoglucanase I. FEBS Lett. 275: 135–138

Morag E, Halevy I, Bayer EA and Lamed R (1991) Isolation and properties of a major cellobiohydrolase from the cellulosome of *Clostridium thermocellum*. J. Bacteriol. 173: 4155–4162

Morpeth FF (1985) Some properties of cellobiose oxidase from the white rot fungus *Sporotrichum pulverulentum*. Biochem. J. 228: 557–564

Mountfort DO (1987) The rumen anaerobic fungi. FEMS Microbiol. Revs. 46: 401–408

Navarro A, Chebrou MC, Béguin P and Aubert JP (1991) Nucleotide sequence of the cellulase gene *celF* of *Clostridium thermocellum*. Res. Microbiol. 142: 927–936

Niku-Paavola ML, Lappalainen A, Enari TM and Nummi M (1985) A new appraisal of the endoglucanases of the fungus *Trichoderma reesei*. Biochem. J. 231: 75–81

Pearce PD and Bauchop T (1985) Glycosidases of the rumen anaerobic fungus *Neocallimastix frontalis* grown on cellulosic substrates. Appl. Environ. Microbiol. 49: 1265–1269

Penttilä M, Lehtovaara P, Nevalainen H, Bhikhabhai R and Knowles JKC (1986) Homology between cellulase genes of *Trichoderma reesei*: complete nucleotide sequence of the endoglucanase I gene. Gene 45: 253–263

Petré D, Millet J, Longin R, Béguin P, Girard H and Aubert JP (1986) Purification and properties of the endoglucanase C of *Clostridium thermocellum* produced in *Escherichia coli*. Biochimie. 68: 687–695

Pilz I, Schwarz E, Kilburn DG, Miller RC Jr, Warren RAJ and Gilkes NR (1990) The tertiary structure of a bacterial cellulase determined by small-angle X-ray-scattering analysis. Biochem. J. 271: 277–280

Puls J and Wood TM (1988) The degradation pattern of cellulose by extracellular cellulases of aerobic and anaerobic microorganisms. Bioresource Technol. 36: 15–20

Reese ET, McGuire AH and Parrish FW (1967) Glucosidase and exo-glucanases. Can. J. Biochem. 46: 25–34

Reese ET, Siu RGH and Levinson HS (1950) Biological degradation of soluble cellulose derivatives. J. Bacteriol. 9: 485–497

Reymond P, Durand R, Hébraud M and Fevre M (1991) Molecular cloning of genes from the rumen anaerobic fungus *Neocallimastix frontalis*: expression during hydrolase induction. FEMS

Microbiol. Lett. 77: 107–112

Rouvinen J, Bergfors T, Pettersson G, Knowles JKC and Jones TA (1989) Crystallographic studies on the core protein of cellobiohydrolase II from *Trichoderma reesei*. First European Workshop on Crystallography of Biological Macromolecules. Como, Italy, May 15–19, 1989

Rouvinen J, Bergfors T, Teeri T, Knowles JKC and Jones TA (1990) Three-dimensional structure of cellobiohydrolase II from *Trichoderma reesei*. Science 249: 380–386

Ryu DDY, Kim C and Mandels M (1984) Competitive adsorption of cellulase components and its significance in a synergistic mechanism. Biotechnol. Bioeng. 26: 488–496

Sadana JC and Patil RV (1985) The purification and properties of cellobiose dehydrogenase from *Sclerotium rolfsii* and its role in cellulolysis. J. Gen. Microbiol. 131: 1917–1923

Sadana JC and Patil RV (1988) Cellobiose dehydrogenase from *Sclerotium rolfsii*. In: WA Wood and ST Kellog (eds) Methods in Enzymology. Vol 160 (pp 448–454). Academic Press, New York

Sagar BF (1985) Mechanism of cellulase action. In: JF Kennedy, GD Phillips, DJ Wedlock and PA Williams (eds) Proc. Cellucon '84 (pp 199–207). Ellis Horwood, Chichester, U.K.

Saloheimo M, Lehtovaara P, Penttilä M, Teeri TT and Stahlberg J (1988) EGIII, a new endoglucanase from *Trichoderma reesei* and the characterization of both gene and enzyme. Gene 63: 11–21

Salovouri I, Makarow M, Rauvala H, Knowles JKC and Kääriänen L (1987) Low molecular weight high mannose type glycans in a secreted protein in the filamentous fungus *Trichoderma reesei*. Bio/Technol. 5: 152–156

Schmuck M, Pilz I, Hayn M and Esterbauer H (1986) Investigation of cellobiohydrolase from *Trichoderma reesei* by small angle X-ray scattering. Biotechnol. Lett. 8: 397–402

Schwarz WH, Gräbnitz F and Staudenbauer WL (1986) Properties of a *Clostridium thermocellum* endoglucanase produced in *Escherichia coli*. Appl. Environ. Microbiol. 51: 1293–1299

Shoemaker S, Schweickaert V, Ladner M, Gelfand D, Kwok S, Myambo K and Innis M (1983) Molecular cloning of exo-cellobiohydrolase I derived from *Trichoderma reesei* strain L27. Bio/Technol. 1: 691–696

Shoseyov O and Doi RH (1990) Essential 170-kDa subunit for degradation of crystalline cellulose by *Clostridium cellulovorans* cellulase. Proc. Natl. Acad. Sci. U.S.A. 87: 2192–2195

Sprey B and Lambert C (1983) Titration curves of cellulases from *Trichoderma reesei*: demonstration of a cellulase-xylanase-β-glucosidase-containing complex. FEMS Microbiol. Lett. 18: 217–222

Ståhlberg J, Johansson G and Pettersson G (1991) A new model for enzymatic hydrolysis of cellulose based on the two-domain structure of cellobiohydrolase I. Bio/Technol. 9: 286–290

Streamer M, Eriksson KE and Pettersson B (1975) Extracellular enzyme system utilized by the fungus *Sporotrichum pulverulentum (Chrysosporium lignorum)* for the breakdown of cellulose. Functional characterization of five endo-1,4-β-glucanase and one exo-β-1,4-glucanase. Eur. J. Biochem. 59: 607–613

Teeri TT, Salovuori I and Knowles JKC (1983) The molecular cloning of the major cellulase gene from *Trichoderma reesei*. Bio/Technol. 1: 696–699

Teeri TT, Kumar V, Lehtovaara P and Knowles JKC (1987a) Construction of cDNA libraries by blunt end ligation: high-frequency cloning of long cDNAs from filamentous fungi. Anal. Biochem. 164: 60–67

Teeri TT, Lehtovaara S, Kauppinen S, Salovuori I and Knowles JKC (1987b) Homologous domains in *Trichoderma reesei* cellulolytic enzymes: gene, sequence and expression of cellobiohydrolase II. Gene 51: 43–52

Teeri TT, Jones A, Kraulis P, Rouvinen J, Penttilä M, Harkki A, Nevelainen H, Vanhanen S, Saloheimo M and Knowles JKC (1990) Engineering *Trichoderma* and its cellulases. In: CP Kubicek, DE Eveleigh, H Esterbauer, W Steiner and EM Kubicek-Pranz (eds) *Trichoderma reesei* Cellulases: Biochemistry, Genetics, Physiology, and Applications (pp 156–167). Royal Society of Chemistry, London

Tomme P and Claeyssens M (1989) Identification of a functionally important carboxyl group in cellobiohydrolase I from *Trichoderma reesei*: a chemical modification study. FEBS Lett. 243: 239–243

Tomme P, Van Tilbeurgh H, Pettersson G, Van Damme J, Vandekerckhove J, Knowles JKC, Teeri TT and Claeyssens M (1988) Studies of the cellulolytic system of *Trichoderma reesei* QM 9414. Eur. J. Biochem. 170: 575–581

Tomme P, Heriban V and Claeyssens M (1990) Adsorption of two cellobiohydrolases from *Trichoderma reesei* to Avicel: evidence for "exo-exo" synergism and possible "loose complex" formation. Biotech. Lett. 121: 525–530

Tomme P, Chauvaux S, Béguin P, Millet J, Aubert JP and Claeyssens M (1991) Identification of a hystidyl residue in the active center of endoglucanase D from *Clostridium thermocellum*. J. Biol. Chem. 266: 10313–10318

Uzcategui E, Ruiz A, Montesino R, Johansson G and Pettersson G (1991) The 1,4-β-D-glucan cellobiohydrolases from *Phanerochaete chrysosporium*. I. A system of synergistically acting enzymes homologous to *Trichoderma reesei*. J. Biotechnol. 19: 271–285

Vaheri MP (1982a) Acidic degradation products of cellulose during enzymatic hydrolysis by *Trichoderma reesei*. J. Appl. Biochem. 4: 153–160

Vaheri MP (1982b) Oxidation as a part of degradation of crystalline cellulose by *Trichoderma reesei*. J. Appl. Biochem. 4: 356–363

Van Arsdell JN, Kwok S, Schweickart VL, Ladner MB, Gelfand DH and Innis MA (1987) Cloning, characterization and expression in *Saccharomyces cerevisiae* of endoglucanase I from *Trichoderma reesei*. Bio/Technol. 4: 60–64

Van Tilbeurgh H and Claeyssens M (1985) Detection and differentiation of cellulase components using low molecular mass fluorogenic substrates. FEBS Lett. 187: 283–288

Van Tilbeurgh H, Claeyssens M and de Bruyne CK (1982) The use of 4-methylumbelliferyl and other chromophoric glycosides in the study of cellulolytic enzymes. FEBS Lett. 149: 152–156

Van Tilbeurgh H, Bhikhabhai R, Pettersson LG and Claeyssens M (1984) Separation of endo- and exo-type cellulases using a new affinity chromatography method. FEBS Lett. 169: 215–218

Van Tilbeurgh H, Pettersson G, Bhikhabhai R and Claeyssens M (1985) Studies of the cellulolytic system of *Trichoderma reesei* QM 9414. Reaction specificity and thermodynamics of the interactions of small substrates and ligands with the 1,4-β-glucan cellobiohydrolase II. Eur. J. Biochem. 148: 329–334

Van Tilbeurgh H, Tomme P, Claeyssens M, Bhikhabhai R and Pettersson G (1986) Limited proteolysis of the cellobiohydrolase I from *Trichoderma reesei*. FEBS Lett. 204: 223–227

Van Tilbeurgh H, Loontiens FG, De Bruyne CK and Claeyssens M (1988) Fluorogenic and chromogenic glycosides as substrates and ligands of carbohydrases. In: WA Wood and ST Kellog (eds) Methods in Enzymology, Vol 160 (pp 45–59). Academic Press, New York

Walker LP and Wilson DB (1991) Enzymatic hydrolysis of cellulose: an overview. Bioresource Technol. 36: 3–14

Warren RAJ, Beck CF, Gilkes NR, Kilburn DG and Langsford M (1986) Sequence conservation and region shuffling in an endoglucanase and an exoglucanase from *Cellulomonas fimi*. Proteins Struct. Funct. Genet. 1: 335–341

Westermark U and Eriksson (1974) Cellobiose: quinone oxidoreductase – a new wood-degrading enzyme from white rot fungi. Acta Chem. Scand. 28: 209–214

White AR and Brown RM (1981) Enzymatic hydrolysis of cellulose: visual characterization of the process. Proc. Nat. Acad. Sci. U.S.A. 78: 1047–1051

Williams AG and Orpin CG (1987) Glycoside hydrolase enzymes present in the zoospore and vegitative growth stages of the rumen fungi *Neocallimastix patriciarum*, *Piromonas communis* and an unidentified isolate, grown on a range of carbohydrates. Can. J. Microbiol. 33: 427–434

Wilson CA and Wood TM (1992a) Studies on the cellulase of the rumen anaerobic fungus *Neocallimastix frontalis*, with special reference to the capacity of the enzyme to degrade crystalline cellulose. Enzyme Microbiol. Technol. 14: 258–264

Wilson CA and Wood TM (1992b) The anaerobic fungus *Neocallimastix frontalis*: isolation and properties of a cellulosome-type enzyme fraction with the capacity to solubilize hydrogen-bond-ordered cellulose. Appl. Microbiol. Biotechnol. 37: 125–129

Wilson DB (1988) Cellulases of *Thermomonospora fusca*. In: WA Wood and ST Kellog (eds) Methods in Enzymology. Vol 160 (pp 314–323). Academic Press, New York

Wilters SG, Dombroski D, Beaven LA, Kilburn DG, Miller RC Jr, Warren RAJ and Gilkes NR

(1986) Direct ^{1}H NMR determination of the stereochemical course of hydrolysis catalyzed by glucanase components of the cellulase complex. Biochem. Biophys. Res. Commun. 139: 487–494

Wood JD and Wood PM (1992) Evidence that cellobiose: quinone oxidoreductase from *Phanerochaete chrysosporium* is a breakdown product of cellobiose oxidase. Biochim. Biophys. Acta. 1119: 90–96

Wood TM (1975) Properties and mode of action of cellulases. Biotechnol. Bioeng. Symp. 5: 111–137

Wood TM (1989) Mechanisms of cellulose degradation by enzymes from aerobic and anaerobic fungi. In: MP Coughlan (ed) Enzyme Systems in Lignocellulose Degradation (pp 17–35). Elsevier Applied Science, London

Wood TM (1991) Fungal cellulases. In: PJ Weimer and CA Hagler (eds) Biosynthesis and Biodegradation of Cellulose (pp 491–534). Marcel Dekker Inc., New York

Wood TM (1992) Fungal cellulases. Biochem. Soc. Trans. 20: 46–53

Wood TM and Garcia-Campayo V (1990) Enzymology of cellulose degradation. Biodegradation. 1: 147–161

Wood TM and McCrae SI (1972) The purification and properties of the C1 component of *Trichoderma koningii* cellulase. Biochem. J. 128: 1183–1192

Wood TM and McCrae SI (1978) The cellulase of *Trichoderma koningii*. Purification and properties of some endoglucanase components with special reference to their action on cellulose when acting alone and in synergism with the cellobiohydrolase. Biochem. J. 171: 61–72

Wood TM and McCrae SI (1979) Synergism between enzymes involved in the solubilization of native cellulose. Adv. Chem. Ser. 181: 181–209

Wood TM and McCrae SI (1986a) Purification and properties of a cellobiohydrolase from *Penicillium pinophilum*. Carbohyd. Res. 148: 331–334

Wood TM and McCrae SI (1986b) The cellulase of *Penicillium pinophilum*. Synergism between enzyme components in solubilizing cellulose with special reference to the involvement of two immunologically-distinct cellobiohydrolases. Biochem. J. 234: 93–99

Wood TM, McCrae SI and MacFarlane CC (1980) The isolation, purification and properties of the cellobiohydrolase component of *Penicillium funiculosum* cellulase. Biochem. J. 189: 51–65

Wood TM, Wilson CA, McCrae SI and Joblin KN (1986) A highly active extracellular cellulase from the anaerobic rumen fungus *Neocallimastix frontalis*. FEMS Microbiol. Lett. 34: 37–40

Wood TM, McCrae SI, Wilson CA, Bhat KM and Gow LA (1988) Aerobic and anaerobic fungal cellulases, with special reference to their mode of attack on crystalline cellulose. In: JP Aubert, P Béguin and J Millet (eds) FEMS Symposium No. 43, Biochemistry and Genetics of Cellulose Degradation (pp 31–52). Academic Press, London

Wood TM, McCrae SI and Bhat KM (1989) The mechanism of fungal cellulase action. Synergism between enzyme components of *Pencillium pinophilum* cellulase in solubilizing hydrogen bond-ordered cellulose. Biochem. J. 260: 37–43

Woodward J (1991) Synergism in cellulase systems. Bioresource Technol. 36: 67–75

Woodward J, Lima M and Lee NL (1988a) The rôle of cellulase concentration in determining the degree of synergism in the hydrolysis of microcrystalline cellulose. Biochem. J. 255: 895–899

Woodward J, Hayes MK and Lee NL (1988b) Hydrolysis of cellulose by saturating and non-saturating concentration of cellulase: implications for synergism. Bio/Technol. 6: 301–304

Wu JDH, Orme-Johnson WH and Demain AL (1988) Two components of an extracellular protein aggregate of *Clostridium thermocellum* together degrade crystalline cellulose. Biochem. 27: 1703–1709

Yablonsky MD, Bartley T, Elliston KO, Kahrs SK, Shalita ZP and Eveleigh DE (1988) Characterization and cloning of the cellulase complex of *Microbispora bispora*. In: Aubert JP, Béguin P and Millet J (eds) FEMS Symposium No. 43, Biochemistry and Genetics of Cellulose Degradation (pp 249–266). Academic Press, London

Yablonsky MD. Elliston KO and Eveleigh DE (1989) The relationship between the endoglucanase gene MbcelA of *Microbispora bispora* and cellulase genes of *Cellulomonas fimi*. In: MP Coughlan (ed) Enzyme System for Lignocellulose Degradation (pp 73–83). Elsevier Applied Science, London

8. Biodegradation of lignin and hemicelluloses

THOMAS W. JEFFRIES

Institute for Microbial and Biochemical Technology, USDA Forest Service, Forest Products Laboratory, One Gifford Pinchot Drive, Madison, WI 53705-12398, U.S.A.

Abbreviations: APPL – acid precipitable polymeric Lignin; CBQase – cellobiose:quinone oxidoreductase; CEL – cellulase-treated enzyme lignin; DDQ – 2,3-dichloro-5,6-dicyano-1,4-benzoquinone; DHP – dehydrogenative polymerizate (synthetic lignin); DMF – dimethylformamide; DNS – dinitrosalicylic acid; DP – degree of polymerization; HPLC – high performance liquid chromatography; LC – lignin-carbohydrate; LCC – lignin-carbohydrate complex; M_r – relative molecular weight; MWEL – milled wood enzyme lignin; NMR – nuclear magnetic resonance; TLC – thin layer chromatography

I. Introduction

The compositions and percentages of lignin and hemicellulose vary from one plant species to another so it is difficult to arrive at generalizations concerning structure and abundance of these polymers. Moreover, composition varies within a single plant (roots, stems, leaves), with age (heartwood versus sapwood), stage of growth (early wood versus late wood in annual rings) and with the conditions under which the plant grows. Study over many decades has elucidated the major structural features of wood hemicelluloses and lignins, along with the biochemical mechanisms for their degradation. These have been the subjects of comprehensive book-length reviews (e.g. Higuchi 1985a). More specifically, the structures (Fengel and Wegner 1984) and degradation (Shoemaker 1990; Higuchi 1990; Kirk and Farrell 1987; Reiser et al. 1989; Leisola and Garcia 1989) of lignin have been the subject of several reviews, as have the structures (Aspinall 1959; Timell 1964, 1965; Wilkie 1979; Lewis and Paice 1989) and degradation (Dekker and Richards 1976; Reilly 1981; Zimmerman 1989; Dekker 1985; Woodward 1984; Biely 1985; Wong et al. 1988; Johnson et al. 1989) of hemicellulose components. Most recently, attention has turned to the molecular characteristics of these enzymes (Gilkes et al. 1991; Kersten and Cullen 1992). It is beyond the scope of the present review to recapitulate these findings in detail. Rather, the focus will be on specialized or recently revealed aspects.

C. Ratledge (ed.), Biochemistry of Microbial Degradation, 233–277.

II. Hemicellulose

Softwoods (gymnosperms), hardwoods (angiosperms), and grasses (graminaceous plants) have evolved separately, and they contain different lignin and hemicellulose constituents. Moreover, their specialized tissues have varying proportions of cellulose, hemicellulose, lignin, pectins, proteins, and extractives. Lignin is deposited during maturation of cell walls, and some carbohydrates become cross linked to it . Because the lignin and hemicellulose constituents differ, the cross links between these polymers differ from plant to plant and from tissue to tissue.

The term hemicellulose refers to a group of homo- and heteropolymers consisting largely of anhydro-β-(1→4)-D-xylopyranose, mannopyranose, glucopyranose, and galactopyranose main chains with a number of substituents. Hemicelluloses are generally found in association with cellulose in the secondary walls of plants, but they are also present in the primary walls. The principal component of hardwood hemicellulose is glucuronoxylan; that of softwoods, glucomannan. The glucuronoxylan of most hardwoods and graminaceous plants consists of a main β-(1→4)-D-xylan backbone with α-(1→2)-4-*O*-methyl-D-glucuronic acid (or α-(1→2)-D-glucuronic acid) substituents on about 10% of the xylose residues (Fig. 1). Softwood xylans and xylans from most graminaceous plants differ from hardwood xylans in that they have arabinofuranose units linked α-(1→3) to the xylan backbone (Fig. 2). The glucomannoxylans (usually referred to as glucomannans and galactomannans) are made up of β-(1→4)-D-glucopyranose and β-D-mannopyranose residues in linear chains (Fig. 3). Hardwood glucomannans consist of β-(1→4)-linked glucose and mannose units forming chains that are slightly branched. The ratio of mannose:glucose is about 1.5:1 or 2:1 in most hardwoods. Softwood glucomannans have occasional galactose side branches linked α-(1→6) to the mannose main chain (Fig. 4). The α-(1→6) linkage of galactose is very sensitive to acid and alkali and may be cleaved during alkaline extraction (Timell 1965). Softwood xylans and xylans from most graminaceous plants have single L-arabinofuranosyl units attached through α linkages to some *O*-3 positions of the main chain (Wilkie 1979). About 60% to 70% of the xylopyranosyl residues of hardwood xylans are acetylated through ester linkages at position 2 or 3 . Grass xylans are acetylated; the galactomannan of softwoods is also acetylated, albeit to a lesser extent (Lindberg et al. 1973a,b).

A. Assay systems

Hemicellulases are most commonly assayed by measuring the rate of reducing group formation under optimum conditions. A suitable polysaccharide substrate is suspended in buffer and mixed with an enzyme solution that is appropriately diluted to yield a linear response over time. Alternately, several successive two-fold dilutions are assayed for a single fixed time (10 to 30 min), and the enzyme titre is calculated from the average of several successive

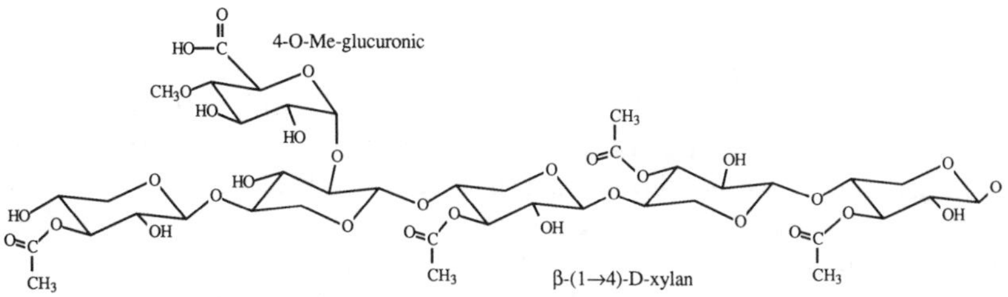

Fig.1. O-Acetyl-4-O-methyl-D-glucuronoxylan from angiosperms (after Dekker 1985).

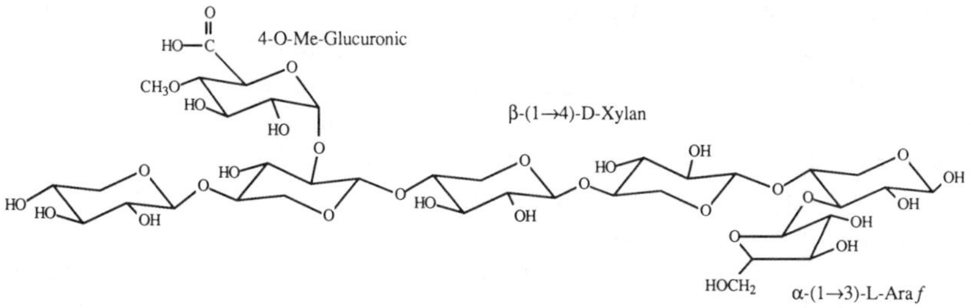

Fig. 2. Arabino-4-O-methylglucuronoxylan from gymnosperms (after Dekker 1985).

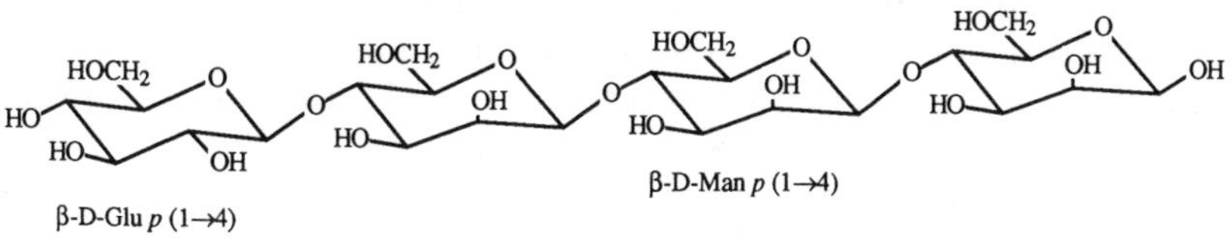

Fig. 3. Glucomannan from angiosperms (after Dekker 1985).

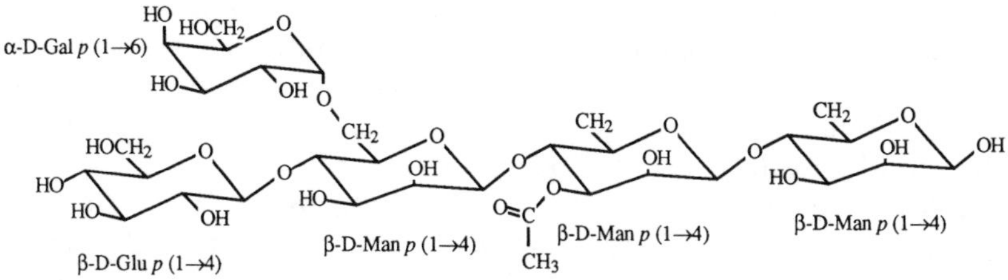

Fig. 4. O-Acetylgalactoglucomannan from gymnosperms (after Dekker 1985).

dilutions that exhibit a consistent enzyme activity. This approach is necessary because at very low dilutions (or long assay times) the substrate is exhausted, and the calculated activity is not representative of the actual value.

The alkaline dinitrosalicylic acid (DNS) method (Miller 1959) or Nelson's modification of the Somogyi method (Nelson 1944; Somogyi 1952) is most commonly used to assay the production of reducing sugars. The DNS method appears to be suitable for the determination of glucose and cellulodextrin reducing sugars in cellulase assays (Miller et al. 1960), but it is known to yield excessively high values when employed with maltodextrins (Roybt and Whelan 1972). It also appears to result in high values when employed for xylan assays as well (Royer and Nakas 1989). The excess reducing group activity appears to be attributable to the degradation of malto- and xylo-dextrins under the alkaline conditions employed. Endoxylanases commonly accumulate xylotriose and higher oligosaccharides, so the DNS assay tends to over-estimate enzymatic activity by three fold or more. Using the DNS assay with various enzyme dilutions and different substrates, Khan et al. (1986) found that the apparent activity of a single enzyme preparation varied by almost seventy-fold.

A number of specialized xylanase assays have been developed that make use of dyed or derivatized substrates. Carboxymethyl xylan will differentiate between xylanases that react with the xylan backbone and those that remove side chains (Khowala et al. 1988). Assays based on dyed substrates are often more sensitive, but the results are also more difficult to interpret from a kinetic perspective (Schmidt and Kebernik 1988). Dyed substrates are particularly useful in developing zymograms from electrophoresis gels (Biely et al. 1985a, 1988). As with all unnatural or substituted substrates, care must be taken to assure that the assay reflects the activity sought. Methods used to assay xylanase activities in various laboratories around the world were summarized by Ghose and Bisaria (1987).

The most exacting method for evaluating hemicellulase activities is to separate and characterize the products. This can be accomplished by paper chromatography (e.g. Biely et al. 1980) or by thin layer chromatography (e.g. Morosoli et al. 1986), but in order to perform quantitative analyses, separation must be accomplished by HPLC. In the past, relatively few HPLC columns were adequate for separating complex oligosaccharides. However, in recent years, Aminex resins such as HPX-42A provided by Bio-Rad (e.g. Kluepfel et al. 1992) or the Carbopac PA1 from Dionex, are capable of separating oligosaccharides with degrees of polymerization (DP) of 5 to 15 or perhaps more. When used in conjunction with a pulsed amperometric detector, highly sensitive and quantitative determinations of action patterns can be obtained with successive samples. The difficulty with this approach is that oligosaccharide standards with DPs in excess of 4 or 5 are hard to obtain. Columns that will separate them are expensive, and the multiple products that show up in such separations are tedious to identify. However, many hemicellulases can be distinguished only by their action patterns or characteristic product profiles, so it is necessary to employ these techniques.

B. Classification

Hemicellulases are classified according to the substrates they act upon, by the bonds they cleave and by their patterns of product formation (Table 1), but greater variety exists among the endo-xylanases and β-glucosidases than is reflected in this simple classification system. One notable distinction is made between endo-1,4-β-xylanase (EC 3.2.1.8) and xylan 1,4-β-xylosidase (EC 3.2.1.37). While the former produces oligosaccharides from random cleavage of xylan, the latter acts principally on xylan oligosaccharides producing xylose. Some endo-xylanases appear to have greater specificity for straight chain substrates, and others appear to be able to accommodate more frequent side chains or branching. Some authors have also described enzymes that remove acetyl, arabinose or 4-*O*-methyl glucuronic acid side chains from xylan backbones (see the following).

Table 1. Recognized hemicellulases.

EC Number	Recommended name	Systematic name
3.2.1.8	Endo-1,4-β-xylanase	1,4-β-D-Xylan xylanohydrolase
3.2.1.37	Xylan 1,4-β-xylosidase	1,4-β-D-Xylan xylohydrolase
3.2.1.32	Xylan endo-1,3-β-xylosidase	1,3-β-D-Xylan xylanohydrolase
3.2.1.72	Xylan 1,3-β-xylosidase	1,3-β-D-Xylan xylohydrolase
3.2.1.55	α-L-Arabinofuranosidase	α-L-Arabinofuranoside arabinofuranosidase
3.2.1.99	Arabinan endo-1,5-α-L-arabinosidase	1,5-α-L-Arabinan 1,5-α-L-arabinohydrolase
3.2.1.78	Mannan endo-1,4-β-mannosidase	1,4-β-D-Mannan mannanohydrolase
3.2.1.100	Mannan 1,4-β-mannobiosidase	1,4-β-D-Mannan mannobiohydrolase
3.2.1.101	Mannan endo-1,6-β-mannosidase	1,6-β-D-Mannan mannanhydrolase

Hemicellulases are usually characterized by their action on defined substrates. In practice, however, most native substrates are relatively complex and bear little similarity to the substrate of commerce. Native substrates (and especially xylans) are often acetylated or otherwise esterified (see the following). This presents a particular problem, because the most common method for hemicellulose recovery – solubilization in alkali – readily removes all ester linkages.

Deacetylation makes native xylans much less soluble in water, which is an observation that causes some consternation to trained organic chemists. Even though acetylation makes the xylan polymer more hydrophobic, it also blocks extensive intrachain hydrogen bonding. When the acetyl groups are removed, hydrogen bonding leads to xylan precipitation. Even though the substrate is less soluble, deactylation generally increases susceptibility of the substrate to enzyme attack. Some acetylesterases, however, show specificity for the native, acetylated substrate (see the following).

Many hemicellulases must accommodate heterogeneous polymers with various degrees of polymerization or side chain branching patterns and substituents. In practice, this means that a given enzyme preparation will exhibit the ability to hydrolyze several different substrates to varying extents. The substrates – like the enzymes that act upon them – are generally difficult to isolate and characterize, and completely homogeneous, defined substrates , as difficult as they are to prepare, may bear little similarity to structures found in the native polymer. Substrate specificities of hemicellulases are, therefore, often ill-defined in the literature.

C. Thermophilic hemicellulases

Thermostable enzymes are often of interest for biotechnological applications, and hemicellulases are no exception. Thermophilic xylanases have been recognized for several years. Ristoph and Humphrey (1985) described an extracellular xylanase produced by the thermophilic actinomycete *Thermonospora* that was stable for approximately 1 month at 55 °C and could withstand up to 80 °C in a 10 min assay. Grüninger and Fiechter (1986) described a highly thermostable xylanase from a thermophilic bacillus that has a catalytic optimum of 78° and a half life of 15 h at 75°. Both enzymes were predominantly endo in activity and produced only trace quantities of xylose after long periods of incubation. The thermophilic fungus *Thermoascus aurantiacus* produces an extracellular endo-xylanase that has a temperature optimum of 80 °C. At that temperature, the half life is 54 min (Tan et al. 1987). These characteristics are fairly similar to thermostable xylanases from *Talaromyces emersonii*, another thermophilic fungus that grows on straws and pulps. *T. emersonii* xylanases are active at an optimal temperature of 80 °C, where they have a half life of 50 to 250 min (Tuohy and Coughlan 1992). In general, xylanases from the actinomycete genus *Actinomadura* exhibit higher temperature stabilities than those of other Actinomycetes. Holtz et al. (1991) examined three strains of *Actinomadura* by cultivating them on xylan and wheat bran. The optimal pH and temperature ranges were from 6.0 to 7.0 and between 70 and 80 °C. The half lives at 75 °C were between 6.5 and 17 h. The record for thermostable xylanases appears to be taken by enzyme from the primitive eubacterium *Thermotoga*. This organism grows optimally at 80 °C. The half life of the enzyme at its optimum pH (5.5) is 90 min at 95 °C. Thermostability can, however, be further improved by immobilization on glass beads and by suspension in 90% sorbitol. Under these conditions, the half life of the enzyme was 60 min at 130 °C (Simpson et al. 1991). Analysis of the molecular basis for thermostable xylanases began with *in vitro* mutagenesis of the xylanase from *Caldocellum saccharolyticum* (Lüthi et al. 1992). The enzyme shows strong sequence homology with other mesophilic xylanases, but the presence of several charged amino acids that could form salt bridges could account for thermostability.

D. Alkaline active xylanases

Most xylanases from fungi have their pH optima in the range of 4.5 to 5.5. Xylanases from many Actinomycetes are most active at pHs between 6.0 and 7.0. A few organisms, however, form xylanases that are active at alkaline pH. Alkaline pH activity could be important for certain applications related to enzymatic treatments of kraft pulps (see the following). The best studied alkaline xylanase is derived from the alkalophilic *Bacillus* sp. C-125 (Honda et al. 1985). This organism produces two xylanases, one (xylanase N) has a pH optimum near neutrality; the other (xylanase A) has a broad pH range showing activity between 5 and 11. The xylanase A gene has been cloned and expressed in *Escherichia coli* (Hamomoto et al. 1987).

An alkaline active enzyme was also purified and characterized from an alkalophilic actinomycete. Tsujibo et al. (1990) purified three endo-xylanases from the culture filtrate of *Nocardiopsis dassonvillei* subsp.*alba* OPC-18. The molecular weights were 23,000, 23,000, and 37,000 for X-I, X-II, and X-III, respectively. Optimum pH levels for all three enzymes was 7.0, whereas, the activities of X-I and X-II fell off to near zero at pH 11; X-III retained about 75% of its activity for pH 8 to 10 and 62% of its initial activity at pH 11.

E. β-Xylosidase

Endo-xylanases are much more common than β-xylosidases, but the latter are necessary in order to produce xylose. Most β-xylosidases are cell bound, and the enzymes are large relative to endo-xylanases. The β-xylosidase present in *Aureobasidium pullulans* is a glycoprotein with an apparent molecular weight of 224,000. On SDS gel electrophoresis, β-xylosidase separates into two subunits of equal molecular mass. β-xylosidase maintains activity over a broad pH range (2.0 to 9.5) and up to 70 °C, but it is present at levels of only 0.35 IU/ml of culture filtrate (Dobberstein and Emeis 1991). The β-xylosidase of *Bacillus sterothermophilus* has a molecular weight of about 150,000 and is stable at up to 70 °C (Nanmori et al. 1990). Utt et al. (1991) described a novel bifunctional β-xylosidase from the ruminal bacterium *Butyrivibrio fibrisolvens* that also possesses α-L-arabinofuranosidase activity. This enzyme was cloned in *E. coli* and shown to consist of a 60,000 M_r protein with dual glycosidase activity.

F. Mannanases and galactanases

Mannanases and galactanases are described far less frequently than xylanases. It is not known whether this is attributable to their lower prevalence in nature or simply because they are sought less often. Bacterial species known to produce mannanases include *Aeromonas hydrophila* (Ratto and Poutanen 1988), *Cellulomonas* sp. (Ide et al. 1983), and *Streptomyces olivochromogenus* (Poutanen et al. 1987). Multiple endo-β-mannanases are found in the extracellular broth of *Polyporous versicolor* (Johnson et al. 1990). They are acidic proteins with molecular weights in the range of 33,900 to 57,500. Multiple endo β-mannanases exhibit maximum activity at pH 5.5 and 65 °C. Torrie et al. (1990) produced β-mannanases by growing *Trichoderma harzianum* E58 on locust bean gum, but the activities were generally low – on the order of less than 1 IU/ml after 6 to 8 days of growth. Araujo and Ward (1990) conducted a systematic search of *Bacillus* species in order to identify strains that will produce large quantities of β-mannanases. One strain, *Bacillus subtilis* NRRL 356, is extraordinary in that it produces 106 IU of mannanase per ml as determined by the DNS assay for reducing sugars using locust bean galactomannanase as the substrate. Araujo and Ward also assayed for β-mannosidase and galactanase activities, but these were produced in much lower titers.

G. Accessory enzymes for hemicellulose utilization

A number of enzymes appear to be critical in the early steps of hemicellulose utilization. These include acetyl xylan esterases, ferulic and *p*-coumaric esterases, α-L-arabinofuranosidases, and α-4-*O*-methyl glucuranosidases.

Acetyl xylan esterase was first described by Biely et al. (1985b) in several species of fungi known to degrade lignocellulose and most especially in *A. pullulans*. Acetyl xylan esterase was subsequently described in a number of different microbes including *Schizophylum commune* (MacKenzie and Bilous 1988), *Aspergillus niger* and *Trichoderma reesei* (Biely et al. 1985b; Poutanen and Sundberg 1988), *Rhodotorula mucilaginosa* (Lee et al. 1987) and *Fibrobacter succinogenes* (McDermid et al. 1990). Acetyl xylan esterase acts in a cooperative manner with endoxylanase to degrade xylans (Biely et al. 1986). This enzyme is not involved in breaking LC bonds because the acetyl esters are terminal groups.

Many different esterase activities have been described, but it has not always been apparent that the assay employed was specific for physiologically important activities. Particular care must be taken in using 4-nitrophenyl acetate as an analog of acetylated xylan, because activities against 4-nitrophenyl acetate and acetylated xylan may show no correlation (Khan et al. 1990). An extracellular esterase from *Aspergillus awamori* (Sundberg et al. 1990) is active against a range of aliphatic and aromatic esters. The native enzyme has an M_r of 150,000 and appears to consist of three equal subunits.

Relatively little is known about enzymes that are capable of releasing aromatic acids from hemicellulose. The substrates are often poorly defined, and most enzymes are obtained only in crude preparations. Ferulic and *p*-coumaric acid esterases were identified in extracellular broths of *S. viridosporus* (Deobald and Crawford 1987), but the activities were not always specific (Donnelly and Crawford 1988). MacKenzie et al. (1987) first assayed for ferulic acid esterase from *Streptomyces flavogrieseus* using native substrate. In this assay, the ferulic acid was identified by HPLC. Ferulic acid esterase is produced along with α-L-arabinofuranosidase and α-4-*O*-methylglucuronidase by cells growing on xylan-containing media. With *S. flavogriseus*, oat spelts xylan was a much stronger inducer for these enzymes than was cellulose. However, not all organisms respond in this manner. *Schizophyllum commune*, for example, produced more xylanase and acetyl xylan esterase when grown on avicel cellulose than when grown on xylan (MacKenzie and Bilous 1988).

The ferulic acid esterase of *S. commune* exhibits specificity for its substrate, and it has been separated from other enzymes. Borneman et al. (1990) assayed feruloyl and *p*-coumaryl esterase activities from culture filtrates of anaerobic fungi using dried cell walls of Bermuda grass (*Cyndon dactylon* [L] Pers) as a substrate. The enzyme preparations released ferulic acid more readily than they released *p*-coumaric acid from plant cell walls. Assays using methyl ferulate or methyl *p*-coumarate as substrates in place of dried cell walls showed the presence of about five times as much enzyme activity. McDermid et al. (1990) employed ethyl esters of *p*-coumarate and ferulate as substrates for these activities.

The use of a realistic model substrate can greatly facilitate purification and kinetic studies. Recently, Hatfield et al. (1991) synthesized 5-*O-trans*-feruloyl-α-L-arabinofuranoside in gram quantities for use as an enzyme substrate. Progress of the reaction can be followed by HPLC or TLC. The substrate is soluble in water and can be readily used for kinetic studies. Much more work needs to be done in this area, particularly with the synthesis of lignin-hemicellulose esters as model substrates.

α-L-Arabinofuranosidase catalyzes the hydrolysis of nonreducing terminal α-L-arabinofuranoside linkages in L-arabinan or from D-xylan. This enzyme has been recognized for a number of years, and the characteristics of α-L-arabinosidases of microbial and plant origin were reviewed (Kaji 1984). Since these enzymes act primarily on terminal arabinose side chains, they are probably not directly involved in removing carbohydrate from lignin. The α-(1→3) linkage is hydrolyzed relatively easily. An enzyme capable of acting on L-arabinofuran substituted at *O*-5 would be of potential interest. Recently, Kormlink et al. (1991a,b) described a 1,4-β-D-arabinoxylan arabino-furanohydrolase produced by *Aspergillus awamori*. This enzyme splits off arabinose side chains from cereal arabinoxylans.

The α-(1→2)-4-*O*-methylglucuronic acid side chain resists hydrolysis by xylanases (Timell 1962), and for that matter by acid. In fact, the presence of the 4-*O*-methylglucuronic acid group stabilizes nearby xylosidic bonds in the xylan main chain, and the substituent must be removed for xylan hydrolysis to proceed. Puls et al. (1987) described α-glucuranosidases from two basidiomycetes, *Agaricus bisporus* and *Pleurotus ostreatus*; the enzyme from *A. bisporus* was partially characterized. This enzyme is evidently not one of the first to attack the xylan polymer because it is relatively large (M_r approximately 450,000) and probably cannot penetrate the microporous structure of lignocelulose. Activity was optimal at pH 3.3 and 52 °C. α-Glucuronidase acts in synergism with xylanases and β-xylosidases to hydrolyze glucuronoxylan. The yield of xylose greatly increases in the presence of this enzyme.

Most filamentous fungi appear to be poor producers of α-methylglucuronidases. Ishihara and Shimizu (1988) systematically screened α-glucuronidase-producing fungi in order to identify other enzymes. Of nine *Trichoderma* and five basidiomycete species (including a strain of *A. bisporus*), *Tyromyces palustris* was the best producer. Concentrated protein precipitates from cell broths were screened against a model substrate of 2-*O*-(4-*O*-methyl-α-D-glucuronopyranose)-D-xylitol. This enabled the detection of reducing group production against a very low background. Even though *T. palustris* produced more α-glucuronidase than did other fungi, total activity amounted to < 0.1 unit/ml. Moreover, the activity was very labile, even in frozen storage. *Streptomyces flavogriesus* and *S. olivochromogenes* also formed α-*O*-methylglucuronidase at low titers (MacKenzie et al. 1987; Johnson et al. 1988).

III. Lignin

Lignin is an aromatic polymer with the substituents connected by both ether and carbon-carbon linkages. It is composed of three principal building blocks: *p*-coumaryl alcohol (*p*-hydroxyphenyl propanol), coniferyl alcohol (guaiacyl propanol), and sinapyl alcohol (syringyl propanol) (Fig. 5). In softwoods, coniferyl alcohol is the principal constituent; the lignin of hardwoods is composed of guaiacyl and syringyl units. Grass lignins contain guiacyl-, syringyl-, and *p*-hydroxyphenyl-units.

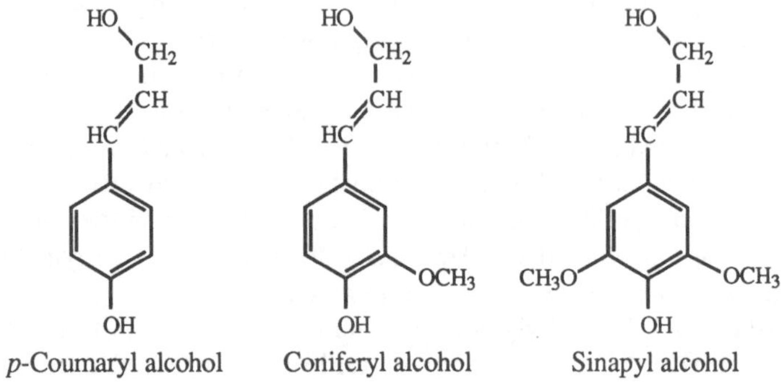

p-Coumaryl alcohol   Coniferyl alcohol   Sinapyl alcohol

Fig. 5. Lignin building blocks.

Lignin inter-monomer linkages are similar in softwood, hardwood, and grass lignins. Biosynthesis of lignin proceeds through a free radical coupling mechanism that was described as dehydrogenative polymerization. Freudenberg (1968) synthesized dehydrogenative polymers (DHP) in the laboratory by reacting coniferyl alcohol with a fungal laccase. In softwoods, the most abundant lignin substructure is the β-aryl ether, which accounts for approximately 40% of the interphenylpropane linkages (Higuchi 1990). This

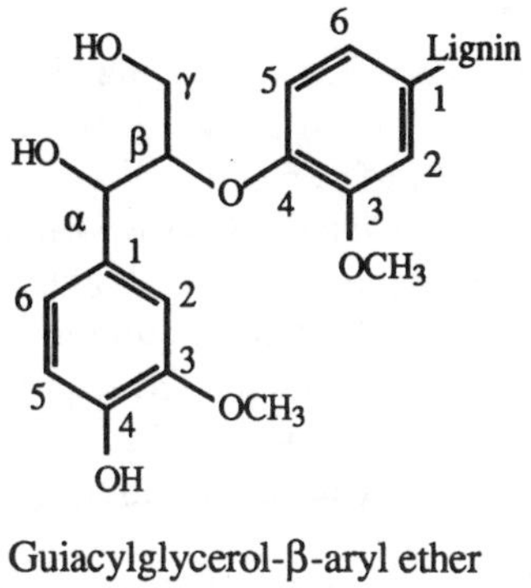

Guiacylglycerol-β-aryl ether

Fig. 6. β-aryl ether lignin model.

linkage is often modeled by guaiacylglycerol-β-guaiacylether dimer (Fig. 6).

The β-arylether substructure has a free benzyl hydroxyl in the α-position and a free phenolic hydroxyl; the remaining hydroxyls are etherified. Both of these hydroxyls were proposed as possible sites for LC bonds. As pointed out by Košíková et al. (1979), nonphenolic benzyl ether bonds are stable in alkaline solution whereas phenolic benzyl ether bonds are easily hydrolysed (Takahashi and Koshijima 1988a). In grasses, *p*-coumaric acid is esterified to the terminal hydroxyl groups of some propyl side chains and comprises 5% to 10% of lignin (Higuchi et al. 1967; Shimada et al. 1971; Scalbert et al. 1985). Ferulic acid is bound via ester linkages to the structural carbohydrates of grasses (Hartley 1973), and diferulic acid can serve to cross link hemicellulose chains (Markwalder and Neukom 1976; Hartley and Jones 1976) (Fig. 7).

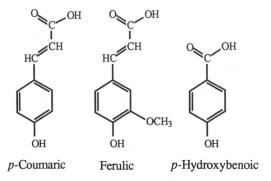

p-Coumaric   Ferulic   *p*-Hydroxybenoic

Fig. 7. *p*-Coumaric, ferulic, and *p*-hydroxybenzoic acids found in lignin-carbohydrate bonds.

A. Lignin-degrading enzymes

Recent years have witnessed an increasing interest in the enzymatic mineralizaiton and depolymerizaiton of lignin. No doubt this stemmed in part from the description of a family of peroxidase-like enzymes from active lignin-degrading cultures of *Phanerochaete chrysosporium* (Tien and Kirk 1983; Glenn et al. 1983). These enzymes, termed lignin peroxidases (LiP), are capable of degrading a number of lignin model compounds – including the β-aryl ether model – and are closely associated with the mineralization of ^{14}C-labelled lignin to CO_2. Of particular note is the ability of enzyme preparations from *P. chrysosporium* to depolymerize methylated spruce lignin (Tien and Kirk 1983).

Lignin degradation could be useful in various pulp bleaching reactions, and a better understanding of the enzymatic mechanisms could facilitate the development of biomimetic catalysts for pulping. From a fundamental perspective, the relative significance of the enzymes found in cultures of fungi that are actively degrading lignin has not been fully resolved, and some controversy exists concerning their roles.

Lignin peroxidases, manganese peroxidases (MnPs), and laccases are three families of enzymes that were implicated in the biological degradation of lignin. LiP production was first shown with *Phanerochaete chrysosporium* (Glenn et al. 1983; Tien and Kirk 1983) and was demonstrated with *Phlebia radiata* (Biswas-Hawkes et al. 1987; Kantelinen et al. 1988b; Niku-Paavola et al. 1988), *Trametes versicolor* (*Coriolus versicolor*) (Bonnarme and Jeffries 1990; Jönsson et al. 1987; Lobarzewski 1987; Waldner et al. 1988), and at least five other white-rot fungi (Bourbonnais and Paice 1988; Fukuzumi 1987; Nerud and Mišurcová 1989; Waldner et al. 1988). Recently, LiPs (veratryl alcohol oxidases) were described in the ascomycete *Chrysonilia sitophila* (Ferrer et al. 1992). MnP was also first shown with *P. chrysosporium* (Glenn and Gold 1985; Huynh and Crawford 1985; Paszczyński et al. 1985; Paszczyński et al. 1986), and it is also produced by *P. radiata*, *P. tremellosa* (Bonnarme and Jeffries 1990), and *T. versicolor*, among other white-rot organisms (Johansson and Nyman 1987). Laccase is not produced by *P. chrysosporium* but is readily demonstrated with various white-rot fungi including *Phlebia* and *Trametes* spp. (Kirk and Farrell 1987; Kirk and Shimada 1985).

LiP and MnP are glycosylated, extracellular heme proteins. Several different LiP isoenzymes are found in the extracellular culture fluids of *P. chrysosporium*. They abstract single electrons from phenolic and nonphenolic aromatic rings – thereby forming cation radicals from the latter (Kersten et al. 1985). Subsequent cleavage reactions (Hammel et al. 1985) partially depolymerize methylated lignins (Tien and Kirk 1983). LiP oxidizes phenolic substrates to phenoxy radicals (Tien and Kirk 1984). *P. chrysosporium* also forms several isoenzymes of MnP. These generate Mn(III) from Mn(II) and H_2O_2. Mn(III), in turn, can oxidize phenols to phenoxy radicals (Gold et al. 1987; Paszczyński et al. 1986). Laccases are extracellular, non-haem, copper-containing proteins that catalyze the one-electron oxidation of phenols to phenoxy radicals. Phenoxy radicals, however formed, can give rise to degradative reactions in certain phenolic model compounds (Kawai et al. 1988; Kirk and Farrell 1987; Kirk and Shimada 1985).

B. Lignin degradation in whole cell cultures

There are essentially two approaches to understanding the roles of these various enzymes in lignin degradation. One approach is to reconstitute a lignin-degrading system *in vitro* using crude, purified, or cloned enzyme constitutents. The other is to regulate the cellular metabolism of lignin-degrading fungi *in vivo* so as to produce either LiP or MnP, and then to observe the cultures for depolymerization and minerilization.

Lignin mineralization can be assayed *in vivo* using ^{14}C-labelled lignins (Crawford and Crawford 1976; Kirk et al. 1978), so it is possible to correlate $^{14}CO_2$ production with the presence or absence of individual enzymes – provided that the enzymatic activities can be regulated differently. Bonnarme and Jeffries (1990) showed that Mn(II) induces MnP while repressing LiP production by *P. chrysosporium* and other white-rot fungi. Mn(II) is therefore a differential effector of these two enzyme families. Manganese is involved in the transcriptional regulation of the MnP gene in *P. chrysosporium* (Brown et al. 1990), and it also regulates expression and lignin degradation in *Dichomitus squalens* (Périé and Gold 1991). We have known for several years that MnO_2 accumulates as black flecks when wood is decayed by several species of white-rot fungi (Blanchette 1984). Also, if the manganese is biologically accessible, the concentrations found in most wood species are sufficient to trigger the regulatory response in *P. chrysosporium* (Bonnarme and Jeffries 1990). Electrochemical and biomimetic studies show that carboxylic acids stabilize Mn(III) in aqueous solutions, and that Mn(III)-α-hydrocarboxylic acid complexes function as diffusible oxidants in lignin model compound oxidations (Cui and Dolphin 1990). Therefore, it appears that manganese could play a role in regulating and carrying out lignin degradation in nature.

Studies by Perez and Jeffries (1990) and Périé and Gold (1991) showed that both the appearance of lignin-degrading enzymes and mineralization of lignin are affected by the presence of manganese. The effects, however, vary with the organisms and test systems employed. In *P. chrysosporium*, elevated Mn concentrations represses LiP while inducing MnP, and in *Phlebia brevispora*, Mn represses LiP while inducing both MnP and laccase. In both cases, high rates of mineralization correlate with the appearance of LiP at low Mn concentrations and not with the appearance of MnP or laccase (Perez and Jeffries 1990). *D. squalens* produces only MnP and laccase, and in this organism, mineralization correlates with the appearance of MnP (Périé and Gold 1991).

More recent studies by Perez and Jeffries (1992) showed that Mn also controls the depolymerization of lignin by whole cell cultures of *P. chrysosporium*.

Here, however, the results are somewhat more complex because both MnP and LiP appear to be responsible in part for depolymerization. At high Mn concentrations, partial depolymerization of lignin occurs, but mineralization is repressed. Partially degraded ^{14}C-labelled lignin accumulates at the cell surfaces. When Mn is present in solution in low amounts (either because of its initial concentration or because it has become insoluble through disproportionation to MnO_2), LiP is induced and mineralization proceeds. When LiP is formed following MnP, depolymerization and mineralization are extensive (Perez and Jeffries 1992).

Investigators studying the biodegradation of lignin have long realized the propensity for fungal mycelia to bind lignin. As often as not, this is viewed as a nuisance. When low concentrations of ^{14}C-labelled lignins from synthetic (Perez and Jeffries 1992) or native (Kern et al. 1989) are added to cultures of *P. chrysosporium*, a significant fraction binds to the mycelium and is essentially nonrecoverable by extraction. This – along with losses attributable to insolubilization or polymerization – plays havoc with material balances. The accumulation of cell-bound lignin is not a static matter. It appears to be a consequence of the physiological state of the cells. Kern et al. (1989) reported that *P. chrysosporium* binds a progressively larger fraction of lignin in the first 3 days of static culture, and then in successive days, the fraction of ^{14}C-lignin decreases while respired $^{14}CO_2$ increases. Therefore, the cell-bound fraction appears to be an intermediate stage in lignin degradation.

Research has not adequately differentiated the cell-bound lignin fraction. The water-soluble and water-insoluble fractions recoverable from culture medium have been characterized by observing changes in molecular mass and chemical features, but the cell-bound fraction – largely because of its adherence to the insoluble cell residue – has not been studied extensively. One suitable approach might be to recover this cell-bound material following digestion of the cells by proteinases and cell wall lytic enzymes.

In some instances, polymerization may preceed depolymerization of lignin in whole-cell cultures. Ritter et al. (1990) examined the degradation of lignosulphonate by immobilized cultures of *P. chrysosporium* and found that with commercial preparations of lignosulphonate, polymerization started immediately, but that depolymerzation followed a few days later. Whether or not polymerization occurs can be indicative of the chemical nature of the lignin.

Michel et al. (1991) used mutants of *P. chrysosporium* in order to dissect the roles of LiP and MnP in the decolorization of lignin-like compounds derived from kraft bleach plant effluents. Mutants blocked in the production of both LiP and MnP were unable to decolorize the effluent, and little decolorization was observed when the wild-type strain was grown in the presence of high nitrogen – a condition that represses both LiP and MnP activity. A LiP mutant of *P. chrysosporium* that produces MnPs but not LiPs exhibited about 80% as much decolorization as did the wild-type strain. When *P. chrysosporium* was grown with high concentrations of Mn, LiP was repressed but MnP was induced, and decolorization proceeded at a very high rate rate. When cells were cultivated with low Mn, LiP production was elevated, MnP was low and decolorization proceeded only slowly. From these results, MnP appears to play an important role in decolorizing kraft bleach plant effluents.

From these data, one can conclude that in whole cultures, MnP and LiP appear to work together, with MnP being responsible for at least some (if not most) of the initial depolymerization steps, and LiP being largely responsible for further degradation of cell-bound oligomeric products. It is not known whether the actions of these two enzymes are synergistic or whether the removal of oligomeric products plays any significant role in the kinetics of depolymerization, but such synergism would be consistent with studies of other extracellular microbial enzyme systems.

C. Degradation by cell-free enzyme systems

Complete catalytic depolymerization of lignin has not been demonstrated *in vitro* with a cell-free enzyme system. Extracellular proteins from *P. chrysosporium*, *Coriolis versicolor*, and *Phlebia radiata* were reported to increase the number of hydroxyl groups, decrease the mean molecular weight, or otherwise chemically alter various lignin preparations (Evans and Palmer 1983; Glenn et al. 1983; Tien and Kirk 1983, 1984; Niku-Paalova 1987). Kern et al. (1989) investigated the action of crude and partially purified LiP from *P. chrysosporium* on the molecular size distribution of ^{14}C-labelled lignins, and he further used pyrolysis GC mass spectrometry to investigate changes in substrate characteristics. They were readily able to reproduce the partial depolymerization of methylated spruce lignin that was initally reported by Tien and Kirk (1983), but the depolymerization was not progressive, and no similar effect was observed with other methylated lignin preparations (Kern et al. 1989).

In contrast to the results with methylated spruce lignin, several studies have shown that LiP will polymerize various lignin preparations (Haemmerli et al. 1986; Odier et al. 1987). Such polymerization is not inhibited by the presence of cellobiose quinone oxidoreductase (Kern and Kirk 1987; Odier et al. 1987), an enzyme long suspected of significance in lignin degradation (Westermark and Ericksson 1974a).

The physical state of the lignin preparation has a strong influence on its reactivity with lignin-degrading enzymes. Kurek et al. (1990) reported that when lignin is finely dispersed in water in a collodial state, it is much more reactive with cell free enzymes than when it is present as reprecipitated lignin. In conventional practice for bioassays using synthetic ^{14}C-labelled lignins, the polymeric model compound is dissolved in dimethylformamide (DMF) and precipitated in medium just prior to inoculation. A fraction of the lignin remains in a collodial state and can be recovered from the supernatant solution following differential centrifugation. The molecular size distributions of the collodial and precipitated lignins do not differ significantly, and the physical state appears to depend on the localized concentration of lignin in the DMF solution and the agitation of the liquid at the time of dispersion. In the studies by Kuerk et al. (1990), H_2O_2 consumption rates in the presence of excess LiP were much greater with collodial lignin than with precipitated lignin. Polymerization predominated in the presence of LiP and H_2O_2 alone, but when veratryl alcohol was added, both polymerization and depolymerization were observed.

Hammel and Moen (1991) showed that lignin peroxidase of *P. chrysosporium* can carry out a partial depolymerization of synthetic lignin *in vitro* in the presence of H_2O_2, veratryl alcohol and 10% DMF. Gel permeation chromatography demonstrated that fragments with molecular weights as low as 170 were formed during the reaction. Approximately 25% to 30% of the original ^{14}C-labelled lignin was converted to volatile products during enzymatic oxidation. Wariishi et al. (1991) also demonstrated depolymerization of ^{14}C-labelled lignins by the MnP of *P. chrysosporium*. In this latter case, the reaction depends on the presence of Mn. One key factor in the success of these studies is the slow addition of enzyme and hydrogen peroxide. If the H_2O_2 is added too quickly, enzyme inactivation and polymerization will predominate. Substrate concentration is also a critical factor, with polymerization predominating at high lignin concentrations and depolymerization being more readily demonstrated at low amounts.

The mechanisms of lignin depolymerization may not be the same in all organisms. Galliano et al. (1991) studied lignin degradation (solubilization) by *Rigidoporous lignosus*, an organism that does not produce LiP, but rather synthesizes MnP and laccase. When the two enzymes were purified and their properties studied *in vitro*, neither enzyme was able to solubilize radioactive lignins. When both enzymes were added to the reaction mixture at the same time, lignin solubilization was extensive; the MnP and the laccase acted synergistically. In addition, glucose oxidase enhanced lignin solubilization by preventing repolymerization of the radicals formed by the two oxidative enzymes.

D. Role of glycosides in lignin degradation

Several recent reports indicate that glycosides might be involved in lignin biodegradation. Kondo and Imamura (1987) first reported that when vanillyl alcohol or veratryl alcohol were included in glucose- or cellobiose-containing media that had been inoculated with wood-rotting fungi, lignin glucosides were formed in the cellulose medium during the early phases of cultivation. Such glucosides could also be formed using a commercial β-glucosidase in place of the culture broth. β-Glucosidase biosynthesized six monomeric glucosides from lignin model compounds and cellobiose (Kondo et al. 1988). These compounds are linked through position *O*-1 of the sugar moiety to alcoholic (but not phenolic) hydroxyls of the lignin model.

In another study by Kondo and Imamura (1989a), three lignin model compounds, 4-*O*-ethylsyringylglycerol-β-syringyl ether, veratryl alcohol, and veratraldehyde, were degraded by *P. chrysosporium* and *Coriolus versicolor* in media containing either monosaccharides or polysaccharides as carbon sources. The authors found that the rate of consumption of the lignin models was much faster in polysaccharide than in monosaccharide media. In media containing xylan or holocellulose, veratryl alcohol was transformed predominantly into veratryl-*O*-β-D-xyloside, which then disappeared rapidly from the medium. Veratraldehyde was first reduced to veratryl alcohol, then glycosylated, and finally consumed.

That carbohydrate is essential for the mineralization of lignin models (Kirk et al. 1976) and that the expression of lignin biodegradation is regulated by carbon catabolite repression (Jeffries et al. 1981) have been known for a long time. However, the role of carbohydrate in the assimilation of lignin degradation products was demonstrated only recently.

One of the first enzymes implicated in lignin biodegradation was cellobiose: quinone oxidoreductase (CBQase). This enzyme catalyzes the reduction of a quinone and the simultaneous oxidation of cellobiose. Westermark and Ericksson (1974a,b) discovered this enzyme and proposed that its role might be to prevent repolymerization of lignin during degradation. Recent studies (Odier et al. 1987) have not borne this out, but the enzyme may be important nonetheless. The CBQase of *P. chrysosporium* binds very tightly to microcrystalline cellulose, but such binding does not block its ability to oxidize cellobiose, indicating that the binding and catalytic sites are in two different domains (Renganathan et al. 1990).

An essential feature of lignin biodegradation is that degradation products resulting from the activity of extracellular enzymes must be taken up by the mycelium; glycosylation by β-glucosidase seems to be an important part of this process. Whether or not sugars attached to lignin in the native substrate by nonglycosidic linkages play a similar role has not been addressed.

Glycosylation could also serve to detoxify lignin degradation products. Veratryl alcohol and vanilyl alcohol, for example, are toxic to the growth of *C. versicolor* and *T. palustris*, whereas the toxicity of the glycosides of these compounds is greatly reduced (Kondo and Imamura 1989b). The presence of a glycosyl group was also shown to prevent polymerization of vanillyl alcohol by phenol oxidase.

Kondo et al. (1990b) recently showed that glycosides can facilitate the depolymerization of DHP by lignin peroxidase of *P. chrysosporium* and decrease repolymerization by laccase III of *C. versicolor*. To clarify the role of glycosylation in lignin degradation, DHP and DHP-glucosides were treated with horseradish peroxidase, commercial laccase, laccase III of *C. versicolor*, and lignin peroxidase of *P. chrysosporium*. During oxidation, DHP changed color and precipitated whereas DHP glucoside only changed color. Moreover, molecular weight distribution studies showed that oxidation of DHP by enzyme preparations containing laccase or peroxidase resulted in polymerization rather than depolymerization. In contrast, enzymatically catalyzed depolymerization of DHP glucoside was observed. DHP-glucoside was also depolymerized more extensively by peroxidase than by laccase.

IV. Lignin-carbohydrate complexes

Lignin-carbohydrate complexes (LCCs) are heterogeneous, poorly defined structures that are found in many plant species. Lignin is directly or indirectly bound covalently to carbohydrate, and the resulting complexes present a barrier to biological degradation (Wallace et al. 1991). LCCs have proven to be highly intractable materials that are in large part responsible for limiting the biodegradation rate of plant materials (Kirk and Chang 1981; Milstein et al. 1983).

Lignin-carbohydrate complexes (LCCs) can be isolated as water-soluble entities (Azuma and Koshijima 1988) and separated by gel filtration into three fractions. The component of lowest molecular weight consists mostly of carbohydrate, the two larger components mostly of lignin. Lignin-carbohydrate bonds are presumed to exist in higher molecular weight lignin fractions that are water insoluble. Softwood LCCs are distinct in that their carbohydrate portions consist of galactomannan, arabino-4-*O*-methylglucuronoxylan, and arabinogalactan linked to lignin at benzyl positions (Azuma et al. 1981; Mukoyoshi et al. 1981). In contrast, carbohydrate portions of hardwood and grass LCCs are composed exclusively of 4-*O*-methylglucuronoxylan and arabino-4-*O*-methylglucuronoxylan, respectively (Azuma and Koshijima 1988). *trans-p*-Coumaric and *p*-hydroxybenzoic acid are esterified to bamboo and poplar lignins, respectively (Shimada et al. 1971), and *trans*-ferulic acid is ether linked to lignin (Scalbert et al. 1985). Many different types of LC bonds have been proposed, but most evidence exists for ether and ester linkages.

LCCs have also been isolated, purified, and characterized following autolysis and steam explosion of *Polulus deltoides* (Johnson and Overend 1991). Four distinct low molecular weight fractions were formed. These were highly acetylated moities containing various amounts of hemicellulsic sugars and lignin. One fraction contained significant amounts of Mn(II). All exhibited micelle formation, indicative of the interaction of hydrophobic lignin and hydrophilic carbohydrate moities in an aqueous environment.

Relatively little attention has been given to enzymes capable of cleaving the chemical linkages between lignin and carbohydrate, the LC bonds. Such bonds occur in low frequency. They are heterogeneous; many are easily disrupted by acid or alkali during isolation and they are still poorly defined. Although many sorts of linkages have been proposed, two have some substantive evidence. They link the a position of the phenyl propane lignin moiety to either carboxyl or free hydroxyls of hemicellulose through ester or ether linkages, respectively. Various chemical and enzymatic procedures have been used to isolate LC complexes, and a few biological systems have been shown to solubilize lignin preparations (see following text).

Problems associated with studying enzymes that will attack LC bonds are similar to those involved in studying lignin and hemicellulose degradation: substrates derived from natural polymers are often poorly defined; model compounds may not reflect the structure of native substrates; the bonds are multifarious; the products can be complex; enzymatic activities can be low; and cofactors and inducers may be required for enzyme activity. For these reasons, no enzymes specific to LC bond cleavage have been described.

Early concepts and evidence about the existence and nature of lignin-polysaccharide complexes were reviewed by Merewether (1957). A more recent review was completed by Koshijima et al. (1989). Higuchi (1985b) reviewed the biosynthesis of LC bonds. Contemporary techniques for isolating milled wood lignin (MWL) and other lignin preparations described here were reviewed and summarized by Obst and Kirk (1988) and techniques for isolating LCCs were described by Azuma and Koshijima (1988). A recent review summarized the effects of LCCs on digestibility of grasses (Wallace et al. 1991).

A. Fractionation of lignin and carbohydrate in wood

The existence and character of LC bonds have long been studied and debated. In 1956, Björkman published techniques for liberating MWL from finely ground wood (Björkman 1956). He also obtained an LCC by exhaustive milling and extraction. Aqueous dioxane was used to extract MWL (Björkman and Person 1957), and dimethylformamide, dimethylsulfoxide, and aqueous acetic acid were used to extract LCC (Björkman 1957). The lignin contents of LCCs are similar to that of the MWLs except that the latter contain less carbohydrate (Azuma et al. 1981). Despite many attempts, Björkman could not separate carbohydrate from LCC without resorting to chemical degradation.

Another method of isolating lignin was published by Pew in 1957. This technique used cellulases and hemicellulases to digest away the carbohydrate from finely ground wood. Materials prepared in this manner are known as milled wood enzyme lignin (MWEL) or cellulase-treated enzyme lignin (CEL). The latter is obtained by extracting the former with aqueous dioxane. MWEL contains nearly all the LC bonds found in wood (Obst 1982). It generally has a carbohydrate content of 10% to 14%, depending on the activity of the enzymes, the wood employed, and the mode of pretreatment. Alkali treatment prior to enzymatic digestion removes ester-linked polysaccharides and results in shorter carbohydrate side chains (Obst 1982). Upon acid hydrolysis, the insoluble MWEL residue liberates all five of the sugars found in wood but it is particularly rich in hemicellulosic sugars and uronic acids.

Chang et al. (1975) compared the lignins prepared by the Björkman and Pew procedures and showed that MWL is adequately representative of the total lignin in wood, but that MWEL is preferable for structural studies, because it is less altered and it can be obtained in good yield with less degradation. A larger fraction of the lignin present can be extracted with dioxane after the milled wood has been digested with cellulases and hemicellulases.

B. Isolation of LCCs

The LCCs are highly heterogeneous and difficult to study. Many conventional chemical treatments, such as alkali, readily disrupt some of the most prevalent LC bonds. The best approaches employ neutral solvents and purified enzyme preparations. Most research has focused on the isolation of water-soluble LCCs, because they can be fractionated, sized, and subjected to spectroscopic study. Larger, insoluble LCCs exist and may even account for the bulk of LC bonds in the substrate, but less is known about them. Most evidence suggests that the chemical linkages are similar to those of the soluble LCCs.

Water-soluble LCCs have been isolated from a large number of plant sources, including sapwood (Azuma et al. 1981) and compression wood (Mukoyoshi et al. 1981) of pine (*Pinus densiflora*), beech (*Fagus crenata*) (Azuma et al. 1985), aspen (*Populus tremula*) (Joseleau and Gancet 1981), Norway spruce (*Picea abies*) (Iversen 1985), and sugar cane bagasse (*Saccharum officinarum*) (Atsushi et al. 1984).

Isolated LCCs show amphiphatic or surface active properties because of the presence of hydrophilic carbohydrates and hydrophobic lignin on the same molecule. As a consequence, LCCs can be isolated by adsorption chromatography (Watanabe et al. 1985). HPLC was used to determine the molecular weight of LCCs (Takahashi and Koshijima 1988a). Enzymatic treatment of water-soluble LCCs showed that lignin moieties were linked together by carbohydrate chains. Molecular weight distribution studies showed that the lignin moieties of beech are 100 times larger but less frequent than those of pine LCC (Takahashi and Koshijima 1988a).

Reports have indicated that the sugars linking hemicellulose to lignin may not be representative of the hemicellulose as a whole. For example, enzymatic digestion of polysaccharides during the isolation of lignin was shown to enrich galactose in the lignin fraction (Iversen et al. 1987). Some researchers had therefore suggested that galactose units were preferentially involved in the LC bond (Eriksson et al. 1980). Iversen et al. (1987) subsequently showed that some commercial enzymes used for such studies are deficient in β-(1→4)-D-galactosidase activity. Care must therefore be taken in interpreting such results, and purified or defined enzyme preparations should be used whenever possible. If ester linkages are sought, esterase activities should be removed from the preparations used in primary digestions.

The LCCs of grasses can differ substantially from those of wood. In Italian rye grass (*Lolium multiflorum*) and alfalfa (*Medicago sativa*), the carbohydrates associated with lignin are rich in arabinose (Kondo et al. 1990a). In pangola grass (*Digitaria decumbens*), most pentosans are linked to lignin through alkali-labile bonds that do not involve uronic acid esters. Presumably, the bulk of these are through ferulic and *p*-coumaric acid. A small percentage of glycosidic linkages also appear to be present in grass LCCs (Ford 1990). The ferulic and *p*-coumaric esters of grasses are discussed elsewhere in this chapter.

C. Chemical characteristics of LC bonds

Existence of chemical bonds between lignin and carbohydrate has been questioned because of the intimate physical integration between the lignin and carbohydrate constituents, the possibility of entrapment or adsorption, and the lability of many linkages. Several different types of LC bonds have been proposed based on knowledge about changes in sugar composition following digestion and about hemicellulose structures. The proposed LC bonds include bonds to xylan, glucomannan, cellulose, various other hemicellulosic sugars, and pectin.

The most labile chemical bonds are ester linkages. Ester linkages occur between the free carboxy group of uronic acids in hemicellulose and the benzyl groups in lignin. Many of these linkages in wood are broken by alkali. Some ester bonds are present as acetyl side groups on hemicellulose, others are between uronic acids and lignin, and still others occur between hemicellulose chains. Monomeric side chains in wood xylans, for example, consist of 4-*O*-methylglucuronic acid units, and some 40% of the uronic acid groups in birch are esterified. In beech, one-third of the glucuronic acid present in LCCs is involved in an ester linkage between lignin and glucuronoxylan (Takahashi and Koshijima 1988b). However, many glucuronic acid groups may be esterified within the xylan polymer (Wang et al. 1967). The question is how does one distinguish among these linkages? Sodium borohydride reduction and alkali will disrupt ester linkages (Ericksson et al. 1980), so loss of bound carbohydrate upon alkali or sodium borohydride treatment is indirect evidence of ester linkages in the LCC.

Alkali-stable and alkali-labile linkages can be distinguished. When Ericksson and Lindgren (1977) prepared LCC from spruce according to the Björkman procedure, they obtained a material that contained 58% carbohydrate (chiefly hemicellulose) by weight. Saccharification of the LCC with cellulases and hemicellulases reduced the carbohydrate content of the insoluble material to 18%, while solubilizing about half the lignin. Sodium borohydride reduction of the enzyme-treated material removed about 50% of the remaining carbohydrate, decreasing the content to about 9%. Further alkali treatment did not remove additional carbohydrate. This evidence suggests that about half the carbohydrate bound to spruce LCC could be attached through ester linkages. The LCCs are obtained only in low yield, and they do not account for all the ester linkages present in wood. The proportion of LC bonds is greater in LCC than in wood as a whole.

Mild alkali treatment also releases about 50% of the carbohydrates from spruce MWL, and xylan is removed to a greater extent than are other carbohydrate constituents (Lundquist et al. 1983). Xylan is the predominant carbohydrate polymer associated with lignin, and about 90% of it can be removed by mild alkaline treatment.

Direct evidence for the chemical nature of ester linkages between lignin and carbohydrate in pine was obtained through the selective oxidation of carbonyls in lignin. Watanabe and Koshijima (1988) used selective cleavage of the methoxybenzyl glucuronate with 2,3-dichloro-5,6-dicyano-1,4-benzoquinone (DDQ) to demonstrate the presence of this bond. They proposed that the 4-O-methylglucuronic acid residue in arabinoglucuronoxylan binds to lignin by an ester linkage in *Pinus densiflora* wood. The linkage position is probably the α- or conjugated γ-position of guaiacylalkane units (Fig. 8). Watanabe et al. (1989) found that mannose, galactose, and glucose are O-6 ether linked and xylose is O-2 or O-3 ether linked to the α-benzyl hydroxyl in a neutral fraction of pine LCC.

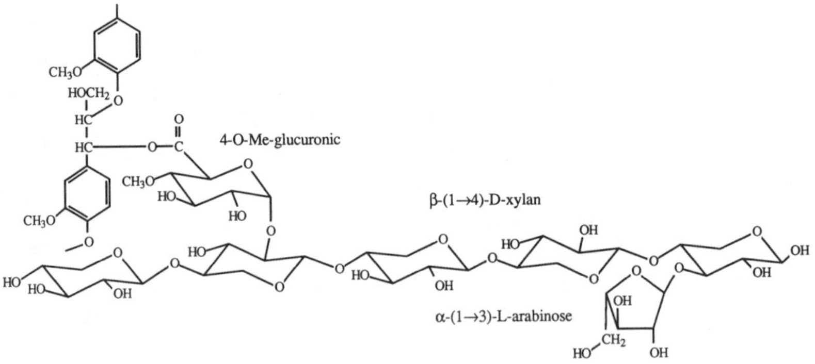

Fig. 8. Proposed structure of ester linkage between lignin and arabino-4-O-methylglucuronoxylan in pine (after Watanabe and Koshijima 1988).

Characterization of the benzyl-carbohydrate bond is complicated by the fact that both benzyl alcoholic and benzyl ether lignin linkages exist in the molecule. Basically, four general structures can be identified. The *p*-hydroxyl and *p*-alkoxy structures can be substituted in the a position with either lignin or sugar moieties. Košíková et al. (1979) studied the alkaline-stabile benzyl ether bonds between lignin and carbohydrate in spruce. Using DDQ and quinone monochloroimide as selective oxidants of total benzyl alcoholic and *p*-hydroxy benzyl alcoholic groups, respectively, combined with selective acid and base hydrolysis, these authors estimated that the lignin-lignin bonds were two to eight times more frequent than were the lignin-saccharidic structures.

Watanabe et al. (1989) also studied alkali-stable LC linkages. In LCCs from normal and compression wood of pine, free hydroxyls on lignin and sugar moieties were first protected by acetylation under mild conditions. DDQ was then used to carry out a selective oxidation of LC bonds at the α- and conjugated γ-positions of *p*-etherified phenyl alkane or alkene units. The newly formed hydroxyl groups were then methylated, and the position of the bonds was determined. The alkali-stable linkages in LCC prepared from *Pinus densiflora* consist of acetyl glucomannan and β-(1→4) galactan bound to the lignin at the O-6 position of the hexoses (Fig. 9). The GC-MS analysis of

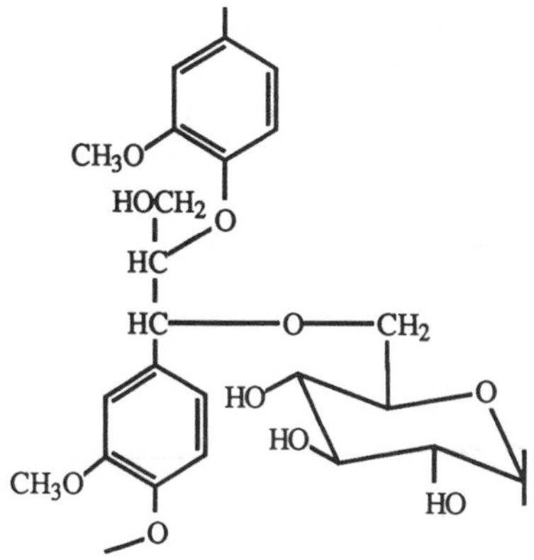

Fig. 9. Proposed structure for ether linkages between lignin and glucomannan of pine (after Watanabe et al. 1989).

methylated sugar derivatives led Watanabe et al. (1989) to conclude that the arabinoglucuronoxylan is bound to the lignin at the O-2 and O-3 positions of the xylose units, and that the linkage position in the lignin subunits is in the α- or conjugated β-positions of phenyl propane or propene units.

Minor (1982) investigated the LC bonds of loblolly pine MWEL using methylation analysis to determine the positions of linkages in carbohydrates. The MWEL, containing about 12% residual carbohydrate anhydride, was reduced by sodium borohydride, and the residue was methylated. Borohydride reduction removed a small fraction of the alkali-sensitive bonds, but the alkali-stable bonds of importance to residual kraft lignin were retained. The carbohydrates exist as oligomeric chains with degrees of polymerization of 7 to 14. Approximately one sugar unit per oligomer chain is bonded to lignin. Hexose units are generally bonded at O-6, and L-arabinose is bonded exclusively at O-5. Galactan and arabinan are structurally of the 1→4 and 1→5 type, respectively, characteristic of the neutral substituents of pectins.

Ericksson et al. (1980) found that in spruce, lignin is linked by ester bonds to 4-O-methylglucuronic acid, and arabinoxylan is linked through ether bonds to the O-2 or O-3 positions of L-arabinose. In softwood xylans, L-arabinofuranose residues exist primarily as terminal side chains. The α-(1→2) linkage is extremely susceptible to acid hydrolysis. The nature of ether linkages through L-arabinose is determined by the conformation of the sugar. L-Arabinose exists primarily as the furan, but the more stable pyranose is sometimes also present (Timell 1965). In the furan form, the O-5 position is a primary hydroxyl and is available to form an ether linkage with lignin. For galactoglucomannan, ether bonds to position 3 of galactose have been indicated. Ericksson et al. (1980) did not find direct evidence for covalent linkages between lignin and cellulose, but the persistence of carbohydrate in enzyme-treated LCC preparations suggested to these authors that some bonds might be present.

A study of the water-soluble LCC from beech showed that xylose, arabinose, and galactose residues are linked to lignin, possibly by benzyl ether bonds at *O*-2 or *O*-3, *O*-5, and *O*-6 positions. Phenol-glycoside linkages are not involved (Takahashi and Koshijima 1987). Esters of *p*-coumaric acid in sugar cane lignin and of *p*-hydroxybenzoic acid in aspen lignins were first demonstrated by Smith (1955) who proposed that the carboxy group of *p*-hydroxybenzoic acid is esterified to lignin in the α-position of the phenyl propane (Fig. 10).

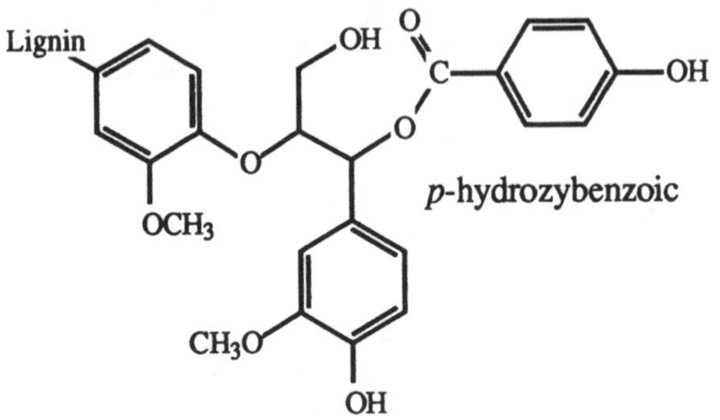

Fig. 10. Proposed structure for lignin-*p*-hydroxybenzoic acid ester in aspen.

Pectic substances may play an important role in binding lignin to the hemicellulose. Pectins consist of poly-α-(1→4)-galactopyranosyluronic acid. As such, they are free to form both ester and ether linkages with lignin. LCC from birch was shown to contain about 7% galacturonic and 4% glucuronic acid, and a small amount of galacturonic acid was found in the LCC of spruce (Meshitsuka et al. 1983). Thus, linkages between lignin and pectin are indicated. When LCC from birch was treated with a purified endopectate lyase, the lignin content of the residual material increased from about 20% to 71%, while the xylose content decreased from 69% to 4%. These results suggested to the authors that pectins may connect portions of lignin to hemicellulose in this herbaceous species (Meshitsuka et al. 1983).

Pectins are abundant in some fiberous plant materials such as mitsumata (*Edgeworthia papyrifera*). Fibre bundles are held together by pectic substances, and the pectins are aggregated with LCCs. Aggregates between lignin and pectins are particularly present in bast fibres, and *endo*-pectin and *endo*-pectate lyases from the soft-rot bacterium *Erwinia carotovora* release pectic fragments from this substrate (Tanabe and Kobayashi 1988). Alkaline presoaking accelerates biochemical pulping of mitsumata by pectinolytic enzymes (Tanabe and Kobayashi 1987). Presoaking releases a LCC rich in arabinose, suggesting that an ester linkage is cleaved, but chemical bonds between pectin and LCC have not been clearly demonstrated (Tanabe and Kobayashi 1986, 1988).

D. *Ferulic and* p-*coumaric ester side chains*

Grass and bamboo lignins differ from those of hardwoods and softwoods in that they are formed not only from guiacyl and syringyl units, but also from *p*-hydroxyphenyl units. Esterified *p*-coumaric acid can comprise 5% to 10% of the total weight of isolated lignin (Shimada et al. 1971). The participation of *p*-hydroxyphenyl glycerol-β-aryl ether structures is of minor importance (Higuchi et al. 1967). Based on methanolysis and spectral studies, Shimada et al. (1971) proposed that the majority of *p*-coumaric acid molecules in bamboo and grass lignins are ester-linked to the terminal γ carbon of the side chain of the lignin molecule (Fig. 11). Ferulic acid is also present in small amounts. The *p*-coumaric ester linkages are extremely stable as they are not removed by methanolysis, thioglycolic acid treatment, or catalytic hydrogenolysis.

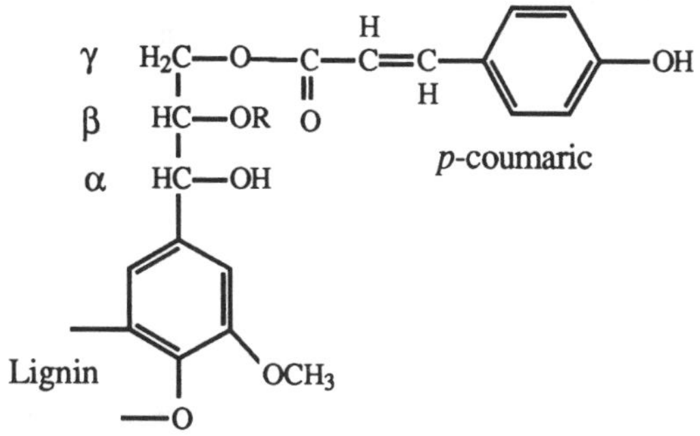

Fig. 11. Proposed *p*-coumaric ester linkage in grass lignins (after Shimada et al. 1971).

Since *p*-coumaric and ferulic acids are bifunctional, they are able to form ester or ether linkages by reaction of their carboxyl or phenolic groups, respectively. *p*-Coumaric is mainly associated with lignin; ferulic acid, on the other hand, is mainly esterified with hemicellulose (Scalbert et al. 1985; Atushi et al. 1984). In wheat straw, at least 93% of the *p*-coumaric acid bound to milled or enzyme lignin fractions is alkali labile and is thus linked by an ester bond. In contrast, 35% to 75% of the recovered ferulic acid is only released by acidolysis in refluxing dioxane and appears to be ether linked. Ferulic acid ethers might form cross links between lignin and hemicelluloses by the simultaneous esterification of their carboxyl group to arabinose substituents of arabinoglucuronxylan and etherification of their hydroxyl group to phenyl hydroxyls of lignin. Evidence for the presence of ferulic acid cross links between lignin and carbohydrate was shown by ^{13}C NMR (Scalbert et al. 1985). This study did not indicate participation of *p*-coumaric acid in such linkages.

Linkage of ferulic acid to lignin can be by ester or ether bonds. The specific components of lignin and the localization of these bonds has not been determined. Linkage of ferulic acids to polysaccharides, on the other hand, is highly specific. It is attached to the C-5 of arabinofuranosyl side chains on arabinoxylans of grasses and in dicotoledonous plants, it is associated with pectins (Hatfield et al. 1991).

Not all of the ferulic acid present serves to link lignin and carbohydrate. Some – perhaps most – of the acid may simply be present as terminal ester-linked groups on arabinoglucuronoxylan. Alkaline hydrolysis of the LCC of rye grass (*Lolium perenne*) releases low molecular weight fragments containing ferulic and *p*-coumaric acids. LCC from Italian rye grass (*L. multiflorum*) likewise releases significant amounts of *p*-coumaric and ferulic acids following alkaline or acid hydrolysis (Kondo et al. 1990a). Purified cellulase releases ferulic acid from grass cell walls in the form of a water-soluble compound with the ferulic acid carboxy group esterified to carbohydrate (Hartley 1973; Hartley et al. 1976). Sodium hydroxide treatment will also release diferulic acid from these same walls (Hartley and Jones 1976). Diferulic acids can be formed by polymerization with peroxidase/water, and studies suggest that these acids can cross link hemicellulosic chains (Markwalder and Neukom 1976; Morrison 1974) (Fig. 12).

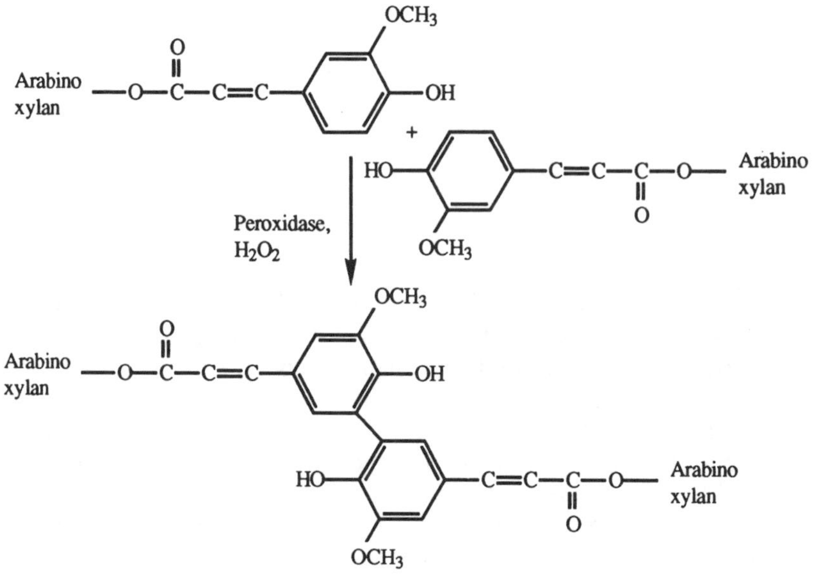

Fig. 12. Formation of diferulic acid in grasses (after Markwalder and Neukom 1976).

Feruloylated arabinoxylans were isolated from the LCC of bagasse (Kato et al. 1987), bamboo shoot cell walls (Ishii and Hiroi 1990), and pangola grass (*Digitaria decumber*) (Ford 1989), following saccharification with purified *endo*-cellulase or xylanase. Several different structures are apparent, but one tetrasaccharide described by Kato et al. (1987) and by Ishii and Hiroi (1990) has

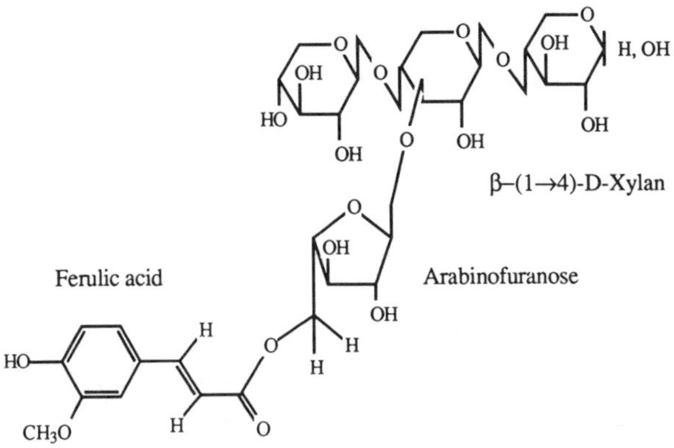

Fig. 13. Ferulic acid ester linkage to grass arabinoxylan (after Kato et al. 1987).

carboxy terminus of ferulic acid esterified to the O-5 of L-arabinose, which in turn is glycosidically bound α-(1→3) to the xylan main chain (Fig. 13).

E. Frequency and stability of LC bonds

The amount of carbohydrate remaining on the lignin can be measured by sugar analysis following acid hydrolysis. Obst (1982) found 10.8% carbohydrate in a MWEL from loblolly pine. A fraction of this amount (11%) was removed by dilute alkali. Sodium borohydride reduction prior to acid hydrolysis allowed the calculation of an average degree of polymerization (DP) of 4 for the alkali-soluble material and a DP of 6 for the carbohydrate attached to the residue. From these data, Obst estimated that there are approximately three LC bonds for every 100 phenylpropane units in lignin. The length of the side chains varies with the method by which MWEL is prepared. If extensive alkaline hydrolysis is performed prior to enzymatic digestion, a larger fraction of the residual carbohydrate is removed.

F. Residual lignin in kraft pulp

Kraft pulping removes large quantities of hemicellulose and lignin and disrupts ester linkages between lignin and carbohydrate and between hemicellulose chains. Even so, significant amounts of carbohydrate remain bound to the residual lignin after kraft pulping. Chemical delignification during pulping may be retarded by the association of lignin with carbohydrates. Lignin and carbohydrate are functionally different. Even though lignin can be solubilized under alkaline conditions, covalent alkali-stable bonds between lignin and cellulose can bind lignin in the matrix. During the phase of bulk delignification, kraft pulping removes lignin without greatly degrading the cellulose. As the reaction progresses, however, residual lignin becomes harder to remove.

Lignin remaining in the kraft pulp cannot be removed without unacceptable large yield losses.

Many explanations have been proposed for the resistance of residual lignin to kraft delignification: (1) lignin does not react with the cooking liquor and is consequently insoluble, (2) lignin undergoes extensive condensation during pulping so that it becomes intractable and unreactive, (3) the molecular weight of lignin is so high that the lignin is trapped within the cell walls, (4) lignin is strongly adsorbed to the carbohydrate, and (5) lignin is covalently bound to the carbohydrate.

In an attempt to resolve some of these possibilities, Yamasaki et al. (1981) characterized the residual lignin in a bleachable-grade kraft pulp prepared from loblolly pine. The pulp was exhaustively digested with cellulases and hemicellulases prior to analysis, leaving an insoluble residue that consisted chiefly of lignin. The residual lignin had a higher molecular weight than did either MWL from pine or lignin solubilized during the kraft process. About 10% to 20% of the residue consisted of hemicellulosic sugars that could not be removed by further digestion. However, the residual lignin was fully soluble in dilute sodium hydroxide or sodium carbonate, so insolubility in cooking liquors could not explain why it had not been removed during the pulping process. The residue was somewhat more condensed than lignin extracted from wood meal, but it was less condensed than kraft lignin that had been solubilized in the pulping process. Therefore, condensation alone does not provide an explanation. The molecular weight (20,000), although higher than that of most material extracted during pulping, was lower than that of the lignin extracted during the final phase of pulping (50,000); therefore, pore diffusion alone would not seem to provide an explanation. Because wood polysaccharides are well known for their ability to adhere to lignin, Yamasaki et al. (1981) prepared holocellulose from kraft pulp and mixed it with the residual lignin. In each case, the components could easily be separated by a simple extraction, so adhesion could not explain the persistence of residual carbohydrate. From these results, the authors concluded that covalent linkages between lignin and carbohydrate were the most probable reasons for lignin to remain in the kraft pulp. More recently, Hortling et al. (1990) arrived at a similar conclusion.

Based on the nature of the residual sugars, Yamasaki et al. (1981) proposed that the most likely bond is an α phenyl ether linkage to the O-6 of galactoglucomannan. Although this structure is in accord with what is known about native LC bonds, the bonds in kraft pulp can be very different. The nature of the covalent linkages in kraft pulp have not been fully characterized because (1) many rearrangements occur during pulping, (2) difficulties are encountered in isolating the residual lignin, and (3) reliable degradation and characterization methods are lacking (Gierer 1970). At least some changes appear to result from the formation of alkali- and acid-stable carbon-carbon bonds between lignin and carbohydrate (Gierer and Wännström 1984). Several rearrangements are possible. Particularly, primary hydroxyls of glucose and mannose can react with the α, β or γ carbons of phenylpropane units to form ether linkages. The reducing-end groups can also react (Iversen and Westermark 1985). Glucose is the most prevalent sugar bound to residual lignins from kraft pulps, and it seems probable that this results from the reaction of cellulose with lignin during the pulping process. This conclusion is supported, but not proven, by the observation that the glucose content of residual lignins from pulps is much higher than the glucose content of lignins from wood (Hortling et al. 1990). Reactions with the β and γ hydroxyls of the phenyl propane are particularly acid- and alkali-stable and may escape detection in methylation analysis (Iversen and Wännström 1986).

Minor (1986) used methylation analysis to determine the characteristics of polysaccharides attached to residual lignin in loblolly pine kraft pulps. The total carbohydrate content of the residual MWEL was only about 8%, as compared to the 12% obtained with MWEL from native wood, but methylation data indicated that the carbohydrate bonding was similar in kraft and native wood. The primary O-6 position was most frequently found for hexans and the primary O-5 for arabinan. Xylan was bonded to lignin at O-3, with a small amount at O-2. The predominant methylated derivatives obtained from galactose and arabinose indicated 1→4 and 1→5 linkages, respectively. The apparent DP ranged from 4 for xylan to almost 13 for galactan. Because of the small differences in methylation patterns between carbohydrates from MWELs of pine and pine kraft pulp, Minor was not able to confirm the possible formation of LC bonds during pulping.

G. Biodegradation of LCCs

Most carbohydrate chains or side groups appear to be attached to lignin through the nonreducing moieties. Because *exo*-splitting enzymes generally attack a substrate from the nonreducing end of a polysaccharide, removing substituents progressively toward the reducing end of the molecule, complete degradation is not possible. Even when carbohydrates are attached to the lignin by the O-1 hydroxyl, a single sugar residue could remain attached even after complete attack by *exo*-splitting glycosidases.

The action of *endo*-splitting glycanases is even more constrained. The binding sites of most *endo*-splitting polysaccharidases have not been well characterized. However, from transferase activities and other kinetic studies, Biely et al. (1981) showed that the substrate binding site of the *endo*-xylanase from *Cryptococcus* has eight to ten subsites for binding pyranose rings and the catalytic groups that make up the active site are located in the center. It is not surprising, therefore, that digestion of LCCs with *endo*-xylanases and *endo*-cellulases leaves residual polysaccharide oligomers with a DP of 4 or more attached to the lignin moiety.

The action of an *endo*-xylanase purified from a commercial preparation from *Myrothecium verrucaria* illustrates the effects of substrate binding (Comtat et al. 1974). This preparation will cleave xylose residues from the nonreducing end of aspen 4-O-methylglucuronoxylan until it leaves a

4-O-methyl-a-D-glucuronic acid substituent linked 1→2 to a β-(1→4)-xylotriose. The enzyme cannot cleave the two β-(1→4)-D-xylan linkages immediately to the right of the 4-O-methylglucuronic acid side chain (in the direction of the reducing end). The residual xylotriose represents that portion of the substrate that is bound but that the enzyme cannot cleave. Presumably, other *endo*-splitting enzymes encounter even more extreme difficulties when carbohydrate is chemically linked adjacent to the lignin polymer because the side chains are generally greater than four residues. This could also be attributable to steric hindrance or to the presence of multiple cross links between the lignin and carbohydrate polymers.

If there is low enzymatic activity for a particular side chain, longer hemicellulosic branches are produced. This was evidenced with the apparent enrichment of galactomannan during enzymatic digestion of Norway spruce (Iversen et al. 1987). The low activity of β-(1→4)-D-galactanase in enzyme preparations from *Trichoderma reesei* and *Aspergillus niger* led to the enrichment of galactomannan in the milled wood enzyme lignin (MWEL). One must be on guard against these possibilities when studying the structures of lignin complexes prepared by enzymatic digestion.

Through selective enzymatic degradation, it should be possible to identify which carbohydrates form LC linkages. Joseleau and Gancet (1981) took such an approach in characterizing residual carbohydrate after enzymatic digestion, alkaline hydrolysis (0.5 M NaOH), and mild acid hydrolysis (0.01 M oxalic) of an LCC isolated from aspen by dimethylsulfoxide extraction. The original LCC contained mostly carbohydrate, and the final material was predominantly lignin. Enzyme treatments and alkali extraction removed the bulk of the xylose and glucose, while increasing the relative portions of arabinose and galactose. Rhamnose was also very persistent; its fraction relative to other sugars increased appreciably, indicating that rhamnose might be involved in an LC bond. Uronic acid likewise was significantly represented in the enzyme-, alkali-, and acid-treated material. It is not known whether uronic acid remains associated because of cross linkages between lignin and carbohydrate or simply because of the stability of glycosidic linkages between uronic acids and xylan.

Recent studies by Ross et al. (1992) have investigated the abilities of β-mannanases, xylanases, and acetyl xylan esterases to hydrolyze LCCs derived from *Populus deltoidies* by steam explosion. Synergism was observed to occur between enzymes, but negative cooperativity was also apparent in some instances. Substantial synergism existed between acety xylan esterase and xylanase. Less synergism was seen with mannanases. Johnson and Overend (1992) found that acetyl xylan esterase was essential for maximum liberation of reducing groups from LCCs of *Populus deltoides*. Prior to enzyme treatment, the LCCs appeared to be homogeneous: material absorbing at 280 nm and material analyzing as carbohydrate co-chromatographed on Sephadex G-50. Following deacetylation and enzyme treatment, the A_{280} (lignin) and carbohydrate fractions of the LCC could be separated. Generally, the A_{280} moiety was of higher molecular weight and the carbohydrate fraction was lower. The A_{280}

fraction was rich in lignin and uronic acid; the carbohydrate-rich fraction had a much lower concentration of lignin. Monomeric sugars were not formed.

1. Residual LC structures after exhaustive enzymatic digestion

The presence of lignin, aromatic acids, and other modifications of hemicellulose clearly retards digestion of cellulose and hemicellulose by ruminants. Phenolic acids associated with forage fibre are known to decrease fibre digestion when they are in the free state. *p*-Coumaric, ferulic, and sinapic acids inhibit the activity of rumen bacteria and anaerobic fungi (Akin and Rigsby 1985). It seems likely, however, that the cross linkages that these acids establish between hemicellulose chains and the lignin polymer are more important than is the toxicity attributable to the monomers themselves. Ruminant digestibility is greatly enhanced by alkali treatment, regardless of whether the phenolic acids are washed out of the residue.

In a comparison of different ruminant feeds, an increase in the extent of substitution at the *O*-5 position of arabinose correlated with the amount of residual material in digested residues (Chesson et al. 1983). The influence of the lignin, however, was out of proportion to its presence in the feed (Chesson 1988). When a cow digests grass, LCCs are solubilized in the rumen. This dissolution accounts for about half the total lignin intake (Gaillard and Richards 1975). At least two studies of LCCs isolated from ruminants have been published (Neilson and Richards 1982; Conchie et al. 1988). The amounts of residual carbohydrate in the LCC isolated from ruminants are relatively low (7.7% from bovine; 5.5% from sheep), and the glycosidic chains are relatively short. Methylation analysis indicates that unlike MWEL, the predominant side chain on rumen LCC is a single glucose residue bound glycosidically to the residual polyphenol. It is unclear how such a structure might arise, but it could result from transglycosylation reactions similar to those observed with the degradation of lignin model compounds (Kondo et al. 1988). Other branched chain structures are also observed, some of which are linked into the polyphenol at more than one location (Fig. 14). As in the case of MWEL, oligoxylosyl residues are also present, some of which are substituted with L-arabinose. Because of the anaerobic conditions that exist in the rumen, relatively little degradation of lignin is believed to occur.

The influence of LCCs on ruminant digestion was studied by examining the solubilization of LCCs using cell-free hemicellulase complexes from the rumen (Brice and Morrison 1982). LCCs from grasses of increasing maturity were isolated and treated with cell-free rumen hemicellulases. As the lignin content increased, the extent of degradation declined, indicating that the lignin content of the LC was the overriding factor in determining its digestibility.

Degradation of lignin does not seem to occur in the anaerobic environment of the rumen, even though substantial solubilization takes place. This observation was recently confirmed using an artificial rumen reactor (Kivaisi et al. 1990). The supernatant solutions of effluents contained lignin-derived compounds that were released by rumen microorganisms.

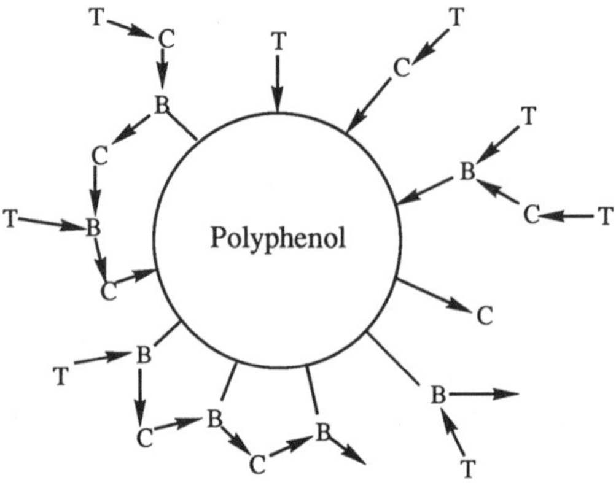

Fig. 14. Proposed structure for residual lignin carbohydrate complex from sheep rumen: T, nonreducing terminal sugar residue; C, chain residue; B, branch point; →, glycosidic linkage (arrow pointing away from C-1); —, ether linkage; →, free reducing group. (After Conchie, et al. 1988.)

2. Solubilization of LCC by microbial activity

In recent years, studies on the solubilization of lignin from grasses or wood labelled with ^{14}C phenylalanine have proliferated (Crawford 1978; Reid et al. 1982; McCarthy et al. 1984). The bulk of the radioactive label is incorporated into the lignin rather than into carbohydrate or protein of the plant, but it is clear that lignin purified from the labelled plant tissue contains significant amounts of carbohydrate. This material is probably closer to the structure of native lignin than is synthetic lignin prepared by *in vitro* dehydrogenative polymerization of coniferyl alcohol (Kirk et al. 1975).

Commonly, biodegradation of the MWL is followed by trapping the $^{14}CO_2$ respired from active cultures. Between 25% and 40% of the ^{14}C added to active cultures can be recovered as $^{14}CO_2$. Given that a fully aerobic organism will respire about half the carbon provided to it while incorporating the other half as cellular material, this represents the metabolism of 50% to 80% of the total lignin present. However, not all the ^{14}C lignin added to cultures is released as $^{14}CO_2$. A significant fraction of the total lignin – as much as 30% – can be recovered from solution as a polymer (Crawford et al. 1983). This material precipitates from culture filtrates following acidification to pH 3 to 5. In this characteristic, it is similar to the LCC solubilized in the rumen of cattle (Gaillard and Richards 1975). The solubilized acid precipitable polymeric lignin (APPL) has an apparent molecular weight of ≥ 20,000 and shows signs of partial degradation. *Phanerochaete chrysosporium* will form water-soluble products from lignin isolated from aspen (Reid et al. 1982) or wheat seedlings (McCarthy et al. 1984), as will a number of actinomycetes. Organisms reported to solubilize grass lignins include *Streptomyces viridosporus*, *S. baddius* (Pometto and Crawford 1986), *S. cyanus*, *Thermomonospora mesophila*, and *Actinomadura* sp. (Mason et al. 1988, 1990; Zimmerman and Broda 1989).

Many different enzymatic activities from these organisms have been reported, including activities of *endo*-glucanase, xylanase, several esterases, and an extracellular peroxidase (Ramachandra et al. 1987). The roles that these enzymes play in lignin solubilization are not entirely clear, but various correlations were made between the appearance of extracellular peroxidase activity and lignin solubilization or mineralization. The streptomycete lignin peroxidase was purified and partially characterized (Ramachandra et al. 1988). It is a heme protein with an apparent molecular weight of 17,800. As such, it is appreciably smaller than the 42,000 lignin peroxidase or the 46,000 manganese peroxidase described in *P. chrysosporium* and other white-rot fungi (Tien and Kirk 1983, 1984; Paszczynski et al. 1985). The streptomycete enzyme was reported to cleave a β-aryl ether model dimer in a manner similar to that observed for *P. chrysosporium* (Ramachandra et al. 1988).

The enzyme (or enzymes) responsible for solubilizing lignin has not been fully characterized. Strains of streptomyces that produce significant quantities of APPL also produce peroxidases and cellulases (Adhi et al. 1989). With *T. mesophila*, the ability to solubilize labelled wheat lignocellulose is extracellular and inducible, but it does not correlate with xylanase or cellulase production (McCarthy et al. 1986).

Zimmerman and Broda (1989) recently reported the solubilization of lignocellulose from undigested ball-milled barley by the extracellular broth from *S. cyanus*, *T. mesophila*, and *Actinomadura* sp. Ground and ball-milled barley straw samples were incubated with cultures of these organisms, and weight losses were recorded. Such solubilization might simply represent the release of low molecular weight ^{14}C-labelled lignin from the carbohydrate complex. This could occur if a portion of the lignin is released as the carbohydrate is removed. Mason et al. (1988), working in the same laboratory, described the production of extracellular proteins in the broths of these cultures. They assayed fractions from a sizing gel and found that the solubilizing activity had an apparent molecular weight of about 20,000 – similar to the peroxidase reported by Ramachandra et al. (1988). Because all the xylanases and cellulases present had an apparent molecular weight of about 45,000, the authors concluded that the solubilization activity is unlikely to be from a cellulase or xylanase. An international patent application for the use of a cell-free enzyme from *S. cyanus* for the solubilization of lignocellulose was filed (Broda et al. 1987).

Esterases from *S. viridosporus* were reported to release *p*-coumaric and vanillic acid into the medium when concentrated, extracellular enzyme was placed on appropriately labelled substrates (Donnelly and Crawford 1988). At least eight different esterases were described. When grown on lignocellulose from corn stover, *S. viridosporus* T7A and *S. badius* 252 produced endoglucanase, xylanase, and lignin peroxidase activity (Adhi et al. 1989). Since the *p*-coumaric acid is known to be esterified essentially only to the lignin, these organisms apparently attack that substrate. In the process, they produce an acid precipitable polymeric lignin (APPL) that can be detected in supernatant solutions.

In summary, research on the solubilization of LCC by microbial activity has periodically shown that cellulases, hemicellulases, esterases, and perhaps peroxidases all correlate with lignin solubilization. The mineralization rates and extents reported for streptomyces are relatively low, and the solubilized lignin is not extensively modified. Lignin mineralization and solubilization could, therefore be attributable to two (or more) different enzymes. These studies require additional rigorous clarification.

V. Enzymatic treatments of pulps

For many applications, residual lignin in kraft pulp must be removed by bleaching. Successive chlorination and alkali extraction remove the remaining lignin to leave a bright, strong pulp suitable for printing papers and other consumer products. Although chlorine bleaching solves the immediate problem of residual lignin, the chlorinated aromatic hydrocarbons produced in the bleaching step are recalcitrant and toxic. These chlorinated products are hard to remove from waste streams, and trace quantities are left in the paper, so other bleaching processes were devised. One approach is to use hemicellulases to facilitate bleaching.

Several different research groups found that the bleaching of hardwood and softwood kraft pulp can be enhanced by xylanase. The xylanase treatment decreases chemical consumption and kappa number and increases brightness. Pine kraft pulp was delignified > 50% following hemicellulase treatment and oxygen bleaching (Kantelinen et al. 1988a). Small amounts of lignin were released by the enzyme treatment alone. Fungal xylanase from *Sporotricum dimorphosum* lowered the lignin content of unbleached softwood and hardwood kraft pulps (Chauvet et al. 1987). Hemicellulases having different specificities for substrate DP and side groups have been used, but more data are needed on the effects of mannanases, cellulases, and other enzymes.

Enzyme-treated paper sheets show slight decreases in interfibre bonding strength. The mechanical strength of fibres is not affected, but interfibre bonding decreases if cellulases are present. This has been confirmed with the pretreatment of pulp with a xylanase from the thermophilic actinomycete *Saccharomonospora viridis* (Roberts et al. 1990). Viscosity decreases with some enzyme treatments, but only limited hemicellulose hydrolysis is necessary to enhance lignin removal. In the absence of cellulase, xylanase treatment increases viscosity. Pulps treated with cloned xylanase from *Bacillus subtilis* retained viscosity and strength properties while lignin removal was facilitated (Jurasek and Paice 1988; Paice et al. 1988a,b).

The mechanism by which xylanases facilitate bleaching is not completely understood. It was hypothesized that the enzymes open up the porosity of the pulp in order to allow the free diffusion of bleaching chemicals or that they sever linkages between the residual lignin and carbohydrate. But more recently, Paice et al. (1992) and Kantelinen et al. (1992) suggested that chromophores generated during the kraft pulping process are trapped on the surfaces of the cellulose fibres when xylan precipitates near the end of the pulping process.

VI. Conclusions

Lignin and hemicellulose are complex polymers occuring in plant materials. Either polymer alone presents a formidable challenge to microbial degradation. In native substrates, however, lignin and hemicellulose are intermeshed and chemically bonded through covalent cross-linkages. As such, they are even more resistant. Covalent lignin-carbohydrate linkages can be divided into two types: ester linkages through the free carboxy terminus of uronic and aromatic acids and ether linkages through sugar hydroxyls. Linkages to sugar hydroxyls can either be through the primary hydroxyl of L-arabinose (*O*-5) or of D-glucose or D-mannose (*O*-6) or through the secondary hydroxyl, as in the case of the *O*-2 or *O*-3 of xylose. Some linkages through the glycosidic hydroxyl (*O*-1) also appear to exist. In all cases, the α-carbon of the phenylpropane subunit in lignin appears to be involved in native lignin-carbohydrate bonds. In kraft pulps, the picture is less clear because many rearrangements occur during pulping. Although some microbial esterases and other enzymes have been shown to attack and solubilize lignin from lignocellulose, their substrate specificities have not been fully characterized. The positive effect of glycosylation on assimilation and degradation of lignin model compounds is supported by long-standing evidence that carbohydrate is necessary for lignin mineralization. The role of carbohydrates linked by nonglycosidic bonds remains to be clarified. It is not clear what role, if any, uronic and aromatic acid esters might play in facilitating or hindering lignin biodegradation. The principal xylanases and lignin-degrading enzymes have been identified from a few species of fungi and bacteria, but the reasons for their multiplicity and the range of their substrate specificities are only poorly understood. The consortia of organisms and enzymes found in nature has not been examined comprehensively *in vitro*, and much remains to be learned about their synergistic interactions. To date, no enzyme has been shown to cleave any bond between polymeric lignin and carbohydrate.

References

Adhi TP, Korus RA and Crawford DL (1989) Production of major extracellular enzymes during lignocellulose degradation by two *Streptomyces* in agitated submerged culture. Appl. Environ. Microbiol. 55: 1165–1168

Akin DE and Rigsby LL (1985) Influence of phenolic acids on rumen fungi. Agronomy J. 77: 180–185

Araujo A and Ward OP (1990) Hemicellulases of *Bacillus* species: preliminary comparative studies on production and properties of mannanases and galactanases. J. Appl. Bacteriol. 68: 253–261

Aspinall GO (1959) Structural chemistry of the hemicelluloses. Adv. Carbohyd. Chem. 14: 429–468

Atsushi K, Azuma J-I and Koshijima T (1984) Lignin-carbohydrate complexes and phenolic acids in bagasse. Holzforschung 38: 141–149

Azuma J-I and Koshijima T (1988) Lignin-carbohydrate complexes from various sources. Methods Enzymology 161: 12–18

Azuma J-I, Takahashi N and Koshijima T (1981) Isolation and characterization of lignin-carbohydrate complexes from the milled-wood lignin fraction of *Pinus densiflora* Sieb et Zucc. Carbohyd. Res. 93: 91–104

Azuma J-I, Takahashi N, Isaka M and Koshijima T (1985) Lignin-carbohydrate complexes extracted with aqueous dioxane from beech wood. Mokuzai Gakkaishi 31: 587–594

Biely P (1985) Microbial xylanaolytic systems. Trends Biotechnol. 3: 286–290

Biely P, Vršanská M and Krátký Z (1980) Xylan-degrading enzymes of the yeast *Cryptococcus albidus*. Identification and cellular localization. Eur. J. Biochem. 108: 313–321

Biely P, Krátky Z and Vršanská M (1981) Substrate-binding site of endo-1,4-β-xylanase of the yeast *Cryptococcus albidus*. Eur. J. Biochem. 119: 559–564

Biely P, Markovič and Mislovičová D (1985a) Sensitive detection of endo-1,4-β-glucanases and endo-1,4-β-xylanases in gels. Anal. Chem. 144: 147–151

Biely P, Puls J and Schneider H (1985b) Acetyl xylan esterases in fungal xylanolytic systems. FEBS 186: 80–84

Biely P, MacKenzie CR, Puls J and Schneider H (1986) Cooperativity of esterases and xylanases in the enzymatic degradation of acetyl xylan. Bio/Technology 4: 731–733

Biely P, Markovič and Toman R (1988) Remazol brilliant blue-xylan: a soluble chromogenic substrate for xylanases. Meth. Enzymol. 160: 536–541

Biswas-Hawkes DA, Dodson PJ, Harvey PJ and Palmer JM (1987) Ligninases from white-rot fungi. In: E Odier (ed) Lignin Enzymic and Microbial Degradation (pp 125–130). INRA Publications, Paris

Björkman A (1956) Studies on finely divided wood. Part 1. Extraction of lignin with neutral solvents. Svensk Papperstid. 59: 477–485

Björkman A (1957) Studies on finely divided wood. Part 3. Extraction of lignin-carbohydrate complexes with neutral solvents. Svensk Papperstid. 60: 243–251

Björkman A and Person B (1957) Studies on finely divided wood. Part 2. The properties of lignins extracted with neutral solvents from softwoods and hardwoods. Svensk Papperstid. 60: 158–169

Blanchette RA (1984) Manganese accumulation in wood decayed by white-rot fungi. Phytopathology 74: 725–730

Bonnarme P and Jeffries TW (1990) Mn(II) regulation of lignin peroxidases and manganese-dependent peroxidases from lignin-degrading white-rot fungi. Appl. Environ. Microbiol. 56: 210–217

Borneman WS, Hartley RD, Morrison WH, Akin DE and Ljungdahl LG (1990) Feruloyl and *p*-coumaroyl esterase from anaerobic fungi in relation to plant cell wall degradation. Appl. Microbiol. Biotechnol. 33: 345–351

Bourbonnais R and Paice MG (1988) Veratryl alcohol oxidases from the lignin-degrading basidiomycete *Pleurotus sajor-caju*. Biochem. J. 255: 445–450

Brice RE and Morrison IM (1982) The degradation of isolated hemicelluloses and lignin-hemicellulose complexes by cell-free rumen hemicellulases. Carbohyd. Res. 101: 93–100

Broda PMA, Mason JC and Zimmerman WK (1987) Decomposition of lignocellulose. International Patent WO 87/06609

Brown JA, Glenn JK and Gold MH (1990) Manganese regulates expression of manganese peroxidase by *Phanerochaete chrysosporium*. J. Bacteriol. 172: 3125–3130

Chang H-M, Cowling EB, Brown W, Adler E and Miksche G (1975) Comparative studies on cellulolytic enzyme lignin and milled wood lignin of sweetgum and spruce. Holzforschung 29: 153–159

Chauvet J-M, Comtat J and Noe P (1987) Assistance in bleaching of never-dried pulps by the use of xylanases: consequences on pulp properties. 4th Intl. Symp. Wood Pulping Chem. (Paris), Poster Presentations Vol 2: 325–327

Chesson A (1988) Lignin-polysaccharide complexes of the plant cell wall and their effect on microbial degradation in the rumen. Animal Feed Sci. Technol. 21: 219–228

Chesson A, Gordon AH and Lomax JA (1983) Substituent groups linked by alkali-labile bonds to arabinose and xylose residues of legume, grass and cereal straw cell walls and their fate during digestion by rumen microorganisms. J. Sci. Food. Agric. 34: 1330–1340

Comtat J, Joseleau J-P, Bosso C and Barnoud (1974) Characterization of structurally similar neutral and acidic tetrasaccharides obtained from the enzymic hydrolysate of a 4-*O*-methyl-D-glucurono-D-xylan. Carbohyd. Res. 38: 217–224

Conchie J, Hay AJ and Lomax JA (1988) Soluble lignin-carbohydrate complexes from sheep rumen fluid: their composition and structural features. Carbohyd. Res. 177: 127–151

Crawford D (1978) Lignocellulose decomposition by selected *Streptomyces* strains. Appl. Environ. Microbiol. 35: 1041–1045

Crawford DL, and Crawford RL (1976) Microbial degradation of lignocellulose: the lignin component. Appl. Environ. Microbiol. 31: 714–717

Crawford DL, Pometto AL and Crawford RL (1983) Lignin degradation by *Streptomyces viridosporus*: isolation and characterization of a new polymeric lignin degradation intermediate. Appl. Environ. Microbiol. 45: 898–904

Cui F and Dolphin D (1990) The role of manganese in model systems related to lignin biodegradation. Holzforschung 44: 279–283

Dekker RF (1985) Biodegradation of the hemicelluloses. In: T Higuchi (ed) Biosynthesis and Biodegradation of Wood Components (pp 505–532). Academic Press, New York

Dekker RFH and Richards GN (1976) Hemicellulases: their occurrence, purification, properties and mode of action. Adv. Carbohyd. Chem. Biochem. 32: 277–352

Deobald LE and Crawford DL (1987) Activities of cellulase and other extracellular enzymes during lignin solubilization by *Streptomyces viridosporus*. Appl. Microbiol. Biotechnol. 26: 158–163

Dobberstein J and Emeis CC (1991) Purification and Characterization of β-xylosidase from *Aureobasidium pullulans*. Appl. Microbiol. Biotechnol. 35: 210–215

Donnelly PK and Crawford DL (1988) Production by *Streptomyces viridosporus* T7A of an enzyme which cleaves aromatic acids from lignocellulose. Appl. Environ. Microbiol. 54: 2237–2244

Ericksson Ö and Lindgren BO (1977) About the linkage between lignin and hemicelluloses in wood. Svensk Papperstind. 80: 59–63

Ericksson Ö, Goring DAI and Lindgren BO (1980) Structural studies on the chemical bonds between lignins and carbohydrates in spruce wood. Wood Sci. Technol. 14: 267–279

Evans CS and Palmer JM (1983) Ligninolytic activity of *Coriolus versicolor*. J. Gen. Microbiol. 129: 2103–2108

Fengel D and Wegner D (1984) Wood Chemistry, Ultrastructure, Reactions. Walter de Gruyter, Berlin, New York

Ferrer I, Esposito E and Durán N (1992) Lignin peroxidase from *Chrysonilia sitophilia*: heat denaturation kinetics and pH stability. Enzyme Microb. Technol. 14: 402–406

Ford CW (1989) A feruloylated arabinoxylan liberated from cell walls of *Digitaria decumbens* (pangola grass) by treatment with borohydride. Carbohyd. Res. 190: 137–144

Ford CW (1990) Borohydride-soluble lignin-carbohydrate complex esters of *p*-coumaric acid from the cell walls of a tropical grass. Carbohyd. Res. 201: 299–309

Freudenberg K (1968) The constitution and biosynthesis of lignin. In: K Freudenberg and AC Neish AC (eds) Constitution and Biosyntheses of Lignin (pp 45–122). Springer-Verlag, Berlin

Fukuzumi T (1987) Ligninolytic enzymes of *Pleurotus sajor-caju*. In: E Odier (ed) Lignin Enzymic and Microbial Degradation (pp 137–142). INRA Publications, Paris

Gaillard BDE and Richards GN (1975) Presence of soluble lignin-carbohydrate complexes in the bovine rumen. Carbohyd. Res. 42: 135–145

Galliano H, Gas G, Seris JL and Boudet AM (1991) Lignin degradation by *Rigidoporous lignosus* involves synergistic action of two oxidizing enzymes: Mn peroxidase and laccase. Enzyme Microb. Technol. 13: 478–482

Ghose TK and Bisaria VS (1987) Measurement of hemicellulase activities. Part 1: Xylanases. Pure and Appl. Chem. 59: 1739–1752

Gierer J (1970) The reactions of lignin during pulping. Svensk Papperstid. 73: 571–596

Gierer J and Wännström S (1984) Formation of alkali-stable C-C-bonds between lignin and carbohydrate fragments during kraft pulping. Holzforschung 38: 181–184

Gilkes NR, Henrissat B, Kilburn DG, Miller RC and Warren RAJ (1991) Domains in microbial β-1,4-glycanases: sequence conservation, function, and enzyme families. Microb. Rev. 55: 303–315

Glenn, JK and Gold MH (1985) Purification and characterization of an extracellular Mn(II)-dependent peroxidase from the lignin-degrading basidiomycete *Phanerochaete chrysosporium*. Arch. Biochem. Biophys. 242: 329–341

Glenn JK, Morgan MA, Mayfield MB, Kuwahara M and Gold MH (1983) An extracellular H_2O_2-requiring enzyme preparation involved in lignin biodegradation by the white-rot basidiomycete *Phanerochaete chrysosporium*. Biochem. Biophys. Res. Comm. 114: 1077–1083

Gold MH, Wariishi H, Akileswaran L, Mino Y, and Loehr TM (1987) Spectral characterization of Mn-peroxidase, an extracellular heme enzyme from *Phanerochaete chrysosporium*. In: E Odier (ed) Lignin Enzymic and Microbial Degradation (pp 113–118). INRA Publications, Paris

Grüninger H and Fiechter A (1986) A novel, highly thermostable D-xylanase. Enzyme Microb. Technol. 8: 309–314

Haemmerli SD, Liesola MSA and Fiechter A (1986) Polymerization of lignins by ligninases from *Phanerochaete chrysosporium*. FEMS Microbiol. Lett. 35: 33–36

Hammel KE and Moen MA (1991) Depolymerization of a synthetic lignin *in vitro* by lignin peroxidase. Enzyme Microb. Technol. 13: 15–18

Hammel KE, Tien M, Kalyanaraman B, and Kirk TK (1985) Mechanism of oxidative C_α-C_β cleavage of a lignin model dimer by *Phanerochaete chrysosporium* ligninase. J. Biol. Chem. 260: 8348–8353

Hamomoto T, Honda H, Kudo T and Horikoshi K (1987) Nucleotide sequence of the xylanase A gene of alkalophilic *Bacillus* sp. strain C-125. Agric Biol. Chem. 51: 953–955

Hartley RD (1973) Carbohydrate esters of ferulic acid as components of cell-walls of *Lolium multiflorum*. Phytochemistry 12: 661–665

Hartley RD and Jones EC (1976) Diferulic acid as a component of cell walls of *Lolium multiflorum*. Phytochemistry 15: 1157–1160

Hartley RD, Jones EC and Wood TM (1976) Carbohydrates and carbohydrate esters of ferulic acid released from cell walls of *Lolium multiflorum* by treatment with cellulolytic enzymes. Phytochemistry 15: 305–307

Hatfield RD, Helm RF and Ralph J (1991) Synthesis of methyl 5-*O*-*trans*-feruloyl-α-L-arabinosuranoside and its use as a substrate to assess feruloyl esterase activity. Anal. Biochem. 194: 25–33

Higuchi T (ed) (1985a) Biosynthesis and Biodegradation of Wood Components. Academic Press, Tokyo

Higuchi T (1985b) Biosynthesis of lignin. In: T Higuchi (ed) Biosynthesis and Biodegradation of Wood Components (pp 141–160). Academic Press, Tokyo

Higuchi T (1990) Lignin biochemistry: biosynthesis and biodegradation. Wood Sci. Technol. 24: 23–63

Higuchi T, Ioto Y, Shimada M and Kawamura I (1967) Chemical properties of milled wood lignin of grasses. Phytochemistry 6: 1551–1556

Holtz C, Kaspari H and Klemme J-H (1991) Production and properties of xylanases from thermophilicactinomycetes. Antonie van Leeuwenhoek 59: 1–7

Honda H, Kudo T, Ikura Y and Horikoshi K (1985) Two types of xylanases of alkalophilic *Bacillus* sp. no. C-125. Can. J. Microbiol. 31: 538–542

Hortling B, Ranua B and Sundquist J (1990) Investigation of the residual lignin in chemical pulps. Part 1. Enzymatic hydrolysis of the pulps and fractionation of the products. Nordic Pulp Paper Res. J. 1: 33–37

Huynh V-B, and Crawford RL (1985) Novel extracellular enzymes (ligninases) of *Phanerochaete chrysosporium*. FEMS Microbiol. Lett. 28: 119–123

Ide JA, Daly JM and Rickards PAD (1983) Production of glycosidase activity by cellulomonas during growth on various carbohydrate substrates. Eur J. Appl. Microbiol. Biotechnol. 18: 100–102

Ishihara M and Shimizu K (1988) α-(1,2)-glucuronidase in the enzymatic saccharification of hardwood xylan I. Screening of a-glucuronidase producing fungi. Mokuzai Gakkaishi 34: 58–64

Ishii T and Hiroi T (1990) Isolation and characterization of feruloylated arabinoxylan oligosaccharides from bamboo shoot cell-walls. Carbohyd. Res. 196: 175–183

Iversen T (1985) Lignin-carbohydrate bonds in a lignin-carbohydrte complex isolated from spruce. Wood Sci. Technol. 19: 243–251

Iversen T and Wännström S (1986) Lignin-carbohydrate bonds in a residual lignin isolated from pine kraft pulp. Holzforschung 40: 19–22

Iversen T and Westermark U (1985) Lignin carbohydrate bonds in pine lignins dissolved during kraft pulping. Cellu. Chem. Technol. 19: 531–536

Iversen T, Westermark U and Samuelsson B (1987) Some comments on the isolation of galactose-containing lignin-carbohydrate complexes. Holzforschung 41: 119–121

Jeffries TW, Choi S and Kirk TK (1981) Nutritional regulation of lignin degradation by *Phanerochaete chrysosporium*. Appl. Environ. Microbiol. 42: 290–296

Johansson T, and Nyman PO (1987) A manganese(II)-dependent extracellular peroxidase from the white-rot fungus *Trametes versicolor*. Acta Chem. Scand. B41: 762–765

Johnson KG and Overend RP (1991) Lignin-carbohydrate complexes from *Populus deltoides* I. Purification and characterization. Holzforschung 45: 469–475

Johnson KG and Overend RP (1992) Lignin-carbohydrate complexes from *Populus deltoides*. II. Effect of hydrolyzing enzymes. Holzforschung 46: 31–37

Johnson KG, Harrison BA, Schneider H, Mackenzie CR and Fontana JD (1988) Xylan-hydrolyzing enzymes from *Streptomyces* spp. Enzyme Microb. Technol. 10: 403–409

Johnson KG, Silva MC MacKenzie CR, Schneider H and Fontana JD (1989) Microbial degradation of hemicellulosic materials. Appl. Biochem. Biotechnol. 20/21: 245–258

Johnson KG, Ross NW and Schneider H (1990) Purification and some properties of β-mannanase from *Polyporous versicolor*. World J. Microbiol. Biotechnol. 6: 245–254

Jönsson L, Johansson T, Sjöström K and Nyman PO (1987) Purification of ligninase isozymes from the white-rot fungus *Trametes versicolor*. Acta Chem. Scand. B41: 766–769

Joseleau J-P, and Gancet C (1981) Selective degradations of the lignin-carbohydrate complex from aspen wood. Svensk Papperstidning 84: R123–R127

Jurasek L and Paice MG (1988) Biological beaching of pulp. Tappi Internat. Pulp Bleach. Conf., Orlando, Florida, pp 11–13

Kaji A (1984) L-Arabinosidases. Advan. Carbohyd. Chem. Biochem. 42: 383–394.

Kantelinen A, Rättö M, Sundquist J, Ranua M, Viikari L and Linko M (1988a) Hemicellulases and their potential role in bleaching. Tappi Internat. Pulp Bleaching Conf., Orlando, Florida, pp 1–9

Kantelinen A, Waldner R, Niku-Paavola M-L, and Leisola MSA (1988b) Comparison of two lignin-degrading fungi: *Phlebia radiata* and *Phanerochaete chrysosporium*. Appl. Microbiol. Biotechnol. 28: 193–198

Kantelinen A, Sundquist J, Linko M and Viikari L (1992) The role of reprecipitated xylan in the enzymatic bleaching of kraft pulp. Proceedings of the Sixth International Symposium on Wood and Paper Chemistry, pp 493–500

Kato A, Azuma JI and Koshijima T (1987) Isolation and identification of a new feruloylated tetrasaccharide from bagasse lignin-carbohydrate complex containing phenolic acid. Agric. Biol. Chem. 51: 1691–1693

Kawai S, Umezawa T and Higuchi T (1988) Degradation mechanisms of phenolic β-1 lignin substructure model compounds by laccase of *Coriolus versicolor*. Arch. Biochem. Biophys. 262: 99–110

Kern HW and Kirk TK (1987) Influence of molecular size and ligninase treatment on degradation of lignins by *Xanthomonas* sp. strain 99. Appl. Environ. Microbiol. 53: 2242–2246

Kern H, Haider K, Pool W, de Leeuw JW and Ernst L (1989) Comparison of the action of *Phanerochaete chrysosporium* and its extracellular enzymes (lignin peroxidases) on lignin preparations. Holzforschung 43: 375–384

Kersten P and Cullen D (1992) Fungal enzymes for lignocellulose degradation In: JR Kinghorn and G Turner (eds) Applied Molecular Genetics of Filamentous Fungi (pp 100–131). Blackie and Sons, London

Kersten PJ, Tien M, Kalyanaraman B, and Kirk TK (1985) The ligninase of *Phanerochaete chrysosporium* generates cation radicals from methoxybenzenes. J. Biol. Chem. 260: 2609–2612

Khan AW, Tremblay D and LeDuy L (1986) Assay of xylanase and xylosidase activities in bacterial and fungal cultures. Enzyme Microb. Technol. 8: 373–377

Khan AW, Lanm KA and Overend RP (1990) Comparison of natural hemicellulose and chemically acetylated xylan as substrates for the determination of acetyl-xylan esterase activity in *Aspergilli*. Enzyme Microb. Technol. 12: 127–131

Khowala S, Mukherjee M and Sengupta S (1988) Carboxymethyl xylan – a specific substrate directly differentiating backbone-hydrolyzing and side chain-reacting β-D-(1,4)-xylanases of the mushroom *Termitomyces clypeatus*. Enzyme Microb. Technol. 10: 563–567

Kirk TK and Chang H-M (1981) Potential applications of biolignolytic systems. Enzyme Microb. Technol. 3: 189–197

Kirk TK and Farrell RL (1987) Enzymatic "combustion": the microbial degradation of lignin. Ann Rev. Microbiol. 41: 465–505

Kirk TK and Shimada M (1985) Lignin biodegradation: the microorganisms involved and the physiology and biochemistry of degradation by white-rot fungi. In: T Higuchi (ed) Biosynthesis and Biodegradation of Wood Components (pp 579–605). Academic Press, San Diego, CA

Kirk TK, Connors WJ, Bleam RD, Hackett WF and Zeikus JG (1975) Preparation and microbial decomposition of synthetic [^{14}C] lignins. Proc. Natl. Acad. Sci. U.S.A. 72: 2515–2519

Kirk TK, Connors WJ and Zeikus JG (1976) Requirement for a growth substrate during lignin decomposition by two wood-rotting fungi. Appl. Environ. Microbiol. 32: 192–194

Kirk TK, Schultz E, Connors WJ, Lorenz LF and Zeikus JG (1978) Influence of culture parameters on lignin metabolism by *Phanerochaete chrysosporium*. Arch. Microbiol. 117: 277–285

Kivaisi AK, Op den Camp HJM, Lubberding HJ, Boon JJ and Vogels GD (1990) Generation of soluble lignin-derived compounds during degradation of barley straw in an artificial rumen reactor. Appl. Microbiol. Biotechnol. 33: 93–98

Kluepfel D, Daigneault N, Morosoli R and Sharek F (1992) Purification and characterization of a new xylanase (xylanase C) produced by *Streptomyces lividans* 66. Appl. Microbiol. Biotechnol. 36: 626–631

Kondo R and Imamura H (1987) The formation of model glycosides by wood-rotting fungi. Lignin enzymatic and microbial degradation. INRA Publications, Paris

Kondo R and Imamura H (1989a) Formation of lignin model xyloside in polysaccharides media by wood-rotting fungi. Mokuzai Gakkaishi 35: 1001–1007

Kondo R and Imamura H (1989b) Model study on the role of the formation of glycosides in the degradation of lignin by wood-rotting fungi. Mokuzai Gakkaishi 35: 1008–1013

Kondo R, Imori T and Imamura H (1988) Enzymatic synthesis of glucosides of monomeric lignin compounds with commercial β-glucosidase. Mokuzai Gakkaishi 34: 724–731

Kondo T, Hiroi T, Mizuno K and Kato T (1990a) Characterization of lignin-carbohydrate complexes of Italian ryegrass and alfalfa. Can. J. Plant Sci. 70: 193–201

Kondo R, Imori T, Imamura H and Kishida T (1990b) Polymerization of DHP and depolymerization of DHP glucoside by lignin oxidizing enzymes. J. Biotechnol. 13: 181–188

Kormlink FJM, Searle-van Leeuwen MJF, Wood TM and Voragen AGJ (1991a) (1,4)-β-D-arabinoxylan arabinofuranohydrolase: a novel enzyme in the bioconversion of arabinoxylan. Appl. Microbiol. Biotechnol. 35: 231–232

Kormlink FJM, Searle-van Leeuwen MJF, Wood TM and Voragen AGJ (1991b) Purification and characterization of a (1,4)-β-D-arabinoxylan arabinofuranohydrolase from *Aspergillus awamori*. Appl. Microbiol. Biotechnol. 35: 753–758

Koshijima T, Watanobe T and Yaku T (1989) Structure and properties of the lignin-carbohydrate complex polymer as an amphipathic substance. In: WG Glasser and S Sarkanen (eds) Lignin Properties and Materials. ACS Symposium Ser. 397 (pp 11–28). American Chemical Society, Washington, DC

Košíková B, Joniak D and Kosáková L (1979) On the properties of benzyl ether bonds in the lignin-saccharidic complex isolated from spruce. Holzforschung 33: 11–14

Kurek B, Monties B and Odier E (1990) Influence of the physical state of lignin on its degradability by the lignin peroxidase of *Phanerochaete chrysosporium*. Enzyme Microb. Technol. 12: 771–777

Lee H, To RJB, Latta RK, Biely P and Schneider H (1987) Some properties of extracellular acetylxylan esterase produced by the yeast *Rhodotorula mucilaginosa*. Appl. Environ. Microbiol. 53: 2831–2834

Leisola MSA and Garcia S (1989) The mechanism of lignin degradation. In: MP Coughlan (ed) Production, Characterization and Application of Cellulose, Hemicellulose and Lignin Degrading Enzyme Systems (pp 89–99). Elsevier, Amsterdam

Lewis NG and Paice MG (eds) (1989) Plant Cell Wall Polymers: Biogenesis and Biodegradation. ACS Symp. Ser. Vol. 399, American Chemical Society, Washington, DC

Lindberg B, Rosell KG and Svensson S (1973a) Positions of the O-acetyl groups in birch xylan. Svensk Papperstid. 76: 30–32

Lindberg B, Rosell KG and Svensson S (1973b) Positions of the O-acetyl groups in pine glucomannan. Svensk Papperstid.76: 383–384

Lobarzewski, J (1987) A lignin-biotransforming peroxidase from *Trametes versicolor* and its use in immobilized form. In: E Odier (ed) Lignin Enzymic and Microbial Degradation (pp 197–202). INRA Publications, Paris

Lundquist K, Simonson R and Tingsvik K (1983) Lignin carbohydrate linkages in milled wood lignin preparations from spruce wood. Svensk Papperstid. R44–R47

Lüthi E, Reif K, Jasmat NB and Bergquist PL (1992) In vitro mutagenesis of a xylanase from the extreme thermophile *Caldocellum saccharolyticum*. Appl. Microbiol. biotechnol. 36: 503–506

MacKenzie CR and Bilous D (1988) Ferulic acid esterase activity from *Schizophyllum commune*. Appl. Environ. Microbiol. 54: 1170–1173

MacKenzie CR, Bilous D, Schneider H and Johnson KG (1987) Induction of cellulolytic and xylanolytic enzyme systems in *Streptomyces* spp. Appl. Environ. Microbiol. 53: 2835–2839

Markwalder HU and Neukom H (1976) Diferulic acid as a possible crosslink in hemicelluloses from wheat germ. Phytochemistry 15: 836–837

Mason JC, Richards M, Zimmerman W and Broda P (1988) Identification of extracellular proteins from actinomycetes responsible for the solubilization of lignocellulose. Appl. Microbiol. Biotechnol. 28: 276–280

Mason JC, Birch OM and Broda P (1990) Preparation of ^{14}C-radiolabelled lignocelluloses from spring barley of differing maturities and their solubilization by *Phanerochaete chrysosporium* and *Streptomyces cyanus*. J. Gen. Microbiol. 136: 227–232

McCarthy AJ, MacDonald MJ, Paterson A and Broda P (1984) Degradation of ^{14}C] lignin-labelled wheat lignocellulose by white-rot fungi. J. Gen. Microbiol. 130: 1023–1030

McCarthy AJ, Paterson A and Broda P (1986) Lignin solubilization by *Thermonospora mesophila*. Appl. Microbiol. Biotechnol. 24: 347–352

McDermid KP, MacKenzie CR and Forsberg CW (1990) Esterase activities of *Fibrobacter succinogenes* subsp *Succinogenes* S85. Appl. Environ. Microbiol. 56: 127–132

Merewether JWT (1957) A lignin-carbohydrate complex of wood. Holzforschung 11: 65–80

Meshitsuka G, Lee ZZ, Nakano J and Eda S (1983) Contribution of pectic substances to lignin-carbohydrate bonding. Int. Symp. Wood Pulping Chem. 1: 149–152

Michel FC, Dass SB, Grulke EA and Reddy CA (1991) Role of manganese peroxidases and lignin peroxidases of *Phanerochaete chrysosporium* in the decolorization of kraft bleach plant effluent. Appl. Environ. Microbiol. 57: 2368–2375

Miller GL (1959) Use of Dinitorsalicylic acid reagent for determination of reducing sugar. Anal. Chem. 31: 426–428

Miller GL, Blum R, Glennon WE and Burton AL (1960) Measurement of carboxymethylcellulose activity. Anal. Biochem. 2: 127–132

Milstein O, Vered Y, Sharma A, Gressel J and Flowers HM (1983) Fungal biodegradation and biotransformation of soluble lignocarbohydrate complexes from straw. Appl. Environ. Microbiol. 46: 55–61

Minor JL (1982) Chemical linkage of pine polysaccharides to lignin. J. Wood Chem. Technol. 2(1): 1–16

Minor JL (1986) Chemical linkage of polysaccharides to residual lignin in loblolly pine kraft pulps. J. Wood. Chem. Technol. 6(2): 185–201

Morrison IM (1974) Structural investigation on the lignin-carbohydrate complexes of *Lolium perene*. Biochem J. 139: 197–204

Morosoli R, Bertrand JL, Mondou F, Sharek F and Kluepfel D (1986) Purification and properties of a xylanase from *Streptomyces lividans*. Biochem. J. 239: 587–592

Mukoyoshi SI, Azuma JI and Koshijima T (1981) Lignin-carbohydrate complexes from compression wood of *Pinus densiflora* Sieb et. Zucc. Holzforschung 35: 233–240

Nanmori T, Watanabe T, Shinke R, Kohno A and Kawamura Y (1990) Purification and properties of thermostable xylanase and β-xylosidase produced by a newly isolated *Bacillus sterothermophilus* strain. J. Bacteriol. 172: 6669–6672

Neilson MJ and Richards GN (1982) Chemical structures in a lignin-carbohydrate complex isolated from bovine rumen. Carbohyd. Chem. 104: 121–138

Nelson N (1944) A photometric adaptation of the Somogyi method for the determination of glucose. J. Biol. Chem. 153: 375–380

Nerud F and Mišurcová Z (1989) Production of ligninolytic peroxidases by the white-rot fungus *Coriolopsis occidentalis*. Biotechnol. Lett. 11: 427–432

Niku-Paavola M-L (1987) Ligninolytic enzymes of the white-rot fungus *Phlebia radiata*. In: E Odier (ed) Lignin Enzymic and Microbial Degradation (pp 119–123). INRA Publications, Paris

Niku-Paavola M-L, Karhunen E, Salola P and Raunio V (1988) Ligninolytic enzymes of the white-rot fungus *Phlebia radiata*. Biochem. J. 254: 877–884

Obst JR (1982) Frequency and alkali resistance of lignin-carbohydrate bonds in wood. Tappi 65(4): 109–112

Obst JR and Kirk TK (1988) Isolation of lignin. Meth. Enzymol. 161: 3–12

Odier E, Mozuch M, Kalyanaraman B and Kirk TK (1987) Cellobiose: quinone oxidoreductase does not prevent oxidative coupling of phenols or polymerization of lignin by ligninase (pp 131–136). INRA Publications, Paris

Paice MG, Bernier R and Jurasek L (1988a) Viscosity-enhancing bleaching of hardwood kraft pulp with xylanase from a cloned gene. Biotechnol. Bioeng. 32: 235–239

Paice MG, Bernier R and Jurasek L (1988b) Bleaching hardwood kraft with enzymes from cloned systems. CPPA Ann. Mtg. (Montreal) preprints 74A: 133–136

Paice MG, Gurnagul N, Page DH and Jurasek L (1992) Mechanism of hemicellulose-directed prebleaching of kraft pulps. Enzyme Microb. Technol. 14: 272–276

Paszczyński A, Huynh V-B and Crawford R (1985) Enzymatic activities of an extracellular manganese-dependent peroxidase from *Phanerochaete chrysosporium*. FEMS Microbiol. Lett. 29: 37–41

Paszczyński, A, Huynh V-Band Crawford R (1986) Comparison of ligninase-1 and peroxidase-M2 from the white-rot fungus *Phanerochaete chrysosporium*. Arch. Biochem. and Biophys. 244: 750–765

Périé FH and Gold MH (1991) Manganese regulation of manganese peroxidase expression and lignin degradation by the white-rot fungus *Dichomitus squalens*. Appl. Environ. Microbiol. 57: 2240–2245

Perez J and Jeffries TW (1990) Mineralization of ^{14}C-ring-labelled synthetic lignin correlates with the production of lignin peroxidase, not of manganese peroxidase or laccase. Appl. Environ. Microbiol. 56: 1806–1812

Perez J and Jeffries TW (1992) The roles of manganese and organic acid chelators in regulating lignin degradation and biosynthesis of peroxidases by *Phanerochaete chrysosporium*. Appl. Environ. Microbiol. 58: 2402–2409

Pew JC (1957) Properties of powdered wood and isolation of lignin by cellulytic enzymes. Tappi 40: 553–558

Pometto AL and Crawford DL (1986) Catabolic fate of *Streptomyces viridosporus* T7A-produced, acid-precipitable polymeric lignin upon incubation with ligninolytic *Streptomyces* species and *Phanerochaete chrysosporium*. Appl. Environ. Microbiol. 51: 171–179

Poutanen K and Sundberg M (1988) An acetyl esterase of *Trichoderma reesei* and its role in the hydrolysis of acetyl xylans. Appl. Microbiol. Biotechnol. 28: 419–424

Poutanen K, Ratto M, Puls J and Viikari L (1987) Evaluation of different microbial xylanaolytic systems. J. Biotechnol. 6: 49–60

Puls J, Schmidt O and Granzow C (1987) α-Glucuronidase in two microbial xylanolytic systems. Enzyme Microb. Technol. 9: 83–88

Ramachandra M, Crawford DL and Pometto AL (1987) Extracellular enzyme activities during lignocellulose degradation by *Streptomyces* spp.: a comparative study of wild-type and genetically manipulated strains. Appl. Environ. Microbiol. 53: 2754–2760

Ramachandra M, Crawford DL and Hertel G (1988) Characterization of an extracellular lignin peroxidase of the lignocellulolytic actinomycete *Streptomyces viridosporus*. Appl. Environ. Microbiol. 54: 3057–3063

Ratto M and Poutanen K (1988) Production of mannan degrading enzymes. Biotechnol. Lett. 10: 661–664

Reid ID, Abrams GD and Pepper JM (1982) Water soluble products from the degradation of aspen lignin by *Phanerochaete chrysosporium* Can. J. Bot. 60: 2357–2364

Reilly PJ (1981) Xylanases: Structure and function. In: A Hollaender (ed) Trends in Biology of Fermentations for Fuels and Chemicals (pp 111–129). Plenum Press, New York

Reiser J, Kälin M, Walther I and Fiechter A (1989) Structure and expression of lignin peroxidase genes. In: MP Coughlan MP (ed) Production, Characterization and Application of Cellulose, Hemicellulose and Lignin Degrading Enzyme Systems (pp 135–146). Elsevier, Amsterdam

Renganathan V, Usha SN and Lindenburg F (1990) Cellobiose-oxidizing enzymes from the lignocellulose-degrading basidiomycete *Phanerochaete chrysosporium*: interaction with microcrystalline cellulose. Appl. Microbiol. Biotechnol. 32: 609–613

Ristoph DL and Humprey AE (1985) Kinetic characterization of the extracellular xylanases of *Thermonospora* sp. Biotechnol. Bioeng. 27: 832–836

Ritter D, Jaklin-Farcher S, Messner K and Stachelberger H (1990) Polymerization and depolymerization of lignosulphonate immobilized on foam. J. Biotechnol. 13: 229–241

Roberts JC, McCarthy AJ, Flynn NJ and Broda P (1990) Modification of paper properties by the pretreatment of pulp with *Saccharomonospora viridis* xylanase. Enzyme Microb. Technol. 12: 210–213

Ross NW, Johnson KG, Braun C, MacKenzie CR and Schneider H (1992) Enzymatic hydrolysis of water soluble lignin-carbohydrate complexes from *Populus deltoides*: effects of combinations of β-mannanases, xylanase, and acetyl xylan esterase. Enzyme Microb. Technol. 14: 90–95

Roybt JF and Whelan WH (1972) Reducing value methods for maltodextrins: I. Chain length dependence of alkaline 3,5-dinitrosalicylate and chain-length independence of alkaline copper. Anal. Biochem. 45: 510–516

Royer JC and Nakas JP (1989) Xylanase production by *Trichoderma longibrachiatum*. Enzyme. Microb. Technol. 11: 405–410

Scalbert A, Monties B, Lallemand JY, Guittet E and Rolando C (1985) Ether linkage between phenolic acids and lignin fractions from wheat straw. Phytochemistry 24: 1359–1362

Schmidt O and Kebernik U (1988) A simple assay with dyed substrates to quantify cellulase and hemicellulase activity of fungi. Biotechnol. Tech. 2: 153–158

Shimada M, Fukuzuka T and Higuchi T (1971) Ester linkages of *p*-coumaric acid in bamboo and grass lignins. Tappi 54: 72–78

Shoemaker HE (1990) On the chemistry of lignin biodegradation. Recl. Trav. Chim. Pays-Bas 109: 255–272

Simpson HD, Haufler UR and Daniel RM (1991) An extremely thermostable xylanase from the thermophilic eubacterium *Thermotoga*. Biochem J. 227: 413–417

Smith DCC (1955) Ester groups in lignin. Nature 176: 267–268

Somogyi M (1952) Notes on sugar determination. J. Biol. Chem. 195: 19–23

Sundberg M, Poutanen K, Markkanen P and Linko M (1990) An extracellular esterase of *Aspergillus awamori*. Biotechnol. Appl. biochem. 12: 670–680

Takahashi N and Koshijima T (1987) Properties of enzyme-unhydrolyzable residue of lignin-carbohydrate complexes isolated from beech wood. Wood Res. 74: 1–11

Takahashi N and Koshijima T (1988a) Molecular properties of lignin-carbohydrate complexes from beech (*Fagus crenata*) and pine *(Pinus densiflora)* woods. Wood Sci. Technol. 22: 177–189

Takahashi N and Koshijima T (1988b) Ester linkages between lignin and glucuronoxylan in a lignin-carbohydrate complex from beech (*Fagus crenata*) wood. Wood Sci. Technol. 22: 231–241

Tan LUL, Mayers P and Saddler JN (1987) Purification and characterization of a thermostable xylanase from a thermophilic fungus *Thermoascus aurantiacus*. Can. J. Microbiol. 33: 689–692

Tanabe H and Kobayashi Y (1986) Enzymatic maceration mechanism in biochemical pulping of mitsumata (*Edgeworthia papyrifera* Sieb. et Zucc.) bast. Agric. Biol. Chem. 50: 2779–2784

Tanabe H and Kobayashi Y (1987) Effect of lignin-carbohydrate complex on maceration of mitsumata (*Edgeworthia papyrifera* Sieb. et Zucc.) bast by pectinolytic enzymes from *Erwinia carotovora*. Holzforschung 41: 395–399

Tanabe H and Kobayashi Y (1988) Aggregate of pectic substances and lignin-carbohydrate complex in mitsumata (*Edgeworthia papyrifera* Sieb. et Zucc.) bast and its degradation by pectinolytic enzymes from *Erwinia cartovora*. Holzforschung 42: 47–52

Tien M and Kirk TK (1983) Lignin-degrading enzyme from hymenomycete *Phanerochaete chrysosporium* Burds. Science 221: 661–663

Tien M and Kirk TK (1984) Lignin-degrading enzyme from *Phanerochaete chrysosporium*: purification, characterization, and catalytic properties of a unique H_2O_2-requiring oxygenase. Proc. Natl. Acad. Sci. U.S.A. 81: 2280–2284

Timell TE (1962) Enzymatic hydrolysis of a 4-*O*-methylglucuronoxylan from the wood of white birch. Holzforschung 11: 436–447

Timell TW (1964) Wood hemicelluloses. Part I. Adv. Carbohyd. Chem. 19: 247–302

Timell TW (1965) Wood hemicelluloses. Part II. Adv. Carbohyd. Chem. 20: 409–493

Torrie JP, Senior DJ and Saddler JN (1990) Production of β-mannanases by *Trichoderma harzianum* E58. Appl. Microbiol. Biotechnol. 34: 303–307

Tsujibo H Sakamoto T, Nishino N, Hasegawa T and Inamori Y (1990) Purification and properties of three types of xylanases produced by an alkalophilic actinomycete. J. Appl. Bacteriol. 69: 398–405

Tuohy MG and Coughlan MP (1992) Production of thermostable xylan-degrading enzymes by *Talaromyces emersonii*. Bioresource Technol. 39: 131–137

Utt EA, Eddy CK, Keshav KF and Ingram LO (1991) Sequencing and expression of the *Butyrivibrio fibrisolvens xylB* gene encoding a novel bifunctional protein with β-D-xylosidase and alpha-L-arabinofuranosidase activities. Appl. Environ. Microbiol. 57: 1227–1234

Waldner R, Leisola MSA and Fiechter A (1988) Comparison of ligninolytic activities of selected white-rot fungi. Appl. Microbiol. Biotechnol. 29: 400–407

Wallace G, Chesson A, Lomax JA and Jarvis MC (1991) Lignin-carbohydrate complexes in graminaceous cell walls in relation to digestibility. Animal Feed Sci. Technol. 32: 193–199

Wang PY, Bolker HI and Purves CB (1967) Uronic acid ester groups in some softwoods and hardwoods. Tappi 50(3): 123–124

Wariishi H, Valli K and Gold MH (1991) *In vitro* depolymerization by manganese peroxidase of *Phanerochaete chrysosporium*. Biochem. Biophys. Res. Comm. 176: 269–275

Watanabe T and Koshijima T (1988) Evidence for an ester linkage between lignin and glucuronic acid in lignin-carbohydrate complexes by DDQ-oxidation. Agric. Biol. Chem. 52: 2953–2955

Watanabe T, Azma J and Koshijima T (1985) Isolation of lignin-carbohydrate complex fragments by adsorption chromatography. Mokuzai Gakkaishi 31: 52–53

Watanabe TJ, Ohnishi Y, Kaizu YS and Koshijima T (1989) Binding site analysis of the ether linkages between lignin and hemicelluloses in lignin-carbohydrate complexes by DDQ-oxidation. Agric. Biol. Chem. 53: 2233–2252

Westermark U and Ericksson KE (1974a) Carbohydrate-dependent enzymic quinone reduction during lignin degradation. Acta Chem. Scand. B 28: 204–208

Westermark U and Ericksson KE (1974b) Cellobiose-quinone oxidoreductase, a new wood-degrading enzyme from white-rot fungi. Acta Chem. Scand. B 28: 209–214

Wilkie KCB (1979) The hemicelluloses of grasses and cereals. Adv. Carbohyd. Chem. Biochem 36: 215–264

Wong KKY, Tan LUL and Saddler JN (1988) Multiplicity of β-1,4-xylanase in microorganisms: functions and applications. Microbiol. Rev. 52: 305–317

Woodward J (1984) Xylanases: Functions, properties and applications. In: A Wiseman (ed) Topics in Enzyme and Fermentation Biotechnology V. 8 (pp 9–30). Wiley, New York

Yamasaki TS, Hosoya CL, Chen JS, Gratzl JS and Chang HM (1981) Characterization of residual lignin in kraft pulp, Vol 2, June 9–12. The Eckman-Days Int. Symp. on Wood and Pulping Chem., Stockholm, Sweden, pp 34–42

Zimmerman W (1989) Hemicellulolytic enzyme systems from Actinomycetes. In: MP Coughlan (ed) Production, Characterization and Application of Cellulose, Hemicellulose and Lignin Degrading Enzyme Systems (pp 161–175). Elsevier, Amsterdam

Zimmerman W and Broda P (1989) Utilization of lignocellulose from barley straw by actinomycetes. Appl. Microbiol. Biotechnol. 30: 103–109

9. Physiology of microbial degradation of chitin and chitosan

GRAHAM W. GOODAY

Department of Molecular and Cell Biology, University of Aberdeen, Marischal College, Aberdeen, AB9 1AS, U.K.

I. Introduction: chitin and chitosan

Chitin, the (1–4)-β-linked homopolymer of *N*-acetyl-D-glucosamine (Fig. 1), is produced in enormous amounts in the biosphere. A recent working estimate for both annual production and steady-state amount is of the order of 10^{10} to 10^{11} tons (Gooday 1990a). Chitin is utilized as a structural component by most species alive today. Its phylogenetic distribution is clearly defined:

(a) Prokaryotes. Despite its chemical similarity to the polysaccharide backbone of peptidoglycan, chitin has only been reported as a possible component of streptomycete spores and of the stalks of some prosthecate bacteria.

(b) Protista. Chitin provides the tough structural material for many protists; in cyst walls of some ciliates and amoebae; in the lorica walls of some ciliates and chrysophyte algae; in the flotation spines of centric diatoms; and in the walls of some chlorophyte algae and oomycete fungi (Gooday 1990a).

(c) Fungi. Chitin appears to be ubiquitous in the fungi (Bartnicki-Garcia and Lippman 1982). Reported exceptions, such as *Schizosaccharomyces*, prove to have small but essential amounts of chitin. *Pneumocystis carinii*, of uncertain affinity, has chitin in the walls of its cysts and trophozoites (Walker et al. 1990).

(d) Animals. Chitin is the characteristic tough material playing a range of structural roles among most invertebrates (Jeuniaux 1963, 1982). It is absent from vertebrates.

(e) Plants. Chitin *sensu stricto* is probably absent from plants, but polymers rich in (1–4)-β-linked *N*-acetylglucosamine have been reported (Benhamou and Asselin 1989).

Chitin occurs in a wide variety of manners. Three hydrogen-bonded crystalline forms have been characterised: α-chitin with antiparallel chains, β-chitin with parallel chains and γ-chitin with a three-chain unit cell, two "up" – one "down" (Blackwell 1988). α-Chitin is by far the most common, being the form found in fungi and most protistan and invertebrate exoskeletons. The importance of physical form to biological function is indicated by squid, *Loligo*,

C. Ratledge (ed.), Biochemistry of Microbial Degradation, 279–312.

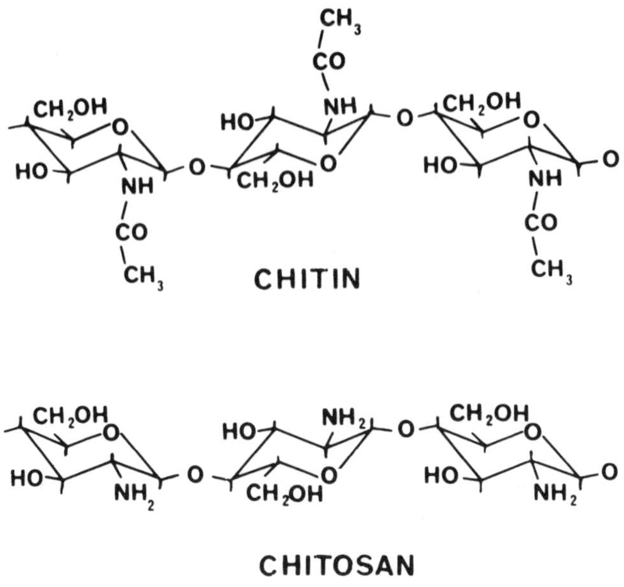

Fig. 1. Structures of chitin and chitosan.

having α-chitin in its tough beak, β-chitin in its rigid pen, and γ-chitin in its flexible stomach lining (Rudall and Kenchington 1973).

With one exception, that of diatom spines, chitin is always found cross-linked to other structural components. In fungal walls it is cross-linked covalently to other wall components notably β-glucans (Sietsma et al. 1986; Surarit et al. 1988). In insects and other invertebrates, the chitin is always associated with specific proteins, with both covalent and noncovalent bonding, to produce ordered structures (Blackwell 1988). There are often also varying degrees of mineralization, in particular calcification, and sclerotization, involving interactions with phenolic and lipid molecules (Poulicek et al. 1986; Peter et al. 1986).

Another modification of chitin is its deacetylation to chitosan, the (1–4)-β-linked polymer of D-glucosamine (Fig. 1). This is mediated by the enzyme chitin deacetylase. In the fungi this occurs in the Mucorales, where chitosan is a major component of the cell wall (Datema et al. 1977; Davis and Bartnicki-Garcia 1984) and in *Saccharomyces cerevisiae*, where it is a major component of ascospore walls. The biological significance of this deacetylation in fungi may be to give them added resistance to lysis by chitinolytic organisms. Deacetylation also occurs in arthropods, where its occurrence seems to be related to chitinous structures that undergo subsequent expansion, such as the abdominal cuticle of physogastric queen termites, and eye-lens cuticles (Aruchami et al. 1986).

With this complexity of chemical and physical form in nature, it is not surprising that a wide range of lytic enzymes are produced, each with activities specific for particular forms of chitins, chitosans, related glucosaminoglycans and their oligomers. Typically a chitinolytic microbe will produce several chitinases and *N*-acetylglucosaminidases, distinguished from each other by their substrate specificities and other properties.

II. Pathways of chitin degradation

The vast annual production of chitin is balanced by an equal rate of recycling. The bulk of this chitin degradation is microbial; in the sea chiefly by bacteria – free-living and in association with animal guts; in the soil chiefly by fungi and bacteria. Their biochemical pathways are reviewed by Davis and Eveleigh (1984). Organisms that degrade chitin solely by hydrolysis of glycosidic bonds are known as chitinolytic; a more general term, not specifying the mechanism, is chitinoclastic.

The best-studied pathway is the action of the chitinolytic system, of hydrolysis of the glycosidic bonds of chitin (Cabib 1987). Exochitinase cleaves diacetylchitobiose units from the non-reducing end of the polysaccharide chain. Endochitinase cleaves glycosidic linkages randomly along the chain, eventually giving diacetylchitibiose as the major product, together with some triacetylchitotriose. There may not always be a clear distinction between these two activities (see also Davis and Eveleigh 1984), as the action of these enzymes is dependant on the nature of the substrate. Thus the pure crystalline β-chitin of diatom spines is degraded only from the ends of the spines by *Streptomyces* chitinase complex, to yield only diacetylchitobiose, whereas colloidal (reprecipitated) chitin is degraded to a mixture of oligomers and diacetylchitobiose (Lindsay and Gooday 1985a). Lysozyme has a low endochitinolytic activity, but can readily be distinguished from chitinases as it hydrolyses *Micrococcus* peptidoglycan whereas they do not. Diacetylchitobiose (often called chitobiose, but beware confusion with the product of chitosanase) is hydrolysed to *N*-acetylglucosamine by β-*N*-acetylglucosaminidase (sometimes called chitobiase but beware confusion with glucosaminidase). Some β-*N*-acetylglucosaminidases can also act weakly as exochitinases, cleaving monosaccharide units from the non-reducing ends of chitin chains. Together, the chitinases and β-*N*-acetylglucosaminidases are known as "the chitinolytic system".

An alternative system for degrading chitin is via deacetylation to chitosan which is hydrolysed by chitosanase to give chitobiose, glucosaminyl-(1–4)-β-glucosaminide, which in turn is hydrolysed by glucosaminidase to glucosamine. This pathway appears to be important in some environments, for example in estuarine sediments, where chitosan is a major organic constituent (Hillman et al. 1989a,b; Gooday et al. 1991). As yet, there are no reports of a third possible pathway, involving deamination of the aminosugars (Davis and Eveleigh 1984).

III. Identification and assay of chitinolytic activities

A ready method for screening for microbial chitinolytic activities is to look for zones of clearing around colonies growing on agar plates containing colloidal or regenerated chitin (e.g. Lindsay and Gooday 1985b; Cody et al. 1990). This, however, only detects production of excreted lytic activities, and not all chitinolytic microbes give such a zone of clearing. Neugebauer et al. (1991), for example, describe the chitinolytic activity of *Streptomyces lividans* when grown in liquid medium that was not readily apparent on solid medium. O'Brien and Colwell (1987) have described a preliminary rapid screen to detect N-acetylglucosaminidase as being a good indicator for chitinolysis, but in a survey of *Bacillus* spp., Cody (1989) reported that many strains negative for endochitinase gave a strong positive response for N-acetylglucosaminidase. Clearing of chitin or glycolchitin agar overlays can also be used to detect chitinase activity in gels, with sensitivity being enhanced by staining with Congo red or Calcofluor White (e.g. Trudel and Asselin 1989; Cole et al. 1989).

There is a wide range of assays for chitinolytic activities in culture media and cell fractions, differing widely in sensitivity, applicability and cost. They fall into two categories: those using macromolecular chitin or its derivatives in various forms, and those using soluble oligomers or their derivatives. In the former category, examples include measurement of release of reducing sugars or N-acetylglucosamine (requiring N-acetylglucosaminidase together with chitinase) (Ulhoa and Peberdy 1991; Vasseur et al. 1990); the use of [^{3}H]- or [^{14}C]-chitin (Molano et al. 1977; Cabib 1988; Rast et al. 1991); viscometric measurements of soluble chitin derivatives (Ohtakara 1988; Lindsay and Gooday 1985b); and release of soluble dye-labelled products from dyed chitin derivatives (Wirth and Wolf 1990; Evrall et al. 1990). In the latter category, chromogenic soluble model substrates have provided the basis for useful assays, notably 3,4-dinitrophenyl tetra-N-acetyl-β chitotetraose (Aribisala and Gooday 1978; Rast et al. 1991). More versatile, however, are assays following the hydrolysis of glycosides of the fluorophore, 4-methylumbelliferone. By using a range of these, comparative activities of N-acetylglucosaminidases, exochitinases and endochitinases can be characterised (Robbins et al. 1988; Watanabe et al. 1990a; Butler et al. 1991; Hood 1991; McCreath and Gooday 1992). The release of the fluorophore can also be used to detect chitinase activity cytochemically in cells (Manson et al. 1992) or in gels after non-denaturing electrophoresis (McNab and Glover 1991).

An important point that should be emphasised is that an enzyme designated as a chitinase by its action in a chitinase assay may not have chitin as its direct natural substrate. Instead, *in vivo* it may act on an as yet unrecognised glucosaminoglycan/mucopolysaccharide/glycoprotein in that tissue. Thus De Jong et al. (1992) described a morphogenetic role for an acidic endochitinase in the development of carrot somatic embryos, in which neither substrate nor product of the enzyme activity have been identified.

IV. Autolytic and morphogenetic chitinolysis

Where investigated in detail, all chitin-containing organisms also produce chitinolytic enzymes. In some cases, such as arthropod moulting, a role is obvious. Microbial examples include the basidiomycete fungi, the inkcaps, *Coprinus* species, and the puff-balls, *Lycoperdon* species, where massive autolysis follows basidiospore maturation (Iten and Matile 1970; Tracey 1955). In the case of *Coprinus*, the basidiospore discharge starts at the outermost edges of the gills which then progressively autolyse upwards so that the spores are always released with only a fraction of a millimetre to fall into the open air for dispersal. Thus, unlike most agarics, precise vertical orientation of the gills is not required. In the case of *Lycoperdon*, the spore-producing gleba autolyses to give a capillitium of long dry springy hyphae packed with dry spores. Raindrops cause the puff-ball to act like bellows, expelling puffs of spores into the open air. Autolytic chitinases must also act in consort with other lytic enzymes to allow plasmogamy during sexual reproduction in fungi, for example to break down the gametangial walls in the Mucorales (Sassen 1965), and to break down septa to allow nuclear migration during dikaryotization in basidiomycetes (Janszen and Wessels 1970). The accumulation of autolytic enzymes in culture filtrates of senescent fungal cultures in well-documented (Reyes et al. 1984, 1989; Isaac and Gokhale 1982) but it is unclear to what extent the chitin is recycled by these mycelia.

Chitinous fungi also produce chitinases during exponential growth. Examples include *Mucor* (Humphreys and Gooday 1984a,b,c; Gooday et al. 1986; Pedraza-Reyes and Lopez-Romero 1989; Rast et al. 1991), *Neurospora crassa* (Zarain-Herzberg and Arroyo-Begovich 1983) and *Candida albicans* (Barrett-Bee and Hamilton 1984). Humphreys and Gooday (1984a,b,c) report that as well as soluble chitinase activities, in *Mucor mucedo* there is also membrane-bound chitinase requiring phospholipids for activity and having properties in common with chitin synthase activities. Similar results for related fungi were reported by Manocha and Balasubramanian (1988), but Dickinson et al. (1991) report that in *C. albicans*, the membrane-associated activity was only 0.3% of the total, and was not associated with any particular membrane fraction.

Possible roles for these soluble and membrane-bound chitinases are discussed by Gooday et al. (1986), Gooday (1990b) and Rast et al. (1991) and they include the following.
(a) Maturation of chitin microfibrils. The form of microfibrils in the wall differs in different fungi and between different life stages in the same fungus (Gow and Gooday 1983). The formation of antiparallel α-chitin microfibrils of particular orientation, length and thickness may require

modelling of the chitin chains by chitinases, both by their lytic activities and their transglycosylase activities (Gooday and Gow 1991). Their transglycosylase activities may also have a role in covalently linking chains with other wall polysaccharides.

(b) Apical growth. The "unitary model" of hyphal growth (Bartnicki-Garcia 1973) envisages a delicate balance between wall synthesis and wall lysis allowing new chitin chains to be continually inserted into the wall, with concomitant lysis of pre-existing chains to allow this. There is much circumstantial evidence for the role of chitinases and other lytic enzymes in this process (Gooday and Gow 1991) but as yet there is no direct evidence. The membrane-bound *Mucor* chitinase studied by Humphreys and Gooday (1984a,b,c) shared with chitin synthase the property of being activatible by trypsin, i.e. being zymogenic, suggesting that the two enzymes could be co-ordinately regulated, as would be required for orderly chitin deposition.

(c) Branching. It is generally accepted that chitinase action will be required to form a branch. The cylindrical wall of a hypha, unlike the apex, is a rigid structure. Its chitin microfibrils are wider, more crystalline, and are cross-linked with other wall components (Wessels 1988). The site of the new branch must be weakened to allow a new apex to be formed, and lytic enzymes are obvious contenders for this process. Rast et al. (1991) presented a detailed speculative scheme for the controlled lysis of chitin during branching, and perhaps during apical growth, through the concerted action of chitinase, β-N-acetylglucosaminidase and chitin synthase. This scheme is based on their observations of a multiplicity of chitinase activities with a range of properties arising during exponential growth of *Mucor rouxii*. The localised outgrowth of a new tip from the hyphal tube is envisaged as involving successive interrelated stages. Co-operation of chitinase molecules in the densely packed chitin of the wall results in a high incidence of transglycosylation events, leading to a slow onset of wall-loosening. As chitinolysis proceeds, the proportion of transglycoslation events will be decreased and the concentration of oligomers and monomer will increase. These will allosterically activate the chitin synthase (cf. Gooday 1977), allowing insertion of chitin into the stretching wall.

(d) Spore germination. Germination of fungal spores, and indeed hatching of protozoal cysts, requires the breaching of the wall. It seems likely that chitinases have a role in this process in at least some cases: for example in sporangiospore germination of *Mucor mucedo* where the initial spherical growth is accompanied by a co-ordinated activation of chitinase and chitin synthase (Gooday et al. 1986). Pedraza-Reyes and Lopez-Romero (1991a,b) presented results of a study of chitinase activities of germinating cells of *M. rouxii*, during spherical growth at four hours, when they found the highest specific activity. This was confirmed by Gooday et al. (1992) who showed that germination was delayed, but not prevented, by treatment with high concentrations of the inhibitor, allosamidin. In a similar way, hatching of eggs of nematodes is also delayed but not prevented by treatment with allosamidin (K. Arnold et al. 1993).

(e) Cell separation in yeasts. In the budding yeast, *Saccharomyces cerevisiae*, chitin is mostly confined to the septum separating the bud from the mother cell, where it is a major component. Elango et al. (1982) showed that chitinase is a periplasmic enzyme in these yeast cells and suggested that it plays a role in cell separation. More direct evidence for this is provided by the findings that treatment with the chitinase inhibitors, allosamidin and demethylallosamidin, inhibits cell separation during budding (Sakuda et al. 1990). Budding yeast cells of *Candida albicans* show the same response with treatment with allosamidin leading to clumps of cells (Gooday et al. 1992). The chitinase of *S. cerevisiae* is a mannoprotein (Correa et al. 1982; Orlean et al. 1991). Its structural gene *CTS1* has been cloned and sequenced by Kuranda and Robbins (1988, 1991). In cultures growing in rich medium, most of this chitinase was secreted to the medium in parallel with growth but a significant amount was also associated with the cell wall through binding of the carboxyl-terminal domain to chitin. Kuranda and Robbins suggested that it is this wall-bound enzyme fraction that is active in cell separation. SDS-polyacrylamide gel electrophoresis showed the enzyme to be a single polypeptide of about 130 kDa, corresponding to the predicted molecular mass of protein of 60 kDa with about 90 short *O*-linked mannose oligosaccharides on its serine and threonine residues. Its size varied between different strains. Different strains provided two chitinase genomic clones, probably allelic variations of a single chitinase locus. Strains were constructed in which the *CTS1* gene was disrupted. Growth was unaffected but the cells could not separate after budding and formed large aggregates attached by their septal regions. Thus chitinase is required for cell separation. Kuranda and Robbins (1991) also studied the secretion of the chitinase by using temperature-sensitive secretory mutants and showed that these accumulated a form of the enzyme that was clearly different to the one that was normally secreted. During studies of chitin synthesis in *S. cerevisiae*, Cabib et al. (1989) showed that deletion of the chitin synthase 1 gene gave yeast cells that grew normally except in acidic conditions when some of the mother cells lysed with leakage of cytoplasm from their bud scars. This damage, which was prevented by allosamidin, led Cabib et al. (1990) to suggest that it was the result of over-action of chitinase during bud separation. Cabib et al. (1992) showed that this is the case as this cell lysis was prevented by disruption of the chitinase gene. The chitin synthase 1 can thus be seen as a repair enzyme, replenishing chitin during cytokinesis, following the action of chitinase. During investigation of the cell-cycle regulated transcription of *ACE2*, a transcriptional activating gene encoding a zinc-finger DNA-binding protein in *S. cerevisiae*, Dohrmann et al. (1992) observed that an *ace2* mutant strain had a clumpy phenotype, similar to that of strains with a disrupted *CTS1* gene. They showed that *CTS1* mRNA was absent from *ace2* strains and concluded that *ACE21* is a major transcriptional activator of *CTS1* in late G_1 phase of cell cycle. Further, from similarity with activation of *HO* mating-type switching gene by the homologous regulator *SW15* they suggested that *ACE2* expression may only activate *CTS1* in the mother cell, which will bear the chitinous bud scar, and not in the daughter cell.

Villagomez-Castro et al. (1992) described a chitinase activity expressed during formation of the chitinous cyst wall by the protozoan, *Entamoeba invadens*. They suggested that it is involved in the orderly deposition of the chitin. Treatment with allosamidin slowed, but did not prevent, the process of encystment.

V. Nutritional chitinolysis

A. Bacteria

Chitinolytic bacteria are widespread in all productive habitats. Chitinases are produced by many genera of Gram-negative and Gram-positive bacteria but not by Archaebacteria (Gooday 1979; Berkeley 1979; Monreal and Reese 1969).

The sea produces vast amounts of chitin, chiefly as carapaces of zooplankton, which are regularly moulted as the animals grow. Most of this chitin is produced near to the surface and studies have shown that its recycling occurs both in the water column and in sediments (reviewed by Gooday 1990a). The rate of degradation will be enhanced by phenomena of adherence of chitinolytic microflora and by passage through animals guts. The importance of these processes is highlighted by the repeated finding of chitinolytic bacteria, principally of the genera *Vibrio* and *Photobacterium*, associated with zooplankton and particulate matter (e.g. Hood and Meyers 1977). Estimations of population densities of chitinolytic bacteria, both as total counts and as percentages of total heterotrophs, have shown considerable variation but consistently higher counts have been reported from marine sediments than from the overlying seawater (Gooday 1990a). Pisano et al. (1992) described the isolation of chitinolytic actinomycetes from marine sediments and comment on the high correlation between chitinolysis and antibiotic production in their isolates. Studies such as that by Helmke and Weyland (1986) conclude that indigenous bacteria are capable of decomposing chitin particles throughout the depth of the Antarctic Ocean, as are chitinases produced in surface waters and transported down by sinking particles.

Estuaries are particularly productive and Reichardt et al. (1983) isolated 103 strains of chitinolytic bacteria from the estuarine upper Chesapeake Bay, Maryland. Of these, 44 were yellow-orange pigmented *Cytophaga*-like bacteria with a range of salt requirements. Others were vibrios, pseudomonads and *Chromobacterium* strains. Chan (1970) presented studies of chitinolytic bacteria from Puget Sound, Washington. Genera identified, in decreasing order of abundance, were *Vibrio, Pseudomonas, Aeromonas, Cytophaga, Streptomyces, Photobacterium, Bacillus* and *Chromobacterium*.

Pel and Gottschal (1986a,b, 1989) and Pel et al. (1989, 1990) have investigated chitinolysis by *Clostridium* strains isolated from sediments and the anoxic intestine of plaice from the Eems-Dollard estuary, The Netherlands. They found that in pure culture, chitin was degraded slowly; diacetylchitobiose accumulated but soon disappeared as *N*-acetylglucosamine accumulated. They suggested that the *Clostridium* strains are specialised utilizers of diacetylchitobiose and that the accumulation of *N*-acetylglucosamine represents non-utilizable monomers appearing during random hydrolysis of chitin oligomers. Chitin degradation was greatly enhanced by co-culture with other bacteria from the sediments. One aspect of this enhancement, they suggest, is the release of stimulatory growth factors, such as a thioredoxin-type compound that maintained the reduced state of essential sulphydryl groups in the chitinolytic system. Interspecies interactions may also play a role for this bacterium if it is exposed to O_2 in the upper layers of sediments, as accumulating mono- and disaccharides could provide substrates for facultative aerobic bacteria which would consume O_2 to render the microenvironment anaerobic again. While investigating the chitinolytic microflora of a solar saltern, Liaw and Mah (1992) isolated a novel, halophilic, anaerobic chitinolytic bacterium, *Haloanaerobacter chitinovorans*. This isolate grew at NaCl concentrations of 0.5 to 5 M and at temperatures between 23 and 50°C. The remarkable ecosystems of the deep-sea thermal vents should be rich areas for the isolation of novel chitinolytic microbes, as their dominant fauna produces chitinous structures such as clam shells, crab carapaces and pogonophoran tubes.

Chitinolytic bacteria are also abundant in freshwaters; characteristic genera in the water column being *Serratia, Chromobacterium, Pseudomonas, Flavobacterium* and *Bacillus*, with *Cytophaga johnsonae* and actinomycetes in sediments (Gooday 1990a).

The soil contains many chitinous animals and fungi as its normal living components. Consequently, chitinolytic bacteria can be isolated readily. The numbers and types reported vary greatly with different soils and methods of isolation but major genera are *Pseudomonas, Aeromonas, Cytophaga johnsonae, Lysobacter, Arthrobacter, Bacillus* and actinomycetes (Gooday 1990a). In a recent survey, Cody (1989) reported that 17 of 52 strains of *Bacillus* were chitinolytic. Recent reports of chitinases from *Streptomyces* species include those by Ueno et al. (1990), Okazaki and Tagawa (1991) and Neugebauer et al. (1991).

When grown in liquid culture, most chitinolytic bacteria secrete chitinases into the medium. *Cytophaga johnsonae*, a ubiquitous soil organism, characteristically binds to chitin as it degrades it. Wolkin and Pate (1985) described a class on non-motile mutants with an interesting pleiotropy: they were all unable to digest and utilize chitin, as well as being resistant to phages that infect the parental strain, and had relatively non-adherent and non-hydrophobic surfaces compared with wild-type strains. The authors concluded that all characteristics associated with this pleiotropy require moving cell surfaces, and that chitin digestion requires some feature of this, presumably involving enzymatic contact between bacterium and substrate. Pel and Gottschal (1986a) illustrated direct contact between cells of the chitinolytic *Clostridium* str. 9.1 and chitin fibrils and, as for cellulolytic *Clostridium* species, this may involve specific enzymatic structures on the cell surface. Particular attention has been paid to adsorption of the pathogenic but also chitinolytic *Vibrio* species. Kaneko and Colwell (1978) described strong adsorption to chitin of *Vibrio parahaemolyticus* isolated from the estuarine Chesapeake Bay. They suggested that this has an ecological as well as digestive significance to the bacteria as the adsorption was decreased by increasing values of salinity and pH from those of the estuary to those of sea-water. This phenomenon would favour retention of the bacteria within the estuary. Bassler et al. (1989 1991a,b) and Yu et al. (1991) presented a detailed study of the utilization of chitin by *Vibrio furnissii*. Adhesion to model substrates was assessed by nixing radio-labelled cells with gel beads that had been covalently coated with carbohydrate residues. Cells of *V. furnissii* adhered to glycosides of *N*-acetylglucosamine and, to a lesser extent, of glucose and mannose. A calcium-requiring lectin was responsible for this binding to the three sugars. Adherent cells continued to divide, and to stay attached, but the population gradually shifted to a large fraction of free swimming cells. Metabolic energy was required for binding but either transient or no adhesion occurred in incomplete growth media. The authors suggested that this active adhesion/de-adhesion process allows the cells to continuously monitor the nutrient status of their environment, enabling them to colonise a suitable chitinous substrate. They suggested that the next step is chemotaxis to chitin hydrolysis products. In a capillary assay, swimming cells of *V. furnissii* showed low level constitutive chemotaxis to *N*-acetylglucosamine (GlcNAc), but induction by prior growth in the presence of GlcNAc greatly increased the effect. No taxis was observed to GlcNAc oligomers by cells grown on lactate, but strong inducible taxis occurred. Bassler et al. (1991a) described the induction of two or more receptors recognizing $(GlcNAc)_n$, n = 2 to 4. Bassler et al. (1991b) described the utilization of chitin oligomers by the cells. They characterized two cell-associated enzymes hydrolysing oligomers that entered the periplasmic space: a membrane-bound chitodextrinase and an *N*-acetylglucosaminidase. Both enzymes were inducible by chitin oligomers, especially *N*,*N*′-diacetylchitobiose $(GlcNAc)_2$.

Where investigated, chitinase production by other bacteria has been shown to be inducible by chitin oligomers and low levels of *N*-acetylglucosamine (Jeuniaux 1963; Monreal and Reese 1969; Kole and Altosaar 1985).

B. Fungi

Chitinolytic fungi are readily isolated from soils where they rival or even exceed the chitinolytic activities of bacteria. Most common are Mucorales, especially *Mortierella* spp., and Deuteromycetes and Ascomycetes, especially the genera *Aspergillus*, *Trichoderma*, *Verticillium*, *Thielavia*, *Penicillium* and *Humicola* (Gooday 1990a). These fungi characteristically have inducible chitinolytic systems (Sivan and Chet 1989). Induction and characterization of an extracellular chitinase from *Trichoderma harzianum* have been described by Ulhoa and Peberdy (1991, 1992). Chitinase production was induced by chitin but repressed by glucose and *N*-acetylglucosamine. Vasseur et al. (1990) have isolated chitinase over-producing mutants of *Aphanocladium album*, by screening for increased clearing zones around colonies on colloidal chitin agar following mutagenesis. One strain showed a 26-fold increase in maximal extracellular chitinase activity in liquid medium with crystalline chitin as sole carbon source, compared to the wild-type strain. McCormack et al. (1991) described the production of a thermostable chitinolytic activity from *Talaromyces emersonii* which was optimally active at 65°C. Baiting of freshwater sites with chitin can yield a range of chitinolytic fungi, interesting members of which are the chytrids, such as *Chytriomyces* species (Reisert and Fuller 1962), and *Karlingia astereocysta*, which has a nutritional requirement for chitin that can only be relieved by *N*-acetylglucosamine; i.e. it is an "obligate chitinophile" (Murray and Lovett 1966). Fungi are rare in the sea, but the sea is rich in chitin, and Kohlmeyer (1972) described a range of fungi degrading the chitinous exoskeletons of hydrozoa. Only one could be identified: the ascomycete *Abyssomyces hydrozoicus*.

C. Slime moulds, protozoa and algae

The Myxomycetes, "true slime moulds", are a rich source of lytic enzymes, and *Physarum polycephalum* produces a complex of extracellular chitinases (Pope and Davies 1979). Soil amoebae, *Hartmanella* and *Schizopyrenus*, produce chitinases. These enzymes must participate in the digestion of chitinous food particles engulfed by the slime mould plasmodium and by the amoebae. Phagocytotic ciliates probably also have the capacity to digest chitin and chitinase activities have been implicated in the unusual feeding strategies of *Ascophrys*, a chitinivorous ectosymbiont of shrimps (Bradbury et al. 1987) and *Grossglockneria*, which feeds by digesting a tiny hole through a fungal hypha and sucking out the cytoplasm (Petz et al. 1986). The colourless heterotrophic diatom, *Nitzschia alba*, is also reported to digest chitin (A.E. Linkens, quoted by Hellebust and Lewin 1977).

VI. Chitinolysis in pathogenesis and symbiosis

Pathogens of chitinous organisms characteristically produce chitinases. These can have two roles; they can aid the penetration of the host; and they can provide nutrients both directly, in the form of amino sugars, and indirectly by exposing other host material to enzymatic digestion. Examples include the oomycete *Aphanomyces astaci*, a pathogen of crayfish (Soderhall and Unestam 1975); the fungus *Paecilomyces lilacius*, a pathogen of nematode eggs (Dackman et al. 1989); the entomopathogenic fungi, *Beauveria bassiana*, *Metarhizium anisopliae*, *Nomuraea rileyi* and *Verticillium lecanii* (Smith and Grula 1983; Coudron et al. 1984; St Leger et al. 1986); mycophilic fungi, *Cladobotryum* species and *Aphanocladium album* (G.W. Gooday unpublished; Zhloba et al. 1980; Srivastava et al. 1985; Kunz et al. 1992); the bacteria *Serratia*, insect pathogens (Lysenko 1976; Flyg and Boman 1988); and a *Photobacterium* species causing exoskeleton lesions of the tanner crab (Baross et al. 1978).

As well as being a component of insect exoskeletons, chitin also has a major structural role in the ephemeral protective lining of insect guts, the peritrophic membrane. Treatment of isolated peritrophic membranes with chitinase leads to their perforation (Brandt et al. 1978; Huber et al. 1991). Addition of exogenous chitinase aids the pathogenesis of insects by *Bacillus thuringiensis* (Smirnoff 1974; Morris 1976), and by a gypsy moth nuclear polyhedrosis baculovirus (Shapiro et al. 1987). There are now several examples where the pathogen's endogenous chitinolytic activities appear to aid penetration of the peritrophic membrane or other chitinous barriers, perhaps aiding eventual release and spread of the pathogen as well as uptake. Gunner et al. (1985) reported a positive correlation between chitinase activity among chitinase-producing strains of *B. thuringiensis* and host mortality. That the chitin of the peritrophic membrane is a site of attack by other insect pathogenic bacteria is suggested by experiments with *Drosophila melanogaster* (Flyg and Boman 1988). Flies with mutations in two genes, *cut* and *miniature*, are more susceptible than the wild type to infection by *Serratia marcescens*. That the *cut* and *miniature* mutations lead to deficiencies in chitin content was demonstrated by showing that pupal shells from the mutant strains were more readily digested by *Serratia* chitinase, and especially by synergistic action of chitinase and protease, than those of other strains. Also a mutant bacterial strain, deficient in chitinase and protease, was much less pathogenic to the flies.

Daoust and Gunner (1979), studying bacterial pathogenesis of larvae of the gypsy moth, showed that the virulence of the chitinolytic bacterium strain 501B was synergistically enhanced by co-feeding the larvae with fermentative nonpathogenic bacteria. They explained this by the acid production by the fermentative bacteria having the effect of lowering the alkaline pH of the larval gut to a value that gave greater activity of the chitinase from 501B, leading to disruption of the peritrophic membrane. The sugar-beet root maggot, however, has turned the chitinolysis by *Serratia* to its advantage by developing a symbiotic relationship with *S. liquefaciens* and *S. marcescens* (Iverson et al. 1984). These bacteria become embedded in the inner puparial surface, and aid the emergence of the adult fly by their digestion of the chitin of the puparium. The symbiotic bacteria are present in all developmental stages, including the eggs. Maternally inherited chitinolytic bacteria are also implicated in susceptibility of tsetse flies to infection with trypanosomes (Maudlin and Welburn 1988). The susceptible flies have infections of "rickettsia-like organisms", which produce chitinase when in culture in insect cells. The resistance of refractory tsetse flies (lacking the bacterial infection) is ascribed to killing of the trypanosomes in the gut mediated by a lectin. Maudlin and Welburn (1988) suggested that bacterial chitinolysis releases amino sugars that inhibit the lectin-trypanosome binding and thus results in survival of the trypanosomes. An alternative explanation is that the chitinolytic bacteria weaken the insect's peritrophic membrane, aiding the penetration of the trypanosomes. Schlein et al. (1991) reported that cultures of the trypanosomatids, *Leishmania* species, produced their own chitinase activities to aid penetration of the insect gut. This needs re-investigating, however, as their culture medium included bovine serum, a rich source of chitinase. They did, however, find activity associated with the *Leishmania* cells. Arnold et al. (1992) detected no chitinase activity in cells or medium of *Trypanosoma brucei* var. *brucei* when cultured in medium depleted of chitinase by affinity adsorption onto chitin. The invasive form of the malarial parasite, *Plasmodium gallinaceum*, is the ookinete, which penetrates the peritrophic membrane of the host mosquito. Huber et al. (1991) reported the formation of chitinase during the maturation process of *Plasmodium* zygotes to ookinetes and implicated its appearance with the invasion of parasites. The filarial nematode, *Brugia malayi*, also has mosquitoes as its vectors between mammalian hosts. Microfilariae, produced during infection of the mammal, are covered by a chitin-rich coat, formed by stretching of the original eggshell. In model infections in gerbils, Fuhrman et al. (1992) have shown that a major antigen of the microfilariae is a nematode chitinase. This is recognized by the monoclonal antibody, MF1, that they had previously shown to be responsible for clearance of the peripheral microfilariae in the gerbils. Sequencing the cDNA of the MF1 antigen showed homologies with known chitinase genes (cf. Table 2). The microfilarial chitinase may play a role in the regulation of stretching of the chitinous sheath, or it may aid the penetration of the mosquito gut peritrophic membrane.

A further example of an insect pathogen producing chitinase is the baculovirus *Autographa californica* nuclear polyhedrosis virus (NPV). This virus is used for biological control of insect pests and by molecular biologists as a system for the expression of heterologous proteins in infected cell cultures. The insect cell cultures produce their own chitinases, at a low activity, but on infection with *A. californica* NPV, an enormous increase in chitinase activity is observed (Hawtin et al. 1993). This is encoded by the virus genome. The amino acid sequence shows very high homology to that for the chitinase A from

Serratia marcescens (Table 2). This suggests that there has been lateral gene transfer relatively recently, especially as *S. marcescens* is itself an insect gut pathogen. The more likely direction is bacterium to virus, as other baculoviruses do not have an homologous gene. A strain of *A. californica* NPV from which the chitinase gene had been deleted was less pathogenic to larvae of the cabbage looper, *Trichoplusia ni*, but the insects still died. A dramatic difference, however, was that after death the insects infected by the chitinolytic virus were totally liquefied, whereas those infected by the mutant strain were dry cadavers. Thus, as with the bacterial, protozoal and microfilarial chitinases, this baculovirus chitinase may aid penetration of the peritrophic membrane of the insect host but its major significance is in aiding release of viruses from the dead host.

Another example of a chitinase activity involved in microbial interactions is that of the yeast killer toxin produced by the yeast *Kluyveromyces lactis* (Butler et al. 1991). This is a trimeric protein, of α, β and γ subunits. Intracellular γ subunit is responsible for killing a susceptible cell of *Saccharomyces cerevisiae*, the β subunit has no known role, and the α subunit has exochitinase activity that is essential for the action of toxin. This is shown by the inhibition of toxin activity by the specific inhibitor of chitinase, allosamidin. The significance of this chitinase activity remains unclear, but may involve the binding of toxin to the susceptible yeast cell surface to facilitate the uptake of the γ subunit. The amino acid sequence of the α subunit has striking homologies to other chitinases in two regions, one corresponding to the catalytic domain of microbial and some acidic plant chitinases (Table 2) and another corresponding to the cysteine-rich chitin-binding domain of some basic plant chitinases and lectins (Butler et al. 1991).

Chitinase production by the entomopathogenic fungi is inducible by chitin oligomers, *N*-acetylglucosamine and glucosamine (Smith and Grula 1983; St Leger et al. 1986, 1991). St Leger et al. (1986) also reported that chitosanase is co-induced with chitinase in *Metarhizium anisopliae*. In insect pathogenesis, any chitinase will be in acting in synergism with proteases and Bidochka and Khachatourians (1988) have suggested that both activities are coordinately regulated. They showed that low levels of *N*-acetylglucosamine will induce a serine protease in *B. bassiana* and suggested that an initial constitutive chitinase attack on the insect cuticle would yield *N*-acetylglucosamine, leading to the coordinate induction of chitinases and proteases. St Leger et al. (1987) questioned the importance of chitinase as in their experiments *M. anisopliae* did not appear to produce chitinase during penetration of cuticle of *Manduca sexta*. Bidochka and Khachatourians (1992) have investigated the growth of *B. bassiana* on cuticular components from the migratory grasshopper. After removal of lipids and protein, the residual chitin (about 30% w/w) was a relatively poor source of nutrients for germination and fungal growth, but their electron micrographs clearly showed penetration of the chitinous material by germ tubes. More positive evidence for the importance of fungal chitinase activities in insect pathogenesis, particularly during spore germination, was provided by El-Sayed et al. (1989) in a comparative study of exo- and endo-chitinase activities of three isolates of *Nomuraea rileyi*. The two virulent isolates had much higher chitinase activities during early growth than an avirulent isolate.

Chitin in fungi and invertebrates comprises a considerable part of the diet of many herbivorous and carnivorous animals. There can be three sources of chitinolytic enzymes in the animal's digestive system: from the animal itself, from endogenous gut microflora or from the ingested food (Gooday 1990a). Most work has been done with fish, where a typical marine fish gut microflora is dominated by chitinolytic strains of *Vibrio*, *Photobacterium* and enterobacteria. However, it is clear that the fish produce their own chitinases which they use as food processing enzymes rather than directly nutritional enzymes. Thus the gut bacteria cannot be regarded as mutualistic symbionts with respect to chitin in the same way that the rumen symbionts are regarded with respect to cellulose degradation (Lindsay et al. 1984; Lindsay and Gooday 1985b; Gooday 1990a). With mammals the situation is less clear: whales have chitinolytic microflora in their stomachs which may contribute to a rumen-type fermentation (Seki and Taga 1965; Herwig et al. 1984); Patton and Chandler (1975) described digestion of chitin by calves and steers implying a chitinolytic rumen flora; and Kuhl et al. (1978) found elevated caecal weights in chitin-fed rats, suggesting participation of intestinal bacteria in chitin digestion.

Among invertebrates, chitin digestion is widespread with or without participation of a microbial chitinolytic flora (Jeuniaux 1963). Borkott and Insam (1990), working with the soil springtail, *Folsomia candida*, concluded that at least in this arthropod there is a mutualistic symbiosis with its gut chitinolytic bacteria, *Xanthomonas* and *Curtobacterium* species. Thus the steady increase in biomass in animals fed every four days with chitin plus yeast extract was prevented by treatment with the antibiotic tetracycline. In a food preference experiment, the animals chose to feed on chitin-agar strips that had been pre-inoculated with the chitinolytic bacteria or the animals' faeces, suggesting that some pre-digestion of the chitin was aiding its utilization by the animal.

VII. Degradation of chitosan

As described earlier, chitosan is a major component of the walls of the common soil fungi, the zygomycetes, and is produced by deacetylation of chitin to form a major organic component of estuarine sediments. Chitosanase was discovered and shown to be widespread among microbes by Monaghan et al.(1973) and Monaghan (1975). It is produced by bacteria such as species of *Myxobacter*, *Sporocytophaga*, *Arthrobacter*, *Bacillus* and *Streptomyces*, and by fungi such as species of *Rhizopus*, *Aspergillus*, *Penicillium*, *Chaetomium* and the basidiomycete that is a very rich source of glucanase, "Basidiomycete sp. QM 806". Davis and Eveleigh (1984) screened soils from barnyard, forest and salt marsh for chitosan-degrading bacteria and found them at 5.9, 1.5 and 7.4% respectively of the total heterotrophic isolates, compared with 1.7, 1.2 and 7.4% chitin-degraders. They investigated chitosanase production by a soil isolate of *Bacillus circulans* in more detail and showed that it was inducible by chitosan but not by chitin or carboxymethylcellulose, and was only active on chitosan. In contrast, the chitosanase from a soil isolate of *Myxobacter* species was active against both chitosan and carboxymethylcellulose (Hedges and Wolfe 1974). Mitsutomi et al. (1990) and Ohtakara et al. (1990) reported the action patterns of chitinases from *Aeromonas hydrophila* and *Streptomyces griseus*, respectively, on partially *N*-acetylated chitosan. In both cases, but especially for *S. griseus*, there was specificity for cleavage of the *N*-acetyl-β-D-glucosaminidic linkages. In contrast, a purified chitosanase from *Nocardia orientalis* attacked 33% acetylated chitosan by hydrolysing between glucosamine and either glucosamine or *N*-acetylglucosamine, but not between *N*-acetylglucosamine and glucosamine (Sakai et al. 1991a). Sakai et al. (1991a) proposed a scheme for the total hydrolysis of partially acetylated chitosans by *N. orientalis* by the cooperative action of chitosanase, β-*N*-acetylhexosaminidase, and a novel exo-β-glucosaminidase characterized by Nanjo et al. (1990). Seino et al. (1991) described the cleavage pattern of a purified *Bacillis* chitosanase on a series of glucosamine oligomers, as measured by HPLC analysis of products, and concluded that the enzyme mainly hydrolyses chitosan in a random fashion.

Pelletier and Sygusch (1990) have purified three chitosanase activities from *Bacillus megaterium* P1. The major activity, chitosanase A, had a high specificity for chitosan, with just a trace of activity against carboxymethyl-cellulose, while chitosanases B and C had much lower activity against chitosan, and also activities against chitin, carboxymethyl-cellulose and cellulase. None had lysozyme activity. These broad specificities shown by enzymes B and C are remarkable and deserve further investigation. Somashekar and Joseph (1992) described a chitosanase activity secreted constitutively into the medium by the yeast, *Rhodotorula gracilis*. This activity was measured by decrease in viscosity of a chitosan solution and yielded a detectable chitosan oligomers. In view of this, and the observation that growth of the yeast was inhibited by even low amounts of chitosan, Somashekar and Joseph (1992) suggested that this enzyme is involved in morphogenesis of the cell wall.

VIII. Biotechnology of chitinases and chitosanases

With chitin and chitosan being an enormous renewable resource, much of which from the shellfish and fungal fermentation industries currently goes to waste, and with their essential roles in fungi and invertebrates, it is not surprising that there is a great deal of current interest in these polysaccharides and in their degradative enzymes (Muzzarelli and Pariser 1978; Hirano and Tokura 1982; Zikakis 1984; Muzzarelli et al. 1986; Deshpande 1986;

Skjak-Braek et al. 1989; Roberts 1992). The use of chitinolytic microbes in the production of single cell protein or ethanol from chitinous wastes has been investigated (Tom and Carroad 1981; Vyas and Deshpande 1991; Cody et al. 1990) but much further work is required to evaluate these ideas.

A. Cloning of chitinase genes

Genes coding for various chitinases from bacteria, fungi and plants have been cloned. Of many bacterial isolates, Monreal and Reese (1969) found *Serratia marcescens* and *Serratia liquefaciens* (*Enterobacter liquefaciens*) to be the most active producers of chitinases. Roberts and Cabib (1982) describe purification of the chitinases and mutant strains with increased production of chitinase have been produced (Kole and Altosaar 1985; Reid and Ogrydziak 1981). Two chitinase genes *chiA* and *chiB* from random cosmid clones of *S. marcescens* have been inserted into *Escherichia coli*, and then into *Pseudomonas fluorescens* and *Pseudomonas putida*, resulting in four strains of genetically manipulated *Pseudomonas* that have considerable chitinase activities (Suslow and Jones 1988). The rationale to this work was to produce chitinolytic rhizosphere bacteria potentially of value for the biocontrol of soil-borne fungal and nematode diseases of crop plants, as chitin is an essential component of fungal walls and nematode egg cases (Gooday 1990d). In another approach using the same genes Jones et al. (1986, 1988), Taylor et al. (1987) and Dunsmuir and Suslow (1989) have obtained expression of *chiA* in transgenic tobacco plants using a range of promoters. These transgenic plants showed increased resistance to the tobacco brown-spot pathogen *Alternaria longipes*. Lund et al. (1989) showed that the *chiA* gene product was secreted by the plant cells in a modified form and suggested that the bacterial signal sequence is functioning in the plant cells and that the chitinase is *N*-glycosylated through the secretory pathway. Lund and Dunsmuir (1992) have investigated the relative effects of plant versus bacterial signal sequences on secretion of *S. marcescens* chitinase A by transgenic tobacco cultures. Only a fraction of the chitinase with the bacterial sequence was secreted and glycosylated, while replacement by a plant signal sequence resulted in efficient glycosylation and secretion. The glycosylation was not, however, essential for secretion as the non-glycosylated protein was also secreted. Fuchs et al. (1986) have characterized five chitinases in *S. marcescens*, and identified clones from a cosmid library encoding for the *chiA* gene. Their aim was biological control of pathogens and pests by enhancing chitinase activities of phylloplane and rhizoplane bacteria. Horwitz et al. (1984) described attempts at cloning the *Serratia* chitinases into *E. coli*, then back into *S. marcescens* on a high copy number plasmid, to produce a bacterium of value for a bioconversion process to treat shellfish waste. They isolated multiple phage clones, encoding both *N*-acetylglucosaminidase and chitinase activity, and suggested that these are linked in a *chi* operon, which was also suggested by Soto-Gil and Zyskind (1984) in their work towards cloning these genes from *Vibrio harveyi* in *E. coli*.

Shapira et al. (1989) have cloned a chitinase gene from *S. marcescens* into *E. coli* and showed that both the *E. coli* containing the appropriate plasmid and enzyme extracts produced by this strain have potential for biological control of fungal diseases of plants under greenhouse conditions.

Streptomyces species are well-known producers of active chitinases (Jeuniaux 1963). A chitinase from *S. erythraeus* has been purified and sequenced: it has 290 amino acid residues, a molecular weight of 30,400 and two disulphide bridges (Hara et al. 1989; Kamei et al. 1989). A chitinase from *S. plicatus* has been cloned from a DNA library and expressed in *Escherichia coli* (Robbins et al. 1988, 1992). The *Streptomyces* chitinase was secreted into the periplasmic space of *E. coli* with its signal sequence having been removed by the *E. coli* signal peptidase. A gene for chitinase from *Vibrio vulnificus* has also been cloned into *E. coli* and was expressed but the protein was not secreted into the medium (Wortman et al. 1986). Similarly, Roffey and Pemberton (1990) expressed a chitinase gene from *Aeromonas hydrophila* in *E. coli* and found the resultant enzyme to be accumulated in the periplasmic space. In contrast, Chen et al. (1991) reported the excretion of an *A. hydrophila* chitinase cloned in *E. coli*. Watanabe et al. (1990b, 1992) described the cloning of chitinase genes from *Bacillus circulans*, the properties of which are described later. A gene for chitinase from *Saccharomyces cerevisiae* has been cloned by transforming the yeast with vector plasmids containing a genomic library and then screening for over-producing transformants (Kuranda and Robbins 1988, 1991) Again this is described later. Fink et al. (1991) reported the cloning of a chitosanase-encoding gene from the actinomycete, *Kitasatosporia*, into *Streptomyces lividans*.

Expression of microbial chitinase genes is typically induced by chitin but repressed by glucose. Delic et al. (1992) described the characterization of promoters for two chitinase genes from *Streptomyces plicatus*. Each one had a pair of perfect 12 base-pair, direct repeat sequences which overlapped the putative RNA polymerase binding site. Similar promoters were also found for chitinase genes for *Streptomyces lividans* (Miyashita et al. 1991).

Plants produce chitinases as major component of their "pathogenesis-related proteins" induced following attack by potential pathogens or treatment with ethylene (Mauch and Staehelin 1989). Some of these plant chitinases have antifungal activity (Mauch et al. 1988; Broekaert et al. 1988) greater than that of some bacterial chitinases (Roberts and Selitrennikoff 1988). Leah et al. (1991) have used a microtitre-well assay to assess the antifungal activity of a purified chitinase from barley seeds. Treatment of both *Trichoderma reesei* and *Fusarium sporotrichiodes* with 375 nM protein resulted in about 50% inhibition of growth but there were strong synergistic inhibitions with either or both of a ribosome-inactivating protein and a glucanase from the barley seeds.

There is now sufficient information to classify the plant chitinases into at least three structural groups: Class I, basic proteins located primarily in the vacuole, sharing amino-terminal sequence homology with wheat germ agglutinin and hevein; Class II, acidic, extracellular, having sequence homology with the catalytic domain of Class I, but without the hevein domain; Class III, acid, extracellular, with no homologies to Classes I or II (Payne et al. 1990; Shinshi et al. 1990). Several genes for plant chitinases have been cloned (e.g. Broglie et al. 1986; Payne et al. 1990) and expressed in other plants (Linthorst et al. 1990) with the aim of increasing the plants' resistance to fungal pathogens.

B. Uses of chitinases and chitosanases

Oligomers of chitin and chitosan have value as fine chemicals and as potential pharmaceuticals (Gooday 1990c). As well as direct hydrolysis of chitin by chitinases, a promising development is the characterization of the transglycosylase activities of these enzymes. Thus Usai et al. (1987, 1990) and Nanjo et al. (1989) described the use of a chitinase from *Nocardia orientalis* for the interconversion of *N*-acetylglucosamine oligomers, especially to produce hexa-*N*-acetylchitohexose, an oligosaccharide with reported anti-tumour activity (Suzuki et al. 1986). The transglycosylase activity is favoured by a high substrate concentration and a lowered water activity, e.g. in increasing concentrations of ammonium sulphate. Takayanagi et al. (1991) described transglycosylase activities of thermostable chitinases produced by a thermophilic strain of *Bacillus licheniformis*. When incubated with a 5% (w/v) solution of $(GlcNAc)_4$ at 50°C, the chitinases produced yields of about 10% $(GlcNAc)_6$ after a few minutes. The production of the disaccharide, N,N'-diacetylchitobiose, from chitin was described by Takiguchi and Shimahara (1988, 1989). They isolated two bacteria, *Vibrio anguillarum* strain E-383a and *Bacillus licheniformis* strain X-Fu, whose growth in chitin-containing medium resulted in the accumulation of 40 and 46%, respectively, conversion of chitin to diacetylchitobiose.

Sakai et al. (1991b) report the use of a column reactor of immobilised chitinase and *N*-acetylhexosaminidase from *Nocardia orientalis* for the continuous production of *N*-acetylglucosamine from soluble chitin oligomers. Pelletier and Sygusch (1990) and Pelletier et al. (1990) described the characterization of chitosanases from *Bacillus megaterium* and their use in the assay of the degree of deacetylation in samples of chitosan. Nanjo et al. (1991) also described the analysis of chitosan using the chitosanase, exo-glucosaminidase and *N*-acetylhexosaminidase activities from *N. orientalis*. A direct medical use has been suggested for chitinases in the therapy of fungal diseases in potentiating the activity of antifungal drugs (Pope and Davies 1979; Orunsi and Trinci 1985). Immunological problems however, probably debar this until anti-iodiotypic antibodies for appropriate chitinases are developed.

Chitinases have extensive uses in the preparation of protoplasts from fungi, a technique of increasing importance in biotechnology (Peberdy 1983). Examples include the chitinases from *Aeromonas hydrophila* subsp. *anaerogenes* (Yabuki et al. 1984) and *Streptomyces* species (Beyer and Diekmann 1985; Tagawa and Okazaki 1991). Chitosanases are required to make protoplasts from species of the Mucorales (Reyes et al. 1985).

C. Uses of chitinolytic organisms in biocontrol

As most fungal and invertebrate pests and pathogens have chitin as an essential structural component (Gooday (1990d), chitinase activity could have an important place in the repertoire of mechanisms for biological control. Thus the strongly chitinolytic fungus, *Trichoderma harzianum*, has good potential for the control of a range of soil-borne plant pathogens (Lynch 1987; Sivan and Chet 1989). Dackman et al. (1989) reported that chitinase activity is required for soil fungi to infect eggs of cyst nematodes. Sneh (1981) discussed the use of rhizophere chitinolytic bacteria for biological control. Inbar and Chet (1991) suggested that rhizosphere colonization by *Aeromonas caviae* gives biocontrol against soil-borne fungal pathogens by increasing the chitinolytic activity of the rhizosphere. They demonstrated chitinolysis around the roots by staining for cleaving in chitin agar with Congo red. Use of genetic manipulation for the development of organisms with enhanced chitinolytic activities for biological control has been discussed earlier. As well as application of the organisms themselves, there have been reports of biological control by addition of chitin to the soil, presumably as this encourages the growth of chitinolytic microbes which then have a better inoculum potential to infect the soil-borne pathogens and pests, but results currently are variable and the procedures need further investigation (Gooday 1990a).

IX. Specific inhibitors of chitinases

Allosamidin is an antibiotic produced by *Streptomyces* strains, discovered independently by Sakuda et al. (1987a) and as metabolite A82516 by Somers et al. (1987) in screens for chitinase inhibitors as potential insecticides. Allosamidin is insecticidal to the silkworm by preventing ecdysis. It does not affect egg hatching of the housefly but prevents development from larvae to pupae. It has an interesting spectrum of activity, strongly inhibiting chitinases from nematodes and fish, less strongly those of insects and fungi, weakly those of bacteria and not inhibiting yam plant chitinase (Gooday 1990a,c). Allosamidin is a pseudo-trisaccharide, being a disaccharide of *N*-acetylallosamine (until now unknown in nature) linked to a novel aminocyclitol derivative, allosamizoline (Sakuda et al. 1987b; Fig. 2). Demethylallosamidin, a minor cometabolite, has similar activity to allosamidin in inhibiting the silkworm chitinase but is more inhibitory to the chitinase from *Saccharomyces cerevisiae* (Isogai et al. 1989; Sakuda et al. 1990). Allosamidin inhibits chitinases from the fungi *Candida albicans* (Dickinson et al. 1989; Milewski et al. 1992), *Neurospora crassa* (McNab and Glover 1991) and *Mucor rouxii* (Pedraza-Reyes and Lopez-Romero 1991a) and the nematode *Onchocerca*

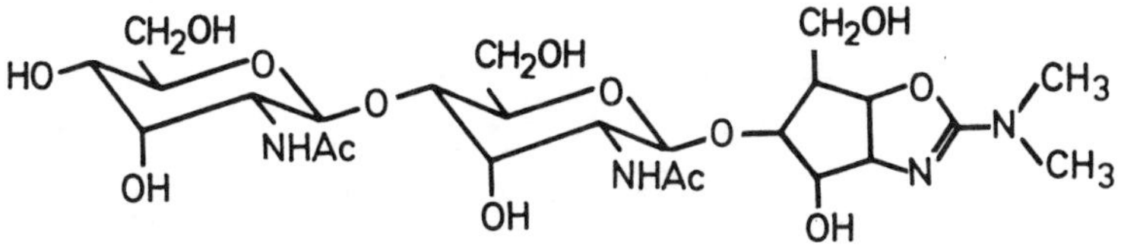

Fig. 2. Structure of allosamidin.

gibsonii (Gooday et al. 1988). *In vivo*, however, reports of its activities are very limited.

As discussed earlier, treatment with allosamidin and demethylallosamidin inhibits cell separation in budding yeasts, such as *S. cerevisiae* and *C. albicans*, and delays germination of spores of *M. rouxii*, hatching of nematode eggs and encystment of *E. invadens*. Nishimoto et al. (1991) described further minor co-metabolites of allosamidin and reported the comparative activities of six allosamidins against chitinase preparations from three fungi; *Candida albicans*, *S. cerevisiae* and *Trichoderma* sp. (Table 1). Distinctly different patterns of inhibition were apparent, with the *S. cerevisiae* activity showing a hundred-fold variation in susceptibility to the different metabolites, while the *C. albicans* and *Trichoderma* activities showed a ten-fold and a two-fold variation, respectively. Mild alkaline hydrolysis of allosamidin and glucoallosamidin A yielded pseudo-disaccharides that retained their inhibition against the *C. albicans* activity but were no longer inhibitory against activities from *S. cerevisiae* and *Trichoderma* sp. Milewski et al. (1992) presented a detailed account of the competitive inhibition of chitinase from *C. albicans* showing that it is strongly pH-dependent, with IC_{50} values of 280 nM at pH 5.0 and 21 nM at pH 7.5. At higher, micromolar concentrations allosamidin inactivates this chitinase in a time- and concentration-dependent manner. Kinetic studies of this inactivation provide evidence for the formation of a reversible complex between allosamidin and chitinase, characterized by $K_{inact} = 5\ \mu M$, followed by irreversible modification of the enzyme consistent with an active site-directed, covalent enzyme modification.

Rast et al. (1991) described the inhibition of a range of chitinase activities of *M. rouxii* by a synthetic analogue of *N,N'*-diacetylchitobiose, *N,N'*-diacetychitobiono-1,5-lactone oxime. This was a competitive inhibitor with a K_i value of around 175 μM, compared to slight inhibition by *N,N'*-diacetylchitobiose (IC_{50} value of about 20 mM).

X. Sequence homologies of chitinases

There is a growing number of amino acid sequences of chitinases. Homologies between them have been classified by Henrissat (1990, 1991, and personal

Table 1. Inhibitory activity of allosamidins and derivatives against chitinase preparations from three fungi.

	IC$_{50}$ (μg/ml)		
	Candida albicans	*Sacharomyces cerevisiae*	*Trichoderma* sp.
Allosamidin	6.2	33.8	0.8
Demethyl allosamidin	0.7	0.3	0.8
Methylallosamidin	8.8	37.2	1.2
Methyl-*N*-demethyl allosamidin	0.6	0.4	1.3
Glucoallosamidin A	3.4	31.3	0.8
Glucoallosamidin B	0.8	0.5	1.6
Hydrolysed allosomidin	1.3	>200	>50
Hydrolysed glucoallosamidin A	5.7	>200	>50

IC$_{50}$ is the concentration causing 50% inhibition.
Compiled from Nishimoto et al. (1991).

communication). The microbial chitinases and the plant acidic chitinases form one group (family 18 in a classification of glycosyl hydrolases) distinct from the plant basic chitinases (family 19). All glycosyl hydrolases were thought to act by an acid catalysis mechanism in which two amino acid residues participate in a displacement reaction. Henrissat's analysis identifies two invariant residues, an aspartate and a glutamate, separated by three amino acids in all chitinases of family 18 examined to date (Table 2). In agreement with this, chemical modification studies of the active centre of the chitinase from *Candida albicans* show specific inactivation by the carboxyl-specific reagent, 1-ethyl-3 (3-dimethylamino-propyl) carbodiimide (EDC), in a single step process (Milewski et al. 1992). In contrast, Verburg et al. (1992) surprisingly reported inactivation of the basic chitinase from maize, *Zea mays*, by reaction of EDC with a tyrosine residue. This residue, however, is conserved in other basic plant chitinases. Table 2 shows homologies of just a short stretch of a range of chitinases and of endo-β-*N*-acetylglucosaminidase H from *Streptomyces plicatus* in the region most likely to contain the active site. The significance of the remarkable sequence homology between the chitinases of the virus *A. californica* NPV and *S. marcescens* has been discussed earlier, as has the chitinase activity of the α-subunit of the toxin from *Kluyveromyces lactis*. Watanabe et al. (1992), Fuhrman et al. (1992) and Kuranda and Robbins (1991) also have discussed homologies at other regions of the chitinase sequences.

Kuranda and Robbins (1991) presented a model of endochitinase encoded by *CTS1* of *S. cerevisiae*, with four functional regions:

Signal – Hydrolytic – Ser,Thr-rich – Chitin-binding

The signal sequence (amino acids 1 to 20) is recognised and cleaved by the usual secretion pathway. The hydrolytic region (amino acids 21 to 237) contains the conserved region shown in Table 2, with the invariant aspartate and glutamate residues, and another conserved region (amino acids 102 to 116). The serine-

Table 2. Alignment of the putative active site region in microbial and plant chitinases.

											*			*						
Gram positive bacteria																				
Bacillus circulans A	(190)	L	R	K	Y	N	F	D	G	V	D	L	D	W	E	Y	P	V	S	(207)
Bacillus circulans D	(290)	I	S	T	Y	G	F	N	G	L	D	I	D	L	E	G	S	S	L	(307)
Flavobacterium sp. (a)	(115)	V	S	K	Y	G	L	D	G	V	D	L	D	D	E	Y	S	D	Y	(132)
Streptomyces plicatus (a)	(161)	V	A	K	Y	G	L	D	G	V	D	F	D	D	E	Y	A	E	Y	(178)
Streptomyces plicatus	(370)	R	W	A	D	V	F	D	G	I	D	L	D	W	E	Y	P	N	A	(387)
Streptomyces erythraeus	(103)	I	D	A	Y	G	L	K	A	I	D	V	D	I	E	A	T	E	F	(120)
Gram negative bacteria																				
Serratia marcescens B	(131)	M	K	D	Y	G	F	D	G	B	D	I	D	W	E	Y	P	Q	A	(148)
Serratia marcescens A	(302)	Q	T	W	K	F	F	D	G	V	D	I	D	W	E	F	P	G	G	(319)
Viruses																				
Autographa californica NPV	(292)	Q	V	W	K	F	F	D	G	V	D	I	D	W	E	F	P	G	G	(309)
Fungi																				
Kluyveromyces lactis (b)	(482)	M	N	K	Y	N	L	D	G	I	D	L	D	W	E	Y	P	G	A	(499)
Saccharomyces cervisiae	(144)	F	D	S	A	V	V	D	G	F	D	F	D	I	E	N	N	N	E	(161)
Plants																				
Cucumis sativis	(139)	L	G	A	A	V	L	D	G	V	D	F	D	I	E	S	G	S	G	(156)
Hevea brasiliensis	(114)	L	G	D	A	V	L	D	G	I	D	F	D	I	E	H	G	S	T	(131)
Arabidopsis thaliana A	(143)	L	G	D	A	V	L	D	G	I	D	F	N	I	E	L	G	S	P	(160)
Nematode																				
Brugia malayi	(135)	L	R	K	N	N	F	D	G	F	D	L	D	W	E	Y	P	V	G	(162)

Asterisks indicate the two invariant aspartate and glutamate residues. Left hand number in parentheses represents position of amino acid from amino-terminus of protein.

(a) *N*-Acetylglucosaminidases; (b) toxin-α-chain.

From Henrissat (1990), with additions: B. Henrissat, personal communication (*B.circulans, K.lactis, H.brasiliensis, A.thaliana, Flavobacterium* sp.); *S.cerevisiae* (Karanda and Robbins, 1991); *A.californica* NPV (Hawtin et al. 1993) *B.malayi* (Fuhrman et al. 1992) *S.plicatus* (Delic et al. 1992).

One letter symbols for amino acids are: A *Ala*, B *Asx*, D *Asp*, E *Glu*, F *Phe*, G *Gly*, H *His*, I *Ile*, K *Lys*, L *Leu*, M *Met*, N *Asn*, P *Pro*, Q *Gln*, R *Arg*, S *Ser*, T *Thr*, V *Val*, W *Trp*, Y *Tyr*.

threonine-rich domain (amino acids 328 to 480) is glycosylated with sugar chains containing from 2 to 5 mannose residues. It may act as a "hinge" region between the catalytic and chitin-binding domains. The high affinity chitin-binding domain (amino acids 481 to 562) has conservation with a cellulose-binding sequence of *Trichoderma reesei* cellulase, with an exact match of a block of 7 amino acids flanked by 2 cysteines. The chitin-binding domain, however, does not display significant affinity for cellulose. Its chitin-binding properties were directly demonstrated in four ways. 1) A carboxyl-terminal deletion product of *CTS1* did not bind to chitin, but retained its catalytic properties. 2) Controlled hydrolysis of wild-type enzyme bound to chitin resulted in an undigested chitin-bound peptide with the sequence starting at amino acid 480. 3) Selective deletion of *CTS1* to remove amino acids 21 to 481 gave direct fusion of the signal sequence to the chitin-binding domain and resulted in secretion of an 18 kDa peptide with high affinity binding to chitin. 4) Expression of a fusion protein between yeast invertase and the chitin-binding domain led to secretion of an invertase that bound efficiently to chitin.

There are strong homologies in the catalytic sites of bacterial chitinases, such as those from *Streptomyces plicatus*, (Robbins et al. 1992), *Streptomyces erythrasus* (Kamei et al. 1989), *Serratia marcescens* (Jones et al. 1986) and *Bacillus circulans* (Watanabe et al. 1990a,b, 1992) (Table 2). *B. circulans* produces at least 6 distinct chitinases. Chitinases A1 has the structure: signal sequence – hydrolytic domain – chitin-binding domain – short carboxyl terminus. The hydrolytic domain, i.e. the *N*-terminal two-thirds of the molecule, has 33% amino acid match to chitinase A from *S. marcescens*. The chitin-binding domain has a tandem repeat of 95-amino acid sequences that are 70% homologous to each other but also have homology to the "type III homology units" of fibronectin, a mammalian cell adhesion molecule (Watanabe et al. 1990b). *S. plicatus* chitinase 63 has a single sequence near the C terminus which is 40% identical to the "type III homology units" of *B. circulans* chitinase A1. The *N*-terminal one-third of *B. circulans* chitinase D shows remarkable similarity to the *C*-terminal one-third of chitinase A and it is immediately upstream of the *ChiA* gene. Watanabe et al. (1992) suggested that this is a result of a complex gene duplication. Thus, the structure of chitinase D contains an *N*-terminal 47 amino acid segment with 62% amino acid match with the *C*-terminus of chitinase A1; a 95 amino acid segment with 63 and 61% matches, respectively, with the "type III homology units" of chitinase 1, and a 73 amino acid segment with the active site with considerable homology to other chitinases (cf. Table 2).

XI. Conclusions

It is clear that the simple definition of chitinase activity, "hydrolysis of N-acetyl-D-glucosaminide (1–4)-β-linkages in chitin and chitodextrins", belies the complexity and diversity of this group of enzymes. There is increasing awareness of the biological roles and importance of chitin and related glucosaminylglycans, both in nature and technology, and we can look forward to major advances in the next few years.

Note added in proof
Further microbial chitinases and their genes that have been characterised are: from the marine bacterium *Alteromonas* sp. Strain 0–7 (Tsujibo H, Orikoshi H, Tanno H, Fujimoto K, Miyamoto K, Imada C, Okami Y and Inamori Y (1993) J. Bacteriol. 175: 176–181); from *Streptomyces lividans* (Fujii T and Miyashita K (1993) J. Gen. Microbiol. 139: 677–686); from the Zygomycete *Rhizopus oligosporus* (Yanai K, Takaya N, Kojima N, Horiuchi H, Ohta A and Takagi M (1992) J. Bacteriol. 174: 7398–7406); and from the Deuteromycete *Aphanocladium album* (Blaiseau P and Lafay J (1992) Gene 120: 243–248).

References

Aribisala OA and Gooday GW (1978) Properties of chitinase from *Vibrio alginolyticus*, as assayed with the chromogenic substrate 3,4-dinitrophenyl tetra-*N*-acetylchitotetraoside. Biochem. Soc. Trans. 6: 568–569

Arnold K, Gooday GW and Chappell LH (1992) Chitinases in Trypanosomatidae: a cautionary note. Parasitol. Today 8: 273

Arnold K, Brydon LJ, Chappell LH and Gooday GW (1993) Chitinolytic activities in *Heligmosomoides polygyrus* and their role in egg hatching. Mol. Biochem. Parasitol. 58: 317–324

Aruchami M, Sundara-Rajulu G and Gowri N (1986) Distribution of deacetylase in arthropoda. In: R Muzzarelli, C Jeuniaux and GW Gooday (eds) Chitin in Nature and Technology (pp 263–265). Plenum Press, New York

Baross JA, Tester PA and Morita RY (1978) Incidence, microscopy and etiology of exoskeleton lesions in the tanner crab *Chionectes tanner*. J. Fish Res. Board Can. 35: 1141–1149

Barrett-Bee K and Hamilton M (1984) The detection and analysis of chitinase activity from the yeast form of *Candida albicans*. J. Gen. Microbiol. 130: 1857–1861

Bartnicki-Garcia S (1973) Fundamental aspects of hyphal morphogenesis. In: JM Ashworth JM and JE Smith (eds) Microbial Differentiation (pp 245–268). Cambridge University Press, Cambridge

Bartnicki-Garcia S and Lippman E (1982) Fungal wall composition. In AJ Laskin and HA Lechevalier (eds) CRC Handbook of Microbiology, 2nd edition, Vol. IV, Microbial Composition: Carbohydrates, Lipids and Minerals (pp 229–252). CRC Press, Boca Raton

Bassler BL, Gibbons P and Roseman S (1989) Chemotaxis to chitin oligosaccharides by *Vibrio furnissii*, a chitinivorous marine bacterium. Biochem. Biophys. Res. Comm. 161: 1172–1176

Bassler BL, Gibbons PJ, Yu C and Roseman S (1991a) Chitin utilization by marine bacteria. Chemotaxis to chitin oligosaccharides by *Vibrio furnissii*. J. Biol. Chem. 266: 24268–24275

Bassler BL, Yu C, Lee YC and Roseman S (1991b) Chitin utilization by marine bacteria. Degradation and catabolism of chitin oligosaccharides by *Vibrio furnissii*. J. Biol. Chem. 266: 24276–24286

Benhamou N and Asselin A (1989) Attempted localization of a substrate for chitinases in plant cells reveals abundant *N*-acetyl-D-glucosamine residues in secondary walls. Biol. Cell 67: 341–350

Berkeley RCW (1979) Chitin, chitosan and their degradative enzymes. In: RCW Berkeley, GW Gooday and DC Ellwood (eds) Microbial Polysaccharides and Polysaccharases (pp 205–236). Academic Press, London

Beyer M and Diekmann H (1985) The chitinase system in *Streptomyces* sp. ATCC 11238 and its significance for fungal cell wall degradation. Appl. Microbiol. Biotech. 23: 14–146

Bidochka MJ and Khachatourians GG (1988) *N*-Acetyl-D glucosamine-mediated regulation of extracellular protease in the entomopathogenic fungus *Beauvaria bassiana*. Appl. Environ. Microbiol. 54: 2699–2704

Bidochka MJ and Khachatourians GG (1992) Growth of the entomopathogenic fungus *Beauveria bassiana* on cuticular components from the migratory grasshopper, *Melanoplus sanguinipes*. J. Invert. Path. 59: 165–173

Blackwell J (1988) Physical methods for the determination of chitin structure and conformation. Meth. Enzymol. 161: 435–442

Borkott H and Insam H (1990) Symbiosis with bacteria enhances the use of chitin by the springtail, *Folsomia candida* (Collembola). Biol. Fert. Soils 9: 126–129

Bradbury P, Deroux G and Campillo A (1987) The feeding of a chitinivorous ciliate. Tissue Cell 19: 351–363

Brandt CR, Adang MJ and Spence KDS (1978) The peritrophic membrane: ultrastructural analysis and function as a mechanical barrier to microbial infection in *Orgyia pseudotsugata*. J. Invert. Pathol. 32: 12–24

Broekaert WF, Van Parijs J, Allen AK and Peumans WJ (1988) Comparison of some molecular, enzymatic and antifungal properties of chitinases from thorn-apple, tobacco and wheat. Physiol. Mol. Pl. Pathol. 33: 319–331

Broglie KE, Gaynor JJ, Broglie RM (1986) Ethylene-regulated gene expression: molecular cloning of the genes encoding an endochitinase from *Phaseolus vulgaris*. Proc. Nat. Acad. Sci. U.S.A. 86: 6820–6824

Butler AR, O'Donnell RW, Martin VJ, Gooday GW and Stark M JR (1991) *Kluyveromyces lactis* toxin has an essential chitinase activity. Eur. J. Biochem. 199: 483–488

Cabib E (1987) The synthesis and degradation of chitin. Adv. Enzymol. 59: 59–101

Cabib E (1988) Assay for chitinase using tritiated chitin. Meth. Enzymol. 161: 424–426

Cabib E, Sburlati A, Bowers B and Silverman SJ (1989) Chitin synthase 1, an auxiliary enzyme for chitin synthesis in *Saccharomyces cerevisiae*. J. Cell Biol. 108: 1667–1672

Cabib E, Silverman SJ, Sburlati A and Slater ML (1990) Chitin synthesis in yeast *Saccharomyces cerevisiae* In: PJ Kuhn, APJ Trinci, MJ Jung, MW Goosey and LG Copping (eds) Biochemistry of Cell Walls and Membranes in Fungi (pp 31–41). Springer-Verlag, Berlin

Cabib E, Silverman SJ and Shaw JA (1992) Chitinase and chitin synthase 1: counter balancing activities in cell separation of *Saccharomyces cerevisiae*. J. Gen. Microbiol. 138: 97–102

Chan JG (1970) The occurrence, taxonomy and activity of chitinolytic bacteria from sediment, water and fauna of puget sound. PhD thesis, University of Washington, Seattle

Chen JP, Nagagdma F and Change MC (1991) Cloning and expression of a chitinase gene from *Aeromonas hydrophila* in *Escherichia coli*. Appl. Environ. Microbiol. 57: 2426–2428

Cody RM (1989) Distribution of chitinase and chitobiase in Bacillus. Curr. Microbiol. 19: 201–205

Cody RM, Davis ND, Lin J and Shaw D (1990) Screening microorganisms for chitin hydrolysis and production of ethanol from amino sugars. Biomass 21: 285–295

Cole TA, Marburger RE and Cabib E (1989) A substrate-included polyacrylamide disc gel-electrophoretic assay for chitinase. In: G Skjak-Braek, T Anthonsen and P Sandford (eds) Chitin and Chitosan (pp 343–351). Elsevier, London

Correa JU, Elango N, Polachek I and Cabib E (1982) Endochitinase, a mannan-associated enzyme from *Saccharomyces cerevisiae*. J. Biol. Chem. 257: 1392–1397

Coudron TA, Kroha MJ and Ignoffo CM (1984) Levels of chitinolytic activity during development of three entomopathogenic fungi. Comp. Biochem. Physiol. B 79: 339–348

Dackman C, Chet I and Nordbring-Hertz B (1989) Fungal parasitism of the cyst nematode *Heterodera schachtii*: infection and enzymatic activity. FEMS Microbiol. Ecol. 62: 201–208

Daoust RA and Gunner HB (1979) Microbial synergists pathogenic to *Lymantria dispar*: chitinolytic and fermentative bacterial interactions. J. Invert. Path. 33: 368–377

Datema R, van den Ende H and Wessels JGH (1977) The hyphal wall of *Mucor mucedo* 2. Hexosamine-containing polymers. Eur. J. Biochem. 80: 621–626

Davis B and Eveleigh DE (1984) Chitosanases: occurrence, production and immobilization. In: JP Zikakis (ed) Chitin, Chitosan and Related Enzymes (pp 161–179). Academic Press, Orlando

Davis LL and Bartnicki-Garcia S (1984) The co-ordination of chitosan and chitin synthesis in *Mucor rouxii*. J. Gen. Microbiol. 130: 2095–2102

De Jong AJ, Cordewener J, Schiavo FL, Terzi M, Vanderckhove J, Van Kammer A and De Vries SC (1992) A carrot somatic embryo mutant is rescued by chitinase. Plant Cell 4: 425–433

Delic I, Robbins P and Westpheling J (1992) Direct repeat sequences are implicated in the regulation of two *Streptomyces* chitinase promoters that are subject to carbon catabolite control. Proc. Nat. Acad. Sci. U.S.A. 89: 1885–1889

Deshpande MV (1986) Enzymatic degradation of chitin and its biological applications. J. Sci. Indust. Res. 45: 273–281

Dickinson K, Keer V, Hitchcock CA and Adams DJ (1989) Chitinase activity from *Candida albicans* and its inhibition by allosamidin. J. Gen. Microbiol. 135: 1417–1421

Dickinson K, Keer V, Hitchcock CA and Adams DJ (1991) Microsomal chitinase activity from *Candida albicans*. Biochim. Biophys. Acta 1073: 177–182

Dohrmann PR, Butler G, Tamai K, Dorland S, Greene JR, Thiele DJ and Sullivan DJ (1992) Parallel pathways of gene regulation: homologous regulators *SWI5* and *ACE2* differentially control transcription of *HO* and chitinase. Genes Devel. 6: 93–104

Dunsmuir P and Suslow T (1989) Structure and regulation of organ- and tissue-specific-genes: chitinase genes in plants. In: J Schell and IK Vaszil (eds) Cell Culture and Somatic Cell Genetics in Plants, Vol. 6 Molecular Biology of Plant Nuclear Genes (pp 215–227). Academic Press, San Diego

El-Sayed GN, Coudron TA, Ignoffo CM and Riba G (1989) Chitinolytic activity and virulence associated with native and mutant isolates of an entomopathogenic fungus *Nomuraea rileyi*: J. Invert. Path. 54: 394–403

Elango N, Correa JV and Cabib E (1982) Secretory nature of yeast chitinase. J. Biol. Chem. 257: 1398–1400

Evrall CC, Attwell RW and Smith CA (1990) A semi-micro quantitative assay for determination of chitinolytic activity in microorganisms. J. Microbiol. Meth. 12: 183–187

Fink D, Boucher I, Denis F and Brezinski R (1991) Cloning and expression in *Streptomyces lividans* of a chitosanase-encoding gene from the actinomycete *Kitasatosporia* N174 isolated from soil. Biotech. Lett. 13: 845–850

Flyg C and Boman HG (1988) *Drosphila* genes *cut* and *miniature* are associated with the susceptibility to infection in *Serratia marcescens*. Genet Res 52: 51–56

Fuchs R, McPherson S and Drahos D (1986) Cloning of a *Serratia marcescens* gene encoding chitinase. Appl. Environ. Microbiol. 51: 504–509

Fuhrman JA, Lane WS, Smith RF, Piessens WF and Perler FB (1992) Transmission-blocking antibodies recognise microfilarial chitinase in brugian lymphatic filariasis. Proc. Nat. Acad. Sci. U.S.A. 89: 1548–1552

Gooday GW (1977) Biosynthesis of the fungal wall-mechanisms and implications. The first Fleming Lecture. J. Gen. Microbiol. 99: 1–11

Gooday GW (1979) A survey of polysaccharase production: a search for phylogenetic implications. In: RCW Berkeley, GW Gooday and DC Ellwood (eds) Microbial Polysaccharides and Polysaccharases (pp 437–460). Academic Press, London

Gooday GW (1990a) The ecology of chitin degradation. In: KC Marshall (ed) Advances in Microbial Ecology, Vol. 11 (pp 387–430). Plenum Press, New York

Gooday GW (1990b) Inhibition of chitin metabolism. In: PJ Kuhn, APJ Trinci, MJ Jung, MW Goosey and LG Copping (eds) Biochemistry of Cell Walls and Membranes in Fungi (pp 61–79). Springer-Verlag, Berlin

Gooday GW (1990c) Chitinases. In: G Leatham and M Himmel (eds) Enzymes in Biomass Conversion (pp 478–485). American Chemical Society Books, Washington

Gooday GW (1990d) Chitin metabolism: a target for antifungal and antiparasitic drugs. Pharmacol. Ther. Suppl. pp 175–185

Gooday GW and Gow NAR (1991) The enzymology of tip growth in fungi. In: IB Heath (ed) Tip

Growth of Plant and Fungal Cells (pp 31–58). Academic Press, New York.

Gooday GW, Humphreys AM and McIntosh WH (1986) Roles of chitinase in fungal growth. In: RAA Muzzarelli, C Jeuniaux and GW Gooday (eds) Chitin in Nature and Technology (pp 83–91). Plenum Press, New York

Gooday GW, Brydon LJ and Chappell LH (1988) Chitinase in female *Onchocerca gibsoni* and its inhibition by allosamidin. Mol. Biochem. Parasitol. 29: 223–225

Gooday GW, Prosser JI, Hillman K and Cross M (1991) Mineralization of chitin in an estuarine sediment: the importance of the chitosan pathway. Biochem. Systematics. Ecol. 19: 395–400

Gooday GW, Zhu W-Y and O'Donnell RW (1992) What are the roles of chitinases in the growing fungus? FEMS Microbiol. Lett. 100: 387–392

Gow NAR and Gooday GW (1983) Ultrastructure of chitin in hyphae of *Candida albicans* and other dimorphic and mycelial fungi. Protoplasma 115: 52–58

Gunner HB, Met MZ and Berger S (1985) Chitinase-producing B T strains. In: DG Grimble and FB Lewis (eds) Microbial Control of Spruce Budworms and Gypsy Moths (pp 102–108). U.S. Forestry Service GTR-NE-100

Hara S, Yamamura Y, Fujii Y, Mega T and Ikenaka T (1989) Purification and characterization of chitinase produced by *Streptomyces erythraeus*. J. Biochem. 105: 484–489

Hawtin RE, Ayres MD, Arnold K, Gooday GW, Chappell LH, Kitts PA, King LA and Possee RD (1993) Liquefaction of insect larvae by a baculovirus-encoded chitinase. (in press)

Hedges A and Wolfe RS (1974) Extracellular enzyme from *Myxobacter AL-1* that exhibits both β-1,4-glucanase and chitosanase activities. J. Bacteriol. 120: 844–853

Hellebust JA and Lewin J (1977) Heterotrophic nutrition. In: D Werner (ed) The Biology of Diatoms (pp 169–197). Blackwell, Oxford

Helmke E and Weyland H (1986) Effect of hydrostatic pressure and temperature on the activity and synthesis of chitinases of Antarctic Ocean bacteria. Mar. Biol. 91: 1–7

Henrissat B (1990) Weak sequence homologies among chitinases detected by clustering analysis. Sequ. Data. Anal. 3: 523–526

Henrissat B (1991) A classification of glycosyl hydrolases based on amino acid sequence similarities. Biochem. J. 280: 309–316

Herwig RP, Staley JT, Nerini MK and Braham HW (1984) Baleen whales: preliminary evidence for forestomach microbial fermentation. Appl. Environ. Microbiol. 47: 421–423

Hillman K, Gooday GW and Prosser JI (1989a) The mineralization of chitin in the sediments of the Ythan Estuary, Aberdeenshire, Scotland. Estuarine Coastal Shelf Sci. 29: 601–612

Hillman K, Gooday GW and Prosser JI (1989b) A simple model system for small scale *in vitro* study of estuarine sediment ecosystems. Lett. Appl. Microbiol. 4: 41–44

Hirano S and Tokura S (1982) Chitin and Chitosan. Japanese Society of Chitin and Chitosan, Tottori

Hood MA (1991) Comparison of four methods for measuring chitinase activity and the applications of the 4-MUF assay in aquatic environments. J. Microbiol. Meth. 13: 151–160

Hood MA and Meyers SP (1977) Microbial and chitinoclastic activities associated with *Panaeus setiferus*. J Oceangraph. Soc. Japan 33: 235–241

Horwitz M, Reid J and Ogaydziak D (1984) Genetic improvements of chitinase production of *Serratia marcescens*. In: J Zikakis (ed) Chitin, Chitosan and Related Enzymes (pp 191–208). Academic Press, Orlando

Huber M, Cabib E and Miller LH (1991) Malaria parasite chitinase and penetration of the mosquito peritrophic membrane. Proc. Nat. Acad. Sci. U.S.A. 88: 2807–2810

Humphreys AM and Gooday GW (1984a) Properties of chitinase activities from *Mucor mucedo*: evidence for a membrane-bound zymogenic form. J. Gen. Microbiol. 130: 1359–1366

Humphreys AM and Gooday GW (1984b) Phospholipid requirement of microsomal chitinase from *Mucor mucedo*. Curr. Microbiol. 11: 187–190

Humphreys AM and Gooday GW (1984c) Chitinase activities from *Mucor mucedo*. In: C Nombela (ed) Microbial Cell Wall Synthesis and Autolysis (pp 269–273). Elsevier, Amsterdam

Inbar J and Chet I (1991) Detection of chitinolytic acitivity in the rhizosphere using image analysis. Soil Biol. Biochem. 3: 239–242

Isaac S and Gokhale AV (1982) Autolysis: a tool for protoplast production from *Aspergillus nidulans*. Trans. Br. Mycol. Soc. 78: 389–394

Isogai A, Sato M, Sakuda S, Nakuyama A (1989) Structure of demethylallosamidin as an insect chitinase inhibitor. Agric. Biol. Chem. 53: 2825–2826

Iten W and Matile P (1970) Role of chitinase and other lysosomal enzymes of *Coprinus lagopus* in the autolysis of fruiting bodies. J. Gen. Microbiol. 61: 301–309

Iverson KL, Bromel MC, Anderson AW and Freeman TP (1984) Bacterial symbionts in the sugar beet root maggot *Tetanops myopaeformis* (von Roder). Appl. Environ. Microbiol. 47: 22–27

Janszen FH and Wessels JGH (1970) Enzymic dissolution of hyphal septa in a Basidiomycete. Antonie van Leewenhoek 36: 255–257

Jeuniaux C (1963) Chitine and Chitinolyse. Masson, Paris

Jeuniaux C (1982) La chitine dans le régne animal. Bull. Soc. Zool. France 107: 363–386

Jones J, Grady K, Suslow T and Bedbrook J (1986) Isolation and characterization of genes encoding two distinct chitinase enzymes from *Serratia marcescens* EMBO J. 5: 467–473

Jones JDG, Dean C, Gidoni D, Gilbert D, Bond-Nutter D, Nedbrook J and Dunsmuir P (1988) Expression of a bacterial chitinase protein in tobacco leaves using two photosynthetic gene promoters. Mol. Gen. Genet. 212: 536–542

Kamei K, Yamamura Y, Hara S and Ikenda T (1989) Amino acid sequence of chitinase from *Streptomyces erythaeus*. J. Biochem. 105: 979–985

Kaneko T and Colwell RR (1978) The annual cycle of *Vibrio parahaemolyticus* in Chesapeake Bay. Microbial Ecol. 4: 135–155

Kohlmeyer J (1972) Marine fungi deteroriating chitin of hydrozoa and keratin-like annelid tubes. Mar. Biol. 12: 277–284

Kole MM and Altosaar I (1985) Increased chitinase production by a non-pigmented mutant of *Serratia marcescens*. FEMS Microbiol. Lett. 26: 265–269

Kuhl J, Nittinger J and Siebert G (1978) Verwertung von Krillschalen in Futterungsversuchen an der Ratte. Arch Fischereiwiss 29: 99–103

Kunz C, Ludwig A, Bertheau Y and Boller T (1992) Evaluation of the antifungal activity of the purified chitinase 1 from the filamentous fungus *Aphanocladium album*. FEMS Microbiol. Lett. 90: 99–103

Kuranda MJ and Robbins PW (1988) Cloning and heterologus expression of glycosidase genes from *Saccharomyces cerevisiae*. Proc. Nat. Acad. Sci. U.S.A. 84: 2585–2589

Kuranda MJ and Robbins PW (1991) Chitinase is required for cell separation during growth of *Saccharomyces cerevisiae*. J. Biol. Chem. 266: 19758–19767

Leah R, Tommerup H, Svendsen I and Mundy J (1991) Biochemical and molecular characterization of three barley seed proteins with antifungal properties. J. Biol. Chem. 266: 1564–1573

Liaw HJ and Mah RA (1992) Isolation and characterization of *Haloanaerobacter chitinovorans* gen. nov., sp. nov., a halophilic, anaerobic, chitinolytic bacterium from a solar saltern. Appl. Environ. Microbiol. 58: 260–266

Lindsay GJH and Gooday GW (1985a) Action of chitinase in spines of the diatom *Thalassiosira fluviatilis*. Carbohydr. Polymers 5: 131–140

Lindsay GJH and Gooday GW (1985b) Chitinolytic enzymes and the bacterial microflora in the digestive tract of cod, *Gadus morhua*. J. Fish Biol. 26: 255–265

Lindsay GJH, Walton MJ, Adron JW, Fletcher TC, Cho CY and Cowey CB (1984) The growth of rainbow trout (*Salmo gairdneri*) given diets containing chitin and its relationships to chitinolytic enzymes and chitin digestibility. Aquaculture 37: 315–334

Linthorst HJM, van Loon LC, van Rossum CMA, Mayer A, Bol JF, van Roekel JSC, Meulenhoff EJS and Cornelissen BJC (1990) Analysis of acid and basic chitinases from tobacco and petunia and their constitutive expression in transgenic tobacco. Mol. Plant-Microbe Interact. 3: 252–258

Lund P and Dunsmuir P (1992) A plant signal sequence enhances the secretion of bacterial ChiA in transgenic tobacco. Plant Mol. Biol. 18: 47–53

Lund P, Lee RY and Dunsmuir P (1989) Bacterial chitinase is modified and secreted in transgenic tobacco. Plant Physiol 91: 130–135

308 G.W. Gooday

Lynch J (1987) *In vitro* identification of *Trichoderma harzianum* as a potential antagonist of plant pathogens. Curr. Microbiol. 16: 49–53

Lysenko O (1976) Chitinase of *Serratia marcescens* and its toxicity to insects, J. Invertebr. Pathol. 27: 385–386

Manocha MS and Balasubramanian R (1988) In vitro regulation of chitinase and chitin synthase activity of two mucoraceous hosts of a mycoparasite. Can. J. Microbiol. 34: 1116–1121

Manson FDC, Fletcher TC and Gooday GW (1992) Distribution of chitinolytic enzymes in blood cells of turbot *Scophthalmus maximus*. J. Fish Biol. 40: 919–927

Mauch F and Staehelin LA (1989) Functional implications of the subcellular localization of ethylene-induced chitinase and β-1,3-glucanase in bean leaves. Plant Cell 1: 447–457

Mauch F, Mauch-Mani B and Boller T (1988) Antifungal hydrolases in pea tissue. Inhibition of fungal growth by combination of chitinase and β-1,3-glucanase. Plant Physiol. 87: 936–942

Maudlin I and Welburn SC (1988) Tsetse immunity and the transmission of trypanosomiasis. Parasit. Today 4: 109–111

McCormack J, Hackett TJ, Tuohy MG and Coughlan MP (1991) Chitinase production by *Taleromyces emersonii*. Biotech. Lett. 13: 677–682

McCreath K and Gooday GW (1992) A rapid and sensitive microassay for determination of chitinolytic activity. J. Microbiol. Meth. 14: 229–237

McNab R and Glover LA (1991) Inhibition of *Neurospora crassa* cytosolic chitinase by allosamidin. FEMS Microbiol. Lett. 82: 79–82

Milewski S, O'Donnell RW and Gooday GW (1993) Chemical modification studies of the active centre of *Candida albicans* chitinase and its inhibition by allosamidin. J. Gen. Microbiol. 138: 2545–2550

Mitsutomi M, Ohtakara A, Fukamizo T and Goto S (1990) Action pattern of *Aeromonas hydrophila* chitinase on partially *N*-acetylated chitinase. Agric. Biol. Chem. 54: 871–877

Miyashita K, Fugii T and Sawada Y (1991) Molecular cloning and characterization of chitinase genes from *Streptomyces lividans* 66. J. Gen. Microbiol. 137: 2065–2072

Molano J, Duran A and Cabib E (1977) A rapid and sensitive assay for chitinase using tritiated chitin. Anal. Biochem. 83: 648–656

Monaghan RL (1975) The discovery, distribution and utilization of chitosanase. PhD thesis, Rutgers University, New Brunswick, NJ

Monaghan RL, Eveleigh DE, Tewari RP and Reese ET (1973) chitosanase, a novel enzyme. Nature (London) New Biol. 245: 78–80

Monreal J and Reese ET (1969) The chitinase of *Serratia marcescens*. Can. J. Microbiol. 15: 689–696

Morris ON (1976) A two-year study of the efficacy of *Bacillus thuringiensis*-chitinase combinations in spruce budworm (*Choristoneura fumiferana*) control. Can. Entomol. 108: 225–233

Murray CL and Lovett JS (1966) Nutritional requirements of the chytrid, *Karlingia asterocysta*, an obligate chitinophile. Am. J. Bot. 53: 469–476

Muzzarelli RAA and Pariser ER (1978) Proceedings of the First International Conference on Chitin/Chitosan. MIT Sea Grant Report MITSG 78-7

Muzzarelli RAA, Jeuniaux C and Gooday GW (1986) Chitin in Nature and Technology. Plenum Press, New York

Nanjo F, Sakai K, Ishikawa M, Isobe K and Usui T (1989) Properties and transglycosylation reaction of an chitinase from *Nocardia orientalis*. Agric. Biol. Chem. 53: 2189–2195

Nanjo F, Katsumi R and Sakai K (1990) Purification and characterization of an exo-β-D-glucosaminidase, a novel type of enzyme, from *Nocardia orientalis*. J. Biol. Chem. 265: 10088–10094

Nanjo F, Katsumi R and Sakai K (1991) Enzymatic method for determination of the degree of deacetylation of chitosan. Anal. Biochem. 193: 164–167

Neugebauer E, Gamuche B, Dery CV and Brzezinski R (1991) Chitinolytic properties of *Streptomyces lividans*. Arch. Microbiol. 156: 192–197

Nishimoto Y, Sakuda S, Takayama S and Yamada Y (1991) Isolation and characterization of new allosamidins. J. Antibiot. 44: 716–722

O'Brien M and Colwell RR (1987) A rapid test for chitinase activity that uses 4-methylumbelliferyl-*N*-acetyl-β-D-glucosaminide. Appl. Environ. Microbiol. 53: 1718–1720

Ohtakara A (1988) Viscosimetric assay of chitinase. Meth. Enzymol. 161: 426–430

Ohtakara A, Matsunaga H and Mitsutomi M (1990) Action pattern of *Streptomyces griseus* chitinase on partially *N*-acetylated chitosan. Agric. Biol. Chem. 54: 3191–3199

Okazaki K and Tagawa K (1991) Purification and properties of chitinase from *Streptomyces cinereoruber*. J. Ferment. Bioeng. 71: 237–241

Orlean P, Kuranda MJ and Albright CF (1991) Analysis of glycoproteins from *Saccharomyces cerevisiae*. Meth. Enzymol. 194: 682–697

Orunsi NA and Trinci APJ (1985) Growth of bacteria on chitin, fungal cell walls and fungal biomass, and the effect of extracellular enzymes produced by these cultures on the antifungal activity of amphotericin B. Microbios 43: 17–30

Patton RS and Chandler PT (1975) *In vivo* digestibility evaluation of chitinous material, J. Dairy Sci. 58: 397–403

Payne S, Ahl P, Moyer M, Harper A, Beck J, Meins F and Ryals J (1990) Isolation of complementary DNA clones encoding pathogenesis-related proteins P and Q, two acid chitinases from tobacco. Proc. Nat. Acad. Sci. U.S.A. 87: 98–102

Peberdy J (1983) Genetic recombination in fungi following protoplast fusion and transformation. In: JE Smith (ed) Fungal Differentiation (pp 559–581). Marcel Dekker, New York

Pedraza-Reyes M and Lopez-Romero E (1989) Purification and some properties of two forms of chitinase from mycelial cells of *Mucor rouxii*. J. Gen. Microbiol. 135: 211–218

Pedraza-Reyes M and Lopez-Romero E (1991a) Chitinase activity in germinating cells of *Mucor rouxii*. Antonie van Leeuwenhoek 59: 183–189

Pedraza-Reyes M and Lopez-Romero E (1991b) Detection of nine chitinase species in germinating cells of *Mucor rouxii*. Curr. Microbiol. 22: 43–46

Pel R and Gottschal JC (1986a) Chitinolytic communities from an anaerobic estuarine environment, In: RAA Muzzarelli, C Jeuniaux and GW Gooday (eds) Chitin in Nature and Technology (pp 539–546). Plenum Press, New York

Pel R and Gottschal JC (1986b) Stimulation of anaerobic chitin degradation in mixed cultures, Antonie van Leeuwenhoek 52: 359–360

Pel R and Gottschal JC (1989) Interspecies interaction based on transfer of a thioredoxin-like compound in anaerobic chitin-degrading mixed cultures. FEMS Microbiol. Ecol. 62: 349–358

Pel R, Wessels G, Aalfs H and Gottschal JC (1989) Chitin degradation by *Clostridium* sp. strain 9.1 in mixed cultures with saccharolytic and sulphate-reducing bacteria. FEMS Microbiol. Ecol. 62: 191–200

Pel R, Van Den Wijngaard AJ, Epping E and Gottschal JC (1990) Comparison of the chitinolytic properties of *Clostridium* sp. strain 9.1 and a chitin degrading bacterium from the intestinal tract of the plaice *Pleuronectes platessa* (*L.*). J. Gen. Microbiol. 136: 695–704

Pelletier A and Sygusch J (1990) Purification and characterization of three chitosanase activities from *Bacillus megaterium* P1. Appl. Environ. Microbiol. 56: 844–848

Pelletier A, Lemire I, Sygusch J, Chornet E and Overend RP (1990) Chitin/chitosan transformation by thermo-mechano-chemical treatment including characterization by enzymatic depolymerisation. Biotech. Bioeng. 36: 310–315

Peter MG, Kegel G and Keller R (1986) Structural studies on sclerotized insect cuticle. In: R Muzzarelli, C Jeuniaux, GW Gooday (eds) Chitin in Nature and Technology (pp 21–28), Plenum Press, New York

Petz W, Foissner W, Wirnsberger E, Kruatgartner WD and Adam H (1986) Mycophagy, a new feeding strategy in autochthonous ciliates. Naturwissenschaften 73: S.560–561

Pisano MA, Sommer MJ and Taras L (1992) Bioactivity of chitinolytic actinomycetes of marine origin. Appl. Microbiol. Biotech. 36: 553–555

Pope AMS and Davies DAL (1979) The influence of carbohydrases on the growth of fungal pathogens *in vitro* and *in vivo*. Postgrad. Med. J. 55: 674–676

Poulicek M, Voss-Foucart MF and Jeuniaux C (1986) Chitinoproteic complexes and mineralization in mollusk skeletal structures. In: R Muzzarelli, C Jeuniaux, GW Gooday (eds)

Chitin in Nature and Technology (pp 7–12). Plenum Press, New York

Rast DM, Horsch M, Funter R and Gooday GW (1991) A complex chitinolytic system in exponentially growing mycelium of *Mucor rouxii*: properties and functions. J. Gen Microbiol 137: 2797–2810

Reichardt W, Gunn B and Colwell RR (1983) Ecology and taxonomy of chitinoclastic *Cytophaga* and related chitin-degrading bacteria isolated from an estuary, Microb. Ecol. 9: 273–294

Reid JD and Ogrydziak DM (1981) Chitinase-overproducing mutant of *Serratia marcescens*. Appl. Environ. Microbiol. 41: 664–669

Reisert PS and Fuller MS (1962) Decomposition of chitin by *Chytridiomyces* species. Mycologia 54: 647–657

Reyes F, Perez-Leblic MI, Martinez MJ and Lahoz R (1984) Protoplast production from filamentous fungi with their own autolytic enzymes. FEMS Microbiol. Lett. 24: 281–283

Reyes F, Lahoz R, Martinez MJ and Alfonso C (1985) Chitosanases in the autolysis of *Mucor rouxii*. Mycopathologia 89: 181–187

Reyes F, Calatayud J, Vazquez C and Martinez MJ (1989) β-*N*-Acetylglucoasminidase from *Aspergillus nidulans* which degrades chitin oligomers during autolysis. FEMS Microbiol. Lett. 65: 83–88

Robbins PW, Albright C and Benfield B (1988) Cloning and expression of a *Streptomyces plicatus* chitinase (chitinase-63) in *Escherichia coli*. J. Biol. Chem. 263: 443–447

Robbins PW, Overbye K, Albright C, Benfield B and Pero J (1992) Cloning and high-level expression of chitinase-encoding gene of *Streptomyces plicatus*. Gene 111: 69–76

Roberts GAF (1992) Chitin Chemistry. Macmillan, Basingstoke

Roberts RL and Cabib E (1982) *Serratia marcescens* chitinase: one-step purification and use for the determination of chitin. Anal. Biochem. 127: 402–412

Roberts WK and Selitrennikoff CP (1988) Plant and bacterial chitinases differ in antifungal activity. J. Gen. Microbiol. 134: 169–176

Roffey PE and Pemberton JM (1990) Cloning and expression of an *Aeromonas hydrophila* chitinase gene in *Escherichia coli*. Curr. Microbiol. 21: 329–337

Rudall KM and Kenchington W (1973) The chitin system. Biol. Rev. 4: 597–636

Sakai K, Katsumi R, Isobe A and Nanjo F (1991a) Purification and hydrolytic action of a chitosanase from *Nocardia orientalis*. Biochim. Biophys. Acta 1079: 65–72

Sakai K, Uchiyama T, Matahira Y and Nanjo F (1991b) Immobilization of chitinolytic enzymes and continuous production of *N*-acetylglucosamine with the immoobolized enzymes. J. Ferment. Bioeng. 72: 168–172

Sakuda S, Isogai A, Matsumoto S and Suzuki A (1987a) Search for microbial insect growth regulators II. Allosamidin, a novel insect chitinase inhibitor. J Antibiot 40: 296–300

Sakuda S, Isogai A, Makita T, Matsumoto S, Koseki K, Kodama H and Suzuki A (1987b) Structures of allosamidins, novel insect chitinase inhibitors, produced by actinomycetes. Agric. Biol. Chem. 51: 3251–3259

Sakuda S, Nishimoto Y, Ohi M, Watanabe M, Takayama J, Isogai A and Yamada Y (1990) Effects of demethylallosamidin, a potent yeast chitinase inhibitor, on the cell division of yeast. Agric. Biol. Chem. 54: 1333–1335

Sassen MMA (1965) Breakdown of the plant cell wall during the cell fusion process. Acta Bot Neerlandica 14: 165–196

Schlein Y, Jacobson RL and Shlomai J (1991) Chitinase secreted by *Leishmania* functions in the sandfly vector. Proc. R. Soc. Lond. B. 245: 121–126

Seino H, Tsukuda K and Shimasue Y (1991) Properties and action pattern of a chitosanase from *Bacillus* sp. P1-75. Agric. Biol. Chem. 55: 2421–2423

Seki H and Taga N (1965) Microbial studies on the decomposition of chitin in marine environment – VI. Chitinoclastic bacteria in the digestive tract of whales from the Antarctic Ocean. J. Oceanogr. Soc. Japan. 20: 272–277

Shapira R, Ordentlich A, Chet I and Oppenheim AB (1989) Control of plant diseases by chitinase expressed from cloned DNA in *Escherichia coli*. Phytopathology 79: 1246–1249

Shapiro M, Preisler HK and Robertson JL (1987) Enhancement of baculovirus activity in gypsy moth (Lepidoptera:Lymantriidae) by chitinase. J. Econ. Ent. 80: 1113–1116

Shinshi H, Beuhaus J, Ryals J and Meins F (1990) Structure of a tobacco chitinase gene: evidence that different chitinase genes can arise by a transposition of sequences encoding a cysteine-rich domain. Plant Mol. Biol. 14: 357–368

Sietsma JH, Vermeulen CA and Wessels JGH (1986) The role of chitin in hyphal morphogenesis. In: R Muzzarelli, C Jeuniaux and GW Gooday (eds) Chitin in Nature and Technology (pp 63–69). Plenum Press, New York

Sivan A and Chet I (1989) Degradation of fungal cell walls by lytic enzymes of *Trichoderma harzianum*. J. Gen. Microbiol. 135: 675–682

Skjak-Braek G, Anthonsen T and Sandford P (1989) Chitin and Chitosan. Elsevier, London

Smirnoff WA (1974) Three years of aerial field experiments with *Bacillus thuringiensis* plus chitinase formulation against the spruce budworm. J. Invert. Path. 24: 344–348

Smith RJ and Grula EA (1983) Chitinase is an inducible enzyme in *Beauvaria bassiana*. J. Invert. Path. 42: 319–326

Sneh B (1981) The use of rhizophere chitinolytic bacteria for biological control. Phytopath Zeitschrift 100: 251–256

Soderhall K and Unestam T (1975) Properties of extracellular enzymes from *Aphanomyces astaci* and their relevance in the penetration process of crayfish cuticle. Physiol. Plant 35: 140–146

Somashekar D and Joseph R (1992) Partial purification and properties of a novel chitosanase secreted by *Rhodotorula gracilis*. Lett. Appl. Microbiol. 14: 1–4

Somers PJB, Yao RC, Doolin LR, McGowan MJ, Fakuda DS and Mynderse JS (1987) Method for the detection and quantitation of chitinase inhibitors in fermentation broths; isolation and insect life cycle. Effect of A82516. J. Antibiot. 40: 1751–1756

Soto-Gil RW and Zyskind JW (1984) Cloning of *Vibrio harveyi* chitinase and chitobiase gene in *Escherichia coli*. In: JP Zikakis (ed) Chitin, Chitosan and Related Enzymes (pp 209–223). Academic Press, Orlando

Srivastava AK, Defago G and Boller T (1985) Secretion of chitinase by *Aphanocladium album*, a hyperparsite of wheat. Experientia 41: 1612–1613

St Leger RJ, Cooper RM and Charnley AK (1986) Cuticle-degrading enzymes of entomopathogenic fungi: regulation of production of chitinolytic enzymes. J. Gen. Microbiol. 132: 1509–1517

St Leger RJ, Cooper RM and Charnley AK (1987) Production of cuticle degrading enzymes by the entomopathogen *Metarhizium anisopliae* during infection of cuticles from *Calliphora vomitaria* and *Manduca sexta*. J. Gen. Microbiol. 133: 1371–1382

St Leger RJ, Cooper RM and Charnley AK (1991) Characterization of chitinase and chitobiase produced by the entomopathogenic fungus *Metarhizium anisopliae*. J. Invert. Path. 58: 415–426

Surarit R, Gopel PK and Shepherd MG (1988) Evidence for a glycosidic linkage between chitin and glucan in the cell wall of *Candida albicans*. J. Gen. Microbiol. 134: 1723–1730

Suslow TV and Jones J (1988) Chitinase-producing bacteria. U.S. Patent No. 4751081

Suzuki K, Mikami T, Okawa Y, Tokora A, Suzuki S and Suzuki M (1986) Antitumour effect of hexa-*N*-acetylchitohexase and chitohexaose. Carb. Res. 151: 403–408

Tagawa K and Okazaki K (1991) Isolation and some cultural conditions of *Streptomyces* species which produce enzymes lysing *Aspergillus niger* cell wall. J. Ferment. Bioeng. 71: 230–236

Takayanagi T, Ajisaka K, Takiguch Y and Shimahara K (1991) Isolation and properties of thermostable chitinases from *Bacillus licheniformis* X-7u. Biochim. Biophys. Acta 1078: 404–410

Takiguchi Y and Shimahara K (1988) *N*,*N'*-Diacetylchitobiose production from chitin by *Vibrio anguillarum* strain E-383a. Lett. Appl. Microbiol. 6 129–131

Takiguchi Y and Shimahara K (1989) Isolation and identification of a thermophilic bacterium producing *N*,*N'*-diacetylchitobiose from chitin. Agric. Biol. Chem. 53: 1537–1541

Taylor JL, Jones JDG, Sandler S, Mueller GM, Bedbrook J and Dunsmuir P (1987) Optimizing the expression of chimeric genes in plant cells. Mol. Gen. Genet. 210: 572–577

Tom RA and Carroad PA (1981) Effect of reaction conditions on hydrolysis of chitin by *Serratia marcescens* QM B1466 chitinase. J. Food Sci. 49: 379–380

Tracey MV (1955) Chitinase in some Basidiomycetes. Biochem. J. 61: 579–589

Trudel J and Asselin A (1989) Detection of chitinase activity after polyacrylamide gel electrophoresis. Anal. Biochem. 178: 362–366

Ueno H, Miyashita K, Sawada Y and Oba Y (1990) Purification and some properties of extracellular chitinases from *Streptomyces* sp. 5-84. J. Gen. Appl. Microbiol. 36: 377–392

Ulhoa CJ and Peberdy JF (1991) Regulation of chitinase synthesis in *Trichoderma harzianum*. J. Gen. Microbiol. 137: 2163–2169

Ulhoa CJ and Peberdy JF (1992) Purification and some properties of the extracellular chitinase produced by *Trichoderma harzianum*. Enzyme Microbiol. Technol. 14: 236–240

Usai T, Hayashi Y, Nanjo F, Sakai K and Ishido Y (1987) Transglycosylation reaction of a chitinase purified from *Nocardia orientalis*. Biochim. Biophys. Acta 923: 302–309

Usai T, Matsui M and Isobe K (1990) Enzymic synthesis of useful chito-oligosaccharides utilizing transglycosylation by chitinolytic enzymes in buffer containing ammonium sulfate. Carb. Res. 203: 65–77

Vasseur V, Arigoni F, Andersen H, Defago G, Bompeix G and Seng J-M (1990) Isolation and characterization of *Aphanocladium album* chitinase-overproducing mutants. J. Gen. Microbiol. 136: 2561–2567

Verburg JG, Smith CE, Lisek CA and Huynh QK (1992) Identification of an essential tyrosine residue in the catalytic site of a chitinase isolated from *Zea mays* that is selectively modified during inactivation with 1-ethyl-3 (3-dimethylamino-propyl)-carbodiimide. J. Biol. Chem. 267: 3886–3893

Villagomez-Castro JC, Calvo-Mendez C and Lopez-Romero E (1992) Chitinase activity in encysting *Entamoeba invadens* and its inhibition by allosamidin. Mol. Biochem. Parasitol. 52: 53–62

Vyas P and Deshpande M (1991) Enzymatic hydrolysis of chitin by *Myrothecium verrucaria* chitinase complex and its utilization to produce SCP. J. Gen. Appl. Microbiol. 37: 267–275

Walker AN, Garner RE and Horst MN (1990) Immunocytochemical detection of chitin in *Pneumocystis carinii*. Infect. Immun. 58: 412–415

Watanabe T, Oyanagi W, Suzuki K and Tanaka H (1990a) Chitinase system of *Bacillus circulans* WL-12 and importance of chitinase A1 in chitin degradation. J. Bact. 172: 4017–4022

Watanabe T, Suzuki K, Oyanagi W, Ohnishi K and Tanaka H (1990b) Gene cloning of chitinase A1 from *Bacillus circulans* WL-12 revealed its evolutionary relationship to *Serratia* chitinase and to the type III homology units of fibronectin. J. Biol. Chem. 265: 15659–15665

Watanabe T, Oyanagi W, Suzuki K, Ohnishi K and Tamaka H (1992) Structure of the gene encoding chitinase D of *Bacillus circulans* WL-12 and possible homology of the enzyme to other prokaryotic chitinases and Class III plant chitinases. J. Bact. 174: 408–414

Wellburn SC, Arnold K, Maudlin I and Gooday GW (1993) Rickettsia-like organisms and chitinase production in relation to transmission of trypanosomes in tsetse flies. Parasitology 107: (in press)

Wessels JGH (1988) A steady state model for apical wall growth. Acta Bot. Neerl. 37: 3–16

Wirth SJ and Wolf GA (1990) Dye-labelled substrates for the assay and detection of chitinase and lysozyme activity. J. Microbiol. Meth. 12: 197–205

Wolkin RH and Pate JKL (1985) Selection for nonadherent or nonhydrophobic mutants co-selects for non-spreading mutants of *Cytophaga johnsonae* and other gliding bacteria. J. Gen. Microbiol. 131: 737–750

Wortman AT, Somerville CC and Colwell RR (1986) Chitinase determinants of *Vibrio vulnificus*: gene cloning and applications of a chitinase probe. Appl. Environ. Microbiol. 52: 142–145

Yabuki M, Kasai Y, Ando A and Fujii T (1984) Rapid method for converting fungal cells into protoplasts with a high regeneration frequency. Exp. Mycol. 8: 386–390

Yu C, Lee AM, Bassler BL and Roseman S (1991) Chitin utilization by marine bacteria. A physiological function for bacterial adhesion to immobilized carbohydrates. J. Biol. Chem. 266: 24260–24267

Zarain-Herzberg A and Arroya-Begovich A (1983) Chitinolytic activity from *Neurospora crassa*. J. Gen. Microbiol. 129: 3319–3326

Zhloba NM, Tiunova NA and Sidorova II (1980) Extracellular hydrolytic enzymes of mycophilic fungi. Mikol Fitopatol 14: 496–499

Zikakis J (1984) Chitin, Chitosan and Related Enzymes. Academic Press, Orlando

10. Biodegradation of starch and α-glycan polymers

MATUR V. RAMESH[1], BADAL C. SAHA[2,3], SAROJ P. MATHUPALA[1],
S. PODKOVYROV[1] and J. GREGORY ZEIKUS[1,2]
[1]*Department of Biochemistry, Michigan State University, East Lansing, MI 48824, U.S.A.;*
[2]*Michigan Biotechnology Institute, 3900 Collins Road, Lansing, MI 48910, U.S.A.;*
[3]*Fermentation Biochemistry Research Unit, USDA-ARS-NCAUR, 1815 North University Street, Peoria, IL 61604, U.S.A.*

I. Introduction

A. *Starch and related polymers*

Glucose is one of the most abundant, readily utilizable carbon sources found in different carbohydrate polymers. Starch and cellulose are the major glucose-containing carbohydrate polymers used as energy sources by many microbes. Of these two polymers, cellulose is linked by β-1,4 glucosidic linkages and is a linear crystalline substrate that is more slowly degraded by microbes. On the other hand, starch is a branched polymer comprised of an α-1,4 linked backbone and α-1,6 linked branches and it is more readily degraded by microbes. Amylosaccharidases, the starch-degrading enzymes, are ubiquitous in nature and they cleave α-1,4 and α-1,6 linkages of starch and related polymers to form a mixture of smaller saccharides ranging from glucose to maltodextrins with an average degree of polymerization of 10 glucose units.

Starch, amylose, amylopectin, glycogen, maltodextrins, pullulan and cyclodextrins are α-glycan-based polymers. Starch is composed of two types of high molecular weight polymers: amylose and amylopectin. Amylose is a linear polymer with α-D-glucose units linked at 1,4 positions. Amylopectin contains branched oligosaccharides with α-1,6 linkages at branch points on a linear glucose polymer backbone comprised of α-1,4 linkages. These two polymers differ considerably in their physical properties. Amylose accounts for about 17–25% of starch, depending on the source. The absorption maximum for the amylose-iodine blue complex is at 650 nm while that of the amylopectin-iodine purple complex is at 550 nm (Whistler and Daniel 1984). The exact structure of amylose in solution is still a source for debate. The different studies on this subject propose that it exists as a helix, interrupted helix, or a random coil in solution (Whistler and Daniel 1984). The average distance between the branch points on the 1,4 glycosidic-linked backbone of amylopectin varies. Amylopectin is described to contain three types of chains: A-chains, linked to the polymer molecule only through the reducing ends; B-chains, also linked through reducing ends, but are branched at α-1,6 position in one or more of the

C. Ratledge (ed.), Biochemistry of Microbial Degradation, 313–346.
© *1994 Kluwer Academic Publishers. Printed in the Netherlands.*

glucose residues; and C-chains which contain the reducing end group. The widely accepted structure of amylopectin is branched, with an A:B chain ratio of 1:1 (Whistler and Daniel 1984).

Glycogen is similar to amylopectin in chemical structure. However, it is more highly branched than amylopectin creating a more complex structure. Amylases vary in their action on this substrate probably because of the differences in accessibility of the substrate to the active site of the enzyme. Pullulan is a different polysaccharide with α-1,4 and 1,6 glucocidic linkages. This polymer is produced by a yeast-like fungus, *Aureobasidium pullulans*. However, there is no true branching in this polymer, although it has α-1,4 and α-1,6 linked glucose residues. Pullulan is a linear polymer of maltotriose units linked at the α-1,6 position. Cyclodextrins are dextrin molecules of 6 to 8 glucose units linked to form cyclic structures. They do not have a reducing end and these molecules are formed by the action of certain amylosaccharidases on starch. Depending on the number of glucose residues in each cyclic molecule, they are classified α (6) β (7) and γ (8) cyclodextrins. Cyclodextrins are very interesting products of enzymatic catalysis since they form inclusion complexes with many organic compounds and have commercial utililty because of this property.

In this review, we present new concepts and information on microbial amylosaccharidases in a consolidated form to give a general working idea to the reader on the physiology, molecular biology, biochemistry and biotechnology of starch degradation. Since the information available on this subject is voluminous, and many reviews on amylosaccharidases have recently been published (Vihinen and Mantsala 1989; Fogarty and Kelly 1983, 1990), we limited our review to only those publications that contribute greatly to our interest and the aim of this article.

B. Microbes degrading α-glycans

Amylosaccharidases are most widely distributed in fungi, yeast, actinomycetes and bacteria (see Table 1). In many cases, this group of enzymes is secreted by the microbe into the microbial environment to break the polymer into smaller oligosaccharides, ranging from glucose to maltodextrins. Aerobic and anaerobic microorganisms are equally efficient in utilizing these carbohydrate sources in the environment.

Among eukaryotic microbes, *Aspergillus*, *Rhizopus* and *Mucor* are best known for their amylase systems; and their amylolytic activities are used for industrial production and in commercial processes (Kvesitadze et al. 1974; Underkofler 1954; Belloc et al. 1975). These fungi are used to produce saccharifying α-amylase and glucoamylase. Other fungi that are reported to produce amylolytic activities include *Basidiomycetes*, *Aureobasidium*, *Fusarium* and *Neurospora* (Federeci 1982; Das et al. 1979; Narayanan and Shanmugasundaram 1967). Yeasts are well studied for amylase production because it is an important factor for starch utilization and reduces the cost of

Table 1. Representative microorganisms producing amylosaccharidases.[a]

Microorganism	Type of enzyme
Bacteria	
Bacillus	α-, β-amylase; α-glucosidase, neopullulanase, amylopullulanase, cyclodextrin glucosyl transferases, cyclodextrinase
Pseudomonas	α-, β-amylase; pullulanase, isoamylase
Clostridium	α-, β-amylase, α-glucosidase; pullulanase, CGTase, cyclodextrinase, glucoamylase
Thermoanaerobacter	amylopullulanase, α-glucosidase
Thermotoga	α-, β- and glucoamylase
Pyrococcus	α-amylase, α-glucosidase
Thermus	pullulanase
Actinomycetes	
Streptomyces	α-, β-amylase; α-glucosidase isoamylase, pullulanase
Thermoactinomyces	α-, β-amylase
Yeasts	
Saccharomyces	α-amylase, isoamylase
Endomycopsis, Candida	glucoamylase, α-glucosidase
Fungi	
Aspergillus, Rhizopus, Trichoderma,	α-, β-amylase, pullulanase, glucoamylase,
Fusarium	α-glucosidase

[a] This is not an inclusive list.

substrate in alcohol production. In general, not many yeasts can grow on starch. Sills and Stewart (1982) studied several yeasts for amylase production. De Mot et al. (1984a,b) studied ascomycetous and non-ascomycetous yeasts for amylase production. Among the ascomycetous yeast, few species of genera *Endomycopsis, Lipomyces, Pichia* and *Schwanniomyces* produce high amylolytic activity in the culture fluid, while others have moderate to no amylolytic activity at all (De Mot et al. 1984a). Among the non-ascomycetous yeast, starch utilization is relatively poor. In genera of *Candida, Cryptococcus* and *Trichosporon*, high extracellular amylolytic activity enabling them to utilize soluble and unmodified starch was described for the first time (De Mot et al. 1984b).

Among Gram-positive bacteria, *Bacillus* is used for industrial production of a liquefying α-amylase. *Pseudomonas* produce maltotetraose forming α-amylase and isoamylase, a debranching amylase. *Klebsiella* produce both pullulanase and cyclodextrin glucosyl transferase (CGTase). Actinomycetes such as *Thermomonospora, Thermoactinomycetes* and *Streptomyces* are well studied for amylase production. Many thermophilic, anaerobic bacteria such as *Clostridium, Thermoanaerobacter, Dictyoglomus, Pyrococcus,* and *Bacteroides* produce amylases during growth on starch.

C. Comparison of amylosaccharidase activities

Amylosaccharidases are classified into several groups based on their activity, substrate specificity, products of hydrolysis and conformation of the products (see Table 2). They are mainly divided into two categories: endo-acting and exo-acting enzymes. Endo-acting amylosaccharidases attack the starch polymer producing a mixture of malto-oligosaccharides. α-Amylases, cyclodextrin glucosyl transferases and pullulanases attack the polymer in this fashion. α-Amylases were proposed to have multiple attack mechanisms, where enzyme attacks the substrate molecule repeatedly (Robyt and French 1967). Single chain attack is where the enzyme binds to one substrate molecule and cleaves it from the reducing end to the non-reducing end. Once the substrate molecule is completely cleaved, the enzyme binds to another molecule. Multi-chain attack is the process in which the enzyme attacks the substrate randomly, cleaving one bond at a time in each molecule. Here one of the two cleaved fragments get separated from the enzyme and the other attached molecule is cleaved at several points before release from the enzyme. α-Amylases were proposed to have a multiple attack mechanism (Robyt and French 1967). α-Amylases cleave only α-1,4 linkages of starch and they can bypass the α-1,6 branch points.

β-Amylases are exo-acting enzymes that cleave only α-1,4 linkages of starch from the non-reducing end of the substrate in such a way that maltose is produced and activity stops at α-1,6 branch points. Amyloglucosidase or

Table 2. Action of different amylosaccharidases and basis for classification.

Enzyme	Endo/Exo action	End products	Configuration of anomeric carbon of products	Bond cleavage	EC number
α-amylase	endo	malto-saccharides, maltose, glucose	α	α-1,4	3.2.1.1
β-amylase	exo	maltose	β	α-1,4	3.2.1.2
amylo-glucosidase	exo	glucose	β	α-1,4 and α-1,6	3.2.1.3
α-glucosidase	exo	glucose	α	α-1,4 and α-1,6	3.2.1.20
pullulanase	endo	maltotriose from pullulan	α	α-1,6	3.2.1.41
isoamylase	endo	malto-dextrin from starch and glycogen	α	α-1,6	3.2.1.68
cyclo-dextrin glucosyl transferase	endo	malto-saccharides cyclo-dextrins	α	α-1,4	2.4.1.19
iso-pullulanase	endo	isopanose from pullulan	α	α-1,4	3.2.1.57
oligo-1,6 glucosidase	exo	glucose from maltose	α	α-1,6	3.2.1.10
cyclodextrinase	–	maltose, glucose from CD	α	α-1,4	3.2.1.54

glucoamylase cleaves glucose units one by one from the non-reducing end to reducing end of the substrate. This exo-acting enzyme can cleave both α-1,4 and α-1,6 linkages in starch and forms exclusively β-D-glucose from starch. α-Glucosidases cleave short chain dextrins like maltose, maltotriose and maltotetraose into glucose and vary in their specificity towards maltodextrins (Fogarty and Kelly 1983).

Pullulanases are another kind of amylosaccharidases that cleave α-1,6 linkages of starch and pullulan but have no action on glycogen. The pullulanases of *K. pneumoniae* and *B. flavocaldarius* (Eisele et al. 1972; Suzuki et al. 1991) cleave specifically α-1,6 linkages and have no action on α-1,4 linkages. Recently several reports appeared on a new kind of amylosaccharidase which can cleave both α-1,4 and α-1,6 linkages of starch or glycogen and α-1,6 linkages of pullulan (Coleman et al. 1987; Melasniemi 1987; Odibo and Obi 1988; Saha et al. 1988, 1989, 1990; Plant et al. 1986, 1987) and these enzymes are referred to here as amylopullulanases.

A unique class of amylosaccharidase is cyclodextrin glucosyl transferase (CGTase), which have α-amylase activity and, in addition, cyclize maltosaccharides of 6 to 8 glucose units to form cyclodextrins. CGTases are not reported so far from any source other than bacteria (Schmid 1989). The ecophysiological relevance of these cyclodextrins (CDs) on cell growth or metabolism is not yet known. Since cyclodextrins are known to form inclusion complexes with a variety of organic substances, and the fact that CDs formed by CGTases comprise only 2 to 8% of the total products, suggests speculation that CDs immobilize compounds toxic to cell growth, by forming inclusion complexes. There are a few reports on cyclodextrinases which specifically cleave CDs (Yoshida et al. 1991; Depinto and Campbell 1968; Kitahata et al. 1983; Bender 1981; Saha and Zeikus 1990a). Some α-amylases were also reported to cleave CDs (Keay 1970) while β-CD acts as an inhibitor for most amylases and amylopullulanases (Saha et al. 1988, 1990a).

Isoamylases are debranching enzymes that cleave α-1,6 linkages of amylopectin and glycogen but do not cleave pullulan at all. They are not very efficient in cleaving 1,6 linkages because they have specificity for α-1,6 side-chains of a certain length. Isoamylases can debranch glycogen completely but cannot hydrolyse 2 or 3 glucose unit side chains of α- or β-limit dextrins. This limits their usage in combination with other amylases in the production of maltose syrup. Isopullulanases are pullulan hydrolysing enzymes that cannot cleave α-1,6 linkage in pullulan instead hydrolyse the adjacent α-1,4 linkages forming isopanose from pullulan. Recently, many new types of amylosaccharidases have been reported with properties intermediate to more than one kind of enzyme. This makes it difficult to classify them based on the mode of action, configuration of anomeric carbon and end products of hydrolysis, the criteria used earlier for their classification. Table 3 summarizes key properties of these new amylosaccharidases.

Neopullulanase is a pullulan hydrolysing enzyme that produces panose from pullulan. However, it was demonstrated that neopullulanase can actually

Table 3. New enzymes with intermediary properties of classified amylosaccharidases.

Enzyme	Source	Type of action	End products	Bond cleavage	Ref.
β-amylase	*Bacillus megaterium*	exo	maltose	α-1,4	Takasaki 1989
α-amylase	*Streptomyces praecox*	endo	maltose	α-1,4	Wako et al. 1978
α-amylase	*Microbacterium imperiale*	endo	maltotriose	α-1,4	Takasaki et al. 1991a
α-amylase	*Pseudomonas stutzeri*	exo, endo	maltotetraose	α-1,4	Zhou et al. 1989
α-amylase glucosyl transferase	*Bacillus circulans* MG-4	exo, endo	maltotetraose	α-1,4	Takasaki et al. 1991b
neopullulanase	*Bacillus stearother- mophilus*	endo	panose from pullulan	α-1,4 and α-1,6	Imanaka and Kuriki 1989
amylopullulanase	*Thermo- anaerobacter* sp. *Bacillus circulans* F-2	endo	maltobiose, maltotriose and malto- tetraose	α-1,4 and α-1,6	Melasniemi et al. 1990; Saha et al., 1990; Saha et al., 1988; 1989

hydrolyse α-1,4 and α-1,6 linkages of pullulan forming panose, maltose and glucose in a ratio of 3:1:1, respectively (Imanaka and Kuriki 1989). These authors proposed a three-step conversion of pullulan to these three products. This enzyme is not an α-amylase with debranching activity since it hydrolyses pullulan more efficiently than starch (Imanaka and Kuriki 1989). Unlike *T. vulgaris* α-amylase which cleaves starch more efficiently than pullulan (Sakano et al. 1982), neopullulanase cleaves α-1,6 linkages of partially hydrolysed pullulan. In terms of repeated action pattern on partly hydrolysed pullulan molecules, neopullulanase may have some similar mechanism to that of multiple attack of α-amylases.

Amylopullulanase hydrolyses α-1,4 and α-1,6 linkages in starch forming maltobiose, maltotriose and maltotetraose; and, it cleaves pullulan into maltotriose (Saha et al. 1988; Saha and Zeikus 1989a,b; Mathupala 1992). Amylopullulanase has a higher activity and affinity for pullulan than starch and cleaves both the α-1,4 and α-1,6 linkages of starch and glycogen.

II. Microbial amylosaccharidase synthesis, regulation, and localization

A. Aerobic fungi and bacteria

Different fermentation conditions and nutrients influencing α-amylase and glucoamylase synthesis by *Rhizopus*, *Aspergillus* and *Bacillus* were reviewed by Fogarty and Kelly (1980) and Frost and Moss (1987). Amylosaccharidase synthesis in fungi occurs during growth; and, the amylosaccharidases are secreted into the extracellular environment. Fungi are known to grow and utilize complex carbohydrates as substrate. They can grow efficiently on wheat bran which contains some starch but is rich in cellulose and hemicellulose. *Aspergillus niger* and *A. oryzae* are used in solid state fermentation (SSF) with wheat bran as substrate for production of glucoamylase and Taka α-amylase, respectively. It is well known that cultural conditions influence the synthesis of amylases in fungi. Higher activities of enzyme are produced in SSF than in submerged fermentation systems and the ratio of α-amylase to glucoamylase synthesized is also influenced, depending on the organism used. For example, *A. niger* produces more glucoamylase and *A. oryzae* produces more α-amylase in solid state fermentation. The higher activities of fungal α-amylase occurring in SSF may be because this system mimics natural environmental conditions, creating local variations in substrate concentrations which makes the organism produce more hydrolysing enzymes. In submerged fermentation systems, the substrate is always distributed throughout the fermentation medium unlike in solid state system (Mudgett 1986).

When *B. licheniformis* is grown under solid state fermentation conditions, high activities of α-amylase occur. In submerged culture, however, end-product repression occurs as the glucose formed from starch hydrolysis represses further synthesis of α-amylase (Ramesh and Lonsane 1991a,b).

There are two modes for regulation of enzyme synthesis: positive and negative, known to regulate inducible and repressible operons in prokaryotes (Goldberger 1979). In the *Escherichia coli* lactose operon model, glucose causes depletion of cAMP thereby halting transcription. In *Bacillus*, glucose also represses α-amylase synthesis. Unlike *E. coli*, *Bacillus* does not contain cAMP under normal growth conditions (Bernlohr et al. 1974). This indicates that catabolite repression of α-amylase synthesis in this genus is mediated by some other means, but not by cAMP.

Laoide et al. (1989) reported that catabolite repression of α-amylase synthesis in *B. licheniformis* occurs at the transcriptional level. The *amyL* gene is subject to catabolite repression when expressed in *B. subtilis* with or without its own promoter on the plasmid, and without its promoter when integrated into *B. subtilis* chromosome. There is very little DNA homology between α-amylase genes of these two *Bacilli*. Studies showed that glucose-mediated repression of the α-amylase gene occurs irrespective of the distance between *amyL* gene and the promoter transcribing it. Laoide and McConnell (1989) identified the *cis* sequences involved in mediating the catabolite repression of the *amyL* gene in *B. subtilis* in the region downstream from the promoter and upstream of the signal sequence cleavage site. This region is 108 bp long with an inverted repeat sequence of TGTTTCAC–20bp–ATGAAACA. It was also shown that mutations which relieve catabolite repression, also relieve repression of α-amylase synthesis in *B. subtilis*. Sporulation resistant strains which show expression of genes that are catabolite repressed in wild-type cells, also expressed the *amyL* gene fused with *lacZ* in *B. subtilis*, indicating that there is a common factor involved in these two processes (Laoide and McConnell 1989).

In *Bacillus*, amylase synthesis is growth associated or initiated in the late growth phase, with initiation of sporulation (Priest 1977, 1985). In general, it was concluded that in *Bacillus*, α-amylase synthesis is constitutive and the enzyme is secreted into the medium. At all times, there are low activities of α-amylase synthesized in the absence of the inducer. α-Amylase synthesis may be triggered by the addition of an inducer and repressed by glucose (Priest 1985). In *B. stearothermophilus*, α-amylase synthesis is induced by maltodextrins and is growth associated (Welker and Campbell 1963).

Expression of the isoamylase gene of *P. amyloderamosa* is not repressed by glucose but it is induced by maltose. Maltose induced transcription of the *iam* gene (Fujita et al. 1989a). In *P. stutzeri*, maltose induces the synthesis of the amylase (*amyP*) which forms maltotetraose and this gene is repressed by glucose (Robyt and Ackerman 1971). Fujita et al. (1990) later showed that the G4-forming amylase of *P. stutzeri* is also induced at the transcriptional level by maltose. However, glucose repression is not at the transcriptional level but at some post-transcriptional stage, unlike in *Bacillus*, where glucose represses transcription of α-amylase synthesis. In *Pseudomonas*, synthesis of the *iam* and *amyP* gene products are regulated differently.

α-Amylase synthesis in *A. oryzae* and *A. awamori* is also repressed by glucose and induced by maltose or starch (Erratt et al. 1984); and starch depresses the glucose catabolite repression of α-amylase synthesis. α-Amylase synthesis in *A. awamori* is controlled at the translational level rather than at transcription (Bhella and Altosaar 1987). The concentration of mRNA inside the cells remain constant during glucose repression. No increases in mRNA were noticed during depression caused by starch, although a four-fold increase in enzyme synthesis was observed.

Pullulanases of *Thermus aquaticus* and *B. thetaiotaomicron* are cell bound (Plant et al. 1986; Smith and Salyers 1989). *B. thetaiotaomicron*, a human colon anaerobe, also degrades and utilizes starch without secreting amylases. Mutants of this organism were developed to show genetic evidence that two types of starch-binding sites, A and M, are involved in binding to large oligomers and small maltodextrins, respectively (Anderson and Salyers 1989). Genes encoding these starch-binding sites and starch-hydrolysing enzymes apparently are under the same regulational control.

B. Thermophilic anaerobic bacteria

Thermophilic anaerobic bacteria are capable of growing very readily on complex carbohydrates such as starch and glycogen. Thermoanaerobes, as they are often known, seem to be more efficient than aerobic bacteria and actinomycetes in terms of production of a wide spectrum of amylosaccharidases for complete degradation of starch. Amylopullulanase and α-glucosidase are produced by many thermophilic *Clostridia*, *Thermoanaerobacter*,

Dictyoglomus, and *Pyrococcus*. Unlike aerobes, thermoanaerobes vary in the types and cellular locations of amylosaccharidases produced. In certain organisms the amylosaccharidases are either cell bound or secreted but their synthesis is growth associated. Amylopullulanase and glucogenic amylases of *C. thermohydrosulfuricum*, are cell-bound activities when starch is in excess as a growth substrate but the amylopullulanase is secreted when starch is a limiting growth substrate (Hyun and Zeikus 1985b). Pullulanase of *Thermoanaerobium* TOK6B-1 (Plant et al. 1987), amylopullulanase of *Thermoanaerobacter* B6A (Saha et al. 1990), pullulanase and α-amylase of *C. thermosulfurogenes* (Spreinat and Antranikian 1990), β-amylase of *C. thermosulfurogenes* (Hyun and Zeikus 1985a), α-amylase of *P. woesei* (Koch et al. 1991) are secreted into the culture medium during growth on starch. The factors that control the cellular location of the amylosaccharidases in thermoanaerobes have not been clarified.

Starch is a high molecular weight polymer and cannot enter the cell's cytoplasm. In order to breakdown this polymer, the microorganism needs at least two enzymes: a solubilizing activity and a glucogenic activity. *C. thermosulfurogenes* EM1 produces an α-amylase, a pullulanase, and an α-glucosidase (Antranikian 1990). Antranikian et al. (1987) observed some remarkable changes in cell envelope structures of *C. thermosulfurogenes* EM1 in starch- and glucose-limited cultures where high activities of these amylosaccharidases were secreted. Under these conditions, vesicles originating from the cell membrane were seen covered with small particles. By immunolabelling studies it was shown that α-amylase and pullulanase were synthesized in active (extracellular) and inactive (intracellular) forms (Specka et al. 1991).

C. thermohydrosulfuricum produces an amylopullulanase which is either cell-bound or secreted depending on the specific growth conditions. The amylopullulanase may be part of a differentiated S-layer which contains starch-solubilizing enzymes and starch receptors that function as an amylosome (S.P. Mathupala and J.G. Zeikus, unpublished observations). In *Clostridia*, the variation in secretion or retention of proteins in a cell-bound form may in part be explained based on the presence or absence of an S-layer on the cell surface. *Clostridia* which contain S-layers for example, are subject to intracellular accumulation of toxins while those without S-layers secrete the toxins during the exponential growth phase (Sleytr and Messner 1983). The S-layer is a glycoprotein complex thought originally to function as a molecular sieve. However, it is possible that the S-layer can differentiate into a hydrolosome structure comprised of substrate binding receptors and glycoprotein hydrolases such as endoxylanases, endo-β-glucanases, or endo-α-glucanases (Lamed et al. 1988; Zeikus et al. 1991). *C. thermocellum* produces a cellulosome that contains cellulase which is cell bound during early growth on cellulose but which is released into the medium late in growth. The cellulosome is a glycoprotein complex that appears associated with the outer wall of the cell (Lamed et al. 1988).

The thermoanaerobic eubacterium, *Thermotoga maritima*, also grows on starch without releasing any amylase activity into the medium (Schumann et al. 1991). In this organism, amylases are not present in the periplasm but are associated with the Toga surface of the cell in such a way that the active site of the enzyme is accessible for substrate binding and catalysis. In contrast, *Pyrococcus furiosus* produces extracellular amylolytic activities and intracellular α-glucosidase activity (Constantino et al. 1990).

The reason for synthesis of dual-active amylopullulanases by a majority of thermoanaerobes and retaining them on the cell surface may be because of energy constraints and limited substrate availability for growth in their natural environments. Compared to either α-amylases or pullulanases, amylopullulanases are more efficient in terms of activity and conservation of energy for their synthesis. Amylopullulanases are higher in molecular weight compared to α-amylases. In our laboratory, amylopullulanases from *C. thermohydrosulfuricum* 39E and *Thermoanaerobacter* B6A-RI have been cloned and expressed in *E. coli* (Mathupala 1992; M.V. Ramesh and J.G. Zeikus, unpublished results). The amylopullulanase gene was deleted from 4.4 kb to 2.9 kb without loss of dual activity or thermostability. This suggests other functions for a major portion of the amylopullulanase such as compartmentalization into the cell surface.

III. Biochemistry of amylosaccharidases

A. General features and thermostability

Microbial amylosaccharidases were among the first industrially produced enzymes mainly because of their utility and broad pH and temperature activity/ stability ranges. Fungal amylosaccharidases are stable at acidic pH and active up to 60 °C, whereas *Bacillus* amylases are stable at near neutral to alkaline pH and are active at above 100 °C. α-Amylase from both *B. licheniformis* (Madsen et al. 1973) and *B. stearothermophilus* are stable and active above 90 °C. Recently isolated pullulanases and amylopullulanases from thermoanaerobic bacteria are as active and stable at high temperature as *Bacillus* α-amylases. α-Amylase, pullulanase, and α-glucosidase activities of *Pyrococcus furiosus*, an archaebacterium are optimally active at 100 °C (Brown et al. 1990). A CGTase, active at pH 4.5 and 95 °C, from *Thermoanaerobacter* sp. was reported to have better stability and liquefaction properties than that of commercial *B. licheniformis* α-amylase (Starnes et al. 1991).

Although the thermostability of amylases can be attributed to the growth temperature of microorganism to some extent, enzymes from certain moderate thermophiles and mesophiles are active at temperatures much higher than their growth temperatures. It is generally accepted that mesophilic microbes evolved from thermophilic forms. The exact chemical basis of enzyme thermostability has been a subject of detailed investigations. Many different mechanisms have been shown to contribute to thermostability of enzymes, like the number of arginine residues, metal ions, disulphide linkages, hydrophobic amino acid content, etc.

The mechanism of irreversible thermal inactivation of enzymes at higher temperatures was studied by Ahern and Klibanov (1985). The causes found for thermal inactivation were deamination of asparagine residues, peptide bond hydrolysis at aspartate residues, disruption of disulphide bonds, and the formation of scrambled structures which are highly dependent on pH. The main cause of thermal inactivation of *B. amyloliquefaciens* α-amylase at pH 8.0 and 90 °C, was formation of a scrambled structure which could be decreased by adding the substrate (Tomazic and Klibanov 1988a). In the presence of the substrate, deamination of asparagine and glutamine were the main factors for irreversible inactivation. Conformational changes are the main cause of inactivation of *B. stearothermophilis* α-amylase, which is more stable than the *B. amyloliquefaciens* enzyme at pH 5.0 and 90 °C. At pH 8.0 and 90 °C, oxidation of cysteine residues and deamination of asparagine and glutamine residues limit the thermostability of the enzyme. Apparently, the rates of these processes are the same in all proteins and cannot be controlled (Tomazic and Klibanov 1988a).

In subsequent studies on determining the possible factors for greater stability of α-amylase from *B. stearothermophilus* versus *B. amyloliquefaciens*, extra electrostatic interactions by salt bridges were identified (Tomazic and Klibanov 1988b). These additional salt bridges in the *B. stearothermophilus* enzyme are provided by one or two lysine residues which are absent at corresponding positions in the *B. amyloliquefaciens* amylase. The *B. licheniformis* α-amylase is apparently more stable than the former two enzymes because of the upper limit of thermostability inherent in its primary structure, a deamination process leads to thermoinactivtion. This enzyme may further be stabilized by substitution of asparagine and glutamine residues near the active site through site-directed mutagenesis (Tomazic and Klibanov 1988b). However, similar thermostability studies have not yet been done with archaebacterial enzymes which are highly stable above 100 °C in part because of the lack of information on their primary structure.

The thermal stability of certain kinds of amylosaccharidases has been attributed in part to the number of proline residues. Suzuki et al. (1991) correlated the linear increase in proline content to thermostability of pullulanases of *Klebsiella pneumoniae*, *B. acidopullulyticus* and *B. flavocaldarius*. The results are consistent with similar observations in oligo-1,6 glucosidases of bacteria where a linear increase in thermostability could be seen with an increasing number of proline residues (Suzuki et al. 1987). Suzuki (1989) has suggested that thermostability of proteins can be enhanced by increasing the frequency of proline residues in the second sites of β-turns which might decrease the entropy of unfolding without a significant change in the secondary or tertiary structure. High thermostability of α-amylase from *B. caldovelox*, however, was attributed to amino acid residues causing more hydrophobic interactions (Bealin-Kelly et al. 1991). This enzyme, unlike other *Bacillus* amylases, does not undergo aggregation during thermoinactivation (Tomazic and Klibanov 1988a,b). *B. licheniformis* α-amylase is more thermostable despite its low proline content (Yuuki et al. 1985) and its high thermostability is reported to be due to glycine and lysine rich regions in the enzyme (Suzuki et al. 1989).

Recently, our laboratory has generated a series of nested deletion mutants of the amylopullulanase gene of *C. thermohydrosulfuricum* strain 39E. The original 4.4 kb gene was deleted to 2.9 kb without loss of dual α-1,4, α-1,6 cleavage activity or thermostability at 90 °C. Larger deletion mutants were created that maintained activity but lost significant thermostability (Mathupala 1992). These results suggested a new amino acid sequence domain separate from the catalytic domain of the protein that is essential for thermostability.

B. *Mechanism of hydrolysis and catalytic residues*

Despite their differences in activity, stability, specificity and molecular weight, all amylases possess similar conserved amino acid sequences which are believed to be involved in substrate binding and catalytic activity. The overall amino acid sequence homology among amylases is low. Based on X-ray crystallographic data analysis of Taka α-amylase, a general acid-base hydrolysis mechanism was proposed for catalysis of amylases, as was proposed for lysozyme which cleaves β-1,4 glycosidic linkages (Matsuura et al. 1984). Glu230 acts as general acid and Asp297 acts as general base during catalysis. His122, Asp206, Lys209, His210, and His296 are proposed to be involved in substrate binding. Based on the X-ray crystallographic data analysis of *A. oryzae* Taka α-amylase A (Matsuura et al. 1984), amylases belong to a super family of proteins with $(\beta/\alpha)_8$ barrel structure. α-Amylases contain three domains, A, B, and C. Domain B is involved in substrate binding site and is linked to domain A by an essential Ca^{2+} ion. The domain C has an immunoglobulin-type fold. Mutation studies on *B. stearothermophilus* α-amylase showed that the domain C is also needed for activity. Nakamura et al. (1992) identified Asp229, Glu257, Asp238 as involved in catalysis in CGTase of an alkalophilic *Bacillus* sp. In case of glucoamylases of *A. niger*, Trp120 stabilizes the substrate transition state and Glu179 and Asp176 act as general acid-base catalysts.

The dual active amylopullulanase (α-amylase-pullulanase) which cleaves both α-1,4 and α-1,6 linkages may also have similar mechanism. Melasniemi et al. (1990) proposed two active centres for such an enzyme. Based on enzyme kinetics and site-directed mutagenesis data, only one active centre is involved in catalysis (Mathupala et al. 1990; Mathupala 1992). A catalytic triad comprised of two Asp and one Glu is required for both α-1,4 and α-1,6 cleavage activity of amylopullulanase or cyclodextrinase produced by *C. thermohydrosulfuricum* strain 39E (S.P. Mathupala, S. Podkovyrov and J.G. Zeikus, in press). Substitution of Asp625, Glu657, and Asp734 with Asn, Gln

Asn, respectively, in separate oligonucleotide directed mutagenesis experiments, resulted in loss of activity.

Kuriki et al. (1991) analysed the active centre of *B. stearothermophilus* neopullulanase through site-directed mutagenesis. The putative Glu357 and Asp424 of neopullulanase involved in catalysis, when substituted by oppositely charged amino acids, the enzyme activity on α-1,4 and 1,6 bonds was lost. Similarly, Asp328 of neopullulanase corresponding to one of substrate binding amino acids (based on Taka amylase structure) when replaced by His or Asn, also lost activity, indicating that it is also needed for activity. It appears that hydrolysis of α-1,4 and α-1,6 linkages is by the same active site mechanism. The specificity of neopullulanase towards α-1,4 and α-1,6 bonds can be changed by site-directed mutagenesis for increasing production of panose (Kuriki et al. 1991).

IV. Molecular biology of amylosaccharidases

A. Gene cloning, expression and gene organization

Amylases from different microbes have been cloned and expressed in different microbes. Fungal Taka α-amylase has been extensively studied and characterized. The amino acid sequence and three dimensional structure of the enzyme have also been studied (Matsuura et al. 1984; Tada et al. 1989). These studies were based on the amino acid sequence derived from the purified protein. Little was known about the DNA until the gene from the parent organism was cloned and expressed. *A. oryzae* was recently reported to contain multiple genes for Taka amylase A (TAA) expression (Gines et al. 1989; Tsukagoshi et al. 1989). cDNA and genomic DNA for two nearly identical Taka amylase genes (TAA), *AmyI* and *AmyII*, were cloned and the DNA sequences showed that only three nucleotides are different, with two pairs of amino acids, 35Arg-Glu and 151Phe-Leu being different in the total molecule. The 3′ untranslated regions (UTR) of cDNA related to *AmyI* are shorter than those of *AmyII* and 3′ UTR of *AmyI* lacks inverted repeated sequences and contains a putative 'AATAAA' polyadenylation signal region (Gines et al. 1989). The TAA gene contains 8 introns unlike *S. fibuligera* amylase (Itoh et al. 1987b). *AmyI* and *AmyII* are not linked on the chromosome of *A. oryzae* and they are separated by at least 6–10 kb on the chromosomal DNA. Tsukagoshi et al. (1989) also reported similar results. Tada et al. (1991) showed by deletion experiments that the −299 to −377 bp region from the start codon of TAA gene is required for high level expression and induction of the gene. Whether these elements in this region are required for regulational control is not yet clear.

Bacterial amylases, especially from the genus *Bacillus*, have been studied in great detail since they are thermostable and produced in large amounts. The *B. licheniformis* α-amylase gene has a restriction map similar to that of *B. coagulans* (Piggott et al. 1984; Cornelis et al. 1982) but these two enzymes differ in their catalytic properties. *B. stearothermophilis* α-amylase is carried on a 26 kb naturally-occurring plasmid (Mielenz 1983; Thudt et al. 1985). The α-amylase gene is located in a 5.4 kb *HindIII* fragment of the plasmid. The host strain does not seem to have a counterpart of the amylase gene on the chromosomal DNA. Thudt et al. (1985) cloned and expressed the α-amylase gene of *B. stearothermophilus* into several *Staphylococcus* species. *S. aureus* secreted half of the recombinant α-amylase produced into the medium but *S. xylosus* retained the plasmid in the absence of selective pressure (antibiotic) even after 60 generations. *B. brevis* 47-5, carrying the plasmid pBAM101 harbouring the *B. stearothermophilus* α-amylase, produced about 100 times more enzyme than the non-recombinant parent organism. The high amounts of enzyme secretion by *B. brevis* was attributed to its unique cell wall structure (Tsukagoshi et al. 1985). The cloned *B. stearothermophilus* α-amylase gene was processed correctly by the recombinant organism.

C. *thermohydrosulfurogenes* EM1 α-amylase (*amyA*) has a 27 amino acid signal peptide. The processed enzyme in *E. coli* had a mol wt of 75,000 which is slightly higher compared to extracellular enzyme (68000) produced by the parent organism (Bahl et al. 1991). The enzyme from the parent organism is secreted only under certain cultural conditions and there may be a membrane anchor sequence in the mature enzyme that is used to maintain the enzyme sticking on the membrane surface. However, hydropathy plots did not show hydrophobic regions in the COOH-terminus that are loose enough to maintain the enzyme on the membrane. The DNA sequence of *amyA* showed a consensus sequence resembling promoter binding sites but whether it regulates gene expression is not clear since there is no unequivocal transcriptional terminator found upstream of the *amyA* gene. The upstream region of the *amyA* gene has a 100% amino acid homology with that of the β-amylase gene of *C. thermosulfurogenes* (Kitamoto et al. 1988). The COOH-terminal end of the ORF for *amyA* encodes a polypeptide (122 amino acid) that has 30% homology to that of the malG protein of *E. coli*. These observations suggest that α-amylase of *C. thermosulfurogenes* is translated from a polycistronic mRNA that is initiated upstream of α-amylase, encoding proteins for starch hydrolysis (Bahl et al. 1991).

The mature α-amylase of *B. stearothermophilus* contains 515 amino acids and that of *B. licheniformis* contains 483. These two genes were cloned and expressed in *E. coli* (Gray et al. 1986). The proteins were processed correctly and 80 and 90% of the enzymes were located in the periplasm and less than 1% was extracellular. Though the NH_2^- terminus of *B. stearothermophilus* DY-5 α-amylase expressed in *E. coli* was confirmed with that of the mature protein from the parent strain, the exact site of translation of mRNA is ambiguous because of the presence of three to four putative ribosome binding sites in the sequence (Ihara et al. 1985). These multiple ribosome binding sites within phase initiation and termination codons are postulated to influence efficient translation of mRNA (Movva et al. 1980). The *B. subtilis* α-amylase gene also contains similar multiple ribosome binding sequences (Emori et al. 1990;

Chung and Friedburg 1980). In contrast, *B. amyloliquefaciens* and *B. licheniformis* α-amylases lack such sequences between the promoter and the structural gene (Ihara et al. 1985). The α-amylase gene in *B. subtilis* (*natto*) IAM 1212 encodes 477 amino acids that is smaller than that of *B. sutilis* strains 168 and NA64 strains (Yamane et al. 1984). This difference is due to deletion of 32 bp in the COOH-terminal of the open reading frame resulting in a frame shift and termination codon near the deleted region. The deleted part of the sequence was attributed to the inefficiency of this enzyme in maltotriose hydrolysis; whereas the amylases of *B. subtilis* strains 168, Marburg and NA64 can produce glucose and maltose from starch. Recently, Emori et al. (1990) cloned and sequenced the *B. subtilis* 2633 α-amylase gene that encodes 477 amino acids but hydrolyses maltotriose. Comparison of the amino acid sequences of the amylases from *B. subtilis* strains (*natto*) IAM 1212 and 2633 showed that they differ in only 5 amino acids with 99% homology. By constructing chimeric α-amylases, it was determined that only one amino acid substitution at position 295 causes the difference in hydrolytic pattern of soluble starch by these two enzymes. Ser at position 295 (in *B. subtilis* strain 2633) enables the enzyme to hydrolyse maltotriose and its replacement with Tyr (in *B. subtilis natto* IAM 1212) prevents this.

The inverted repeat structure in front of the structural gene (*amyE*) of *B. subtilis* α-amylase acts as transcription terminator. Deletion of this inverted repeat sequence results in a read-through transcript of the gene and decreasing amounts of enzyme synthesis (Kallio et al. 1986). The presence of this inverted repeat sequence prevents synthesis of read through mRNA thus allowing high levels of α-amylase synthesis in *B. subtilis*. However, this palindromic sequence does not enhance the synthesis of α-amylase of *B. subtilis*. Although Takano et al. (1987) reported that this palindromic sequence preceding *amyE* of *B. subtilis* enhances enzyme synthesis by 6-fold, their studies did not include the deletion experiments of the gene preceding this palindromic sequence and it appears that the results of Kallio et al. (1986) are more conclusive.

B. polymyxa produces a pre-amylase that is cleaved by proteases into α-amylase and β-amylase (Uozumi et al. 1989). The mature α-amylase (1161 aa) from a 3588 bp gene is divided into two portions by a direct repeat sequence in the middle of the gene. The 5′ region upstream from the direct repeat sequence encodes β-amylase while the 3′ region downstream of this sequence encodes for α-amylase. This might be the first report in prokaryotes where a single protein precursor gives rise to two enzymes. This protein may be a result of fusion of two genes at the direct repeat sequences. The proteases involved in processing of this bifunctional protein in *B. polymyxa* were identified to be a neutral protease and an intracellular protease. Though they differ in their catalytic properties, these two proteases process the precursor molecule into a separate α- and β-amylase enzyme activity (Takekawa et al. 1991). The two large domains of the α- and β-amylase precursor protein may be folded in such a way that the junction region (DNA direct repeat sequence) of 200 amino acids is exposed to the surface and accessible for protease digestion.

The *amyl* gene of *S. limosus* encodes 566 amino acid sequence with a mol wt of 59 670 and a signal peptide of 28 amino acids (Long et al. 1987). The enzyme is synthesized from monocistronic mRNA originating from a unique promoter. The nucleotide sequence shows two inverted repeats after 7 bp from the transcription initiation site before the translation-start codon. The function of this sequence is unknown. *S. hygroscopicus* α-amylase could be expressed in *E. coli* only after fusing the gene to the *lac* promoter while it was expressed with its own promoter in *S. lividans* (Hoshiko et al. 1987). The 30-mer leader sequence was similar to those of other prokaryotic enzymes. Using mung bean nuclease for mapping, two additional transcriptional initiation sequences were detected upstream of the 5' region of the transcription initiation site in *S. hygroscopicus* α-amylase. Two direct repeats in the COOH-terminal region and a palindromic sequence after the stop codon may be involved in transcription regulation and termination. α-Amylase of *S. griseus* IMRU3570 is intracellularly processed when cloned and expressed in *S. lividans*, into two proteins of 57 and 50 kDa and these two enzymes are secreted (Garcia-Gonzalez et al. 1991). *S. griseus* α-amylase, when cloned and expressed in *S. lividans*, is susceptible to the proteolytic activity of the latter organism. This susceptibility of *S. griseus* α-amylase to proteases appears to be an inherent nature of the protein, as no such proteolytic processing of α-amylases of *S. hygroscopicus* or *Thermomonospora curvata* was observed when expressed in *S. lividans* (Hoshiko et al. 1987; Petricek et al. 1992). That there was no consensus promoter sequence identified in *Streptomyces* suggests that there may be different RNA polymerases involved in recognition of these sequences.

Relatively little is known about cloning and sequencing of β-amylases when compared to α-amylases. β-Amylase of *C. thermohydrosulfurogenes* is encoded by 1653 bp open-reading frame that contains a signal peptide of 32 residues. The protein is translated from a monocistronic mRNA (Kitamoto et al. 1988). The gene was expressed in *B. subtilis* and the protein was secreted into the medium. The enzyme showed more hydrophobic amino acid and cys residues than *B. polymyxa* β-amylase. The β-amylase of *B. polymyxa* is synthesized as a precursor of 130 kDa protein that is cleaved into active α-amylase and β-amylase.

Among *Klebsiella* species, only *K. oxytoca* M5a1 (formerly *K. pneumoniae* M5a1) produces CGTase extracellularly and this organism can grow on starch. The CGTase gene (*cst* gene) of this organism was cloned and expressed in *K. pneumoniae* KAY 2026 and in *E. coli* (Binder et al. 1986). This enzyme has a 30 amino acid leader sequence and is secreted by these two host organisms. A transcription start site -CCTTAT- was identified about 150 bp upstream from the initiation codon; and, 90 bp upstream of the transcription start site an *E. coli* catabolite activator protein-binding motif, TGTGA was detected. Between the ribosome binding site and the initiation codon was a six-fold TTGTAG repeat sequence, the function of which is unknown. The *cst* gene contains a low GC content of 36% against 52 to 56% of the total chromosomal DNA of the organism. This unusual nature of DNA within the *cst* gene indicates that it may be acquired by lateral transfer from some other organism (Binder et al. 1986).

B. ohbensis, an alkalophilic bacterium, has a 20 bp inverted repeat preceding the *cst* gene forming secondary structures in mRNA, thereby impeding its expression in *E. coli* (Sin et al. 1991). The *cst* gene could be expressed in *E. coli* only when the inverted repeat preceding the gene was deleted and fused to the *lac* promoter of pUC19. Substitution of TTG codon with ATG doubled gene expression. The expressed protein was processed the same way as in the parent organism and was secreted to the periplasmic space. The *B. macerans cst* gene has a ribosome binding sequence of GAGGAGAGG and the promoter binding sequence was predicted to be TTTTT (Takano et al. 1986). The CGTase enzyme is a dimer of 74,000 Da protein. For some reason, the *B. circulans cst* gene with its own promoter failed to express in *B. subtilis* (Paloheimo et al. 1992). The gene was expressed only when the promoter was changed to that of *B. amyloliquefaciens* α-amylase. Kaneko et al. (1988) cloned CGTase gene of alkalophilic *Bacillus* sp. no. 38-2 in *E. coli* and the putative ribosome binding site was identified. The CGTase gene of *B. circulans* (Nitschke et al. 1990) has two initiation codons 11 to 14 bp after the ribosome binding site, separated by 6 bp. The first ATG codon is the likely true initiator.

Pullulanase of *K. pneumoniae* is regulated positively in *E. coli* by *malT*, the regulator gene of the maltose region. Adjacent to the *pulA* promoter, an additional *malT* controlled promoter *malX* was identified to be oriented in the opposite direction (Chapon and Raibaud 1985). These two promoters are separated by 25 bp GC-rich sequences with almost all purines on one strand and pyrimidines on the other strand. The consensus sequence of these two promoters is GGAT/GGA and occurs − 35 bp upstream from the transcription initiation site. These two hexanucleotides are believed to be simultaneously involved in promoter function.

It would be interesting to know if pullulanases of *Clostridia* spp. are also regulated in an operon. However, such studies have not yet been done in detail. *B. thetaiotaomicron* contains two allelic *pul* genes producing two enzymes. The presence of two genes in this anaerobe was demonstrated by transposon insertional mutagenesis (Smith and Salyers 1989). The bacterium which was carrying the inserted transposon in the *pul* gene was capable of growing on pullulan. However, there appears to be no problem with protease action on the expressed pullulanase enzyme in *E. coli*. A common problem with anaerobic bacterial α-amylases or pullulanases expressed in *E. coli*, is appearance of multiple activity bands on SDS-PAGE. This is mainly due to protease action on the expressed pullulanase (Coleman et al. 1987; Melasniemi and Paloheimo 1989; Mathupala 1992; M.V. Ramesh and J.G. Zeikus, unpublished results).

d'Enfert et al. (1987) reported cloning and expression in *E. coli* of the pullulanase gene *pulA* of *K. pneumoniae* UNF5023. The cloned enzyme was secreted as a lipoprotein. In the parent organism, the enzyme was shown to be expressed even though all other maltose-regulated genes are repressed and the protein is neither exposed on the surface nor secreted. Pullulanase is synthesized as a precursor protein with a signal peptide and is then fatty acylated at the amino terminal cysteines in *K. pneumoniaea* (Pugsley et al. 1986). It was earlier reported that the cloned pullulanase of *K. pneumoniae*, though processed correctly as a lipoprotein in *E. coli*, did not reach the proper location on outer membrane and was not secreted (d'Enfert et al. 1987). However, when the *pulA* gene of *K. pneumoniae* was cloned into *E. coli* in an 18.8 kb fragment together with secretion genes, the enzyme was secreted into the medium. *E. coli* carrying this chromosomal DNA fragment was induced by maltose and the expressed pullulanase was secreted (d'Enfert et al. 1987) indicating that the secretion genes located on both sides of *pulA* were part of a maltose regulon and coexpressed. In a similar study, Tokizawa et al. (1991) cloned and expressed in *E. coli* the complete pullulanase gene secretion system in a 22 kb chromosomal DNA fragment of *K. pneumoniae* W70. Here expression, induction and secretion of the enzyme in response to maltose and soluble starch was the same as in the parent strain. There are at least four genes, *pulB*, *pulC*, *pulK* and *pulL* which were responsible for secretion of pullulanase in *K. pneumoniae* though the exact role of each gene is not clear yet. The cloned pullulanases in *E. coli* are also processed the same way as in the parent organism, between gly and cys residues which is a common site for processing of bacterial pre-lipoprotein (Katsuragi et al. 1987).

Pseudomonas amyloderamosa produces an extracellular isoamylase with a 26 amino acid signal peptide. The gene was expressed in *E. coli* and the protein seems to be translated from monocistronic mRNA (Amemura et al. 1988). The *amyP* of *P. stutzeri* encoding an amylase that forms maltotetraose is also transcribed as monocistronic mRNA. Expression of the gene was poor in *E. coli* because its RNA polymerases may not be efficient, or their recognition sites are different from that of *Pseudomonas* enzymes (Fujita et al. 1990). Another maltotetraose-forming amylase from *P. saccharophila* was expressed in *E. coli* (Zhou et al. 1989). In this case, the nucleotide sequences of *Pseudomonas amyP* were similar to the consensus sequences of *E. coli* and functioned as promoters for gene expression.

Cyclodextrinases (CDase) are a new class of amylosaccharidases with specificity for cyclodextrins as substrates. CDase of *C. thermohydrosulfuricum* 39E is encoded by a 2 kb gene and its amino acid sequence showed similar conserved catalytic-domain regions to other amylases. The CDase is most homologous to amylopullulanases in the putative conserved catalytic domain sequences (Podkovyrov and Zeikus 1992). Oligo-1,6-glucosidase of *B. cereus* was expressed in *E. coli* and the 2.2 kb gene was sequenced (Watanabe et al. 1990). The cloned and native proteins have no antigenic similarity with that of *B. thermoglucosidasius* oligo-1,6-glucosidase but the amino acid sequence showed 72% homology.

B. Homology among amylosaccharidases

Amylases from different sources have low total homology in their overall amino acid sequences but they share high homology in catalytic-domain sequences of three to four regions, depending on the type of enzyme. α-Amylases of *B. stearothermophilus* (*amyS*) and *B. licheniformis* (*amyL*) have overall 59% DNA sequence homology and 62% sequence homology with different signal peptides. The COOH-terminal end of *amyS* has an extended stretch of 32 amino acids more than in *amyL* (Gray et al. 1986). However, *amyL* is more closely related in structure to *amyA* of *B. amyloliquefaciens* than to *amyS*. *B. subtilis* (*natto*) and *B. subtilis* α-amylases are structurally similar with 99% DNA sequence homology yet they differ in substrate hydrolysing activity because of one amino acid replacement (Emori et al. 1990). The NH$_2$-terminal part of *B. amyloliquefaciens* and *B. licheniformis* α-amylases are highly homologous. 60 out of 75 amino acids are identical with an overall 67% amino acid homology between these two enzymes (Ihara et al. 1985).

α-Amylase of *Streptomyces limosus* has considerable sequence similarity to those of mouse pancreas, *Drosophila melanogaster*, human pancreas and pig pancreatic α-amylases (Long et al. 1987) while little homology exists with *Bacillus*, *A. oryzae*, and barley α-amylases. *S. limosus* α-amylase is also inhibited by a mammalian α-amylase inhibitor, further indicating the structural similarity to mammalian enzymes. *Bacillus* and *A. oryzae* α-amylases are not affected by this inhibitor. It is not clear why such a unique similarity of actinomycetes enzyme with mammalian enzymes exists. *Thermomonospora curvata* α-amylase has 46 to 48% amino acid sequence homology with those of *S. hygroscopicus*, *S. limosus* and *S. venezuela*. All these amylases have four conserved regions of amino acids, i.e., 89–99, 117–124, 184–192 and 342–349, in addition to homologous regions of other amylases (Petricek et al. 1992). The COOH terminal part of *S. limosus*, *S. venezuelae*, and *T. curvata* α-amylases have homology to COOH terminal region of *A. niger* glucoamylase and *C. thermohydrosulfurogenes* β-amylase, which is believed to be involved in binding to raw starch (Belshaw and Williamson 1990). The repeated sequence of 10–12 amino acids at the COOH terminus of *S. hygroscopicus* amylase is not seen in other amylases (Hoshiko et al. 1987).

The CGTase of *B. ohbensis* and an alkalophilic *Bacillus* sp. are 80% homologous at amino acid level and 60% identical with CGTases of other *Bacillus* sp. (Sin et al. 1991) irrespective of their specificity of product formation (i.e., α, β, or γ CD). Like other amylases, CGTases also have four homologous conserved regions in their primary structure. CGTases have NH$_2$- and COOH-terminal regions that are slightly homologous to *B. subtilis* α-amylase and *A. niger* glucoamylase (Schmid 1989). *B. ohbensis* CGTase has 58% homology to α-amylase of *B. circulans* that produces maltohexaose from starch (Sin et al. 1991). These observations strongly suggest that α-amylases and CGTases have evolved from a common ancestral gene. CGTase from

K. oxytoca M5a1 (formerly *K. pneumoniae*) has similar conserved amino acids in the active site region. However, homology with other amylases is limited in regions 1 and 3 of the four conserved regions recognized in α-amylases (Binder et al. 1986).

The maltohexaose-producing amylase of an alkalophilic *Bacillus* sp. 707 has no amino acid homology with saccharifying α-amylases of *B. subtilis* or *B. subtilis* (*natto*) (Tsukamoto et al. 1988) while it has about 66% amino acid sequence homology to the liquefying amylase of *B. licheniformis*, *B. amyloliquefaciens* and *B. stearothermophilus* indicating that it is a liquefying-type amylase. The maltotetraose-forming amylase of *P. stutzeri* MO-19 (Fujita et al. 1989b) has similar conserved catalytic amino acid regions but a limited similarity in amino acid sequence to those of *Bacillus* α-amylases (Yamane et al. 1990). By deletion of 32 bp of the *B. subtilis* α-amylase gene, a stop codon was created downstream of the direct-repeat sequence, causing a deletion of 200 amino acids of the COOH terminus of enzyme. This protein was still active and the number of amino acids was similar to that of *B. subtilis* (*natto*) α-amylase (Yamane et al. 1984). This observation indicates that the deleted region is not required for catalytic activity.

Amylopullulanase from significantly different *C. thermohydrosulfuricum* strains have more than 90% DNA sequence homology (Mathupala 1992; Melasniemi et al. 1990). Amylopullulanase of *Thermoanaerobacter* B6A-RI has an amino acid sequence homology of about 75% with these *C. thermohydrosulfuricum* enzymes. This observation and the fact that this enzyme is less thermostable than the other two, implies that the *Thermoanaerobacter* enzyme evolved divergently from these two other enzymes (M.V. Ramesh and J.G. Zeikus, unpublished results). Part of the 4.4 kb *apu* gene of *C. thermohydrosulfuricum* was deleted to give a 2.9 kb mutant which still expressed an active enzyme (S. P. Mathupala 1992). Figure 1 shows the similarity in conserved sequence homology regions of amylopullulanases to α-amylases, pullulanase and glucoamylase. Figure 2 shows Hydrophobic cluster analysis of similarity between alpha amylase, cyclodextrinase and other alpha glycanases, (Podkovyrov et al. 1993).

The amino acid sequence homology in conserved regions of *B. fibrisolvens* amylase is restricted to the middle from NH₂-terminus region. The other part of the enzyme at the COOH-terminus does not have any amino acid sequence homology with other known amylases (Rumbak et al. 1991). This enzyme is still active even after deleting 40% of the total gene from COOH-terminus. Recently, corrections for many errors in the DNA sequence of pullulanase of *K. aerogenes*, by Katsuragi et al. (1987), were reported (Katsuragi et al. 1992). Since any change in DNA sequence would also show up in the amino acid sequence, at this point it is difficult to discuss the comparative results of the DNA or amino acid sequences of this enzyme with those of amylopullulanases.

β-Amylase of *C. thermosulfurogenes* and *B. polymyxa* have 54% sequence homology and structurally they are similar. However, β-amylase of *C*

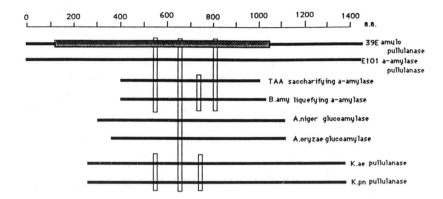

Fig. 1. Overall alignment of the deduced sequence of amylopullulanase of *C. thermohydrosulfuricum* 39E with amylases from microbial and fungal origin. 39E = *C. thermohydrosulfuricum* 39E; E 101 = *C. thermohydrosulfuricum* E 101; TAA = *Aspergillus oryzae*; B. amy = *Bacillus amyloliquefaciens*; K. ae = *Klebsiella aerogenes*; K. pn = *K. pneumoniae*. The open boxes represent regions putatively identified on all sequences based on the four conserved regions of α-amylase of *A. oryzae*. The 100 kDa thermostable amylopullulanase fusion protein encoded by the 2.9 kb *apu* gene fragment, is shown by the shaded area on amylopullulanase.

E101 = *C. thermohydrosulfuricum* (Melasniemi et al. 1990); TAA = *Aspergillus oryzae* (Tada et al. 1989); B. amy = *Bacillus amyloliquefaciens* (Nakajima et al. 1986); K. ae and K. pn = *K. aerogenes* and *K. pneumoniae* (Katsuragi et al. 1987, 1992).

thermosulfurogenes organism is more thermostable than that of the latter (Kitamoto et al. 1988). *C. thermosulfurogenes* β-amylase contains seven cysteine residues while that of *B. polymyxa* contains only three. These two enzymes, moreover, are more homologous at their NH₂- than COOH-termini.

A novel amylase digesting raw starch and showing CGTase activity has been isolated from a *Bacillus* sp. and shown to have a COOH-terminal with high homology to that of *A. niger* glucoamylase and that of the maltotetraose-forming *P. stutzeri* amylase (Itkor et al. 1990). CGTase of *K. pneumoniae* M5a1 has about 30% homology with those of *Bacillus* CGTases (Nitschke et al. 1990). *B. circulans* CGTase has 74% amino acid sequence homology to that of *B. circulans* F-2. *B. ohbensis* CGTase has about 80% identify with that of *Bacillus* sp. 1-1 (Sin et al. 1991). *B. circulans* α-amylase has 58% amino acid sequence homology with that of *B. ohbensis*. The former enzyme produces only maltohexaose from starch (Sin et al. 1991). The *P. saccharophila* amylase sequence has a little similarity (less than 9%) with other α-amylases, CGTases, maltohexaose forming amylases, pullulanases, and isoamylases (Zhou et al. 1989). This enzyme still has three highly conserved sequence regions similar to those identified in catalysis in other amylosaccharidases.

Glucoamylases from yeast and fungi have considerable similarity in their sequences of amino acids. Fungal glucoamylases show little sequence similarity with the conserved amino acid regions of bacterial amylases. Of all microbial amylases, glucoamylases and β-amylases show considerable deviation from the consensus, conserved sequences of other amylosaccharidases. In glucoamylases, four highly conserved sequence regions can be identified in yeast and fungal enzymes (Itoh et al. 1987a). *S. fibuligera* glucoamylase has a fifth conserved region similar to those of *Aspergillus* and *Rhizopus* glucoamylases. However, this fifth region is absent in *S. diastaticus* and *S. cerevisiae* and probably is not required for catalytic function of the enzyme (Itoh et al. 1987a). *S. diastaticus* glucoamylase apparently is evolved from a fusion *S. cerevisiae* glucoamylase gene and a fragment of DNA of the same organism which encodes a protein for binding raw starch (Yamashita et al. 1987). Glucoamylases of *Aspergillus* species have high sequence homology, i.e., above 90%. However, *A. niger* and *A. oryzae* glucoamylases show considerable variation. *A. oryzae* glucoamylase has 67 and 30% overall amino acid sequence homology with those of *A. niger* and *Rhizopus oryzae* glucoamylases, respectively (Hata et al. 1991). However, the homology in the conserved amino acid regions was 84% for glucoamylases of *A. niger* and *A. oryzae*. Interestingly, *A. oryzae* glucoamylase lacks the threonine-serine rich region (TS-region) which is responsible for binding to raw starch. Correspondingly, *A. oryzae* glucoamylase has less affinity for raw starch (Hata et al. 1991). The function of the TS-rich region in binding to raw starch was demonstrated by Hayashida et al. (1989).

V. Biotechnology of amylosaccharidases

Amylosaccharidases have a broad base for biotechnology applications. Three types of enzymes are involved in the production of sugars from starch: 1) endo-amylase (α-amylase), 2) exo-amylase (β-amylase, glucoamylase); and, 3) debranching enzymes (pullulanase, isoamylase). Starch bioprocessing usually involves two steps: liquefaction and saccharification. First, an aqueous slurry of starch (30 to 40% w/v dissolved solids) is gelatinized (105 °C, 5 to 7 min) and partially hydrolysed (95 °C, 2 h) by highly thermostable α-amylase to a dextrose equivalent (DE) of 5 to 10. The optimum pH for the reaction is 6.0–6.5 and Ca^{2+} is required. Then, in the saccharification step, a saccharifying enzyme or a combination of two saccharifying enzymes with or without a debranching enzyme is added depending on the type of sugar syrup desired.

→

Fig. 2. HCA plots of the CDase from *T. ethanolicus* 39E, Taka-amylase A, neopullulanase from *B. stearothermophilus*, and CGTase from *B. circulans*. Abbreviations of the enzymes are the same as on Fig. 1. Numbering starts from the first amino acid of the mature protein for CDX and NPL. The first amino acid of the plot for AMY (arginine, R) corresponds to the 110-th amino acid of the mature protein, and the first amino acid of the plot for CGTase (leucine, L) is 121-st amino acid of the mature protein. Proline is symbolized by ★, glycine by ◆, serine by ⊡ and threonine by □. Proposed catalytic residues are ringed with circles. For the parts of the sequences containing catalytic residues the correspondences between hydrophobic clusters (segments I–VI) are shown by vertical lines.

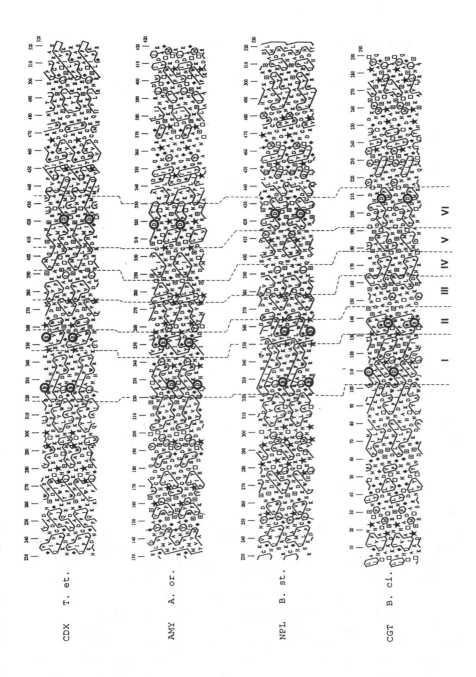

CDX   T. et.

AMY   A. or.

NPL   B. st.

CGT   B. ci.

A. Amylodextrin production

Various dextrin products are widely used as viscosity improver, filler or ingredient of food. Dextrin is generally prepared from starch by liquefaction by thermostable α-amylase, followed by refining the liquefied starch solution and spray-drying (Taji 1988). These liquefied starches are generally referred to as maltodextrins (DE < 20).

B. Maltose production

Maltose syrups are used in the brewing, baking, soft drink, canning, and confectionery industries. Among the important functional characteristics of high maltose syrups are low hygroscopicity, low viscosity in solution, resistance to crystallization, low sweetness, reduced browning capacity and good heat stability. Thus, maltose syrups can be used as moisture conditioners, crystallization inhibitors, stabilizers, carriers and bulking agents (Saha and Zeikus 1987). There are two types of high maltose syrups: high maltose syrups and extreme high maltose syrups. The high maltose syrups have DE of about 35 to 50 and a maltose content of 45 to 50%; the extreme high maltose syrups have DE 45 to 60 and a maltose content about 70 to 85%. High maltose syrups are produced from liquefied starch by the use of a maltogenic amylase such as plant or microbial β-amylase or fungal α-amylase at pH 5.0 to 5.5 and 50 to 55 °C. In order to produce extreme high maltose syrups it is necessary also to employ a debranching enzyme (Saha and Zeikus 1989a,b).

Saha and Zeikus (1989b) developed an improved method for producing various high maltose containing syrups from liquefied and raw starch by saccharifying the starch at higher temperatures than presently used with the thermostable α-amylase from a thermoanaerobe and other thermostable enzymes.

C. High conversion syrup production

The high conversion syrups have been widely applied in many industries such as brewing, baking, confectionery, canning and soft drink. These syrups are characterized by high content of glucose (35 to 45%) and maltose (30 to 37%) (fermentable sugars, above 85% w/v). High conversion syrups have high DE value, 60 to 70, but are stable enough to resist crystallization below 4 °C and 80 to 83% DS (Tegge et al. 1986). The glucose content in these syrups should not exceed 43% (w/v). This type of syrup is produced from liquefied starch (DE ~40) by saccharifying it with β-amylase or fungal α-amylase and glucoamylase (Saha and Zeikus 1987). After completion of saccharification (48 to 72 h), the syrup is heated to destroy the enzyme action to stop further glucose formation. Recently, Saha and Zeikus (1989a, 1990b) demonstrated the preparation of high conversion syrups by using thermostable amylases from thermoanaerobes on raw or liquefied starch at 75 °C and pH 5.5.

D. Dextrose production

High dextrose (glucose) syrups contain about 95 to 98% (w/v) glucose. They are produced by the saccharification of liquefied starch with a glucoamylase and a debranching enzyme such as pullulanase at pH 4.5 and 60 °C. These syrups are widely used as starting material for production of high-fructose, corn syrups and as substrate for fermentation, production of fine powdered glucose, and crystalline glucose.

E. Cyclodextrin production

Cyclodextrins (CDs) are cyclic oligosaccharides composed of six or more α-1,4 linked glucose units. CDs of 6,7 and 8 α-D-glucose residues are called, α-, β-, and γ-CD, respectively. CDs possess the unique ability to form inclusion complexes with various chemicals and are useful in the food, cosmetic, pharmaceutical and plastics industries as emulsifiers, antioxidants and stabilizing agents. CDs are produced from starch by cyclodextrin glycosyltransferase (CGTase). CGTase first detaches linear short-chain length oligosaccharides from starch and then links the two ends to form a cyclic molecule. Various methods for the production of a specific CD including the enhancement in the production of cyclodextrins by the presence of organic solvents such as acetone, propanol and ethanol have been developed (Bender 1986).

Branched CDs are produced by the reverse action of hydrolytic enzymes. Maltosyl cyclodextrins can be prepared from maltose and a CD by the reverse reaction of pullulanase (Sakano et al. 1985).

F. Other product uses

Amylases are used in animal feeds as digestives for animals as de-sizing agents in the textile industry, for making speciality saccharides and substrates for alcohol fermentation. During the past 30 years great advances have been made in the development of enzyme-based, starch conversion technologies. Future work is likely to be focused towards improving the existing technologies. For example, industry is looking to improve the biotechnology of amylosaccharidases by developing an improved saccharifying α-amylase that works at high temperature and low pH with a low Ca^{2+} requirement. Also, a highly thermostable and pH compatible pullulanase is required. Future directions of research also include new approaches in starch processing technology using amylopullulanase and other newly discovered α-glycan degrading enzymes with unique properties.

VI. Future directions

Although amylases have been extensively used in the starch industry, there are certain areas that have not been paid much attention by researchers. For example, bacterial α-glucosidases have not yet been studied at the molecular level. We could not trace any report on cloning and sequencing of bacterial α-glucosidase gene. This may be mainly due to the properties of the enzyme being unsuitable for industrial application. Nevertheless, α-glucosidase is one of the key enzymes in starch utilization by many microbes. The maltosaccharides produced by the action of endo-acting amylases have to be further hydrolysed in order to be transported inside the cell. In most cases, α-glucosidases are cell-associated and are not secreted. This enzyme is also probably under the glucose repression effect. In microbes which utilize starch without releasing amylolytic activity into the medium, there should be some mechanism by which they bind to the starch granules and hydrolyse them. It would be quite interesting to investigate such mechanisms since it is not understood so far why few species of the same genus, sometimes the same species but often different strains, differ in their secretion of amylases. Probably in Gram-negative bacteria, secretion factors may play a major role. It is also possible that microbes growing in environments where substrate availability is scarce and in very low concentrations, dictate secretion of their own enzymes. Hence, it is always important to look at the site from where a microbe is isolated before pre-judging the utilization of starch by the organism. Understanding the cellular localization and juxtaposition of α-glycan-degrading enzyme and starch-binding receptors in relation to the overall rate of starch degradation by microbes will be of fundamental and applied significance.

Another interesting area, though not new, is hydrolysis of raw starch by certain amylases. It is apparent that such enzymes have two sites, one for binding to raw starch granules and the other one for catalysis (Hayashida et al. 1990). Development of a more efficient hybrid enzyme by gene fusion between such an enzyme and a known stable amylase is another possibility if the raw starch binding site is separate and clearly identified. However, development of hybrid enzymes through gene fusion may not be cost-effective for starch liquefaction because of the process economics and enzyme cost. Such approaches are good if used for producing speciality chemicals. High amounts of recombinant amylases can be produced using B. brevis as host (Schmid 1989). Production of cyclodextrins with a raw starch-digesting CGTase will be promising. Cyclodextrinases are a recent addition to the amylase industry for producing maltohexaose, maltoheptaose or maltooctaose. Bacterial glucoamylases, active under the same conditions of bacterial α-amylase will also be interesting if identified and characterized completely (Ohnishi et al. 1991).

Nature continues to be a good source for new microbes with new enzyme systems. In this regard, one can anticipate both new types of α-glycan-degrading activities and highly thermostable amylosaccharidases from new hyper-thermophilic archaebacterial sources that have been recently discovered. Finally, protein engineering is another approach that can be used for development of hybrid enzymes that have improved activity or stability. Based on current knowledge, these enhancements in amylosaccharidase performance may be achieved by site-directed mutagenesis to improve catalytic efficiency and by protein fusion to achieve thermostability.

Acknowledgements

This work was supported in part by a grant from the United States Department of Agriculture, Bioprocessing for Utilization of Agricultural Raw Materials, No. 90-34189-5014.

References

Ahern TJ and Klibanov AM (1985) The mechanism of irreversible enzyme inactivation at 100 °C. Science 228: 1280–1284

Amemura A, Chakraborty R, Fujita M, Noumi T and Futai M (1988) Cloning and nucleotide sequence of the isoamylase gene from *Pseudomonas amyloderamosa* SB-15. J. Biol. Chem. 263: 9271–9275

Anderson KL and Salyers AA (1989) Genetic evidence that outer membrane binding of starch is required for starch utilization by *Bacteroides thetaiotaomicron*. J. Bacteriol. 171: 3199–3204

Antranikian G (1990) Physiology and enzymology of thermophilic anaerobic bacteria degrading starch. FEMS Microbiol. Rev. 75: 201–218

Antranikian G, Herzberg C, Mayer F and Gottschalk G (1987) Changes in cell envelope structure of *Clostridium* sp. strain EM1 during massive production of α-amylase and pullulanase. FEMS Microbiol. Lett. 41: 193–197

Bahl H, Burchhardt G, Spreinat A and Antranikian G (1991) α-Amylase of *Clostridium thermohydrosulfurogenes* EM1: nucleotide sequence of the gene, processing of the enzyme, and comparison to other α-amylases. Appl. Env. Microbiol. 57: 1554–1559

Bealin-Kelly F, Kelly CT and Fogarty WM (1991) Studies on the thermostability of the α-amylase of *Bacillus caldovelox*. Appl. Microbiol. Biotechnol. 36: 332–336

Belloc A, Florent J, Mancy D and Verrier J (1975) Novel amylase by fermentation of a *Penicillium* strain. U.S. Patent 3906113

Belshaw NJ and Williamson G (1990) Production and purification of a granular-starch-binding domain of glucoamylase from *Aspergillus niger*. FEBS Lett. 269: 350–353

Bender H (1981) A bacterial glucoamylase degrading cyclodextrins: partial purification and properties of the enzyme from a *Flavobacterium* species. Eur. J. Biochem. 115: 287–291

Bender H (1986) Production, characterization and application of cyclodextrins. Adv. Biotechnol. Proc. 6: 31–71

Bernlohr RW, Maddock NK and Goldberg ND (1974) Cyclic guanosine 3′:5′ monophosphate in *Escherichia coli* and *Bacillus licheniformis*. J. Biol. Chem. 249: 4329–4331

Bhella RH and Altosaar I (1987) Transcriptional control of α-amylase gene expression in *Aspergillus awamori*. Biotechnol. Appl. Biochem. 9: 287–293

Binder F, Huber O and Bock A (1986) Cyclodextrin glucosyltransferase from *Klebsiella pneumoniae* M5a1: cloning, nucleotide sequence and expression. Gene 47: 269–277

Brown SH, Constantino HR and Kelly RM (1990) Characterization of amylolytic enzyme activities associated with the hyperthermophilic archaebacterium *Pyrococcus furiosus*. Appl. Environ. Microbiol. 57: 1985–1991

Chapon C and Raibaud O (1985) Structure of two divergent promoters located in front of the gene encoding pullulanase of *Klebsiella pneumoniae* and positively regulated by the *malT* product. J. Bacteriol. 164: 639–645

Chung H and Friedburg F (1980) Sequence of N-terminal half of *Bacillus amyloliquefaciens* α-amylase. Biochem. J. 185: 387–395

Coleman RD, Yang SS, McAlister MP (1987) Cloning of the debranching enzyme gene from *Thermoanaerobacter brockii* into *Escherichia coli* and *Bacillus subtilis*. J. Bacteriol. 169: 4302–4307

Constantino HR, Brown SH and Kelly RM (1990) Purification and characterization of α-glucosidase from a hyperthermophilic archaebacterium, *Pyrococcus furiosus*, exhibiting a temperature optimum of 105 to 115 °C. J. Bacteriol. 172: 3654–3660

Cornelis P, Digneffe C and Willemot K (1982) Cloning and expression of *Bacillus coagulans* amylase gene in *Escherichia coli*. Mol. Gen. Genet. 186: 507–511

Das A, Chatterjee M and Roy A (1979) Enzymes of some higher fungi. Mycologia. 71: 530–536

De Mot R, Andries K and Verachtert H (1984a) Coomparative study of starch degradation and amylase production by ascomycetous yeast species. Syst. Appl. Microbiol. 5: 106–111

De Mot R, Andries K and Verachtert H (1984b) Comparative study of starch degradation and amylase production by non-ascomycetous yeast species. Syst. Appl. Microbiol. 5: 421–426

d'Enfert C, Ryter A and Pugsley AP (1987) Cloning and expression in *Escherichia coli* of the *Klebsiella pneumoniae* genes for production, surface localization and secretion of pullulanase. EMBO J. 6: 3531–3538

Depinto JA and Campbell LL (1968) Purification and properties of cyclodextrinase of *Bacillus macerans*. Biochemistry 7: 121–125

Eisele B, Rasched R and Wallenfels K (1972) Molecular characterization of pullulanase from *Aerobacter aerogenes*. Eur. J. Biochem. 26: 62–67

Emori EM, Takagi M, Maruo B and Yano K (1990) Molecular cloning, nucleotide sequencing and expression of the *Bacillus subtilis* (natto) IAM1212 α-amylase gene, which encodes an α-amylase structurally similar to but enzymatically distinct from that of *Bacillus subtilis* 2633. J. Bacteriol. 172: 4901–4908

Erratt JA, Douglas PE, Mpraneli F and Seligy VL (1984) The induction of α-amylase by starch in *Aspergillus oryzae*: evidence for controlled mRNA expression. Can. J. Biochem. 62: 678–690

Federeci F (1982) Extracellular enzymatic activities in *Aureobasidium pullulans*. Mycologia 74: 738–743

Fogarty WM and Kelly CT (1980) Amylases, amyloglucosidase and related glucanases. In: AH Rose (ed) Economic Microbiology 5: Microbial Enzymes and Bioconversions (pp 116–170). Academic Press, London

Fogarty WM and Kelly CT (1983) Microbial amylases. In: Microbial Enzymes and Biotechnology (pp 1–92). Elsevier Applied Science Publishers, London

Fogarty WM and Kelly CT (1990) Recent advances in microbial amylases. In: Microbial Enzymes and Biotechnology (pp 71–131). Elsevier Applied Science Publishers, London

Frost GM and Moss DA (1987) Production of enzymes by fermentation. In: HJ Rehm and GE Reed (eds) Biotechnology 7a: Enzyme Technology (pp 65–211). VCH, Weinheim, Germany

Fujita M, Amemura A and Futai A (1989a) Transcription of the isoamylase gene (*iam*) in *Pseudomonas amyloderamosa* SB-15. J. Bacteriol. 171: 4320–4325

Fujita M, Torigoe K, Nakada T, Tsukai K, Kubota M, Sakai S and Tsujisaka Y (1989b) Cloning and nucleotide sequence of the gene (*amyP*) for maltotetraose forming amylase from *Pseudomonas stutzeri* MO 19. J. Bacteriol. 171: 1333–1339

Fujita M, Futi M and Amemura A (1990) In vivo expression of *Pseudomonas stutzeri* forming amylase gene (*amyP*). J. Bacteriol. 172: 1595–1599

Garcia-Gonzalez MD, Martin JF, Vigal T and Liras P (1991) Characterization, expression in *Streptomyces lividans* and processing amylase of *Streptomyces griseus* IMRU 3570: two different amylases are derived from the same gene by an intracellular processing mechanism. J. Bacteriol. 173: 2451–2458

Gines MJ, Dove MJ and Seligy VL (1989) *Aspergillus oryzae* has two nearly identical Taka-amylase genes, each containing eight introns. Gene 79: 107–117

Goldberger RF (1979) Biological Regulation and Development 1. Plenum Press, New York

Gray GL, Mainzer SE, Rey MW, Lamsa M, Kinle KL, Carmona C and Requadt C (1986) Structural genes encoding the thermophilic α-amylase of *Bacillus licheniformis*. J. Bacteriol. 166: 635–643

Hata Y, Kitamoto K, Gomi K, Kumagai C, Tamura G and Hara S (1991) The glucoamylase of cDNA from *Aspergillus oryzae*: Its cloning, nucleotide sequencing and expression in *Saccharomyces cerevisiae*. Agric. Biol. Chem. 55: 941–949

Hayashida S, Nakamura K, Kanlayakrit W, Hara T and Teramoto Y (1989) Characteristics and function of raw-starch-affinity site on *Aspergillus awamore* var. *kawachi* glucoamylase I molecule. Agric. Biol. Chem. 53: 143–149

Hayashida S, Teramoto Y, Inoue T and Mitsuiki S (1990) Occurrence of an affinity site apart from the active site on the raw starch digesting but non-raw starch-absorbable *Bacillus subtilis* 65 α-amylase. Appl. Environ. Microbiol. 56: 2584–2586

Hoshiko H, Makabe O, Nojiri C, Katsumura K, Satoh E and Nagaoka K (1987) Molecular cloning and characterization of *Streptomyces hygroscopicus* α-amylase gene. J. Bacteriol. 169: 1029–1036

Hyun HH and Zeikus JG (1985a) General biochemical characterization of thermostable extracellular α-amylase from *Clostridium thermosulfurogenes*. Appl. Environ. Microbiol. 49: 1162–1167

Hyun HH and Zeikus JG (1985b) General biochemical characterization of thermostable pullulanase and glucoamylase from *Clostridium thermohydrosulfuricum*. Appl. Environ. Microbiol. 49: 1168–1173

Ihara H, Sasaki T, Tsuboi A, Yamagita H, Tsukagoshi N and Udaka S (1985) Complete nucleotide sequencing of a thermophilic α-amylase gene: homology between prokaryotic and eukaryotic α-amylases at the active sites. J. Biochem. 98: 95–103

Imanaka T and Kuriki T (1989) Pattern of action of *Bacillus stearothermophilus* neopullulanase on pullulan. J. Bacteriol. 171: 369–374

Itkor P, Tsukagoshi N and Udaka S (1990) Nucleotide sequence of the raw starch digesting amylase gene from *Bacillus* sp. B1018 and its strong homology to the cyclodextrin glucanotransferase genes. Biochem. Biophys. Res. Comm. 166: 630–636

Itoh T, Ohtsuki I, Yamashita I and Fukui S (1987a) Nucleotide sequence of the glucoamylase gene *GLU 1* in the yeast *Saccharomycapsis fibuligera*. J. Bacteriol. 169: 4171–4176

Itoh T, Yamashita I and Fukui S (1987b) Nucleotide sequence of the α-amylase gene *ALP 1* in yeast *Saccharomycapsis fibuligera*. FEBS Lett. 219: 339–342

Kallio P, Ulmanen I and Palva I (1986) Isolation and characterization of a 2.2 kb operon preceding the α-amylase gene of *Bacillus amyloliquefaciens*. Eur. J. Biochem. 158: 497–504

Kaneko T, Hamamoto T and Horikoshi K (1988) Molecular cloning and nucleotide sequence of the cyclomaltodextrin glucanotransferase gene from the alkalophilic *Bacillus* sp. strain no. 38-2. J. Gen. Microbiol. 134: 97–105

Katsuragi N, Takizawa N and Murooka Y (1987) Entire nucleotide sequence of the pullulanase gene of *Klebsiella aerogenes* W70. J. Bacteriol. 169: 2301–2306

Katsuragi N, Takizawa N and Murooka Y (1992) Erratum. J. Bacteriol. 174: 3095

Keay L (1970) The action of *Bacillus subtilis* saccharifying amylase on starch and α-cyclodextrin. Starch. 22: 153–157

Kitahata S, Taniguchi M, Beltran SD, Sugimoto T and Okada S (1983) Purification and some properties of cyclodextrinase from *Bacillus coagulans*. Agric. Biol. Chem. 47: 1441–1447

Kitamoto N, Yamagata H, Kato H, Tsukagoshi N and Udaka S (1988) Cloning and sequencing of the gene encoding thermophilic β-amylase of *Clostridium thermosulfurogenes*. J. Bacteriol. 170: 5848–5854

Koch R, Spreinat A, Lamke K and Antranikian G (1991) Purification and properties of a hyperthermoactive α-amylase from the archaebacterium *Pyrococcus woesei*. Arch. Microbiol. 155: 572–578

Kuriki T, Takada H, Okada S and Imanaka T (1991) Analysis of the active center of *Bacillus stearothermophilus* neopullulanase. J. Bacteriol. 173: 6147–6152

Kvesitadze GI, Kokonashvili GN and Fenixova RV (1974) α-amylase. U.S. Patent 3 826 716

Lamed R, Bayer S, Saha BC and Zeikus JG (1988) Biotechnological potential of enzymes from unique extreme thermophiles In: G Durand, L Bobichon, and J Florent (eds) Proc. 8th Intl. Biotech. Symposium (pp 371–383). Société Française de Microbiologie, Paris

Laoide BM and McConnell DJ (1989) cis Sequences involved in modulating expresison of Bacillus licheniformic amyL in Bacillus subtilis: effect of sporulation mutations and catabolite repression resistance mutations on expression. J. Bacteriol. 171: 2443–2450

Laoide BM, Chambliss GH and McConnell DJ (1989) Bacillus licheniformis α-amylase gene amyLis subject to promoter independent catabolite repression in Bacillus subtilis. J. Bacteriol. 171: 2435–2442

Long CM, Virolle M-J, Chang S-Y, Chang S and Bibb MJ (1987) α-Amylase gene of Streptomyces limosus: nucleotide sequence, expression motifs and amino acid sequence homology to mammalian and invertebrate α-amylases. J. Bacteriol. 169: 5745–5754

Madsen GB, Norman BE and Slott S (1973) A new heat stable bacterial amylase and its use in high temperature liquefaction. Starch/Starke. 25: 304–308

Mathupala SP (1992) Biochemical characterization of amylopullulanase from Clostridium thermohydrosulfuricum 39E. PhD thesis, Michigan State University, U.S.A.

Mathupala SP, Saha BC and Zeikus JG (1990) Substrate competition and specificity at the active site of amylopullulanase from Clostridium thermohydrosulfuricum 39E. Biochem. Biophys. Res. Comm. 166: 126–132

Matsuura Y, Kusunoki M, Harada W and Kukudo M (1984) Structure and possible catalytic residues of Taka amylase A. J. Biochem. (Tokyo) 95: 697–702

Melasniemi H (1987) Characterization of α-amylase-pullulanase activities of Clostridium thermohydrosulfuricum. Evidence for novel thermostable amylase. Biochem. J. 250: 813–818

Melasniemi H and Paloheimo M (1989) Cloning and expression of the Clostridium thermohydrosulfuricum α-amylase-pullulanase gene in Escherichia coli. J. Gen. Microbiol. 135: 1755–1762

Melasniemi H, Paloheimo M and Hemio L (1990) Nucleotide sequence of the α-amylase-pullulanase gene from Clostridium thermohydrosulfuricum. J. Gen Microbiol. 136: 447–454

Mielenz JR (1983) Bacillus stearothermophilus contains a plasmid-borne gene for α-amylase. Proc. Natl. Acad. Sci. U.S.A. 80: 5875–5979

Movva NR, Nakamura K and Inouye M (1980) Regulatory region of the gene for the ompA protein, a major outer membrane protein of Escherichia coli. Proc. Natl. Acad. Sci. U.S.A. 77: 3845–3849

Mudgett RE (1986) Solid state fermentations: In: AL Demain and NA Solomen (eds) Manual of Industrial Microbiology and Biotechnology (pp 66–83). American Society for Microbiology, Washington, DC

Nakajima R, Imanaka T and Aiba S (1986) Comparison of amino acid sequences of eleven different α-amylases. Appl. Microbiol. Biotechnol. 23: 355–360

Nakamura A, Haga A, Ogawa S, Kuwano K, Kimura K and Yamane K (1992) Functional relationships between cyclodextrin glucano transferase from an alkalophilic Bacillus and α-amylases: site-directed mutagenesis of the conserved two Asp and one Glu residues. FEBS Lett. 296: 37–40

Narayanan AS and Shanmugasundaram ERB (1967) Studies on amylase from Fusarium vasinfectum. Archiv. Biochem. Biophys. 118: 317–322

Nitschke L, Heeger K, Bender H and Schulz GE (1990) Molecular cloning, nucleotide sequencing and expression in Escherichia coli of β-cyclodextrin glucosyl transferase gene from Bacillus circulans strain no. 8. Appl. Microbiol. Biotechnol. 33: 542–546

Odibo FJC and Obi SKC (1988) Purification and characterization of a thermostable pullulanase from Thermoactinomyces thalpophilus. J. Ind. Microbiol. 3: 343–350

Ohnishi H, Sakai H and Ohta T (1991) Purification and some properties of a glucoamylase from Clostridium sp. G0005. Agric. Biol. Chem. 55: 1901–1902

Paloheimo M, Haglund D, Aho S and Korhola M (1992) Production of cyclomaltodextrin glucanotransferase of Bacillus circulans var. alkalophilus ATCC 21783 in Bacillus subtilis.

Appl. Microbiol. Biotechnol. 36: 584–591

Petricek M, Tichy P and Kuncova M (1992) Characterization of α-amylase encoding gene from *Thermomonospora curvata*. Gene 112: 77–83

Piggott RP, Rossiter A, Ortlepp SA, Pembroke JT and Ollington JF (1984) Cloning in *Bacillus subtilis* of an extremely thermostable alpha amylase: comparison with othger cloned heat stable alpha amylases. Biochem. Biophys. Res. Comm. 122: 175–183

Plant AR, Morgan HW and Daniel RM (1986) A highly stable pullulanase from *Thermus aquaticus* YT-1. Enz. Microbial. Technol. 8: 668–672

Plant AR, Clemens RM, Daniel RM and Morgan HW (1987) Purification and preliminary characterization of an extracellular pullulanase from *Thermoanaerobium* TOKB1. Appl. Microbiol. Biotechnol. 26: 427–433

Podkovyrov SM and Zeikus JG (1992) Structure of the gene encoding cyclomaltodextrinase from *Clostridium thermohydrosulfuricum* 39E and characterization of the enzyme purified from *Escherichia coli*. J. Bacteriol. 174: 5400–5405

Podkovyrov SM, Burdette D and Zeikus JG (1993) Analysis of the catalytic center of cyclomaltodextrinase from *Thermoanaerobacter ethanolicus* 39E. FEBS Lett. 317: 259–262

Priest FG (1977) Extracellular enzyme synthesis in the genus *Bacillus*. Bacteriol. Rev. 41: 711–753

Priest FG (1985) Extracellular enzymes. In: CL Cooney and AE Humphrey (eds) Comprehensive Biotechnology: Principles, Applications and Regulations of Biotechnology in Industry, Agriculture and Medicine (pp 587–604). Pergamon Press, Oxford

Pugsley AP, Chapon C and Schwartz M (1986) Extracellular pullulanase of *Klebsiella pneumoniae* is a lipoprotein. J. Bacteriol. 166: 1083–1088

Ramesh MV and Lonsane BK (1991a) Ability of solid state fermentation technique to significantly minimize catabolite repression of α-amylase production by *Bacillus licheniformis* M27. Appl. Microbiol. Biotechnol. 35: 591–593

Ramesh MV and Lonsane BK (1991b) Regulation of α-amylase production in *Bacillus licheniformis* M27 by enzyme end-products in submerged fermentation and its overcoming in solid state fermentation system. Biotechnol. Lett. 13: 355–360

Robyt JF and Ackerman RJ (1971) Isolation, purification and characterization of a maltotetraose producing amylase from *Pseudomonas stutzeri*. Arch. Biochem. Biophys. 145: 105–114

Robyt JF and French D (1967) Multiple attack hypothesis of α-amylase action: action of porcine pancreatic, human salivary and *Aspergillus oryzae* α-amylases. Arch. Biochem. Biophys. 122: 8–16

Rumbak E, Rawlings DE, Lindsey GG and Woods DR (1991) Cloning, nucleotide sequencing and enzymatic characterization of an α-amylase from the ruminal bacterium *Butyrivibrio fibrisolvens* H17C. J. Bacteriol. 173: 4203–4211

Saha BC and Zeikus JG (1987) Biotechnology of maltose production. Proc. Biochem. 22: 78–82

Saha BC and Zeikus JG (1989a). Novel highly thermostable pullulanase from thermophiles. Trends in Biotechnol. 7: 234–239

Saha BC and Zeikus JG (1989b) Improved method for preparing high maltose conversion syrups. Biotechnol. Bioeng. 34: 299–303

Saha BC and Zeikus JG (1990a) Characterization of thermostable cyclodextrinase from *Clostridium thermohydrosulfuricum* 39E. Appl. Environ. Microbiol. 56: 2941–2943

Saha BC and Zeikus JG (1990b) Preparation of high conversion syrups by using thermostable amylases from thermoanaerobes. Enz. Microb. Technol. 12: 229–231

Saha BC, Mathupala S and Zeikus JG (1988) Purification and characterization of highly thermostable pullulanase from *Clostridium thermohydrosulfuricum*. Biochem J. 252: 343–348

Saha BC, Shen G-J, Srivastava KC, LeCureux LW and Zeikus JG (1989) New thermostable α-amylase-like pullulanase from thermophilic *Bacillus* sp. 3183. Enz. Microb. Technol. 11: 760–764

Saha BC, Lamed R, Lee C-Y, Mathupala SP and Zeikus JG (1990) Characterization of an endo-acting amylopullulanase from *Thermoanaerobacter* strain B6A. Appl. Environ. Microbiol. 56: 881–886

Sakano Y, Haraiwa S and Fukushima J (1982) Enzymatic properties and action patterns of *Thermoactinomyces vulgaris* α-amylase. Agric. Biol. Chem. 46: 1121–1129

Sakano Y, Sano M and Kobayashi T (1985) Preparation of enzymatic hydrolysis of maltosyl-α-cyclodextrin. Agric. Biol. Chem. 49: 3391–3398

Schmid G (1989) Cyclodextrin glycosyl transferase production: yield enhancement by overexpression of cloned genes. Trends Biotechnol. 7: 244–248

Schumann J, Wraba A, Jaenicke R and Stetter KO (1991) Topographical enzymatic characterization of amylases from extremely thermophilic eubacterium *Thermotoga maritima*. FEBS Lett. 282: 122–126

Sills AM and Stewart AM (1982) Production of amylolytic enzymes by several yeast species. J. Inst. Brew. 88: 313–318

Sin K-A, Nakamura A, Kobayashi K, Masaki H and Uozumi T (1991) Cloning and sequencing of a cyclodextrin glucanotransferase from *Bacillus ohbensis* and its expression in *Escherichia coli*. Appl. Microbiol. Biotechnol. 35: 600–605

Sleytr UB and Messner P (1983) Crystalline surface layers on bacteria. Annu. Rev. Microbiol. 37: 311–339

Smith KA and Salyers AA (1989) Cell-associated pullulanase from *Bacteroides thetaiotaomicron*: cloning, characterization and insertional mutagenesis to determine role in pullulan utilization. J. Bacteriol. 171: 2116–2123

Specka U, Spreinat A, Antranikian G and Mayer F (1991) Immunochemical identification and localiation of active and inactive α-amylase and pullulanase in cells of *Clostridium thermosulfurogenes* EM1. Appl. Environ. Microbiol. 57: 1062–1069

Spreinat A and Antranikian G (1990) Purification and properties of a thermostable pullulanase from *Clostridium thermosulfurogenes* EM1 which hydrolyses both α-1,6 and α-1,4 glycosidic bonds. Appl. Microbiol. Biotechnol. 33: 511–518

Starnes RL, Hoffman CL, Flint VM, Trackamn PC, Duhart DJ and Katkocin DM (1991) Starch liquefaction with a highly thermostable cyclodextrin glycosyl transferase from *Thermoanaerobacter* species. In: GF Leatham and ME Himmel (eds) Enzymes in Biomass Conversion (pp 384–393). ACS Symposium Series 460. American Chemical Society, Washington, DC

Suzuki Y (1989) A general principle of increasing protein thermostability. Proc. Japan. Acad. 65(B): 146–148

Suzuki Y, Oishi K, Nakano H and Nagayama T (1987) A strong correlation between the increase in number of proline residues and the rise in thermostability of five *Bacillus* oligo-1,6 glucosidases. Appl. Microbiol. Biotechnol. 26: 546–551

Suzuki Y, Ito N, Yuuki T, Yamgat H and Udaka S (1989) Amino acid residues stabilizing a *Bacillus* α-amylase against irreversible thermoinactivation. J. Biol. Chem. 264: 18933–18938

Suzuki Y, Hatagaki K and Oda H (1991) A hyperthermostable pullulanase produced by an extreme thermophile, *Bacillus flavocaldarius* KP 1228 and evidence for proline theory of increasing protein thermostability. Appl. Microbiol. Biotechnol. 34: 707–714

Tada S, Imura Y, Gomi K, Takahashi K, Hara S and Yoshizawa K (1989) Cloning and nucleotide sequencing of the genomic Taka amylase A gene of *Aspergillus oryzae*. Agri. Biol. Chem. 53: 593–599

Tada S, Gomi K, Kitamoto K, Kitagai K, Tamura G and Hara S (1991) Identification of the promoter region of the Taka amylase A gene required for starch induction. Agri. Biol. Chem. 55: 1939–1945

Taji N (1988) Industrial utilization α-amylase. In: The Amylase Research Society of Japan (ed) Handbook of Amylases and Related Enzymes: Their Sources, Isolation Methods, Properties and Applications (pp 196–198). Pergamon Press, Oxford

Takano T, Fukuda M, Monma M, Kobayashi S, Kainuma K and Yamane K (1986) Molecular cloning, DNA nucleotide sequencing and expression in *Bacillus subtilis* cells of the *Bacillus macerans* cyclodextrin glucanotransferase gene. J. Bacteriol. 166: 1118–1122

Takano J, Kinoshita T and Yamane K (1987) Modulation of *Bacillus subtilis* α-amylase promoter activity by the presence of a palindromic sequence in front of the gene. Biochem. Biophys. Res. Comm. 146: 73–79

Takasaki Y (1989) Novel maltose producing amylase from *Bacillus megaterium* G-2. Agric. Biol. Chem. 53: 341–347

Takasaki Y, Katajima M, Tsuruta T, Nonoguchi M, Hayashi S and Imada K (1991a) Maltotriose-producing amylase from *Microbacterium imperiale*. Agric. Biol. Chem. 55: 687–692

Takasaki Y, Shinohara H, Tsuruhisa M, Hayashi S and Imada K (1991b) Maltotetraose-producing amylase from *Bacillus* sp. MG-4. Agric. Biol. Chem. 55: 1715–1720

Takekawa S, Uozumi N, Tsukagoshi N and Udaka S (1991) Proteases involved in generation of β- and α-amylases from a large amylase precursor in *Bacillus polymyxa*. J. Bacteriol. 173: 6820–6825

Tegge G, Richter G, and Richter G (1986) Optimization of the production of maltose syrups by different enzyme combinations. Starch/Starke 38: 61–67

Thudt K, Schleiffer KH and Gotz F (1985) Cloning and expression of α-amylase gene from *Bacillus stearothermophilus* in several staphylococcal species. Gene 37: 163–169

Tokizawa N, Shiro H, Hatta T, Nagao K and Kiyohara H (1991) Extracellular production of *Klebsiella* pullulanase by *Escherichia coli* that carries the pullulanse secretion genes. Agri. Biol. Chem. 55: 1467–1473

Tomazic SJ and Klibanov AM (1988a) Mechanism of irreversible thermal inactivation of *Bacillus* α-amylases. J. Biol. Chem. 263: 3086–3091

Tomazic SJ and Klibanov AM (1988b) Why is one α-amylase more resistant against irreversible thermoinactivation than another? J. Biol. Chem. 228: 3092–3096

Tsukagoshi N, Iritani S, Sasaki T, Takemura T, Ihara H, Idota Y, Yamagata H and Udaka S (1985) Efficient synthesis and secretion of a thermophilic α-amylase by protein producing *Bacillus brevis* 47 carrying the *Bacillus stearothermophilic* amylase gene. J. Bacteriol. 164: 1182–1187

Tsukamoto A, Kimura K, Ishii Y, Takano T and Yamane K (1988) Nucleotide sequence of the maltohexaose-producing amylase gene from an alkalophilic *Bacillus* sp. 707 and structural similarity to liquefying type α-amylases. Biochem. Biophys. Res. Comm. 151: 25–31

Tsukagoshi N, Furukawa M, Nagaba H, Kirita N, Tsuboi A and Udaka S (1989) Isolation of a cDNA encoding *Aspergillus oryzae* Taka-amylase A: evidence for multiple related genes. Gene. 84: 319–327

Underkofler LA (1954) Fungal amylolytic enzymes. In: LA Underkofler and RJ Hickey (eds) Industrial Fermentations 2 (pp 97–121). Chemical Publishing, New York

Uozumi N, Sakurai K, Saaki T, Takekawa S, Yamagata H, Tsukagoshi N and Udaka S (1989) A single gene directs the synthesis of a precursor protein with β and α-amylase activities in *Bacillus polymyxa*. J. Bacteriol. 171: 375–382

Vihinen M and Mantsala P (1989) Microbial amylolytic enzymes. Crit. Rev. Biochem. Mol. Biol. 24: 329–418

Wako K, Takahashi C, Hashimoto S and Kanaeda J (1978) Studies on maltotriose and maltose-forming amylases from *Streptomyces*. J. Jpn. Soc. Starch Sci. 25: 155–160

Watanabe K, Kitamura K, Iha H and Suzuki Y (1990) Primary structure of the oligo-1,6 glucosidase of *Bacillus cereus* ATCC 7064 deduced from the nucleotide sequence of the cloned gene. Eur. J. Biochem. 192: 609–620

Welker NE and Campbell LL (1963) Induction of α-amylase of *Bacillus stearothermophilus* by maltodextrins. J. Bacteriol. 86: 687–691

Whistler RL and Daniel JR (1984) Molecular structure of starch. In: RL Whistler, JN Bemiller and EP Paschall (eds) Starch: Chemistry and Technology (pp 153–182). Academic Press, Inc., New York

Yamane K, Kirata Y, Furusato T, Yamazaki H and Nakayama A (1984) Changes in the properties and molecular weights of *Bacillus subtilis* M-type and N-type α-amylase resulting from a spontaneous deletion. J. Biochem. 96: 1849–1858

Yamane K, Nakamura A, Kimura K, Takano T and Kobayashi S (1990) Starch degrading enzymes of *Bacilli*: cloning and analyses of the gene and gene products. In: H Heslot, J Davies, J Florent, L Bobichon, G Durand and L Penasse (eds) 6th International Symposium on Genetics of Industrial Microorganisms. Proceedings Vol II (pp 923–934). Strasbourg, France

Yamashita I, Nakamura M and Fukui S (1987) Gene fusion is a possible mechanism underlying the evolution of STA1. J. Bacteriol. 169: 2142–2149

Yoshida A, Iwasaki Y, Akiba T and Horikoshi B (1991) Purification and properties of cyclomaltodextrinase from alkalophilic *Bacillus* sp. J. Ferment. Bioeng. 71: 226–229

Yuuki T, Nomura T, Tezuko H, Tsuboi A, Yamagata M, Tsukagoshi N and Udaka S (1985) Complete nucleotide sequence of a gene coding for heat and pH-stable α-amylase of *Bacillus licheniformis*: comparison of the amino acid sequences of three bacterial liquefying α-amylases deduced from DNA sequences. J. Biochem. 98: 1147–1156

Zeikus JG, Lee C, Lee Y-E, and Saha BC (1991) Thermostable saccharidases: New sources, uses and biodesigns In: GF Leatham and ME Himmel (eds) ACS Symposium Series 460, Enzymes in Biomass Conversion (pp 36–51). American Chemical Society, Washington, DC

Zhou J, Baba T, Takano T, Kobayashi S and Arai Y (1989) Nucleotide sequence of the maltotetraohydrolase gene from *Pseudomonas saccharophila*. FEBS Lett. 255: 37–41

11. The physiology of aromatic hydrocarbon degrading bacteria

MARK R. SMITH

Division of Industrial Microbiology, Agricultural University, P.O. Box 8129, 6700 EV Wageningen, The Netherlands (present address: Department of Microbiology, NIZO, P.O. Box 20, 6710 BA Ede, The Netherlands)

I. Introduction

Aromatic hydrocarbons are ubiquitous in nature. Indeed, next to glucosyl residues, the benzene ring is the most widely distributed unit of chemical structure in nature (Dagley 1981). The amount and variety of aromatic hydrocarbons generated by commercial and industrial activities has continued to increase over the years. The occurrence of such compounds is a cause for great concern due to their potential hazard to the well being of both plants and animals. Benzene, toluene, ethylbenzene, styrene and the xylenes are among the 50 largest-volume industrial chemicals produced, with production figures of the order of millions of tonnnes per year. These compounds are widely used as fuels and industrial solvents. In addition, they and the polynuclear aromatic compounds provide the starting materials for the production of pharmaceuticals, agrochemicals, polymers, explosives and many other everyday products (Smith 1990). The use of man-made aromatic hydrocarbons has inevitably led to their release (either accidental or otherwise) into the environment and this problem is still escalating in spite of governmental intervention.

The biodegradation of aromatic compounds can be considered, on the one hand as part of the normal process of the carbon cycle and as the removal of man-made pollutants from the environment, on the other. Over the last decades this topic has been extensively reviewed (Hopper 1978; Cripps and Watkinson 1978; Gibson and Subramanian 1984; Cerniglia 1984; Dagley 1986; Smith 1990, as examples). The majority of the recent advances made have consolidated previous findings, tying up some of the loose ends.

The purpose of this current review is to examine the present status of our understanding of the physiology of aromatic hydrocarbon degradation, using examples from a selected range of key compounds (benzene, alkylbenzenes, alkenylbenzenes, tetralin, biphenyl, naphthalene and polycyclic aromatic hydrocarbons – see Fig. 1).

Some of the salient physical and chemical properties of certain aromatic hydrocarbons are given in Table 1.

C. Ratledge (ed.), Biochemistry of Microbial Degradation, 347–378.

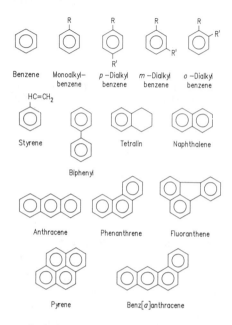

Fig. 1. Structures of principal aromatic hydrocarbons discussed in this chapter.

Table 1. Some physical properties of selected aromatic hydrocarbons.

Compound	C atoms	Mol.wt.	m.p. (°C)	b.p. (°C)	Solubility in in water (mg/L)
Benzene	6	78.1	5.5	80.1	1790.0
Toluene	7	92.1	−95.1	11.0	579.0
Ethylbenzene	8	106.2	−95.0	136.0	135.0
p-Xylene	8	106.2	13.0	138.0	221.0
m-Xylene	8	106.2	−50.0	139.0	260.0
o-Xylene	8	106.2	−25.0	144.0	215.0
iso-Propyl-benzene	9	120.2	−96.0	152.0	48.3
n-Propyl-benzene	9	120.2	−100.0	159.0	52.2
t-Butylbenzene	10	134.2	−58.0	169.0	29.5
iso-Butyl-benzene	10	134.2	−51.6	173.0	10.1
n-Butylbenzene	10	134.2	−81.0	183.0	13.8
1,2,3-Tri-methylbenzene	9	120.2	−15.0	176.0	48.2
n-Pentyl-benzene	11	148.3	–	295.0	10.5
Naphthalene	10	128.2	80.2	218.0	30.6
Tetralin	10	132.2	−35.0	207.0	15.2
Biphenyl	12	154.2	70.0	256.0	7.0
Acenaphthalene	12	154.2	96.0	278.0	3.9
Phenanthrene	14	178.2	100.0	339.0	1.2
Anthracene	14	178.2	217.0	340.0	0.7

Data from Eastcott et al. (1988)

Attention in this review will focus on both the modes of degradation and the effects that the aromatic hydrocarbons have on the physiology of the bacteria. The uptake of aromatic compounds, the mechanisms of toxicity, the consequences of mixed substrates and the possibilities of anaerobic degradation of aromatic hydrocarbons, are all relatively new aspects that will be discussed here, though the latter area in only briefly discussed as much greater detail is given in the chapter by Fuchs et al. (Chapter 16). Although many recent advances have centered on the genetics of these microorganisms this aspect will not be covered here.

The most important criterion for inclusion in this review is that the aromatic hydrocarbon undergoes defined ring-cleavage. Here I have been somewhat lenient in some cases, where ring-cleavage is only implied, although essential, for the observed growth of the bacteria. Compounds such as long chain alkylbenzenes have not been included as often the alkyl moiety itself is sufficient to support growth.

Bacterial growth on water-insoluble aromatic hydrocarbons is often accompanished by physiological alterations and many such events will be dealt with here. It is however impossible to cover all aspects of this multidicipline subject and what follows is in many ways a personal view of the recent advances. One key area not discussed here is the formation and role of surfactants produced by the bacteria to solubilize the hydrocarbons, this topic is covered in depth elsewhere in this volume (Hommel, Chapter 3).

II. Benzene

There have been few reports on the bacterial biodegradation of benzene in the literature over the last five years. The excellent studies carried out in the previous three decades (Marr and Stone 1961; Gibson et al. 1968, 1970; Högn and Jaenicke 1972; Axell and Geary 1975) elucidated the pathways involved, identified the intermediates and characterized the enzyme systems. The two divergent pathways employed both share the same initial mode of attack resulting in the formation of catechol which is further catabolized by either catechol 1,2-dioxygenase (the so called *ortho-* or intradiol-cleavage) and subsequently via the β-ketoadipate pathway or catechol 2,3-dioxygenase (the so called *meta-* or extradiol-cleavage). Both routes have been described in different benzene utilizing strains. Figure 2 illustrates these two key enzyme ring cleavage steps.

Although there have been reports of the isolation of new bacterial strains which can grow on benzene (Shirai 1986; van den Tweel et al. 1988; Winstanley et al. 1987, as examples) the biodegradation routes were, not surprisingly, the same as those outlined above.

Growth of *Pseudomonas* sp. 50 on benzene dissolved in an organic solvent has been investigated (Rezessy-Szabó et al. 1987). The organic phase (dibutyl phthalate) was not attacked by the bacterium but did serve as a reservoir for the

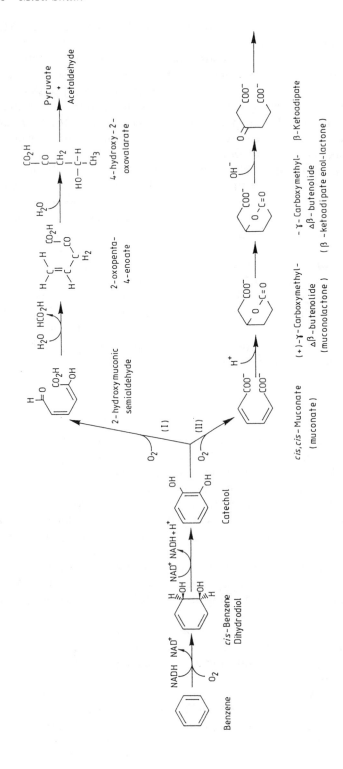

Fig. 2. The biodegradative routes of benzene. (I) *meta*-cleavage route. (II) *ortho*-cleavage route. (Smith 1990.)

poorly water-soluble, cytotoxic benzene. Thus it was possible to cultivate *Pseudomonas* sp. 50 at much higher benzene concentrations than in the absence of solvent. These authors also claimed growth of this bacterium in a chemostat with benzene supplied to the cells via dibutyl phthalate. The toxicity of aromatic hydrocarbons to the bacteria degrading them is discussed below.

Winstanley et al. (1987) described a new benzene-utilizing bacterium, *Acinectobacter calcoaceticus* RJE74, which carries a large plasmid (pWW174) encoding the enzymes for the catabolism of benzene via the β-ketoadipate pathway (that involving catechol 1,2-dioxygenase). This was the first report of the *ortho* pathway being plasmid encoded and was one of very few citations of plasmids in this genus.

III. Arenes

The introduction of a substituent group(s) onto the benzene ring opens up the possibility of alternative modes of biodegradation: either side-chain attack or ring attack. Indeed with the longer chain length alkylbenzenes the oxidation of the side chain is sufficient to support growth and the organisms may not be able to degrade the aromatic moiety. Such compounds may be regarded as substituted alkanes rather than substituted benzenes. For the purposes of this review I will only discuss those cases where the benzene ring undergoes ring-cleavage (either before or after side chain modifications).

A. Mono-alkylbenzenes

Toluene, the simplest of these substituted benzenes, has been shown to be biodegraded by both ring attack and methyl-group hydroxylation. The evidence for these routes is well established (Smith 1990).

Recently, a novel pathway of toluene catabolism has been postulated (Shields et al. 1989). Toluene was degraded by an unidentified bacterium strain (G4) via hydroxylation in the 2-position by a monooxygenase thus producing *o*-cresol. This same enzyme was also thought to be responsible for the further oxidation to give 3-methylcatechol which then served as the substrate for a *meta*-type ring cleavage enzyme. Kukor and Olsen (1990) also quote a fourth degradation route for toluene in which the molecule is also initially hydroxylated by a monooxygenase this time to *p*-cresol. This is further catabolized via sequential oxidation of the methyl group to give *p*-hydroxybenzoate. This is subsequently oxidized to protocatechuate which undergoes *ortho*-cleavage. This pathway has also been found in *Achromobacter xylosoxidans* by G.M. Whited (unpublished data).

When reviewing the literature concerning toluene over the last few years one finds an enormous wealth of references regarding plasmid-encoded biodegradation. Williams and his colleagues at the University of Wales (Bangor, UK) have shown that toluene (and *m*-, *p*-xylenes, etc.) is degraded via catechol and subsequently the *meta* pathway by several strains of *Pseudomonas* by enzymes encoded on plasmids, designated TOL (Bayly and Barbour 1984). These plasmids often contain two catabolic operons (Nakazawa et al. 1980; Franklin et al. 1981). The "upper" pathway operon encodes enzymes for the successive oxidation of the hydrocarbons to the corresponding alcohol, aldehyde and carboxylic acid derivatives. The "lower", or *meta*-cleavage pathway operon encodes enzymes for the conversion of the carboxylic acids to catechols, whose aromatic rings are then cleaved˙(*meta*-fission) to produce corresponding semialdehydes, which are then further catabolized through the TCA cycle (Ramos et al. 1987). Figure 3 outlines the TOL plasmid encoded pathway.

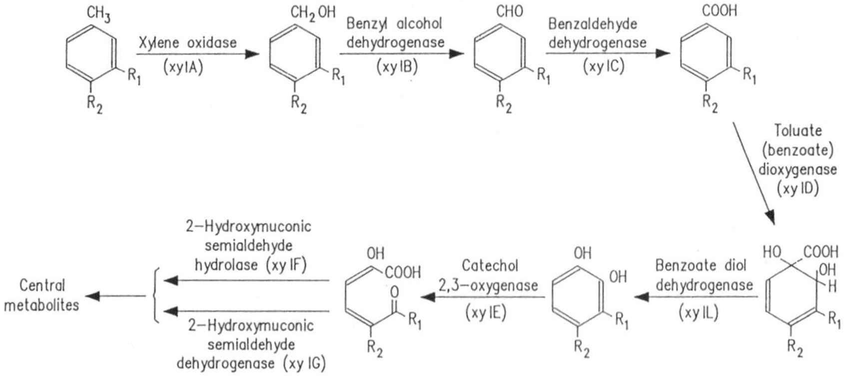

Fig. 3. The early enzymes of the TOL plasmid degradative pathway. The primary metabolites of the pathway are: toluene ($R_1 = R_2 = H$), *m*-xylene ($R_1 = CH_3$, $R_2 = H$), *p*-xylene ($R_1 = H$, $R_2 = CH_3$) and 1,2,4-trimethylbenzene ($R_1 = CH_3$, $R_2 = CH_3$). Adapted from Keil and Williams (1985).

To cover the huge amount of data published in this area would require more space than is available in such a review and therefore the reader is advised to consult an excellent review by Assinder and Williams (1990).

An interesting recent study in this field has studied the DNA sequence of the (pWWO) *xyl GFJ* genes and has proposed their involvement in the evolution of aromatic catabolism (Horn et al 1991). The *xylG* gene (specifying 2-hydroxymuconic semialdehyde dehydrogenase) encodes for a 51.7 kDa protein with significant homology to (eukaryotic) aldehyde dehydrogenases. The *xylG* (specifying 2-hydroxymuconic semialdehyde hydrolase) was delineated by *N*-terminal sequence analysis of the purified gene product and is shown to encode a protein of 30.6 kDa. Homology analysis revealed sequence similar to serine hydrolase.

The physiological stability of the TOL plasmid has also been investigated. Early studies showed that growth of TOL-plasmid harbouring strains on benzoate caused them to lose their toluene (and hence xylene) degrading ability (see Assinder and Williams 1990). This was shown to be caused by the loss of 39 kb region rather than the total plasmid (Williams et al. 1988). Additionally, Duetz and van Andel (1991) have studied TOL plasmid (pWWO) stability under non-selective conditions in continuous culture. Under succinate-, sulphate-, ammonium-, or phosphate- limitations, the TOL$^-$ cells were also shown to be as a result of partially deleted plasmids, lacking the catabolic genes. The TOL$^-$ cells had a growth advantage (under non-selective conditions) over the TOL$^+$ cells and this was most likely caused by the so called "metabolic burden".

A thermotolerant *Bacillus* sp. that grew on toluene at 50 °C was isolated by Simpson et al. (1987). Biodegradation (Simpson et al. 1987) was via *cis*-toluene dihydrodiol, 3-methylcatechol and *meta*-cleavage. The *cis*-toluene dihydrodiol dehydrogenase from this organism was purified and shown to possess different properties to those previously reported, most notable was a temperature optimum of 80 °C (Simpson et al. 1987).

Vecht et al. (1988) recently reported on the growth of *Pseudomonas putida* in chemostat cultures with toluene as the sole source of carbon and energy (after initial growth on *m*-toluic acid). Steady states were maintained for several months with maximal biomass concentrations of 3.2 g cell dry weight/l. The maximum specific growth rate was 0.13 h^{-1}, with a cellular yield of 1.05 g cell dry weight/g toluene utilized.

The kinetics of concentration-dependent toluene metabolism in various pseudomonads were examined by Robertson and Button (1987). The metabolism kinetics were characterized by very small Michaelis constants (ranging from approximately 0.005 μM to 0.5 μM). Induction of cells had little effect on these values but did cause a marked increase in the specific affinity (0.03 to 320 l/g cells/h). The observed cell yields (0.10 to 0.17 g cells/g toluene) were low due to occurrence of extracellular intermediates (dihydrodiol, methylcatechol, *meta* ring fission product, acetate, formate and pyruvate).

The anaerobic biodegradation of toluene has recently been reported by independent groups. Many of these reports have isolated denitrifying bacteria (mostly pseudomonads) using nitrate as the terminal electron acceptor (Zeyer et al. 1990). In addition, Lovley and Lonergan (1990) reported on the isolation of an unidentified bacterial strain (GS-15) which was demonstrated to couple the oxidation of aromatic compounds (including toluene) to the reduction of Fe(III). The biochemistry of anaerobic hydrocarbon degradation is dealt with in great depth elsewhere in this volume (Fuchs et al., Chapter 16).

Inoue and Horikoshi (1989) have isolated a variant strain of *Pseudomonas putida* which is able to grow in the presence of up to 50% (v/v) toluene (and high concentrations of other aromatic hydrocarbons) in the medium, although the organism could not utilize aromatic hydrocarbons. Most organisms that can grow on toluene can only tolerate concentrations of around 0.3% (v/v), indeed, toluene is commonly used to lyse microbial cells. Their studies indicated that the newly isolated strain possessed a unique cell surface and went on to isolate a 10 kb fragment from the chromosome which coded for solvent (toluene etc.) tolerance. The subject of solvent tolerance and toxicity is further discussed later in this review.

The initial oxidation and subsequent ring-cleavage of the aromatic moiety of alkylbenzenes, without prior attack of the alkyl side chain, has been demonstrated for many alkylbenzenes of chain length C_2–C_7 (including branched chain species – Smith 1990).

We recently re-examined the biodegradation of alkylbenzenes in an attempt to investigate what chain length a single strain could tolerate before the compound became, in effect, a substituted alkane on which the organism would not grow (Smith and Ratledge 1989a). *Pseudomonas* sp. NCIB 10643 grew on a range of n-alkylbenzenes (C_2 to C_7) and on several branched species within this chain size (isopropylbenzene, isobutylbenzene, *sec*-butylbenzene, *tert*-butylbenzene and *tert*-amylbenzene). The organism could not grow on n-alkanes of any chain length. All of the alkylbenzenes were catabolized via ring attack, rather than side chain attack, proceeding via initial dioxygenase activity resulting in the corresponding 2,3-dihydro-2,3-dihydroxyalkylbenzene which underwent reduction to the corresponding 2,3-dihydroxyl-intermediate (3-alkylsubstituted catechols). The 3-substituted catechols were ring-cleaved by an extra-diol type enzyme between C_1 and C_2 resulting in characteristic *meta* ring-fission products. Further catabolism was by hydrolytic attack to give alkyl-chain dependent carboxylic acids and, presumably, 2-oxopenta-4-enoate.

A general pathway for the complete catabolism of these mono-alkylbenzenes is given in Fig. 4.

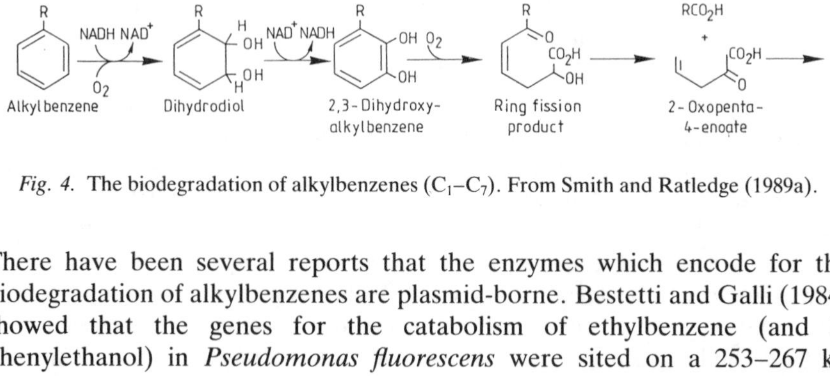

Fig. 4. The biodegradation of alkylbenzenes (C_1–C_7). From Smith and Ratledge (1989a).

There have been several reports that the enzymes which encode for the biodegradation of alkylbenzenes are plasmid-borne. Bestetti and Galli (1984) showed that the genes for the catabolism of ethylbenzene (and 1-phenylethanol) in *Pseudomonas fluorescens* were sited on a 253–267 kb plasmid. Eaton and Timmis (1986) have demonstrated that the catabolism of isopropylbenzene by *Pseudomonas putida* RE 204 is also plasmid-encoded. The pathway was shown to be identical to that outlined above (see Fig. 4). Our own studies with *Pseudomonas* sp. NCIB 10643 suggested that the genes of alkylbenzene (and biphenyl) biodegradation were chromosomal (Smith and Ratledge 1989b).

When the alkyl chain length exceeds C_7 the preferred route seems to be by attack on the alkyl chain but this will only occur in those organisms possessing an α-alkane hydroxylase (see Chapter 1).

B. Di-alkylbenzenes

Until recently only the *m*- and *p*- isomers of xylene had been shown to be biodegraded by bacteria. Both compounds are biodegraded by certain strains of *Pseudomonas* (notably those containing the TOL plasmid – see above) by initial oxidation of one of the methyl groups to the corresponding methyl benzylalcohols, tolualdehydes, toluic acids and methyl catechol (Davey and Gibson 1974; Davis et al. 1968). The biodegradation of *m*- and *p*-xylenes to their corresponding methylcatechols is shown in Fig. 5. The resultant catechols then undergo *meta*-cleavage. The ring-fission products of the two different methyl catechols (3-methylcatechol from *m*-xylene; 4-methylcatechol from *p*-xylene) are catabolized by different enzyme systems to 2-oxopenta-4-enoate (Duggleby and Williams 1986). The product from 3-methylcatechol cleavage is further biodegraded by a single hydrolase type enzyme (Duggleby and Williams 1986; Smith and Ratledge 1989a), whereas the product from 4-methylcatechol, an aldehyde, is converted via the enzymes of the 4-oxocrotonate branch (Sala-Trepat et al. 1972; Wigmore et al. 1974). These pathways are illustrated in Fig. 6.

An alternative mode of attack of *p*- and *m*-xylenes proceeds via direct dioxygenase attack of the aromatic moiety. This was shown to be via the corresponding *cis*-dihydrodiol and subsequent conversion to substituted catechols (3,6-dimethylcatechol from *p*-xylene; 3,5-dimethylcatechol from *m*-xylene) by dehydrogenase type enzymes (Gibson et al. 1974). However, although this is often cited as an alternative pathway for the degradation of xylenes (Baggi et al. 1987, for example) the resultant catechols are not further degraded and this should be regarded as a biotransformation reaction.

Members of the genus *Nocardia* have been demonstrated to co-metabolize all three of the isomers of xylene (see Gibson and Subramanian 1984 for review). It is noteworthy that *p*- and *m*- xylenes were co-metabolized (when hexadecane was used as the principal substrate) via *ortho* cleavage whereas *o*-xylene was attacked by *meta*-cleavage.

The first reports of the complete biodegradation of *o*-xylene as sole source of carbon and energy by pure cultures were provided by Baggi et al. (1987) and Schraa et al. (1987). A strain of *Pseudomonas stutzeri* was isolated from activated sludge of a waste water treatment plant. Initial studies suggested that *o*-xylene was catabolized via 3,4-dimethylcatechol and subsequent *meta*-cleavage (Baggi et al. 1987). Independently, Schraa et al. (1987) reported on the characterization of *Corynebacterium* strain C125 able to grow on *o*-xylene as the sole source of carbon and energy. The proposed pathway was the same as that suggested by Baggi et al. (1987) and confirmed 2-dihydroxy-5-methyl-6-oxo-2,4-heptadienoate as the ring-fission product. The pathway is illustrated in Fig. 7.

Most recently a report has appeared which showed all isomers of xylene to be degraded by a single (unidentified) bacterium, F199 (Fredrickson et al. 1991). The organism was also capable of growth on a very wide range of other

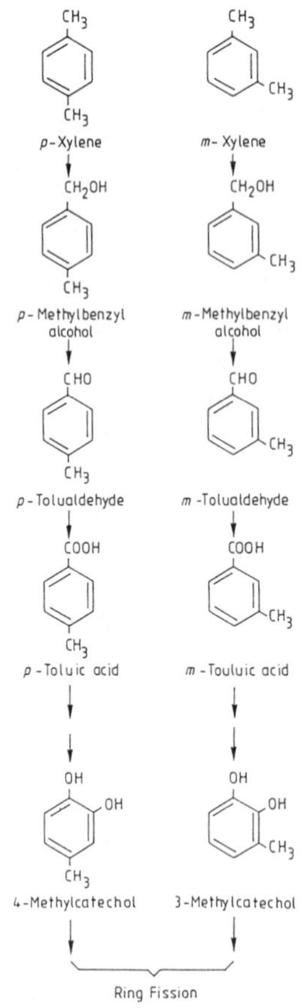

Fig. 5. The initial reactions in the biodegradation of *m*- and *p*-xylenes (Smith 1990).

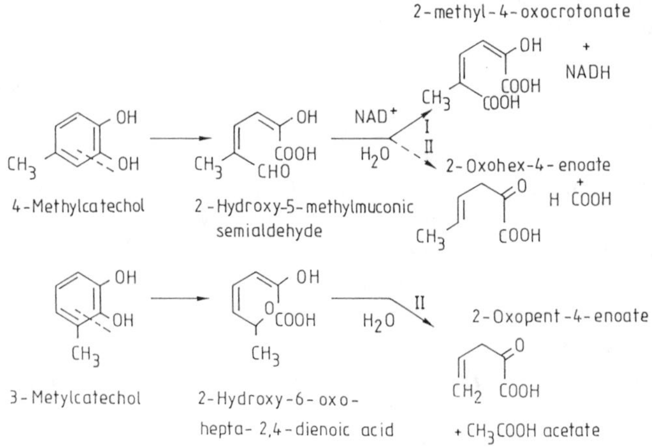

Fig. 6. The alternative pathways for the catabolism of 4- and 3-methylcatechols. (I) 4-Oxocrotonate branch. (II) Hydrolytic branch. (Smith 1990.)

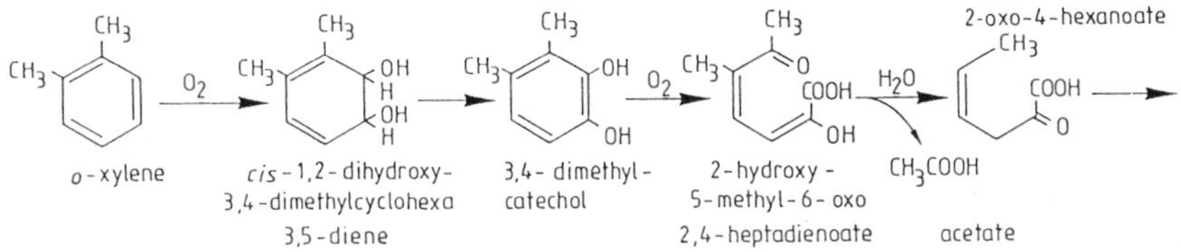

Fig. 7. The biodegradative pathway of *o*-xylene. Adapted from Schraa et al. (1987) and Baggi et al. (1983).

aromatic compounds (toluene, naphthalene, dibenzothiophene, salicylate, benzoate, *p*-cresol), however, to date no pathways have been elucidated.

With regard to other dialkylsubstituted benzenes, there have been no recent advances; our knowledge is limited to the biodegradation routes of *p*-cymene (DeFrank and Ribbons 1976; DeFrank and Ribbons 1977a,b) and 3-ethyltoluene (Jigami et al. 1974; Kunz and Chapman 1981a). In both these cases, attack begins on the smaller substituent of the ring (i.e. the methyl group), however, there is too little evidence available to judge whether this is always the case. The *o*-xylene degrading strain isolated by Schraa et al. (1987) (*Corynebacterium* strain C125) degrades 1,2-diethylbenzene via *meta*-ring cleavage (J. Sikkema personal communication), although the exact nature of the degradation remains to be clarified.

C. Tri-alkylbenzenes

Evidence of the degradation of multiple substituted benzenes is limited. Kunz and Chapman (1981a) were the first to recognize the versatility of the function of the TOL plasmid demonstrating the catabolism of 1,2,4-trimethylbenzene (pseudocumene) by *Pseudomonas putida* mt-2 (pWWO baring strain). Catabolism proceeds through the formation of 3,4-dimethylcatechol which is ring cleaved via extra-diol oxidation to the corresponding characteristic semialdehyde (Fig. 3). This same pathway was also confirmed to be present in *Pseudomonas putida* strain HS1 (Kunz and Chapman 1981b), also to be plasmid encoded (TOL, pDK1). A strain of *Pseudomonas putida* (TMB) was subsequently isolated by Bestetti and Galli (1987) and shown to resemble *Pseudomonas putida* mt-2 in its ability to degrade 1,2,4-trimethylbenzene (toluene, *m*-xylene, and *p*-xylene). However, these authors suggested a different regulatory model. Their strain was selected using trimethylbenzene as the sole source of carbon and energy and carried an 85 kb plasmid (pGB). This plasmid was initially thought to encode the enzymes for the degradation of 1,2,4-trimethylbenzene (Bestetti and Galli 1987) but this was later refuted by the same group (Polissi et al. 1990), demonstrating the enzymes to be on the chromosome. The strain has one regulatory gene (induced by the hydrocarbon, alcohol and carboxylic acid) unlike the TOL-pWWO plasmid which possess two (see above). These chromosomal genes of *Pseudomonas putida* strain TMB showed homology with the pWWO *xyl* operons.

D. Alkenylbenzenes

Reports on the biodegradation of alkenylbenzenes are relatively scarce. This is somewhat surprising when one considers the enormous quantities styrene (the simplest member of this series) produced by the petrochemical industry (3.6 million tonnes in the United States in 1987 – Hartmans et al. 1989). There have however been reports of bacteria able to grow on styrene (Baggi et al. 1983; Shirai and Hisatsuka 1979; van den Tweel et al. 1986) and methylstyrenes (Omori et al. 1974; Dzhusupova et al. 1985) as sole sources of carbon and energy and although most have concentrated on the initial reactions there has been one study showing the degradation beyond ring-cleavage (Baggi et al. 1983). Figure 8 shows the various modes of initial attack and includes the pathway postulated for a pseudomonad (Baggi et al. 1983) involving the formation of phenylacetic acid (by an unspecified route) followed by two mono-oxygenase steps to 2,5-dihydroxyphenylacetate and subsequent *meta*-cleavage (2,5-dihydroxyphenylacetate 1,2-dioxygenase) resulting in maleylacetoacetic acid. The ring-cleavage product was further catabolized via acetoacetate and fumarate. This pathway was later shown to be inducible and encoded on a 37 kb plasmid (Bestetti et al. 1984).

The initial degradation of styrene by *Xanthobacter* strain 124X (a strain isolated by enrichment on styrene by van den Tweel et al. 1986) was recently reported by Hartmans et al. (1989). Although styrene oxide and 2-phenylethanol have been implicated in the degradation by other bacteria (Shirai and Hisatsuka 1979), and were shown to be oxidized by cells grown on

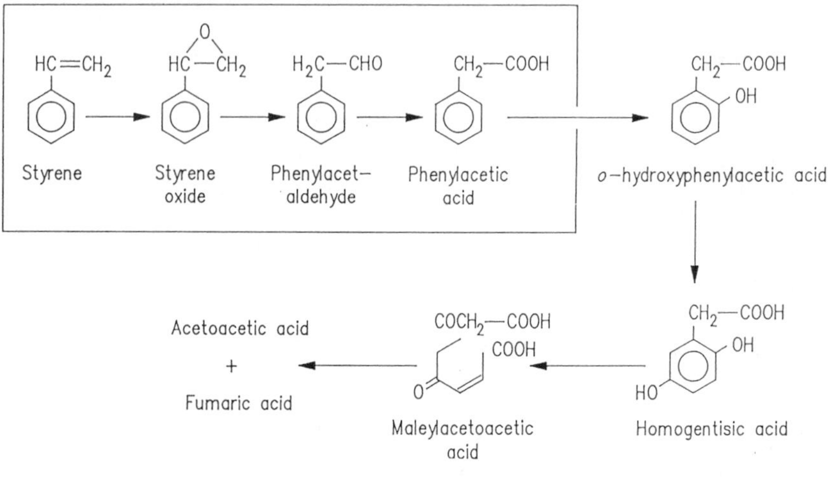

Fig. 8. The biodegradation of styrene. The initial reactions (to phenylacetate) have been confirmed by an unidentified Gram-positive strain (Hartmans et al. 1990); conversion of styrene to phenylacetate and further catabolism to central metabolites has been demonstrated for a pseudomonad (Baggi et al. 1983).

styrene, these authors concluded that the initial step in styrene metabolism is O_2-dependent and probably involves oxidation of the aromatic nucleus. Previously, this organism was also shown to grow on ethylbenzene and toluene (van den Tweel et al. 1986).

Subsequently, this same group (Hartmans et al. 1990) isolated 14 strains of bacteria able to grow on styrene as the sole source of carbon and energy. One of these was studied in more detail (S5) and the initial reactions characterized. Styrene was converted to styrene oxide by a novel flavin mono-oxygenase. Further biodegradation proceeded via phenylacetaldehyde and phenylacetic acid. The conversion of phenylacetic acid to central metabolites was not investigated but one may postulate that this is further metabolized as described for *Pseudomonas fluorescens* by Baggi et al. (1983). This is supported by the observation that both strains grow on 2- and 3-hydroxyphenylacetates.

There have been several preliminary reports of the involvement of various *meta*-cleavage of alkenylbenzenes. Sielicki et al. (1978) observed the development of yellow culture fluids during stationary phase of growth of a mixed culture with styrene as the sole source of carbon and energy (the appearance of yellow culture fluids often occurs during the degradation of aromatic compounds and is caused by the accumulation *meta*-cleavage products). This observation was not persued further. Hartmans et al. (1989) observed a transient accumulation of a yellow product during growth of *Xanthobacter* strain 124X on styrene and 1-phenylethanol. They reported that the compound had different spectral properties to those reported for the ring-cleavage product of the catechol of 1-phenylethanol (2,7-dihydroxy-6-oxoocta-2,4-dienoate) (Cripps et al. 1978), but did not identify the product. In contrast, Dzhusupova et al. (1985) showed induced levels of protocatechuate 3,4-dioxygenase (an intra-diol cleaving enzyme) in strains of *Pseudomonas* grown on α-methylstyrene, suggesting a novel pathway, sadly without explanation.

The biodegradation of other members of this class of aromatic hydrocarbons has received very little attention. The biotransformation of styrenes by a strain of *Pseudomonas putida* capable of growth on α-methylstyrene has been recorded (Bestetti et al. 1989). The growth substrate was catabolized via 2-phenyl-2-propen-1-ol and 1,2-dihydroxy-3-isopropenyl-3-cyclohexene, implying a pathway different to that previously reported [via (−)-*cis*-2,3-dihydroxy-1-isopropenyl-6-cyclohexene (Omori et al. 1974)]. Furthermore, the strain was also able to biotransform styrene to 1,2-dihydroxy-3-ethenyl-3-cyclohexene.

Clearly from the foregoing discussion alkenylbenzenes are degraded by bacteria and it seems certain, from the evidence available, that *meta* ring-cleavage is the preferred route of attack on the aromatic ring.

IV. Biphenyl

Biphenyl is a unique chemical compound which may be considered as substituted benzene even though the substituent is in fact benzene itself; biphenyl is not a polycyclic aromatic hydrocarbon.

Biphenyl is a toxicologically important compound since it is used as a fungistat in the storage and transport of certain fruits (Cerniglia 1981). It is also the parent hydrocarbon of polychlorinated biphenyls (PCBs),[1] which are established as worldwide pollutants. The catabolism of biphenyl has therefore received much attention recently due to the increasing concern over the fate of such compounds.

The pathway for the biodegradation of biphenyl is given in Fig. 9. Initial attack of biphenyl proceeds via 2,3-dihydro-2,3-dihydroxybiphenyl (Catelani et al. 1971; Catelani et al. 1973; Gibson et al. 1973) and 2,3-dihydroxybiphenyl (Catelani et al. 1973; Smith and Ratledge 1989b). The catechol-type compound is then ring cleaved between carbon atoms 1 and 2 to form 2-hydroxy-6-oxo-6-phenylhexa-2,4-dienoate (Catelani et al. 1973; Catelani and Colombi 1974; Ishigooka et al. 1986; Smith and Ratledge 1989b). Further catabolism of this ring fission-product is via 2-oxo-penta-4-enoate and benzoate (Smith and Ratledge 1989b). This pathway is common to many species of bacteria.

Contemporary investigators have revealed some subtle differences between different bacteria. For example, we found that benzoate was a dead-end product in the biodegradation by *Pseudomonas* sp. NCIB 10643, whereas *Nocardia* sp. NCIB 10503 catabolized the benzoate via oxidative-decarboxylation to catechol which was subsequently degraded via the β-ketoadipate pathway (Smith and Ratledge 1989b). Omori et al. (1986) demonstrated the NADPH dependent reduction of the ring-cleavage product to 2,6-dioxo-6-phenylhexanoic acid in biphenyl-grown cells of *Pseudomonas cruciviae*, although no physiological significance of this enzyme step was postulated. 3-Hydroxybenzoate and cinnamic acid were also reported as intermediates in the plasmid encoded catabolism of biphenyl by a strain of *Pseudomonas putida* (Starovoitov et al. 1986). No scheme of dissimilation was proposed to account for the production of cinnamic acid. Furukawa and Suzuki (1988) transferred the plasmid encoding for the degradation of biphenyl in *Pseudomonas pseudoalcaligenes* into *Pseudomonas aeruginosa* and subsequently detected 2,3,2′,3′-tetrahydroxybiphenyl, suggesting that the initial two enzymes of biphenyl degradation had a broader substrate specificity than hitherto considered. Khan and Walia (1990) have recently cloned the genes encoding for two of the key enzymes of biphenyl biodegradation (3-phenylcatechol dioxygenase and 2-hydroxy-6-oxo-6-phenylhexa-2,4-dienoate hydrolase) from *Pseudomonas putida* into *Escherichia coli* succesfully expressing the transferred genes.

Our own investigation with *Nocardia* sp. NCIB 10503 (Smith and Ratledge

---

[1]Biphenyl forms a significant proportion of commercial PCB mixtures.

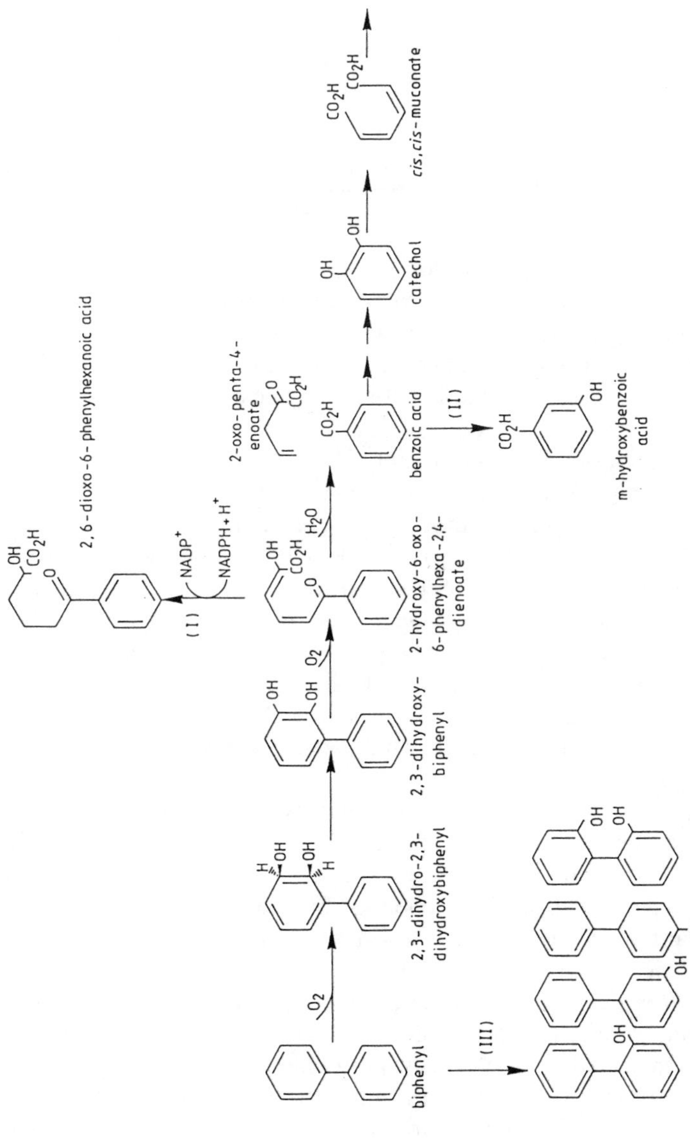

Fig. 9. The major pathway for the biodegradation of biphenyl and some alternative side reactions.

1989b) was the first report of the complete degradation of biphenyl by an actinomycete. Previously (Schwartz 1981), growth has been reported, but the only intermediates identified were mono-hydroxybiphenyls and 2,2'-dihydroxybiphenyl (Fig. 9). It remains to be seen if other members of the actinomycetes biodegrade biphenyl via mono-oxygenase type enzymes rather than the more conventional dioxygenases.

V. Tetralin

1,2,3,4-Tetrahydronaphthalene (Tetralin) consists of an aromatic ring fused to an alicyclic ring, resulting in a hydrocarbon exhibiting both aromatic and alicyclic properties. Tetralin is widely used in a diverse number of applications, as a solvent (for oils, fats, waxes, resins, asphalt and rubber), as a substitute for turpentine and as a larvicide (Sikkema and de Bont 1991a). Until recently little was known about the bacterial catabolism of this bulk chemical. As a result, tetralin metabolism has not been covered in any recent reviews of aromatic (or aliphatic) hydrocarbon biodegradation. The few early reports of bacterial tetralin utilization all showed either degradation or transformation initiated by hydroxylation of the alicyclic moiety (Jamison et al. 1971). A co-substrate (Soli and Bens 1972; Kappeler and Wuhrmann 1978) or mixed culture (Strawinski and Stone 1940) have long thought to be prerequisites for utilization. Schreiber and Winkler (1983) did though provide evidence of tetralin being metabolized, very slowly, as the sole source of carbon and energy by a *Pseudomonas* sp. Once again catabolism was via initial hydroxylation of the alicyclic ring. Recently, however, there has been a series of papers on tetralin metabolism (Sikkema and de Bont 1991a,b; Sikkema et al. 1992) which have demonstrated the major problem of isolating tetralin degraders; cell toxicity, developed a protocol to over come this, isolated bacteria capable of growth and described a biodegradation route.

Eight pure bacterial strains which grew on tetralin as the sole source of carbon and energy were selected in the work of Sikkema and colleagues. Of these, four had been previously isolated using alternative aromatic hydrocarbons (two using *o*-xylene – Schraa et al. 1987; one on styrene – Hartmans et al. 1990; one on 1,3,5-trimethylbenzene). The four original isolates belong to the genera *Acinetobacter*, *Arthrobacter* (two strains) and *Moraxella* (i.e. both Gram-negative and Gram-positive bacteria). These isolations were achieved (and growth of all of the bacteria) by either supplying the substrate to the cells via the vapour phase or using an organic solvent/water two-phase system. Tetralin was found to be toxic to the cells when very low concentrations (15μl/l) were presented directly to the medium (Sikkema and de Bont 1991a). The nature of the toxicity of tetralin was subsequently investigated (Sikkema et al. 1992) using a model membrane system (proteoliposomes in which beef heart cytochrome *c* oxidase was reconstituted as a proton motive force generating mechanism) and both tetralin degraders

Acinetobacter T5, *Arthrobacter* T2 – Sikkema and de Bont 1991a; *Corynebacterium* C125 – Schraa et al. 1987) and non-degraders *Escherichia coli* and *Bacillus subtilis*) were investigated. Tetralin was shown to partition into the lipid membranes causing expansion and impairment of membrane functions (especially the proton motive force). It was postulated that would in turn impair other physiological process and lead to low or zero growth rates. Thus demonstrating for the first time that the key to the toxicity of many (aromatic) hydrophobic compounds lies in their chronic effects caused by their solubility in cell membranes, which then disrupt the key physiological barrier.

The biodegradation route(s) of tetralin by these bacterial strains has yet to be confirmed. Initial data have shown that one of the strains originally tested (Sikkema and de Bont 1991a) initiate attack on the aromatic moiety via the classical dioxygenase-dehydrogenase scheme. *Corynebacterium* C125 starts by attacking the aromatic moiety to form a *cis*-dihydrodiol (*cis*-1,2,5,6,7,8-hexahydro-1,2-naphthalene diol; Sikkema and de Bont 1991b) which then presumably undergoes reaction with a dehydrogenase. The resultant dihydroxyl tetralin then undergoes *meta* ring cleavage; the pathway appears to then follow that presented for naphthalene (J. Sikkema personal communication).

VI. Fused-ring aromatic compounds

A. Naphthalene

The bacterial degradation of naphthalene has frequently been reported over the last thirty years. The biodegradative route employed by the vast majority of micro-organisms is given in Fig. 10 and evidence to support these pathways can be found in previous review articles (Cerniglia and Heitkamp 1989; Smith 1990). The catabolic divergence in the catabolism of salicylate illustrated in Fig. 10, may be somewhat misleading as it has only been reported in *Pseudomonas fluorescens*; the norm being the oxidative decarboxylation of the salicylate to yield catechol (Gibson and Subramanian 1984).

This pathway has recently been confirmed for a *Mycobacterium* sp. (Kelley et al. 1990). There was also evidence of a side reaction leading to the formation of *trans*-naphthalene 1,2-dihydrodiol (the ratio of *cis*-diol to *trans*-diol was 25:1) via a cytochrome P-450 monooxygenase which formed an epoxide (naphthalene 1,2-oxide) which was converted to the *trans*-diol by an epoxide hydrolase enzyme. This reaction scheme is common in the eukaryotic naphthalene catabolism but this was the first report in bacteria. A second example of the role of cytochrome P-450 monooxygenase in the removal of naphthalene was provided by Kulish and Vilker (1991). These authors showed that it was possible to use the camphor P-450 enzyme from *Pseudomonas putida* PpG 786 to remove traces of naphthalene from aqueous solutions at high rates.

The catabolism of naphthalene by pseudomonads has frequently been

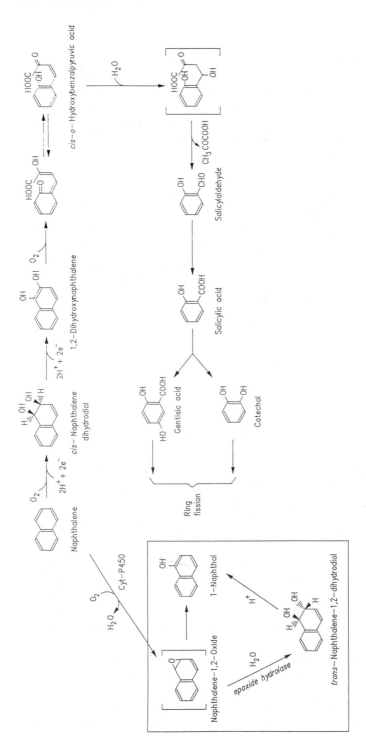

Fig. 10. The initial reactions of naphthalene biodegradation.
The involvement of the cytochrome P-450 monooxygenase reaction has only been confirmed in a *Mycobacterium* sp. (Kelley et al. 1990). The catabolic divergence in the catabolism of salicylate has only been reported in *Pseudomonas fluorescens* (Gibson and Subramanian 1984).

demonstrated to be plasmid-encoded as originally reported by Dunn and Gunsalus (1973). Subsequent studies have demonstrated that the genes are located on two operons (*nah* and *sal*) cited on plasmids of about 80 kb. The *nah* operon encodes for the genes for the conversion of naphthalene to salicylate, whilst the *sal* operon encodes for the conversion of salicylate to central metabolites (You et al. 1988). Recent reports on the genetic basis of naphthalene biodegradation have dissected the structure and function of the individual genes (Schell 1986; Schell and Wender 1986; You and Gunsalus 1986; You et al. 1988).

The association and uptake of naphthalene in *Pseudomonas putida* was recently reported (Bateman et al. 1986). Neither an energized membrane or ATP were essential for the association or uptake of naphthalene. The evidence for the a non-specific naphthalene cellular association was based on two important findings. Firstly, the absence of structure specific carrier protein as evidenced by the lack of saturation of association, the lack of competitive inhibition by protein inhibitors and the observation that naphthalene was able to pass through membranes even in the presence of iodacetamide. Secondly, there was no requirement for an intact membrane potential or for ATP as concluded by the absence of inhibition of association in the presence of an ionophore and electron transport inhibitors, lack of stimulation of association by phenazine methosulphate and ascorbate and intracellular formation of the *cis*-naphthalene dihydrodiol in the presence of an ionophore and inhibitors of electron transport or under conditions which depleted intracellular ATP. It was also recorded that association was not influenced by the catabolic plasmid.

The growth kinetics of a soil pseudomonad capable of growing on naphthalene, phenanthrene and anthracene has recently been reported (Volkering et al. 1992). These authors demonstrated growth on crystalline or absorbed polycyclic aromatic hydrocarbons resulting in increased biomass concentrations. Using a simple mathematical model they showed that mass transfer from the solid phase to the liquid phase was rate limiting to growth.

B. Polycyclic aromatic hydrocarbons

There is currently great concern about the amounts of polycyclic hydrocarbons in the environment. The presence in the environment of large quantities may be attributed to both petrogenic and pyrogenic sources (Laflamme and Hite 1978) and these compounds are considered extremely undesirable as most are potential carcinogens, mutagens and tetragens.

There is consequently much interest in the potential use of biodegradation of these compounds as this is the major route through which polycyclic aromatic compounds could be removed from the environment.

1. Anthracene and phenanthrene
The biodegradative routes for phenanthrene and anthracene have been proposed (Figs. 11 and 12 respectively) and the reader is directed to previous

review articles (Gibson and Subramanian 1984; Cerniglia and Heitkamp 1989). There still remain many unanswered questions regarding the biodegradation of these compounds; to date most of the studies have focused on the initial attack of these compounds and much of the pathways remain predicted rater than proven. It has been shown that both compounds undergo *meta* ring-fission after classical dihydroxylation (Cerniglia and Heitkamp 1989).

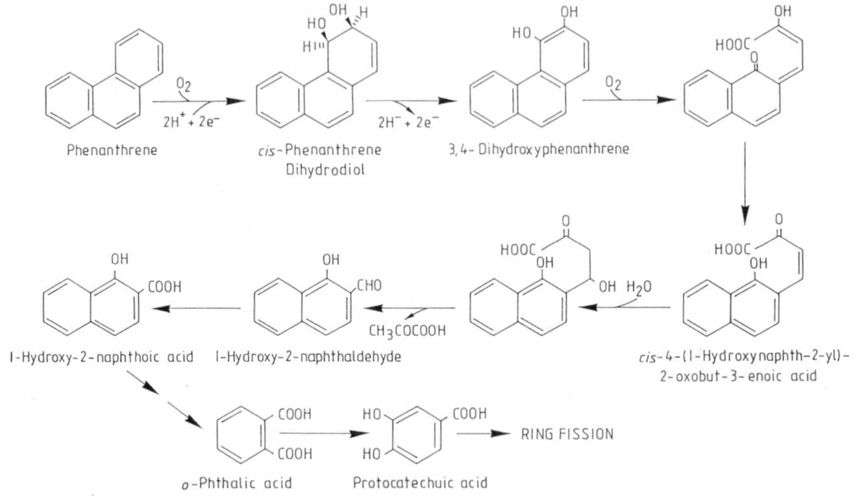

Fig. 11. Various proposed steps in the biodegradation of phenanthrene. From Gibson and Subramanian (1984).

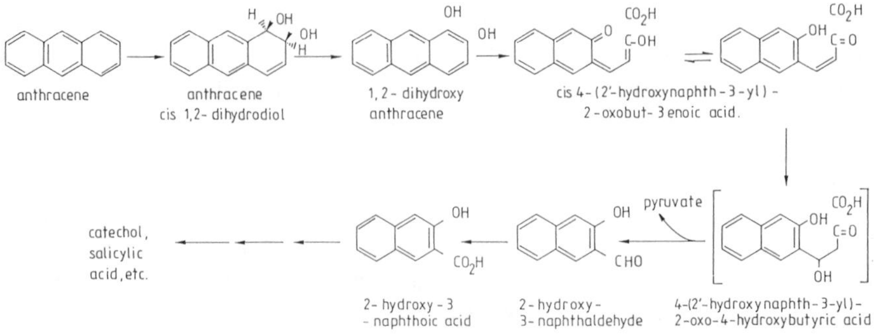

Fig. 12. Various proposed steps in the biodegradation of anthracene. From Gibson and Subramanian (1984).

Foght et al. (1990) recently reported on the complete biodegradation of [^{14}C]-phenanthrene in mineral oil by various bacteria; the use of radio-labelled substrates offers conclusive evidence that the compound is biodegraded. This same technique has also been applied by other workers in this field (Keuth and

Rehm 1991; Heitkamp and Cerniglia 1989) who also showed mineralization of phenanthrene.

2. *Other polycyclic aromatic hydrocarbons*

The biodegradation of higher molecular weight polycyclic aromatic compounds is one of the least well understood aspects of aromatic hydrocarbon degrading bacteria. There have been a few reports on the co-metabolism of these compounds (see Cerniglia and Heitkamp 1989). Until recently the only reports concerned the initial oxidation of high molecular weight polycyclic compounds by bacteria grown on alternative aromatic compounds and reflected the substrate specificity of the enzymes, rather than novel modes of biodegradation. However, there have been several recent advances in this area.

Heitkamp and Cerniglia (1988) isolated a Gram-positive bacterium capable of degrading naphthalene, phenanthrene, fluoranthene, and pyrene. The percentage mineralization of these compounds to CO_2 was 60%, 51%, 90% and 63% respectively, and were so encouraging that the authors suggested that this isolate may be an attractive alternative to physico-chemical methods for the remediation of environments contaminated with polycyclic hydrocarbons. In follow-up studies they identified not only the bacterium (*Mycobacterium* sp.) but also the initial ring oxidation and the ring fission products. This was the first report on the enzymatic mechanisms of the metabolism of pyrene (Heitkamp et al. 1988a,b). Further, they also showed the involvement of both dioxygenase and monooxygenase in the initial attack (analogous to the situation outlined above with naphthalene – Kelley et al. 1990). The structures of the compounds identified in the metabolism of pyrene by *Mycobacterium* sp. are given in Fig. 13.

Walter et al. (1991) isolated a strain of *Rhodococcus* sp. (UW1) which utilized pyrene as sole source of carbon and energy (72% mineralization in 14 days; maximum degradation rate 0.08 mg/ml per day; doubling time 30 h). During growth a metabolite was detected in the culture fluid and shown to result from the recyclization of the *meta* ring fission product of pyrene, assuming dihydroxylation in either the 1,2- or 4,5-positions. This is outlined in Fig. 13.

The complete biodegradation of fluoranthene by pure cultures of *Pseudomonas paucimobilis* was demonstrated by Mueller et al. (1990). This was partially attributable to the use of Tween 80 in the medium to increase the bioavailability of the substrate. Fluoranthene-grown cells were able to oxidize a wide range of aromatic compounds (anthraquinone, benzo[*b*]fluorene, biphenyl, chrysene and pyrene. These authors suggested that this was due to novel biodegradative routes but no route(s) have yet been postulated.

Weissenfels et al. (1990) isolated various pure cultures capable of degrading different polycyclic aromatic compounds; *Pseudomonas paucimobilis* grew on phenanthrene, *Pseudomonas vesicularis* could degrade fluorene and *Alcaligenes denitrificans* grew on fluoranthene. The last of these bacteria was

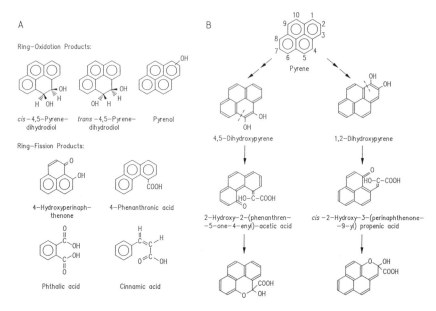

Fig. 13. The metabolism of pyrene. (A) structures of the compounds identified in the metabolism of pyrene by *Mycobacterium* sp. (Heitkamp et al. 1988b). (B) *meta* Ring fission pathway proposed by Walter et al. (1991) for a strain of *Rhodococcus* sp. (UW1).

then studied in detail (Weissenfels et al. 1991). The strain showed a maximum degradation rate of 0.3 mg fluoranthene/ml per day. Several intermediates (also shown to be further degraded) were identified in the culture fluids. These and a tentative biodegradation route are given in Fig. 14.

The versatile *Mycobacterium* sp. of Cerniglia's group has also been studied for the mineralization of fluoranthene (Kelley and Cerniglia 1991; Kelley et al. 1991). Initial studies used [3-^{14}C]fluoranthene, demonstrating up to 78% conversion to [14]CO$_2$ in 5 days (Kelley and Cerniglia 1991). Then 9-fluorenone-1-carboxylic acid (Fig. 14) was identified as an intermediate (Kelley et al. 1991).

With regard to the more complex members of the polycyclic aromatic hydrocarbon family, there have been no reports of complete biodegradation. This is not surprising when one considers their solubility in water. Naphthalene is relatively soluble in water (approximately 30 mg/l). The addition of a third aromatic ring decreases the solubility by between 30 and 700 times and as more aromatic rings are added so the solubility approaches truly insoluble. Couple to this the extreme chemical stability of such compounds and one can understand the difficulty microorganisms are faced with. There is, however, now increasing evidence that complex, polycyclic aromatic compounds can be (partially) metabolized, beyond ring fission, by bacterial strains pre- induced by

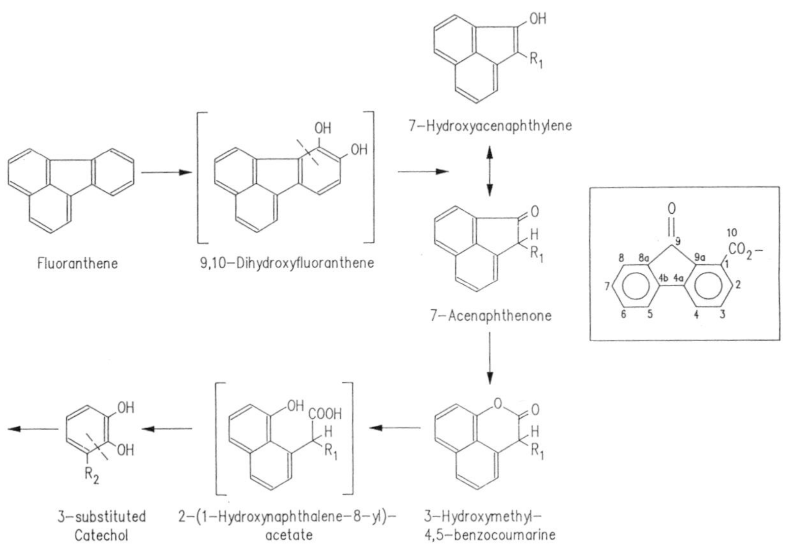

Fig. 14. Pathway for the degradation of fluorene by *Alcaligenes denitrificans* (WW1). Re-drawn from Weissenfels et al. (1991).

growth on an alternative aromatic substrate. Initial studies (Gibson et al. 1975) showed that a mutant strain of *Beijerinckia* sp. (B8/36) oxidized benzo[a]anthracene to three distinct diols (Fig. 15) after induction by growth on biphenyl. Mahaffey et al. (1988) then went on to show that the wild type of this bacterium (*Beijerinckia* sp. B1) metabolized benzo[a]anthracene to a mixture of *o*-hydroxy(polyaromatic) acids after induction on biphenyl, *m*-xylene or salicylate. In this metabolic process this carcinogenic, genotoxic, five-ringed compound was oxidized to hydroxylated substituted three-ring carboxylic acids, which are neither carcinogenic nor tetragenic, thus presenting the first ever evidence of the ring cleavage of such a polycyclic aromatic hydrocarbon. Additionally this latter study noted the partitioning of the benzo[a]anthracene into the cell membrane due to its hydrophobic nature.

The number and complexity of polycyclic aromatic hydrocarbons now known to be dissimilated by pure bacterial strains is rapidly increasing. As the problems, such as water solubility, are overcome, so the list is expected to grow. It is also hoped that the promising reports on the isolation of bacteria capable of growing on these compounds are followed up by detailed studies into the exact nature of the pathways employed.

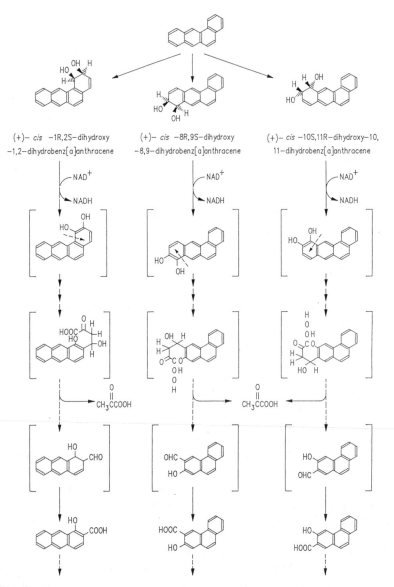

Fig. 15. Metabolism of benzo[*a*]anthracene by *Beijerinckia* sp. From Gibson et al. (1975) and Mahaffey et al. (1988).

VII. Mixed substrates

In the environment it is unlikely that the micro-organisms are confronted with single aromatic hydrocarbons; mixtures are much more likely to occur. Much more information concerning the biodegradation kinetics of mixtures of common industrial chemicals, such as (aromatic) solvents, needs to be gathered (Bitzi et al. 1991). It is well established that certain mixtures are more rapidly biodegraded than when the compounds are present singularly (McCarty et al. 1984). However, these studies are mainly based on work with non-growth-supporting (secondary) substrates, with biomass being created by growth on one or more easily degraded primary substrate, present in high concentrations. Little is known about substrate interactions among biodegradable aromatic hydrocarbons present in growth supporting concentrations. Recently, Bauer and Capone (1988) investigated the biodegradation of mixtures of polycyclic aromatic hydrocarbons. Amongst their findings they showed that naphthalene stimulated the biodegradation of phenanthrene but not that of anthracene. Arvin et al. (1989) demonstrated the interaction of aromatic substrates during the biodegradion of benzene. The presence of toluene or xylene stimulated the degradation of benzene but that toluene and xylene had an antagonistic effect on the utilization benzene. Similarly, Alvarez and Vogel (1991) showed that toluene enhanced the degradation of both benzene and *p*-xylene in an *Arthrobacter* sp. whilst detrimental substrate interactions included retardation in benzene and toluene degradation by the presence of *p*-xylene in *Pseudomonas* sp. An important conclusion from this study was that catabolic deversity precludes generalizations about stimulation or inhibition of growth by mixed substrates. We (Smith et al. 1990) observed complete inhibition of growth of *Pseudomonas* sp. when presented with a mixture of biphenyl and ethylbenzene; both compounds being readily degraded when present singularly (by the same enzymes).

None of these groups could explain their data and further research is required to establish if this is a commonly occurring phenomenon and to elucidate the mechanisms involved. Law and Button (1986) investigated the metabolic competition between hydrocarbons in a bacterium (*Pseudomonas* sp. strain T2). This strain was shown to have a low Michaelis constant for toluene, however, no competitive inhibition of toluene metabolism was seen with benzene, naphthalene, xylene, dodecane or amino acids; there was neither stimulation nor inhibition. The authors suggest that this might be due to the fact that the different (aromatic) compounds are degraded by different metabolic routes, hence no competition for common active-sites. The converse (competitive inhibition) could be the result of shared metabolic routes (or uptake) of structurally similar (aromatic) hydrocarbons meaning that when faced with mixtures of compounds at low (environmentally significant) concentrations bacterial growth would be reduced. This theory was first formulated back in the late 1960s (Gibson 1968) and it remains to be seen if the combination of very low Michaelis constants and shared pathways and/or uptake systems is the answer to the increasing number of observations of mixed substrate inhibition.

VIII. Conclusions

When reviewing a subject as broadly studied as that of bacterial aromatic hydrocarbon degradation one is immediately impressed by the wealth of knowledge which has been accumulated over the years. In addition, current developments in this field continue at a vigorous pace which does not compromise the quality of the end result. In general, we now know which compounds can be degraded and which routes are employed. However, preparing such a manuscript does allow the author to stand back a little and wonder at some of the fundamental gaps which still exist in our understanding of the biodegradation of aromatic hydrocarbons by bacteria.

One of the most overlooked areas of research in this field is that of association and uptake. There have been but a handful of reports of the kinetics of these process.

All aromatic hydrocarbons are to a lesser or greater degree hydrophobic compounds, many are extremely toxic to cells of all types. What makes an aromatic hydrocarbon utilizing strain different to the norm? Preliminary studies have highlighted the hydrophobic nature of the aromatic compound, but not all strains of the same bacteria respond in the same manner to the same compounds, let alone all bacteria to one compound. The key here must lie in the structure of the outer barriers of the cells, yet very little has been studied in the specific area of aromatic degrading bacteria.

The recent discovery of pure bacterial strains capable of anaerobic growth on benzene, toluene, ethylbenzene and xylenes (see Chapter 16 – Fuchs et al.) seems to have opened up a new exciting field for researchers. It will be intriguing to see what the mechanisms of dissimilation are and what the other aromatic hydrocarbons can be so degraded. It is also of immense interest that certain pseudomonads are able to degrade toluene and other aromatic hydrocarbons by both aerobic dissimilation and anaerobic denitrifying conditions (Kukor and Olsen 1990; M.R. Smith, unpublished data). These microorganisms will surely provide new insights into the catabolism of aromatic hydrocarbons and the physiological control mechanisms involved.

In the last 30 years or more there has been a wealth of knowledge acquired regarding the biodegradation of aromatic hydrocarbons. The studies have gone along way to showing us how the bacteria tackle these hydrophobic pollutants.In some cases the genetic nature and control of the enzymes have been dissected. Let us now hope that future studies will consolidate our understanding of the full impact that this diverse group of organic compounds has on the physiology of the bacteria which degrade them.

Acknowledgements

I am much indebted to the many research groups who have sent reprints and pre-prints to me during the preparation of this chapter. Further I would like to thank Jan Sikkema for supplying many useful references and for his comments and support during the preparation of the manuscript.

References

Alvarez PJJ and Vogel TM (1991) Substrate interactions of benzene, toluene and *para*-xylene during microbial degradation by pure cultures and mixed culture aquifer slurries. Appl. Environ. Microbiol. 57: 2981–2985

Arvin E, Jensen BK and Gundersen AT (1989) Substrate interactions during the areobic degradation of benzene. Appl. Environ. Microbiol. 55: 3221–3225

Assinder SJ and Williams PA (1990) The TOL plasmids: Determinants of the catbolism of toluene and the xylenes. In: AH Rose and DW Tempest (eds) Advances in Microbial Physiology, Vol 31 (pp 1–69). Academic Press Limited, London, UK

Axell BC and Geary PJ (1975) The metabolism of benzene by bacteria. Biochem. J. 136: 927–934

Baggi G, Boga MM, Catelani D, Galli E and Trecccani V (1983) Styrene catabolism by a strain of *Pseudomonas fluorescens*. Syst. Appl Microbiol. 4: 141–147

Baggi G, Barbieri P, Galli E and Tollari S (1987) Isolation of a *Pseudomonas stutzeri* strain that degrades o-xylene. Appl. Environ. Microbiol. 53: 2129–2132

Bateman JN, Speers B, Feduik L and Hartline RA (1986) Naphthalene association and uptake in *Pseudomonas putida*. J. Bacteriol. 166: 155–161

Bauer JE and Capone DG (1988) Effects of co-occurring aromatic hydrocarbons on the degradation of individual polycyclic aromatic hydrocarbons in marine sediment slurries. Appl. Environ. Microbiol. 54: 1649–1655

Bayly RC and Barbour MG (1984) The degradation of aromatic compounds by the *meta* and gentisate pathways. In: DT Gibson (ed) Microbial Degradation of Organic Compounds (pp 253–294). Marcel Dekker, New York

Bestetti G and Galli E (1984) Plasmid-coded degradation of ethylbenzene and 1-phenylethanol in *Pseudomonas fluoresens*. FEMS Microbiol. Letts. 21: 165–168

Bestetti G and Galli E (1987) Characterization of a novel TOL-like plasmid from *Pseudomonas putida* involved in 1,2,4-trimethylbenzene degradation. Appl. Environ. Microbiol. 169: 1780–1783

Bestetti G, Galli E, Ruzzi M, Baldacci G, Zennaro E and Frontali L. (1984) Molecular characterization of a plasmid involved in styrene degradation. Plasmid 12: 181–188

Bestetti G, Galli E, Benigini C, Orsini F and Pelizzoni F (1989) Biotransformation of styrenes by a *Pseudomonas putida*. Appl. Microbiol. Biotechnol. 30: 252–256

Bitzi U, Egli T and Hamer G (1991) The biodegradation of mixtures of organic solvents by mixed and monocultures. Biotech. Bioeng. 37: 1037–1042

Catelani D and Colombi A (1974) Metabolism of biphenyl. Structure and physical properties of 2-hydroxy-6-oxo-6-phenylhexa-2,4-dienoate, the *meta* cleavage product from 2,3-dihydroxybiphenyl by *Pseudomonas putida*. Biochem. J. 143: 431–434

Catelani D, Sorlini C and Treccani V (1971) The metabolism of biphenyl by *Pseudomonas putida*. Experientia 27: 1173–1174

Catelani D, Colombi A, Sorlini C and Treccani V (1973) Metabolism of biphenyl. 2-Hydroxy-6-oxo-6-phenylhexa-2,4-dienoate: the *meta* cleavage product from 2,3-dihydroxybiphenyl by *Pseudomonas putida*. Biochem. J. 134: 1063–1066

Cerniglia CE (1981) Aromatic hydrocarbons: metabolism of bacteria, fungi and algae. In: E Hodgson, JR Bend and RM Philpot (eds) Biochemical Toxicology, Vol 3 (pp 321–361). Elsevier/North Holland, New York

Cerniglia CE (1984) Microbial metabolism of polycyclic aromatic compounds. Adv. Appl. Microbiol. 30: 31–71

Cerniglia CE and Heitkamp MA (1989) Microbial degradation of polycyclic compounds (PAH) in the aquatic environment In: V Varanasi (ed) Metabolism of PAHS in the Aquatic Environment (pp 41–68). CRC Press Inc., Boca Raton, Florida

Cripps RE and Watkinson RJ (1978) Polycyclic aromatic hydrocarbons: Metabolism and environmental aspects. In: RJ Watkinson (ed) Developments in the Biodegradation of Hydrocarbons (pp 113–134). Applied Science Publishers, London

Cripps RE, Trudgill PW and Wheatley JG (1978) The metabolism of 1-phenylethanol and actophenone by *Nocardia* T5 and an *Arthrobacter* species. Eur. J. Biochem. 86: 175–186

Dagley S (1981) New perspectives in aromatic catabolism. In: T Leisinger, AM Cook, R Hütter and J Nüesch (eds) Microbial Degradation of Xenobiotics and Recalcitrant Compounds (pp 181–186). Academic Press, New York

Dagley S (1986) Biochemistry of aromatic hydrocarbon degradation in Pseudomonads. In: JR Sokatch (ed) The Bacteria, Vol 10 (pp 527–555). Academic Press, New York

Davey JF and Gibson DT (1974) Bacterial metabolism of *p*- and *m*- xylene: oxidation of the methyl substituent. J. Bacteriol. 119: 923–929

Davis RS, Hossler FE and Stone RW (1968) Metabolism of *p*- and *m*-xylene by species of *Pseudomonas*. Can. J. Microbiol 27: 1005–1009

DeFrank JJ and Ribbons DW (1976) The *p*-cymene pathway in *Pseudomonas putida* PL: isolation of a dihydrodiol accumulated by a mutant. Biochem. Biophys. Res. Commun. 70: 1129–1135

DeFrank JJ and Ribbons DW (1977a) *p*-Cymene pathway in *Pseudomonas putida*: initial reactions. J. Bacteriol. 129: 1356–1364

DeFrank JJ and Ribbons DW (1977b) *p*-Cymene pathway in *Pseudomonas putida*: ring cleavage of 2,3-dihydroxy-p-cumate and subsequent reactions. J. Bacteriol. 129: 1365–1375

Duetz WA and van Andel JG (1991) Stability of TOL plasmid pWWO in *Pseudomonas putida* mt-2 under non-selective conditions in continuous culture. J. Gen. Microbiol. 137: 1369–1374

Duggleby CJ and Williams PA (1986) Purification and some properties of the 2-hydroxy-6-oxohepta-2,4-dienoate hydrolase (2- hydroxymuconic semialdehyde hydrolase) encoded by the TOL plasmid pWWO from *Pseudomonas putida* mt 2. J Gen Microbiol 132: 717–726

Dunn NW and Gunsalus IC (1973) Transmissible plasmid coding for the early enzymes of naphthalene oxidation in *Pseudomonas putida*. J. Bacteriol. 114: 974–979

Dzhusupova DB, Baskunov BP, Golovleva LA, Alieva RM and Ilyaletdinov AN (1985) Peculiarities of the oxidation of α-methylstyrene by bacteria of the genus *Pseudomonas*. Mikrobiologiya 54: 136–140

Eastcott L, Shiu WY and Mackay D (1988) Environmentally relevant physical-chemical properties of hydrocarbons: a review of data and development of simple correlations. Oil and Chemical Pollution 4: 191–216

Eaton RW and Timmis KN (1986) Characterization of a plasmid-specified pathway for catabolism of isopropylbenzene in *Pseudomonas putida* RE 204. J. Bacteriol. 168: 123–131

Foght JM, Fedorak PM and Westlake DWS (1990) Mineralization of [^{14}C]hexadecane and [^{14}C]phenanthrene in crude oil: specificity amoung bacterial isolates. Can. J. Microbiol 36: 169–175

Franklin FCH, Bagdasarian M, Bagdasarian MM and Timmis KN (1981) Molecular and functional analysis of TOL plasmid pWWO from Pseudomonas putida and cloning of the entire regulated aromatic ring *meta* cleavage pathway. Proc. Natl. Acad. Sci. U.S.A. 78: 7458–7462

Fredrickson JK, Brockman FJ, Workman DJ, Li SW and Stevens (1991) Isolation and characterization of a subsurface bacterium capable of growth on tolunene, naphthalene, and other aromatic compounds. Appl. Environ. Microbiol. 57: 796–803

Furukawa K and Suzuki H (1988) Gene manipulation of catabolic activities for production of intermediates of various biphenyl compounds. Appl. Microbiol. Biotechnol. 29: 363–369

Gibson DT (1968) The microbial degradation of aromatic compounds. Science 161: 1093–1097

Gibson DT and Subramanian V (1984) Microbial degradation of aromatic hydrocarbons. In: DT Gibson (ed) Microbial Degradation of Organic Compounds (pp 361–369). Marcel Dekker, New York

Gibson DT, Koch JR and Kallio RE (1968) Oxidative degradation of aromatic hydrocarbons by micro-organisms. I. Enzymatic formation of catechol from benzene. Biochemistry 7: 2643–2656

Gibson DT, Cardini GE, Maesels FC and Kallio RE (1970) Incorporation of ^{18}O into benzene. Biochemistry 9: 1631–1635

Gibson DT, Roberts RL, Wells MC and Kobal VM (1973) Oxidation of biphenyl by a *Beijerinckia* species. Biochem. Biophys. Res. Commun. 50: 211–219

Gibson DT, Mahadaven V and Davey JF (1974) Bacterial metabolism of *p*- and *m*- xylene: oxidation of the aromatic ring. J. Bactriol. 119: 930–936

Gibson DT, Mahadaven V, Jerina DM, Yagi H and Yeh H (1975) Oxidations of the carcinogens benzo[*a*]pyrene and benzo[*a*]anthracene to dihydrodiols by bacterium. Science 189: 295–297

Hartmans S, Smits JP, van der Werf MJ, Volkering F and de Bont JAM (1989) Metabolism of styrene oxide ande 2-phenylethanol in the styrene-degrading *Xanthobacter* strain 124X. Appl. Environ. Microbiol. 55: 2850–2855

Hartmans S, van der Werf MJ and de Bont JAM (1990) Bacterial degradation of styrene involving a novel flavin adenine dinucleotide-dependent styrene monooxygenase. Appl. Environ. Microbiol. 56: 1347–1351

Heitkamp MA and Cerniglia CE (1988) Mineralization of polycyclic aromatic hydrocarbons by a bacterium isolated from sediment below an oil field. Appl. Environ. Microbiol. 54: 1612–1614

Heitkamp MA and Cerniglia CE (1989) Polycyclic aromatic hydrocarbon degradtion by a *Mycobacterium* sp. in microcosms containing sediment and water from a pristine ecosystem. Appl. Environ. Microbiol. 55: 1968–1973

Heitkamp MA, Franklin W and Cerniglia CE (1988a) Microbial metabolism of polycyclic aromatic hydrocarbons: isolation and characterization of a pyrene-degrading bacterium. Appl. Environ. Microbiol. 54: 2549–2555

Heitkamp MA, Freeman JP, Miller DW and Cerniglia CE (1988b) Pyrene degradation by a *Mycobacterium* sp.: identification of ring oxidation and ring fission products. Appl. Environ. Microbiol. 54: 2556–2565

Högn T and Jaenicke L (1972) Benzene metabolism of *Moraxella* sp. Eur. J. Biochem. 30: 369–375

Hopper DJ (1978) Microbial degradation of aromatic hydrocarbons. In: RJ Watkinson (ed) Developments in the Biodegradation of Hydrocarbons (pp 85–112). Applied Science Publishers, London

Horn JM, Harayama S and Timmis KN (1991) DNA sequence determination on the TOL plasmid (pWWO) *xylGFJ* genes of *Pseudomonas putida*: implications for the evolution of aromatic catabolism. Molecular Microbiol. 5: 2459–2474

Inoue A and Horikoshi K (1989) A *Pseudomonas* thrives in high concentations of toluene. Nature 338: 264–266

Ishigooka H, Yashida Y, Omori T and Minoda Y (1986) Enzymatic dioxygenation of biphenyl 2,3-diol and 3-isopropylcatechol. Agric. Biol. Chem. 50: 1045–1046

Jamison VW, Raymond RL and Hudson JO (1971) Hydrocarbon co-oxidation by *Nocardia corallina* strain V49. Dev. Ind. Microbiol. 12: 99–105

Jigami Y, Omori T, Minoda Y and Yamada K (1974) Formation of of 3-ethylsalicylic acid from 3-ethyltoluene by *Pseudomonas ovalis*. Agric. Biol. Chem. 38: 467–469

Kappeler T and Wuhrmann K (1978) Microbial degradation of the water-soluble fraction of gas oil. II Bioassay with pure strains. Water Res. 12: 335–342

Keil H and Williams PA (1985) A new class of TOL plasmid deletion mutants in *Pseudomonas putida* MT15 and their reversion by tandem gene amplification. J. Gen. Microbiol.131: 1023–1033

Kelley I and Cerniglia CE (1991) The metabolism of fluoranthene by a species of *Mycobacterium*. J Industrial Microbiol. 7: 19–26

Kelley I, Freeman JP and Cerniglia CE (1990) Identification of metabolites from degradation of naphthalene by a *Mycobacterium* sp. Biodegradation 1: 283–290

Kelley I, Freeman JP, Evans FE and Cerniglia CE (1991) Identification of a carboxylic acid metabolite from the catabolism of fluoranthene by a *Mycobacterium* sp. Appl. Environ. Microbiol. 57: 636–641

Keuth S and Rehm H-J (1991) Biodegradation of phenanthrene by *Arthrobacter polychromogenes* isolated from a contaminated soil. Appl. Microbiol. Biotechnol. 34: 804–808

Khan AA and Walia SK (1990) Identification and localization of 3-phenylcatechol dioxygenase and 2-hydroxy-6-oxo-6-phenylhexa-2,4-dienoate hydrolase genes of *Pseudomonas putida* and expression in *Eschericha coli*. Appl. Environ. Microbiol. 56: 956–962

Kukor JJ and Olsen RH (1990) Diversity of toluene degradation following long term exposure to BTEX *in situ*. In: D Kamely, A Chakrabarty and GS Omenn (eds) Biotechnology and Biodegradation (pp 405–421). Gulf Publishing Company, Houston

Kulisch GP and Vilker VL (1991) Application of *Pseudomonas putida* PpG 786 containing P-450 cytochrome monooxygenase for the removal of trace naphthalene concentrations. Biotechnol. Prog. 7: 93–98

Kunz DA and Chapman PJ (1981a) Catabolism of pseudocumene and 3-ethyltoluene by *Pseudomonas putida (arvilla)* mt-2: Evidence for new functions of the TOL (pWWO) plasmid. J. Bacteriol. 146: 179–191

Kunz DA and Chapman PJ (1981b) Isolation and characterization of spontaneously occurring TOL plasmid mutants of *Pseudomonas putida* HS1. plasmid. J. Bacteriol. 146: 952–964

Laflamme RE and Hite RA (1978) The global distribution of polycyclic aromatic hydrocarbons in recent sediments. Geochim. Cosmomchim. Acta. 42: 289–303

Law AT and Button DK (1986) Modulation of affinity of a marine pseudomonad for toluene and benzene by hydrocarbon exposure. Appl. Environ. Microbiol. 51: 469–476

Lovley DR and Lonergan DJ (1990) Anaerobic oxidation of toluene, phenol and *p*-cresol by the dissimilatory iron-reducing organism, GS-15. Appl. Environ Microbiol. 56: 1858–1864

Mahaffey WR, Gibson DT and Cerniglia CE (1988) Bacterial oxidation of chemical carcinogens: formation of polycyclic aromatic acids from Benz[*a*]anthracene. Appl. Microbiol. Microbiol. 54: 2415–2423

Marr EK and Stone RW (1961) Bacterial oxidation of benzene. J. Bacteriol. 85: 425–430

McCarty PL, Rittmann BE and Bouwer EJ (1984) Microbial processes affecting chemical transformations in groundwater. In: G Bitton and CP Gerba (eds) Groundwater Pollution Micobiology (pp 89–115). John Wiley and Sons, New York

Mueller JG, Chapman PJ, Blattmann BO and Pritchard PH (1990) Isolation and characterization of a fluoranthene-utilizing strain of *Pseudomonas paucimobilis*. Appl Environ. Microbiol. 56: 1079–1086

Nakazawa T, Inouye S and Nakazawa A (1980) Physical and functional analysis of RP4-TOL plasmid recombinants: analysis of insertion and deletion mutants. J. Bacteriol. 144: 222–231

Omori T, Jigami Y and Minoda Y (1974) Microbial oxidation of α-methylstyrene and β-methylstyrene. Agric. Biol. Chem. 38: 409–415

Omori T, Ishigooka H and Minoda Y (1986) Purification and some properties of 2-hydroxy-6-oxo-6-phenylhexa-2,4-dienoic acid (HODPA) reducing enzyme from *Pseudomonas cruciviae* S93B1, involved in the degradation of biphenyl. Agric. Biol. Chem. 50: 1513–1518

Polissi A, Bestetti G, Bertoni G, Galli E and Dehò G (1990) Genetic analysis of chromosomal operons involved in degradation of aromatic hydrocarbons in *Pseudomonas putida* TMB. J. Bacteriol. 172: 6355–6362

Ramos JL, Mermod N and Timmis KN (1987) Regulatory circuits controlling transcription of TOL plasmid operon encoding *meta*-cleavage pathway for degradation of alkylbenzoates by *Pseudomonas*. Mol. Microbiol. 1: 293–300

Rezessy-Szabó JM, Huijberts GNM and de Bont JAM (1987) Potential of organic solvents in cultivating micro-organisms on toxic water-insoluble compounds. In: C Laane, J Tramper and MD Lilly (eds) Biocatalysis in Organic Media, proceedings of an international symposium held at Wageningen, The Netherlands, 7–10 December 1986 (pp 295–302). Elsevier Science Publishers, Amsterdam

Robertson BR and Button (1987) Toluene induction and uptake kinetics and their inclusion in the specific-affinity relationship for describing rates of hydrocarbon metabolism. Appl. Environ. Microbiol. 53: 2193–2205

Sala-Trepat JM, Murray K and Williams PA (1972) The metabolic divergence in the *meta* cleavage pathway of catechols by *Pseudomonas putida* NCIB 10015. Eur. J. Biochem. 28: 347–356

Schell MA (1986) Homology between nucleotide sequences of promoter regions of *nah* and *sal* operons of NAH7 plasmid of *Pseudomonas putida*. Proc. Natl. Acad. Sci. U.S.A. 83: 369–373

Schell MA and Wender PE (1986) Indentification of the *nahR* gene product and nucleotide sequence required for its activation of the *sal* operon. J. Bacteriol. 166: 9–14

Schraa G, Bethe BM, van Neerven ARW, van den Tweel WJJ, van der Wende E and Zehnder AJB (1987) Degradation 1,2-dimethylbenzene by *Corynebacterium* strain C125. Antonie van Leewenhoek 53: 159–170

Schreiber AF Winkler UK (1983) Transformation of tetralin by whole cells of *Pseudomonas stutzeri* AS39. Eur. J. Appl. Microbiol. Biotechnol. 18: 6–10

Schwartz RD (1981) A novel reaction: *meta* hydroxylation of biphenyl by an actinomycete. Enzyme Microbial Technol. 3: 158–159

Shields MS, Montgomery SO, Chapman PJ, Cuskey SM and Pritchard PH (1989). Novel pathway of toluene catabolism in the trichloroethylene-degrading bacterium G4. Appl. Environ. Microbiol. 55: 1624–1629

Shirai K (1986) Screening microorganisms for catechol production from benzene. Agric. Biol. Chem. 50: 2875–2880

Shiraiso K and Hisatsuka K (1979) Isolation and identification of styrene assimiliating bacteria. Agric. Biol. Chem. 43: 1595–1596

Simpson HD, Green J and Dalton H (1987) Purification and some properties of a novel heat-stable *cis*-toluene dihydrodiol dehydrogenase. Biochem. J. 244: 585–590

Sielicki M, Focht DD and Martin JP (1978) Microbial transformations of styrene and [^{14}C]styrene in soil and enrichment cultures. Appl. Environ. Microbiol. 35: 124–128

Sikkema J and de Bont JAM (1991a) Isolation and initial characterization of bacteria growing on tetralin. Biodegradation 2: 15–23

Sikkema J and de Bont JAM (1991b) Biocatalytic production of hydroxylated aromatic and alicyclic compounds: products derived from tetralin. Recl. Trav. Chim. Pays-Bas 110: 189–194

Sikkema J, Poolman B, Konings WN and de Bont JAM (1993) The membrane action of tetralin: Effects on the functional and structural properties of artificial and bacterial membranes (In press)

Smith MR (1990) The biodegradation of aromatic hydrocarbons by bacteria. Biodegradation 1: 191–206

Smith MR and Ratledge C (1989a) Catabolism of alkylbenzenes by *Pseudomonas* sp. NCIB 10643. Appl. Microbiol. Biotechnol. 32: 68–75

Smith MR and Ratledge C (1989b) Catabolism of biphenyl by *Pseudomonas* sp. NCIB 10643 and *Nocardia* sp. NCIB 10503. Appl. Microbiol. Biotechnol. 30: 395–401

Smith MR, Ewing M and Ratledge C (1990) The interactions of various aromatic substrates degraded by *Pseudomonas* sp. NCIB 10643: synergistic inhibition of growth by two compounds which serve as growth substrates. Appl. Microbiol. Biotechnol. 34: 536–538

Soli G and Bens EM (1972) Bacteria which attack petroleum hydrocarbons in a saline medium. Biotechnol. Bioeng. 14: 319–330

Starovoitov II, Selfionov SA, Nefedova MY and Adanin VM (1986) Catabolism of diphenyl by *Pseudomonas putida* strain BS893 containing biodegradation plasmid pBS241. Microbiology (Engl. transl.) 54: 726–727

Strawinski RJ and Stone RW (1940) The utilization of hydrocarbons by bacteria. J. Bacteriol. 40: 461–462

van den Tweel WJJ, Janssens and de Bont JAM (1986) Degradation of 4-hydroxyphenylacetate by *Xanthobacter* 124X. Antonie van Leeuwenhoek 52: 309–318

van den Tweel WJJ, Vorage MJAW, Marsman EH, Koppejan J, Tramper J and de Bont JAM (1988) Continuous production of *cis*-1,2-dihydroxycyclohexa-3,5-diene (*cis*-benzeneglycol) from benzene by a mutant of a benzene degrading *Pseudomonas* sp. Enzyme Microbial Technol. 10: 134–142

Vecht SE, Platt MW, Er-El Z and Goldberg I (1988) The growth of *Pseudomonas putida* on *m*-toluic acid and on toluene in batch and chemostat cultures. Appl. Microbiol. Biotechnol. 27: 587–592

Volkering F, Breure AM, Sterkenburg A and van Andel JG (1992) Microbial degradation of polycyclic aromatic hydrocarbons: effect of substrate availability on bacterial growth kinetics. Appl. Microbiol. Biotechnol. 36: 548–552

Walter U, Beyer M, Klein J and Rehm H-J (1991) Degradation of pyrene by *Rhodococcus* sp. UW1. Appl. Microbiol. 34: 671–676

Weissenfels WD, Beyer M and Klein J (1990) Degradation of fluorene and fluoranthene by pure bacterial cultures. Appl. Microbiol. Biotechnol. 32: 479–484

Weissenfels WD, Beyer M, Klein J and Rehm H-J (1991) Microbial metabolism of fluorene: isolation and identification of ring fission products. Appl. Microbiol. Biotechnol. 34: 528–535

Wigmore GJ, Bayley RC and DiBerardino D (1974) *Pseudomonas putida* mutants defective in the catabolism of the products of *meta* fission of catechol and its methyl analogues. J. Bacteriol. 120: 31–37

Williams PA, Taylor SD and Gibb AE (1988) Loss of the toluene-xylene catabolic plasmid pWWO during growth of *Pseudomonas putida* on benzoate is due to a selective growth advantge of "cured" segregants. J. Gen. Microbiol. 134: 2039–2048

Winstanley C, Taylor SC and Williams PA (1987) pWW174: A large plasmid from *Acinetobacter calcoaceticus* encoding benzene catabolism by the β-ketoadipate pathway. Mol. Microbiol. 1: 219–227

You IS and Gunsalus IC (1986) Regulation of the *nah* and *sal* operons of plasmid NAH7: evidence for a new function in *nahR*. Biochem. Biophys. Res. Commun. 141: 986–992

You IS, Ghosal D and Gunsalus IC (1988) Nucleotide sequence of plasmid NAH7 gene *nahR* and DNA binding of the *nahR* product. J. Bacteriol. 170: 5409–5415

Zeyer J, Eicher P, Dolfing J and Schwarzenbach RP (1990) Anaerobic degradation of aromatic compounds. In: D Kamely, A Chakrabarty and GS Omenn (eds) Biotechnology and Biodegradation (pp 33–40). Gulf Publishing Company, Houston

12. Microbial dehalogenation of haloaliphatic compounds

J. HOWARD SLATER

School of Pure and Applied Biology, University of Wales, Cardiff, Wales CF1 3TL, U.K.

I. Introduction

Microorganisms remove halogens from aliphatic compounds by the activity of enzymes known as dehalogenases. Microbes synthesizing these enzymes are widely distributed and easily isolated (Slater and Bull 1982; Hardman et al. 1988; Hardman 1991) because many naturally-occurring halogenated compounds are present throughout the biosphere (Marais 1944; Bracken 1954; Petty 1961; Fowden 1968; Murray and Riley 1973; Suida and DeBernardis 1973; Lovelock 1975; King 1986, 1988; Vogel et al. 1987; Symonds et al. 1988). Microbes have evolved dehalogenating mechanisms for two main reasons: firstly, to use halogenated compounds as growth nutrients and, secondly, as detoxification mechanisms since many of these compounds are potent metabolic inhibitors.

In recent times – over the last 50 years or thereabouts – another important factor has impinged on microbial dehalogenases. The chemical industry has produced large quantities of thousands of novel halogenated compounds that have been deliberately released into the biosphere (Bollag 1974; Hill 1978; Alexander 1981; Slater and Bull 1982; Motosugi and Soda 1983; Dagley 1984; Muller and Lingens 1986). The active ingredients of many pesticides, for example, are halogenated compounds and a plethora of difficulties accompany their use, such as: distribution, bioaccumulation, persistence, toxicity and human health issues. Indisputable these issues would be much more serious were it not for microbial populations adapting to degrade these xenobiotic compounds. The evolution of dehalogenase-producing microbes using novel halo-compounds is scientifically interesting and practically important.

Microbial dehalogenases are intrinsically interesting because several enzyme classes with fundamentally different catalytic mechanisms have been discovered and, within each class, isoenzymes readily characterized (Clarke and Slater 1986). Comparative enzyme studies, particularly as a result of molecular studies, are now revealing exciting details of evolutionary relationships.

Microbial catabolism of halogenated compounds has a special place in the

C. Ratledge (ed.), Biochemistry of Microbial Degradation, 379–421.
© 1994 *Kluwer Academic Publishers. Printed in the Netherlands.*

history of microbial physiology and biochemistry. Den Dooren de Jong (1926) made the then startling observation that some bacteria dehalogenated certain compounds, such as bromopropionate and bromosuccinate. This opened the door to countless biodegradation studies and led Kluyver (1931) to realize that enzymes did not necessarily exhibit extreme substrate specificity, an observation which in turn led to enzyme adaptation and evolution studies (Clarke 1978, 1982).

II. Dehalogenation of Haloalkanoic Acids (HAAs)

A. Some basic considerations

Dehalogenation of substituted alkanoic acids results in the formation of molecules which are either direct intermediates of central metabolism or can be easily converted to intermediary metabolites. Normally, dehalogenase-producing microorganisms are selected for their ability to use halogenated alkanoic acids (HAAs, especially 2HAAs) or haloalkanes as carbon and energy sources. However, in some instances, dehalogenase activities yield products not used by dehalogenase-synthesizing organisms. In these cases this results in the stoichiometric conversion of substrate to a product which accumulates or supports the growth of non-dehalogenase-producing microorganisms (Slater and Bull 1982). The enrichment and selection of such co-metabolic reactions is more difficult and it is likely that they are more widespread in nature than is fully appreciated. Many enrichment strategies have been used to isolate dehalogenating microorganisms and these are described elsewhere (Slater and Bull 1982; Slater and Hardman 1982).

There is another reason why dehalogenases are advantageous to microorganisms. Many halogenated compounds are toxic (Peters 1952) and dehalogenation mechanisms may be physiologically more important as detoxification mechanisms aiding microbial survival in hostile environments than as mechanisms for releasing a halogenated compound as a carbon and energy source for growth.

Jensen (1951, 1957, 1959, 1960) made a thorough study of soil microorganisms mineralizing and growing on a number of halogenated compounds. He first suggested the name dehalogenase for enzymes involved in dehalogenation processes (Jensen 1963). This is now widely accepted as the generic name for the different enzymes and mechanisms involved in dehalogenating HAAs (Hardman 1991; Goldman et al. 1968; Little and Williams 1971; Janssen et al. 1985).

Jensen's work showed, in general terms, what has now been confirmed by later studies that many microorganisms produce dehalogenases with the following general features (Clarke and Slater 1986; Slater 1988).

● They are normally inducible, less frequently constitutive, and the number of inducing compounds for a particular enzyme is often greater than the number of the substrates of that enzyme.

- The enzymes are soluble and readily purified.
- They have comparatively poor substrate specificities, normally in the mM range for K_m values, and each enzyme has activity towards a wide range of substrates.
- They show broad pH profiles with optimum values in the range pH 9.0 to 10.0.
- Often more than one dehalogenase is synthesized by a single organism, and the regulation of synthesis is complex, different for each enzyme, and highly dependent on environmental conditions, especially factors such as growth rate and growth substrate.
- Dehalogenases with stereospecific catalytic properties have been described, with all possible types known; namely enzymes which dehalogenate both L- and D-isomers, enzymes which attack only L-isomers, and enzymes specific for D-isomers.
- On the basis of experimental techniques such as relative mobility in polyacrylamide electrophoresis, the same enzymes are isolated from independently isolated microorganisms.
- The genes for dehalogenases frequently are plasmid encoded and recent evidence indicates that dehalogenase genes can be associated with transposons.

Dehalogenases normally catalyse the key initial step in the catabolism of halogenated compounds. Halogen removal is the initial step and the efficiency of dehalogenation depends in part on which halogen is involved. The stability of the carbon-halogen bond increases with increasing electronegitivity of the halogen (Hoffman 1950). For example, the C-F bond energy (in CH_3-F it is 109 kcal mol^{-1}) is twice the C-I bond energy (in CH_3-I it is 56 kcal mol^{-1}). So it is simpler to isolate microbes which dehalogenate chloro-substituted compounds than fluoro-compounds.

Three other factors affect microbial dehalogenation mechanisms, namely: compound stereospecificity, the number of halogens per molecule, and the position of the halogen(s) on the substituted molecule. Both L- and D-specific dehalogenases have been described (see later). It is easier to dehalogenate compounds substituted in the C2 position than in other positions, such as the C3 position. Dehalogenases with activity towards 2-halo-substituted compounds do not dehalogenate 3-halo-substituted compounds (Slater et al. 1979). However, microbes with the ability to attack 3-monochloropropionic acid (3MCPA) have been isolated (Bollag and Alexander 1971; Hughes 1988) (Section II.J.3). Generally, the rate of dehalogenation declines with increasing number of halogen substitutions. Also increasing chain length of the substituted compound decreases dehalogenase reactivity.

B. Dehalogenating microbes

Many different genera of dehalogenase-synthesizing microorganism have been isolated. The most commonly reported are *Pseudomonas* species (Davies and

Evans 1962; Goldman 1965, 1972; Goldman et al. 1968; Little and Williams 1971; Senior et al. 1976; Slater et al. 1979; Kawasaki et al. 1981a; Hardman and Slater 1981a; Hardman 1982; Motosugi et al. 1982a,b; Klages et al. 1983; Tsang et al. 1988; Smith et al. 1989). However, bacteria such as *Moraxella* (Kawasaki et al. 1981b), *Rhizobium* (Allison et al. 1983), *Alcaligenes* (Kohler-Staub and Kohler 1989), *Arthrobacter* (Kearney et al. 1964), *Xanthobacter* (Janssen et al. 1985), *Bacillus* and *Microbacterium* (D.J. Hardman, personal communication), and fungi, such as *Trichoderma viride* (Jensen 1959) are known to produce 2HAA dehalogenases. This list is undoubtedly far from complete, especially since plasmid-encoded dehalogenase genes ensure widespread dissemination.

C. Early studies on the catabolism of 2HAAs by 2HAA dehalogenases

Some forty years ago, Jensen proved the existence of these dehalogenases, demonstrating that they were easily assayed in crude cell-free extracts provided that the cells were grown on an appropriate 2HAA (Jensen 1951, 1957, 1960). The expected products, namely hydroxy- and oxo-alkanoic acids, were identified (Section II.F) and it was shown that different enzymes were synthesized in response to different growth substrates. For example, *Trichoderma viride* produced two isoenzymes, one induced by the presence of monochloroacetic acid (MCA) and dechlorinated only this substrate, and the second induced by dichloroacetic acid (DCA) which could dehalogenate both MCA and DCA (Jensen 1959). Jensen classified the enzymes on the basis of substrate specificities, although now these groupings are no longer sustainable. The principle was established that one organism produced different enzymes under different environmental conditions. At the time Jensen's work was remarkably advanced, coming at the time that induction and repression mechanisms were being discovered. Now, since environmentally directed mutations can significantly influence the expression of dehalogenases in *Pseudomonas putida*, these early observations are particularly interesting (Section II.I).

Thirty years ago interest in dechlorination mechanisms was stimulated by the introduction of the herbicide Dalapon by the Dow Chemical Company. This successful herbicide was widely applied and was particularly effective in controlling rye-grasses. Many studies showed that Dalapon's active ingredient, 2,2-dichloropropionic acid (22DCPA), was quickly degraded in soils (Magee and Colmer 1959; Hirsch and Alexander 1960; Macgregor 1963; Kearney et al. 1965; Kearney 1966; Burge 1969; Foy 1975). Microbes dehalogenating 22DCPA as the sole carbon and energy source were isolated and pure cultures obtained. For example, Kearney et al. (1964) isolated an *Arthrobacter* species which removed both halogens from 22DCPA in cell-free extracts and partially purified the enzyme by ammonium sulphate fractionation.

D. *Monofluoroacetic acid dehalogenases sensitive to thiol reagents*

Another important early reason for research centred on the toxic tricarboxylic acid cycle inhibitor, monofluoroacetic acid (MFA) (Goldman 1965; Kelly 1965; Tonomura et al. 1965; Goldman and Milne 1966; Lien et al. 1979). Goldman (1965) isolated a soil pseudomonad which grew on MFA as its carbon and energy source, and demonstrated that the key reaction was catalysed by a hydrolase. Goldman and his colleagues showed in experiments with ^{18}O-labelled water that the fluorine was replaced by the hydroxyl of water (Goldman et al. 1968; Goldman and Milne 1966; Goldman 1972). The product of the reaction, glycolate, was fully characterized and the enzyme partially purified by ion exchange chromatography. The first conclusive evidence of wide substrate specificity was obtained since the MFA dehalogenase also cleaved MCA (at less than 20% of the MFA hydrolysis rate), with the defluoro- and dechloro-activities co-eluting during protein purification. Moreover, the activities showed the same heat denaturation kinetics and it was rightly concluded that the same protein was responsible for both activities. Indeed, growth experiments showed that this was the only dehalogenase produced by the MFA-degrading microorganism. The dehalogenase had a K_m for MFA of 2.4 mM, a tenth of its affinity for MCA, and it was unable to dehalogenate substituted propionates, such as 2-monofluoropropionic acid (2MFPA) and D-2-monochloropropionic acid (D-2MCPA). Unfortunately some of the more commonly used 2HAAs in contemporary studies were not evaluated, such as DL-2-monochloropropionic acid (DL-2MCPA), DCA and 22DCPA. However, from the standpoint of determining the types of mechanisms used by dehalogenases, one important observation was made: the MFA dehalogenase was powerfully inhibited by thiol reagents, such as N-ethylmaleimide, iodoacetamide, and p-chloromercuribenzoate sulphonate. It was proposed that the enzyme's active site contained a sulphydryl group and that the enzyme-bound intermediate during hydrolysis was a thioether compound (Fig. 1). Similar observations were made by Davies and Evans (1962) showing that the dehalogenase synthesized by *Pseudomonas dehalogens* NCIMB 9062 was strongly inhibited by p-chloromercuribenzoate.

Goldman and his colleagues distinguished the MFA dehalogenase from others on the basis of its high specificity towards substituted acetic acids and no activity for halogenated propionic acids. Indeed, this resulted in its own enzyme classification number, EC 3.8.1.3, to distinguish it from other dehalogenases (Section II.E). However, to this author, this seems an artificial distinction since it is based on the legacy of Jensen's attempts to classify dehalogenases on the basis of substrate specificities only, without regard to mechanistic differences or similarities. Furthermore, sub-dividing this type of dehalogenase into two groups, depending only on whether or not they defluorinated MFA, seems equally artificial. It seems more likely, although not proved, that the different substrate specificities simply reflect different evolutionary history of the same basic mechanism to produce "specialist"

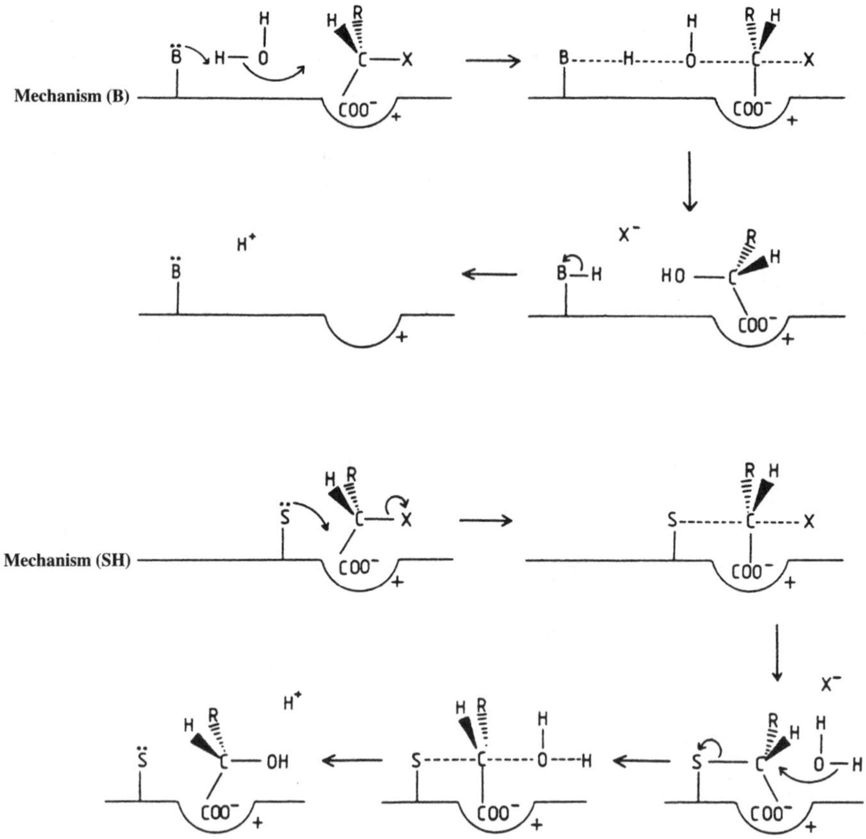

Fig. 1. Mechanisms for the dehalogenation of 2-monochloropropionic acid and other halogenated alkanoic acids by DehI (mechanism B) and DehII (mechanism SH) of *Pseudomonas putida* PP3 (from Weightman et al. 1982).

enzymes with varying substrate profiles. The enzyme isolated by Goldman and his colleagues is discussed here purely because of its historical significance and it should be considered as one of the 2HAA dehalogenases (Section II.E).

E.   Three 2HAA dehalogenases insensitive to thiol reagents

Goldman et al. (1968) isolated a DCA-degrading Pseudomonad that grew on a wide range of substrates and synthesized two dehalogenases, named halidohydrolase I and halidohydrolase II. These enzymes had wider substrate ranges than the MFA dehalogenase (Section II.D), and differed from each other in their capability to dehalogenate HAAs. Halidohydrolase I used MCA as its principle substrate, whilst halidohydrolase II's main substrate was DCA and it dehalogenated this compound at twice the rate halidohydrolase I dechlorinated MCA. Both enzymes dechlorinated L-2MCPA, yielding D-lactate as the product, but only halidohydrolase II used DL-2-bromopropionic acid (2MBPA). Neither utilized D-2MCPA.

A significant difference between the MFA dehalogenase described in Section II.D and these two enzymes was that the latter were unaffected by sulphydryl-blocking reagents. This suggested that these two dehalogenases operated by a different mechanism, accounting for the thiol reagent insensitivity and for the inversion of configuration between a stereospecific substrate (L-2MCPA) and its product (D-lactate) (Section II.F). However, largely on the basis of its ability to dehalogenate substituted propionic acids, these enzymes were classified as EC 3.8.1.3.

In another important study, Little and Williams (1971) re-examined one of Jensen's original isolates, *Pseudomonas dehalogens* NCIMB 9061, purifying one dehalogenase which was similar to the halidohydrolase I isolated by Goldman et al. (1968). Like Goldman's 2HAA dehalogenases, this enzyme showed narrow substrate stereospecificity since it was unable to dechlorinate D-2MCPA, and was similarly unaffected by thiol reagents.

F. Hydrolytic 2HAA dehalogenase mechanisms

Although this chapter has yet to discuss other 2HAA dehalogenases that have been characterized, it is convenient now to outline the dehalogenating mechanisms. All studies to date suggest that 2HAA dehalogenases depend on a hydrolytic mechanism of the general form for mono-substituted alkanoic acids:

$$R - \underset{\underset{Cl}{|}}{CH} - COOH + H_2O \;\rightarrow\; R - \underset{\underset{OH}{|}}{CH} - COOH + H^+ + Cl^-$$

2-chloroalkanoic acid   2-hydroxyalkanoic acid

and for di-substituted alkanoic acids:

$$R - \underset{\underset{Cl}{|}}{\overset{\overset{Cl}{|}}{C}} - COOH + H_2O \;\rightarrow\; R - \underset{\underset{OH}{|}}{\overset{\overset{Cl}{|}}{C}} - COOH + H^+ + Cl^-$$

2,2-dichloroalkanoic acid

$$\rightarrow R - \underset{\underset{O}{\|}}{C} - COOH + H^+ + Cl^-$$

2-oxoalkanoic acid

Other possibilities have been considered. For example, Bollag (1974) suggested that dehalogenases might catalyse a reductive dehalogenation leading to the formation of the unsubstituted alkanoic acid. He also proposed a mechanism based on dehydro-dehalogenation reactions, leading to the introduction of a double bond into the product. There is no evidence for this, although Bollag and Alexander (1971) thought that the dehalogenation of 3MCPA might involve the formation of acrylic acid. At other times different mechanisms have been proposed, but none have been confirmed. For example, Kearney et al. (1964) and Kearney (1966) suggested that 2,2-dichloropropionic acid (22DCPA) was dehalogenated by an oxidative mechanism involving 2-chloroacrylate and 2-chloro 2-hydroxy-propionic acid, followed by spontaneous dechlorination.

The two detailed mechanisms which currently explain all observations are those shown in Fig. 1. The generalized base catalysis mechanism (B mechanism) (Fig. 1a) might involve a histidine residue (Little and Williams 1971), and applies to dehalogenases which are unaffected by thiol reagents and invert substrate-product configuration. The three dehalogenases described in Section II.E (Goldman et al. 1968; Little and Williams 1971) are B mechanism dehalogenases. Dehalogenases which show sensitivity towards sulphydryl-blocking reagents and retain substrate-product configurations (through a double inversion) are examples of reactions which involve an -SH group in the enzymes' active site (SH mechanism) (Fig. 1b).

As noted in Section II.D, too much has been made of comparatively minor differences in substrate specificities, leading to much unnecessary confusion with regard to the types of dehalogenases found in nature. It seems more appropriate at present to consider just two classes of dehalogenase based on the mechanisms given above, until such time as molecular techniques indicate otherwise.

G. The diversity of 2HAA dehalogenases

1. Pseudomonas putida PP3
The two 2HAA dehalogenases of a strain of *Pseudomonas putida* isolated from a stable microbial community growing on the herbicide Dalapon (22DCPA) has been extensively studied (Slater et al. 1976; Senior et al. 1976; Senior 1977; Slater and Bull 1978; Slater et al. 1979). *P. putida* PP3 was a strain which evolved during the growth of the community and was of particular interest since it appeared to acquire dehalogenating capabilities. *P putida* PP3 synthesized two dehalogenases, termed DehI (previously Fraction I) and DehII (previously Fraction II), which dehalogenated a similar range of 2HAAs (Slater et al. 1979; Weightman et al. 1979a,b). Originally, it was thought that the two dehalogenases were closely related, having evolved slightly different substrate specificities following gene duplication of an ancestral gene. This proved not to be the case since the two enzymes had markedly different properties. Their molecular sizes were different: 46 kDa for DehI and 50 kDa for DehII (A.J. Topping, A.J. Weightman and J.H. Slater, unpublished observations).

P. putida PP3 grew on DL-2MCPA and the chloride release data showed that both enantiomers were dechlorinated by both DehI and DehII, each enzyme dechlorinating L-2MCPA at about 80% of the V_{max} for D-2MCPA (Weightman 1981; Weightman et al. 1982). However, DehI retained the same product configuration as the substrate, whilst DehII inverted the product's configuration (Table 1). In addition DehI was strongly inhibited by sulphydryl-blocking reagents whereas DehII was unaffected. Thus a single organism synthesized two dehalogenases, one showing the B mechanism (DehII) and the other the SH mechanism (DehI) (Fig. 1). Recently Barth et al. (1992) showed that the two dehalogenases in *Pseudomonas putida* AJ1 were unrelated at the level of their DNA sequences, although both enzymes were of the B mechanism type (Section II.G.8). Conversely, Schneider et al. (1991) showed that one organism synthesized two closely related enzymes (Section II.G.6).

Table 1. Stereospecificity of 2-monochloropropionic acid dehalogenation by dehalogenases DehI and DehII of *Pseudomonas putida* PP3. The production of L-lactate from DL-, D- and L-2-monochloropropionic acid (from Weightman et al. 1982).

	Fraction I dehalogenase			Fraction II dehalogenase		
	DL-2MCPA	D-2MCPA	L-2MCPA	DL-2MCPA	D-2MCPA	L-2MCPA
Chloride release during lactate preparation (mM)	23.20 ±0.83	27.63 ±1.00	26.02 ±0.36	13.00 ±0.17	6.02 ±0.08	5.42 ±0.14
Amount of L-lactate formed (mM)	12.13 ±0.27	0.62 ±0.03	25.39 ±0.12	6.30 ±0.10	5.83 ±0.06	0.49 ±0.05
Percentage of dechlorinated 2MCPA yielding L-lactate	52.2	2.2	97.7	48.5	96.7	9.3

The dehalogenases were incubated with each substrate (50 mM starting concentration) for 12 h (DehI) or 1 h (DehII). The errors quoted are standard deviations of three determinations for the chloride release values and two determinations for the L-lactate concentrations.

Thomas (1990) sequenced a 2561 bp *Sma*I fragment known to carry part of DehI and its regulator gene DehR$_I$ which was part of the transposon *DEH* (Thomas et al. 1992a,b). It was shown that this fragment failed to yield a functional DehI but recently the complete sequence has been obtained from another fragment (A.J. Topping, A.J. Weightman and J.H. Slater, unpublished observations). There is no significant homology with any other sequenced dehalogenase. DehI has not been purified adequately for the N-terminal amino acid sequence to be determined. However, Murdiyatmo (1991) purified an enzyme termed HdlV from laboratory isolate K37 and determined the first 13 amino acids. Inspection of the predicted amino acid sequence of DehI showed that this corresponded exactly with the expected sequence starting from the second methionine residue in the putative DehI open reading frame. Within the sequence between the first and second methionine residues was a strong Shine-Dalgarno region separated by 8 bases from the initiation codon, a distance considered to be optimal for transcription (Gold 1988). However, another possible Shine-Dalgarno sequence 10 bases up-stream from the first methionine (initiation) codon was also found. Accordingly, the actual polypeptide sequence of DehI awaits confirmation.

A putative promotor of the $-12/-24$ type was identified upstream from the initiation codon. These promotors are recognized by sigma-54 RNA polymerases, normally considered to be associated with nitrogen-fixing genes (Thony and Hennecke 1989; Dixon 1986; Merrick and Stewart 1985). Expression of DehI was dependent on the presence of a functional sigma-54 polymerase since it was not expressed in *rpoN* mutants of *P. putida* (Thomas et al. 1992b). At the moment this is the only dehalogenase known to depend on a sigma-54 dependent promotor.

A further relationship with nitrogen-fixing genes was shown for the regulatory gene *dehR$_I$*. From the same fragment used to determine the *dehI* sequence, another open reading frame, on the opposite DNA strand and reading in the opposite direction, was found (Thomas 1990; A.W. Thomas, A.W. Topping, A.J. Weightman and J.H. Slater, unpublished observations). This sequence showed significant similarity with other positive regulatory genes, including *nifA* and *ntrC* of *Klebsiella pneumoniae*, *nifA* of *Rhizobium meliloti*, and *tyrR* of *Escherichia coli* and *xylR* of *P. putida*. Three of the four domains of the proteins, regions B, C and D, were substantially homologous, and variations in the remaining domain, region A, presumably accounted for the regulators' specificity.

2. Pseudomonas *species 113*

Motosugi et al. (1982a,b) isolated many soil microbes growing on DL-2MCPA and found two groups of isolates. Group 1 organisms grew only on L-2MCPA and could not dehalogenate or grow on D-2MCPA. Group 2 organisms grew separately on either L-2MCPA or D-2MCPA. One strain which grew on both D-2MCPA and L-2MCPA, *Pseudomonas* species 113, had a single dehalogenase which, unlike DehI or DehII of *P. putida* PP3, dechlorinated the L-isomer faster than the D-isomer (72% of the L-isomer rate). A dimeric enzyme with a molecular size of 68 kDa was purified, shown to be unaffected by sulphydryl-blocking reagents and inverted the configuration between substrate and product. On the basis of these properties, this enzyme was similar to DehII of *P. putida* PP3.

Recently this enzyme has been used to catalyse the dehalogenation of long-chain haloalkanoic acids and aromatic substituted haloalkanoic acids using lyophilized preparations of the purified dehalogenase dissolved in organic solvents (Hasan et al. 1991). The enzyme was found to be stable and catalytically active in solvents such as anhydrous dimethyl sulphoxide (DMSO) and toluene, dehalogenating compounds such as 2-chloropentanoic acid, 2-bromooctanoic acid, 2-bromohexadecanoic acid and 2-bromo-2-phenylacetic acid to yield the corresponding inverted 2-hydroxyalkanoic acids, such as 2-hydroxypentanoic acid, 2-hydroxyoctanoic acid, 2-hydroxyhexadecanoic acid, and 2-hydroxy-3-phenylpropanoic acid, respectively. This is probably the first enzyme found to be active in anhydrous DMSO. The rates of dechlorination for some substrates were almost three times the rate of L-2MCPA dehalogenation in aqueous solutions: for example, 2-bromohexadecanoic acid was not dechlorinated in aqueous solution, but was dechlorinated at 2.7 times the aqueous L-2MCPA rate. L-2MCPA itself was only dehalogenated at 6% of the aqueous rate. It was postulated that the enzyme's activity on higher chain length haloalkanoic acids was due to their greater solubility in organic solvents.

3. Rhizobium *species*

Skinner and his colleagues (Berry et al. 1976, 1979; Allison et al. 1983; Leigh et al. 1988) isolated a fast-growing, Gram-negative microorganism, identified as a *Rhizobium* species (although there is some doubt as to the exact classification – D.J. Hardman, personal communication), growing on 22DCPA and 2MCPA. Many other 2HAAs served as enzyme inducers but most were not substrates. Unusually MCA was dehalogenated by crude cell-free extracts but it did not act as an enzyme inducer. Originally, two dehalogenases were identified (Allison et al. 1983) but later it was found to synthesize a third dehalogenase (Leigh et al. 1988). Dehalogenase I showed activity only to L-2MCPA and DCA, inverted substrate-product configuration and had a molecular size of 93 kDa. Leigh et al. (1988) considered the enzyme to be similar to the B mechanism dehalogenases of Goldman et al. (1968) and *Pseudomonas dehalogens* (Little and Williams 1971). Dehalogenase II degraded both D-2MCPA and L-2MCPA at similar rates and inverted the substrate-product configuration and was probably similar to DehII of *Pseudomonas putida* PP3 and the dehalogenase of *Pseudomonas* species 113. However, unexpectedly it showed some sensitivity towards thiol reagents and so there must be doubts about its exact position in the general classification of dehalogenases.

Dehalogenase III was a B mechanism enzyme with the unusual property of only catalysing the dechlorination of D-2MPCA, but not L-2MCPA, and inverting the configuration of the product (L- lactate). In this respect, it was similar to one of the dehalogenases described in *Pseudomonas putida* AJ1 (Section II.G.8).

4. Moraxella *species B*

Kawasaki and associates (Kawasaki et al. 1981b,c,d) described two dehalogenases, H1 and H2, from an organism enriched with MFA as the sole carbon and energy source. Dehalogenase H1 dechlorinated both MFA and MCA, whilst H2 only degraded MCA. H2 was purified and found to be sensitive towards sulphydryl-blocking reagents. Unfortunately, its response towards stereospecific compounds was not investigated.

In this study mutants unable to grow on MFA and MCA were obtained by treatment with mitomycin C, a compound used to prevent plasmid replication yielding strains without plasmids (Kawasaki et al. 1981c) (Section VI).

5. Pseudomonas *species A*

Another isolate was obtained by Kawasaki et al. (1981a) by growth on MFA and shown to have an H1-type dehalogenase. It is possible that this strain evolved as a result of plasmid transfer involving the *Moraxella* species B.

6. Pseudomonas *species CBS3*

This Pseudomonad was isolated on 4-chlorobenzoate as the sole carbon and energy source (Klages and Lingens 1980). As well as mechanisms for dehalogenating chlorinated aromatic compounds, two HAA dehalogenases (DehCI and DehCII) which dehalogenated MCA and 2MCPA were synthesized. Both enzymes dehalogenated L-2MCPA to yield D-lactate, but were unable to attack D-2MCPA (Klages et al. 1983; Morsberger et al. 1991). In this respect they were similar to the dehalogenases of *Xanthobacter autotrophicus* GJ10 (Section II.G.9) and *Pseudomonas cepacia* MBA4 (Section II.G.7), but unlike the dehalogenase of *Pseudomonas putida* AJ1 (Section II.G.8). The molecular sizes of the active proteins were estimated to be 42 kDa (Klages et al. 1983) and, since SDS-PAGE estimated the molecular sizes of both DehCI and DehCII at 28 kDa, these dehalogenases were evidently dimeric proteins.

Recently, Schneider et al. (1991) cloned and sequenced the genes for DehCI (*dehCI*) and DehCII (*dehCII*) from small fragments (a 1.1 kb *Sma*I - *Sst*I fragment for *dehCI* and a 1.0 kb *Bam*HI fragment for *dehCII*). In various vectors carrying these fragments, dehalogenase expression was constitutive and, since both dehalogenases carried their own promotors and were not under the control of the vector's promotor, this indicated a negative regulatory mechanism.

The *dehCI* sequence had an open reading frame producing a protein of 227 amino acids with a molecular size of 25,401 Da which corresponded well with the value determined from the protein (M_v = 28 kDa). The first 23 amino acids predicted from the DNA sequence corresponded exactly with the N-terminal amino acid sequence from DehCI. A Shine-Dalgarno ribosome binding region was found close to the start of the open reading frame, and a promotor of the archetypal $-10/-35$ consensus sequence located.

For the *dehCII* sequence the first eight amino acids predicted corresponded with the observed N-terminal amino acid sequence and overall coded for a protein of 229 amino acids with a molecular size of 25,683 Da. Again, suitable Shine-Dalgarno and $-10/-35$ promotor regions were found.

The two genes showed 45% homology at the base sequence level, 37.5% at the identical amino acid level, and more than 70% at the similar amino acid level. Clearly these two proteins were closely related. Moreover comparison with the amino acid sequence predicted for dehalogenase DehIVa showed a close relationship (Murdiyatmo et al. 1992). For DehIVa and DehCI there was 67% matching at the identical amino acid level, and 81% matching at the similar amino acid level. For DehIVa and DehCII the values were 37% and 56% respectively.

7. Pseudomonas cepacia *MBA4*

This organism, a batch culture isolate growing on monobromoacetic acid (MBA), had two dehalogenases (Tsang et al. 1988). Dehalogenase III was not studied in detail but dehalogenase IVa (DehIVa) showed substrate stereospecificity for L-2MCPA, producing D-lactate, and was insensitive to thiol reagents, indicating that it was a B mechanism enzyme (Murdiyatmo, 1991; Asmara 1991; Murdiyatmo et al. 1992; Asmara et al. 1992).

The structural gene for DehIVa (*hdlIVa*) was isolated from a 1600 bp genomic DNA fragment and cloned onto vector pBTac1 under the control of the *tac* promoter (Murdiyatmo 1991; Murdiyatmo et al. 1992). Analysis of the DNA sequence located an open reading frame coding for a protein of 231 amino acids and a molecular size of 25.9 kDa. SDS-PAGE had previously suggested a molecular size of 23 kDa, whilst gel filtration suggested that the value was about 45 kDa, leading to the conclusion that the active protein was a dimer (Tsang et al. 1988). This compared well with the information from *Pseudomonas* strain A (Section II.G.5), *Pseudomonas* species CBS3 (Section II.G.6), and *Moraxella* (Section II.G.4). Comparison with the known N-terminal sequence from the purified protein fixed the start of the coding sequence. A Shine-Dalgarno ribosome binding region was found close to the start of the structural gene. Two promotor regions of the $-10/-35$ consensus sequence type characteriztic of the sigma-70 RNA polymerase (*rpoD*) were found and expression was under positive control.

Recently Asmara et al. (1992) in experiments involving chemical modification of DehIVa, and random and site-directed mutagenesis identified two amino acid residues, His-20 and Arg-42, as the key residues for catalytic activity. Asp-18 was also implicated possibly by positioning the correct tautomer of His-20 in the enzyme-substrate complex. Interestingly similar amino acid residues seem to be involved in the active site of a hydrolytic haloalkane dehalogenase (Janssen et al. 1989; Franken et al. 1991).

8. Pseudomonas putida *AJ1*

This organism, isolated from soil pre-enriched with 2MCPA, contained two dehalogenases (encoded by genes *hadD* and *hadL*) (Taylor 1985, 1988; Barth 1988; Smith et al. 1989a,b, 1990; Barth et al. 1992). Uniquely, both enzymes were stereospecific, one for D-isomers and the second for L-isomers. The D-2MCPA specific dehalogenase was purified and shown to invert the product configuration (L-lactate). This was an unusual enzyme since the molecular size was much greater (135 kDa) than for other dehalogenases previously reported and was composed of four identical sub-units (Smith et al. 1990).

Barth et al. (1992) cloned the genes for both dehalogenases (*hadD* and *hadL*) by shot-gun cloning *P. putida* AJ1 genomic DNA into vectors pTB107 and pTB244 which were transformed into *E coli* C600. Eight clones were selected with the expected plasmid markers, and for the presence of the D-2MCPA specific dehalogenase gene (*hadD*) using small oligonucleotides as specific *hadD* probes. Plasmids were mobilized or transformed into two strains of *P. putida* or *E coli* J53 and transconjugants selected for growth on L-2MCPA and D-2MCPA. Both substrates were used but since only *hadD* had been screened, this strongly suggested that *hadD* and *hadL* were closely positioned on the genome. For some transconjugants carrying inserts of 4 kb, the dehalogenase activities were constitutive. For other transconjugants carrying inserts of either 10 kb or 13 kb, the activities were induced by MCA as in the parent strain. This suggested that the larger fragments carried a functional regulatory system in *P. putida* AJ1.

A sub-cloning procedure separated the *hadD* and *hadL* genes. Analysis of plasmid pTB316 showed that it expressed the D-2MCPA specific dehalogenase, and SDS polyacrylamide gel analysis found that strains containing this plasmid synthesized a new protein with a molecular size of about 30 kDa. This plasmid contained an insert of 1752 bp corresponding to the complete *hadD* gene and about half the *hadL* gene. A suitable open reading frame gave a DNA sequence corresponding exactly to the first 21 amino acid from the N-terminal amino acid sequence of the purified D-2MCPA dehalogenase. The nucleotide sequence predicted a protein of about 300 amino acids, a molecular size of 33,601 and an overall amino acid composition which was consistent with that of the purified protein. Upstream of *hadD* were two open reading frames, one of which contained a Shine-Dalgarno region. This region, named *hadA*, did not have a -35/-10 sequence characteriztic of a sigma-70 promoter region (Section II.G.7). This D-specific dehalogenase did not have any sequence homology with any other dehalogenase.

9.   Xanthobacter autotrophicus *GJ10*

This bacterium was isolated on 1,2-dichloroethane (DCE) as the sole carbon and energy source (Janssen et al. 1985) but, in addition to a haloalkane dehalogenase (Section III.B.1), it also synthesized a 2HAA dehalogenase. The enzyme was active against halogenated acetates, propionates and butyrates, dehalogenating L-2MCPA to D-lactate and was not inactivated by thiol reagents. The molecular size of the protein was estimated to be between 28 and 36 kDa and from gene (*dhlB*) sequence data to be 27.3 kDa (van der Ploeg et al. 1991).

10.   Alcaligenes *strain CC1*

This microorganism was isolated from an activated sewage sample on *trans* 3-chlorocrotonic acid, and shown to synthesize a constitutive 2HAA dehalogenase that did not act on 3-chloroalkanoic acids (Kohler-Staub and Kohler 1989). The organism grew on various 2HAAs including 2-chlorobutyric acid, 2-monochloropropionic acid and monochloroacetate with growth rates varying between 0.05 and 0.11 h^{-1}. However, the interest in this organism is more concerned with its ability to dehalogenate four carbon 3-haloalkanoic acids (Section II.J.1).

H. Environmental influences on the physiology of 2HAA dehalogenases

1. Growth rate and growth substrates

Previous sections have shown the range of characterized 2HAA dehalogenases and demonstrated that the expression of dehalogenases can be variable, dependent on the growth conditions and the specific growth rate of the organism. More dehalogenases will undoubtedly be discovered if greater attention is paid to important ecological and physiological conditions. Hardman and Slater (1981a) found dehalogenase isoenzyme variation on the basis of electrophoretic mobility in activity gels of 16 independently isolated Gram-negative soil microorganisms (Fig. 2). This study, although it provided no details of the type of mechanisms used by the different dehalogenases, showed that over 75% of the isolates synthesized at least two dehalogenases following batch culture on MCA or 2MCPA. One isolate, *Pseudomonas* species E6, produced at least four different dehalogenases. Moreover, later experiments showed that the dehalogenase activity in one strain, *Pseudomonas* species E4, depended on whether the organism was grown in open or closed culture systems, the nature of the growth-limiting substrate and the specific growth rate of the population (Hardman and Slater 1981b). Continuous-flow culture revealed that a third dehalogenase, not seen in batch grown cells, was synthesized by this organism (Fig. 3). Other isolates showed the same propensity to synthesize additional dehalogenases under different growth conditions (Hardman 1982, 1991). These studies did not elucidate the physiological significance of the variations in expression, and it may be that the variations were due to factors such as post-translational modifications (Hardman 1991). Dehalogenase III of *Pseudomonas cepacia* MBA4 was only synthesized during continuous-flow culture (Tsang et al. 1988). The specific growth rate also influenced the level of expression of DehI and DehII activity in *Pseudomonas putida* PP3 grown in continuous-flow culture (Slater et al. 1979; Weightman and Slater 1980) and, since this is a general physiological response, it undoubtedly will occur in other strains (Clarke and Lilly 1969).

2. Inhibitory HAAs

Many HAAs are toxic and uptake into cells causes cessation of growth and cell death. For example, addition of DCA at the normal growth concentration of 0.5 g carbon l^{-1} to a population of succinate-growing *P. putida* PP3 caused an immediate cessation of growth followed by the induction of DehI and DehII. Once the dehalogenases were synthesized, the DCA concentration decreased, eventually reaching a low enough concentration for growth to resume (Slater et al. 1979). Little is known about the mechanisms of HAA assimilation but the process is probably an active one involving specific permeases (Slater et al. 1985). For cells which were already induced for dehalogenase activity, for example by growth on 2MCPA, the addition of a toxic compound, for example 2-monobromobutanoic acid (2MBBA), was more traumatic than for non-induced cells, presumably because the compound entered the induced cells

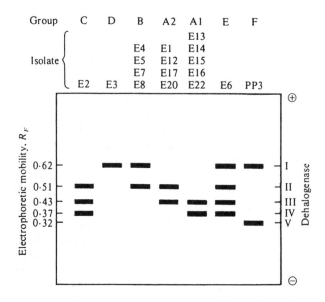

Fig. 2. Distribution of five dehalogenases from 16 soil bacteria and *Pseudomonas putida* PP3 after separation by polyacrylamide gel electrophoresis. The bars indicate that activity was detected at any level without reference to specific enzyme substrates (from Hardman and Slater 1981a).

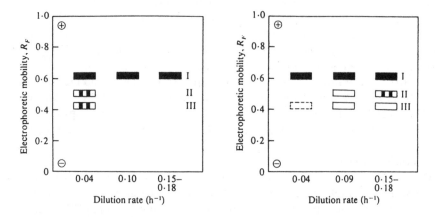

Fig. 3. Patterns of dehalogenases I, II and III activities towards a mixture of monochloroacetic acid and dichloroacetic acid for *Pseudomonas* species E4 grown at different dilution rates on: (a) monochloroacetic acid and (b) 2-monochloropropionic acid as the growth-limiting substrates. The bars represent an assessment of enzyme activity towards monochloroacetic acid and dichloroacetic acid ranging from slight to great in the sequence (from Hardman and Slater 1981b).

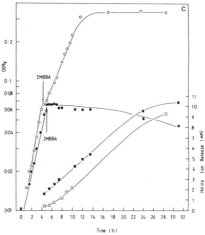

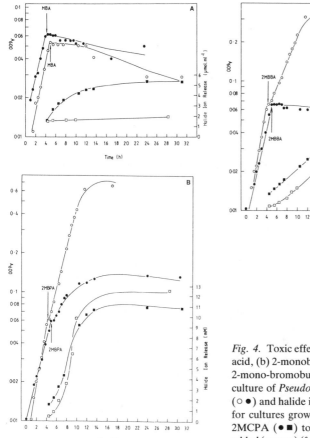

Fig. 4. Toxic effects of (a) monobromoacetic acid, (b) 2-monobromopropionic acid and (c) 2-mono-bromobutanoic acid on growing culture of *Pseudomonas putida* PP3. Growth (○ ●) and halide ion concentration measured for cultures growing on succinate (○ □) and 2MCPA (● ■) to which MBA and 2MBBA added (arrows) (from Weightman et al. 1985).

much more rapidly (Fig. 4) (Weightman et al. 1985). Addition of 2MBBA to non-induced cells resulted in little growth perturbation since induction of the dehalogenase system concomitantly with the uptake system ensured that there was no intracellular accumulation of 2MBBA.

It is possible to select mutants of dehalogenase-producing microorganisms that are resistant to toxic levels of HAAs (HAA^R mutants) (Weightman 1981; Slater et al. 1985; Weightman et al. 1985). In some cases, and in particular with the HAA^R mutants of *P. putida* PP3, the frequency of HAA^R mutation is unusually high. The significance of these observations are discussed in Section II.I. The mutations all involved attenuation in the rate of uptake of HAAs. In other instances, for example MBA^R strains of *Xanthobacter autotrophicus* GJ10, the frequency of mutation was lower and resistance was due to elevated dehalogenase levels, presumably allowing more rapid degradation of the toxic compound once it had entered the cell (van der Ploeg et al. 1991). Conversely it was observed for mutants of *P. putida* PP3 producing 10-fold greater amounts of DehI, that the sensitivity to toxic HAAs was increased. In this case, it was argued that although the rate of dehalogenation was increased, the rate of uptake of HAAs was also increased and the balance between uptake and dehalogenation was such that the net effect was greater sensitivity to HAAs (Weightman and Slater 1980).

I. HAA^R mutants of Pseudomonas putida PP3

Mutants of *P. putida* PP3 resistant to between 20 to 100 mM MCA and DCA were selected at high frequencies. For example, the mutation frequencies were 3.2 x 10^{-1} for MCA and 5.3 x 10^{-3} for DCA (both at 42 mM) after 10 days incubation (Slater et al. 1985). Time of incubation greatly affected the calculated frequencies since mutants appeared after about two days' incubation and new mutants continued to appear up to 14 days' incubation. The mutation frequency was increased to 0.01 after 9 days' starvation in basal medium lacking a carbon source. When 20 mM TCA (a good dehalogenase inducer but a very poor dehalogenase substrate and a non-growth compound) was included under the same starvation conditions, the frequency of DCA^R mutation was 1.0.

Analysis of the mutants revealed that there were five classes of DCA^R and MCA^R mutant (Table 2) (Slater et al. 1985). All the mutants had decreased capability to assimilate HAAs compared with *P. putida* PP3 and five classes were distinguished by their ability to synthesize DehI and DehII. For example, the PP411 class synthesized only DehI, the PP412 class produced only DehII, and the PP40 class synthesized neither dehalogenase. In view of the mutation frequencies, it was postulated that the mutant classes were due to the loss of one or more of three transposons carrying the two dehalogenases genes and one of the uptake systems (Slater et al. 1985). However, subsequent analysis of the two dehalogenases showed that only DehI was located on a mobile DNA

Table 2. Characteriztics of various dichloroacetic acid resistant mutants derived from *Pseudomonas putida* PP3 and comparison with *P. putida* PP1 and PP3 (from Slater et al. 1985).

Strain or mutant class	Growth rate on 2MCPA (h^{-1})[a]	Response to DCA (85 mM)	Dehalogenase complement[b]		Permease activity[c]	% Total mutants
			I	II		
PP1	0	R	–	–	ND	...
PP3	0.31 ± 0.02	S	+	+	8,000	...
PP40	0	R	–	–	100	16
PP411	0.14 ± 0.02	R	+	–	1,900	54
PP412	0.23 ± 0.02	R	–	+	2,100	7
PP4120	0.04 ± 0.01	R	–	+	100	3
PP42	0.26 ± 0.03	R	+	+	657	20

ND = not determined.

[a] Based on $n = 3$ independent determinations, except for PP412 ($n = 4$) and PP4120 ($n = 6$).

[b] Determined by rapid gel electrophoresis.

[c] Determined from the initial rate of uptake of (I-^{14}C)-MCA. Units are counts per minute of radioactivity take up per minute of incubation by 1 mg of cell protein.

element (Beeching et al. 1983; Thomas 1990; Thomas et al. 1992a,b).

Most recently, it was demonstrated that the transposition of DehI was under environmental control since favourable mutations involving DehI (and DehII) occurred in response to specific environmental conditions, namely, starvation, the presence of toxic HAAs and the presence of metabolisable HAAs (Thomas et al. 1992c). Fluctuation tests demonstrated that the movement of DehI, to produce the various classes of mutants described above, was in response to the nature of the environment. The mutations involved complex rearrangements of DNA associated with DehI resulting in the switching off of DehI and/or DehII synthesis to protect cells against the toxic effects of HAAs. Under different conditions where toxic compounds were absent, dehalogenase synthesis was restored to enable organisms to utilize suitable HAAs as growth substrates.

J. Catabolism of other HAAs

1. Alcaligenes *species CC1*
This organism, isolated on 5 mM *trans* 3-chlorocrotonic acid, utilized both *cis* 3-chlorocrotonic acid and 3-chlorobutyric acid (Kohler-Staub and Kohler 1989). Under induced conditions, cell-free extracts dechlorinated the 3-substituted acids in a reaction which was strictly dependent on CoA, ATP and Mg^{2+}. Although CoA esters were not detected, these authors proposed the catabolic pathways shown in Fig. 5. The initial reaction was not a hydrolytic dehalogenation, which seems the ubiquitous mechanism for 2HAA dehalogenation, but was more akin to mechanisms proposed for anaerobic dehalogenations (Section IV).

2. An unidentified coryneform strain FG41
Van Hylckama Vlieg and Janssen (1991) isolated a coryneform bacterium strain 41 by growth on *trans* 3-chloroacrylic acid (3CAA). The isolate grew on *cis* 3CAA and crotonic acid but could not grow on a variety of other halogenated compounds. The organisms synthesized two dehalogenases, one specific for *cis* 3CAA and the other specific for *trans* 3CAA, and both yielded hydroxymuconic semialdehyde as the product. Both enzymes were induced by *trans* 3CAA, *cis* 3CAA and 3-chlorocrotonic acid and had pH optima in the range pH 7.5 to 8.0.

Despite the similar, restricted catalytic activity, analysis of the two proteins showed that they were not closely related (cf. DehI and DehII in *Pseudomonas putida* PP3 – Section II.G.1). The *cis* 3CAA dehalogenase had a molecular size of 38 kDa and was composed to two identical sub-units of 16.2 kDa. The *trans* 3CAA dehalogenase had a molecular size of 50 kDa, contained two unequal sub-units of 8.7 and 7.4 kDa, and was probably a hexameric protein. *N*-terminal amino acid analysis confirmed the lack of polypeptide relationship, Yet again this study serves to illustrate the wide diversity of hydrolytic dehalogenases, even amongst those which have evolved to attack very similar substrates.

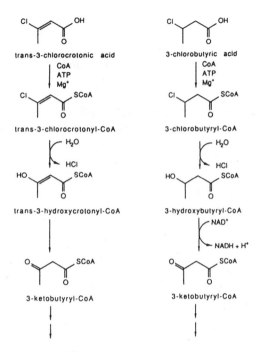

Fig. 5. Proposed pathway for the catabolism of *trans*-3-chlorocrotonate and 3-chlorobutyrate by *Alcaligenes* species CC1 (from Kohler-Staub and Kohler 1989).

3. 3MCPA-degrading microorganisms

Hughes (1988) isolated a variety of 3MCPA-utilizing yeasts and bacteria from both batch and continuous culture enrichments. Most of the isolates grew slowly, but one, *Pseudomonas pickettii* strain SH1, had a growth rate of $0.13\,h^{-1}$ in batch culture. None of the isolates grew on 2HAAs suggesting a different dehalogenating mechanism. Enzyme activity was not visualized in activity gels suggesting a more complex reaction requiring specific cofactors. Whole cell assays showed that O_2 was required for 3MCPA dechlorination. O_2 uptake experiments indicated that *trans* 3-chloroacrylic acid (*trans* 3CAA) and acrylate could stimulate 3MCPA degradation. The 3MCPA degradation product was identified as 3-hydroxypropionate. On the basis of these results Hughes (1988) proposed that the pathway was one of two mechanisms: either there was an initial dehydrogenase reaction which simultaneously removed the chloride, yielding acrylic acid which was subsequently hydrated; alternatively, there was an initial dehydrogenase reaction which did not result in the removal of the halide, yielding *trans* 3CAA which in turn was hydrated with simultaneous chloride removal to produce 3-hydroxypropionate also. Clearly the similarity with the coryneform FG41 (Section II.J.2) is notable.

III. Dehalogenation of haloalkanes under aerobic conditions

A. Some basic considerations

Haloalkanes are important environmental contaminants because of their toxicity and apparent recalcitrance. Natural sources of haloalkanes are known (Gschwend et al. 1985) but most occur in the biosphere through the activity of man (Keith and Teilliard 1979). Thus understanding the biodegradation mechanisms are important for the development of practical treatment systems (Galli and Leisinger 1985). Many of the general considerations that apply to haloalkanoic acids (Section II.A) also apply to haloalkanes. However detailed studies on the microbiology and biochemistry of haloalkane catabolism is less complete because of the comparatively greater difficulties in isolating pure cultures of microbes with the required capabilities. In part this may be due to the importance of co-metabolic processes, as is the case under anaerobic conditions (Section IV) and, in part, because these substrates are significantly more toxic even to those microbes with the potential to degrade haloalkanes. Consequently it was supposed that haloalkanes were non-biodegradable until Omori and Alexander (1978b) found that about 1% of soil samples contained microorganisms with the ability to transform 1,9-dichlorononane. Later, Brunner et al. (1980) demonstrated the microbial degradation of dichloromethane (DCM). Wilson and Wilson (1985) showed that trichloroethylene was degraded to CO_2 by soil columns exposed to 0.6% natural gas (methane) in air under aerobic conditions. Murphy and Perry (1983, 1984) found that three fungi, *Cunninghamella elegans*, *Penicillium zonatum* and *Candida lipolytica*, used 1-chlorohexadecane and 1-chlorooctadecane as sole carbon and energy sources. In addition, the fatty acids of these organisms were heavily (60 to 70% of the fatty acids were halogenated) halogenated. Recently, we have isolated mesophilic and thermophilic microbes able to use a range of chloro-propanes, -pentanes and -hexanes, but the enrichments have been long and slow probably because the microbes required a period of adaptation to grow in the presence of toxic compounds (Endang Soetarto, A.T. Bull, D.J. Hardman and J.H. Slater, unpublished observations). In this regard, it will be very interesting to see if haloalkane-utilizing microorganisms exhibit the same properties of resistant phenotypes seen with some of the haloalkanoic acid utilizers (Section II.I).

There is a greater diversity of enzyme mechanisms involved in haloalkane degradation, including NADH-linked mechanisms (Omori and Alexander 1978a,b) oxygenases (Hartmans et al. 1985, 1986; Yokota et al. 1986), glutathione (GSH)-dependent dehalogenases and hydrolytic dehalogenases (Kohler-Staub and Leisinger 1985; Janssen et al. 1985; Yokota et al. 1986). In general the pH profiles are broad but the optimum pH value is slightly lower than for 2HAA dehalogenases. In common with 2HAA dehalogenases, haloalkane dehalogenases have wide substrate activity ranges.

B. Hydrolytic haloalkane dehalogenases

1. Xanthobacter autotrophicus *GJ10*

This organism has already been partly considered since it produced a 2HAA dehalogenase (Section II.G.9) (Janssen et al. 1985). However, the organism was isolated as a dichloroethane (DCE) utilizer and crude cell-free extracts contained two dehalogenases. One was heat labile with a pH optimum about 7.5 and the second was heat stable with a pH optimum of about 9.0. The former was found to act on various C_1 to C_4 substituted haloalkanes, such as DCE, bromoethane, 1-chloropropane, 1-chlorobutane, 1,3-dichloropropane and 3-chloropropene, yielding various corresponding alcohols, such as ethanol, 1-propanol and 1-butanol, as products of the reactions. This enzyme also dehalogenated similar bromo- and iodo-alkanes. In *Xanthobacter autotrophicus* GJ10 the enzyme was constitutively expressed and represented about 2 to 3% of the total soluble protein. The latter enzyme was a HAA dehalogenase discussed previously (Section II.G.9).

On the basis of these results Janssen and his colleagues proposed that the two dehalogenases were part a simple pathway involving two separate dehalogenation reactions as follows:

$$\underset{(a)}{CH_2ClCH_2Cl \xrightarrow[]{H_2O \quad HCl}} CH_2ClCH_2OH \underset{(b)}{\xrightarrow[]{X \quad XH_2}}$$

$$\underset{(c)}{CH_2ClCHO \xrightarrow[]{H_2O + Y \quad YH_2}} CH_2ClOOH \underset{(d)}{\xrightarrow[]{H_2O \quad HCl}} CH_2OHCOOH$$

where reaction (a) is catalysed by the haloalkane dehalogenase; (b) by a specific alcohol dehydrogenase (Janssen et al. 1987a); (c) by a specific aldehyde dehydrogenase (Janssen et al. 1987a); and (d) by a 2HAA dehalogenase.

The purified hydrolytic haloalkane dehalogenase had a molecular size of 36 kDa and did not require any cofactors (Keuning et al. 1985). The enzyme had a limited substrate range, dehalogenating terminally substituted haloalkanes with up to four carbon atoms in the case of 1-chloroalkanes and up to at least 10 carbon atoms for 1-bromoalkanes (Janssen et al. 1989). The enzyme-substrate affinities were not very high: for example, the K_m for DCE was 1.1 mM. The enzyme had a pH optimum of 8.2. Janssen et al. (1988) proposed that the enzyme catalysed a nucleophilic substitution reaction with water and showed that the reaction was strongly inhibited by thiol reagents (Keuning et al. 1985). Despite some superficial similarities to 2HAA dehalogenases, Keuning et al. (1985) found that there was no immunological cross reaction with the 2HAA dehalogenase from the same organism, suggesting that these two enzymes probably did not share a common ancestral gene.

Using the cosmid vector pLAFR1, a gene bank was prepared from which the *dhlA* gene was sub-cloned. It efficiently expressed from its own promotor in strains of *Pseudomonas* and *Escherichia coli*. The nucleotide sequence was determined (Janssen et al. 1989) and the enzyme found to have 310 amino acid residues and a molecular size of 35 kDa. Originally, it was proposed that one of its four cysteine residues was involved in the reaction centre in view of the inhibitory effect of thiol reagents. However, the detailed protein crystal structure of the enzyme has recently been determined (Rozeboom et al. 1988; Franken et al. 1991). The three dimensional structure showed that the monomeric enzyme was spherical with two main domains. Domain I had an α/β structure with a central eight-stranded parallel β sheet. Overarching domain I was a second domain II consisting of α helices connected by loops. The active site was located within an internal hydrophobic cavity and it was proposed that Asp-124 was the nucleophilic residue essential for catalysis. This residue interacted with His-289 and a hydrogen-bonded partner Asp-260. Franken et al. (1991) noted the similarity of the overall structure to a dienelactone hydrolase (Pathak et al. 1988; Pathak and Ollis 1990) although the total sequence homology was poor. However, strikingly homologous stretches were found in the putative active sites. For example, dehalogenase residues Asp-260 and His289 were at equivalent positions in the hydrolase at Asp-171 and His-202.

Another *Xanthobacter autotrophicus* strain GJ11 with a similar capability to grow on DCE is discussed in Section III.B.6.

2. Xanthobacter autotrophicus *GJ70*

This organism, originally identified as an *Acinetobacter* species (Janssen et al. 1987b), was isolated from activated sludge by growth on 1,6-dichlorohexane. It grew on a range of compounds such as 1-chlorobutane, 1-chloropentane 1-bromopentane, ethyl bromide, 1-iodopropane and 1,9-dichlorononane. Under inducing conditions a monomeric protein, comprising about 4% of the total cellular protein, was identified as the active haloalkane dehalogenase with a molecular size of 28 kDa. The enzyme contrasted significantly with the dehalogenase from strain GJ10 by virtue of its very wide substrate range (Table 3), particularly an ability to dehalogenate chloroalkanes with a chain length greater than C_5. The enzyme had no activity against HAAs but had significant activities against haloalcohols, haloethers, and haloalkanes substituted elsewhere on the molecule, such as 1-phenyl 2-bromopropane. It showed a broad pH optimum over pH 8.0 to 9.0 and K_m values in the range 0.1 to 0.9 mM at least for the short chain haloalkanes. The enzyme was slightly inhibited by sulphydryl-blocking reagents (10 to 30%). For chiral substrates, such as 2-bromobutane, the dehalogenase reaction inverted the product configuration. Janssen et al. (1988) proposed that this enzyme functioned either via a nucleophilic attack involving a carboxyl residue or via a general base reaction. This enzyme was markedly different from the *Xanthobacter autotrophicus* GJ10 enzyme, again illustrating the diversity of enzyme types involved in the dehalogenation of haloorganic compounds.

Table 3. Substrate range of the dehalogenase from *Xanthobacter autotrophicus* GJ70 (from Janssen et al. 1988).

Substrate	Rate
	%
Halomethanes	
Methylbromide	143
Methyliodide	75
Dibromomethane	13
Bromochloromethane	5
n-*Haloalkanes*	
Bromoethane	143
Iodoethane	93
l-Chloropropane	15
l-Bromopropane	100
l-Iodopropane	66
3-Chloropropene	139
l-Chlorobutane	66
l-Chloropentane	65
Secondary haloalkanes	
2-Bromopropane	97
2-Bromobutane	56
2-Bromopentane	38
Bromocyclohexane	16
α,ω-*Dihaloalkanes*	
1,2-Dichloroethane	13
1,2-Dibromoethane	172
1,3-Dichloropropane	102
1,3-Dichloropropene	133
1,6-Dichlorohexane	67
1,9-Dichlorononane	26
Halogenated alcohols	
2-Bromoethanol	55
3-Bromopropanol	123
l-Chloro 6-hexanol	60
l-Bromo 6-hexanol	68
Halogenated ethers	
2-Chloroethylvinylether	13
Bis(2-chloroethyl)ether	30
Haloalkylbenzenes	
l-Phenyl 2-bromopropane	18
2-*Substituted l-haloalkanes*	
1,2-Dibromopropane	148
l-Bromo 2-methylpropane	39

Janssen et al. (1987b) utilized the substrate toxicity, especially towards 1,2-dibromoethane, to isolate mutants resistant to the substrates and found that resistant strains lacked different dehalogenase activities (cf. *Pseudomonas putida* PP3 – Section II.I). Mutants, such as strain GJ70M16, isolated in the presence of 3 mM 1,2-dibromoethane lacked only the initial dehalogenase, could not grow on dibromomethane, but could still utilize bromoalcohols. Other dehalogenase-minus mutants, such as GJ70M4, lacked other dehalogenases which prevent growth on bromoalcohols as well.

3.Corynebacterium *species m15-3 and other unidentified bacteria*
Yokota et al. (1986) evaluated several unidentified strains isolated on substrates such as methane, various *n*-alkanes, 1-chlorobutane and 1,2-dichlorononane, for their ability to utilize 22 mono- and di-substituted chloroalkanes. Five strains in particular had a broad ability to dechlorinate these compounds, with the activity profile for strain m2C-32 shown in Table 4. This study demonstrated clearly that chlorine atoms from any position in the molecule were removed and suggested that two different types of reaction were involved, namely hydrolase and oxygenase reactions. Strain H2 failed to dehalogenate DCE under anaerobic conditions, and the ^{18}O-incorporation profile clearly implicated an oxygenase-type reaction (Section III.C). The hydrolytic dehalogenase activities were different from the 2HAA dehalogenases since they would not attack HAAs, but did attack 3-substituted compounds. Yokota et al. (1986) suggested that the enzymes of β-oxidation systems were able to utilize chlorinated substrates and were responsible for their dehalogenation. This study also showed that some of the isolates, including *Corynebacterium* species m15-3, were   to   dehalogenate chloroalcohols.

The dehalogenase of strain m2C-32 has so far not been purified but the dehalogenase of *Corynebacterium* species m15-3 was purified and characterized (Yokota et al. 1987). The enzyme hydrolysed C_2 to C_{12} mono- and di-halogenated alkanes, some haloalcohols, including 3-chloropropanol and 3-chlorobutanol, and perhaps some HAAs. The later group must be viewed with caution since the rates are low and definitive controls were not quoted. As Hardman (1991) has pointed out, the rates of dehalogenation for HAAs were not much above the expected spontaneous dehalogenation rates. The enzyme had a molecular size of between 33 to 36 kDa and was composed of a single polypeptide chain. The K_m for 1-chlorobutane (its enrichment substrate) was 0.18 mM. The enzyme appeared was similar to the hydrolytic haloalkane dehalogenase of *Xanthobacter autotrophicus* strain GJ10 (Keuning et al. 1985) (Section III.B.1), but seemed to be more closely related to the haloalkane dehalogenases from *Arthrobacter* species HA1 (Section III.B.4) and *Rhodococcus erythropolis* Y2 (Section III.B.5).

Table 4. Dehalogenation activity of cell-free extracts of strain m2C-32 on haloalkanes (from Yokota et al. 1986).

Substrate	Relative activity	Substrate	Relative activity
Monohalogenated alkanes		1,3-Dichloropropane	154
		1,4-Dichlorobutane	131
Bromoethane	209	1,5-Dichloropentane	91
1-Chloropropane	48	1,6-Dichlorohexane	119
1-Chlorobutane	100	1,7-Dichloroheptane	115
1-Chloropentane	83	1,8-Dichlorooctane	85
1-Chlorohexane	106	1,9-Dichlorononane	85
1-Chloroheptane	100	1,1-Dichloroethane	<1
1-Chlorooctane	69	1,1-Dichlorobutane	14
1-Chlorononane	74	1,2-Dichlorobutane	17
1-Bromobutane	89	1,3-Dichlorobutane	41
1-Iodobutane	90		
		Multihalogenated alkanes	
Dihalogenated alkanes			
		1,2,3-Trichloropropane	14
1,2-Dichloroethane	10	1,1,2,2-Tetrachloroethane	<1

Relative to 1-chlorobutane = 100.

4. Arthrobacter *species HA1*

In enrichments with 1-chlorobutane, 1-chloropentane and 1-chlorohexane as the growth substrates, four coryneform bacteria were isolated, one of which was identified as *Arthrobacter* species HA1 (Scholtz et al. 1987a). This microbe used 18 different 1-chloro-, 1-bromo-, and 1-iodoalkanes for growth, but not 1-fluoroalkanes. Under inducing conditions with 1-chlorobutane, cell-free extracts released halide form a wide range of substrates (29 in total) by mechanisms which were not dependent on O_2, any co-factors or the presence of membranes, yielding *n*-alcohols as the products. Scholtz et al. (1987a) demonstrated that chloro-substituted C_1 to C_8 alkanes, and most of the equivalent monobromo- and iodo-substituted compounds, were biodegradable.

Subsequently, Scholtz et al. (1988a) showed that *Arthrobacter* species HA1 synthesized three dehalogenases. One was an inducible 1-bromoalkane debrominase. The second was an inducible 1-chlorohexane halidohydrolase with an exceptionally wide substrate profile of over 50 substrates including C_1 and C_3 substituted alkanes. The third enzyme was induced by substituted compounds of a chain length greater than C_9, such as 1-chloro- or 1-bromohexadecane, and could only use compounds of C_4 to C_{16} carbon chain length. All enzymes operated under anaerobic conditions and it was concluded that the mechanisms involved were hydrolytic.

The inducible 1-chlorohexane halidohydrolase had previously been purified (Scholtz et al. 1987b) and shown to be a monomeric protein with a molecular size of 37 kDa. It had a broad pH optimum around pH 9.5. As might have been expected from its growth substrate range, this enzyme was extremely versatile, dehalogenating over 50 compounds, including C_3 to C_{10} 1-chloroalkanes, C_1 to C_9 1-bromoalkanes C_1 to C_7 1-iodoalkanes. The K_m values ranged from 0.02 mM for 1-bromopropane to 6.4 mM for 1-iodomethane. In addition, sub-terminal, branched chain and non-saturated haloalkanes were dehalogenated as were some halogenated aromatic compounds such as bromobenzene and benzyl bromide.

This study neatly showed the wide range of capabilities of haloalkane dehalogenases and it seems that the enzyme diversity found in nature is only limited by the effort made to isolate and characterize different microorganisms. Little weight should be given to differences such as substrate profiles since these features are unlikely to reflect major differences.

5. Rhodococcus erythropolis *Y2*
This organism was isolated from soil previously exposed to haloalkanes by enrichment on 1-chlorobutane, but grew on several other compounds, such as 1-chloro-substituted propane, pentane, hexane, dodecane, tetradecane, hexadecane and octadecane (Sallis et al. 1990). Originally a single inducible dehalogenase was characterized (Section III.B.1) and, as is normally the case, shown that it dehalogenated a much wider range of compounds than supported growth (Table 5). The enzyme was better at dehalogenating terminally substituted compounds than mid-chain substitutions. In addition some chloroalcohols were dehalogenated (Table 5) but not HAAs. The K_m for 1-chlorobutane was 0.26 mM. The reaction was independent of O_2 and the products were alcohols, for example, 1-chlorobutane producing *n*-butanol. The purified enzyme was a monomeric protein with a molecular size of 34 to 36 kDa. Thiol reagents were effective inhibitors.

This enzyme closely resembled the dehalogenases of *Corynebacterium* species m15-3 (Section III.B.3) and *Arthrobacter* species HA1 (Section III.B.4). As well as the properties described here, these three enzymes also had similar isoelectric points and activation energies. Furthermore, the first 18 N-terminal amino acid residues were identical. This was particularly interesting since these two organisms were independently isolated from geographically well separated environments. There was no amino acid sequence homology between these three enzymes and the dehalogenase from *Xanthobacter autotrophicus* GJ10 (Keuning et al. 1985; Janssen et al. 1989). Other features of the dehalogenating systems of this organism are discussed in Section III.C.ii

6. Ancylobacter aquaticus *AD20, AD25 and AD27*
Van den Wijngaard et al. (1992) isolated a number of facultative methylotrophs, including these three strains of *Ancylobacter aquaticus*, able to grow on DCE by the same pathway shown in *Xanthobacter autotrophicus* GJ10 (Janssen et al. 1985). From the same enrichment another strain of *X. autotrophicus*, strain GJ11, was also isolated. All four isolates synthesized a constitutive haloalkane (DCE) dehalogenase but there were significant

Table 5. Relative activity of the crude haloalkane halidohydrolase of *Rhodococcus erythropolis* Y2 towards halogenated aliphatic acids (from Sallis et al. 1990).

Substrate	Relative activity	Substrate	Relative activity
l-Chloropropane	8	1,2-Dibromoethane	802
l-Chlorobutane	100	1,3-Dichloropropane	202
l-Bromobutane	100	1,4-Dichlorobutane	232
l-Chloropentane	132	1,6-Dichlorohexane	168
l-Chlorohexane	107	1,9-Dichlorononane	61
l-Bromohexane	66	1,10-Dichlorodecane	61
l-Chloroheptane	104	2-Chlorobutane	3
l-Chlorooctane	72	2-Bromobutane	67
l-Chlorononane	85	1,2-Dichlorobutane	4
l-Chlorododecane	53	1,2-Dibromopropane	132
l-Chlorotetradecane	10	Chloroacetate	0
l-Chlorohexadecane	8	2-Chloropropionate	0
l-Chlorooctadecane	0	2-Chloroethanol	15
Dichloromethane	0	3-Chloropropan-1-ol	18
Dibromomethane	3	4-Chlorobutan-1-ol	121
1,2-Dichloroethane	6	bis(2-Chloroethyl)ether	60

Enzyme assays contained 5 ml 100 mM-glycine/NaOH buffer, pH 9·1, 50 μl of cell-free extract, and 4 μl of one of the substrates (or 40 μl of a 10%, w/v, neutral solution of the substrate for the haloacids). Relative activities were determined by comparison with the activity towards l-chlorobutane [0·012 U (mg protein)$^{-1}$ standardized to 100].

differences in the individual specific activities: for example strain AD25 produced 15 to 20 times more than strain GJ11, which corresponded to about 30 to 40% of the total cellular protein in this single enzyme. The product of DCE dehalogenation, 2-chloroethanol was converted to chloroacetaldehyde by a PMS-linked alcohol dehydrogenase. In strains AD20 and GJ11, this enzyme was induced by growth on DCE but in strain AD25 high activities were observed even in cells grown on glucose or methanol. In all strains, an inducible NAD$^+$-dependent chloroacetaldehyde dehydrogenase was detected. And finally the product of this reaction, MCA, was converted to glycolate by a constitutive dehalogenase synthesized to varying levels in the different strains.

Apart from the different expression levels seen, the most interesting feature was the identical nature of the DCE dehalogenase from all the strains. Sequence analysis of the first 34 to 40 N-terminal amino acids from the purified enzyme from strains AD20 and AD25 were identical. Moreover, these sequences were also identical with the DCE dehalogenase from *Xanthobacter autotrophicus* GJ10 (Section III.B.1). Not surprisingly, the PCR-amplified sequences of the putative *dhlA* genes from strains AD20, AD25 and GJ11 were identical and the same as the sequence for strain GJ10. This strongly suggests that the DCE catabolic capability seen in different microorganisms and strains was due to horizontal gene transfer. Indeed, recent evidence suggested that the DCE dehalogenase is plasmid-encoded.

C. Oxygenase-type haloalkane dehalogenases

1. Methane-oxidizing bacteria

It has been known for many years that microorganisms capable of growth on methane were also able to transform many other substituted alkanes including halogenated compounds (Dalton 1980; Anthony 1986). Methane monooxygenases which normally catalyse the conversion of methane to methanol via NADH and O_2 dependent reactions, generally show poor substrate specificities and can therefore transform many substituted alkanes (Colby et al. 1977; Higgins et al. 1979; Patel et al. 1982). Generally these substrates are co-metabolised and do not therefore act as sole carbon and energy sources. Several different mechanisms for their degradation have been proposed (Stirling and Dalton 1980; Bouwer and McCarty 1985; Imai et al. 1986). Co-metabolic transformation of longer chain length haloalkanes has also been suggested (Omori and Alexander 1978b; Yokota et al. 1986) but in general the potential for bacterial monooxygenases to dehalogenate equivalent halosubstituted alkanes has not been explored fully.

2. Rhodococcus erythropolis Y2

Further examination of this bacterium (Section III.B.5) revealed that it synthesized a second haloalkane dehalogenase of the oxygenase type (Armfield et al. 1992). This enzyme was induced by C_7 to C_{16} 1-haloalkanes and *n*-alkanes, compared with the C_3 to C_6 1-haloalkane induction for the hydrolytic haloalkane dehalogenase (Section III.B.5). Resting cell suspensions of organism, induced by prior growth on hexadecane, 1-chlorotetradecane, 1-chlorohexadecane or 1-chlorooctadecane, possessed a dehalogenase which was strictly aerobic. The activity was sensitive to treatment with Triton X-100 or toluene suggesting that the activity was membrane associated, and was inhibited by PMS, 2,4-DNP and azide. The activity was less stable than the hydrolytic dehalogenase activity, had a pH optimum of about 9.2 and activity over a wide pH range.

A full characterization of this oxygenase is awaited with interest. There are two important features of this system that may have a more general importance. Firstly, this is the first organism reported to have both a hydrolytic haloalkane dehalogenase as well as an oxygenase-type dehalogenase. Secondly it seems that the hydrolytic enzyme operates with the greatest efficiency on short chain length compounds whilst the oxygenase-type functions on the long chain length molecules. Whether this is fortuitous or not remains to be determined but it opens the way for some interesting comparative studies.

D. Glutathione (GSH)-dependent dehalogenases

The first indication of another mechanism for dehalogenation of haloalkanes was provided by Brunner et al. (1980) who isolated a facultative bacterium strain DM1 that grew with a specific growth rate of 0.11 h^{-1} on 2 to 5 mM dichloromethane (DCM). The dehalogenation mechanism was unknown since, under aerobic conditions, resting cell suspensions (but not cell-free extracts) dechlorinated DCM to formaldehyde. The activity was powerfully induced by over 500-fold compared with methanol grown cells. In retrospect, it is likely that this was an example of GSH-dependent dehalogenation.

1. Hyphomicrobium *species DM2 and related isolates*

Stucki et al. (1981) isolated a *Hyphomicrobium* species DM1 growing on 10 mM DCM with a specific growth rate of 0.07 h^{-1}. In cell-free extracts the conversion of DCM to formaldehyde and free chloride was strictly dependent on the catalytic effects of GSH (with no other thiols able to substitute). It was noted that this system was analogous to the well characterized GSH S-transferase (EC 2.5.1.18) found in the liver cytosol (Heppel and Porterfield 1948; Ahmed and Anders 1976, 1978). These enzymes converted one mol DCM to one mol of formaldehyde and two HCl by nucleophilic substitution of the halogens with hydroxy groups. S-Chloromethyl GSH conjugate was an intermediate produced by the GSH S-transferase which was non-enzymatically converted to S-hydroxymethyl GSH, in turn, cleaved to yield formaldehyde and GSH. The enzyme had a 10-fold greater specific activity in *Hyphomicrobium* species compared with the liver enzyme.

The DM2 enzyme was purified by Kohler-Staub and Leisinger (1985) and found to be a hexameric protein with an overall molecular size of 195 kDa and subunits of 33 kDa. The enzyme was highly specific for four substrates, namely: DCM, bromochloromethane, dibromomethane and diiodomethane with K_m values ranging from 5 to 30 μM. Despite these affinities, the enzyme's low turnover rate (33 mol substrate mol enzyme^{-1} min^{-1}) resulted in a requirement for high enzyme concentrations (up to 16% of the total protein composed of this single protein) in order to sustain a reasonable growth rate on these substrates. This suggested that the enzyme had recently evolved and its catalytic activity was still in the process of evolving.

Subsequently, Kohler-Staub et al. (1986) compared the GSH-dependent dehalogenases from four facultative methylotrophs: *Methylobacterium* (formerly *Pseudomonas*) DM1 and DM4, and *Hyphomicrobium* DM2 and GJ21. All synthesized a similar GSH-dependent dehalogenase with a subunit molecular size of 33 kDa and which was immunologically identical. The N-terminal amino acid sequences were identical for the first 15 amino acid residues. It was concluded that these enzymes must have recently evolved, although the possibility of horizontal gene transfer by plasmid-mediated mechanisms was not fully excluded. Indeed, Galli and Leisinger (1988) showed that *Methylobacterium* DM4 contained three plasmids with sizes about 120 kb, 40 kb and 8 kb and that curing experiments indicated that the DCM dehalogenase gene (*dcm*) was encoded by the larger plasmid. However, cloning of the *dcm* gene from a 21 kb *Hind*III fragment and its subsequent use as a probe suggested that the *dcm* gene was located on the chromosome since the probe hybridised with genomic DNA but not plasmid DNA. These experiments were not sufficiently rigorous, however, to exclude one or more of the following possibilities that: an unidentified plasmid with a copy of the *dcm* gene was not detected; regulatory functions were associated with the 120 kb plasmid; or multiple copies of the *dcm* gene were both plasmid and genomically located.

La Roche and Leisinger (1990) sequenced the DCM dehalogenase gene (*dcmA*) from *Methylobacterium* DM4 carried on a 2.8 kb *Bam*H1-*Pst*I fragment and found an open reading frame coding for 287 amino acids and a molecular size of 37,430 which agreed with the previous amino acid sequence data, overall amino acid composition and molecular size. An appropriate -10/-35 promotor region was located and the strong DCM induction was abolished by deleting a 1.3 kb region upstream of the *dcmA* gene. This presumably indicated the position of an associated regulatory gene that functioned via a negative mechanism (cf. *dehI* and *dehR_I* Section II.G.1). An interesting conclusion from this study concerned the relationship with GSH S-transferases from eukaryotic sources. A comparison of the amino acid sequences of GSH S-transferases from maize, a helminth, rat and man showed that there were three highly conserved regions (amino acid positions 64 to 106, 171 to 182 and 205 to 220). The first region coded for a heptadecapeptide which putatively represented the GSH-binding region: the arginine residue at position 83 was thought to be the GSH-binding site in the bacterial enzyme. The bacterial enzyme was as closely related to the eukaryotic enzymes as the eukaryotic enzymes were related to each other. Even though the eukaryotic enzymes were dimeric and the bacterial protein was hexameric, these enzymes clearly formed a supragene family.

Scholtz et al. (1988b) isolated a fast-growing DCM-utilizing bacterium strain DM11 which synthesized an enzyme that was immunologically distinct from the previously characterized enzyme with a different N-terminal amino acid sequence. Strain DM11 produced substantially less protein as a fraction of the total protein, but the specific activity was over two times greater. So yet again, there are clear indications of great enzyme diversity and at least two classes of enzyme involved in DCM catabolism.

IV. Dehalogenation of haloalkanes under anaerobic conditions

It has been well established that haloalkanes are readily transformed by biological systems under anaerobic conditions. For example, Bouwer et al. (1981) and Bouwer and McCarty (1983) showed that methanogenic bacteria were able to degrade a wide range of haloalkanes from starting concentrations as high as 100 μg ml^{-1}, achieving over 90% degradation after two days' incubation. Compounds such as chloroform, carbon tetrachloride and DCE were mineralized to CO_2. More heavily substituted compounds, such as 1,1,2,2-tetrachloroethane, were reductively dechlorinated to 1,1,2-trichloroethane.

Belay and Daniels (1987) demonstrated that some halogenated alkanes were transformed by methanogenic bacteria, although they were toxic at comparatively low concentrations. For example, *Methanococcus thermolithotrophicus*, *M. deltae*, and *Methanobacterium thermoautotrophicum* transformed compounds such as bromoethane, dibromoethane, dichloroethane and 1,2-dibromoethylene to yield ethane, ethylene and acetylene. However, the brominated compounds were much more toxic than chlorinated compounds, being growth inhibitory at about 1 μM.

Vogel and McCarty (1985) suggested a series of reductive dehalogenations to completely mineralize tetrachloroethylene (PCE) and trichloroethylene (TCE) via dichloroethylene (DCE) and vinyl chloride (VC) to CO_2. The obligate anaerobe DCB-1 and two strains of *Methanosarcina* dechlorinated PCE and various halomethanes to yield tricarboxylic acid cycle intermediates (Fathepure et al. 1987; Mikesell and Boyd 1990). *Methanosarcina* DCM and *M. mazei* S6 converted chloroform, carbon tetrachloride and bromoform to CO_2. During the degradation of chloroform, 70% accumulated transiently as DCM. Fogel et al. (1986) found that the toxic chlorinated ethanes, such as TCE, were anaerobically converted to 50% CO_2 and 50% biomass. Both the *cis* and *trans* compounds were degraded by a pathway which involved VC and vinylidene chloride as intermediates.

Fathepure and Boyd (1988) proposed that electrons released during normal methanogenic reduction of CO_2 to methane were instead transferred to the halogenated alkanes. Interestingly, Fathepure et al. (1987) observed that active methanogenic consortia increased the rate of PCE degradation, suggesting that microbial activity involving electron transfers (Slater and Lovatt 1984) might be more effective in reductive dehalogenation processes. Egli et al. (1988) proposed for *Acetobacterium woodii* that the anaerobic, reductive dehalogenation of tetrachloromethane (TCM) was always associated with the acetyl-coenzyme A pathway and the production of acetate. This would necessitate the direct conversion of TCM to CO_2.

V. Dehalogenation of haloalcohols

Much less is known about the catabolism of haloalcohols. Castro and Bartnicki (1965) demonstrated that 3-bromopropanol was biodegradable. Later the same researchers isolated a *Flavobacterium* species that grew on 2,3-dibromopropan-1-ol and enzymatically converted halohydrins to epoxides (Castro and Bartnicki 1968; Bartinicki and Castro 1969). Stucki and Leisinger (1983) isolated an aerobic, Gram-negative soil microbe strain CE1r which grew on 2-chloroethanol, demonstrating that the pathway involved an alcohol dehydrogenase specific for 2-chloroethanol and producing 2-chloroacetaldehyde. In turn, this was the substrate for an aldehyde dehydrogenase yielding MCA that was dehalogenated by a constitutive 2HAA dehalogenase.

A. Pseudomonas *species AD1 and* Arthrobacter *species AD2 and AD3*

Van den Wijngaard et al. (1989) isolated *Pseudomonas* strain AD1 and *Arthrobacter* strains AD2 and AD3 from freshwater sediments growing on epichlorohydrin as the sole carbon and energy source. In strains AD1 and AD3, 3-chloro 1,2-propanediol was identified as an intermediate, produced as the result of the activity of epoxide hydrolase, an enzyme which also hydrolysed epibromohydrin, glycidol and propylene oxide. There was no evidence for the direct dehalogenation of epichlorohydrin to glycidol in these three strains, although it had previously been shown that the haloalkane dehalogenase of *X. autotrophicus* converted epichlorohydrin to glycidol at 12% of the rate of dehalogenation of 1,2-dichloroethane (Janssen et al. 1985; Keuning et al. 1985). Thus the actual dehalogenation step was catalysed by a halohydrin dehalogenase yielding glycidol from 3-chloro 1,2-propanediol. In these strains the halohydrin dehalogenase was able to dehalogenate eight halohydrins and was inducible by epichlorohydrin. A mutant of strain AD2 which could not grow on epichlorohydrin did not synthesize the dehalogenase. Strain AD2 differed significantly from the other two strains by an inability to synthesize an epioxide hydrolase and so the strain was dependent on the spontaneous hydrolysis of epichlorohydrin to 3-chloro 1,2-propanediol.

Van der Wijngaard et al. (1989) proposed the following pathway for epichlorohydrin catabolism:

$$CH_2 - CH - CH_2Cl \xrightarrow[\text{(a)}]{H_2O} CH_2OH - CHOH - CH_2Cl \xrightarrow[\text{(b)}]{HCl}$$

$$CH_2OH - CH - CH_2 \xrightarrow[\text{(c)}]{H_2O} CH_2OH - CHOH - CH_2OH$$

where reaction (a) is catalysed by the epoxide hydrolase; (b) by the halohydrin dehalogenase; and (c) by the same epoxide hydrolase.

More recently, van den Wijngaard et al. (1991) showed that *Arthrobacter* strain AD2 synthesized an inducible haloalcohol dehalogenase with activity against 1,3-dichloro 2-propanol, 3-chloro 1,2-propanediol, 1-chloro 2-propanol and their brominated analogues, as well as 2-bromoethanol, chloroacetone and 1,3-dichloroacetone. The enzyme was not active against epichlorohydrin or 2,3-dichloro 1-propanol. The reaction products were epoxides and so, in combination with the pathway given above, organisms with this haloalcohol dehalogenase were able to grow on haloalcohols by conversion to glycerol. The enzyme was a dimer with an molecular size of 69 kDa and the subunits each of 29 kDa. The K_m values were in the range 8.5 to 48 mM and the dehalogenase had a broad pH profile with an optimum about pH 8.5. MCA was a competitive inhibitor. Analysis of the first 34 residues of N terminal, indicated no homology with any other known protein sequence.

B. Pseudomonas putida *US2*

Pseudomonas putida US2 was isolated by growth on 2-chloroethanol (Strotmann et al. 1990). An inducible pathway converting the growth substrate to glycolate via 2-chloroacetaldehyde and MCA was found. The final dehalogenase is therefore a 2HAA dehalogenase, with a molecular size of 28 kDa, and the mechanism of dehalogenation is similar to that described by Stucki and Leisinger (1983).

VI. Plasmids encoding genes for haloaliphatic dehalogenase

A great deal of information suggests that dehalogenases are often encoded by genes located on plasmids since these enzymes catalyse exactly the type of catabolic function associated with catabolic plasmids (Slater and Bull 1982; Reanney et al. 1983). Kawasaki and his colleagues (Kawasaki et al. 1981b,c,d, 1982, 1983a,b, 1984, 1985) showed that MFA- and MCA-utilizing isolates treated with mitomycin C to prevent plasmid replication, lost the ability to grow on halogenated compounds. With *Moraxella* species B, two classes of mutant were produced: type I mutants, such as strain 86, lost the ability to grow on MCA but not MFA; and type II mutants, such as strain 123, could not grow on either substrate. Wild-type strain B contained a single large plasmid (pUO1 – size 66 kb) encoding for dehalogenase H1 and H2 (Section II.G.4), strain 86 a smaller plasmid encoding only dehalogenase H1 (H2 having been lost as a result of the decrease in plasmid size associated with DNA loss), and strain 123 did not have a plasmid and so no H1 or H2 activity. In loosing the gene for H2, plasmid pUO1 (or a more stable derivative, plasmid pUO2, sized at 64.5 kb, and carrying both H1 and H2 genes) lost a discrete DNA sequence of 5.4 kb (Kawasaki et al. 1983a,b, 1984). It was suggested, but not proved, that this element might represent a H2 dehalogenase transposon (Kawasaki et al. 1985), a conclusion which was subsequently demonstrated conclusively for the *dehI* gene and the *DEH* transposon from *Pseudomonas putida* PP3 (Thomas et al. 1992a,b) (Sections II.G.1 and II.I). It was shown that the H2 gene could be transferred to another plasmid, such as RP4 (Kawasaki et al. 1985).

Hardman (1982) showed that almost all novel HAA-utilizing soil bacteria contained large plasmids of varying size (Hardman et al. 1986). From strains of *Pseudomonas* and *Alcaligenes* species, plasmids ranging in size from about 150 kb to 300 kb were isolated which, if lost either spontaneously or as the result of plasmid curing, produced strains without the ability to grow on 2HAAs, suggesting that the dehalogenase genes were plasmid-located.

Although *Pseudomonas putida* PP3 did not have a plasmid (at least by the time the reference strain came to be analysed), Beeching (1985) showed that it was possible to mobilize the *dehI* gene on to suitable target plasmids, such as RP4 and RP45 (Beeching et al. 1983). Subsequently, it has been shown that some dehalogenases are encoded on transposons and it therefore likely that many more systems will be described in which dehalogenase genes move between plasmids and chromosomes, and so between different plasmids.

VII. Conclusions

The general picture of the diversity of microbial dehalogenases is now clear, not only in terms of the taxonomic distribution but also the basic mechanisms of enzymatic activity. Research interest will now centre increasingly on the comparative studies at the molecular level and attempts will be made to describe accurately the evolutionary history of the dehalogenase genes. Furthermore at the applied level, the ability to manipulate dehalogenases to breakdown even more recalcitrant molecules by modern molecular methods will be of particular interest. The role of dehalogenases as detoxifying catalysts deserves greater investigation, especially in the light of the high mutation rates which appear to be under environmental regulation (Thomas et al. 1992c). A wider search for novel catabolic functions should also prove interesting. For example, a dehalogenase called a dechlorinase has been identified as an important component in the control of differentiation in the slime mould *Dictyostelium discoideum* (Traynor and Kay 1991; Insall et al. 1992; Nayler et al. 1992). DIF-1, a chlorinated alkyl phenone – [1-[(3,5-dichloro 2,6-dihydroxy 4-methoxy)phenyl] hexan 1-one] – induces differentiation in *D. discoideum*. The DIF-1, 3(5)-dechlorinase, removes one chlorine atom from DIF-1 to yield DIF-3 by a mechanism which involves GSH and the addition of a proton in place of the chlorine atom. Thus, it seems unlikely that this enzyme is closely related to any of the dehalogenases described in this chapter. However, it does illustrate the point that novel dehalogenating mechanisms continue to be described.

Acknowledgements

I am very grateful for the friendship with and the contributions made by my research associates over many years: Alan Bull, Eric Senior, Andrew Weightman, Alison Weightman, Barry Hall, David Hardman, Peter Gowland, John Beeching, Steve Hope, Simon Bale, Sian Hughes, Andrew Thomas, Andrew Topping and Endang Soetarto. I acknowledge the kindness in providing unpublished information and manuscripts in press by Peter Barth, David Hardman, Dick Janssen, Rob Kay, Rudi Muller, and Steven Taylor. This chapter is dedicated to the memory of Ashley Skinner – a Gower Street friend.

References

Ahmed AE and Anders MW (1976) Metabolism of dihalomethanes to formaldehyde and inorganic halide. I. *In vitro* studies. Drug Metab. Disp. 4: 357–361

Ahmed AE and Anders MW (1978) Metabolism of dihalomethanes to formaldehyde and inorganic halide. II. Studies on the mechanism of the reaction. Biochem. Pharmacol. 27: 2021–2025

Alexander M (1981) Biodegradation of chemicals of environmental concern. Science 211: 132–138

Allison N, Skinner AJ and Cooper RA (1983) The dehalogenases of a 2,2-dichloropropionate-degrading bacterium. J. Gen. Microbiol. 129: 1283–1293

Anthony C (1986) Bacterial oxidation of methane and methanol. Adv. Microb. Physiol. 27: 113–210

Armfield SJ, Sallis PJ, Baker PB, Bull AT and Hardman DJ (1993) Biodegradation of haloalkanes by *Rhodococcus erythropolis* Y2: the presence of an oxygenase-type dehalogenase complements a halidohydrolase activity. J. Gen. Microbiol. (in press)

Asmara W (1991) Molecular biology of two 2-haloacid halidohydrolases. PhD thesis, University of Kent at Canterbury, U.K.

Asmara W, Murdiyatmo U, Baines AJ, Bull AT and Hardman DJ (1993) Protein engineering of the 2-haloacid halidohydrolase IVa from *Pseudomonas cepacia* MBA4. J. Gen. Microbiol. (in press)

Barth PT (1988) Genetic stability and expression. In: M Sussman, CH Collins, FA Skinner and DD Stewart-Tull (eds) The Release of Genetically Engineered Microorganisms (pp 239–240). Academic Press, London

Barth PT, Bolton L and Thomson JC (1992) Cloning and partial sequencing of an operon encoding two *Pseudomonas putida* haloalkanoate dehalogenases of opposite stereospecificity. J. Bacteriol. 174: 2612–2619

Bartinicki EW and Castro CE (1969) Biodehalogenation. The pathway for transhalogenation and the stereochemistry of epoxide formation from halohydrins. Biochemistry 8: 4677–4680

Beeching JR (1985) The transfer and stability of the dehalogenase I gene of *Pseudomonas putida* PP3. PhD thesis, University of Warwick, Coventry, U.K.

Beeching JR, Weightman AJ and Slater JH (1983) The formation of an R-prime carrying the fraction I dehalogenase gene from *Pseudomonas putida* PP3 using the Inc P plasmid R68-44. J. Gen. Microbiol. 129: 2071–2078

Belay N and Daniels L (1987) Production of ethane, ethylene and acetylene from halogenated hydrocarbons by methanogenic bacteria. Appl. Environ. Microbiol. 53: 1604–1610

Berry EKM, Skinner AJ and Cooper RA (1976) The bacterial degradation of Dalapon. Proc. Soc. Gen. Microbiol. 4: 38–39

Berry EKM, Allison N and Skinner AJ (1979) Degradation of the selective herbicide 22DCPA (Dalapon) by a soil bacterium. J. Gen. Microbiol. 110: 39–45

Bollag JM (1974) Microbial transformations of pesticides. Adv. Appl. Microbiol. 18: 75–130

Bollag JM and Alexander M (1971) Bacterial dehalogenation of chlorinated aliphatic acids. Soil Biol. Biochem. 3: 241–243

Bouwer EJ and McCarty PL (1983) transformation of 1- and 2-carbon halogenated aliphatic organic compounds under methanogenic conditions. Appl. Environ. Microbiol. 45: 1286–1294

Bouwer EJ and McCarty PL (1985) Ethylene dibromide transformation under methanogenic conditions. Appl. Environ. Microbiol. 50: 527–528

Bouwer EJ, Rittman BE and McCarty PL (1981) Anaerobic degradation of halogenated 1- and 2-carbon organic compounds. Environ. Sci. Technol. 15: 596–599

Bracken A (1954) Naturally-occurring chlorine-containing organic substances. Manuf. Chemistry 25: 533–538

Brunner W, Staub D and Leisinger T (1980) Bacterial degradation of dichloromethane. Appl. Environ. Microbiol. 40: 950–958

Burge WD (1969) Populations of Dalapon-decomposing bacteria in soil as influenced by additions of Dalapon or other carbon sources. Appl. Microbiol. 17: 545–550

Castro CE and Bartnicki EW (1965) Biological cleavage of carbon-halogen bonds in the metabolism of 3-bromopropanol. Biochim. Biophys. Acta 100: 384–392

Castro CE and Bartnicki EW (1968) Biodehalogenation. Epoxidation of halohydrins, epoxide opening, and transdehalogenation by a *Flavobacterium* species. Biochemistry 7: 3213–3218

Clarke PH (1978) Experiments in microbial evolution. In: LN Ornston and JR Sokatch (eds) The Bacteria, Vol VI (pp 137–218). Academic Press, New York

Clarke PH (1982) The metabolic versatility of pseudomonads. Antonie van Leeuwenhoek 48: 105–130

Clarke PH and Lilly MD (1969) The regulation of enzyme synthesis during growth. In: PM Meadows and SJ Pirt (eds) Microbial Growth (pp 113–159). Cambridge University Press, Cambridge

Clarke PH and Slater JH (1986) Evolution of enzyme structure and function in *Pseudomonas*. In: JR Sokatch (ed) The Bacteria, Vol X (pp 71–144). Academic Press, London

Colby J, Stirling DI and Dalton H (1977) The soluble methane monooxygenase of *Methylococcus capsulatus* (Bath), its ability to oxygenate *n*-alkanes, *n*-alkenes, ethers and alicyclic aromatic and heterocyclic compounds. Biochem. J. 165: 395–402

Dagley S (1984) Introduction. In: DT Gibson (ed) Microbial Degradation of Organic Compounds (pp 1–10). Marcel Dekker, New York

Dalton H (1980) Oxidation of hydrocarbons by methane monooxygenases from a variety of microbes. Adv. Appl. Microbiol. 26: 71–87

Davies JI and Evans WC (1962) The elimination of halide ions from aliphatic halogen-substituted organic acids by an enzyme preparation from *Pseudomonas dehalogens*. Proc. Biochem. Soc. 82: 50P

Den Dooren de Jong LE (1926) Bijdrage tot de kennis van het mineralisatieproces. Nijgh van Ditmar, Rotterdam

Dixon RA (1986) The *xyl*ABC promotor from the *Pseudomonas putida* TOL plasmid is activated by nitrogen regulatory genes in *Escherichia coli*. Mol. Gen. Genet. 203: 129–136

Egli C, Tschan T, Scholtz R, Cook AM and Leisinger T (1988) Transformation of tetrachloromethane to dichloromethane and carbon dioxide by *Acetobacter woodii*. Appl. Environ. Microbiol. 54: 2819–2824

Fathepure BZ and Boyd SA (1988) Dependence of tetrachloroethylene dechlorination on methanogenic substrate consumption by *Methanosarcina sp.* strain DCM. Appl. Environ. Microbiol. 54: 2976–2980

Fathepure BZ, Nengu JP and Boyd SA (1987) Anaerobic bacteria that dechlorinate perchloroethylene. Appl. Environ. Microbiol. 53: 2671–2674

Fogel MM, Taddeo AR and Fogel S (1986) Biodegradation of chlorinated ethanes by a methane-utilizing mixed culture. Appl. Environ. Microbiol. 51: 720–724

Fowden L (1968) The occurrence and metabolism of carbon-halogen compounds. Proc. R. Soc. London Ser. B 171: 5–18

Foy CL (1975) The chlorinated aliphatic acids. In: PC Kearney and DD Kaufman (eds) Herbicides, Chemistry, Degradation and Mode of Action (pp 399–452). Marcel Dekker, New York

Franken SM, Rozeboom HJ, Kalk KH and Dijkstra BW (1991) Crystal structure of haloalkane dehalogenase: an enzyme to detoxify halogenated alkanes. EMBO J. 10: 1297–1302

Galli R and Leisinger T (1985) Specialized bacterial strains for the removal of dichloromethane from industrial waste. Conservation and Recycling 8: 91–100

Galli R and Leisinger T (1988) Plasmid analysis and cloning of the dichloromethane utilization genes of *Methylobacterium* species DM4. J. Gen. Microbiol. 134: 943–952

Gold L (1988) Post transcriptional regulatory mechanisms in *Escherichia coli*. Annu. Rev. Biochem. 57: 199–233

Goldman P (1965) The enzymic cleavage of C-F bond in fluoroacetate. J. Biol. Chem. 240: 3434–3438

Goldman P (1972) Enzymology of carbon-halogen bonds In: The Degradation of Synthetic Organic Molecules in the Biosphere (pp 147–165). Nat. Acad. Sci. U.S.A., Washington

Goldman P and Milne GWA (1966) Carbon-fluorine bond cleavage. J. Biol. Chem. 241: 5557–5559

Goldman P, Milne GWA and Keister DB (1968) Carbon-halogen bond cleavage. III. Studies on bacterial halidohydrolases. J. Biol. Chem. 243: 428–434

Gschwend PM, MacFarlane JK and Newman KA (1985) Volatile halogenated organic compounds released to seawater from temperate marine macroalgae. Science 227: 1033–1035

Hardman DJ (1982) Dehalogenases in soil bacteria. PhD thesis, University of Warwick, Coventry, U.K.

Hardman DJ (1991) Biotransformation of halogenated compounds. Crit. Rev. Biotechnol. 11: 1–40

Hardman DJ and Slater JH (1981a) Dehalogenases in soil bacteria. J. Gen. Microbiol. 123: 117–128

Hardman DJ and Slater JH (1981b) The dehalogenase complement of a soil pseudomonad grown in closed and open cultures on haloalkanoic acids. J. Gen. Microbiol. 127: 399–405

Hardman DJ, Gowland PC and Slater JH (1986) Plasmid-encoded dehalogenase genes in *Pseudomonas* species. Appl. Environ. Microbiol. 51: 44–51

Hardman DJ, Slater JH and Marks T (1988) Biotransformation of halogenated compounds (pp 1–83). Laboratory of the Government Chemist, Department of Trade and Industry, London

Hartmans S, de Bont JAM, Tramper J and Luyben KCAM (1985) Bacterial degradation of vinyl chloride. Biotechnol. Lett. 7: 383–388

Hartmans S, Schmuckle A, Cook AM and Leisinger T (1986) Methyl chloride naturally occurring toxicant and C1 growth substrate. J. Gen. Microbiol. 132: 1139–1142

Hasan AKMQ, Takada H, Esaki N and Soda K (1991) Catalytic action of L2-halo acid dehalogenase on long-chain L2-haloalkanoic acids in organic solvents. Biotechnol. Bioeng. 38: 1114–1117

Heppel LA and Porterfield VA (1948) Enzymatic dehalogenation of certain brominated and chlorinated compounds. J. Biol. Chem. 176: 763–769

Higgins IJ, Hammond RC, Sariaslani FS, Best D, Davies MM, Tryhorn SE and Taylor F (1979) Biotransformation of hydrocarbons and related compounds by whole organism suspension of methane-grown *Methylosinus trichosporium* OB3b. Biochem. Biophys. Res. Comm. 89: 671–677

Hill IR (1978) Microbial transformation of pesticides. In: IR Hill and SJL Wright (eds) Pesticide Microbiology (pp 137–202). Academic Press, London

Hirsch P and Alexander M (1960) Microbial decomposition of halogenated propionic and acetic acids. Can. J. Microbiol. 6: 241–249

Hoffman FW (1950) Aliphatic fluorides. II. 1-halogeno-co-fluoroalkanes. J. Org. Chem. 15: 425–434

Hughes S (1988) Microbial growth on 3-chloropropionic acid. PhD thesis, University of Wales, Cardiff, U.K.

Imai T, Takikawa H, Nakagawa S, Shen G-J, Kodarma T and Minoda Y (1986) Microbial oxidation of hydrocarbons and related compounds by whole cell suspensions of methane-oxidizing bacterium H-2. Appl. Environ. Microbiol. 52: 1403–1406

Insall R, Nayler O and Kay RR (1992) DIF-1 induces its own breakdown in *Dictyostelium*. EMBO J. 11: 2849–2854

Janssen DB, Scheper A, Dijkhuizen L and Witholt B (1985) Degradation of halogenated aliphatic compounds by *Xanthobacter autotrophicus* GJ10. Appl. Environ. Microbiol. 49: 673–677

Janssen DB, Keuning S and Witholt B (1987a) Involvement of a quinoprotein alcohol dehydrogenase and an NADdependent aldehyde dehydrogenase in 2-chloroethanol metabolism in *Xanthobacter autotrophicus* GJ10. J. Gen. Microbiol. 133: 85–92

Janssen DB, Jager D and Witholt B (1987b) Degradation of *n*-haloalkanes and α,ω-dihaloalkanes by wild-type and mutants of *Acientobacter* species strain GJ70. Appl. Environ. Microbiol. 53: 561–566

Janssen DB, Gerritse J, Brackman J, Kalk C, Jager D and Witholt B (1988) Purification and characterization of a bacterial dehalogenase with activity toward halogenated alkanes, alcohols and ethers. Eur. J. Biochem. 171: 67–72

Janssen DB, Pries F, van de Ploeg J, Kazemier B, Terpstra P and Witholt B (1989) Cloning of 1,2-dichloroethane degradation genes of *Xanthobacter autotrophicus* GJ10 and expression and sequencing of the *dhlA* gene. J. Bacteriol. 171: 6791–6799

Jensen HL (1951) Decomposition of chlorosubstituted aliphatic acids by soil bacteria. Can. J. Microbiol. 3: 151–164

Jensen HL (1957) Decomposition of chloroorganic acids by fungi. Nature, London 180: 1416

Jensen HL (1959) Decomposition of chlorine-substituted organic acids by fungi. Acta Agri. Scand. 9: 421–434

Jensen HL (1960) Decomposition of chloroacetates and chloropropionates by bacteria. Acta Agri. Scand. 10: 83–103

Jensen HL (1963) Carbon nutrition of some microorganisms decomposing halogen-substituted aliphatic acids. Acta Agri. Scand. 13: 404–412

Kawasaki H, Miyoshi K and Tonomura K (1981a) Purification, crystallisation and properties of haloacetate halidohydrolase from *Pseudomonas* sp. Agri. Biol. Chem. 45: 543–544

Kawasaki H, Tone H and Tonomura K (1981b) Plasmid determined dehalogenation of haloacetates in *Moraxella* species. Agri. Biol. Chem. 45: 29–34

Kawasaki H, Tone H and Tonomura K (1981c) Purification and properties of haloacetate halidohydrolase specified by plasmid from *Moraxella* species strain B. Agri. Biol. Chem. 45: 35–42

Kawasaki H, Yahara H and Tonomura K (1981d) Isolation and characterization of plasmid pUO1 mediating dehalogenation of haloacetate and mercury resistance in *Moraxella* species B. Agri. Biol. Chem. 45: 1477–1481

Kawasaki H, Hayashi S, Yahara H, Minami F and Tonomura K (1982) Plasmid pUO2 determining haloacetate dehalogenase and mercury resistance in *Pseudomonas* species. J. Fermentation Technol. 60: 5–11

Kawasaki H, Yahara H and Tonomura K (1983a) Cleavage maps of dehalogenation plasmid pUO1 and its deletion derivative harboured in *Moraxella* species. Agri. Biol. Chem. 47: 1639–1641

Kawasaki H, Yanase N, Yahara H and Tonomura K (1983b) Molecular modification of a dehalogenation plasmid originating from a *Moraxella* species in a foreign host. Agri. Biol. Chem. 47: 1643–1645

Kawasaki H, Yahara H and Tonomura K (1984) Cloning and expression in *Escherichia coli* of the haloacetate dehalogenation genes from *Moraxella* plasmid pUO1. Agri. Biol. Chem. 48: 2627–2632

Kawasaki H, Takao M, Koiso A and Tonomura K (1985) Genetic rearrangement of plasmids: *in vivo* recombination between a dehalogenation plasmid and multiple-resistance plasmid RP4 in *Pseudomonas* species. Appl. Environ. Microbiol. 49: 1544–1546

Kearney PC (1966) Metabolism of herbicides in soils. Adv. Chem. Ser. 60: 250–262

Kearney PC, Kaufman DD and Beall ML (1964) Enzymatic dehalogenation of 2,2-dichloropropionate. Biochem. Biophys. Res. Comm. 14: 29–33

Kearney PC, Harris CI, Kaufman DD and Sheets TJ (1965) Behaviour and fate of chlorinated aliphatic acids in soils. Adv. Pest Control Res. 6: 1–30

Keith LH and Teilliard WA (1979) Priority pollutants I: a perspective view. Environ. Sci. Technol. 13: 416–423

Kelly M (1965) Isolation of bacteria able to metabolise fluoroacetate or fluoroacetamide. Nature, London 208: 809–810

Keuning S, Janssen DB and Witholt B (1985) Purification and characterization of hydrolytic haloalkane dehalogenase from *Xanthobacter autotrophicus* GJ10. J. Bacteriol. 163: 635–639

King GM (1986) Inhibition of microbial activity in marine sediments by bromophenol from a hemichordate. Nature, London 232: 257–259

King GM (1988) Dehalogenation in marine sediments containing natural sources of halophenols. Appl. Environ. Microbiol. 54: 3079–3085

Klages U and Lingens F (1980) Degradation of 4-chlorobenzoic acid by a *Pseudomonas* species. Zentralbl. Bakteriol. Hyg. I. Abt. Orig. C 1: 215–223

Klages U, Krauss S and Lingens F (1983) 2-haloacid dehalogenase from a 4-chlorobenzoate-degrading *Pseudomonas* species CBS3. Hoppe-Seylers Z. Physiol. Chem. 364: 529–535

Kluyver AJ (1931) The Chemical Activities of Microorganisms. University of London Press, London

Kohler-Staub D and Kohler HPE (1989) Microbial degradation of β-chlorinated four-carbon aliphatic acids. J. Bacteriol. 171: 1428–1434

Kohler-Staub D and Leisinger T (1985) Dichloromethane dehalogenase of *Hyphomicrobium* species strain DM2. J. Bacteriol. 162: 676–681

Kohler-Staub D, Hartmans S, Galli R, Suter F and Leisinger T (1986) Evidence for identical dichloromethane dehalogenases in different methylotrophic bacteria. J. Gen. Microbiol. 132: 2837–2843

La Roche SD and Leisinger T (1990) Sequence analysis and expression of the bacterial dichloromethane dehalogenase structural gene, a member of the glutathione *S*-transferase supragene family. J. Bacteriol. 172: 164–171

Leigh JA, Skinner AJ and Cooper RA (1988) Partial purification, stereospecificity and stoichiometry of three dehalogenases from a *Rhizobium* species. FEMS Microbiol. Lett. 49: 353–356

Lien BC, Cole ALJ, Walker JRL and Peters JA (1979) Effect of sodium fluoroacetate ("Compound 1080") on the soil microflora. Soil Biol. Biochem. 11: 13–18

Little M and Williams PA (1971) A bacterial halidohydrolase. Its purification, some properties and its modification by specific amino acid reagents. Eur. J. Biochem. 21: 99–109

Lovelock JE (1975) Natural halocarbons in the air and in the sea. Nature, London 256: 193–194

Macgregor AN (1963) The decomposition of dichloropropionate by soil microorganisms. J. Gen. Microbiol. 30: 497–501

Magee LA and Colmer AR (1959) Decomposition of 2,2-dichloropropionic acid by soil bacteria. Can. J. Microbiol. 5: 255–260

Marais JSC (1944) Monofluoroacetic acid, the toxic principle of "gifblaar in *Dichapetalum cymosum*" (Hook) Engl. Onderstepoort. J. Vet. Sci. 20: 208–218

Merrick MJ and Stewart WDP (1985) Studies on the regulation and function of the *Klebsiella pneumoniae ntr* A gene. Gene 35: 297–303

Mikesell MD and Boyd SA (1990) Dechlorination of chloroform by *Methanosarcina* strains. Appl. Environ. Microbiol. 56: 1198–1201

Morsberger F-M, Muller R, Otto MK, Lingens F and Klube KD (1991) Purification and characterization of 2-halocarboxylic acid dehalogenase II from *Pseudomonas* species CBS3. Biol. Chem. Hoppe-Seyler 372: 915–922

Motosugi K and Soda K (1983) Microbial degradation of synthetic organochlorine compounds. Experientia 39: 1214–1220

Motosugi K, Esaki N and Soda K (1982a) Bacterial assimilation of D and L2-chloropropionates and occurrence of a new dehalogenase. Arch. Microbiol. 131: 179–183

Motosugi K, Esaki N and Soda K (1982b) Purification and properties of a new enzyme, DL2-haloacid dehalogenase from *Pseudomonas* sp. J. Bacteriol. 150: 522–527

Muller R and Lingens F (1986) Microbial degradation of halogenated hydrocarbons: a biological solution to the pollution problem? Angew. Chem. Int. Ed. Engl. 25: 779–789

Murdiyatmo U (1991) Molecular genetic analysis of a 2-haloacid halidohydrolase structural gene. PhD thesis, University of Kent at Canterbury, Canterbury, U.K.

Murdiyatmo U, Asmara W, Tsang JSH, Baines AJ, Bull AT and Hardman DJ (1993) Molecular biology of the 2-haloacid halidohydrolase from *Pseudomonas cepacia* MBA4. Biochem. J. (in press)

Murphy GL and Perry JJ (1983) Incorporation of chlorinated alkanes into fatty acids of hydrocarbon-utilizing mycobacteria. J. Bacteriol. 156: 1158–1164

Murphy GL and Perry JJ (1984) Assimilation of chlorinated alkanes and fatty acids. Appl. Environ. Microbiol. 160: 1171–1174

Murray AJ and Riley JP (1973) Occurrence of some chlorinated aliphatic hydrocarbons in the environment. Nature 242: 37–38

Nayler O, Insall R and Kay RR (1992) Differentiation-inducing-factor dechlorinase, a novel cytosolic dechlorinating enzyme from *Dictyostelium discoideum*. Eur. J. Biochem. 208: 531–536

Omori T and Alexander M (1978a) Bacterial and spontaneous dehalogenation of organic compounds. Appl. Environ. Microbiol. 35: 512–516

Omori T and Alexander M (1978b) Bacterial dehalogenation of halogenated alkanes and fatty acids. Appl. Environ. Microbiol. 35: 867–871

Patel RN, Hou CT, Laskin AJ and Felix A (1982) Microbial oxidation of hydrocarbons: properties of a soluble methane monooxygenase from a facultative methane-utilizing organism, *Methylobacterium* species strain CRL-26. Appl. Environ. Microbiol. 44: 1130–1137

Pathak D and Ollis D (1990) Refined structure of dienelactone hydrolase at 1.8Å. J. Mol. Biol. 214: 497–525

Pathak D, Ngai KL and Ollis D (1988) X-ray crystallographic structure of dienelactone hydrolase at 2.8Å. J. Mol. Biol. 204: 435–445

Peters RA (1952) Lethal synthesis. Proc. R. Soc. London Ser. B 139: 143–167

Petty MA (1961) An introduction to the origin and biochemistry of microbial halometabolites. Bacteriol. Rev. 25: 111–130

Reanney DC, Gowland PC and Slater JH (1983) Genetic interactions among microbial communities. In: JH Slater, R Whittenbury and JWT Wimpenny (eds) Microbes in Their Natural Environments (pp 379–421). Cambridge University Press, Cambridge

Rozeboom HJ, Kingma J, Janssen DB and Dijkstra B (1988) Crystallization of haloalkane dehalogenase from *Xanthobacter autotrophicus* GJ10. J. Mol. Biol. 200: 611–612

Sallis PJ, Armfield SJ, Bull AT and Hardman DJ (1990) Isolation and characterization of a haloalkane halidohydrolase from *Rhodococcus erythropolis* Y2. J. Gen. Microbiol. 136: 115–120

Schneider B, Muller R, Frank R and Lingens F (1991) Complete nucleotide sequences and comparison of the structural genes of two 2-haloalkanoic acid dehalogenases from *Pseudomonas* species strain CBS3. J. Bacteriol. 173: 1530–1535

Scholtz R, Schmuckle A, Cook AM and Leisinger T (1987a) Degradation of eighteen 1-monohaloalkanes by *Arthrobacter* species strain HA1. J. Gen. Microbiol. 133: 267–274

Scholtz R, Leisinger T, Suter F and Cook AM (1987b) Characterization of 1-chlorohexane halidohydrolase, a dehalogenase of wide substrate range from an *Arthrobacter* species. J. Bacteriol. 169: 5016–5021

Scholtz R, Messi F, Leisinger T and Cook AM (1988a) Three dehalogenases and physiological restraints in the biodegradation of haloalkanes by *Arthrobacter* species strain HA1. Appl. Environ. Microbiol. 54: 3034–3038

Scholtz R, Wackett LP, Egli C, Cook AM and Leisinger T (1988b) Dichloromethane dehalogenase with improved catalytic activity isolated from a fast-growing dichloromethane-utilizing bacterium. J. Bacteriol. 170: 5698–5704

Senior E (1977) Characterization of a microbial community growing on the herbicide Dalapon. PhD thesis, University of Kent at Canterbury, U.K.

Senior E, Bull AT and Slater JH (1976) Enzyme evolution in a microbial community growing on the herbicide Dalapon. Nature, London 263: 476–479

Slater JH (1988) Biotransformation of halogenated alkanoic acids. In: DJ Hardman (ed) Biotransformation of Halogenated Compounds (pp 6–34). Lab. Gov. Chem., Dept. Trade and Industry, London

Slater JH and Bull AT (1978) Biochemical basis of microbial interactions. Annal. Appl. Biol. 89: 149–150

Slater JH and Bull AT (1982) Environmental microbiology: biodegradation. Philos. Trans. R. Soc. London 297: 515–597

Slater JH and Hardman DJ (1982) Isolation and studies *in vitro* of microbial communities. In: RG Burns and JH Slater (eds) Experimental Microbial Ecology (pp 255–274). Blackwell Scientific Publications, Oxford, U.K.

Slater JH and Lovatt D (1984) Biodegradation and the significance of microbial communities. In: DT Gibson (ed) Biochemistry of Microbial Degradation (pp 439–485). Marcel Dekker and Sons, New York

Slater JH, Weightman AJ, Senior E and Bull AT (1976) The dehalogenases from *Pseudomonas putida*. Proc. Soc. Gen. Microbiol. 3: 103

Slater JH, Lovatt D, Weightman AJ, Senior E and Bull AT (1979) The growth of *Pseudomonas putida* on chlorinated aliphatic acids and its dehalogenase activity. J. Gen. Microbiol. 114: 125–136

Slater JH, Weightman AJ and Hall BG (1985) Dehalogenase genes of *Pseudomonas putida* PP3 on chromosomally located transposable elements. Mol. Biol. Evol. 2: 557–567

Smith JM, Harrison K and Colby J (1989a) Purification and characterization of D2-haloacid dehalogenase from *Pseudomonas putida* AJ1/23. J. Gen. Microbiol. 136: 881–886

Smith JM, Harrison K, Colby J and Taylor SC (1989b) Determination of D2-halopropionate dehalogenase activity from *Pseudomonas putida* strain AJ1/23 by ion chromatography. FEMS Microbiol. Lett. 57: 71–74

Smith JM, Harrison K and Colby J (1990) Purification and characterization of D2-haloacid dehalogenase from *Pseudomonas putida* strain AJ1/23. J. Gen. Microbiol. 136: 881–886

Stirling DI and Dalton H (1980) Oxidation of dimethyl ether, methyl formate and bromoethane by *Methylococcus capsulatus* (Bath). J. Gen. Microbiol. 116: 277–283

Strotmann UJ, Pentenga and Janssen DB (1990) Degradation of 2-chloroethanol by wild type and mutants of *Pseudomonas putida* US2. Arch. Microbiol. 154: 294–300

Stucki GR and Leisinger T (1983) Bacterial degradation of 2-chloroethanol proceeds via chloroacetic acid. FEMS Microbiol. Lett. 16: 123–126

Stucki GR, Galli R, Ebersold H-R and Leisinger T (1981) Dehalogenation of dichloromethane by cell extracts of *Hyphomicrobium* DM2. Arch. Microbiol. 130: 366–371

Suida JF and DeBernardis JF (1973) Naturally-occurring halogenated organic compounds. Lloydia 36: 107–143

Symonds RB, Rose WI and Reed MH (1988) Contribution of chlorine and fluorine bearing gases to the atmosphere of volcanoes. Nature 334: 415–418

Taylor SC (1985) D2-Haloalkanoic acid halidohydrolase. European Patent No. 179603

Taylor SC (1988) D2-Haloalkanoic acid halidohydrolase. U.S.A. Patent No. 4758518

Thomas AW (1990) Analysis of a mobile genetic element from *Pseudomonas putida* which encodes dehalogenase functions. PhD thesis, University of Wales, Cardiff, U.K.

Thomas AW, Slater JH and Weightman AJ (1992a) The dehalogenase gene *dehI* from *Pseudomonas putida* PP3 is carried on an unusual mobile genetic element designated *DEH*. J. Bacteriol. 174: 1932–1940

Thomas AW, Topping AW, Slater JH, and Weightman AJ (1992b) Localization and functional analysis of structural and regulatory dehalogenase genes carried on *DEH* from *Pseudomonas putida* PP3. J. Bacteriol. 174: 1941–1947

Thomas AW, Lewington J, Hope S, Topping AW, Weightman AJ, and Slater JH (1992c) Environmentally directed mutations in the dehalogenase system of *Pseudomonas putida* strain PP3. Arch. Microbiol. 158: 176–182

Thony B and Hennecke H (1989) The −24/−12 promotor comes of age. FEMS Microbiol. Rev. 63: 341–358

Tonomura N, Futsi F, Tanabe O and Yamaoka T (1965) Defluorination of monofluoroacetate by bacteria. I. Isolation of bacteria and their activity of defluorination. Agri. Biol. Biochem. 29: 124–128

Traynor D and Kay RR (1991) The DIF-1 signalling system in *Dictyostelium*: metabolism of the signal. J. Biol. Chem. 266: 717–719

Tsang JSH, Sallis PJ, Bull AT and Hardman DJ (1988) A monobromoacetate dehalogenase from *Pseudomonas cepacia* MBA4. Arch. Microbiol. 150: 441–446

van den Wijngaard AJ, Janssen D and Witholt B (1989) Degradation of epichlorohydrin and halohydrins by bacterial cultures isolated from freshwater sediments. J. Gen. Microbiol. 135: 2199–2208

van den Wijngaard AJ, Reuvekamp PTW and Janssen DB (1991) Purification and characterization of haloalcohol dehalogenase from *Arthrobacter* sp. strain AD2. J. Bacteriol. 173: 124–129 van den Wijngaard AJ, van der Kamp KWHJ, van der Ploeg J, Pries F, Kazemier B and Janssen DB (1993) Degradation of 1,2-dichloroethane by *Ancylobacter aquaticus* and other facultative methylotrophs. Appl. Environ. Microbiol. (in press)

van der Ploeg J, van Hall G and Janssen DB (1991) Characterization of the haloacid dehalogenase from *Xanthobacter autotrophicus* GJ10 and sequencing of the *dhlB* gene. J. Bacteriol. 173: 7925–7933

van Hylckama Vlieg JET and Janssen DB (1991) Bacterial degradation of 3-chloroacrylic acid and the characterization of *cis*- and *trans*-specific dehalogenases. Biodegradation 2: 25–31

Vogel TM and McCarty PL (1985) Biotransformation of tetrachloroethylene, dichloroethylene, vinylchloride and carbon dioxide under methanogenic conditions. Appl. Environ. Microbiol. 49: 1080–1083

Vogel TM, Criddle CS and McCarty PL (1987) Transformations of halogenated aliphatic compounds. Environ. Sci. Technol. 21: 722–736

Weightman AJ (1981) The catabolism of halogenated alkanoic acids by *Pseudomonas putida* strains: characterization of dehalogenase enzymes and associated functions. PhD thesis, University of Warwick, Coventry, U.K.

Weightman AJ and Slater JH (1980) Selection of *Pseudomonas putida* strains with elevated dehalogenase activities by continuous culture growth on chlorinated alkanoic acids. J. Gen. Microbiol. 121: 187–193

Weightman AJ, Slater JH and Bull AT (1979a) Cleavage of the carbon-chlorine bond by *Pseudomonas putida*. Soc. Gen. Microbiol. Q. 6: 76–77

Weightman AJ, Slater JH and Bull AT (1979b) The partial purification of two dehalogenases from *Pseudomonas putida* PP3. FEMS Microbiol. Lett. 6: 231–234

Weightman AJ, Weightman AL and Slater JH (1982) Stereospecificity of 2-monochloropropionate dehalogenation by the two dehalogenases of *Pseudomonas putida* PP3: evidence of two different dehalogenation mechanisms. J. Gen. Microbiol. 131: 1755–1762

Weightman AJ, Weightman AL and Slater JH (1985) Toxic effects of chlorinated and brominated alkanoic acids on *Pseudomonas putida* PP3: selection at high frequencies of mutations in genes encoding dehalogenases. Appl. Environ. Microbiol. 49: 1494–1501

Wilson JT and Wilson BH (1985) Biotransformation of trichloroethylene in soil. Appl. Environ. Microbiol. 49: 242–243

Yokota T, Fuse H, Omori T and Minoda Y (1986) Microbial dehalogenation of haloalkanes by oxygenase or halidohydrolase. Agri. Biol. Chem. 50: 453–460

Yokota T, Omori T and Kodama T (1987) Purification and properties of haloalkane dehalogenase from *Corynebacterium* species strain m15-3. J. Bacteriol. 160: 4049–4054

13. Biodegradation of halogenated aromatic compounds

LAETITIA C.M. COMMANDEUR and JOHN R. PARSONS
Department of Environmental and Toxicological Chemistry, University of Amsterdam, Nieuwe Achtergracht 166, 1018 WV Amsterdam, The Netherlands

Abbreviations: CBz – chlorobenzene; DCBz – dichlorobenzene; TrCBz – trichlorobenzene; TCBz – tetrachlorobenzene; PCBz – pentachlorobenzene; HCBz – hexachlorobenzene; CBA – chlorobenzoic acid; BBA – bromobenzoic acid; FBA – fluorobenzoic acid; IBA – iodobenzoic acid; CP – chlorophenol; CA – chloroaniline; 2,4-D –2,4-dichlorophenoxyacetic acid; 2,4,5-T –2,4,5-trichlorophenoxyacetic acid; PCBs – polychlorinated biphenyls; CB – chlorobiphenyl; PCDDs – polychlorinated dibenzo-*p*-dioxins; PCDFs – polychlorinated dibenzofurans

I. Introduction

Chlorinated aromatic compounds have been widely used as pesticides (e.g. 2,4-D, 2,4,5-T, chlorophenols) or for industrial applications (e.g. PCBs in electrical equipment and as hydraulic fluids). Others, such as PCDDs and PCDFs, are produced unintentionally as trace contaminants during the industrial production of chlorinated compounds and incineration of chlorine-containing wastes. Brominated aromatic compounds have found use as flame retardants, whereas fluorinated and iodinated aromatic compounds have pharmaceutical applications. The chemical inertness and hydrophobicity of many of these compounds has resulted in them becoming widely distributed in the environment; in particular accumulating in terrestial and aquatic organisms. This, together with their toxicity, has given rise to concern about their fate and effects in the environment.

Despite the fact that naturally occurring halogenated aromatic compounds are rare, many bacteria have been isolated which are able to degrade such chemicals. However, these bacteria are often unable to grow on these compounds, but they degrade them while growing on other compounds, such as their non-halogenated analogues. This process is referred to as co-metabolism. The pathways by which halogenated aromatic compounds are degraded by microorganisms are, in general, similar to those used for the degradation of other aromatic compounds. Under aerobic conditions aromatic

423

C. Ratledge (ed.), Biochemistry of Microbial Degradation, 423–458.
© 1994 *Kluwer Academic Publishers. Printed in the Netherlands.*

compounds are transformed by mono- and di-oxygenase catalysed reactions into dihydroxylated derivatives, which are then subjected to ring cleavage. Under anaerobic conditions, degradation follows reductive pathways leading to loss of aromaticity before ring cleavage takes place. The degradative pathways for aromatic compounds in general are described in more detail elsewhere in this publication by Smith (Chapter 11). We here present a survey of the physiology and genetics of the biodegradation of halogenated aromatic compounds, covering the literature up to the end of 1991.

II. Dehalogenation of aromatic compounds

In general, halogenated organic compounds are less readily degraded by microorganisms than their non-halogenated counterparts. Consequently, the removal of the halogen substituents is a key step in their biodegradation. As is described below, in most cases dehalogenation of aromatic compounds takes place after the aromatic ring system has been cleaved. See, for example, the dehalogenation of the ring-cleavage products of halocatechols during a lactonization reaction (Fig. 4). However, some bacteria can dehalogenate aromatic compounds directly as one of the initial steps in their degradation. Three classes of such dehalogenation reactions are known.

1. *Oxidative dehalogenation*, in which the halogen substituent is lost fortuitously during oxygenation of the ring (Fig. 1). An example of such a reaction is the formation of catechol by 1,2-dioxygenation of 2-halogenated benzoates (see below).
2. *Hydrolytic dehalogenation*, in which a halogen substituent is specifically replaced by a hydroxyl group (Fig. 2). In this case the source of the oxygen atom in the hydroxyl group is water instead of oxygen. This reaction can occur under both aerobic and denitrifying conditions. An example of such reactions is the dehalogenation of 4-chlorobenzoate to form 4-hydroxybenzoate by *Arthrobacter* and *Pseudomonas* strains able to grow on 4-CBA (see below).
3. *Reductive dehalogenation*, in which a halogen is replaced by a hydrogen (Fig. 3). This reaction occurs almost exclusively under sulphate-reducing and methanogenic conditions. It has been proposed that the bacteria carrying out this reaction use the halogenated aromatic compounds as terminal electron acceptors. In the case of the 3-CBA-dechlorinating bacterium *Desulfominile tiedje* it has been shown that reductive dehalogenation is coupled to growth and to ATP formation (see below).

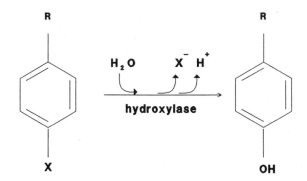

Fig. 1. Oxidative dehalogenation of haloaromatic compounds. R = e.g. COOH, H, NH$_2$; X = F, Cl, Br, I.

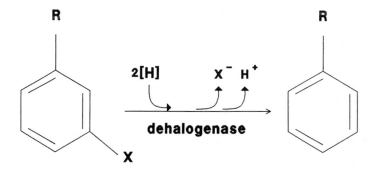

Fig. 2. Hydrolytic dehalogenation of haloaromatic compounds. R = e.g. COOH, OH, NH$_2$; X = F, Cl, Br, I.

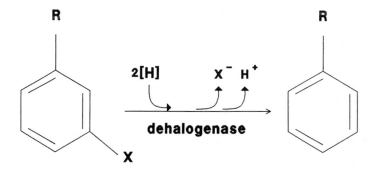

Fig. 3. Reductive dehalogenation of haloaromatic compounds. R = COOH, H, OH, NH$_2$, C$_6$H$_5$; X = F, Cl, Br, I.

III. Biodegradation of halogenated benzoic acids

A. Aerobic biodegradation

Apart from their significance as environmental pollutants through the use of herbicides, such as dicamba (3,6-dichloro-2-methoxybenzoate) and 2,3,6-trichlorobenzoate, and their production by pharmaceutical industries (mainly fluoro- and iodobenzoates), chlorinated benzoates have been identified as the major accumulating intermediates in the aerobic bacterial co-metabolism of polychlorinated biphenyls. Furthermore, halogenated benzoates have been used extensively as model compounds to study the biodegradation of halogenated aromatic compounds. The biodegradation of halogenated monoaromatic compounds has been reviewed by Reineke (1984) and more recently by Reineke and Knackmuss (1988) and Häggblom (1990).

Aerobically, halogenated benzoates are mainly degraded by initial dioxygenation of the aromatic ring to yield halocatechols (Fig. 4). *Ortho*-substituted benzoates give catechols and 3-halocatechols, *meta* substitution gives 3- or 4-halocatechols and *para* substitution gives 4-halocatechols. Ring cleavage of these compounds takes place most efficiently by the *ortho*, or β-ketoadipate, route yielding halo-*cis*,*cis*-muconates (Fig. 5). Many bacterial strains capable of degrading halobenzoates have been described. They are members of the genera *Pseudomonas* (Reineke and Knackmuss 1980; Schreiber et al. 1980; Chatterjee et al. 1981; Schmidt and Knackmuss 1984; Focht and Shelton 1987; Wyndham and Straus 1988; Vora et al. 1988; Hartmann et al. 1979, 1989; Schlömann et al. 1990), *Alcaligenes* (Schmidt and Knackmuss 1984; Wyndham and Straus 1988; Schlömann et al. 1990), *Nocardia* (Cain et al. 1968; Spokes and Walker 1974) and *Azotobacter* (Walker and Harris 1970). Strain FLB 300 which is able to degrade all three monofluorobenzoates was assigned to the *Agrobacterium-Rhizobium* group (Engesser et al. 1990).

Bacteria growing on these halobenzoates have catechol-1,2-dioxygenases with high activities towards halogen-substituted catechols, referred to as type II pyrocatechases, and degrade these compounds by the modified *ortho*

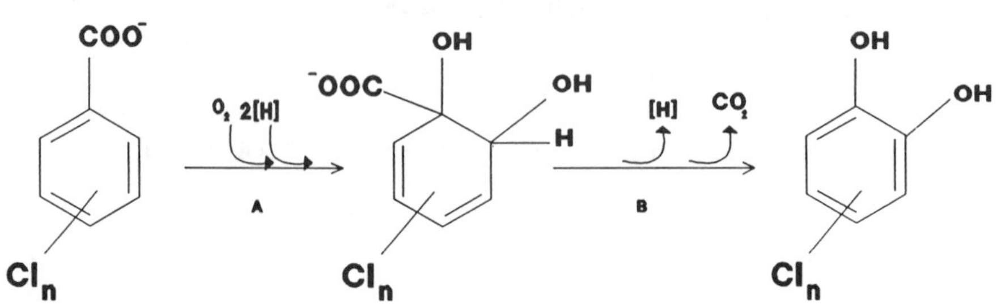

Fig. 4. Aerobic degradation of benzoate to catechol. A = benzoate dioxygenase; B = benzoate dihydrodioldehydrogenase.

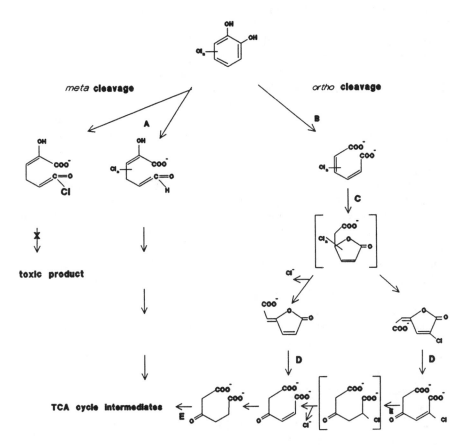

Fig. 5. Metabolic route for aerobic degradation of chlorocatechols. A = catechol-2,3-dioxygenase; B = catechol-1,2-dioxygenase; C = muconate cycloisomerase; D = dienelactone hydrolase; E = maleylacetate reductase (Schlömann et al. 1990).

cleavage pathway. The halogen substituents are eliminated from halo-*cis*,*cis*-muconates, halomuconolactones or halomaleylacetates (Fig. 4).

Genes for the complete mineralization of 3-chlorobenzoate via the modified *ortho* cleavage pathway are known to be located on the plasmid pWR1 (pB13) from *Pseudomonas* sp. strain B13 (Chatterjee and Chakrabarty 1983), clustered on a fragment no larger than 11 kilobases (Weisshaar et al. 1987), on the plasmids pAC25 or pAC27 from *Pseudomonas putida* AC (Chatterjee et al. 1981) and on the plasmid pJP4 from *Alcaligenes eutrophus* JMP134 (Don and Pemberton 1981). The plasmids pWR1 and pAC25 are almost identical, whereas pJP4 shows differences in the regulation of these genes (Ghosal et al. 1985). The *tfdCDEF* genes from pJP4 and *clcABD* genes from pAC27 show high homology and both code for the degradation of chlorocatechols (Ghosal and You 1988, 1989). *Alcaligenes* sp. BR60 was shown to also contain a plasmid (pBR60) specifying 3-CBA degradation. Remarkably, there is little or no homology between pAC27 and pBR60 (Wyndham et al. 1988).

Meta, or extradiol, cleavage of halocatechols is performed by catechol-2,3-dioxygenases. Only *meta* cleavage of 4-halocatechols yields degradable metabolites,   e.g.   5-halo-2-formyl-2-hydroxy-6-oxohexa-2,4-dienoates (Horvath 1970). 3-Halocatechols in particular seem to be critical metabolites. *Meta* cleavage of 3-halocatechols gives, apart from productive halogenated semialdehydes, 5-chloroformyl-2-hydroxypenta-2,4-dienoates. These latter compounds probably bind irreversibly to basic groups of the dioxygenase, which results in inactivation of the enzyme (Bartels et al. 1984). A study of *ortho* and *meta* cleavage enzymes of several organisms resulted in the identification of four groups of organisms with distinctive phenotypes regarding 3-chlorobenzoate degradation (Taeger et al. 1988).

An alternative catabolic route for 3-chlorobenzoate is via 5-chlorosalicylate and 2,3-dihydroxybenzoate. This route is found in *Bacillus* sp. (Spokes and Walker 1974) and has been succesfully transferred to *Pseudomonas* WR1 by genetic manipulation (Lehrbach et al. 1984).

In some cases, dehalogenation of halobenzoates takes place in the first step in the degradation. Halogen substitution in the *ortho* position of benzoates can lead to fortuitous oxidative dehalogenation by benzoate dioxygenases. Oxidative dehalogenation of 2-fluorobenzoate has been described for *Pseudomonas* spp. (Goldman et al. 1967; Milne et al. 1968; Vora et al. 1988), for *Acinetobacter calcoaceticus* (Clarke et al. 1975) and for *Pseudomonas* B13 (Schreiber et al. 1980). 2-Chlorobenzoate was oxidatively dehalogenated by *Pseudomonas* B300 (Sylvestre et al. 1989). *Pseudomonas putida* CLB250 oxidatively dehalogenated 2-fluoro-, 2-bromo- and 2-chlorobenzoates (Engesser and Schulte 1989). In these cases, both 1,2- and 1,6-dioxygenation took place. 1,2-Dioxygenation of 2-halobenzoates leads to catechol and 1,6-dioxygenation to 3-halocatechols. During growth the metabolic route continues via *ortho* cleavage (Fig. 5).

2-CBA can serve as a growth substrate for *Pseudomonas* JB2. This organism is suggested to possess a specific halobenzoate dioxygenase in addition to the benzoate dioxygenase (Hickey and Focht 1990). A apparently constitutive benzoate dioxygenase activity for all singly *ortho*-substituted mono- and dichlorobenzoates is found in *Pseudomonas putida* P111. Growth of this latter strain on 3- and 4-CBA induces a specific dihydrodiol dioxygenase which enables co-metabolism of 3,5-dichlorobenzoate (3,5-DCBA) (Hernandez et al. 1991). Two strains of *Alcaligenes denitrificans*, designated BRI 3010 and BRI 6011 could grow on 2-CBA but not on 3- and 4-CBA. Both strains degraded 2,5- and 2,3-DCBA but not 2,6-DCBA. In contrast to growth on benzoate and 2-CBA, growth on 2,5-DCBA induced a type II catechol-1,2-dioxygenase. Only strain BRI 6011 could degrade 2,4-DCBA (Miguez et al. 1990).

A mechanism in which chlorine is eliminated in the first step by a benzoate-2,3-dioxygenase-catalysed reaction, giving 2,3- dihydroxybenzoate as product was found in *Pseudomonas* sp. 2-CBA during growth on 2-CBA (Fetzner et al. 1989).

Based on the observed growth on salicylate, Higson and Focht (1990) suggested hydrolytic dehalogenation yielding salicylate as the first step in 2-halobenzoate degradation by *Pseudomonas aeruginosa* 2-BBZA. However, most reports on hydrolytic dehalogenation involve the displacement of *para* halogen substituents by a hydroxyl group. The metabolic route proceeds from the 4-halobenzoate to 4-hydroxybenzoate and protocatechuate. 4-Halobenzoate dehalogenation was found to be independent of benzoate degradation. When grown on 4-CBA, *Arthrobacter* sp. DSM 20407 carried out *meta* cleavage of protocatechuate, while when grown on benzoate the *ortho* cleavage route was induced (Ruisinger et al. 1976). Furthermore, a mutant strain of *Pseudomonas* sp. CBS3, which had lost the ability to grow on 4-CBA, could still grow on benzoate (Keil et al. 1981). Markus et al. (1984) proposed an oxygen- and NADH-dependent 3,4-dioxygenase-catalysed reaction for the conversion of 4-chlorophenylacetate to yield protocatechuate in *Pseudomonas* sp. CBS3. This suggests an oxidative dehalogenation mechanism. Subsequently, genes coding for the first enzyme in the degradation of 4-CBA in this bacterium were identified and H_2O was shown to be the hydroxyl donor. Labelling experiments also showed that the oxygen incorporated originated from H_2O, not from O_2 (Marks et al. 1984b; Müller et al. 1984).

The genes of *Pseudomonas* sp. CBS3 specifying 4-chlorobenzoate dehalogenase were cloned in *Pseudomonas putida* KT2440. A 9.5 kilobase-pair fragment inserted in a plasmid conferred on this latter strain the ability to grow on 4-CBA but did not complement mutants unable to grow on 4-HBA (Savard et al. 1986). The 4-CBA dehalogenase enzyme of *Pseudomonas* CBS3 showed higher activities in alcohol than in water (Thiele et al. 1988a) and was also able to dehalogenate 4-chlorodinitrobenzoates and 4-chlorodinitrophenols (Thiele et al. 1988b). The 4-chlorobenzoate dehalogenase genes of *Pseudomonas* sp. CBS3 contain a 4.5-kilobase chromosomal DNA fragment, which codes for a three polypeptide component enzyme system. Magnesium ATP could be identified as a co-substrate and CoA as a cofactor. The dehalogenase activity arises from a 4-chlorobenzoate: CoA ligase-dehalogenase (two components) and a thioesterase (one component) (Scholten et al. 1991).

Dehalogenation by hydroxylation seems very specific for the *para* position of halobenzoates. However, hydroxylation of 3-CBA was reported by Johnston et al. (1972). Furthermore, *Acinetobacter* sp. 4CB1 showed both *para* and *meta* dehalogenation during co-metabolism of 3,4-dichlorobenzoate. However, this compound did not serve as growth substrate (Adriaens et al. 1989). In contrast to growth on 4-CBA, the latter bacterium did not induce dioxygenase activity for the co-metabolism of 3,4-DCBA when grown on benzoate. Although the addition of 3-chloro-4-hydroxybenzoate stimulated oxygen and NADH consumption in cell extracts, this compound was not mono-oxygenated during oxygen uptake experiments. During growth on 4-chlorobenzoate, 3-chloro-4-hydroxybenzoate was oxidized to 4-carboxy-1,2-

benzoquinone. Under anaerobic conditions this dehalogenation rate was nearly 2.5-fold higher than in the presence of oxygen (Adriaens and Focht 1991). These authors proposed that both hydroxyl groups may be derived from water. *Para* hydroxylation of 4-CBA has also been described for *Alcaligenes denitrificans* NTB-1 (van den Tweel et al. 1987). Groenewegen et al. (1990) later characterized this bacterium as a coryneform bacterium NTB-1 and showed that oxygen, and thus ATP, was necessary for the transport of 4-chlorobenzoate through the cell membrane.

Most bacteria capable of 4-CBA dehalogenation also dehalogenate 4-BBA and 4-IBA but not 4-FBA. These reactions have been described for *Arthrobacter* sp. SU DSM 20407 (Müller et al. 1988), for *Pseudomonas* sp. CBS3 (Thiele et al. 1987) and for *Alcaligenes denitrificans* NTB-1 (van den Tweel et al. 1986, 1987). Remarkably *Aureobacterium* sp. RHO25 dehalogenates 4-FBA but not 4-CBA (Oltmanns et al. 1989). Furthermore, Marks et al. (1984a) described a 4-chlorobenzoate dehalogenase with activity for 4-CBA, 4-BBA and 4-FBA.

Reductive dechlorination under aerobic conditions has been described for *Alcaligenes denitrificans* NTB-1 (van den Tweel et al. 1987). Prior to *para* hydroxylation, 2,4-dichlorobenzoate was reductively dehalogenated in the *ortho* position.

B. Anaerobic biodegradation

Under anaerobic conditions halobenzoates are reductively dehalogenated. The first reports on this subject were from Horowitz et al. (1983) and Suflita et al. (1983). They described the reductive dehalogenation of iodo-, bromo- and chlorobenzoates by a methanogenic bacterial consortium isolated from sewage sludge. The dechlorination of chlorobenzoates seemed specific for *meta* substituents whereas dehalogenation of iodo- and bromo-benzoates occurred in all three positions. Fluorobenzoates were never found to be reductively dehalogenated. In this consortium, a 3-CBA-dehalogenating organism (DCB-1) could be detected (Shelton and Tiedje 1984). With DCB-1 as the key dehalogenating bacterium, a defined three-membered consortium was constructed which was able to use 3-CBA as sole source of carbon and energy. In this consortium, consisting of the 3-chlorobenzoate reducer (DCB-1), a benzoate metabolizer (BZ-2) and a methanogenic *Methanospirillum* (PM-1), the reducing equivalents needed for 3-CBA dehalogenation were obtained as hydrogen from the anaerobic oxidation of benzoate (Dolfing and Tiedje 1986, 1987). Benzoate oxidation was shown to be stimulated by addition of 3-CBA to this consortium, which reduces the hydrogen concentration. (Dolfing and Tiedje 1991a).

Several attempts were made to isolate and characterize strain DCB-1 (Shelton and Tiedje 1984; Stevens et al. 1988; Stevens and Tiedje 1988; Linkfield and Tiedje 1990; Mohn et al. 1990). Eventually these attempts were successful and this strain was named *Desulfomonile tiedjei*. This strain is now catalogued as ATCC 49306 (DeWeerd et al. 1990). This bacterium is a sulphate reducer (Linkfield and Tiedje 1990; Mohn and Tiedje 1990), which is capable of reductive dehalogenation of halobenzoates and chloroethenes (Fathepure et al. 1987). Apart from dechlorination of 3-chlorobenzoate, the consortium and *D. tiedjei* are capable of complete dechlorination of 3,5-DCBA, whereas 2,5-DCBA is not dechlorinated further than to 2-CBA (Fig. 6). Additionally, this consortium and *D. tiedjei* are also capable of demethoxylation of methoxybenzoates (Dolfing and Tiedje 1991b,c). Reductive dehalogenation by *D. tiedjei* is inhibited by sulphite and thiosulphate and by the same respiratory inhibitors that inhibit sulphite reduction (DeWeerd et al. 1991). The dehalogenation activity is inducible by its substrate and is membrane associated. Hydrogen and formate are electron donors for dehalogenation, but it is not yet known which endogenous reducing equivalents are involved (DeWeerd and Suflita 1990). The reduction of 3-chlorobenzoate is coupled to ATP production (Dolfing 1990), which results in an increase of growth yield (Mohn and Tiedje 1990). Reductive dehalogenation causes an increase in proton translocation which supports a proton-motive force, which in turn supports ATP synthesis via a proton-driven ATP-ase. Such a process would be a novel mode of anaerobic respiration (Mohn and Tiedje 1991).

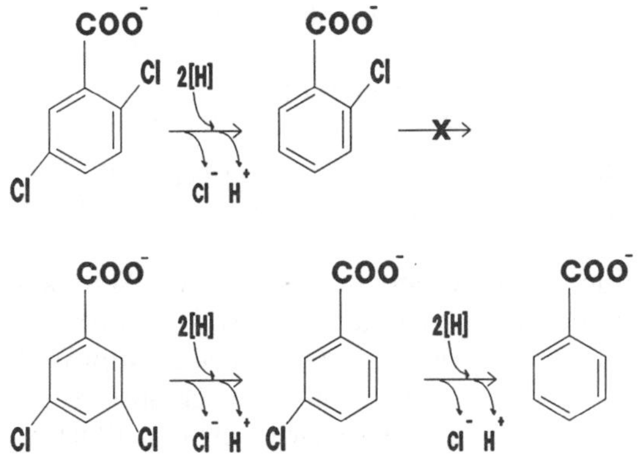

Fig. 6. Anaerobic dehalogenation of 2,5- and 3,5-dichlorobenzoate by *Desulfomonile tiedjei* (Dolfing and Tiedje 1991b).

No relationship was found between reductive dehalogenation and other substituent removal reactions (DeWeerd et al. 1986). Attempts to relate the molecular structure of several substituted benzoates to experimental degradation rates yield no satisfactory correlation. This implies that aryl substituent removal cannot be described as either simple nucleophilic or electrophilic substitution (Dolfing and Tiedje 1991c).

A denitrifying consortium was also found to dehalogenate chlorobenzoates (Sharak Genthner et al. 1989a). The dehalogenation occurred in the presence of nitrate. Under these conditions, the degradative pathways followed are similar to those for anaerobic benzoate degradation. A denitrifying *Pseudomonas* could grow on 2-FBA as the sole source of carbon and energy and showed activity towards 3- and 4-FBA but not towards 2-, 3- and 4-CBA (Schennen et al. 1985).

A mixed phototrophic culture was capable of mineralizing 3-CBA in presence of light and absence of oxygen. As was the case for denitrifying conditions, the metabolic route was similar to that for anaerobic benzoate degradation (Kamal and Wyndham 1990).

IV. Biodegradation of halogenated benzenes

A. Aerobic biodegradation

Despite the fact that halogenated benzenes are the simplest halogenated aromatic compounds and as such can be considered as model compounds for the whole class, relatively little is known about their biodegradation. The use of chlorobenzenes in particular as solvents, fumigants and as intermediates in the production of dyes and pesticides has resulted in widespread environmental contamination by these compounds. Marinucci and Bartha (1979) described the degradation of 1,2,3- and 1,2,4-trichlorobenzenes (1,2,3- and 1,2,4-TrCBz) to CO_2 in soil and media inoculated with soil. They identified 3,4,5-tri-, 2,6-di- and 2,3-dichlorophenols as metabolites of 1,2,3-TrCBz and 2,4-, 2,5- and 3,4-dichlorophenols in incubations with 1,2,4-TrCBz. Similarly, Ballschmiter and Scholz (1981) isolated 2,3-, 3,4- and 2,6-dichlorophenols from incubations of three *Pseudomonas putida* strains with 1,2-DCBz, and 2,4,6-dichlorophenol from 1,3,5-TrCBz.

In contrast, the chlorobenzene-utilizing strain WR1306 was shown to degrade this compound via dioxygenation to form chlorocatechol (Fig. 7), which was degraded further by the *ortho* cleavage pathway (Fig. 4) (Reineke and Knackmuss 1984). Reineke and Knackmuss suggested that the chlorophenols isolated by other workers were in fact artifacts produced by acid-catalysed dehydration of the *cis*-dihydrodiols formed by the dioxygenation reaction. In common with other bacteria which are able to grow on chloroaromatic compounds, this strain contains ring cleavage enzymes showing high activity towards chlorinated substrates.

Similar pathways are active in 1,4-DCBz-utilizing *Pseudomonas* and *Alcaligenes* strains (Oltmanns et al. 1988), by two *Alcaligenes* strains which are able to grow on 1,3- and 1,4-DCBz, chlorobenzene and benzene (De Bont et al. 1986; Schraa et al. 1986), by a 1,2-DCBz-utilizing *Pseudomonas* strain (Haigler et al. 1988), by *Pseudomonas* strain P51 which can grow on 1,2,4-TrCBz and the three dichlorobenzenes (van der Meer et al. 1987) and by *Pseudomonas* strains able to grow on 1,2,4,5-tetrachlorobenzene (1,2,4,5-TCBz), 1,2,4-TrCBz and di- and monochlorobenzenes (Sander et al. 1991).

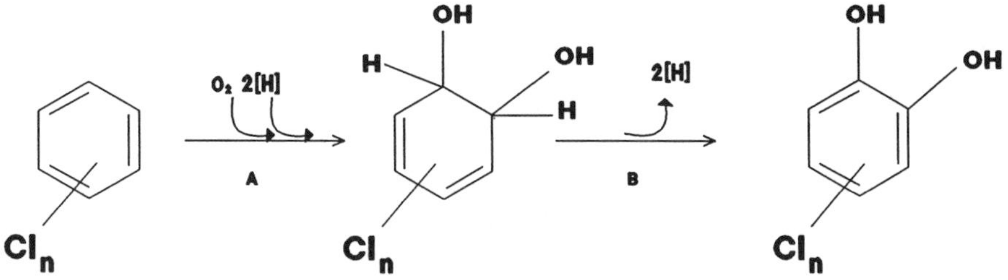

Fig. 7. Aerobic degradation of benzene to catechol. A = benzene dioxygenase; B = benzene dihydrodioldehydrogenase.

Pyrocatechases (catechol-1,2-dioxygenases) with high activities towards chlorinated substrates also appear to be induced in such strains grown on chlorobenzenes.

The genes involved in the degradation of chlorobenzenes in strain P51 are located in two clusters on a 110 kb plasmid (van der Meer et al. 1991b). The *tcbA* and *tcbB* genes, encoding the degradation to chlorocatechols, lie on a transposable element (van der Meer et al. 1991a). The second cluster consists of the *tcbC*, *tcbD* and *tcbE* genes, coding for type II catechol-1,2-dioxygenase, cycloisomerase and hydrolase enzymes. The transposon may have played a role in the transfer of the first two genes and the evolution of this catabolic pathway.

A hybrid chlorobenzene-utilizing strain was obtained by mating the benzene-degrading *Pseudomonas putida* strain F1 with the 3-chlorobenzoate-utilizing *Pseudomonas* strain B13 (Oltmanns et al. 1988). Strain B13 possesses a type II catechol-1,2-dioxygenase which enables it to degrade chlorocatechols efficiently via the modified *ortho* cleavage pathway.

Kröckel and Focht (1987) isolated a chlorobenzene-utilizing recombinant *Pseudomonas putida* strain from mixed cultures of toluene-grown *Pseudomonas putida* and benzoate-grown *Pseudomonas alcaligenes* strains exposed to chlorobenzene. Chromosomal DNA from the *Pseudomonas alcaligenes* strain had been transferred and integrated in a TOL-like plasmid of the *Pseudomonas putida*. During insertion a 24 kB fragment was lost from the plasmid, which resulted in the loss of the ability to grow on xylene and methylbenzoates. This fragment coded for a catechol-2,3-dioxygenase with low specificity and high activity (Carney and Leary 1989; Carney et al. 1989).

Pseudomonas strain JS6 is able to degrade all three dichlorobenzenes, chlorobenzene and toluene (Spain and Nishino 1987; Haigler and Spain 1989). In common with other strains able to grow on chlorinated benzenes, strain JS6 degrades these compounds via the modified *ortho* cleavage pathway. Although methyl-substituted benzenes are usually degraded via the *meta* cleavage pathway, strain JS6 also degrades 4-chlorotoluene via the modified *ortho* cleavage pathway (Haigler and Spain 1989). Unusually, this strain also possesses a *meta* cleavage pathway, but this is inhibited in the presence of

3-chlorocatechol (Pettigrew et al. 1991). In cultures grown on a mixture of chlorobenzene and toluene, some of the toluene is degraded via the modified *ortho* cleavage pathway.

Recently, bromobenzene-utilizing *Pseudomonas* strains have been isolated from chemostat cultures exposed to increasing concentrations of bromobenzene (Sperl and Harvey 1988). Bromocatechols appeared to be intermediates in the degradation of this compound.

B. Anaerobic biodegradation

The dominant biodegradative reaction for chlorobenzenes under anaerobic conditions is reductive dehalogenation. Hexachlorobenzene was dechlorinated by two pathways in anaerobic sewage sludge (Fathepure et al. 1988). The major route gave pentachlorobenzene, 1,2,3,5-TCBz and 1,3,5-TrCBz whereas the minor route yielded PCBz, 1,2,4,5-TCBz, 1,2,4-TrCBz and dichlorobenzenes. There was no evidence for further reduction of 1,3,5-TrCBz. In contrast, Bosma et al. (1988) observed reductive dechlorination of all tri- and dichlorobenzene isomers in anaerobic sediment columns. Fathepure and Vogel (1991) recently reported the reductive dechlorination of hexachlorobenzene in anaerobic biofilms. In this case 1,2,3-TrCBz and 1,2-DCBz were the major products formed, along with trace amounts of 1,2,4-TrCBz and 1,3-DCBz (Fig. 8). A two-stage anaerobic-aerobic biofilm reactor gave a 94% conversion of HCBz to CO_2.

V. Biodegradation of halogenated phenols

A. Aerobic biodegradation

Chlorinated phenols are used on large scale as wood preservatives, fungicides, herbicides and general biocides. Bacteria able to grow on pentachlorophenol, the most widely used chlorophenol, were first described in the early 1970s (Chu and Kirsch 1972; Watanabe 1973). The biodegradation of chlorophenols takes place by three main pathways. In general, mono- and di-chlorophenols are converted into chlorocatechols by monooxygenase-catalysed reactions, whereas the higher chlorinated phenols are hydroxylated to form chlorinated hydroquinones. Under anaerobic conditions, chlorophenols undergo initial reductive dechlorination.

The 3-chlorobenzoate-utilizing *Pseudomonas* B13 is also able to grow on phenol and 4-chlorophenol (4-CP) and to degrade other mono- and di-chlorophenols (Knackmuss and Hellwig 1978). The chlorophenol-degrading cultures of this strain show high activity of a type II catechol 1,2-dioxygenase. The pathway proposed for the degradation of chlorophenols consists of initial monooxygenation to form chlorocatechols, which undergo *ortho* ring cleavage to chloromuconic acids, lactonization with loss of chloride and further

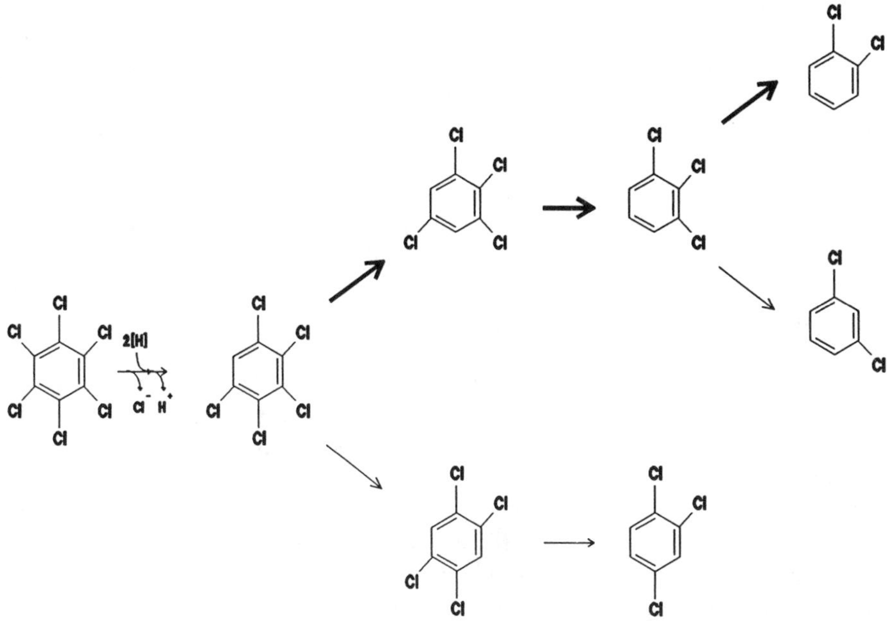

Fig. 8. Anaerobic reductive dechlorination of hexachlorobenzene in a continuous-flow anaerobic biofilm reactor (Fathepure and Vogel 1991).

degradation (via the modified *ortho* or β-ketoadipate pathway) (Fig. 5). Co-metabolism of 2-, 3- and 4-chloro- and 2,4- and 3,4-di-chlorophenols by *Nocardia* sp. DSM 43251 follows a similar mechanism (Engelhardt et al. 1979).

Degradation of monochlorophenols was poor in defined, mixed bacterial cultures containing *Pseudomonas* and *Alcaligenes* strains (Schmidt et al. 1983), due to the accumulation of toxic metabolites formed by the *meta* cleavage of chlorocatechols. In the presence of *Pseudomonas* sp. B13, the chlorophenols were degraded much more quickly with stoichiometric release of chloride. Hybrid strains were isolated from such mixed cultures which were able to grow on all three monochlorophenols, for example *Alcaligenes* strain A 7-2 which had acquired the modified *ortho* cleavage pathway from strain B13 (Schwien and Schmidt 1982).

Rhodococcus strains An 117 and An 213 co-metabolize monochlorophenols via the β-ketoadipate pathway (Janke et al. 1988b). In these strains ring cleavage is catalysed by "ordinary" catechol-1,2-dioxygenases with low activity towards chlorocatechols. Degradation of 3- and 4-CPs, but not of 2-CP, is stimulated in the presence of glucose as an extra source of energy and reducing equivalents (Janke et al. 1988a).

The 2,4,5-trichlorophenoxyacetic acid-degrading *Pseudomonas cepacia* strain AC1100 is also able to degrade a range of di-, tri-, tetra- and penta-chlorophenols (Karns et al. 1983b). These chlorophenols are wholly or partly dechlorinated. Dehalogenation was also observed of 2,4-dibromo-, 2,4,6-tribromo- and penta-bromophenol, but not of 2,4,6-triiodophenol. The enzymes responsible for dechlorination of 2,4,5-T, 2,4,5-trichlorophenol and PCP are induced by 2,4,5-TrCP, whereas those which convert 2,4,5-T to 2,3,4-TrCP are constitutive (Karns et al. 1983a). The mechanism by which 2,4,5-TrCP is degraded by this strain has been identified as conversion of 2,4,5-TrCP to 2,5-dichlorohydroquinone, which undergoes a dehalogenation to 5-chloro-2-hydroxyhydroquinone and subsequent ring cleavage (Sangodkar et al. 1989).

A pathway involving initial hydrolytic dechlorination in the *para* position, to form tetrachlorohydroquinone, and further reductive dechlorinations was proposed for the degradation of pentachlorophenol (PCP) by an aerobic *Flavobacterium* strain isolated from PCP-contaminated soil (Steiert and Crawford 1986). This strain is also able to degrade and dechlorinate a range of di-, tri- and tetra-chlorophenols (Steiert et al. 1987). Chlorophenols with chlorine substituents in both *ortho* (2 and 6) positions were degraded most readily. Of these, 2,4,6-TrCP, 2,3,5,6-TCP and PCP were inducers of the complete PCP degradation pathway. Recently, a periplasmic protein involved in the dechlorination of PCP by this strain was purified (Xun and Orser 1991). The *pcpA* gene coding for this protein was cloned and sequenced but does not show strong homology with any other genes characterized.

The PCP-utilizing *Rhodococcus chlorophenolicus* also attacks tri-, tetra- and penta-chlorophenols by *para*-hydroxylation to produce chlorinated hydroquinones (Fig. 9) (Apajalahti and Salkinoja-Salonen 1987a). Originally, these authors concluded on the basis of labelling experiments that this reaction was a hydrolytic substitution in which the hydroxyl group is derived from a water molecule (see above). However, subsequent work indicated that this reaction is catalysed by a membrane-bound enzyme which is inhibited by different inhibitors of cytochrome P-450 and had a carbon monoxide-dependent absorbance peak at 457 nm (Uotila et al. 1991). This reaction is therefore probably a cytochrome P-450-catalysed monooxygenation.

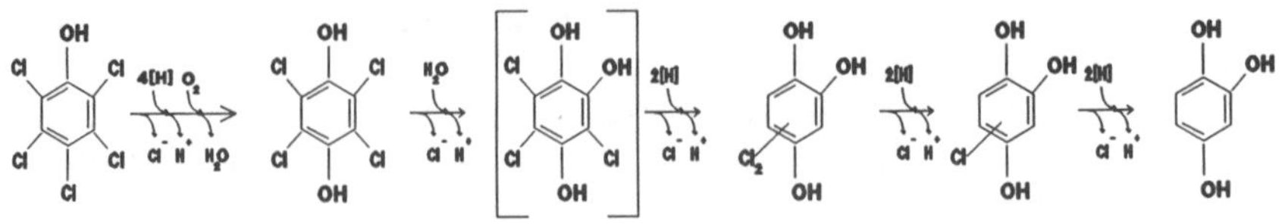

Fig. 9. Metabolic route for aerobic PCP degradation by *Rhodococcus chlorophenolicus* PCP-1 (Apajalathi and Salkinoja-Salonen 1987a,b; Uotila et al. 1991).

The tetrachlorohydroquinone formed from PCP is subsequently converted to a dichlorotrihydroxybenzene by a reaction sequence involving both hydrolytic and reductive dechlorinations (Apajalahti and Salkinoja-Salonen 1987b). Two further reductive dechlorinations then give 1,2,4-trihydroxybenzene. Trichlorohydroquinone is degraded very slowly, suggesting that it is not an intermediate in this pathway. Similar results have been found for *Rhodococcus* strain CP-2 (Häggblom et al. 1988, 1989b).

Schenk et al. (1990) used $H_2^{18}O$ and $^{18}O_2$ to investigate the mechanism by which *Arthrobacter* strain ATCC 33790 converts PCP to tetrachlorohydroquinone. Labelled product was only formed in experiments with $H_2^{18}O$. However, since they also found that unlabelled tetrachlorohydroquinone became labelled in incubations with the enzyme fraction in $H_2^{18}O$, they were not able to distinguish between the hydrolytic and oxidative mechanisms. The requirement for oxygen and NADPH indicated that this reaction is an oxygenation. These results also explain the formation of labelled tetrachlorohydroquinone in $H_2^{18}O$ observed by other investigators. *Azotobacter* sp. strain GP1 grows on both 2,4,6-trichlorophenol and phenol, but by different pathways (Li et al. 1991). 2,6-Dichlorohydroquinone accumulated in cultures of mutants blocked in 2,4,6-TrCP degradation, whereas phenol-deficient mutants accumulated catechol.

Many microorganisms, including several strains of *Rhodococcus*, *Acinetobacter* and *Pseudomonas*, are able to O-methylate halophenols, but are not capable of further degradation (Allard et al. 1985; Neilson et al. 1988; Häggblom et al. 1989a).

B. Anaerobic biodegradation

The anaerobic biodegradation of halogenated phenols is receiving increasing interest. Reductive dechlorination of mono-, di- and penta-chlorophenols and pentabromophenol takes place in anaerobic sewage sludges (Fig. 10) (Boyd and Shelton 1984; Mikesell and Boyd 1986). Which positions are dechlorinated most rapidly depends on which monochlorophenol the sludges are adapted to. Complete reductive dehalogenation of PCP and PBP and mineralization to methane and CO_2 was observed in sludges adapted to all three monochlorophenols. An anaerobic mixed culture enriched from sewage sludge was able to dehalogenate 2-chloro-, 2-bromo- and 2,6-dichloro-phenols but could only remove the *ortho* substituent from 2,4-dichlorophenol (Dietrich and Winter 1990). The culture lost its dehalogenation ability if a spirochaete-like organism was lost from it.

Reductive dehalogenation of 2- and 3-CP, 2,4-DCP, 2,4-DBP and 2,4,6-TrBP has also been observed in anaerobic consortia enriched from aquatic

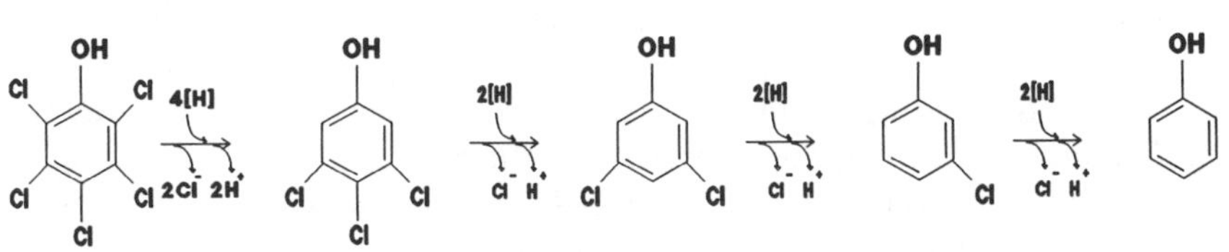

Fig. 10. Anaerobic dehalogenation of pentachlorophenol by a mixture of anaerobic sludges (Mikesell and Boyd 1986).

sediments (King 1988; Sharak Genthner et al. 1989a,b). Again, adaptation of anaerobic sediment slurries influenced the specificity of the dechlorination of PCP (Bryant et al. 1991). In unadapted slurries, dechlorination required a lag period of at least 40 days. Slurries adapted to 2,4-dichlorophenol dechlorinated PCP preferentially in the *ortho* positions without a lag phase, whereas those adapted to 3,4-dichlorophenol initially removed the *para* chlorine.

The anaerobic dechlorination of chlorophenols is inhibited in the presence of sulphate and other sulphuroxy anions (Kohring et al. 1989; Madsen and Aamand 1991). In experiments with a culture derived from sewage sludge, the inhibition was reduced on addition of H_2. The inhibition may therefore be caused by competion for H_2 between sulphate reduction and dechlorination (Madsen and Aamand 1991). In contrast, Häggblom and Young (1990) reported dechlorination of 2,4- and 2,6-dichlorophenol, and oxidation of the resulting monochlorophenols, under sulphate-reducing conditions, by cultures derived from estuarine sediments and an anaerobic bioreactor used for treating pulp-bleaching effluents.

VI. Biodegradation of halogenated anilines

A. Aerobic biodegradation

Chlorinated anilines are used as intermediates in the synthesis of pesticides and enter the environment in industrial wastes. In many cases, degradation of the pesticides after application in the field also yields chlorinated anilines. Few bacteria are known that can mineralize halogenated anilines. *Moraxella* sp. strain G is able to use aniline, 4-fluoro-, 2-chloro-, 3-chloro-, 4-chloro- and 4-bromo-anilines, but not 4-iodoaniline, as sole carbon and nitrogen source (Zeyer and Kearney 1982a; Zeyer et al. 1985). This strain is also able to co-metabolize 2,4-dichloroaniline (2,4-DCA) (Zeyer and Kearney 1982b; Zeyer et al. 1985). Degradation of these compounds proceeds by initial dioxygenation to form halocatechols, catalysed by an aniline oxygenase with a broad substrate specificity. Further degradation is via the modified *ortho* cleavage pathway, involving a catechol-1,2-oxidase with high activity towards substituted catechols, as is the case with other bacteria able to grow on halogenated aromatic compounds (Zeyer et al. 1985).

Similar pathways are involved in the co-metabolism of monochloroanilines by two *Rhodococcus* sp. strains (Janke et al. 1988a,b) and 3,4-DCA by a *Pseudomonas putida* strain (You and Bartha 1982). A *Pseudomonas* strain which degrades aniline via the *meta*-cleavage pathway is not able to degrade chlorinated anilines, although these compounds do induce the enzymes for aniline oxidation (Konopka et al. 1989). *Pseudomonas acidovorens* strains able to grow on aniline and monochloroanilines were isolated by enrichment in soil columns (Loidl et al. 1990). Oxygen uptake experiments indicated that these strains also possessed the modified *ortho* cleavage pathway, although minor amounts of the *meta* cleavage product were produced from 4-CA.

B. Anaerobic biodegradation

In the presence of nitrate-reducing bacteria, chlorinated anilines undergo condensations to chlorinated azobenzenes, triazenes and biphenyls (Minard et al. 1977; Corke et al. 1979). However, these appear to be chemical reactions of diazonium cations derived from the chloroanilines. The role of the bacteria is reduction of nitrate to nitrite, which reacts with chloroanilines to form diazonium cations. Such reactions take place for a variety of substituted anilines, including mono- and dichloroanilines, but not trichloroanilines, and for monobromoanilines, but not for 2-fluoroaniline (Lammerding et al. 1982).

Reductive dehalogenation of chloroanilines takes place under anaerobic conditions. Dechlorination of 2,4-DCA and 3,4-DCA was observed in pond sediment slurries (Struijs and Rogers 1989). In samples from methanogenic, but not sulphate-reducing, aquifers sequential *para* and *ortho* dehalogenation of 2,3,4,5-TCA yielded 2,3,5-TrCA and 3,5-DCA (Kuhn and Suflita 1989). 3,4-DCA was dechlorinated to 3-CA. No further dehalogenations were detected. In the presence of butyrate, 2,3,4,5-TCA was also dehalogenated in the *meta* and *para* positions (Kuhn et al. 1990). In the presence of sulphate dechlorination rates were lower and the latter pathway was preferred.

VII. Biodegradation of halogenated phenoxyacetic acids

A. Aerobic biodegradation

The herbicides 2,4-dichlorophenoxyacetic acid (2,4-D) and 2,4,5-trichlorophenoxyacetic acid (2,4,5-T) have been used for more than 40 years. Several strains are known which can degrade these compounds. In general, biodegradation of 2,4-D takes place via initial cleavage of the ether bond, followed by hydroxylation of the resulting dichlorophenol to chlorocatechols (e.g. Bollag et al. 1968a,b; Evans et al. 1971; Tiedje and Alexander 1969).

One of the best characterized 2,4-D-degrading microorganisms is *Alcaligenes eutrophus* JMP 134. This strain also degrades 4-chloro-2-methylphenoxyacetic acid, 2-methylphenoxyacetic acid and phenoxyacetic acid (Pieper et al. 1988). The ether bond cleavage of these compounds is apparently catalysed by a monooxygenase with a wide substrate specificity (Fig. 11). The chlorocatechol intermediates formed from the chlorinated compounds are degraded by the *ortho* cleavage pathway. Unusually, for the nonchlorinated compounds, both *ortho* and *meta* ring cleavage mechanisms are induced. The genes encoding this pathway are located on a 80 kb plasmid, known as pJP4, which also codes for the degradation of 3-CBA and mercury resistance (Don and Pemberton 1981; Don et al. 1985). The genes for 2,4-D degradation are organized in three operons: the *tfdA* gene coding for

2,4-dichlorophenoxyacetic acid monooxygenase, the *tfdB* gene coding for 2,4-dichlorophenol hydroxylase and the *tfdCDEF* genes coding for for the modified *ortho* cleavage pathway. In addition two regulatory genes have been identified: *tdfR* and *tdfS* (Kaphammer and Olsen 1990). Transfer of pJP4 to *Pseudomonas aeruginosa* and *putida* strains failed to give these the ability to grow on 2,4-D (Kukor et al. 1989). 2-Chloromaleylacetate accumulated in the culture medium. However, transfer of chromosomal genes for maleylacetate reductase, cloned from a *Pseudomonas* sp., in recombinant plasmids together with pJP4 did result in strains able to grow on 2,4-D.

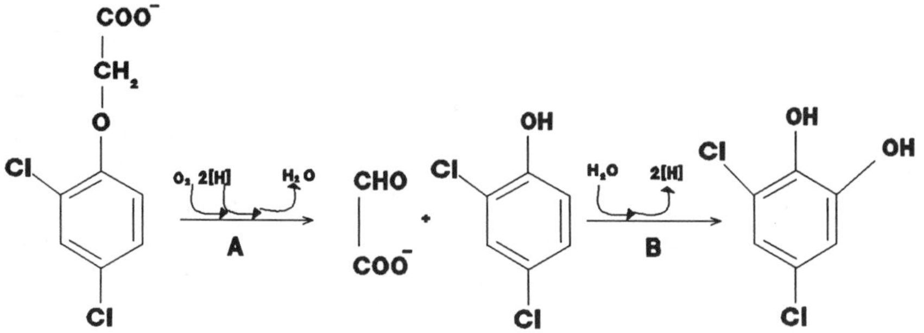

Fig. 11. Aerobic   degradation   of   2,4,5-trichlorophenoxyacetate.   A   =   2,4,5-trichlorophenoxyacetate monooxygenase; B = 2,4-dichlorophenol hydroxylase (Bollag et al. 1968a,b).

Several other *Acinetobacter*, *Arthrobacter*, *Corynebacterium*, *Flavobacterium*, and *Pseudomonas* strains are known which are able to degrade 2,4-D although the degradative pathways are not always known. *Flavobacterium* strain MH degrades 2,4-D by a pathway similar to that of strain JMP134 as well as a range of other 2,4-dichlorophenoxyalkanoic acids (Horvath et al. 1990). Most of the 30 2,4-D-utilizing bacteria isolated from sewage sludge and freshwater samples appeared to be *Alcaligenes* strains (Amy et al. 1985). At least 14 of these strains contained one or more plasmids, but in general there was little similarity between these.

As mentioned above, the 2,4,5-T-utilizing *Pseudomonas cepacia* AC1100 initially converts this compound to 2,4,5-TrCP (Karns et al. 1983b), which is then dechlorinated to form 2,5-dichlorohydroquinone (Sangodkar et al. 1989). This strain degrades 2,4-D to chlorohydroquinone, which accumulates and inhibits 2,4,5-T degradation (Haugland et al. 1990). The 2,4-D degrading *Alcaligenes eutrophus* strain JMP134 does not degrade 2,4,5-T. In mixed cell suspensions of strains AC1100 and JMP134 exposed to both 2,4-D and 2,4,5-T chlorohydroquinone and chlorophenols accumulate, similarly to that in pure suspensions of AC1100. Presumably, the 2,4-D-degradation pathway of JMP134 cannot compete with that of AC1100. Conjugative transfer of the plasmid coding for 2,4-D degradation in JMP1344 to AC1100 gave a constructed strain able to degrade 2,4-D and 2,4,5-T simultaneously.

Nocardioides simplex strain 3E is able to grow on 2,4,5-T (Golovleva et al. 1990). This strain degrades this compound by two, and possibly three, different pathways. Evidence was found for initial cleavage of 2,4,5-T to 2,4,5-trichlorophenol, reductive dehalogenation to 2,4-D and hydrolytic dechlorination to produce a dichlorohydroxyphenoxyacetic acid.

B. Anaerobic biodegradation

Reductive dehalogenation of chlorophenoxyacetic acids appears to be the most important degradative reaction under anaerobic conditions. 2,4,5-T was dehalogenated in methanogenic aquifer samples to form 2,4- and 2,5-dichlorophenoxyacetic acids (Gibson and Suflita 1990). Further degradation resulted in the formation of monochlorophenoxyacetic acids, chlorophenols and phenol. These reactions were inhibited by added sulphate, but stimulated by organic substrates.

VIII. Biodegradation of halogenated biphenyls

A. Aerobic biodegradation

Polychlorinated biphenyls (PCBs) were used for many years, mainly as nonflammable heat transfer fluids, as dielectric fluids in capacitors and transformers, as components of hydraulic fluids and as plasticizers in paints. The biodegradation of chlorinated biphenyls has been reviewed by Furukawa (1982), Parsons et al. (1983) and Safe (1984) and more recently by Abramowicz (1990). The ability to degrade polychlorinated biphenyls (PCBs) is found in several genera of both Gram positive and Gram negative aerobic bacteria (Ohmori et al. 1973; Walia et al. 1988; Unterman et al. 1988). These are mostly members of the genera *Pseudomonas*, *Alcaligenes*, *Arthrobacter* and *Acinetobacter*.

There are many reports of the mineralization (i.e complete degradation to CO_2, often measured by the formation of $^{14}CO_2$) of individual chlorinated biphenyls by bacterial consortia (Shiaris and Sayler 1982; Kong and Sayler 1983; Bailey et al. 1983; Fries and Marrow 1984; Brunner et al. 1985) brominated biphenyls (Kong and Sayler 1983) and commercial PCB mixtures (Hankin and Sawhney 1984; Baxter et al. 1975; Brunner et al. 1985; Fava et al. 1991). Mineralization rates are enhanced by sunlight (Kong and Sayler 1983) and moderately aerobic conditions (Pardue et al. 1988). Also, substrate enrichment and inoculation with PCB-degrading bacteria enhanced mineralization in soil (Brunner et al. 1985).

Mineralization capabilities vary in different bacterial consortia as a result of the differing metabolic abilities of the component strains of the populations (Hiramoto et al. 1989; Pettigrew et al. 1990). In most cases, chlorobenzoates accumulate as endproducts of chlorobiphenyl degradation. One possible explanation for this metabolic variety is that biphenyls are initially degraded by *meta* cleavage whereas *meta* cleavage of the resulting chlorocatechols can lead to toxic end products (see above). Only when enzymes for the *ortho* cleavage route of chlorocatechols are also induced will complete mineralization be possible. However, *Alcaligenes* sp. strain JB1 (originally tentatively identified as a *Pseudomonas* strain) cometabolizes both chlorobiphenyls and chlorobenzoates when grown on benzoate (Parsons et al. 1988). Mokross et al. (1990) constructed a strain capable of growth on monochlorobiphenyls by transconjugation of genes coding for pyrocatechase II (a catechol 1,2-dioxygenase with high specificity for chlorocatechols) from *Pseudomonas* B13 to *Pseudomonas putida* strain BN10 (a biphenyl degrader). By co-culturing chlorobiphenyl and chlorobenzoate degraders, Havel and Reineke (1991) obtained stable co-cultures which could mineralize 4-chlorobiphenyl (4-CB). When genes coding for chlorocatechol degradation from *Pseudomonas* sp. B13 were transferred to *Pseudomonas* sp. strain WR 401, which contains a benzoate dioxygenase with broad substrate specificity, the hybrid strain *Pseudomonas* JH230 was constructed. Transfer of the latter genes to *Pseudomonas* JHR, a biphenyl-degrading strain, resulted in derivative strains capable of growth on 2-, 3- and 4-chlorobiphenyls and 2,4- and 3,5-dichlorobiphenyls (2,4- and 3,5-DCB) (Havel and Reineke 1991).

Complete mineralization was also found for 4-chlorobiphenyl by a two membered culture of *Pseudomonas* CBS3 and a facultative anaerobic strain B-206 (Sylvestre et al. 1985). A co-culture of *Acinetobacter* sp. strain P6 and *Acinetobacter* strain 4CB1 also mineralizes 3,4- and 4,4'-DCB and 3,3',4,4'-TCB via hydrolytic dehalogenation of the chlorobenzoate intermediates (Adriaens et al. 1989; Adriaens and Focht 1990).

In all the strains described above, the major metabolic pathway for PCBs under aerobic conditions consists of 2,3-dioxygenation of the less chlorinated aromatic ring, followed by *meta* cleavage and further degradation to chlorobenzoates. However, 3,4-dioxygenation has been proposed as an alternative route for PCB degradation in some strains (Fig. 12) (Yagi and Sudo 1980; Massé et al. 1984; Bedard et al. 1986). Another possible reaction is nitration, which was reported for 4-chlorobiphenyl degradation by strain B-206 (Sylvestre et al. 1982). Mass spectrometric analysis also indicated the formation of a nitrogen-containing metabolite of 4-CB by *Pseudomonas putida* DA2 (Ahmad et al. 1991b).

The elimination of chlorine is thought to be a fortuitous event, which occurs in later metabolic steps. However, oxidative dehalogenation has been proposed to occur during the initial dioxygenation of PCBs in *Pseudomonas testosteroni* B-356. During 2,3-dioxygenation by biphenyl-2,3-dioxygenase the *ortho* position of 2,2'-DCB and the *para* position of 4,4'-DCB were dehalogenated. The latter dehalogenation is particularly remarkable because this involves a shift of the chlorine substituent (Ahmad et al. 1991a).

Based on studies of PCB metabolism by *Alcaligenes* sp. Y46 and

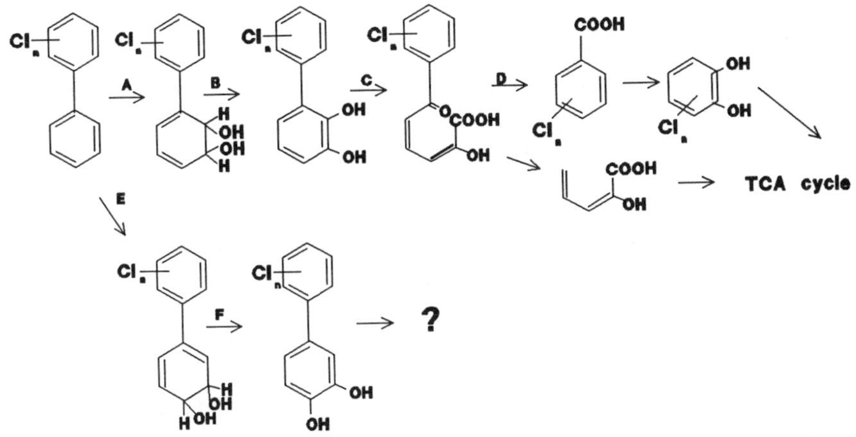

Fig. 12. Metabolic routes for aerobic degradation of PCBs. A = biphenyl-2,3-dioxygenase; B = biphenyl-2,3-dihydrodiol dehydrogenase; C = 2,3-dihydroxybiphenyl dioxygenase; D = 2-hydroxy-6-oxo-6-phenylhexa-2,4-dienate hydrolase; E = biphenyl-3,4-dioxygenase; F = biphenyl-3,4-dihydrodiol dehydrogenase.

Acinetobacter sp. P6, Furukawa (1982) proposed some general relationships between PCB structure and biodegradability. To date the following relationships are still valid for the aerobic degradation.
1. The less chlorinated the biphenyl, the faster degradation takes place.
2. Dioxygenation takes place on the ring with the least chlorine substituents.
3. PCBs with chlorine substituents on both rings are more recalcitrant than isomers containing an unchlorinated ring.

Alcaligenes strain JB1 shows fast degradation of 2,2′,3,3′-tetrachlorobiphenyl and relatively slow degradation of 3,3′,4,4′-TCB (Parsons and Sijm 1988). Also cultures of *Pseudomonas* sp. KKS102 to which supernatant from a culture of the related *Pseudomonas* sp. KKS101 had been added showed relatively fast degradation of 2,2′,3,3′-TCB and slow degradation of 2,2′6,6′-TCB and 3,3′,4,4′-TCB (Kimbara et al. 1988). Similar results were described for *Corynebacterium* sp. MB1 (Bedard et al. 1986).

However, in contrast to these results, *Alcaligenes eutrophus* H850 and *Pseudomonas putida* LB400 metabolize 2,2′,5,5′-TCB completely and even degrade 2,2′,4,4′,6,6′-HCB. Furthermore, *Alcaligenes eutrophus* H850 does not degrade 2,2′,3,3′-TCB whereas *Pseudomonas putida* LB400 does degrade this latter compound but does not metabolize 2,2′,6,6′-TCB (Bedard et al. 1986, 1987b; Bopp 1986). Dioxygenation at the 3,4-positions was proposed as the initial step for the degradation of those congeners containing a 3-, 2,5- or 2,4,5-chlorophenyl ring by *Pseudomonas* sp. LB400 and *Alcaligenes eutrophus* H850 (Bedard and Haberl 1990). All the *meta* cleavage pathway enzymes were detected in strain H850 (Unterman et al. 1988). A new metabolite, 2′,4′,5′-trichloroacetophenone was detected in incubations of *Alcaligenes eutrophus*

H850 with 2,2',4,4',5,5'-HCB (Bedard et al. 1987a). Although there was no direct evidence, the authors suggested the formation of chloroacetophenones to be the result of 3,4-dioxygenation. However, Baxter and Sutherland (1984) proposed the formation of 2'-chloroacetophenone to be the result of photochemical degradation of 3-chloro-2-hydroxy-6-oxo-2'-chlorophenylhexa-2,4-dienoate, the *meta* cleavage product of 2,4'-dichloro-2,3-dihydroxybiphenyl.

The formation of 3,4-dihydro-3,4-dihydroxybiphenyl could be identified in some bacterial cultures grown in the presence of chlorobiphenyls (Massé et al. 1989). Extensive mass spectrometric analysis of the metabolites of 4-chlorobiphenyl produced by a pseudomonad yielded, besides 4'-chloroacetophenone, 2-hydroxy-2-[4'-chlorophenyl]ethane and 2-oxo-2-[4'chlorophenyl]ethanol (Barton and Crawford 1988).

After studying the PCB congener degradation spectrum of eight PCB-degrading strains, Bedard and Haberl (1990) suggest a classification of PCB degrading strains into four classes:

Class I   The sequence of degradation rates of the reacting phenyl ring is 3- ≥ 4- >> 2-chlorophenyl. The degradation rate is largely determined by the chlorination pattern of the nonreacting ring.

Class II   The sequence of degradation rates is identical to Class I but there is no influence of the chlorine substitution of the nonreacting ring.

Class III   The sequence of reactivity of the reacting chlorophenyl ring is 4- ≥ 2- > 3-chlorophenyl. The chlorination pattern of the nonreacting ring has a strong influence.

Class IV   The sequence of reactivity is 2- > 3- >> 4-chlorophenyl. The chlorination pattern of the nonreacting ring has influence.

The few biochemical and genetic data available are in accordance with this four-class distinction. Yates and Mondello (1989) demonstrated that the two members of Class IV, strain LB400 and H850 are genetically similar and distinct from bacteria from the other classes.

The enzymes involved in the degradation of chlorinated biphenyls to benzoates are encoded by genes located both on chromosomes and on plasmids. At least four genes are involved in the degradation of chlorobiphenyls to chlorobenzoates. Gene *bphA* (or alternatively *cbpA*) codes for the biphenyl dioxygenase, gene *bphB* (or *cbpB*) encodes the dihydrodiol dehydrogenase, gene *bphC* (or *cbpC*) the 2,3-dihydroxybiphenyl dioxygenase and gene *bphD* (or *cbpD*) codes for the 2-hydroxy-6-oxo-6-phenylhexa-2,4-dienoate hydrolase (Fig. 12) (Hayase et al. 1990; Kahn and Walia 1991).

The first report that a plasmid (pKF1) was involved in the degradation of 4-CBP was from Furukawa and Chakrabarty (1982), detected in *Acinetobacter* sp. P6 and *Arthrobacter* sp. M5. From colony hybridization of *Alcaligenes* , *Acinetobacter* and *Achromobacter* spp., all isolated from the same sediment, another plasmid (pSS50) involved in PCB degradation was isolated (Shields et al. 1985). After transfer of pSS50 to various other strains several pSS50-like plasmids could be isolated (Pettigrew et al. 1990). In contrast, the *bphABCD genes of Pseudomonas testosteroni* B-356 are located on the chromosome. Using a broad host range plasmid, these genes could be transferred succesfully to *Pseudomonas putida* KT2440 (Ahmad et al. 1990).

The size of the *bphABCD* gene cluster is appoximately 10kb. A 8 kb DNA fragment of *Pseudomonas putida* OU83 was shown to contain the four genes. Further analysis revealed that a 2.3 kb fragment contains the *cbpBCD* genes and the *cbpA* gene is located on a 2.8 kb fragment (Khan and Walia 1991). The first three genes (*bphABC*) are clustered in a operon in *Pseudomonas putida* KF715. This operon is separated by two open reading frames from the *bphD* gene (Hayase et al. 1990). The order of the *cbp* genes from *Pseudomonas putida* OU83 cloned in plasmid pAW6194 is *cbpADCB*, where the *cbpA* gene is seperated by 3 kb from *cbpBC*.

Hybridization of a *bphABCD* DNA-probe with genomic DNA of several different PCB-degrading strains indicated that many PCB-degrading pathways have a common phylogenetic origin (Walia et al. 1990; Ahmad et al. 1990). Comparison of the 2,3-dihydroxybiphenyl dioxygenase gene products of *bphC* of pseudomonads isolated in Japan and America reveal similar molecular weights, subunits and substrate specificities. However, the nucleotide sequence homology between the *bphC* genes of these strains is rather low (38%). Comparison of these nucleotide sequences between strains isolated in Japan show much higher homology (91%). Furthermore, the latter sequence shows homology with the catechol-2,3-dioxygenase-encoding gene. This indicates similar enzymes for both *meta* cleavage reactions. The *bphA* gene product, biphenyl dioxygenase is most likely a multicomponent enzyme like the benzoate, toluene and naphthalene dioxygenases (Furukawa et al. 1990).

B. Anaerobic biodegradation

The first evidence for anaerobic degradation of PCBs came from the analysis of sediment samples from the Hudson River. Compared to the Aroclor mixtures originally discharged, lower relative concentrations of highly chlorinated biphenyls were detected in these sediments. Comparison of changes in PCB concentrations in sterile and nonsterile sediment incubations demonstrated that bacteria were indeed responsible for the reductive dehalogenation of PCBs (Brown et al. 1987a,b; Brown and Wagner 1990; Quensen, III et al. 1988, 1990). Differences in congener specificity for dechlorination were detected for consortia from different locations. However, in general, the final products of anaerobic dechlorination are the *ortho*-substituted congeners, which can be degraded aerobically. Thus, total degradation of PCBs seems possible by sequential anaerobic and aerobic degradation. However, Quensen, III et al. (1990) showed that dehalogenation only occurs with biphenyls with up to seven chlorines, mainly in the *meta* and *para* positions. More highly chlorinated congeners were not dehalogenated.

The addition of several organic substrates to PCB-dechlorinating consortia resulted in different rates of dechlorination but yielded similar dechlorination patterns. No significant dechlorination could be detected in batches receiving no additional organic substrate (Nies and Vogel 1990). In the latter study, *meta* and *para* dechlorination again predominated, resulting in accumulation of *ortho*-substituted products. In contrast to the above results, *ortho* dechlorination has been detected in incubations with methanogenic pond sediment (Fig. 13) (Van Dort and Bedard 1991).

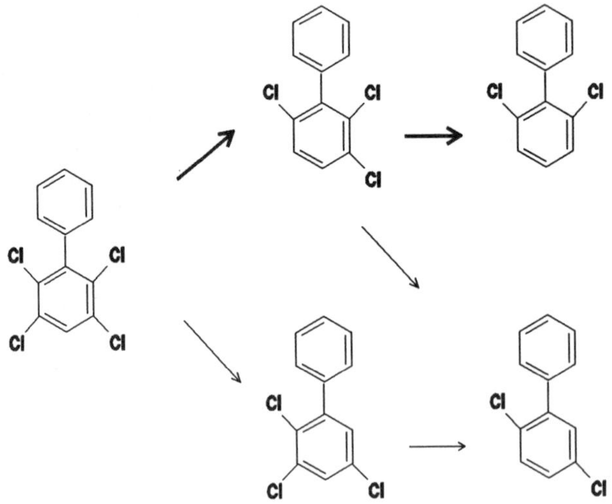

Fig. 13. Anaerobic dehalogenation of PCBs in methanogenic pond sediment (Van Dort and Bedard 1991).

From experiments with deuterated water it was concluded that the proton donor for reductive dechlorination is water (Nies and Vogel 1991). On the basis of the reductive dechlorination of 4,4'-dichlorobiphenyl by consortia isolated from aerobic activated sludge, Mavoungou et al. (1991) suggest that reductive dehalogenation is not a strictly anaerobic process.

IX. Biodegradation of halogenated dibenzo-*p*-dioxins and dibenzofurans

A. Aerobic biodegradation

Polychlorinated dibenzo-*p*-dioxins and dibenzofurans have not been produced intentionally on a large scale by the chemical industry, but are formed as trace contaminants during the incineration of chlorine-containing wastes and the industrial synthesis of organochlorine compounds. The extremely high toxicity of some of these compounds has stimulated interest in their environmental fate. However, very little is known of their biodegradation. The co-metabolism of mono-, di- and tri-chlorodioxins by a biphenyl-utilizing *Beijerinckia* strain has been described by Klečka and Gibson (1980). *cis*-1,2-Dihydrodiols were isolated as the products of dioxygenation of 1-chloro- and 2-chlorodioxins. Acid-catalysed dehydration of these compounds gives 2-hydroxylated compounds. Further metabolism of the dihydrodiols produces 1,2-dihydroxylated derivatives, but there was no evidence for ring cleavage of these compounds. In fact, the 1,2-dihydroxylated derivatives appear to inhibit the ring cleavage enzymes in this strain. The biphenyl-utilizing *Alcaligenes* sp. strain JB1 co-metabolizes mono-, di- and trichlorinated dioxins by a similar mechanism (Parsons and Storms 1989). It is at the moment unclear whether this strain is able to cleave the dioxin ring system.

Very slow oxidative degradation of 2,3,7,8-tetrachlorodibenzo-*p*-dioxin (2,3,7,8-TCDD) has been reported for a number of microorganisms, including *Pseudomonas testosteroni*, *Bacillus megaterium* and *Nocardiopsis* strains (Philippi et al. 1982; Quensen, III and Matsumura 1983). Traces of polar metabolites, probably hydroxylated derivatives, are formed by these strains.

To date no evidence for ring cleavage of chlorinated dibenzo-*p*-dioxins has been reported. Recently, however, oxidative ring cleavage of dibenzo-*p*-dioxin by a dibenzofuran-degrading *Pseudomonas* strain was described (Harms et al. 1990). A 2-phenoxy derivative of muconic acid and minor amounts of 1-hydroxydibenzo-*p*-dioxin and catechol were formed (Fig. 14).

There are very few reports of the biodegradation of halogenated dibenzofurans. The degradation of unchlorinated dibenzofuran by *Pseudomonas* and *Brevibacterium* strains has been described (Fortnagel et al. 1990; Strubel et al. 1991). The degradation was proposed to proceed via initial angular dioxygenation to produce an unstable hemiacetal. This compound then undergoes ether bridge cleavage to produce 2,2′,3-trihydroxybiphenyl, which is degraded further by a pathway analogous to that for biphenyl and chlorinated biphenyls to give salicylate (Fig. 15).

Alcaligenes sp. strain JB1 has been reported to degrade 2-CDF and 2,8-DCDF (Parsons et al. 1990). 5-Chlorosalicylate was identified as a metabolite of 2-CDF, in agreement with the degradation pathway proposed for dibenzofuran. Recently, *Pseudomonas* sp. strain HH69 was also reported to degrade 3-CDF to a mixture of 4-chlorosalicylate and salicylate, presumably as a result of a nonselective initial attack on 3-CDF (Harms et al. 1991).

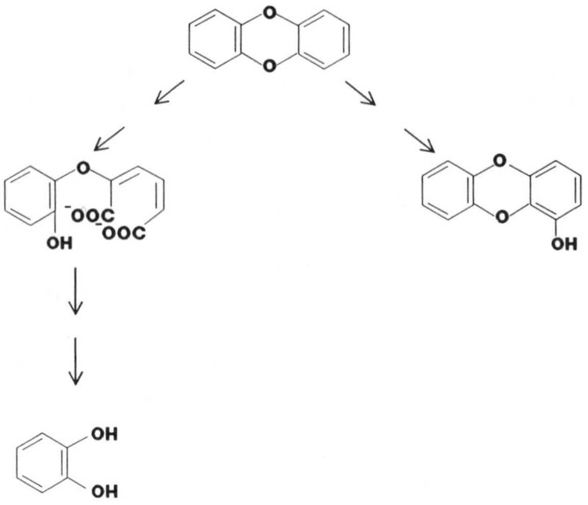

Fig. 14. Aerobic degradation of dibenzo-p-dioxin by Pseudomonas strains (Harms et al. 1990).

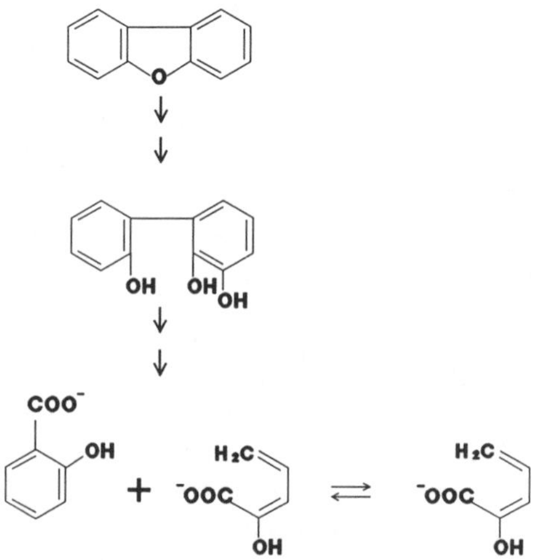

Fig. 15. Aerobic degradation of dibenzofuran by Brevibacterium and Pseudomonas strains (Fortnagel et al. 1990; Strubel et al. 1991).

References

Abramowicz DA (1990) Aerobic and anaerobic biodegradation of PCBs: A review. Crit. Rev. Biotechnol. 10: 241–251

Adriaens P, Kohler H-PE, Kohler-Staub D and Focht DD (1989) Bacterial dehalogenation of chlorobenzoates and coculture biodegradation of 4,4′-dichlorobiphenyl. Appl. Environ. Microbiol. 55: 887–892

Adriaens P and Focht DD (1990) Continuous coculture degradation of selected polychlorinated biphenyl congeners by *Acinetobacter* spp. in an aerobic reactor system. Environ. Sci. Technol. 24: 1042–1049

Adriaens P and Focht DD (1991) Cometabolism of 3,4-dichlorobenzoate by Acinetobacter sp. strain 4-CB1. Appl. Environ. Microbiol. 57: 173–179

Ahmad D, Massé R and Sylvestre M (1990) Cloning and expression of genes involved in 4-chlorobiphenyl transformation by *Pseudomonas testosteroni*: homology to polychlorobiphenyl degrading genes in other bacteria. Gene 86: 53–61

Ahmad D, Sylvestre M and Sondossi M (1991a) Subcloning of *bph* genes from *Pseudomonas testosteroni* B-356 in *Pseudomonas putida* and *Escherichia coli*: evidence for dehalogenation during initial attack on chlorobiphenyls. Appl. Environ. Microbiol. 57: 2880–2887

Ahmad D, Sylvestre M, Sondossi M and Massé R (1991b) Bioconversion of 2-hydroxy-6-oxo-(4'-chlorophenyl)hexa-2,4-dienoic acid, the *meta*-cleavage product of 4-chlorobiphenyl. J. Gen. Microbiol. 137: 1375–1385

Allard A-S, Remberger M and Neilson AH (1985) Bacterial O-methylation of chloroguaiacols: effect of substrate concentration, cell density, and growth conditions. Appl. Environ. Microbiol. 49: 279–288

Amy PS, Schulke JW, Frazier LM and Seidler RJ (1985) Characterization of aquatic bacteria and cloning of genes specifying partial degradation of 2,4-dichlorophenoxyacetic acid. Appl. Environ. Microbiol. 49: 1237–1245

Apajalahti JHA and Salkinoja-Salonen MS (1987a) Dechlorination and para-hydroxylation of polychlorinated phenols by *Rhodococcus chlorophenolicus*. J. Bacteriol. 169: 675–681

Apajalahti JHA and Salkinoja-Salonen MS (1987b) Complete dechlorination of tetrachlorohydroquinone by cell extracts of pentachlorophenol-induced *Rhodococcus chlorophenolicus*. J. Bacteriol. 169: 5125–5130

Bailey RE, Gonsior SJ and Rhinehart WL (1983) Biodegradation of the monochlorobiphenyls and biphenyl in river water. Environ. Sci. Technol. 17: 617–624

Ballschmiter K and Scholz C (1981) Primärschritte der Umwandlung von Chlorbenzol-Derivaten durch *Pseudomonas putida*. Angew. Chem. 93: 1026–1027

Bartels I, Knackmuss H-J, Reineke W (1984) Suicide inactivation of catechol-2,3-dioxygenase from *Pseudomonas putida* mt-2 by 3-halocatechols. Appl. Environ. Microbiol. 47: 500–505

Barton MR and Crawford (1988) Novel biotransformations of 4-chlorobiphenyl by a *Pseudomonas* sp. Appl. Environ. Microbiol. 54: 594–595

Baxter RA, Gilbert PE, Lidgett RA, Mainprize JH and Vodden HA (1975) The degradation of polychlorinated biphenyls by microorganisms. Sci. Total Environ. 4: 53–61

Baxter RM and Sutherland DA (1984) Biochemical and photochemical processes in the degradation of chlorinated biphenyls. Environ. Sci. Technol. 18: 608–610

Bedard DL and Haberl ML (1990) Influence of chlorine substitution pattern on the degradation of polychlorinated biphenyls by eight bacterial strains. Microb. Ecol. 20: 87–102

Bedard DL, Unterman R, Bopp LH, Brennan MJ, Haberl ML and Johnson C (1986) Rapid assay for screening and characterizing microorganisms for the ability to degrade polychlorinated biphenyls. Appl. Environ. Microbiol. 51: 761–768

Bedard DL, Haberl ML, May RJ and Brennan MJ (1987a) Evidence for novel mechanisms of polychlorinated biphenyl metabolism in *Alcaligenes eutrophus* H850. Appl. Environ. Microbiol. 53: 1103–1112

Bedard DL, Wagner RE, Brennan MJ, Haberl ME and Brown, Jr JF (1987b) Extensive degradation of Arochlors and environmentally transformed polychlorinated biphenyls by *Alcaligenes eutrophus* H850. Appl. Environ. Microbiol. 53: 1094–1102

Bollag JM, Briggs GG, Dawson JE and Alexander M (1968a) 2,4-D Metabolism: Enzymatic degradation of chlorocatechols. J. Agric. Food Chem. 16: 829–833

Bollag JM, Helling CS and Alexander M (1968b) 2,4-D metabolism: enzymatic hydroxylation of chlorinated phenols. J. Agric. Food Chem. 16: 826–828

Bopp LH (1986) Degradation of highly chlorinated PCBs by *Pseudomonas* strain LB400. J. Ind. Microbiol. 1: 23–29

Bosma TNP, Van der Meer JR, Schraa G, Tros ME and Zehnder AJB (1988) Reductive dechlorination of all trichloro- and dichlorobenzene isomers. FEMS Microbiol. Ecol. 53: 223–229

Boyd SA and Shelton DR (1984) Anaerobic biodegradation of chlorophenols in fresh and acclimated sludge. Appl. Environ. Microbiol. 47: 272–277

Brown, Jr JF and Wagner RE (1990) PCB movement, dechlorination, and detoxification in the Acushnet estuary. Environ. Toxicol. Chem. 9: 1215–1233

Brown, Jr JF, Bedard DL, Brennan MJ, Carnahan JC, Feng H and Wagner RE (1987a) Polychlorinated biphenyl dechlorination in aquatic sediments. Science 236: 709–712

Brown, Jr JF, Wagner RE, Feng H, Bedard DL, Brennan MJ, Carnahan JC and May RJ (1987b) Environmental dechlorination of PCBs. Environ. Toxicol. Chem. 6: 579–593

Brunner W, Sutherland FH and Focht DD (1985) Enhanced biodegradation of polychlorinated biphenyls in soil by analog enrichment and bacterial inoculation. J. Environ. Qual. 14: 324–328

Bryant FO, Hale DD and Rogers JE (1991) Regiospecific dechlorination of pentachlorophenol by dichlorophenol-adapted microorganisms in freshwater, anaerobic sediment slurries. Appl. Environ. Microbiol. 57: 2293–2301

Cain RB, Trantner EK and Darrah JA (1968) The utilization of some halogenated aromatic acids by Nocardia: oxidation and metabolism. Biochem. J. 106: 211–227

Carney BF and Leary JV (1989) Novel alterations in plasmid DNA associated with aromatic hydrocarbon utilization by Pseudomonas putida R5-3. Appl. Environ. Microbiol. 55: 1523–1530

Carney BF, Kröckel L, Leary JV and Focht DD (1989) Identification of Pseudomonas alcaligenes chromosomal DNA in the plasmid DNA of the chlorobenzene-degrading recombinant Pseudomonas putida strain CB1-9. Appl. Environ. Microbiol. 55: 1037–1039

Chatterjee DK and Chakrabarty AM (1983) Genetic homology between independently isolated chlorobenzoate-degradative plasmids. J. Bacteriol. 153: 532–534

Chatterjee DK, Kellogg ST, Hamada S and Chakrabarty AM (1981) Plasmid specifying total degradation of 3-chlorobenzoate by a modified ortho pathway. J. Bacteriol. 146: 639–646

Chu JP and Kirsch EJ (1972) Metabolism of pentachlorophenol by an axenic bacterial culture. Appl. Environ. Microbiol. 23: 1033–1035

Clarke KF, Callely AG, Livingstone A and Fewson CA (1975) Metabolism of monofluorobenzoates by Acinetobacter calcoaceticus N.C.I.B. 8250: formation of monofluorocatechols. Biochim. Biophys. Acta 404: 169–179

Corke CT, Bunce NJ, Beaumont AL and Merrick RL (1979) Diazonium cations as intermediates in the microbial transformation of chloroanilines to chlorinated biphenyls, azo compounds, and triazenes. J. Agric. Food Chem. 27: 644–646

de Bont JAM, Vorage MJAW, Hartmans S and van den Tweel WJJ (1986) Microbial degradation of 1,3-dichlorobenzene. Appl. Environ. Microbiol. 52: 677–680

DeWeerd KA and Suflita JM (1990) Anaerobic reductive dehalogenation of halobenzoates by cell extracts of "Desulfomonile tiedjei". Appl. Environ. Microbiol. 56: 2999–3005

DeWeerd KA, Suflita JM, Linkfield T, Tiedje JM and Pritchard PH (1986) The relationship between reductive dehalogenation and other aryl substituent removal reactions catalyzed by anaerobes. FEMS Microbiol. Ecol. 38: 331–339

DeWeerd KA, Mandelco L, Tanner RS, Woese CR and Suflita JM (1990) Desulfomonile tiedjei gen. nov. and sp. nov., a novel anaerobic dehalogenating sulfate-reducing bacterium. Arch. Microbiol. 154: 23–30

DeWeerd KA, Concannon F and Suflita JM (1991) Relationship between hydrogen consumption, dehalogenation, and the reduction of sulfur oxyanions by Desulfomonile tiedjei. Appl. Environ. Microbiol. 57: 1929–1934

Dietrich G and Winter J (1990) Anaerobic degradation of chlorophenol by an enrichment culture. Appl. Microbiol. Biotechnol. 34: 252–258

Dolfing J (1990) Reductive dechlorination of 3-chlorobenzoate is coupled to ATP production and growth in an anaerobic bacterium, strain DCB-1. Arch. Microbiol. 153: 264–266

Dolfing J and Tiedje JM (1986) Hydrogen cycling in a three-tiered food web growing on the methanogenic conversion of 3-chlorobenzoate. FEMS Microbiol. Ecol. 38: 293–298

Dolfing J and Tiedje JM (1987) Growth yield increase linked to reductive dechlorination in a defined 3-chlorobenzoate degrading methanogenic coculture. Arch. Microbiol. 149: 102–105

Dolfing J and Tiedje JM (1991a) Kinetics of two complementary hydrogen sink reactions in a defined 3-chlorobenzoate degrading methanogenic co-culture. FEMS Microbiol. Ecol. 86: 25–32

Dolfing J and Tiedje JM (1991b) Acetate as source of reducing equivalents in the reductive dehalogenation of 2,5-dichlorobenzoate. Arch. Microbiol. 156: 356–361

Dolfing J and Tiedje JM (1991c) Influence of substituents on reductive dehalogenation of 3-chlorobenzoate analogs. Appl. Environ. Microbiol. 57: 820–824

Don RH and Pemberton JM (1981) Properties of six pesticide degradation plasmids isolated from *Alcaligenes paradoxus* and *Alcaligenes eutrophus*. J. Bacteriol. 145: 681–686

Don RH, Weightman AJ, Knackmuss H-J and Timmis KN (1985) Transposon mutagenesis and cloning analysis of the pathways for degradation of 2,4-dichlorophenoxyacetic acid and 3-chlorobenzoate in *Alcaligenes eutrophus* JMP134 (pJP4). J. Bacteriol. 161: 85–90

Engelhardt G, Rast HG and Wallnöfer PR (1979) Cometabolism of phenol and substituted phenols by *Nocardia* spec. DSM 43251. FEMS Microbiol. Lett. 5: 377–383

Engesser K-H and Schulte P (1989) Degradation of 2-bromo-, 2-chloro- and 2-fluorobenzoate by *Pseudomonas putida* CLB 250. FEMS Microbiol. Lett. 60: 143–148

Engesser K-H, Auling G, Busse J and Knackmuss H-J (1990) 3-fluorobenzoate enriched bacterial strain FLB 300 degrades benzoate and all three isomeric monofluorobenzoates. Arch. Microbiol. 153: 193–199

Evans WC, Smith BSW, Fernley HN and Davies JI (1971) Bacterial metabolism of 2,4-dichlorophenoxyacetate. Biochem. J. 122: 543–552

Fathepure BZ and Vogel TM (1991) Complete degradation of polychlorinated hydrocarbons by a two-stage biofilm reactor. Appl. Environ. Microbiol. 57: 3418–3422

Fathepure BZ, Nengu JP and Boyd SA (1987) Anaerobic bacteria that dechlorinate perchloroethene. Appl. Environ. Microbiol. 53: 2671–2674

Fathepure BZ, Tiedje JM and Boyd SA (1988) Reductive dechlorination of hexachlorobenzene to tri- and dichlorobenzenes in anaerobic sewage sludge. Appl. Environ. Microbiol. 54: 327–330

Fava F, Zappoli S, Marchetti L and Morselli L (1991) Biodegradation of chlorinated biphenyls (Fenclor 42) in batch cultures with mixed and pure cultures. Chemosphere 22: 3–14

Fetzner S, Müller R and Lingens F (1989) A novel metabolite in the microbial degradation of 2-chlorobenzoate. Biochem. Biophys. Res. Commun. 161: 700–705

Focht DD and Shelton D (1987) Growth kinetics of *Pseudomonas alcaligenes* C-0 relative to inoculation and 3-chlorobenzoate metabolism in soil. Appl. Environ. Microbiol. 53: 1846–1849

Fortnagel P, Harms H, Wittich R-M, Krohn S, Meyer H, Sinnwell V, Wilkes H and Francke W (1990) Metabolism of dibenzofuran by *Pseudomonas* sp. strain HH69 and the mixed culture HH27. Appl. Environ. Microbiol. 56: 1148–1156

Fries GF and Marrow GS (1984) Metabolism of chlorobiphenyls in soil. Bull. Environ. Contam. Toxicol. 33: 6–12

Furukawa K (1982) Microbial degradation of polychlorinated biphenyls (PCBs). In: AM Chakrabarty (ed) Biodegradation and Detoxification of Environmental Pollutants (pp 33–57). CRC, Boca Raton, Florida

Furukawa K and Chakrabarty AM (1982) Involvement of plasmids in total degradation of chlorinated biphenyls. Appl. Environ. Microbiol. 44: 619–626

Furukawa K, Hayase N and Taira K (1990) Biphenyl/polychlorinated biphenyl catabolic gene (*bph* operon): organization, function, and molecular relationship in various pseudomonads. In: S Silver, AM Chakrabarty, B Iglewski and S Kaplan (eds) Pseudomonas: Biotransformations, Pathogenesis and Evolving Biotechnology (pp 111–120). ASM, Washington, DC

Ghosal D and You I-S (1988) Nucleotide homology and organization of chlorocatechol oxidation genes of plasmids pJP4 and pAC27. Mol. Gen. Genet. 211: 113–120

Ghosal D and You I-S (1989) Operon structure and nucleotide homology of the chlorocatechol oxidation genes of plasmids pAC27 and pJp4. Gene 83: 225–232

Ghosal D, You I-S, Chatterjee DK and Chakrabarty AM (1985) Genes specifying degradation of 3-chlorobenzoic acid in plasmids pAC27 and pJP4. Proc. Natl. Acad. Sci. U.S.A. 82: 1638–1642

Gibson SA and Suflita JM (1990) Anaerobic degradation of 2,4,5-trichlorophenoxyacetic acid in samples from methanogenic aquifer: stimulation by short-chain organic acids and alcohols. Appl. Environ. Microbiol. 56: 1825–1832

Goldman P, Milne GWA and Pignataro MT (1967) Fluorine containing metabolites formed from 2-fluorobenzoic acid by *Pseudomonas* species. Arch. Biochem. Biophys. 118: 178–184

Golovleva LA, Pertsova RN, Evtushenko LI and Baskunov BP (1990) Degradation of 2,4,5-trichlorophenoxyacetic acid by a *Nocardioides simplex* culture. Biodegradation 1: 263–271

Groenewegen PEJ, Driessen AJM, Konings WN and de Bont JAM (1990) Energy-dependent uptake of 4-chlorobenzoate in the Coryneform bacterium NTB-1. J. Bacteriol. 172: 419–423

Häggblom M (1990) Mechanisms of bacterial degradation and transformation of chlorinated monoaromatic compounds. J. Basic Microbiol. 30: 115–141

Häggblom MM and Young LY (1990) Chlorophenol degradation coupled to sulfate reduction. Appl. Environ. Microbiol. 56: 3255–3260

Häggblom MM, Nohynek LJ and Salkinoja-Salonen MS (1988) Degradation and O-methylation of chlorinated phenolic compounds by *Rhodococcus* and *Mycobacterium* strains. Appl. Environ. Microbiol. 54: 3043–3052

Häggblom MM, Janke D, Middeldorp PJM and Salkinoja-Salonen MS (1989a) O-Methylation of chlorinated phenols in the genus *Rhodococcus*. Arch. Microbiol. 152: 6–9

Häggblom MM, Janke D and Salkinoja-Salonen MS (1989b) Hydroxylation and dechlorination of tertrachlorohydroquinone by *Rhodococcus* sp. strain CP-2 cell extracts. Appl. Environ. Microbiol. 55: 516–519

Haigler BE and Spain JC (1989) Degradation of *p*-chlorotoluene by a mutant of *Pseudomonas* sp. strain JS6. Appl. Environ. Microbiol. 55: 372–379

Haigler BE, Nishino SF and Spain JC (1988) Degradation of 1,2-dichlorobenzene by a *Pseudomonas* sp. Appl. Environ. Microbiol. 54: 294–301

Hankin L and Sawhney BL (1984) Microbial degradation of polychlorinated biphenyls in soil. Soil Sci. 137: 401–407

Harms H, Wittich R-M, Sinnwell V, Meyer H, Fortnagel P and Francke W (1990) Transformation of dibenzo-*p*-dioxin by *Pseudomonas* sp. strain HH69. Appl. Environ. Microbiol. 56: 1157–1159

Harms H, Wilkes H, Sinnwell V, Wittich R-M, Figge K, Francke W and Fortnagel P (1991) Transformation of 3-chlorodibenzofuran by *Pseudomonas* sp. HH69. FEMS Microbiol. Lett. 81: 25–30

Hartmann J, Reineke W and Knackmuss H-J (1979) Metabolism of 3-chloro-, 4-chloro-, and 3,5-dichlorobenzoate by a pseudomonad. Appl. Environ. Microbiol. 37: 421–428

Hartmann J, Engelberts K, Nordhaus B, Schmidt E and Reineke W (1989) Degradation of 2-chlorobenzoate by in vivo constructed hybrid pseudomonads. FEMS Microbiol. Lett. 61: 17–22

Haugland RA, Schlemm DJ, Lyons, III RP, Sferra PR and Chakrabarty AM (1990) Degradation of the chlorinated phenoxyacetate herbicides 2,4-dichlorophenoxyacetic acid and 2,4,5-trichlorophenoxyacetic acid by pure and mixed bacterial cultures. Appl. Environ. Microbiol. 56: 1357–1362

Havel J and Reineke W (1991) Total degradation of various chlorobiphenyls by cocultures and in vivo constructed hybrid pseudomonads. FEMS Microbiol. Lett. 78: 163–170

Hayase N, Taira K and Furukawa K (1990) *Pseudomonas putida* KF715 *bphABCD* operon encoding biphenyl and polychlorinated biphenyl degradation: cloning analysis and expression in soil bacteria. J.Bacteriol. 172: 1160–1164

Hernandez BS, Higson FK, Kondrat and Focht DD (1991) Metabolism and inhibition by chlorobenzoates in *Psuedomonas putida* P111. Appl. Environ. Microbiol. 57: 3361–3366

Hickey WJ and Focht DD (1990) Degradation of mono-, di-, and trihalogenated benzoic acids by *Pseudomonas aeruginosa* JB2. Appl. Environ. Microbiol. 56: 3842–3850

Higson FK and Focht DD (1990) Degradation of 2-bromobenzoic acid by a strain of *Pseudomonas aeruginosa*. Appl. Environ. Microbiol. 56: 1615–1619

Hiramoto M, Ohtake H and Toda K (1989) A kinetic study on total degradation of 4-chlorobiphenyl by a two-step culture of *Arthrobacter* and *Pseudomonas* strains. J. Fermentation Bioeng. 1: 68–70

Horowitz A, Suflita JM and Tiedje JM (1983) Reductive dehalogenations of halobenzoates by anaerobic lake sediment microorganisms. Appl. Environ. Microbiol. 45: 1459–1461

Horvath RS (1970) Co-metabolism of methyl- and chloro-substituted catechols by an *Achromobacter* sp. possessing a new *meta*-cleaving oxygenase. Biochem. J. 119: 871–876

Horvath M, Ditzelmüller G, Loidl M and Streichsbier F (1990) Isolation and characterization of a 2-(2,4-dichlorophenoxy)propionic acid-degrading soil bacterium. Appl. Microbiol. Biotechnol. 33: 213–216

Janke D, Al-Mofarji T and Schukat B (1988a) Critical steps in degradation of chloroaromatics by rhodococci. II. Whole-cell turnover of different monochloroaromatic non-growth substrates by *Rhodococcus* sp. An 117 and An 213 in the absence/presence of glucose. J. Basic Microbiol. 8: 519–528

Janke D, Al-Mofarji T, Straube G, Schumann P and Prauser H (1988b) Critical steps in the degradation of chloroaromatics by rhodococci. I. Initial enzyme reactions involved in catabolism of aniline, phenol and benzoate by *Rhodococcus* sp. An 117 and An 213. J. Basic Microbiol. 8: 509–518

Johnston HW, Briggs GG and Alexander M (1972) Metabolism of 3-chlorobenzoic acid by a pseudomonad. Soil. Biol. Biochem. 4: 187–190

Kamal VS and Wyndham RC (1990) Anaerobic phototrophic metabolism of 3-chlorobenzoate by *Rhodopseudomonas palustris* WS17. Appl. Environ. Microbiol. 56: 3871–3873

Kaphammer B and Olsen RH (1990) Cloning and characterization of *tfdS*, the repressor-activator gene of *tdfB*, from the 2,4-dichlorophenoxyacetic acid catabolic plasmid pJP4. J. Bacteriol. 172: 5856–5862

Karns JS, Duttagupta S and Chakrabarty AM (1983a) Regulation of 2,4,5-trichlorophenoxyacetic acid and chlorophenol metabolism in *Pseudomonas cepacia* AC 1100. Appl. Environ. Microbiol. 46: 1182–1186

Karns JS, Kilbane JJ, Duttagupta S and Chakrabarty AM (1983b) Metabolism of halophenols by 2,4,5-trichlorophenoxyacetic acid-degrading *Pseudomonas cepacia*. Appl. Environ. Microbiol. 46: 1176–1181

Keil H, Klages U and Lingens F (1981) Degradation of 4-chlorobenzoate by *Pseudomonas* sp. CBS3: induction of catabolic enzymes. FEMS Microbiol. Lett. 10: 213–215

Khan AA and Walia SK (1991) Expression, localization, and function analysis of polychlorinated biphenyl degradation genes *cbpABCD* of *Pseudomonas putida*. Appl. Microbiol. Biotechnol. 57: 1325–1332

Kimbara K, Hashimoto T, Fukuda M, Koana T, Takagi M, Oishi M and Yano K (1988) Isolation and characterization of a mixed culture that degrades polychlorinated biphenyls. Agric. Biol. Chem. 52: 2885–2891

King GM (1988) Dehalogenation in marine sediments containing natural sources of halophenols. Appl. Environ. Microbiol. 54: 3079–3085

Klečka GM and Gibson DT (1980) Metabolism of dibenzo-*p*-dioxin and chlorinated dibenzo-*p*-dioxins by a *Beijerinckia* species. Appl. Environ. Microbiol. 39: 288–296

Knackmuss H-J and Hellwig M (1978) Utilization and cooxidation of chlorinated phenols by *Pseudomonas* sp. B13. Arch. Microbiol. 117: 1–7

Kohring GW, Zhang X and Wiegel J (1989) Anaerobic dechlorination of 2,4-dichlorophenol in freshwater sediments in the presence of sulfate. Appl. Environ. Microbiol. 55: 2735–2737

Kong H-Y and Sayler GS (1983) Degradation and total mineralization of monohalogenated biphenyls in natural sediment and mixed microbial culture. Appl. Environ. Microbiol. 46: 666–672

Konopka A, Knight D and Turco RF (1989) Characterization of a *Pseudomonas* sp. capable of aniline degradation in the presence of secondary carbon sources. Appl. Environ. Microbiol. 55: 385–389

Kröckel L and Focht DD (1987) Construction of chlorobenzene-utilizing recombinants by progressive manifestation of a rare event. Appl. Environ. Microbiol. 53: 2470–2475

Kuhn EP and Suflita JM (1989) Sequential reductive dehalogenation of chloroanilines by microorganisms from a methanogenic aquifer. Environ. Sci. Technol. 23: 848–852

Kuhn EP, Townsend GT and Suflita JM (1990) Effect of sulfate and organic carbon supplements on reductive dehalogenation of chloroanilines in anaerobic aquifer slurries. Appl. Environ. Microbiol. 56: 2630–2637

Kukor JJ, Olsen RH and Siak J-S (1989) Recruitment of a chromosomally encoded maleylacetate reductase for degradation of 2,4-dichlorophenoxyacetic acid by plasmid pJP4, J. Bacteriol. 171: 3385–3390

Lammerding AM, Bunce NJ, Merrick RL and Corke CT (1982) Structural effects on the microbial diazotization of anilines. J. Agric. Food Chem. 30: 644–647

Lehrbach RP, Zeyer J, Reineke W, Knackmuss H-J and Timmis KN (1984) Enzyme recruitment in vitro: use of cloned genes to extend the range of haloaromatics degraded by Pseudomonas sp. strain B13. J. Bacteriol. 158: 1025–1032

Li D-Y, Eberspächer J, Wagner B, Kuntzer J and Lingens F (1991) Degradation of 2,4,6-trichlorophenol by Azotobacter sp. strain GP1. Appl. Environ. Microbiol. 57: 1920–1928

Linkfield TG and Tiedje JM (1990) Characterization of the requirements and substrates for reductive dehalogenation by strain DCB-1. J. Ind. Microbiol. 5: 9–16

Loidl M, Hinteregger C, Ditzelmüller G, Ferschl A and Streichbier F (1990) Degradation of aniline and monochlorinated anilines by soil-born Pseudomonas acidovorens strains. Arch. Microbiol. 155: 56–61

Madsen T and Aamand J (1991) Effects of sulfuroxy anions on degradation of pentachlorophenol by a methanogenic enrichment culture. Appl. Environ. Microbiol. 57: 2453–2458

Marinucci AC and Bartha R (1979) Biodegradation of 1,2,3- and 1,2,4-trichlorobenzene in soil and in liquid enrichment culture. Appl. Environ. Microbiol. 38: 811–817

Marks TS, Smith ARW and Quirk AV (1984a) Degradation of 4-chlorobenzoic acid by Arthrobacter sp. Appl. Environ. Microbiol. 48: 1020–1025

Marks TS, Wait R, Smith ARW and Quirk AV (1984b) The origin of the oxygen incorporated during the dehalogenation/hydroxylation of 4-chlorobenzoic acid by an Arthrobacter sp. Biochem. Biophys. Res. Commun. 124: 669–674

Markus A, Klages U, Krauss S and Lingens F (1984) Oxidation and dehalogenation of 4-chlorophenylacetate by a two component enzyme system from Pseudomonas sp. strain CBS3. J. Bacteriol. 160: 618–621

Massé R, Messier F, Peloquin L, Ayotte C and Sylvestre M (1984) Microbial degradation of 4-chlorobiphenyl, a model compound of chlorinated biphenyls. Appl. Environ. Microbiol. 47: 947–951

Massé R, Messier F, Ayotte C, Lévesque MF and Sylvestre M (1989) A comprehensive gas chromatographic/mass spectrometric analysis of 4-chlorobiphenyl bacterial degradation products. Biomed. Environ. Mass Spectr. 18: 27–47

Mavoungou R, Massé R and Sylvestre M (1991) Microbial dehalogenation of 4,4'-dichlorobiphenyl under anaerobic conditions. Sci. Total Environ. 101: 263–268

Miguez CB, Greer CW and Ingram JM (1990) Degradation of mono- and dichlorobenzoic acid isomers by two natural isolates of Alcaligenes denitrificans. Arch. Microbiol. 154: 139–143

Mikesell MD and Boyd SA (1986) Complete reductive dechlorination and mineralization of pentachlorophenol by anaerobic microorganisms. Appl. Environ. Microbiol. 52: 861–865

Milne GWA, Goldman P and Holtzman JL (1968) The metabolism of 2-fluorobenzoic acid: studies with $^{18}O_2$. J. Biol. Chem. 243: 5374–5376

Minard RD, Russel S and Bollag JM (1977) Chemical transformation of 4-chloroaniline to a triazene in a bacterial culture medium. J. Agric. Food Chem. 25: 841–844

Mohn WW and Tiedje JM (1990) Strain DCB-1 conserves energy for growth from reductive dechlorination coupled to formate oxidation. Arch. Microbiol. 153: 267–271

Mohn WW and Tiedje JM (1991) Evidence for chemiosmotic coupling of reductive dechlorination and ATP synthesis in Desulfomonile tiedjei. Arch. Microbiol. 157: 1–6

Mohn WW, Linkfield TG, Pankratz HS and Tiedje JM (1990) Involvement of a collar structure in polar growth and cell division of strain DCB-1. Appl. Environ. Microbiol. 56: 1206–1211

Mokross H, Schmidt E and Reineke W (1990) Degradation of 3-chlorobiphenyl by in vivo constructed hybrid pseudomonads. FEMS Microbiol. Lett. 71: 179–186

Müller R, Thiele J, Klages U and Lingens F (1984) Incorporation of [^{18}O H$_2$O] water into 4-hydroxybenzoic acid in the reaction of 4-chlorobenzoate dehalogenase from *Pseudomonas* spec. CBS3. Biochem. Biophys. Res. Commun. 124: 178–182

Müller R, Oltmans RH and Lingens F (1988) Enzymic dehalogenation of 4-chlorobenzoate by extracts from *Arthrobacter* sp. SU DSM 20407. Biol. Chem. Hoppe-Seyler 369: 567–571

Neilson AH, Lindgren C, Hynning P-A and Remberger M (1988) Methylation of halogenated phenols and thiophenols by cell extracts of Gram-positive and Gram-negative bacteria. Appl. Environ. Microbiol. 54: 524–530

Nies L and Vogel TM (1990) Effects of organic substrates on dechlorination of Arochlor 1242 in anaerobic sediments. Appl. Environ. Microbiol. 56: 2612–2617

Nies L and Vogel TM (1991) Identification of the proton source for the microbial reductive dechlorination of 2,3,4,5,6-pentachlorobiphenyl. Appl. Environ. Microbiol. 57: 2771–2774

Ohmori T, Ikai T, Minoda Y and Yamada K (1973) Utilization of hydrocarbons by microorganisms. XXV: Utilization of polyphenyl and polyphenyl-related compounds by microorganimsms. Agric. Biol. Chem. 37: 1599–1605

Oltmanns RH, Rast HG and Reineke W (1988) Degradation of 1,4-dichlorobenzene by enriched and constructed bacteria. Appl. Microbiol. Biotechnol. 28: 609–616

Oltmanns RH, Müller R, Otto MK and Lingens F (1989) Evidence for a new pathway in the bacterial degradation of 4-fluorobenzoate. Appl. Environ. Microbiol. 55: 2499–2504

Pardue JH, Delaune RD and Patrick Jr. WH (1988) Effect of sediment pH and oxidation-reduction potential on PCB mineralization. Water Air Soil Pollut. 37: 439–447

Parsons JR and Sijm DTHM (1988) Biodegradation kinetics of polychlorinated biphenyls in continuous cultures of a *Pseudomonas* strain. Chemosphere 17: 1755–1766

Parsons JR and Storms MCM (1989) Biodegradation of chlorinated dibenzo-*p*-dioxins in batch and continuous cultures of strain JB1. Chemosphere 19: 1297–1308

Parsons J, Veerkamp W and Hutzinger O (1983) Microbial metabolism of chlorobiphenyls. Toxicol. Environ. Chem. 6: 327–350

Parsons JR, Sijm DTHM, Van Laar A and Hutzinger O (1988) Biodegradation of chlorinated biphenyls and benzoic acids by a *Pseudomonas* strain. Appl. Microbiol. Biotechnol. 29: 81–84

Parsons JR, Ratsak C and Siekerman C (1990) Biodegradation of chlorinated dibenzofurans by an *Alcaligenes* strain. In: O Hutzinger and H Fiedler (eds) Organohalogen Compounds. Proc. Dioxin '90 – EPRI Seminar, Sept. 10–14, 1990, Bayreuth, Vol 1 (pp 377–380). Ecoinforma Press, Bayreuth, Germany

Pettigrew CA, Breen A, Corcoran C and Sayler GS (1990) Chlorinated biphenyl mineralization by individual populations and consortia of freshwater bacteria. Appl. Environ. Microbiol. 56: 2036–2045

Pettigrew CA, Haigler BE and Spain JC (1991) Simultaneous biodegradation of chlorobenzene and toluene by a *Pseudomonas* strain. Appl. Environ. Microbiol. 57: 157–162

Philippi M, Schmid J, Wipf HK and Hütter RA (1982) A microbial metabolite of TCDD. Experientia 38: 659–661

Pieper DH, Reineke W, Engesser K-H and Knackmuss H-J (1988) Metabolism of 2,4-dichlorophenoxyacetic acid, 4-chloro-2-methylphenoxyacetic acid and 2-methylphenoxyacetic acid by *Alcaligenes eutrophus* JMP 134. Arch. Microbiol. 150: 95–102

Quensen, III JF and Matsumura F (1983) Oxidative degradation of 2,3,7,8-tetrachlorodibenzo-p-dioxin by microorganisms. Environ. Toxicol. Chem. 2: 261–268

Quensen, III JF, Tiedje JM and Boyd SA (1988) Reductive dechlorination of polychlorinated biphenyls by anaerobic microorganisms from sediments. Science 242: 752–754

Quensen, III JF, Boyd SA and Tiedje JM (1990) Dechlorination of four commercial polychlorinated biphenyl mixtures (Aroclors) by anaerobic microorganisms from sediments. Appl. Environ. Microbiol. 56: 2360–2369

Reineke W (1984) Microbial degradation of halogenated aromatic compounds. Microbiol. Ser. 13: 319–360

Reineke W and Knackmuss H-J (1980) Hybrid pathway for chlorobenzoate metabolism in *Pseudomonas* sp. B13 derivatives. J. Bacteriol. 142: 467–473

Reineke W and Knackmuss H-J (1984) Microbial metabolism of haloaromatics. Isolation and properties of a chlorobenzene-degrading bacterium. Appl. Environ. Microbiol. 47: 395–402

Reineke W and Knackmuss H-J (1988) Microbial degradation of haloaromatics. Ann. Rev. Microbiol. 42: 263–287

Ruisinger S, Klages U and Lingens F (1976) Abbau der 4-Chlorobenzoesaure durch eine *Arthrobacter* species. Arch. Microbiol. 110: 253–256

Safe SH (1984) Microbial degradation of polychlorinated biphenyls. Microbiol. Ser. 13: 361–369

Sander P, Wittich R-M, Fortnagel P, Wilkes H and Francke W (1991) Degradation of 1,2,4-trichloro- and 1,2,4,5-tetrachlorobenzene by *Pseudomonas* strains. Appl. Environ. Microbiol. 57: 1430–1440

Sangodkar UMX, Aldrich TL, Haugland RA, Johnson J, Rothmel RK, Chapman PJ and Chakrabarty AM (1989) Molecular basis of biodegradation of chloroaromatic compounds. Acta Biotechnol. 9: 301–316

Savard P, Péloquin L and Sylvestre M (1986) Cloning of *Pseudomonas* strain CBS3 genes specifying dehalogenation of 4-chlorobenzoate. J. Bacteriol. 168: 81–85

Schenk T, Müller R and Lingens F (1990) Mechanism of enzymatic dehalogenation of pentachlorophenol by *Arthrobacter* sp. strain ATCC 3370. J. Bacteriol. 172: 7272–7274

Schennen U, Braun K and Knackmuss H-J (1985) Anaerobic degradation of 2-fluorobenzoate by benzoate degrading denitrifying bacteria. J. Bacteriol. 161: 321–325

Schlömann M, Pieper DH and Knackmuss H-J (1990).Enzymes of haloaromatics degradation: variations of *Alcaligenes* on a theme by *Pseudomonas*. In: S Silver, AM Chakrabarty, B Iglewski and S Kaplan S (eds) Pseudomonas: Biotransformations, Pathogenesis and Evolving Biotechnology (pp 111–120). ASM, Washington, DC

Schmidt E and Knackmuss H-J (1984) Production of *cis,cis*-muconate from benzoate and 2-fluoro-*cis,cis*-muconate from 3-fluorobenzoate by 3-chlorobenzoate degrading bacteria. Appl. Microbiol. Biotechnol. 20: 351–355

Schmidt E, Hellwig M and Knackmuss H-J (1983) Degradation of chlorophenols by a defined mixed microbial community. Appl. Environ. Microbiol. 46: 1038–1044

Scholten JD, Chang K-H, Babbitt PC, Charest H, Sylvestre M and Dunaway-Mariano D (1991) Novel enzymic hydrolytic dehalogenation of a chlorinated aromatic. Science 253: 182–185

Schraa G, Boone ML, Jetten MSM, Van Neerven ARW, Colberg PJ and Zehnder AJB (1986) Degradation of 1,4-dichlorobenzene by *Alcaligenes* sp. strain A175. Appl. Environ. Microbiol. 52: 1374–1381

Schreiber A, Hellwig M, Dorn E, Reineke W and Knackmuss H–J (1980) Critical reactions in fluorobenzoic acid degradation by *Pseudomonas* sp B13. Appl. Environ. Microbiol. 39: 58–67

Schwien U and Schmidt E (1982) Improved degradation of monochlorophenols by a constucted strain. Appl. Environ. Microbiol. 44: 33–39

Sharak Genthner BR, Price, II WA and Pritchard PH (1989a) Anaerobic degradation of chloroaromatic compounds in aquatic sediments under a variety of enrichment conditions. Appl. Environ. Microbiol. 55: 1466–1471

Sharak Genthner BR, Price, II WA and Pritchard PH (1989b) Characterization of anaerobic dechlorinating consortia derived from aquatic sediments. Appl. Environ. Microbiol. 55: 1472–1476

Shelton DR and Tiedje JM (1984) Isolation and partial characterization of bacteria in an anaerobic consortium that mineralizes 3-chlorobenzoic acid. Appl. Environ. Microbiol. 48: 840–848

Shiaris MP and Sayler GS (1982) Biotransformation of PCBs by natural assemblages of freshwater microorganisms. Environ. Sci. Technol. 16: 367–369

Shields MS, Hooper SW and Sayler (1985) Plasmid-mediated mineralization of 4-chlorobiphenyl. J. Bacteriol. 163: 882–889

Spain JC and Nishino SF (1987) Degradation of 1,4-dichlorobenzene by a *Pseudomonas* sp. Appl. Environ. Microbiol. 53: 1010–1019

Sperl GT and Harvey GJ (1988) Microbial adaptation to bromobenzene in a chemostat. Curr. Microbiol. 17: 99–103

Spokes JR and Walker N (1974) Chlorophenol and chlorobenzoic acid co-metabolism by different genera of soil bacteria. Arch. Microbiol. 96: 125–134

Steiert JG and Crawford RL (1986) Catabolism of pentachlorophenol by a *Flavobacterium* bacterium. Biochem. Biophys. Res. Commun. 141: 825–830

Steiert JG, Pignatello JJ and Crawford RL (1987) Degradation of chlorinated phenols by a pentachlorophenol-degrading bacterium. Appl. Environ. Microbiol. 53: 907–910

Stevens TO and Tiedje (1988) Carbon dioxide fixation and mixotrophic metabolism by strain DCB-1, a dehalogenating anaerobic bacterium. Appl. Environ. Microbiol. 54: 2944–2948

Stevens TO, Linkfield TG and Tiedje JM (1988) Physiological characterization of strain DCB-1, a unique sulfidogenic bacterium. Appl. Environ. Microbiol. 54: 2938–2943

Strubel V, Engesser K-H, Fischer P and Knackmuss H-J (1991) 3-(2-Hydroxyphenyl)catechol as substrate for proximal *meta* ring cleavage in dibenzofuran degradation by *Brevibacterium* sp. strain DPO 1361. J. Bacteriol. 173: 1932–1937

Struijs J and Rogers JE (1989) Reductive dehalogenation of dichloroanilines by anaerobic microorganisms in fresh and dichlorophenol-acclimated pond sediment. Appl. Environ. Microbiol. 55: 2527–2531

Suflita JM, Robinson JA and Tiedje JM (1983) Kinetics of microbial dehalogenation of haloaromatic substrates in methanogenic environments. Appl. Environ. Microbiol. 45: 1466–1473

Sylvestre M, Massé R, Messier F, Fauteux J, Bisaillon J-G and Beaudet R (1982) Bacterial nitration of 4-chlorobiphenyl. Appl. Environ. Microbiol. 44: 871–877

Sylvestre M, Massé R, Ayotte C, Messier F and Fauteux J (1985) Total biodegradation of 4-chlorobiphenyl (4-CB) by a two-membered bacterial culture. Appl. Microbiol. Biotechnol. 21: 192–195

Sylvestre M, Mailhiot K, Ahmad D and Massé R (1989) Isolation and preliminary characterization of a 2-chlorobenzoate degrading *Pseudomonas*. Can. J. Microbiol. 35: 439–443

Taeger K, Knackmuss H-J and Schmidt E (1988) Biodegradability of mixtures of chloro- and methylsubstituted aromatics: simultaneous degradation of 3-chlorobenzoate and 3-methylbenzoate. Appl. Microbiol. Biotechnol. 28: 603–608

Thiele J, Müller R and Lingens F (1987) Initial characterization of 4-chlorobenzoate dehalogenase from *Pseudomonas* sp. CBS3. FEMS Microbiol. Lett. 41: 115–119

Thiele J, Müller R and Lingens F (1988a) Enzymatic dehalogenation of 4-chlorobenzoate by 4-chlorobenzoate dehalogenase from *Pseudomonas* sp. CBS3 in organic solvents. Appl. Microbiol. Biotechnol. 27: 577–580

Thiele J, Müller R and Lingens F (1988b) Enzymatic dehalogenation of chlorinated nitroaromatic compounds. Appl. Environ. Microbiol. 54: 1199–1202

Tiedje JM and Alexander M (1969) Enzymatic cleavage of the ether bond of 2,4-dichlorophenoxyacetate. J. Agric. Food Chem. 17: 1080–1084

Unterman R, Bedard DL, Brennan MJ, Bopp LH, Mondello FJ, Brooks RE, Mobley DP, McDermott JB, Schwartz CC and Dietrich DK (1988) Biological approaches for polychlorinated biphenyl degradation. Basic Life Sciences 45: 253–269

Uotila JS, Salkinoja-Salonen MS and Apajalahti JHA (1991) Dechlorination of pentachlorophenol by membrane bound enzymes of *Rhodococcus chlorophenolicus* PCP-I. Biodegradation 2: 25–31

van den Tweel WJJ, Ter Burg N, Kok JB and De Bont JAM (1986) Bioformation of 4-hydroxybenzoate from 4-chlorobenzoate by *Alcaligenes denitrificans* NTB-1. Appl. Microbiol. Biotechnol. 25: 289–294

van den Tweel WJJ, Kok JB and De Bont JAM (1987) Reductive dechlorination of 2,4-dichlorobenzoate to 4-chlorobenzoate and hydrolytic dehalogenation of 4-chloro-, 4-bromo-, and 4-iodobenzoate by *Alcaligenes denitrificans* NTB-1. Appl. Environ. Microbiol. 53: 810–815

van der Meer JR, Roelofsen W, Schraa G and Zehnder AJB (1987) Degradation of low concentrations of dichlorobenzenes and 1,2,4-trichlorobenzene by *Pseudomonas* sp. P51 in nonsterile soil columns. FEMS Microbiol. Ecol. 45: 333–341

van der Meer JR, Van Neerven ARW, De Vries EJ, De Vos WM and Zehnder AJB (1991a) Cloning and characterization of plasmid-encoded genes for the degradation of 1,2-dichloro-,

1,4-dichloro-, and 1,2,4-trichlorobenzene of *Pseudomonas* sp. strain P51. J. Bacteriol. 173: 6–15

van der Meer JR, Zehnder AJB and De Vos WM (1991b) Identification of a novel composite transposable element, Tn*5280*, carrying chlorobenzene dioxygenase genes of *Pseudomonas* sp. strain P51. J. Bacteriol. 173: 7077–7083

Van Dort HM and Bedard DL (1991) Reductive *ortho* and *meta* dechlorination of a polychlorinated biphenyl congener by anaerobic microorganisms. Appl. Environ. Microbiol. 57: 1576–1578

Vora KA, Singh C and Modi VV (1988) Degradation of 2-fluorobenzoate by a pseudomonad. Curr. Microbiol. 17: 249–254

Walia S, Tewari R, Brieger G, Thimm V and McGuire T (1988) Biochemical and genetic characterization of soil bacteria degrading polychlorinated biphenyl. In: R Abbou (ed) Hazardous Waste: Detection Control Treatment (pp 1621–1632). Elsevier, Amsterdam

Walia S, Khan A and Rosenthal N (1990) Construction and applications of DNA probes for detection of polychlorinated biphenyl-degrading genotypes in toxic organic-contaminated soil environments. Appl. Environ. Microbiol. 56: 254–259

Walker N and Harris D (1970) Metabolism of 3-chlorobenzoic acid by *Azotobacter* species. Soil Biol. Biochem. 2: 27–32

Watanabe I (1973) Isolation of pentachlorophenol decomposing bacteria from soil. Soil Sci. Plant Nutr. 19: 109–116

Weisshaar M-P, Franklin FCH and Reineke W (1987) Molecular cloning and expression of the 3-chlorobenzoate degrading genes from *Pseudomonas* sp. strain B13. J. Bacteriol. 169: 394–402

Wyndham RC and Straus NA (1988) Chlorobenzoate catabolism and interaction between *Alcaligenes* and *Pseudomonas* species from Bloody Run Creek. Arch. Microbiol. 150: 230–236

Wyndham RC, Singh RK and Straus NA (1988) Catabolic instability, plasmid gene deletion and recombination in *Alcaligenes* sp. BR60. Arch. Microbiol. 150: 237–243

Xun L and Orser CS (1991) Purification of a *Flavobacterium* pentachlorophenol-induced periplasmic protein (pcpA) and nucleotide sequence of the corresponding gene (*pcpA*). J. Bacteriol. 173: 2920–2926

Yagi O and Sudo R (1980) Microbial degradation of polychlorinated biphenyls by microorganisms. J. Water Poll. Contr. Fed. 52: 1035–1043

Yates JR and Mondello FJ (1989) Sequence similarities in the genes encoding polychlorinated biphenyl degradation by *Pseudomonas* sp. LB400 and *Alcaligenes eutrophus* H850. J. Bacteriol. 171: 1733–1735

You I-S and Bartha R (1982) Cometabolism of 3,4-dichloroaniline by *Pseudomonas putida*. J. Agric. Food Chem. 30: 274–277

Zeyer J and Kearney PC (1982a) Microbial degradation of para-chloroaniline as sole carbon and nitrogen source. Pesticide Biochem. Physiol. 17: 215–223

Zeyer J and Kearney PC (1982b) Microbial metabolism of propanil and 3,4-dichloroaniline. Pesticide Biochem. Physiol. 17: 224–231

Zeyer J, Wasserfallen A and Timmis KN (1985) Microbial mineralization of ring-substituted anilines through an *ortho*-cleavage pathway. Appl. Environ. Microbiol. 50: 447–453

14. Bacterial degradation of N-heterocyclic compounds

GERHILD SCHWARZ and FRANZ LINGENS
Institute for Microbiology, University of Hohenheim, Postfach 70 05 62, D-70593 Stuttgart, Germany

I. Introduction

The nitrogen-containing heterocyclic compounds are a large and very heterogeneous group of biologically most important substances. They play essential roles in the fundamental processes of life and are therefore ubiquitous. The purines and pyrimidines, the amino acids histidine, hydroxyproline, proline and tryptophan, several vitamins and other cofactors, the alkaloids and various other natural metabolites are N-heterocyclic. A great part of the N-heterocyclic compounds occurring in the environment are of anthropogenic origin. In the wastes from the coal and crude oil refining industries and in coal and tar products, compounds like pyridine, quinoline and their derivatives are found. There are also numerous N-heterocyclic compounds which are synthetic products developed in the chemical, agrochemical and pharmaceutical industries.

Both naturally occurring as well as synthetic N-heterocyclics are applied as, or used for, the synthesis of drugs, pesticides, detergents, dyes, plastic polymers and various other commercial products. Some of them serve as industrial solvents. Considering the widespread use and occurrence of all these compounds in our daily life and their polluting activity, the question of their microbial degradability arises. This especially refers to the man-made N-heterocyclic compounds because of their often "non-biological" structure which can make them rather difficult for microorganisms to metabolize.

Most N-heterocyclic compounds are aromatic, but there are also partly or completely saturated ring systems. The aromatic N-heterocyclics are predominantly π-electron deficient systems. The electronegative nitrogen atom leads to a polarization of the molecule, resulting in the ring carbons having a positive net charge. Thus, they are more amenable to nucleophilic attack, preferably in C-2- or C-4-position. Several N-heterocyclic systems are π-electron excessive, for example, pyrrole, and thus are susceptible to electrophilic substitution. These physico-chemical properties of the N-heterocyclic compounds are relevant for their biodegradability.

C. Ratledge (ed.), Biochemistry of Microbial Degradation, 459–486.

A lot of work has been done concerning the microbiological breakdown of N-heterocyclic substances. Callely (1978) published a detailed review of the microbial degradation of the N-, O-, and S-heterocyclic compounds. Many N-heterocyclics, especially the natural ones, were taken into consideration. A detailed report on the biodegradation of purines and pyrimidines, which will not be considered here, was provided by Vogels and van der Drift (1976). Because of the heterogeneity within the N-heterocyclic group, we shall, for the purpose of this review, confine ourselves to the biodegradation of the aromatic heterocyclics with one nitrogen atom and thereby emphasize pyridine, quinoline, indole and their derivatives. Within the pyridines and quinolines, a large number of xenobiotic compounds are found, the biodegradation of which is of special interest.

II. Degradation of pyridine and derivatives

A. Pyridine and alkylpyridines

Bacterial strains with the ability to degrade the unsubstituted pyridine ring are widespread but, although a number of investigations have been made, its catabolism in microorganisms is still not well understood (Stafford and Callely 1970; Houghton and Cain 1972; Watson and Cain 1975; Korosteleva et al. 1981; Shukla and Kaul 1974, 1975, 1986; Sims et al. 1986; Ronen and Bollag 1991). Watson and Cain (1975) studied pyridine catabolism with a *Nocardia* strain and a *Bacillus* strain and the results of these studies are the most comprehensive ones. Pyridine seems to be metabolized via two different pathways (Fig. 1a and b). Unlike its homocyclic analogue benzene, pyridine is reduced in both organisms in an initial step to give 1,4-dihydropyridine. In *Nocardia* sp. this seems to undergo a hydrolytic ring cleavage following deamination to glutaric dialdehyde. Glutaric acid semialdehyde was identified as a metabolite in the same organism. Glutarate and glutaryl-CoA were then proposed as further intermediates. With *Bacillus* sp., succinic acid semialdehyde and formate could be detected suggesting an oxygenative cleavage of 1,4-dihydropyridine and subsequent hydrolysis. Succinate, the putative end product of degradation, can then be metabolized by central metabolic processes. The latter pathway also operates in *Corynebacterium* sp., *Brevibacterium* sp. and *Micrococcus luteus* (Shukla and Kaul 1974, 1975; Sims et al. 1986). In no case were hydroxylated pyridines found as intermediates. One exception is provided by Korosteleva et al. (1981) with *Nocardia* sp. KM-2. They proposed a degradative pathway for pyridine involving 3-hydroxypyridine which was tentatively identified as the first degradation product. *Nocardia* sp. KM-2, however, was unable to grow with 3-hydroxypyridine which was ascribed to impermeability of the cells to this substrate. Pyridine also is degraded under anaerobic conditions, yet no metabolic pathway is known thus far (Ronen and Bollag 1991).

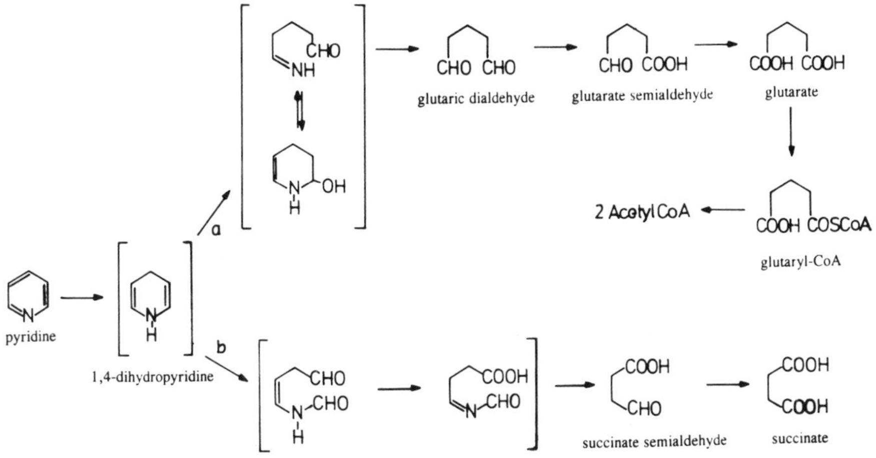

Fig. 1. Degradation pathways of pyridine.

Alkylpyridines are produced mainly during fossil fuel processing and are those pyridines that are most often found as pollutants in the environment (Sims and O'Loughlin 1989). Little is known about their microbial breakdown although. An *Arthrobacter* species that degrades 2-methylpyridine (2-picoline) via a reductive pathway similar to pyridine has been described (Shukla 1974). Various *Mycobacteria* and *Nocardia* strains oxidize 3-methylpyridine (3-picoline) co-metabolically to a high extent to nicotinic acid. 2- and 4-Methyl- and 5-ethylpyridine are also oxidized to the corresponding carboxylic acids (Skryabin and Golovleva 1971; Golovleva et al. 1974; Golovlev 1976; Golovlev et al. 1978). Working with a 3-methylpyridine-utilizing *Pseudomonas* species, Korosteleva et al. (1981) postulated a degradative pathway via 3-hydroxymethylpyridine and nicotinic acid. Both compounds were isolated from the culture fluid. The authors suggest that degradation proceeds via the maleamate pathway (see Fig. 2).

B. Hydroxypyridines

The degradation of 2-hydroxypyridine (2-pyridone), 3-hydroxypyridine and 4-hydroxypyridine (4-pyridone) has been studied thoroughly with different bacterial strains (Houghton and Cain 1972; Cain et al. 1974; Watson et al. 1974a,b; Gupta and Shukla 1975; Khanna and Shukla 1977). Common to all hydroxylated pyridines is the introduction of a second hydroxyl group and subsequent cleavage of the pyridine ring. Houghton and Cain (1972) and Cain et al. (1974) investigated the metabolism of 2-hydroxypyridine and 3-hydroxypyridine with three bacterial strains, then identified as *Achromobacter* species. Their results suggest that both compounds are metabolized through the maleamate pathway (Fig. 2). In both cases, the first hydroxylation leads to

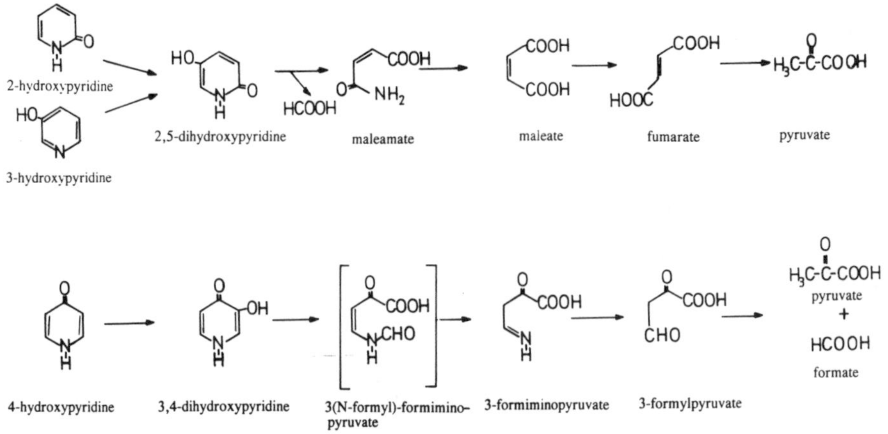

Fig. 2. Bacterial degradation of 2-hydroxypyridine, 3-hydroxypyridine and 4-hydroxypyridine.

2,5-dihydroxypyridine. Respiration experiments provided evidence that the additional oxygen atom in the diol is derived from water and not from O_2. The hydroxylation reaction is therefore not monooxygenase-mediated. 2,5-Dihydroxypyridine is cleaved to maleamate and formate by a pyridine-2,5-diol-dioxygenase. N-Formylmaleamate, an intermediate expected as the primary cleavage product, could not be detected. Maleamate is deaminated to maleate with release of NH_3. *Arthrobacter crystallopoietes*, *Arthrobacter pyridinolis* and other *Arthrobacter* species, able to degrade 2-hydroxypyridine, produce blue pigments during growth on this substrate (Ensign and Rittenberg 1965; Kolenbrander et al. 1976; Kolenbrander and Weinberger 1977). These azaquinone pigments are derived from pyridine diols and triols and are also formed during the degradation of nicotinic acid, isonicotinic acid and nicotine (Knackmuss 1973).

The degradation of 4-hydroxypyridine by *Agrobacterium* sp. involves the formation of 3,4-dihydroxypyridine. In contrast to the enzymes hydroxylating the other two monohydroxypyridines, this reaction is catalysed by a monooxygenase deriving the oxygen that is incorporated from air (Houghton and Cain 1972). The diol is cleaved by a pyridine-3,4-diol dioxygenase to formate and formiminopyruvate. With the release of NH_3, formylpyruvate is formed which is hydrolysed to pyruvate and formate (Watson et al. 1974a,b) (Fig. 2).

C. Pyridoxine and pyridoxamine

Snell and coworkers isolated microorganisms, predominantly pseudomonads, with the ability to grow with pyridoxine, pyridoxamine or pyridoxal – the three forms of vitamin B_6 (Rodwell et al. 1958; Ikawa et al. 1958; Burg et al. 1960; Sundaram and Snell 1969; Burg and Snell 1969; Sparrow et al. 1969; Nyns et al.

1969; Nelson and Snell 1986). Detailed studies with pyridoxine-degrading organisms revealed that there are two slightly different degradative pathways (Fig. 3-I and 3-II). Before ring cleavage, one or both hydroxymethyl side chains of the molecule are successively oxidized. In pathway I, operating in *Pseudomonas* sp. MA-1, pyridoxine is thus converted to 3-hydroxy-2-methylpyridine-5-carboxylic acid which is then cleaved by an FAD-dependent oxygenase to α-(*N*-acetylamino-methylene)succinic acid. With *Arthrobacter* sp. Cr-7 and *Pseudomonas* sp. IA, pyridoxine is oxidized to 5-pyridoxic acid, which is cleaved by a similar oxygenase to α-hydroxymethyl-α'-(*N*-acetylaminomethylene)succinic acid. *Pseudomonas* sp. MA-1 is also able to grow on pyridoxamine which is converted to pyridoxal by transamination with pyruvate (Wada and Snell 1962).

D. Pyridine carboxylic acids

1. Pyridine 2-carboxylic acid (α-picolinic acid)
α-Picolinic acid and α-picolinamide are decomposition products of the bipyridylium herbizide Diquat. Several bacterial strains and the yeast *Rhodotorula* have been isolated with the ability to degrade α-picolinic acid (Dagley and Johnson 1963; Shukla and Kaul 1973; Shukla et al. 1977; Tate and Ensign 1974a,b). The first reaction in the degradation pathway is a hydroxylation to 6-hydroxypicolinic acid. With *Arthrobacter picolinophilus*, Tate and Ensign (1974a,b) showed that the hydroxyl group of this first intermediate is derived from water and is incorporated into the molecule by the enzyme picolinic acid dehydrogenase (picolinic acid hydroxylase), which probably contains molybdenum (Siegmund et al. 1990). 6-Hydroxypicolinic acid is converted to 2,5-dihydroxypyridine via 3,6-dihydroxypicolinic acid and further metabolized via the maleamate pathway. The same degradation pathway was postulated for a 2-picolinate-degrading *Bacillus* sp. and a Gram-negative coccus (Shukla and Kaul 1973; Shukla et al. 1977). Orpin et al. (1972) showed that an unidentified Gram-negative bacillus converts picolinamide to picolinic acid by a deamidase and then degrades it, like picolinic acid, through the maleamate way.

2. Pyridine 3-carboxylic acid (nicotinic acid)
The vitamin nicotinic acid can be degraded both aerobically and anaerobically. In each case 6-hydroxynicotinic acid was identified as the first intermediate. The oxygen atom required for this reaction is, in this case, also derived from water and not from O_2 (Hughes 1954; Hunt et al. 1958; Ensign and Rittenberg 1964; Hirschberg and Ensign 1971) as shown by [18]O experiments. Nagel and Andreesen (1989, 1990) proved molybdenum dependence for the enzyme catalysing this first hydroxylation in the *Bacillus* sp. of Ensign and Rittenberg (1964), which was recently classified as *Bacillus niacini* (Nagel and Andreesen 1991).

During aerobic degradation of 6-hydroxynicotinic acid in *Pseudomonas*

464   *G. Schwarz and F. Lingens*

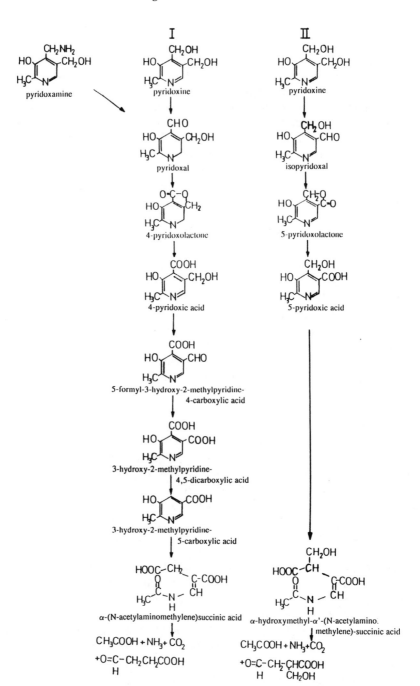

Fig. 3. Degradation pathways of pyridoxine and pyridoxamine: (I) in *Pseudomonas* sp. MA-1; (II) in *Arthrobacter* sp. Cr-7 and *Pseudomonas* sp. IA.

putida (Fig. 4b), an oxidative decarboxylation leads to a para-diol, 2,5-dihydroxypyridine. This is converted to formate and maleamate by a dioxygenase which apparently catalyses both cleavage of the pyridine ring by insertion of two atoms of O_2 and hydrolytic cleavage of the resultant N-formyl group. N-Formylmaleamate, postulated as the ring cleavage product, could not be detected nor could the enzyme attack this compound (Behrman and Stanier 1957; Gauthier and Rittenberg 1971a,b).

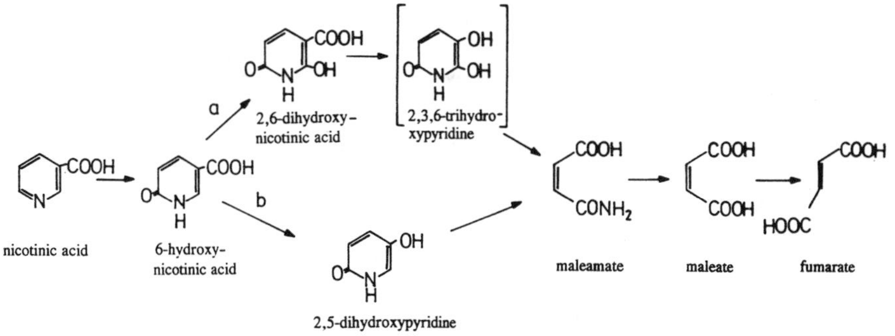

Fig. 4. Aerobic degradation of nicotinic acid via two routes.

There is an alternative oxidative pathway for nicotinic acid elucidated with *Bacillus niacini* (Ensign and Rittenberg 1964) (Fig. 4a). 6-Hydroxynicotinic acid is hydroxylated in 2-position to give 2,6-dihydroxynicotinic acid. A molybdo-enzyme, 6-hydroxynicotinate dehydrogenase, is involved in the reaction (Nagel and Andreesen 1990). 2,6-Dihydroxynicotinic acid is then supposed to be oxidatively decarboxylated to a tripyridol, 2,3,6-trihydroxypyridine. The authors suggest that degradation continues via maleamate.

Clostridium barkeri ferments nicotinic acid to propionic acid, acetic acid, CO_2 and NH_3 (Pastan et al. 1964; Tsai et al. 1966; Holcenberg and Stadtman 1969; Holcenberg and Tsai 1969; Kung et al. 1970; Kung and Stadtman 1971; Kung and Tsai 1971; Stadtman et al. 1972). Studies with ^{14}C-labeled nicotinic acid were helpful for the identification of the intermediates. The initial reaction leads to 6-hydroxynicotinic acid as during the aerobic degradation. This is reduced to 6-oxo-1,4,5,6-tetrahydronicotinic acid which is converted via several unknown steps to α-methyleneglutarate. The formation of methylitaconate from α-methyleneglutarate is catalyzed by the vitamin B_{12}-coenzyme-dependent enzyme, α-methyleneglutarate mutase. A second isomerisation leads to dimethylmaleate which is further converted to the fermentation end products (Fig. 5). *Desulfococcus niacini*, a sulphate-reducing bacterium, is able to use nicotinate as electron donor and carbon source. However, intermediates have not been identified (Imhoff-Stuckle and Pfennig 1983).

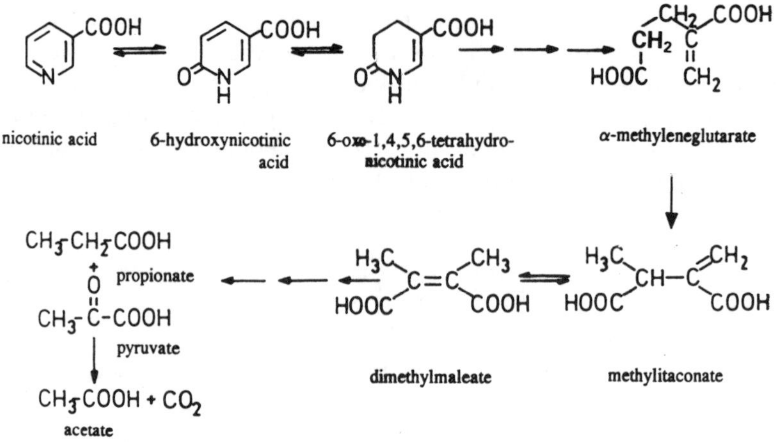

Fig. 5. Anaerobic degradation of nicotinic acid.

3. 5-Chloro-2-hydroxynicotinic acid

A *Mycobacterium* species was found able to grow on 5-chloro-2-hydroxynicotinic acid, a dead-end product of the degradation of 3-chloroquinoline-8-carboxylic acid by *Pseudomonas* sp. 5-Chloro-2,6-dihydroxynicotinic acid is the first metabolite in this degradation sequence. The organisms released chloride from 5-chloro-2-hydroxynicotinic acid, chloromaleic acid and also chlorofumaric acid which made the authors propose a maleamate pathway with chlorinated compounds (Fig. 6) (Tibbles et al. 1989a).

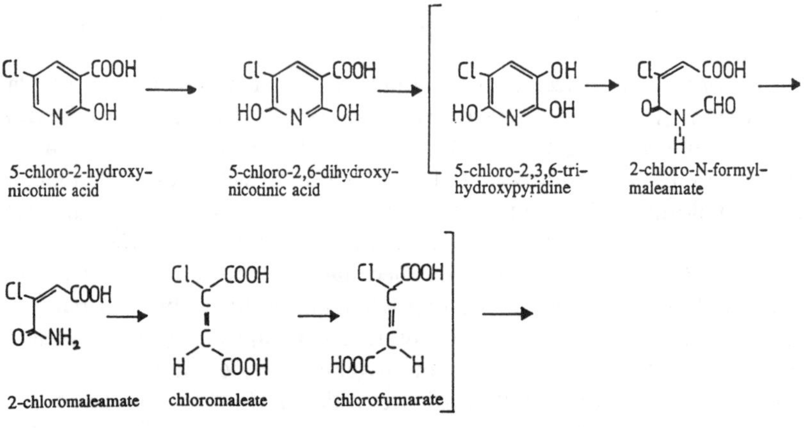

Fig. 6. Degradation of 5-chloro-2-hydroxynicotinic acid.

4. Pyridine 4-carboxylic acid (isonicotinic acid)

There also seem to be different degradation pathways for isonicotinic acid in different organisms. A *Pseudomonas* species accumulated 2,6-dihydroxyisonicotinic acid (citrazinic acid) during growth on isonicotinic acid and, derived from this observation, 2-hydroxyisonicotinic acid was postulated as the first oxidation product (Ensign and Rittenberg 1965). The formation of a soluble blue pigment, probably an azaquinone, led to the assumption that 2,5,6-trihydroxyisonicotinic acid was a metabolite in the degradation pathway. *Sarcina* sp. accumulated 2-hydroxyisonicotinic acid when cultivated with isonicotinic acid (Gupta and Shukla 1978, 1979). *Bacillus brevis* degrades isonicotinic acid via succinic semialdehyde and thus here the pathway for degradation resembles that of pyridine catabolism (Singh and Shukla 1986). Recently, a *Mycobacterium* species was isolated which oxidizes isonicotinic acid via a novel pathway (Fig. 7). 2-Hydroxyisonicotinic acid and citrazinic acid were identified as the first two intermediates. Both dehydrogenases involved in the formation of these compounds are apparently molybdenum-dependent. CoA-activated citrazinate and its reduced form were proposed as intermediates because the respective enzymes were found in crude cell extracts (Kretzer and Andreesen 1991).

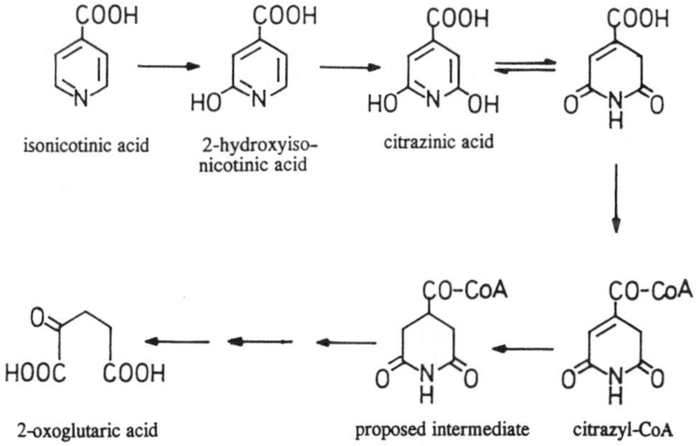

Fig. 7. Biodegradation of isonicotinic acid.

5. Pyridine 2,6-dicarboxylic acid (dipicolinic acid)

Dipicolinic acid is found almost exclusively, and in high amounts, in the *Bacillus* spore and is released into the environment during its germination. The aerobic degradation of dipicolinic acid has been studied by Arima and Kobayashi with a Gram-negative bacterial strain, which at the time was designated as an *Achromobacter* sp. (Arima and Kobayashi 1962; Kobayashi and Arima 1962). They postulated 3-hydroxydipicolinic acid as the first intermediate. Ring cleavage between C2 and C3 atoms, and subsequent hydrolysis, would yield oxalic acid and α-ketoglutaric acid, both identified as metabolites. Both hydroxylation as well as ring cleavage therefore seem to be catalysed by oxygenases.

Seyfried and Schink (1990) isolated a co-culture of strictly anaerobic bacteria able to utilize dipicolinic acid as sole source of carbon, energy and nitrogen. Dipicolinic acid was fermented to acetate, propionate, ammonia and CO_2. The degradation route is still unclear. The authors postulated a reductive step as the first reaction in the sequence, leading to 1,4-dihydrodipicolinic acid.

E. Nicotine

In studies on the bacterial degradation of nicotine various intermediates have been isolated or postulated and several degradative routes have been proposed (Tabuchi 1955; Wada 1957; Decker et al. 1960, 1961; Eberwein et al. 1961; Gries et al. 1961a,b; Hochstein and Rittenberg 1958a,b, 1959; Richardson and Rittenberg 1961a,b; Gherna and Rittenberg 1962; Gherna et al. 1965; Hochstein and Dalton 1965; Holmes and Rittenberg 1972a,b). There are at least two pathways that can be considered to be "established" because the data have been confirmed by different authors. In both cases, the pyrrolidine ring of nicotine is cleaved before the pyridine ring. In one pathway (Fig. 8A), elucidated with different *Pseudomonas* species, nicotine is thus converted to 3-succinoylpyridine. The pyridine ring is then hydroxylated and the resulting 6-hydroxy-3-succinoylpyridine is cleaved at the 3'-position with the formation of succinate and 2,5-dihydroxypyridine (Tabuchi 1955; Gherna and Rittenberg 1962). In *Arthrobacter oxydans*, isolated by Rittenberg and coworkers (Hochstein and Rittenberg 1958a,b; Gherna et al. 1965), the pyridine ring is hydroxylated at carbon 6 before the pyrrolidine ring is oxidized and cleaved to 6-hydroxypseudooxynicotine (Fig. 8B). A further oxidation leads to 2,6-dihydroxypseudooxynicotine from which the metabolically inactive compound 2,6-dihydroxy-N-methylmyosmine is non-enzymatically formed. In the degradation sequence, 2,6-dihydroxypseudooxynicotine is cleaved to γ-methylaminobutyric acid and 2,6-dihydroxypyridine which is further oxidized to 2,3,6-trihydroxypyridine. From the latter two intermediates, blue pigments are derived that form during growth of the bacteria on nicotine. They were identified as diazadiphenoquinones (nicotine blue I and II) (Fig. 9) (Knackmuss 1973; Knackmuss and Beckmann 1973). This second degradation sequence for nicotine has been partly confirmed by Decker et al. (1960, 1961).

III. Degradation of quinoline and derivatives

Quinoline and its derivatives occur in coal tar, mineral oil and bone oil. The quinoline and the isoquinoline skeletons are also found in alkaloids from animals, plants, fungi and bacteria. Furthermore, quinolines are used as reagents in the chemical industry. In recent years, the microbial degradation of these compounds has been thoroughly investigated.

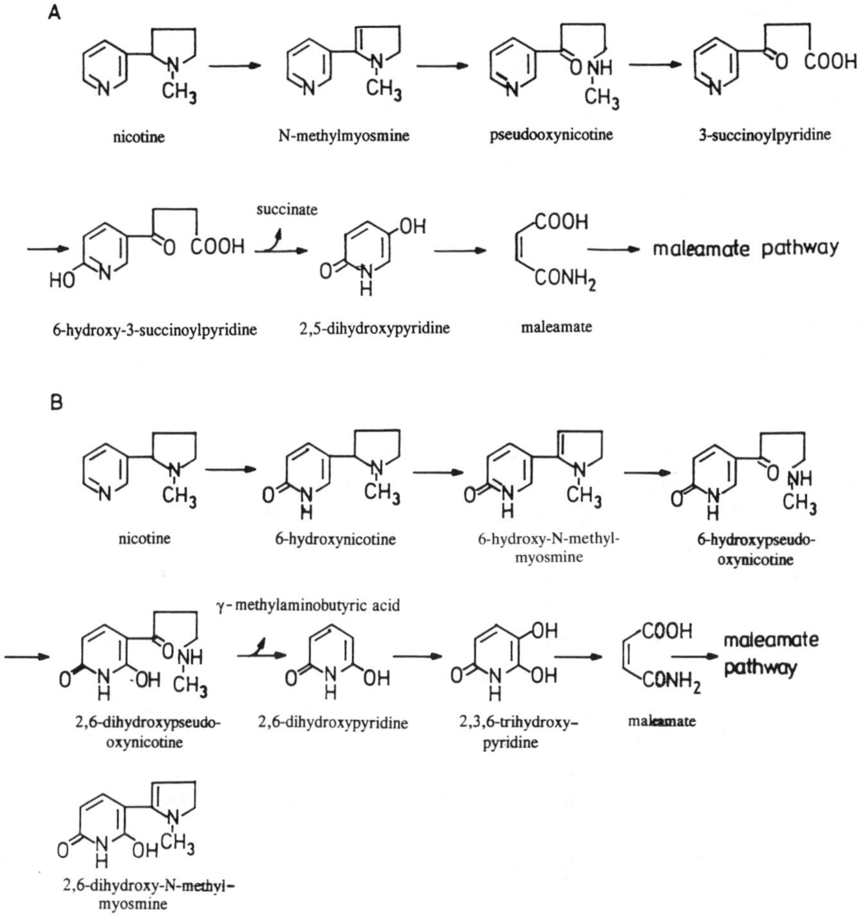

A

nicotine   N-methylmyosmine   pseudooxynicotine   3-succinoylpyridine

6-hydroxy-3-succinoylpyridine   2,5-dihydroxypyridine   maleamate

B

nicotine   6-hydroxynicotine   6-hydroxy-N-methyl-myosmine   6-hydroxypseudo-oxynicotine

2,6-dihydroxypseudo-oxynicotine   2,6-dihydroxypyridine   2,3,6-trihydroxy-pyridine   maleamate

2,6-dihydroxy-N-methyl-myosmine

Fig. 8. Degradation of nicotine: (A) in *Pseudomonas* sp.; (B) in *Arthrobacter oxydans*.

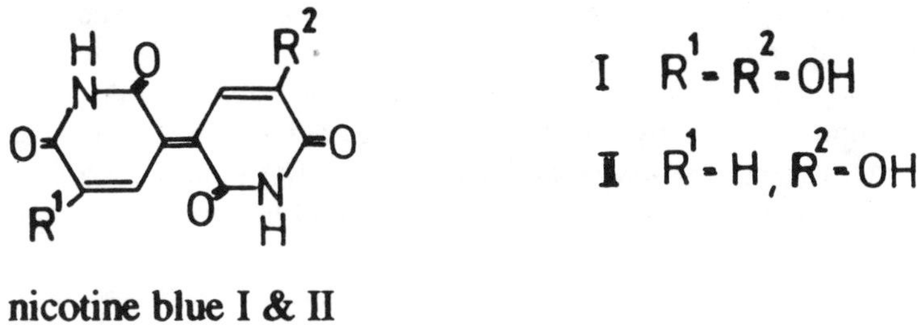

nicotine blue I & II

$$I \quad R^1 = R^2 = OH$$

$$II \quad R^1 = H, R^2 = OH$$

Fig. 9. Nicotine blue I and II from *Arthrobacter oxydans*.

A. Quinoline

Bacteria with the ability to utilize quinoline as sole source of carbon, nitrogen and energy are relatively easy to isolate, and there have been several reports dealing with such cultures and their degradative abilities (Grant and Al-Najjar 1976; Kucher et al. 1980; Dzumedzei et al. 1982; Bennett et al. 1985; Shukla 1986, 1989; Boyd et al. 1987; Pereira et al. 1988; Schwarz et al. 1988; Brockman et al. 1989; Schwarz et al. 1989; Aislabie and Atlas 1990; Aislabie et al. 1990; Brauckhoff et al. 1991; Blaschke et al. 1991; Bott and Lingens 1991). Quinoline is degraded via at least two different pathways in different organisms. In both cases, 2-oxo-1,2-dihydroquinoline is the first intermediate. The quinoline oxidoreductases from *Pseudomonas putida* 86 and *Rhodococcus* sp. B1, catalysing the incorporation of a hydroxyl group derived from water, were shown to be molybdo-iron/sulphur-flavoproteins (Bauder et al. 1990; Hettrich et al. 1991; Hettrich and Lingens 1991; Peschke and Lingens 1991). As was shown before, other enzymes hydroxylating heterocyclic compounds in a position adjacent to the hetero-atom also contain molybdenum. With different *Pseudomonas* species (Shukla 1986; Schwarz et al. 1989; Shukla 1989) degradation continues via 8-hydroxy-2-oxo-1,2-dihydroquinoline, 8-hydroxycoumarin and 2,3-dihydroxyphenylpropionic acid (Fig. 10, pathway I). It is still unknown how the pyridine moiety of the quinoline molecule is cleaved and how the nitrogen atom is eliminated. Also, nothing is known about how 2,3-dihydroxyphenylpropionic acid is further degraded. In the alternative metabolic route (pathway II) found in *Rhodococcus* sp. the benzene moiety of quinoline is cleaved before the pyridine ring (Fig. 11). Another difference to pathway I is that the second hydroxylation takes place at carbon 6. 6-Hydroxy-2-oxo-1,2-dihydroquinoline is supposed to be further hydroxylated at position 5 and then undergoes meta cleavage. The red coloured meta cleavage product was isolated from the medium. It is partly converted non-enzymatically to an azacoumarin, 2-H-pyrano-2-one-[3,2b]-5H-6-pyridone, that accumulates in the culture medium (Schwarz et al. 1989). Further metabolites were not isolated. Because of the similarities to the degradation pathway of kynurenic acid (see below), it is conceivable that the degradation proceeds in an analogous way. An indication for a third degradation route for quinoline was given by Boyd et al. (1987). Besides identifying stable cis-dihydrodiols, 3-hydroxyquinoline and 8-hydroxyquinoline, they detected anthranilic acid with a quinoline-utilizing *Pseudomonas putida* strain. The latter compound is also an intermediate in the degradation of various other heterocyclic substances.

B. Methylquinolines

Bacteria able to attack methylquinolines have been isolated by different workers, yet in most cases nothing is known about the degradation pathway (Aislabie and Atlas 1990; Aislabie et al. 1989; Brauckhoff et al. 1991) The degradation of quinaldine has been studied by Hund et al. (1990) with an

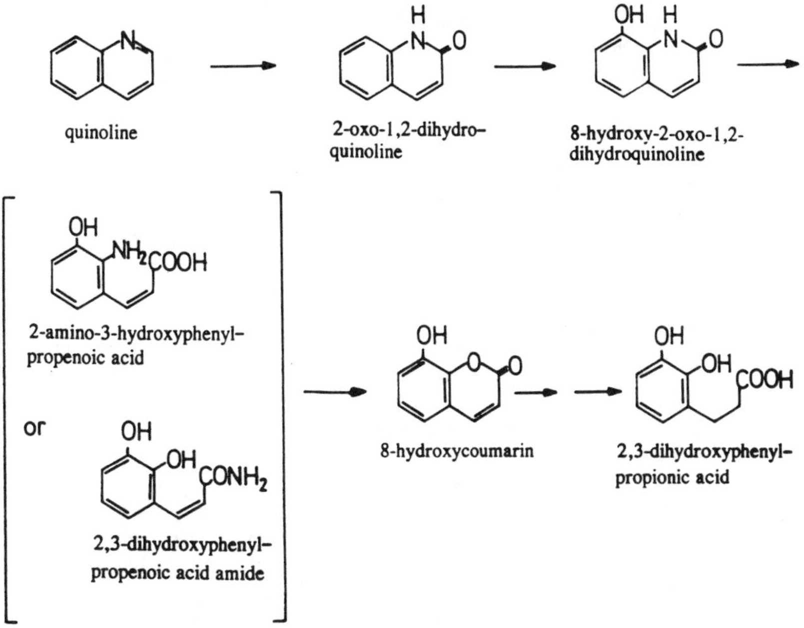

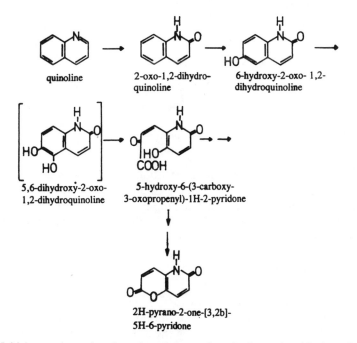

Fig. 10. Initial reactions in the degradation of quinoline via 8-hydroxy-2-oxo-1,2-dihydroxyquinoline (pathway I).

Fig. 11. Initial reactions in the degradation of quinoline via 6-hydroxy-2-oxo-1,2-dihydroquinoline (pathway II).

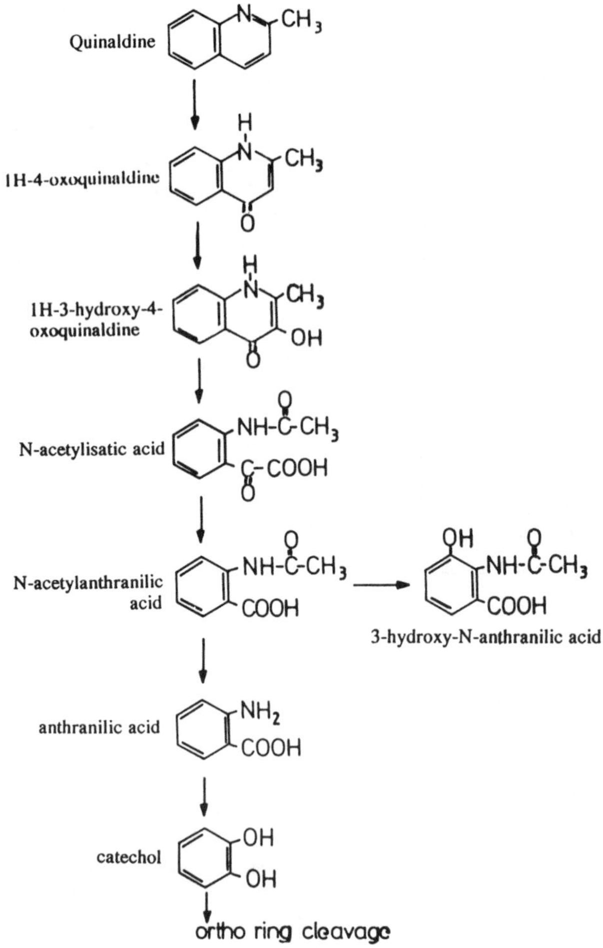

Fig. 12. Degradation pathway of quinaldine.

Arthrobacter species. The first hydroxylation, catalysed by a molybdenum-dependent dehydrogenase (De Beyer, unpublished results), yields 1H-4-oxoquinaldine. The pathway is outlined in Fig. 12. Catechol is metabolized via β-ketoadipate to acetate and succinate. 3-Hydroxy-N-acetylanthranilic acid is considered a by-product of degradation.

C. Hydroxyquinolines

1. 2-Oxo-1,2-dihydroquinoline (2-hydroxyquinoline)

2-Oxo-1,2-dihydroquinoline (2-hydroxyquinoline) is the first intermediate during quinoline degradation in most of the hitherto known quinoline-utilizing bacteria. It is therefore degraded by all of them via one of the two pathways of quinoline degradation.

2. 1H-4-Oxo-quinoline (4-hydroxyquinoline)

Pseudomonas putida 33/1 is able to grow with 1H-4-oxo-quinoline (4-hydroxyquinoline) as sole source of carbon, nitrogen and energy. The degradation scheme proposed by Bott et al. (1990) is illustrated in Fig. 13. The analogy to the degradation pathway of quinaldine is obvious. The first hydroxylation takes place at position 3 and yields the ortho-diol, 1H-3-hydroxy-4-oxoquinoline. This reaction is catalysed by the enzyme 1H-4-oxoquinoline monooxygenase. Compared to other quinoline-oxidizing enzymes, this monooxygenase is not molybdenum-dependent and introduces oxygen into the molecule that is derived from O_2 (Bott et al. 1990; Block and Lingens 1992). In this connection, it is interesting that the initial hydroxylation of 4-hydroxypyridine, leading to 3,4-dihydroxypyridine, also occurs from molecular oxygen and is monooxygenase-catalysed. In the degradation sequence, 1H-3-hydroxy-4-oxoquinoline is cleaved to hypothetical *N*-formylisatic acid. The following intermediates are *N*-formylanthranilic acid, anthranilic acid and catechol. The latter is metabolized via the β-ketoadipate way. A different pathway is proposed by Al-Najjar et al. (1976), who postulate 7,8-dihydroxy-1H-4-oxoquinoline and a red meta-cleavage product as intermediates.

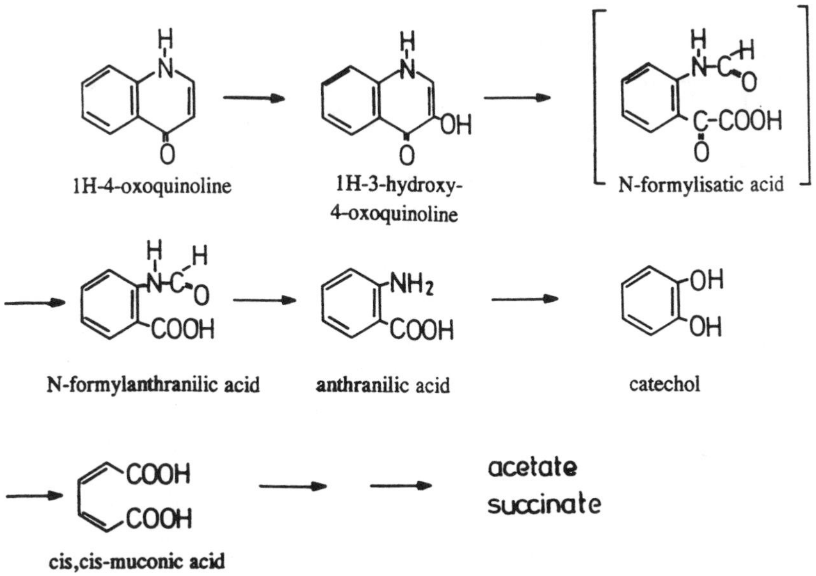

Fig. 13. Degradative route of 1H-4-oxoquinoline.

3. 6-Hydroxyquinoline

6-Hydroxyquinoline is oxidized by *Pseudomonas diminuta* and *Bacillus circulans* to 6-hydroxy-2-oxo-1,2-dihydroquinoline (Bott and Lingens 1991). Further metabolites could not be detected but it is assumed that the following degradation steps are the same as for quinoline. Both strains are able to utilize quinoline and 2-oxo-1,2-dihydroquinoline. Some *Rhodococcus* spp. degrading quinoline via 6-hydroxy-2-oxo-1,2-dihydroquinoline (pathway II) are also able to utilize 6-hydroxyquinoline (Schwarz et al. 1988).

D. Quinoline carboxylic acids

1. Quinoline 2-carboxylic acid (quinaldic acid)

Several bacterial strains capable of utilizing quinaldic acid (quinoline 2-carboxylic acid) as the sole source of carbon and energy were isolated by Dembek and Lingens (1988). With a mutant strain of *Azotobacter* sp. a yellow meta-cleavage product, 6-carboxy-3-(3'-carboxy-1'-hydroxy-3'-oxopropenyl)-1H-2-pyridone (Fig. 14), was isolated. This suggests that the benzene moiety of the quinoline skeleton is primarily attacked.

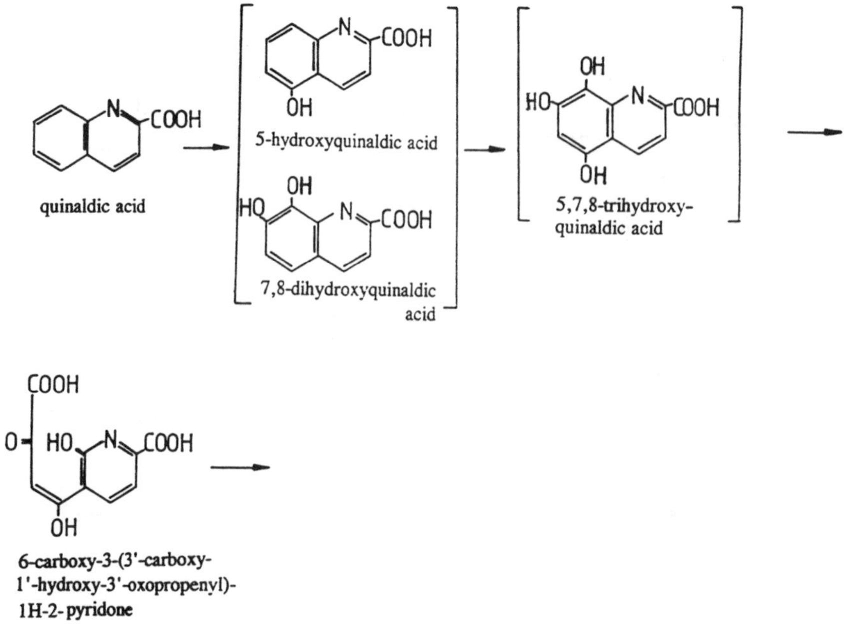

Fig. 14. Initial reactions in the degradation of quinaldic acid by *Azotobacter* sp.

2. Quinoline 4-carboxylic acid

Quinoline 4-carboxylic acid is metabolized by *Microbacterium* sp., *Agrobacterium* sp. and *Pimelobacter simplex* via a set of reactions that is similar to that of quinoline degradation (Röger and Lingens 1989, Schmidt et al. 1991). The degradation route is shown in Fig. 15. The first oxidation product is 2-oxo-1,2-dihydroquinoline-4-carboxylic acid. In all three bacterial strains the quinoline 4-carboxylic acid oxidoreductases catalysing this first reaction were shown to be molybdo-enzymes (G. Bauer, unpublished results). The authors suggest that 8-hydroxycoumarin-4-carboxylic acid is formed via 8-hydroxy-2-oxo-1,2-dihydroquinoline-4-carboxylic acid, which could, however, not be detected. 2,3-Dihydroxyphenylsuccinic acid is the cleavage product of the coumarin derivative. 2-Oxo-1,2,3,4-tetrahydroquinoline-4-carboxylic acid was isolated from cultures of *Agrobacterium* sp., but its role in the degradation scheme is still uncertain.

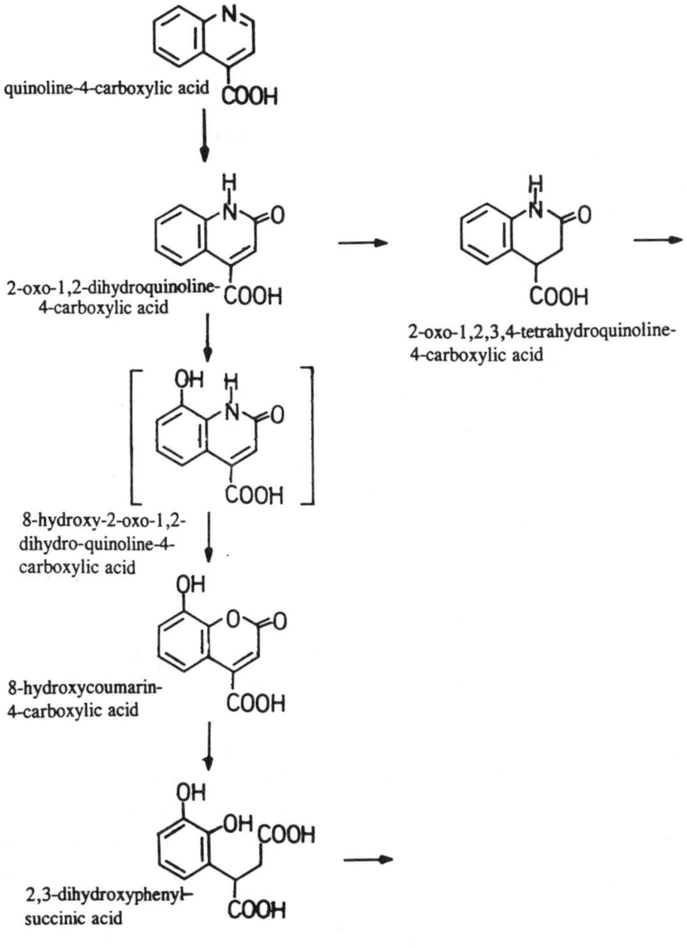

Fig. 15. Degradation pathway of quinoline 4-carboxylic acid.

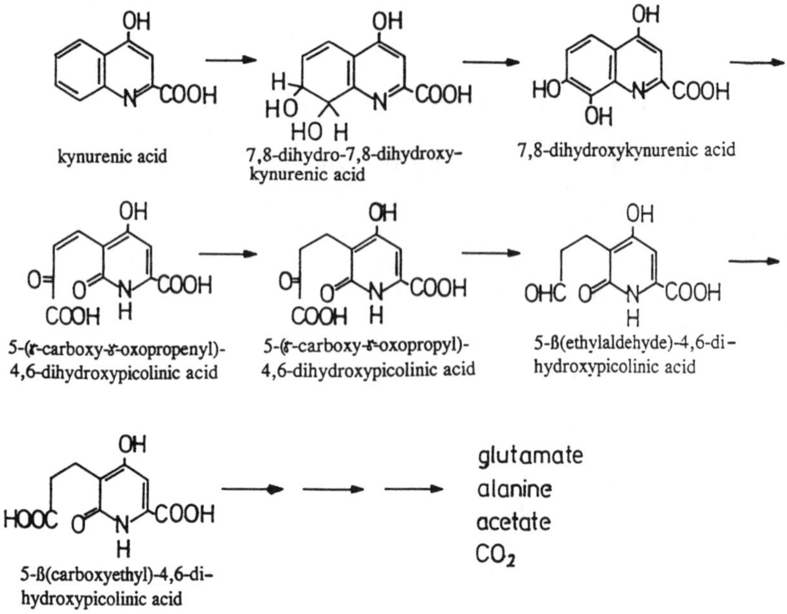

Fig. 16. Biodegradation of kynurenic acid.

3. 4-Hydroxyquinoline 2-carboxylic acid (kynurenic acid)

Kynurenic acid (4-hydroxyquinoline 2-carboxylic acid) is a metabolite of tryptophan catabolism. Its bacterial degradation was studied intensively by Hayaishi and colleagues (Hayaishi et al. 1961; Horibata et al. 1961; Kuno et al. 1961; Taniuchi and Hayaishi 1963) with *Pseudomonas fluorescens*. The pathway of metabolism is illustrated in Fig. 16. The two substituents of the pyridine ring enforce the cleavage of the benzene ring to give a picolinic acid derivative.

4. 3-Chloroquinoline 8-carboxylic acid

Some halogenated quinolines show herbicidal activity and may be used in agriculture. During the degradation of 3-chloroquinoline 8-carboxylic acid, pyridine- but not benzene-derivatives occur as intermediates. Tibbles et al. (1989b) isolated a *Pseudomonas* species capable of using this substrate as carbon and energy source. Fig. 17 illustrates the catabolic route proposed. The degradation proceeds via a meta-cleavage product, 3-(γ-carboxy-γ-oxopropenyl)-2-hydroxy-5-chloropyridine, isolated from a resting cells broth. A second product was isolated and identified as 5-chloro-2-hydroxynicotinic acid which was not further metabolized by the *Pseudomonas* sp. The degradation of this compound by a *Mycobacterium* species is mentioned elsewhere in this review.

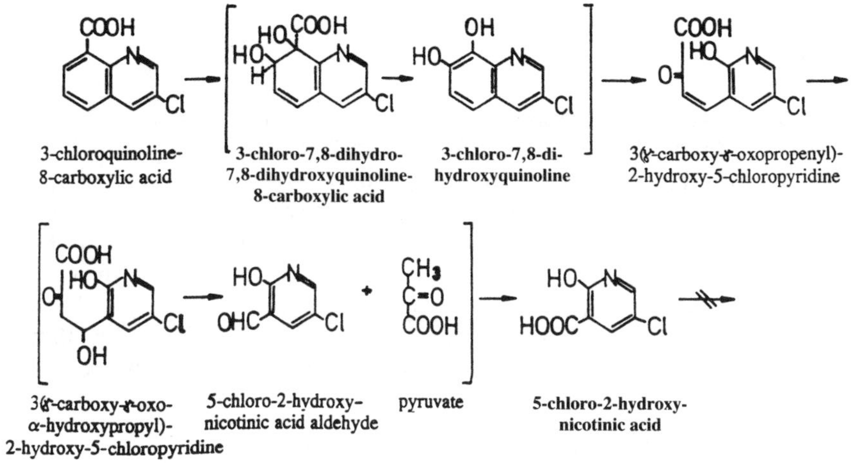

3-chloroquinoline-8-carboxylic acid

3-chloro-7,8-dihydro-7,8-dihydroxyquinoline-8-carboxylic acid

3-chloro-7,8-di-hydroxyquinoline

3(ß-carboxy-ɣ-oxopropenyl)-2-hydroxy-5-chloropyridine

3(ß-carboxy-ɣ-oxo-α-hydroxypropyl)-2-hydroxy-5-chloropyridine

5-chloro-2-hydroxy-nicotinic acid aldehyde

pyruvate

5-chloro-2-hydroxy-nicotinic acid

Fig. 17. Proposed degradation pathway of 3-chloro-quinoline-8-carboxylic acid.

IV. Isoquinoline

Little is known about the catabolism of isoquinoline. 1-Oxo-1,2-dihydroisoquinoline was detected as the first degradation product by Aislabie et al. (1989) with a mixed microbial culture and by Röger et al. (1990) with *Alcaligenes faecalis* and *Pseudomonas diminuta*.

Besides its occurrence in coal tar and fossil fuel, the isoquinoline skeleton is a constituent of many plant alkaloids. The isoquinoline derivative, papaverine, (Fig. 18) is utilized by a *Nocardia* species as sole source of carbon and nitrogen. The isoquinoline moiety of the molecule is attacked primarily, leading to a cleavage of the pyridine ring (Haase-Aschoff and Lingens 1979).

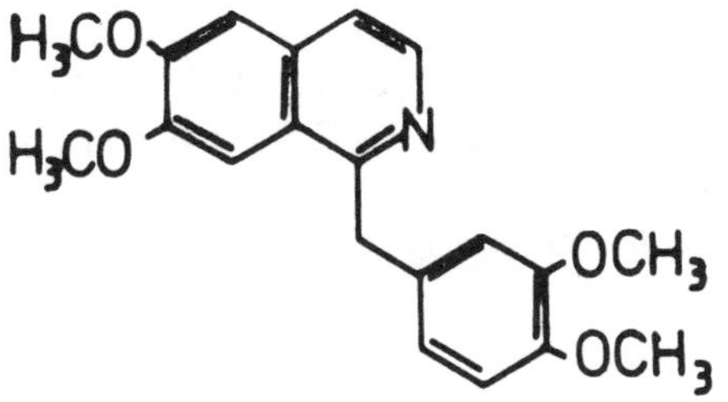

papaverine

Fig. 18. Papaverine.

V. The degradation of indole and derivatives

A. Indole

The benzopyrrole ring system of indole is biodegraded under aerobic and anaerobic conditions. Fujioka and Wada (1967) isolated an aerobic, Gram-positive coccus that converted indole to anthranilic acid. The authors proposed a pathway with indole being oxidized to 2,3-dihydroxyindole by the action of a dioxygenase. 2,3-Dihydroxyindole undergoes ortho cleavage following conversion to anthranilic acid via an unknown cleavage product. An anaerobic bacterial consortium was found to degrade indole to methane. Indole is converted to 1,3-dihydro-2H-indole-2-one (oxindole) and further to anthranilic acid which is finally mineralized to methane and CO_2 (Berry et al. 1987).

B. Tryptophan

Several bacterial degradation routes exist for the amino acid tryptophan. Both D- and L-tryptophan can be metabolized through the "aromatic pathway" via anthranilic acid or through the "quinoline pathway" via kynurenic acid. Both pathways were discovered in *Pseudomonas* species (Stanier et al. 1951; Tashiro et al. 1961; Behrman 1962). The first two reactions are the same in both pathways and involve the cleavage of the pyrrole ring, catalysed by the enzyme tryptophan dioxygenase. The reaction product N-formylkynurenine is cleaved to kynurenine and formate. At this point the pathways diverge. On the one hand, kynurenine can be degraded through the aromatic pathway via anthranilic acid and catechol – a way that resembles the catabolism of many homocyclic aromatic compounds (Gibson and Subramanian 1984). On the other hand, it can be metabolized via kynurenic acid (quinoline pathway) the degradation of which is described in this review (see above). The initial reactions of tryptophan catabolism in bacteria are shown in Fig. 19.

Bacterial degradation pathways involving the alanine side-chain of tryptophan, and leading to indole or indole derivatives, have also been reported (DeMoss and Moser 1969; Roberts and Rosenfeld 1977; Narumiya et al. 1979; Lübbe et al. 1983). Balba and Evans (1980) propose a methanogenic fermentation pathway for tryptophan studied with a microbial consortium. Indole was tentatively identified as an intermediate.

VI. Concluding remarks

A great deal of knowledge has been gained about the mechanisms of degradation of the N-heterocyclic compounds, especially under aerobic conditions. The degradation of many compounds mentioned in this review is initiated by a hydroxylation catalysed by a dehydrogenase with oxygen derived

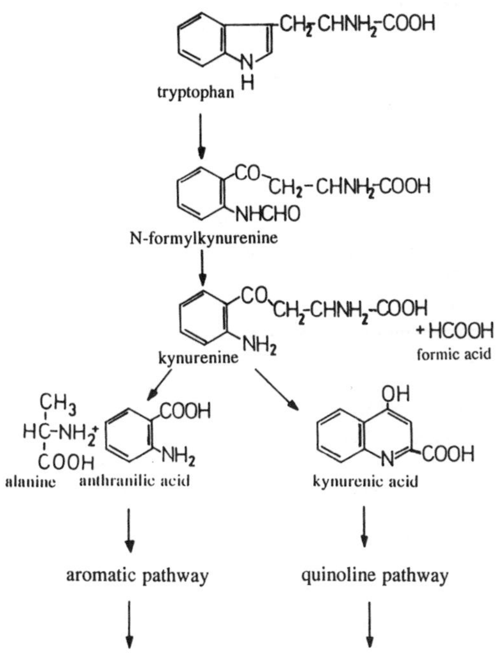

Fig. 19. Initial reactions in the degradation pathways of tryptophan.

from water. Several pyridine derivatives are metabolized via the maleamate pathway, a set of reactions that have been intensively studied. The bicyclic quinolines are degraded via various routes after the initial hydroxylation that, with very few exceptions, involves the pyridine ring. The initial reactions in the degradation of indole and tryptophan are, on the contrary, catalysed by dioxygenases, analogous to the homocyclic aromatic compounds. Nevertheless, there still remain reactions that are only poorly understood. Pyridine, for example, is degraded via an aerobic reductive pathway the single steps of which are still hypothetical. Also, only a little information exists thus far about the catabolism of alkylpyridines.

Most of the studies dealing with the microbial degradation of organic compounds are performed with pure bacterial cultures under defined conditions in the laboratories. They do, however, only partly document the situation occurring in the natural environment. Mixed microbial cultures have to cope with organic, and polluting, compounds under complex and often less-ideal conditions compared with those in the laboratory. So, when a compound is found to be biodegradable under artificial conditions, it may not necessarily be degraded the same way or at the same rate in nature. Future laboratory research with microbial communities can probably provide a better understanding of the ultimate fate of persistent xenobiotics in the biosphere. Pure cultures, however, facilitate the biochemical analysis of the single steps of a degradative pathway. Furthermore, they are necessary for comparative studies of the enzymes and their respective genes involved in the catabolism of one certain compound in different species or genera, an important question under evolutionary aspects.

References

Aislabie J and Atlas RM (1990) Microbial upgrading of Stuart shale oil: removal of heterocyclic nitrogen compounds. Fuel 69: 1155–1157

Aislabie J, Rothenburger S and Atlas RM (1989) Isolation of microorganisms capable of degrading isoquinoline under aerobic conditions. Appl. Environ. Microbiol. 55: 3247–3249

Aislabie J, Bej AK, Hurst H, Rothenburger S and Atlas RM (1990) Microbial degradation of quinoline and methylquinolines. Appl. Environ. Microbiol. 56: 345–351

Al-Najjar TR, Grout RJ and Grant DJW (1976) Degradation of 4-hydroxyquinoline (kynurine) by a soil pseudomonad. Microbios. Lett. 1: 157–163

Arima K and Kobayashi Y (1962) Bacterial oxidation of dipicolinic acid. I. Isolation of microorganisms, their culture conditions and end products. J. Bacteriol. 84: 759–764

Balba MT and Evans WC (1980) Methanogenic fermentation of the naturally occurring aromatic amino acids by a microbial consortium. Biochem. Soc. Trans. 8: 625–627

Bauder R, Tshisuaka B, Lingens F (1990) Microbial metabolism of quinoline and related compounds. VII. Quinoline oxidoreductase from *Pseudomonas putida*: a molybdenum-containing enzyme. Biol. Chem. Hoppe-Seyler 371: 1137–1144

Bauer G and Lingens F (1992) Microbial metabolism of quinoline and related compounds. XV. Quinoline-4-carboxylic acid oxidoreductase from Agrobacterium spec. 1B: A molybdenum-containing enzyme. Biol. Chem. Hoppe-Seyler 373: 699–705

Behrman EJ (1962) Tryptophan metabolism in *Pseudomonas*. Nature 196: 150–152

Behrman EJ and Stanier RY (1957) The bacterial oxidation of nicotinic acid. J. Biol. Chem. 228: 923–945

Bennett JL, Updegraff DM, Pereira WE and Rostad CE (1985) Isolation and identificaton of four species of quinoline-degrading pseudomonads from a creosote-contaminated site at Pensacola, Florida. Microbios. Lett. 29: 147–154

Berry DF, Madsen EL and Bollag J-M (1987) Conversion of indole to oxindole under methanogenic conditions. Appl. Environ. Microbiol. 53: 180–182

Blaschke M, Kretzer A, Schäfer C, Nagel M and Andreesen JR (1991) Molybdenum-dependent degradation of quinoline by *Pseudomonas putida* Chin IK and other aerobic bacteria. Arch. Microbiol. 155: 164–169

Block DW and Lingens F (1992a) Microbial metabolism of quinoline and related compounds. XIII. Purification and properties of 1H-4-oxoquinoline monooxygenase from *Pseudomonas putida* strain 33/1. Biol. Chem. Hoppe-Seyler 373: 249–254

Block DW and Lingens F (1992b) Microbial metabolism of quinoline and related compounds. XIV. Purification and properties of 1H-3-hydroxy-4-oxyquinoline oxygenase, a new extradiol cleavage enzyme from Pseudomonas putida strain 33/1. Biol. Chem. Hoppe-Seyler 373: 343–349

Bott G and Lingens F (1991) Microbial metabolism of quinoline and related compounds. IX. Degradation of 6-hydroxyquinoline and quinoline by *Pseudononas diminuta* 31/1 and *Bacillus circulans* 31/2 A1. Biol. Chem. Hoppe-Seyler 372: 381–383

Bott G, Schmidt M, Rommel TO and Lingens F (1990) Microbial Metabolism of quinoline and related compounds. V. Degradation of 1H-4-oxoquinoline by *Pseudomonas putida* 33/1. Biol. Chem. Hoppe-Seyler 371: 999–1003

Boyd DR, Austin R, McMordie S, Porter HP, Dalton H, Jenkins RO and Howarth OW (1987) Metabolism of bicyclic aza-arenes by *Pseudomonas putida* to yield vicinal cis-dihydrodiols and phenols. J. Chem. Soc. Chem. Commun. 22: 1722–1724

Brauckhoff S, Schacht S and Klein J (1991) Charakterisierung mikrobieller Verwerter N-heterozyklischer Verbindungen. gwf Das Gas- und Wasserfach, Wasser Abwasser 132: 191–192

Brockman FJ, Denocan BA, Hicks RJ and Fredrickson JK (1989) Isolation and characterization of quinoline-degrading bacteria from subsurface sediments. Appl. Environ. Microbiol. 55: 1029–1032

Burg RW and Snell EE (1969) The bacterial oxidation of vitamin B_6. VI. Pyridoxal dehydrogenase and 4-pyridoxolactonase. J. Biol. Chem. 244: 2585–2589

Burg RW, Rodwell CW and Snell EE (1960) Bacterial oxidation of vitamin B_6. III. Metabolites of pyridoxamine. J. Biol. Chem. 235: 1164–1169

Cain RB, Houghton C and Wright KA (1974) Microbial metabolism of the pyridine ring. Biochem. J. 140: 293–300

Callely AG (1978) The microbial degradation of heterocyclic compounds. Progr. Ind. Microbiol. 14: 205–281

Dagley S and Johnson PA (1963) Microbial oxidation of kynurenic, xanthurenic and picolinic acids. Biochim. Biophys. Acta 78: 577–587

De Beyer A and Lingens F (1993) Microbial metabolism of quinoline and related compounds. XVI. Quinaldine oxidoreductase from Arthrobacter spec. Rü 61a: A molybdenum-containing enzyme catalysing the hydroxylation at C-4 of the heterocycle. Biol. Chem. Hoppe-Seyler 374: 101–110

Decker K, Eberwein H, Gries FA and Brühmüller M (1960) Über den Abbau des Nicotins durch Bakterienenzyme. Hoppe Seyler's Z Physiol. Chem. 319: 279–282

Decker K, Gries FA and Brühmüller M (1961) Über den Abbau des Nicotins durch Bakterienenzyme. III. Stoffwechselstudien an zellfreien Extrakten. Hoppe Seyler's Z Physiol. Chem. 323: 249–263

Dembek G and Lingens F (1988) Isolation and characterization of a meta-cleavage product in the degradation of quinaldic acid by Azotobacter sp. FEMS Microbiol Lett 56: 261–264

DeMoss RD and Moser K (1969) Tryptophanase in diverse bacterial species. J. Bacteriol. 98: 167–171

Dzumedzei NV, Shevchenko AG, Turovskii AA and Starovoitov II (1982) Effects of iron ions on the microbial transformation of quinoline. Microbiology (russ. engl. transl) 52: 157–160

Eberwein H, Gries FA and Decker K (1961) Über den Abbau des Nicotins durch Bakterienenzyme. II. Isolierung und Charakterisierung eines nikotinabbauenden Bodenbakteriums. Hoppe-Seyler's Z Physiol. Chem. 323: 236–248

Ensign JC and Rittenberg SC (1964) The pathway of nicotinic acid oxidation by a Bacillus species. J. Biol. Chem. 239: 2285–2291

Ensign JC and Rittenberg SC (1965) The formation of a blue pigment in the bacterial oxidation of isonicotinic acid. Arch. Microbiol. 51: 384–392

Fetzner S and Lingens F (1993) Microbial metabolism of quinoline and related compounds. XVIII. Purification and some properties of the molybdenum- and iron-containing quinaldic acid 4-oxidoreductase from Serratia marcescens 2CC-1. Biol. Chem. Hoppe-Seyler 374: 363–376

Fujioka M and Wada H (1967) The bacterial oxidation cof indole. Biochim. Biophys. Acta 158: 70–78

Gauthier JJ and Rittenberg SC (1971a) The metabolism of nicotinic acid. I. Purification and properties of 2,3-dihydroxypyridine oxygenase from Pseudomonas putida N-9. J. Biol. Chem. 246: 3737–3742

Gauthier JJ and Rittenberg SC (1971b) The metabolism of nicotinic acid. II. 2,5-Dihydroxypyridine oxidation, product formation, and oxygen incorporation. J. Biol. Chem. 246: 3743–3748

Gherna RL and Rittenberg SC (1962) Alternate pathways in nicotine degradation. Bact. Proceed. 62: 107

Gherna RL, Richardson SH, and Rittenberg SC (1965) The bacterial oxidation of nicotine. VI. The metabolism of 2,6-dihydroxypseudooxynicotine. J. Biol. Chem. 24: 3669–3674

Gibson DT and Subramanian V (1984) Microbial degradation of aromatic hydrocarbons. In: DT Gibson (ed) Microbial Degradation of Organic Compounds (pp 181–252). Marcel Dekker Inc, New York, Basel

Golovleva LA, Golovlev EL, Skryabin GK, Sadyrina GA and Ananeva TI (1974) Picolinic acid. Chem. Abstr. 81: 103241

Golovlev EL (1976) Characteristics of the regulation of the microbiological transformation of organic compounds. Chem. Abstr. 85: 156250x

Golovlev EL, Golovleva LA, Eroshina NV and Skryabin GK (1978) Microbiological transformation of xenobiotics by *Nocardia*. In: M Mordarski, W Kurylowicz and J Jeljaszewicz (eds) *Nocardia* and *Streptomyces* (pp 269–283). Gustav Fischer Verlag, Stuttgart, New York

Grant DJW and Al-Najjar TR (1976) Degradation of quinoline by a soil bacterium. Microbios. 15: 177–189

Gries FA, Decker K and Brühmüller M (1961a) Über den Abbau des Nicotins durch Bakterienenzyme. V. Der Abbau des L-6-Hydroxynicotins zu [γ-Methylamino-prophyl]-[6-hydroxy-pyridyl-(3)]-keton. Hoppe-Seyler's Z Physiol. Chem. 325: 229–241

Gries FA, Decker K, Eberwein H and Brühmüller M (1961b) Über den Abbau des Nicotins durch Bakterienenzyme. VI. Die enzymatische Unwandlung des [γ-Methylaminopropyl]-[6-hydroxypyridyl-(3)]-ketons. Biochem. Z 335: 285–302

Gupta RC and Shukla OP (1975) Microbial metabolism of 2-hydroxypyridine. Indian J. Biochem. Biophys. 12: 296–298

Gupta RC and Shukla OP (1978) 2-Hydroxyisonicotinic acid – an intermediate in metabolism of isonicotinic acid hydrazide and isonicotinic acid by *Sarcina* sp. Indian. J. Biochem. Biophys. 15: 492–493

Gupta RC and Shukla OP (1979) Isonicotinic and 2-hydroxy-isonicotinic acid hydroxylases of *Sarcina sp.* Indian. J. Biochem. Biophys. 16: 72–75

Haase-Aschoff K and Lingens F (1979) Mikrobieller Abbau von Papaverin. Hoppe-Seyler's Z Physiol. Chem. 360: 621–632

Hayaishi O, Taniuchi H, Tashiro M and Kuno S (1961) Studies on the metabolism of kynurenic acid. I. The formation of L-glutamic acid, D- and L-alanine, and acetic acid from kynurenic acid by *Pseudomonas* extracts. J. Biol. Chem. 236: 2492–2497

Hettrich D and Lingens F (1991) Microbial metabolism of quinoline and related compounds. VIII. Xanthine dehydrogenase from a quinoline utilizing *Pseudomonas putida* strain. Biol. Chem. Hoppe-Seyler 2: 203–211

Hettrich D, Peschke B, Tshisuaka B and Lingens F 1991) Microbial metabolism of quinoline and related compounds. X. The molybdopterin cofactors of quinoline oxidoreductases from *Pseudomonas putida* 86 and *Rhodococcus spec.* B1 and of Xanthine dehydrogenase from *Pseudomonas putida* 86. Biol. Chem. Hoppe-Seyler 372: 513–517

Hirschberg R and Ensign JC (1971) Oxidation of nicotinic acid by a *Bacillus* species: source of oxygen atoms for the hydroxylation of nicotinic acid and 6-hydroxynicotinic acid. J. Bacteriol. 108: 757–759

Hochstein LI and Dalton BP (1965) The hydroxylation of nicotine: the origin of the hydroxyl oxygen. Biochem. Biophys. Res. Commun. 21: 644–648

Hochstein LI and Rittenberg SC (1958a) The bacterial oxidation of nicotine. I. Nicotine oxidation by cell-free preparations. J. Biol. Chem. 234: 151–156

Hochstein LI and Rittenberg SC (1958b) The bacterial oxidation of nicotine. II. The isolation of the first oxidative product and its identification as (1)-6-hydroxynicotine. J. Biol. Chem. 234: 156–160

Hochstein LI and Rittenberg SC (1959) The bacterial oxidation of nicotine. III. The isolation and identification of 6-hydroxypseudooxynicotine. J. Biol. Chem. 235: 795–799

Holcenberg JS and Stadtman ER (1969) Nicotinic acid metabolism. III. Purification and properties of a nicotinic acid hydroxylase. J. Biol. Chem. 244: 1194–1203

Holcenberg JS and Tsai L (1969) Nicotinic acid metabolism. IV. Ferredoxin-dependent reduction of 6-hydroxynicotinic acid to 6-oxo-1,4,5,6-tetrahydronicotinic acid. J. Biol. Chem. 244: 1204–1211

Holmes PE and Rittenberg SC (1972a) The bacterial oxidation of nicotine. VII. Partial purification and properties of 2,6-dihydroxypyridine oxidase. J. Biol. Chem. 247: 7622–7627

Holmes PE and Rittenberg SC (1972b) The bacterial oxidation of nicotine. VIII. Synthesis of

2,3,6-trihydroxy-pyridine and accumulation and partial characterization of the product of 2,6-dihydroxypyridine oxidation. J. Biol. Chem. 247: 7628–7633

Horibata K, Taniuchi H, Tashiro M, Kuno S and Hayaishi O (1961) The metabolism of kynurenic acid. II. Tracer experiments in the mechanism of kynurenic acid degradation and glutamic acid synthesis by *Pseudomonas extracts*. J. Biol. Chem. 236: 2991–2995

Houghton C and Cain RB (1972) Microbial metabolism of the pyridine ring. Biochem. J. 130: 879–893

Hughes DE (1954) 6-Hydroxynicotinic acid as an intermediate in the oxidation of nicotinic acid by *Pseudomonas fluorescens*. Biochem. J. 60: 303–310

Hund HK, De Beyer A and Lingens F (1990) Microbial metabolism of quinoline and related compounds. VI. Degradation of quinaldine by *Arthrobacter sp*. Biol. Chem. Hoppe-Seyler 371: 1005–1008

Hunt AL, Hughes DE and Löwenstein JM (1958) The hydroxylation of nicotinic acid by *Pseudomonas fluorescens*. Biochem. J. 69: 170–173

Ikawa M, Rodwell VW and Snell E (1958) Bacterial oxidation of vitamin B_6 II. Structure of "260 Compound". J. Biol. Chem. 233: 1555–1559

Imhoff-Stuckle D and Pfennig N (1983) Isolation and characterization of a nicotinic acid-degrading sulfate-reducing bacterium, *Desulfococcus niacini* sp. nov. Arch. Microbiol. 136: 194–198

Khanna M and Shukla OP (1977) Microbial metabolism of 3-hydroxypyridine. Indian J. Biochem. Biophys. 14: 301–302

Knackmuss HJ (1973) Zur Chemie und Biochemie der Azachinone. Angew Chem. 85: 163–169

Knackmuss HJ and Beckmann W (1973) The structure of nicotine blue from *Arthrobacter oxidans*. Arch Mikrobiol 90: 167–169

Kobayashi Y and Arima K (1962) Bacterial oxidation of dipicolinic acid. II. Identification of α-ketoglutaric acid and 3-hydroxypicolinic acid and some properties of cell-free extracts. J. Bacteriol. 84: 765–771

Kolenbrander PE and Weinberger M (1977) 2-Hydroxypyridine metabolism and pigment formation on three *Arthrobacter species*. J. Bacteriol. 132: 51–59

Kolenbrander PE, Lotong N and Ensign JC (1976) Growth and pigment production by *Arthrobacter pyridinolis* n. sp. Arch. Microbiol. 110: 239–245

Korosteleva LA, Kost AN, Vorob'eva LI, Modyanova LV, Terent'ev PB and Kulikov NS (1981) Microbiological degradation of pyridine and 3-methylpyridine. Appl. Biochem. Microbiol. 17: 276–283

Kretzer A and Andreesen JR (1991) A new pathway for isonicotinate degradation by *Mycobacterium sp*. INA1. J. Gen. Microbiol. 137: 1037–1080

Kucher RV, Turovsky AA, Dzumedzei NV and Shevchenko AG (1980) Microbial transformation of quinoline by *Pseudomonas putida* bacteria. Mikrobiol Zh (Kiev) 42: 284–287

Kung H-F and Stadtman TC (1971) Nicotinic acid metabolism. VI. Purification and properties of α-methyleneglutarate mutase (B_{12}-dependent) and methylitaconate isomerase. J. Biol. Chem. 246: 3378–3388

Kung H-F and Tsai L (1971) Nicotinic acid metabolism. VII. Mechanisms of action of clostridial α-methyleneglutarate mutase (B_{12}-dependent) and methylitaconate isomerase. J. Biol. Chem. 246: 6436–6443

Kung H-F, Cederbaum S, Tsai L and Stadtman TC (1970) Nicotinic acid metabolism. V. A cobamide coenzyme-dependent conversion of α-methyleneglutaric acid to dimethylmaleic acid. Proc. Nat. Acad. Sci. U.S.A. 65: 978–984

Kuno S, Tashiro M, Taniuchi H, Horibata K and Hayaishi O (1961) Enzymatic degradation of kynurenic acid. Fed. Proc. 20: 3

Lübbe C, van Pée K-H, Salcher O and Lingens F (1983) The metabolism of tryptophan and 7-chlorotryptophan in *Pseudomonas pyrrocinia* and *Pseudomonas aureofaciens*. Hoppe Seyler's Z Physiol. Chem. 364: 447–453

Nagel M and Andreesen JR (1989) Molybdenum-dependent degradation of nicotinic acid by *Bacillus sp*. DSM 2923. FEMS Microbiol. Lett. 59: 147–152

Nagel M and Andreesen JR (1990) Purification and characterization of the molybdoenzymes nicotinate dehydrogenase and 6-hydroxynicotinate dehydrogenase from *Bacillus niacini*. Arch. Microbiol. 154: 605–613

Nagel M and Andreesen JR (1991) *Bacillus niacini* sp. nov., a nicotinate-metabolizing mesophile isolated from soil. Int. J. Syst. Bacteriol. 41: 134–139

Narumiya S, Katsuji T, Tokuyama T, Noda Y, Ushiro H and Hayaishi O (1979) A new pathway of tryptophan initiated by tryptophan side chain oxidase. J. Biol. Chem. 254: 7007–7015

Nelson MJK and Snell EE (1986) Enzymes of vitamin B_6 degradation. J. Biol. Chem. 261: 15115–15117

Nyns EJ, Zach D and Snell EE (1969) The bacterial oxidation of vitamin B_6. VIII. Enzymatic breakdown of α-(N-acetylaminomethylene)succinic acid. J. Biol. Chem. 244: 2601–2605

Orpin CG, Knight M and Evans WC (1972) The bacterial oxidation of picolinamide, a photolytic product of Diquat. Biochem. J. 127: 819–831

Pastan I, Tsai L and Stadtman ER (1964) Nicotinic acid metabolism. I. Distribution of isotope in fermentation products of labeled nicotinic acid. J. Biol. Chem. 239: 902–906

Pereira WE, Rostad CE, Leiker TJ, Updegraff DM and Bennett JL (1988) Microbial hydroxylation of quinoline in contaminated groundwater: evidence for incorporation of the oxygen atom of water. Appl. Environ. Microbiol. 54: 827–829

Peschke B and Lingens F (1991) Microbial metabolism of quinoline and related compounds. XII. Isolation and characterization of the quinoline oxidoreductase from *Rhodococcus* sp. B1 compared with the quinoline oxidoreductase from *Pseudomonas putida* 86. Biol. Chem. Hoppe-Seyler 372: 1081–1088

Richardson SH and Rittenberg SC (1961a) The bacterial oxidation of nicotine. IV. The isolation and identification of 2,6-dihydroxy-N-methylmyosmine. J. Biol. Chem. 236: 959–963

Richardson SH and Rittenberg SC (1961b) The bacterial oxidation of nicotine. V. Identification of 2,6-dihydroxypseudooxynicotine as the third oxidative product. J. Biol. Chem. 236: 964–967

Roberts J and Rosenfeld HJ (1977) Isolation, crystallization, and properties of indolyl-3-alkane α-hydroxylase. A novel tryptophan-metabolizing enzyme. J. Biol. Chem. 252: 2640–2647

Rodwell VW, Volcani BE, Ikawa M and Snell EE (1958) Bacterial oxidation of vitamin B_6. I. Isopyridoxal and 5-pyridoxic acid. J. Biol. Chem. 233: 1548–1554

Röger P and Lingens F (1989) Degradation of quinoline-4-carboxylic acid by *Microbacterium* sp. FEMS Microbiol. Lett. 57: 279–282

Röger P, Erben A and Lingens F (1990) Microbial metabolism of quinoline and related compounds. IV. Degradation of isoquinoline by *Alcaligenes faecalis* Pa and *Pseudomonas diminuta* 7. Biol. Chem. Hoppe-Seyler 371: 511–513

Ronen Z and Bollag J-M (1991) Pyridine metabolism by a denitrifying bacterium. Can. J. Microbiol .37: 725–729

Rüger A, Schwarz G and Lingens F (1993) Microbial metabolism of quinoline and related compounds. XIX. Degradation of 4-methylquinoline and quinoline by Pseudomonas putida Kl. Biol. Chem. Hoppe-Seyler 374: 479–488

Schach S, Schwarz G, Fetzner S and Lingens F (1993) Microbial metabolism of quinoline and related compounds. XVII. Degradation of 3-methylquinoline by Comamonas testosteroni 63. Biol. Chem. Hoppe-Seyler 374: 175–181

Schmidt M, Röger P and Lingens F (1991) Microbial metabolism of quinoline and related compounds. XI. Degradation of quinoline-4-carboxylic acid by *Microbacterium* sp. H2, *Agrobacterium* sp. 1B and *Pimelobacter simplex* 4B and 5B. Biol. Chem. Hoppe-Seyler 372: 1015–1020

Schwarz G, Senghas E, Erben A, Schäfer B, Lingens F and Höke H (1988) Microbial metabolism of quinoline and related compounds. I. Isolation and characterization of quinoline-degrading bacteria. System Appl. Microbiol. 10: 185–190

Schwarz G, Bauder R, Speer M, Rommel TO and Lingens F (1989) Microbial metabolism of quinoline and related compounds. II. Degradation of quinoline by *Pseudomonas fluorescens* 3, *Pseudomonas putida* 86 and *Rhodococcus* sp. B1. Biol. Chem. Hoppe-Seyler 370: 1183–1189

Seyfried B and Schink B (1990) Fermentative degradation of dipicolinic acid (pyridine-2,6-dicarboxylic acid) by a defined coculture of strictly anaerobic bacteria. Biodegradation 1: 1–7

Shukla OP (1974) Microbial decomposition of α-picoline. Ind. J. Biochem. Biophys. 11: 192–200

Shukla OP (1986) Microbial transformation of quinoline by a *Pseudomonas* sp. Appl. Environ. Microbiol. 51: 1332–1342

Shukla OP (1989) Microbiological degradation of quinoline by *Pseudomonas stutzeri*: the coumarin pathway of quinoline catabolism. Microbios 59: 47–63

Shukla OP and Kaul SM (1973) Microbial Transformation of α-picolinate by *Bacillus sp*. Ind. J. Biochem. Biophys. 10: 176–178

Shukla OP and Kaul SM (1974) A constitutive pyridine degrading system in *Corynebacterium* sp. Ind. J. Biochem. Biophys. 11: 201–207

Shukla OP and Kaul SM (1975) Succinate semialdehyde, an intermediate in the degradation of pyridine by *Brevibacterium* sp. Ind. J. Biochem. Biophys. 12: 326–330

Shukla OP and Kaul SM (1986) Microbiological transformation of pyridine N-oxide and pyridine by *Nocardia* sp. Can. J. Microbiol. 32: 330–341

Shukla OP, Kaul SM and Khanna M (1977) Microbial transformation of pyridine derivatives: α-picolinate metabolism by a Gram-negative coccus. Ind. J. Biochem. Biophys. 14: 292–295

Siegmund I, Koenig K and Andreesen JR (1990) Molybdenum involvement in aerobic degradation of picolinic acid by *Arthrobacter picolinophilus*. FEMS Microbiol. Lett. 67: 281–284

Sims GK and O'Loughlin EJ (1989) Degradation of pyridines in the environment. CRC Crit. Rev. Environ. Control. 19: 309–340

Sims GK, Sommers LE and Konopka A (1986) Degradation of pyridine by *Micrococcus luteus* isolated from soil. Appl. Environ. Microbiol. 51: 963–968

Singh RP and Shukla OP (1986) Isolation, characterization, and metabolic activities of *Bacillus brevis* degrading isonicotinic acid. J. Ferment. Technol. 64: 109–117

Skryabin GK and Golovleva LA (1971) Nicotinic acid. Chem. Abstr. 75: 98452v

Sparrow LG, Ho PPK, Sundaram TK, Zach D, Nyns EJ and Snell EE (1969) The bacterial oxidation of vitamin B₆. VII. Purification, properties, and mechanism of action of an oxygenase which cleaves the 3-hydroxypyridine ring. J. Biol. Chem. 244: 2590–2600

Stafford DA and Callely AG (1970) Properties of a pyridine-degrading organism. J. Gen. Microbiol. 63: XIVP

Stadtman ER, Stadtman TC, Pastan I and Smith LD (1972) *Clostridium barkeri* sp.n. J. Bacteriol. 110: 758–760

Stanier RY, Hayaishi and Tsuchida M (1951) The bacterial oxidation of tryptophan I. A general survey of the pathways. J. Bacteriol. 62: 355–366

Sundaram TK and Snell EE (1969) The bacterial oxidation of vitamin B₆. V. The enzymatic formation of pyridoxal from pyridoxine. J. Biol. Chem. 244: 2577–2584

Tabuchi T (1955) Microbial degradation of nicotine and nicotinic acid. Part 2: Degradation of nicotine. J. Agric. Chem. Soc. Japan 29: 219–225

Taniuchi H and Hayaishi O (1963) Studies on the metabolism of kynurenic acid III. Enzymatic formation of 7,8-dihydroxykynurenic acid from kynurenic acid. J. Biol. Chem. 238: 283–293

Tashiro M, Tsukada K, Kobayashi S and Hayaishi O (1961) A new pathway of D-tryptophan metabolism: enzymic formation of kynurenic acid via D-kynurenine. Biochem. Biophys. Res. Com. 6: 155–160

Tate RL and Ensign JC (1974a) A new species of *Arthrobacter* which degrades picolinic acid. Can. J. Microbiol. 20: 691–694

Tate RL and Ensign JC (1974b) Picolinic acid hydroxylase of *Arthrobacter picolinophilus*. Can. J. Microbiol. 20: 695–702

Tibbles PE, Müller R and Lingens F (1989a) Degradation of 5-chloro-2-hydroxynicotinic acid by *Mycobacterium* sp. BA. Biol. Chem. Hoppe-Seyler 370: 601–606

Tibbles PE, Müller R and Lingens F (1989b) Microbial metabolism of quinoline and related compounds. III. Degradation of 3-chloroquinoline-8-carboxylic acid by *Pseudomonas spec.* EK III. Biol. Chem. Hoppe-Seyler 370: 1191–1196

Tsai L, Pastan I and Stadtman ER (1966) Nicotinic acid metabolism. II. The isolation and characterization of intermediates in the fermentation of nicotinic acid. J. Biol. Chem. 241: 1807–1813

Vogels GD and van der Drift C (1976) Degradation of purines and pyrimidines by microorganisms. Bacteriol. Rev. 40: 403–468

Wada E (1957) Microbial degradation of the tobacco alkaloids, and some related compounds. Arch. Biochem. Biophys. 72: 145–162

Wada H and Snell EE (1962) Enzymatic transamination of pyridoxamine II. Crystalline pyridoxamine-pyruvate transaminase. J. Biol. Chem. 237: 133–137

Watson GK and Cain RB (1975) Microbial metabolism of the pyridine ring. Biochem. J. 146: 157–172

Watson GK, Houghton C and Cain RB (1974a) Microbial metabolism of the pyridine ring. Biochem. J. 140: 265–276

Watson GK, Houghton C and Cain RB (1974b) Microbial metabolism of the pyridine ring. Biochem. J. 140: 277–292

15. Biodegradation of inorganic nitrogen compounds

J.A. COLE

School of Biochemistry, University of Birmingham, Birmingham B15 2TT, U.K.

I. Introduction

Three molecular species, ammonia, nitrate and N_2, dominate any simplified diagram of the Biological Nitrogen Cycle (Fig. 1). As seen in the preceding chapter in this volume, the nitrogen atoms of most organic nitrogen compounds are ultimately released as ammonia by the processes summarized briefly in the first section of this review. Not only is ammonia extremely toxic to mammals but it is also rapidly oxidized in any well-aerated environment by nitrifying bacteria which convert it first to nitrite and ultimately to nitrate. In poorly aerated soils and sediments, nitrate is then reduced by denitrifying bacteria to N_2, intermediates in this process being nitrite, nitric oxide and nitrous oxide. The processes of putrefaction, nitrification and denitrification have provided the basis of by far the largest (if not the most glamorous) biotechnologies, the safe disposal of sewage and the supply of potable water.

The urea cycle, which catalyses the removal of toxic ammonia from mammalian tissue, was the first biochemical cycle to be discovered (Krebs and Henseleit 1932). The biochemistry of the urea cycle and most other aspects of the degradation of organic nitrogen compounds are well established and succinctly summarized in basic text-books (see, for example, Stryer 1990; Bender 1978). The preceding chapter describes the degradation of heterocyclic nitrogen compounds, including purines and pyrimidines, so only a brief summary of how ammonia is released from other major organic nitrogen compounds will be presented here. This review will therefore focus on the loss of nitrogen from the biosphere by the sequential action of the nitrifying and denitrifying bacteria. It will highlight recent developments which should provide the basis for new biotechnological processes or explain why some traditional processes sometimes fail. General principles will be suggested by consideration of specific examples, so the reader must be aware of the danger – indeed, often the fallacy – of extrapolating such information.

C. Ratledge (ed.), Biochemistry of Microbial Degradation, 487–512.

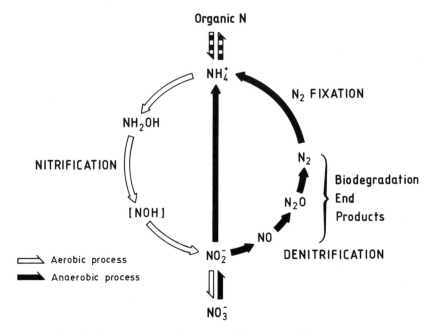

Fig. 1. Biodegradation end products of the Biological Nitrogen Cycle.

II. Degradation of organic nitrogen compounds by animals

A. Ammonia as the ultimate degradation product of animals

Members of the Animal Kingdom are loosely described as ammoniotelic, uricotelic (purinotelic) or ureotelic. The classification is based on whether the major excreted product of the degradation of organic nitrogen compounds is ammonia (as in fish; also tadpoles – but not frogs!), uric acid (birds and terrestrial invertebrates) or urea (man and other mammals). Animals which consume a protein-rich diet must release the nitrogen, usually as ammonia, before the carbon skeleton can be used for catabolism, energy conservation and biosynthesis. As ammonia is toxic to most animals, its concentration must be maintained below a critical limit by ensuring that it is safely removed from body fluids. This is achieved in the liver by a condensation reaction with CO_2 resulting in the ATP-dependent synthesis of carbamoyl phosphate (Fig. 2). In the urea cycle, carbamoyl phosphate and aspartate are converted by a cyclic series of reactions to fumarate and urea with ornithine, citrulline, argininosuccinate and arginine as the regenerated intermediates. Fumarate is readily oxidized to oxaloacetate by tricarboxylic acid cycle enzymes and aspartate is the transamination product of oxaloacetate, for which glutamate is an excellent amino-group donor. As α-ketoglutarate is the preferred amino-group acceptor for most of the other amino acids and glutamate dehydrogenase catalyses the oxidative deamination of glutamate to α-ketoglutarate and ammonia, both of the N atoms of urea can be derived from glutamate. There are, however, other pathways in mammals for the generation of NH_4^+. For example, in the liver and kidneys, ammonia is released by glutaminase from glutamine to generate glutamate. More examples will follow.

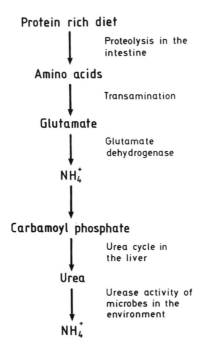

Fig. 2. Catabolism of a protein-rich diet in mammals.

Three instructive points follow from the above scheme. First, although mammals release urea and not ammonia into the environment, microorganisms are a prolific source of urease which rapidly releases the nitrogen atoms of urea as ammonia. Secondly, the abundance of pyridoxal phosphate-dependent transaminases in mammalian tissues enables the nitrogen atoms of other organic nitrogen compounds to be transferred easily to glutamate. Finally, as most animal species synthesize an active glutamate dehydrogenase, glutamate dehydrogenase-catalysed deamination provides the major route in mammals for the degradation of organic nitrogen compounds (Fig. 2).

Despite the importance of the urea cycle, other pathways for the release of nitrogen into the environment are significant both in man and in other animals. The initial step in pyridoxal-phosphate dependent reactions of amino acids is the formation of an aldimine Schiff base at the catalytic site of the enzyme. In transamination reactions (Fig. 3), the next step is a tautomerization of the aldimine to the ketimine form from which the ketoacid is released. This tautomerization in turn results from the withdrawal of electrons from the α-C-H bond which is perpendicular to the plane of the pyridoxylidine conjugated system. In the Schiff base formed with amino acids such as serine and threonine, it is the carboxyl group that is perpendicular to the conjugated system. Consequently, the amino acid is decarboxylated to the corresponding primary amine (Fig. 3). Finally, when the amino acid side chain is in the perpendicular position, as in the threonine aldolase and serine hydroxymethytransferase reactions, it is the side chain that is removed to give glycine and acetaldehyde from threonine, and glycine plus tetrahydrofolate from serine. This then raises the question of how ammonia is released from glycine or primary amines.

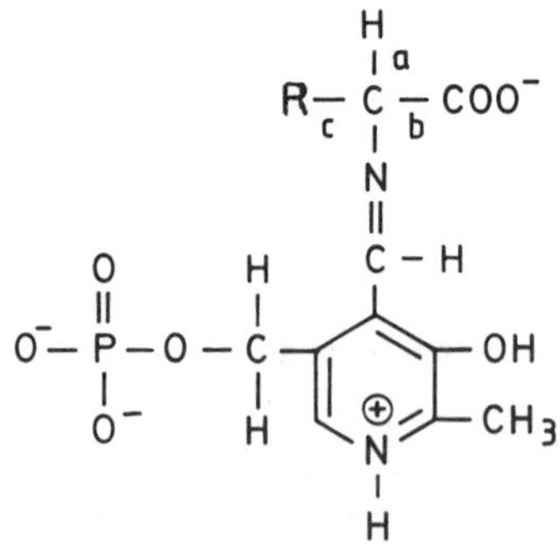

Fig. 3. Structure of pyridoxal phosphate. The type of reaction catalysed is determined mainly by the group which is perpendicular to the plane of the conjugated system in the enzyme active site: in (a), this is the C-H bond, resulting in tautomerization and transamination; (b) it is the carboxyl group, resulting in decarboxylation; and in (c) it is the amino acid side chain, as in the threonine aldolase and serine hydroxymethyltransferase reactions.

The glycine cleavage system provides a second major route for the elimination of nitrogen atoms from organic nitrogen compounds in mammals. In this pyridoxal phosphate-dependent pathway, two molecules of glycine are converted to one molecule each of serine, ammonia and carbon dioxide with the formation of one molecule of methylene tetrahydrofolate from tetrahydrofolate and the reduction of one molecule of NAD^+. Glycine is then regenerated from serine by the action of serine hydroxymethyltransferase.

The amino acid oxidases are another group of extremely active enzymes in both animals and microorganisms. Glyoxylate is an excellent substrate for amino-transfer, resulting yet again in the formation of glycine which can be oxidatively deaminated to ammonia and glyoxylate. This sequence of reactions thus leads to the indirect oxidative deamination of other amino acids. While the potential importance of L-amino acid oxidases is obvious, less clear is the function of the equally active and equally ubiquitous D-amino acid oxidases:

they occur in mammals presumably as a defence against toxic D-amino acids from bacterial contamination or infection. Finally, primary amines are readily deaminated by a group of primary amine oxidases which show variable substrate specificity.

B. Degradation of more problematic organic nitrogen compounds

Mammals seem to have evolved a variety of mechanisms to degrade – and hence release ammonia from – almost any naturally-occurring organic nitrogen compound, but other nitrogen compounds present a greater challenge. Two specific recent examples will illustrate some of the points of principle. Note that in neither case has ammonia been clearly identified as the end product of the biodegradation process.

Although many bacteria synthesize or scavenge polyamines, their physiological roles are far from clear. It is known that they associate with acidic cellular components such as DNA, RNA and ribosomal proteins; it has even been suggested that they can function as osmo-protectants (Tabor and Tabor 1984; Munro et al. 1974). *Escherichia coli* synthesizes two different lysine decarboxylases for the synthesis of cadaverine from lysine. The more active of the two is formed only under specific growth conditions, when lysine is available in excess during anaerobic growth at acid pH. This enzyme catalyses the first step in the degradation of lysine, its decarboxylation to cadaverine – but the cadaverine formed is simply excreted into the environment. Contiguous with the structural gene, *cadA*, for this pH-regulated lysine decarboxylase is a gene, *cadB*, which encodes a hydrophobic membrane-spanning protein homologous with the *Pseudomonas aeruginosa* anti-porter uptake system for arginine and ornithine. It is therefore suggested that CadB might function as a lysine/cadaverine anti-porter with the primary physiological role of raising the pH of the medium during anaerobic growth when excess acid is formed by fermentation (Fig. 4; Meng and Bennett 1992). If so, this example illustrates several principles which should be remembered by environmental biotechnologists interested in achieving a successful biodegradation process. First, even relatively simple, naturally occurring nitrogen compounds might be only partially degraded by microorganisms to intermediates that are more difficult to metabolize than the original compound. Secondly, the synthesis of some microbial biodegradative enzymes occurs only under highly specialized environmental conditions to fulfil a physiological role that is not immediately apparent or relevant to biodegradation. Finally, a microbial consortium, rather than a single species, is often essential for the complete degradation of the starting compound.

The second problem area is the biodegradation of xenobiotic compounds, for example, the N-containing heterocyclic pesticides and herbicides such as linuron or metamitron (Fig. 5; Roberts et al. 1991; Parekh et al. 1992). Bacterial cultures able to degrade linuron were isolated by sequential enrichment culture in a minimal salts medium. The result was a stable mixed

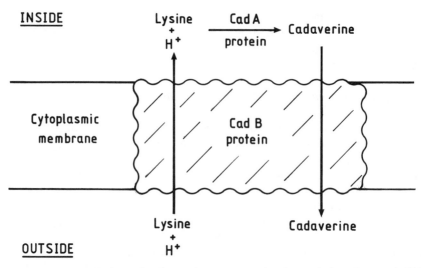

Fig. 4. The proposed lysine-cadaverine antiporter system for the regulation of external pH by *Escherichia coli*. Based on Meng and Bannett (1992).

culture from which pure microbial cultures could be isolated – none of which appeared able to metabolize the pesticide (Roberts et al. 1991). Attempts to isolate a metamitron-degrading organism were more successful, resulting in the purification of a *Rhodococcus* sp. which totally degraded the herbicide in less than 24 h in minimal salts medium supplemented with ammonia. In neither of these examples, however, were either the biochemical pathway or the final degradation products identified: further studies in this important research area are clearly required.

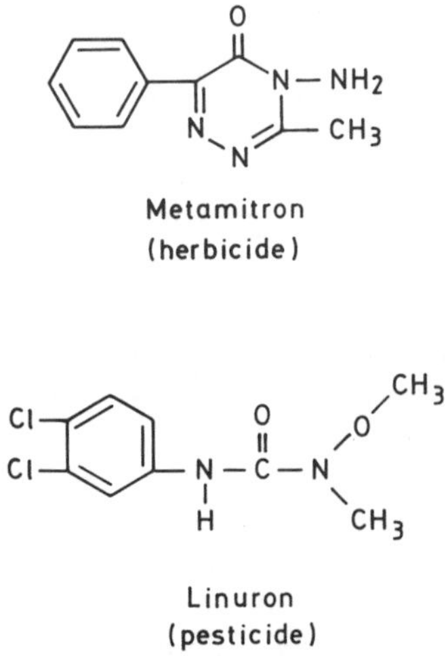

Metamitron
(herbicide)

Linuron
(pesticide)

Fig. 5. Structures of the herbicide, metamitron, and the pesticide, linuron.

III. Nitrification in soil and water

A. *Relative merits of ammonia and nitrate as agricultural N-sources*

The presence of ammonia in agricultural soils is a valuable resource which the farmer seeks to conserve; its absence necessitates the application of expensive nitrogenous fertiliser. In contrast, ammonia in fresh water represents a threat because it is a primary cause of eutrophication, it is toxic to animals and its smell is offensive. In any well-aerated soil or water, nitrifying bacteria rapidly oxidize ammonia to nitrate, a process which is beneficial in waste water treatment plants. Although nitrate is also a perfectly acceptable source of nitrogen for plant growth, it is less acceptable to the farmer than ammonia for at least two reasons. First, before the nitrogen can be assimilated into the pool of amino acids. plants must provide reducing equivalents, which otherwise would be used for CO_2 fixation, to reduce nitrate back to ammonia. Hence, biomass production under these conditions is less than when ammonia is used as the nitrogen source. Secondly, in contrast to ammonia which is retained in the topsoil by binding to anionic soil particles, nitrate is readily leached by rainwater into the unproductive anaerobic layers of soil where it is either lost from the biosphere by denitrification or transferred into the underground aquefers.

B. *The nitrifying bacteria*

Although specialized groups of chemolithotrophic bacteria are responsible for the vast majority of nitrification, heterotrophs and even animals are also known to oxidize reduced nitrogen compounds to nitrate. One of the many uncertainties about nitrification is whether the alternative processes contribute 0.1%, 1% or even 10% to the global oxidation of ammonia (Prosser 1986, 1989).

Nitrification is a two-stage process requiring the combined action of two groups of bacteria which oxidize ammonia to nitrite or nitrite to nitrate, respectively (Aleem 1977). *Nitrosomonas europeae* is by far the best studied of the ammonia-oxidizing bacteria, primarily because it can readily be isolated from most soils, marine and fresh water supplies. The most widespread genus in soils, however, is *Nitrosolobus* (Macdonald 1986); the least common is *Nitrosovibrio*. *Nitrosospira* also have been isolated from both acidic soils and fresh water (Walker and Wickramasinghe 1979; Hall 1986), but *Nitrosococcus* appears to be more important in marine nitrification (Wood 1986, 1988). Similarly, *Nitrobacter* has dominated studies of nitrite oxidation to nitrate, though three other genera are recognised to catalyse the same process. These are *Nitrospira*, *Nitrococcus* and *Nitrospina*; all three grow optimally at high salt concentrations and are found in marine environments, though one *Nitrospira* strain has been isolated from soil (Bock et al. 1986).

The chemolithotrophic nitrifying bacteria must overcome three major biochemical problems in order to survive. First, they are essentially obligate autotrophs which gain little benefit and probably no energy from the oxidation of organic compounds. The synthesis of all organic cell constituents from CO_2 inevitably drains the cell of reducing equivalents, particularly reduced pyridine nucleotides which, in other organisms with less demanding life-styles, would be available for energy conservation. Secondly, the mid-point redox potentials of both the NO_2^-/NH_4^+ and the NO_3^-/NO_2^- couples are too high for the reduction of NAD^+ to be thermodynamically favourable. The generation of reduced pyridine nucleotides required for CO_2 fixation and biosynthesis is therefore an energy-requiring process which places a severe demand on the proton motive force. Finally, far less useful energy is released during nitrite or ammonia oxidation by O_2 than during the oxidation of most other inorganic electron donors such as sulphide, sulphur or H_2.

In summary, the result of all three of these factors is that, although autotrophic growth on CO_2 plus NH_4^+ or NO_2^- requires MORE energy than heterotrophic growth, the only energy-conserving reactions available to nitrifying bacteria deliver LESS energy than almost any other lifestyle.

In the nitrifying bacteria, these problems are solved in three ways. First, extensive intracytoplasmic membranes are synthesized, apparently to optimize the rates of ammonia or nitrite oxidation. Secondly, because NADH is generated by reversed electron flow driven by the proton motive force, it cannot routinely be used as an electron donor for an electrogenic electron transfer chain. The third point is a direct consequence of the second one. In mammals and other heterotrophic organisms, the tricarboxylic acid fulfils two roles: first it generates NADH from the oxidation of reduced organic compounds which are the primary foodstuffs; secondly, it generates intermediates required for the biosynthesis of compounds such as glutamate and its derivatives. As NADH is not oxidized by the electron transfer chain of chemolithotrophs, the tricarboxylic acid cycle functions mainly as a biosynthetic pathway with glutamate as the end product. Little α-keto-glutarate dehydrogenase is synthesized, so the cycle is incomplete. A further consequence of these metabolic limitations is that growth is extremely slow and yields of biomass per mole of electron donor consumed are extremely low.

C. Oxidation of ammonia to nitrite

The ammonia-oxidizing bacteria face additional problems to those described above because the first reaction in the pathway, the conversion of ammonia to hydroxylamine, is catalysed by a mixed function oxidase which therefore consumes electrons (Fig. 6):

$$NH_3 + O_2 + 2e^- + 2H^+ \rightarrow NH_2OH + H_2O$$

Half of the electrons released during hydroxylamine oxidation to nitrite must therefore be used to regenerate the reduced form of cytochrome P460, the co-reductant for ammonia oxidation. Only the remaining two electrons are available for transfer through a conventional, electrogenic electron transfer chain which includes *c*-type cytochromes and a copper-containing cytochrome oxidase (Jones 1982). The substrate for the mono-oxygenase appears to be NH_3 rather than NH_4^+, so ammonia oxidation proceeds rapidly only at neutral or slightly alkaline pH which favours the formation of the unionized ammonia species. The alternative possibility of synthesizing an energy-consuming transport system for ammonia is a luxury these bacteria cannot afford.

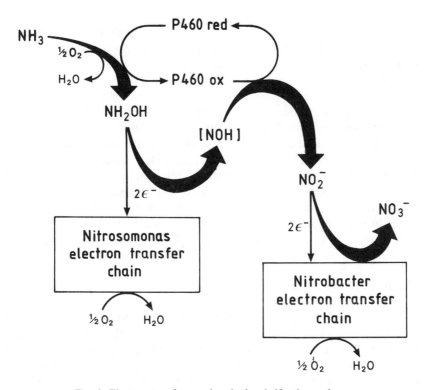

Fig. 6. Electron transfer reactions in the nitrification pathway.

D. *Enzymology of ammonia oxidation to nitrite*

The first stage of ammonia oxidation is catalysed by ammonia mono-oxygenase which is a membrane-bound enzyme with a 28 kDa subunit which also binds suicide substrates (substrates that irreversibly inactivate the enzyme) such as acetylene. The reaction catalysed by ammonia mono-oxygenase is superficially similar to that catalysed by the oligomeric iron-sulphur protein, methane mono-oxygenase, from methanotrophs which oxidizes methane to methanol. Both enzymes are located in the cytoplasmic membrane; both oxidize methane and ammonia due to their broad substrate specificity and are sensitive to the same inhibitors. Other substrates for ammonia mono-oxygenase include propene, benzene (Drozd 1980), cyclohexane, phenol and methanol (Wood 1988). Although methanotrophs can oxidize ammonia at a slower rate than nitrifying bacteria, they are unable to grow with ammonia as the sole energy source.

Two lines of evidence indicate that ammonia mono-oxygenase is a copper-containing enzyme: chelating agents with a high affinity for cuprous ions are potent inhibitors of ammonia oxidation; nitrification is also sensitive to copper deprivation.

The co-reductant for the oxidation of ammonia is the b-type cytochrome, P460, so-called because of the Soret-like absorption maximum at 460 nm in the reduced, Fe^{2+} form of the protein (Fig. 6). Results of Raman resonance spectroscopy suggest that P460 contains a novel haem group with lower symmetry than protoporphyrin IX or chlorins but possibly similar to chlorophylls and isobacteriochlorins (Andersson et al. 1991).

The best characterized enzyme of the nitrification pathway is hydroxylamine oxidoreductase which catalyses the two-stage oxidation of hydroxylamine to nitrite with an enzyme-bound intermediate (formally written as the highly unstable NOH) in which the nitrogen atom is at the oxidation level +1. A c-type cytochrome transfers the 2 electrons released from hydroxylamine to conventional cytochrome oxidases such as cytochromes aa_3 or o and hence to O_2 with the concomitant generation of a proton motive force: this is the only reaction in which *Nitrosomonas* can conserve useful energy. The two electrons released from the oxidation of NOH to nitrite are used to regenerate the reduced form of P460 so that further ammonia molecules can be oxidized (Fig. 6).

E. Nitrite oxidation to nitrate by Nitrobacter

There is no thermodynamic barrier for the direct oxidation of nitrite to nitrate (E_O' +430 mV) by O_2 (E_O' for the O_2/H_2O couple = +820 mV), so the energy released in this reaction could potentially be conseved as a proton motive force. This, indeed, occurs but the favorite model to explain how this is achieved is unusually complex in order to accommodate the following observations:

(i) ^{18}O labelling studies clearly established that the third oxygen atom added to nitrite to form nitrate is derived not from O_2 but from H_2O;

(ii) compounds which dissipate the membrane potential inhibit rather than accelerate nitrite oxidation (as expected by analogy with the effect of uncouplers on mitochondrial electron transfer);

(iii) chemicals which collapse the membrane potential promote the oxidation of a c-type cytochrome and the reduction of the heam a associated with nitrite oxidase;

(iv) the electron transfer chain appears to be unusually complex, including two high-potential cytochromes of the a_1 type,(E_O = +140 and +350 mV), a molybdenum centre (E_O = +340 mV), cytochrome oxidase aa_3 (E_O = +240 and +400 mV) as well as several non-haem iron-sulphur proteins with E_O values in the range + 60 to + 320 mV.

Figure 7 shows the ingenious (but essentially unproven) scheme proposed by Cobley (1976) to accommodate these observations. It envisages the energy-dependent transfer of a hydride ion abstracted from H_2O across the membrane to cytochrome c where the charge is separated and a proton is released into the periplasm. The two electrons then flow through cytochrome oxidase to molecular oxygen which is reduced to H_2O in the cytoplasm. The proton motive force is generated by this final step in the process (Jones 1982).

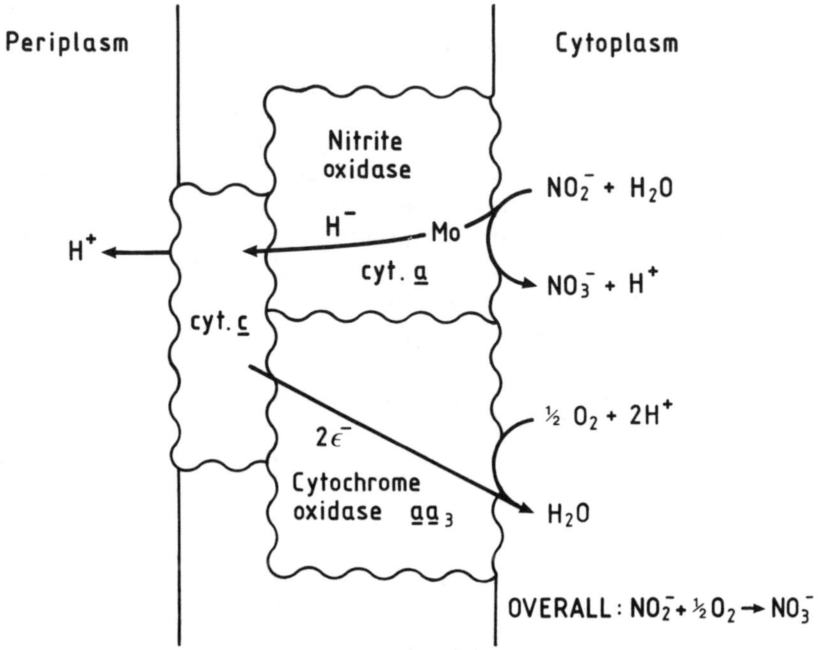

Fig. 7. Cartoon of the scheme postulated by Colby (1976) for the generation of a proton motive force during nitrite oxidation by *Nitrobacter*.

F. Heterotrophic nitrification

Although the most intensively studied nitrifying bacteria are obligate autotrophs, various heterotrophs also oxidize organic nitrogen compounds to nitrite or nitrate (reviewed by Focht and Verstraete 1977; Killham 1986; and summarized by Prosser 1989). A typical example is *Thiosphaera pantotropha*, an organism with many metabolic features indistinguishable from *Paracoccus denitrificans* (indeed, are they identical?). Enzymes responsible for heterotrophic nitrification have not been adequately characterized, so there remains doubt whether ammonia is first released and then oxidized by the same mechanisms as those used by the conventional nitrifying bacteria;

alternatively, it has been suggested that organic derivatives such as R. NHOH, R.NO or $R.NO_2$ might be generated as intermediates from which the corresponding inorganic nitrogen compounds, NH_2OH, (NOH) and NO_2^-, are released. Far more work is required before definitive statements can be made.

An interesting interaction between two different species was reported by Rho (1986) who found that a mixture of both an *Arthrobacter* and a *Corynebacterium*, isolated from an esturine sediment, rapidly oxidized acetaldoxime to nitrate. Neither bacterium alone catalysed this reaction at a significant rate.

Far more rapid progress has been made in characterizing a remarkable "nitrification" process in mammals in which nitric oxide is formed from arginine. Nitric oxide has been implicated in a number of diseases and disorders ranging from septic shock and diabetes to Alzheimer's and Huntington's diseases (Lancaster 1992; Moncada et al. 1991). In healthy individuals, NO is vital for the control of blood pressure, as an intermediary messenger in glutamate-dependent neurotransmission, in the immune surveillance and destruction of cancer cells and, perhaps most importantly, of penile erection (Bredt and Snyder 1992; Ignarro 1991; Lancaster 1992). The discovery of these extremely important roles for NO in mammals came from the observation that animals can excrete more nitrate than they ingest. Subsequently, it was realized that macrophages convert arginine into nitric oxide; the reaction is catalysed by the enzyme, nitric oxide synthase, the gene for which has been cloned (Xie et al. 1992). As the structure of nitric oxide is a free radical, it is very reactive and binds readily to many metalloproteins which, in the presence of O_2, oxidize it to nitrite and nitrate.

IV. Denitrification

A. The bacterial denitrification pathway

There are essentially three, physiologically distinct types of microbial nitrate reduction: denitrification, nitrate assimilation and the dissimilatory reduction of nitrate to ammonia. Cole (1989, 1990) has summarized the various subdivisions of each of these processes according to the types of enzyme involved, their genetic regulation, whether energy is conserved as a proton motive force and the end products. As nitrate assimilation clearly competes with nitrate "degradation" (or removal from the biosphere) and nitrate dissimilation to ammonia essentially reverses nitrification (Cole 1990; Page et al. 1990), these processes will not be considered further.

The formal definition of denitrification is any process which results in the removal of nitrogen atoms from the pool of biologically active nitrogen compounds that can be used for growth. According to this broad definition, three types of bacteria (and also some lower eukaryotes) qualify as denitrifiers because they convert nitrite to N_2O. By far the most significant group includes the most active and best known denitrifying bacteria such as the pseudomonads, *Paracoccus denitrificans*, *Hyphomicrobium* and *Alcaligenes* as well as members of up to 100 other genera (Knowles 1982). These bacteria use nitrate, nitrite, NO and N₂O as alternative electron acceptors to oxygen during anaerobic growth (some minor exceptions will be mentioned subsequently) and conserve some of the free energy released as a proton motive force which can drive ATP synthesis and other forms of useful work. One of the two less conventional groups of microbes that match the broadest definition of denitrifiers are those such as the enterics in which nitrite is reduced slowly to N₂O, probably in a gratuitous side-reaction catalysed by nitrate reductase (Smith 1982, 1983; reviewed by Cole 1988). More recently it has been recognized that enteric bacteria can also release nitric oxide from nitrite under appropriate conditions, so the limits of nitrate and nitrite reduction by these bacteria are still being reassessed (Ji and Hollocher 1988a,b). The second unusual organism is the *Rhizobium* strain HC NTI described by Casella et al. (1986, 1988) which, although it reduces nitrite to N₂O, it apparently lacks the ability to conserve useful energy from the process. Although both of the exceptional processes might have local environmental significance, neither is believed to contribute significantly to the degradation of inorganic nitrogen compounds on a global scale.

The reduction of nitrate to N₂ by denitrifying bacteria proceeds in four stages, the chemical intermediates being nitrite, nitric oxide and nitrous oxide (Fig. 8). Although all four enzymes implicated in the pathway have been characterized and the genes encoding them have been cloned and sequenced, it was only recently that Braun and Zumft (1991) provided overwhelming genetic evidence that NO is an essential, free intermediate in the pathway. A full account of the basis of the long-running controversy is beyond the scope of this chapter, so only a summary will be presented. Braun and Zumft (1991) summarized the main alternative hypotheses which were that nitrite is reduced directly to N₂O by a single enzyme; that nitrite reductase catalyses a two-stage reaction in which the unstable intermediate can release or be formed from free NO; or that NO released from the transient intermediate in nitrite reduction is reduced by a second enzyme, NO reductase, which is not an obligatory intermediate in denitrification. The most significant points in the controversy were:

(i) some nitrite reductases, for example those from *Achromobacter cycloclastes* or *P. aureofaciens*, appeared to produce either NO or N₂O, depending on the artificial electron donor used (Zumft et al. 1987; Iwasaki and Matsubara 1971);

(ii) isotopic data revealed that no ^{15}NO accumulated in experiments with *Pseudomonas aeruginosa* nitrite reductase in which ^{15}NO₂⁻ was mixed with a large excess of ^{14}NO; this was interpreted to mean that NO was not formed as a free imntermediate during nitrite reduction (Garber and Hollocher 1981; Kim and Hollocher 1983);

(iii) unlike N₂O, NO is rarely detected as a major denitrification end product (Payne 1973).

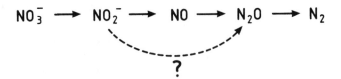

Fig. 8. Chemical intermediates of the denitrification pathway.

Carr et al. (1989) provided plausible explanations for the above observations which were compatible with NO as a free intermediate: one essential point in their argument is that high concentrations of NO inhibit both nitrite and nitrous oxide reduction, so the design of the earlier experiments was flawed. They also showed that NO was the sole product of nitrite reduction by a highly purified sample of *Paracoccus denitrificans* nitrite reductase (Carr et al. 1989). Heiss et al. (1989) and Goretski and Hollocher (1988, 1990) have also provided overwhelming biochemical evidence for the contrary view: NO is a free intermediate and NO reductase is an essential enzyme for denitrification. The final unanswered question is therefore whether there are any as yet uncharacterized nitrite reductases in other bacteria that reduce nitrite directly to N_2O. This would seem to be unlikely. There is, though, a real possiblilty that the nitrite and nitric oxide reductases of some organisms might be physically associated as a bi-functional complex which channels NO formed from nitrite reduction directly to the active site of NO reductase. Such metabolic channelling would clearly require the two reductases to be located close to each other, unlike the situation in *Paracoccus denitrificans* in which nitrite reductase, like nitrous oxide reductase, is located in the periplasm, but NO reductase is embedded in the cytoplasmic membrane (Carr et al. 1989).

By isolating mutants that were devoid of NO reductase activity, unable to reduce NO_2^- to N_2O and accumulating toxic concentrations of NO, Braun and Zumft (1991) provided the much-needed genetic evidence that NO reductase is an essential enzyme in the denitrification pathway. The dashed arrow in Fig. 8 is retained because some doubt remains whether there is a very low rate of N_2O production when the nitrite reductase, cytochrome cd_1, reduces nitrite to NO, and also to accommodate the gratuitous reduction of nitrite to N_2O by nitrate reductase (see above).

Even amongst the conventional denitrifying bacteria there are wide variations in – and different reasons for – the end products of nitrate and nitrite reduction:

(i)   in some but not all bacteria, nitrate inhibits, represses or competes for electrons with nitrite, nitric oxide and nitrous oxide reductases, so nitrite accumulates in some but not all cultures. Similarly, NO inhibits nitrite and nitrous oxide reductases of *Paracoccus denitrificans* resulting in a transient accumulation of NO or N_2O under some conditions (Carr et al. 1989).

(ii) Some bacteria lack genes for one or more of the denitrification enzymes (Payne 1973). For example, some bacteria are unable to reduce N_2O which is therefore the end product of denitrification.

(iii) The four reductases differ in pH optima, sensitivity to oxygen inactivation or repression and to various other inhibitors, so environmental factors play a critical role in determining whether denitrification is partial or complete.

(iv) Genes for some or all of the denitrification enzymes can occur on unstable plasmids: the products of nitrate reduction therefore depend on the proportion of plasmid-deficient bacteria in the population (Romermann and Friedrich 1985).

(v) Even under optimal growth conditions, some bacteria produce far more of one enzyme than another, resulting in the accumulation of the intermediate which is the substrate for the least active enzyme.

B. Characterization of enzymes involved in denitrification: nitrate reductase

There are surprisingly few recent reports of the purification and characterization of nitrate reductase from denitrifying bacteria. Some of the very early reports should be treated with caution because it is not clear whether the enzyme studied was primarily an assimilatory or a dissimilatory nitrate reductase, or whether enzyme sub-units had been lost during the purification procedures.

The enzyme from *Paracoccus denitrificans* is apparently similar to the respiratory nitrate reductase of *Escherichia coli* (Craske and Ferguson 1986). Both enzymes contain three types of subunit in an overall structure of $\alpha_2\beta_2\gamma_4$. Particularly significant was the demonstration that the *P. denitrificans* enzyme contains the third, cytochrome b subunit and that two-subunit derivatives lacking the b-type cytochrome can readily be isolated unless appropriate precautions are taken (Craske and Ferguson 1986). Presumably the two-subunit enzymes isolated from *Pseudomonas aeruginosa* and *Bacillus licheniformis* fall into this category, but confirmation that this is so is still awaited (Carlson et al. 1982; Van't Riet et al. 1979). All nitrate reductases so far isolated are molybdo-proteins with a molybdo-pterin cofactor attached to the largest, α subunit. The other two subunits, which bind the enzyme to the cytoplasmic membrane, are involved in charge separation and the transfer of electrons from the cytochrome b-quinone region of the electron transfer chain to another b-type cytochrome attached to the γ-subunit. It is highly unlikely that the nitrate reductases from all of the other denitrifying bacteria will have identical structures.

Chlorate is a gratuitous substrate for enzymes which contain the molybdenum cofactor. As nitrate reductase is the most abundant molybdo-protein in some bacteria during anaerobic growth, mutants defective in nitrate reductase genes can be isolated amongst the pool of chlorate-resistant mutants that are unable to synthesize the molybdenum cofactor. This provides a convenient approach to the cloning of genes for nitrate reduction and is successful as long as the strain concerned does not synthesize other very active molybdoproteins such as trimethylamine-N-oxide reductase or formate dehydrogenase (Glaser and DeMoss 1972).

C. Nitrite reductases

The nitrite reductases of denitrifying bacteria are essentially of two types, copper proteins or cytochrome cd_1. There are, however, significant differences between the various copper-containing nitrite reductases. Furthermore, they should not be confused with the copper protein, azurin, which has been used as an electron donor for studies of the cytochrome cd_1 nitrite reductase from *P. aeruginosa*.

The nitrite reductase from *Alcaligenes spheroides* is a dimer of 37 kDa subunits which contains only type I copper (Masuko et al. 1984; Meyer and Cusanovich 1989). The *P. aureofaciens* enzyme is a slightly larger dimer of 43 kDa subunits: it contains two type I copper atoms (Zumft et al. 1987). Two further types of copper-containing nitrite reductase both accept electrons from a 12 kDa blue copper protein, cupredoxin (Liu et al. 1986). These are the 69 kDa *Achromobacter cycloclastes* enzyme which contains three copper centres of both types I and II; and nitrite reductases from *Nitrosomonas europaea*, *Rhodopseudomonas palustris* and *Alcaligenes faecalis* which are all 120 kDa proteins (but apparently with different sized sub-units and hence different quaternary structures) and four copper atoms of both types I and II.

Cytochrome cd_1 nitrite reductases are widely distributed amongst different genera. It was first reported in *P. aeruginosa* but was mistakenly identified as a cytochrome oxidase because of its ability to reduce O_2 (Horio 1958). It is a dimer of two identical 60 kDa subunits, both of which contain haem c and d_1 The DNA sequence encoding the enzyme includes an N-terminal amino acid leader sequence, consistent with its periplasmic location (Silvestrini et al. 1989). Its synthesis is repressed by O_2 and nitrate, but induced by nitrite (Unden et al. 1980). The haem d_1 prosthetic group is an unusual dioxo-isobacteriochlorin with a reduced porphyrin ring of the chlorin type (Chang and Wu 1986; Chang et al. 1986). Both types of haem group are reduced by ascorbate, consistent with their positive mid-point redox potentials. This is in marked contrast to the very negative mid-point redox potentials of the hexa-haem nitrite reductases involved in the electrogenic reduction of nitrite to ammonia (Fujita and Sato 1967; Steenkamp and Peck 1981). The natural electron donor to cytochrome cd_1 appears to be another c-type cytochrome, cytochrome c_{551} (Timkovich 1986). How electrons are transferred from physiological substrates such as NADH or succinate to nitrite reductase is not entirely clear. Stopped flow experiments indicated the reduction of cytochrome cd_1 by azurin to be triphasic, the slowest step being the reduction of haem d_1 by an internal electron transfer from haem c (Parr et al. 1977; Silvestrini et al. 1990).

D. Nitric and nitrous oxide reductases

Nitric oxide reductase has been purified from *Paracoccus denitrificans* and *Pseudomonas stutzeri* (Carr et al. 1989; Heiss et al. 1989; Carr and Ferguson 1988, 1990; Dermastasia et al. 1991). In both cases the enzyme consists of two haemoproteins, a 17.5 kDa cytochrome *c* and a 38 kDa cytochrome *b*, both of which are firmly attached to the cytoplasmic membrane.

Two types of nitrous oxide reductase have been purified. Conventional denitrifiers such as *Pseudomonas stutzeri*, *Rhodobacter capsulata*, *Paracoccus denitrificans* and *Achromobacter cycloclastes* synthesize a 77 kDa N_2O reductase with 4 Cu atoms but no haem *c*. In contrast, an 88 kDa nitrous oxide reductase with 4 Cu atoms and haem *c* has been purified from *Wolinella succinogenes* which, although unable to reduce nitrate or nitrite to N_2O, can grow anaerobically and conserve energy during N_2O reduction to N_2. The 88 kDa enzyme from *Wolinella* is far more stable than the 77 kDa N_2O reductases and alone can use dithionite as an electron donor for N_2O reduction (Zhang et al. 1991).

E. Nitrification and denitrification in waste water treatment plants: "Catch 22"

If nitrogen is to be removed from waste water, for example, sewage contaminated with ammonia from piggeries or poultry farms, it seems obvious that the process should begin with an aerobic nitrification step to ensure that all of the nitrogen is available for disposal as N_2 by the denitrifying bacteria. Alas, life is never so simple!

A low but finite concentration of nitrate nitrogen is legally acceptable in water discharged from treatment plants. As traditional nitrification processes are aerobic, reduced carbon compounds abundant in domestic sewage are also oxidized to CO_2, thus reducing the biological oxygen demand of the final product. The catch is that the same reduced carbon compounds could supply the five electrons required for any subsequent denitrification stage for each nitrate ion reduced to N_2. Many water authorities, especially in Europe, therefore use an anaerobic denitrification step as the first, or even the only, process for water purification. This simultaneously removes much of the organic carbon and all of the nitrate without incurring the expense of adding substrates such as methanol for denitrification or the risk of adding an excess of reducing agent which would increase the BOD of the water still more. If the water is unacceptably rich in ammonia, this and any remaining organic carbon is removed in a subsequent nitrification stage, resulting in potable water with little organic carbon, but retaining the low but measurable concentration of nitrate generated from ammonia. If the nitrate concentration is too high, the water can be recycled until the problem is solved.

Clearly, the design of domestic waste water treatment plants is dictated more by economics and the standards set by regulatory authorities for nitrate levels in water than by biochemical logic. Full-scale denitrification plants have been in operation for many years in countries such as France and the USA, but the following two examples of more recent pilot plant experiments illustrate how these principles apply in practice.

In the production of pectin for the food industry, fruit peels are extracted with nitric acid which is subsequently neutralized with ammonia.The wastewater from Copenhagen Pectin Ltd. consequently contains an average of 600 mg of total nitrogen /l (Dalentoft 1991). Pilot plant studies designed to meet increasingly stringent regulatory requirements demonstrated that this could be achieved without the addition of an extra carbon source using a two-stage process with a denitrification zone before the nitrification zone in each stage (Fig. 9). With appropriate ratios of recycling to release at each stage, it was possible to achieve nitrate- and ammonia-N concentrations below 5 and 2 mg/l, respectively. The pilot process was surprisingly resistant to sudden influxes of high concentrations of carbon compounds or ammonia and allowed the company to release water with less than the statutory 8 mg total N /l (Dalentoft 1991).

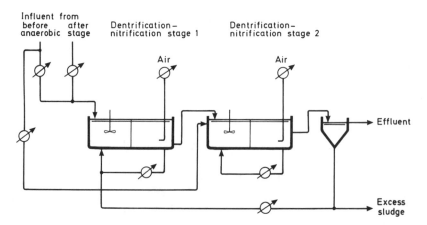

Fig. 9. Design of a two-stage pilot plant for the removal of high concentrations of nitrogen from waste water generated during the extraction of pectin from fruit peel. Each of the two stages incorporates nitrification and denitrification compartments. Based on Delentoft et al. (1991).

In the Blankaart region of Belgium, intense farming activity results in very high concentrations of nitrate in the reservoirs during winter, but not during summer. In contrast to the limit of 25 mg/l recommended by the EC Drinking Water Directive 80/778, throughout the 1980s, special dispensation had been granted on a yearly basis for the 50 mg/l statutory limit to be exceeded. When legislation was passed for the introduction of more stringent limits, Germonpre et al. (1991) investigated both the economics and scientific feasibility of adding a methanol-driven denitrification stage to the water purification process. Their scientific data are summarized in Table 1.

Table 1. Effect of different methanol concentrations on the efficiency of nitrate removal from a water supply. The data are from Germonpre et al. 1991.

Limiting substrate	Input conc. (mg l^{-1})		Conc. (mg.l^{-1}) in effluent		
	Nitrate	Methanol	Nitrate	Nitrate	Methanol
Methanol	66.0	29.3	30.0	4.9	0.0
Nitrate	65.5	53.0	1.0	0.0	2.5
Neither	66.5	50.1	2.0	1.8	0.1

Both the stability of the process and the need to achieve the virtually complete removal of nitrate dictated that the process should operate under conditions of excess methanol; consequently, a further stage to remove methanol, which is toxic to man, would be required together with on-line continuous monitoring to ensure that the methanol concentration in drinking water was never above 0.25 mg/l.

F. Future developments to avoid "Catch 22"

Broda (1977) suggested that it should be possible for bacteria with the appropriate enzyme complement to use ammonia as the electron donor for denitrification in the absence of oxygen, the overall reaction being thermodynamically favourable:

$$5 NH_4^+ + 3 NO_3^- \rightarrow 4 N_2 + 9 H_2O + 2 H^+; G^{0'} = -1483.5 \, kJ/ \, reaction$$

Although it is not yet certain that chemolithotrophic bacteria are involved, they have recently been used in the patented "Anamox" process (van der Graaf et al. 1990). Active biomass was shown to be essential for Anamox activity. Two lines of evidence which indicated that contaminating, low concentrations of O_2 were not required were that the rate of ammonia removal increased with increasing biomass (which is inconsistent with a slow leakage of oxygen into the bioreactor) and was totally dependent upon the addition of nitrate. When nitrate was added, ammonia disappeared and N_2 was formed. Isotopic labelling experiments established that $^{15}NH_4^+$ was converted to N_2. The significance of these observations is that, as O_2 is clearly not essential for nitrification, it should be possible to operate a totally anaerobic, waste-water treatment process capable of achieving nitrification and denitrification simultaneously. If the ratio of nitrate to ammonia is unfavourably low, the balance could presumably be restored by limited aeration to convert some of the ammonia to nitrate by aerobic nitrification.

G. Denitrification during aerobic nitrification

Some nitrifying bacteria also contain genes for denitrification which are regulated by the availability of O_2 (Ritchie and Nicholas 1972; Poth 1986).

Although denitrification is normally repressed by excess O_2, in some bacteria denitrification continues even when the dissolved O_2 concentration is sufficient to saturate cytochrome oxidase (Robertson and Kuenen 1984). As ammonia mono-oxygenase is also inhibited or inactivated by excess O_2, nitrification, like denitrification, is also more rapid under conditions of O_2 limitation. Consequently, some nitrifying bacteria can simultaneously denitrify either nitrate or nitrite to N_2O or N_2, but only when the supply of O_2 is carefully regulated. There are clearly considerable problems in operating a treatment plant which relies solely on this fine balancing act to achieve one-step removal of ammonia as N_2: too frequently the process is only partially successful, resulting in the production of the so-called "Greenhouse gases", NO and N_2O, which are also major contributing factors in the production of "acid rain". It has even been suggested – and vigorously refuted – that poorly-managed waste water treatment plants are currently major contributors to global warming.

V. Cyanide production and metabolism

A. Cyanide production and metabolism by plants and microorganisms

Unlike most of the other inorganic nitrogen compounds, cyanide is both produced and metabolized by eukaryotic as well as prokaryotic microorganisms (comprehensively reviewed by Knowles and Bunch 1986). The most prolific fungal sources of cyanide are probably the cyanogenic glycosides produced by the plants that host phytopathogenic Basidiomycetes, Ascomycetes and Zygomycetes (Colotelo and Ward 1961). The pathogenic fungi synthesize β-glucosidases which release cyanide from α-hydroxynitriles produced as intermediates in glycoside degradation. Other fungi generate cyanide from their own metabolic precursors (Ward and Thorn 1961).

Small quantities of cyanide are also generated by algae; for example, *Chlorella vulgaris* generates cyanide from histidine and other aromatic amino acids in the presence of Mn^{2+} ions and O_2 (Pistorius et al. 1977a). Subsequently it was realized that this is a general reaction catalysed by both D- and L-amino acid oxidoreductases in the presence of O_2 and either appropriate divalent metal ions or peroxidase (Pistorius et al. 1977b). A second pathway has also been found in *C. vulgaris* which involves a chemical reaction between hydroxylamine and glyoxylic acid to form glyoxylic acid oxime followed by the enzymic production of cyanide (Solomonson and Spehar 1979).

The best-studied cyanogenic bacteria are *Chromobacterium violaceum* and various Pseudomonads, the major difference between the two being that *C. violaceum* also degrades cyanide to β-cyanoalanine (Knowles and Bunch 1986). The source of cyanide is glycine, though optimal conditions vary considerably. Cyanogenesis occurs late during the growth cycle, typical of secondary metabolite production. It is critically affected by the dissolved O_2 concentration, primarily because the membrane-bound enzyme complex that generates HCN from glycine is O_2-labile and is therefore inactivated when the O_2 concentration increases as the culture enters the stationary phase of growth. Both types of bacteria synthesize cyanide synthase during growth with glutamate alone but the enzyme is far more active in media supplemented with both glycine and an activator such as methionine. Isotopic data establish that cyanide synthase catalyses the oxidative decarboxylation of glycine, transferring electrons to the electron transfer chain according to the reaction:

$$NH_2CH_2COOH \rightarrow HCN + CO_2 + 4[H]$$

Although O_2 is probably the natural oxidant, it can be replaced by fumarate or artificial electron acceptors such as dichlorophenol-indophenol or phenazonium methosulphate.

B. Cyanide degradation

Three enzymes able to degrade cyanide have been identified in *C. violaceum*. The most significant appears to be the soluble, cytoplasmic β-cyanoalanine synthase which catalyses the reaction:

$$CN^- + \text{cysteine (or O-acetyl serine)} \rightarrow \text{β-cyanoalanine} + H_2S$$
$$\text{(or acetate)}$$

The soluble, cytoplasmic enzyme has been purified to homogeneity (Macadam and Knowles 1984). It is a pyridoxal phosphate-dependent enzyme with two identical 37 kDa subunits. Although its properties are similar to those of cysteine synthases of other bacteria, it is specific for synthesis of cyanoalanine from cyanide (Knowles and Bunch 1986).

The two other enzymes from *C. violaceum* which can degrade cyanide are rhodanese (Rodgers and Knowles 1978):

$$S_2O_3^{2-} + CN^- \rightarrow SO_3^{2-} + SCN^-$$

and γ-cyano-α-aminobutyric acid synthase:

$$\text{Homocystine} + CN^- \rightarrow \text{homocysteine} + SCN^- + \text{γ-cyanoaminobutyrate}$$

Non-cyanogenic phytopathogenic fungi degrade cyanide by hydrolysis to formamide which can then be further metabolized to NH_4^+ and formate (which in turn is decarboxylated to CO_2). The enzyme, cyanide hydratase, catalyses the simple reaction:

$$HCN + H_2O \rightarrow HCO.NH_2$$

Cyanide is an important toxic waste, being produced in large quantities in the electoplating and related industries. The availability of microbes that can degrade it rapidly to non-toxic end products would therefore be of considerable significance in environmental biotechnology. Some of the many reports of such cultures were summarized by Knowles and Bunch (1986). An oxygenase activity that allowed some *P. fluorescens* isolates to grow aerobically with cyanide as the sole nitrogen source has been partially purified and characterized (Harris and Knowles 1983). Addition of 1 mM cyanide inhibited growth of cyanide-adapted cultures, but growth resumed as soon as the cyanide had been degraded to CO_2 and NH_4^+ (possibly via cyanate: Knowles and Bunch 1986).

A cautionary note about cyanide utilization was sounded recently by Hope and Knowles (1991) who, during a search for anaerobes capable of degrading cyanide, isolated a *Klebsiella planticola* that could use cyanide anaerobically as a sole nitrogen source. The conversion of HCN to NH_4^+ was shown convincingly to be due to an abiotic, chemical reaction between HCN and the reducing sugar added as a carbon source for microbial growth. This raises the question whether some of the earlier, less meticulous work on cyanide metabolism might have been based on chemical rather than enzyme catalysed reactions.

VI. Summary and conclusions

Clearly the title of this contribution 'Biodegradation of Inorganic Nitrogen Compounds' needed to be interpreted in context. All of the molecular species considered are intermediates in the biodegradation of organic nitrogen compounds released by plants, animals and microbes. The only molecule which cannot be "degraded" further and which is re-assimilated only by specialized prokaryotes is N_2. Consequently, N_2 is reasonably considered to be the end product of the biodegradation of inorganic nitrogen compounds. We have seen that N_2 is formed by denitrifying bacteria which reduce oxidized nitrogen substrates generated by the nitrifying bacteria and, to a lesser extent, by heterotrophic nitrification. Less abundant compounds and ions such as cyanide can enter the cycle by a variety of routes, but their ultimate fate is the same as that of amino acids, purines and pyrimidines which are metabolized by the routes outlined at the start of the chapter.

Finally, three dominant themes have emerged. First, the metabolism of each inorganic nitrogen compound depends first on its production by other members of the Biosphere – hence the concept of the Biological Nitrogen Cycle which is completed by the N_2 fixing bacteria and by plants and microorganisms that reduce nitrate directly to ammonia (see Cole 1988, 1989, 1990; Fig. 1). Secondly, each step in the process is predominantly the result of the redox reactions exploited by bacteria to generate their life-supporting membrane potential – their so-called energy yielding reactions. Finally, much that has been reviewed is based on the study of relatively few bacterial species, some of which are not dominant members of the relevant natural flora. Consequently, many of the statements made will in due time be shown to be over-simplifications based on too few data.

References

Aleem MIH (1977) Coupling of energy with electron transfer reactions in chemolithotrophic bacteria. In: BA Haddock and WA Hamilton (eds) Microbial Energetics, Symposium 27 of the Society for General Microbiology (pp 351–381). Cambridge University Press, Cambridge

Andersson KK, Babcock GJ and Hooper AB (1991) P460 of hydroxylamine oxidoreductase of *Nitrosomonas europaea*: Soret resonance Raman evidence for a novel heme-like structure. Biochem. Biophys. Res. Commun. 174: 358–363

Bender DA (1978) Amino Acid Metabolism. John Wiley and Sons, Chichester, U.K.

Bock E, Koops H-P and Harms H (1986) Cell Biology of nitrifying bacteria. In: JI Prosser (ed) Nitrification. Special Publications of the Society for General Microbiology, Vol 20 (pp 17–38). IRL Press, Oxford, U.K.

Braun C and Zumft WG (1991) Marker exchange of the structural genes for nitric oxide reductase blocks the denitrification pathway of *Pseudomonas stutzeri* at nitric oxide. J. Biol. Chem. 266: 22785–22788

Bredt DS and Snyder SH (1992) Nitric oxide, a novel neuronal messenger. Neuron 8: 3–11

Broda E (1977) Two kinds of lithotrophs missing in nature. Z. Allg. Microbiol. 17: 491–493

Carlson CA, Ferguson LP and Ingraham JL (1982) Properties of dissimilatory nitrate reductase purified from the denitrifier *Pseudomonas aeruginosa*. J. Bacteriol. 151: 162–171

Carr GJ and Ferguson SJ (1988) Nitric oxide reductase of *Paracoccus denitrificans*. Biochem. Soc. Trans. 16: 187–188

Carr GJ and Ferguson SJ (1990) The nitric oxide reductase of *Paracoccus denitrificans*. Biochem. J. 269: 423–429

Carr GJ, Page D and Ferguson SJ (1989) The energy-conserving nitric-oxide-reductase system in *Paracoccus denitrificans*. Eur. J. Biochem. 179: 683–692

Casella S, Shapleigh JP and Payne WJ (1986) Nitrite reduction in *Pseudomonas "hedysari"* strain HCNT1. Arch. Microbiol. 146: 233–238

Casella S, Shapleigh JP, Lupi F and Payne WJ (1988) Nitrite reduction in bacteroids of *Pseudomonas "hedysari"* strain HCNT1. Arch. Microbiol. 149: 384–388

Chang CK and Wu W (1986) The porphinedione structure of heme d_1. J. Biol. Chem. 261: 8593–8596

Chang CK, Timkovich R and Wu W (1986) Evidence that heme d_1 is a 1,3-porphyrindione. Biochemistry 25: 8447–8453

Cobley JG (1976) Reduction of cytochromes by nitrite in electron-transport particles from *Nitrobacter winogradski*. Biochem. J. 156: 493–498

Cole JA (1988) Assimilatory and dissimilatory reduction of nitrate to ammonia. In: JA Cole and SJ Ferguson (eds) The Nitrogen and Sulphur Cycles, Symposium 40 of the Society for General Microbiology (pp 281–330). Cambridge University Press

Cole JA (1989) Physiology, biochemistry and genetics of nitrite reduction by *Escherichia coli*. In: JR Wray and JR Kinghorn (eds) Molecular and Genetic Aspects of Nitrate Assimilation (pp 229–243). Oxford University Press, U.K.

Cole JA (1990) Physiology, biochemistry and genetics of nitrate dissimilation to ammonia. In: NP Revsbech and J Sorensen (eds) Denitrification in Soil and Sediment. Plenum Press, New York and London

Colotelo N and Ward EWB (1961) β-glycosidase activity and cyanogenesis in the susceptibility of Alfalfa to winter crown rot. Nature (London) 189: 242–243

Craske AL and Ferguson SJ (1986) The respiratory nitrate reductase from *Paracoccus denitrificans*. Eur. J. Biochem. 158: 429–436

Dalentoft E (1991) Biological treatment of a high strength nitrogen wastewater. In: H Verachtert and W Verstraete (eds) Environmental Biotechnology 1991 (pp 45–48). Koninklijke Vlaamse Ingenieursvereniging vzw, Antwerp

Dermastasia M, Turk T and Hollocher TC (1991) Nitric oxide reductase. J. Biol. Chem. 266: 10899–10905

Drozd JW (1980) Respiration in the ammonia oxidizing chemoautotrophic bacteria. In: CJ Knowles (ed) Diversity of Bacterial Respiratory Systems, Vol 2 (pp 87–111). CRC Press, Boca Raton, Florida

Focht DD and Verstraete W (1977) Biochemical ecology of nitrification and denitrification. Adv. in Microbiol. Ecol. 1: 135–214

Fujita T and Sato R (1967) Studies on soluble cytochromes in Enterobacteriaceae V. Nitrite-dependent gas evolution in cells containing cytochrome c_{552}. J. Biochem. 62: 230–240

Garber EAE and Hollocher TC (1981) ^{15}N tracer studies on the role of NO in denitrification. J Biol. Chem. 265: 5459–5465

Germonpre R, Liessens J and Verstraete W (1991) Fluidised bed denitrification of drinking water with methanol: pilot plant experience. In: H Verachtert and W Verstraete (eds) Environmental Biotechnology 1991 (pp 49–52). Koninklijke Vlaamse Ingenieursvereniging vzw, Antwerp

Glaser JH and DeMoss JA (1972) Comparison of nitrate reductase mutants selected by alternative procedures. Mol. gen. genet. 116: 1–10

Goretski J and Hollocher TC (1988) Trapping of nitric oxide produced during denitrification by extracellular hemoglobin. J. Biol. Chem. 263: 2316–2323

Goretski J and Hollocher TC (1990) The kinetic and isotopic competence of nitric oxide as an intermediate in denitrification. J. Biol. Chem. 265: 889–895

Hall GH (1986) Nitrification in lakes. In: JI Prosser (ed) Nitrification, Special Publications of the Society for General Microbiology, Vol 20 (pp 127–156). IRL Press, Oxford, U.K.

Harris R and Knowles CJ (1983) Isolation and growth of a *Pseudomonas* species that utilizes cyanide as a source of nitrogen. J. Gen. Microbiol. 129: 1005–1011

Heiss B, Frunzke K and Zumft WG (1989) Formation of the N-N bond from nitric oxide by a membrane-bound cytochrome *bc* complex of nitrate-respiring (denitrifying) *Pseudomonas stutzeri*. J. Bacteriol. 171: 3288–3297

Hope KM and Knowles CJ (1991) The anaerobic utilisation of cyanide in the presence of sugars by microbial cultures can involve an abiotic process. FEMS Microbiol. Lett. 80: 217–220

Horio T (1958) Terminal oxidation systems in bacteria. 1. Purification of cytochromes from *Pseudomonas aeruginosa*. J. Biochem. 45: 195–205

Ignarro LJ (1991) Signal transduction mechanisms involving nitric oxide. Biochem. Pharmacol. 41: 485–490

Iwasaki H and Matsubara T (1971) Cytochrome c-557 (551) and cytochrome *cd* of *Alcaligenes faecalis*. J. Biochem. 69: 847–857

Ji X-B and Hollocher TC (1988a) Mechanism of nitrosation of 2,3-aminonaphthalene by *Escherichia coli*: enzymatic production of NO followed by O_2-dependent chemical nitrosation. Appl. Environm. Microbiol. 54: 1791–1794

Ji X-B and Hollocher TC (1988b) Reduction of nitrite to nitric oxide by enteric bacteria. Biochem. Biophys. Res. Comm. 157: 106–108

Jones CW (1982) Bacterial respiration and photosynthesis. In: JA Cole and CJ Knowles (eds) Aspects of Microbiology 5. Thomas Nelson and Sons, Surrey, U.K.

Killham K (1986) Heterotrophic nitrification. In: JI Prosser (ed) Nitrification (pp 117–126). IRL Press, Oxford, U.K.

Kim C-H and Hollocher TC (1983) ^{15}N tracer studies on the reduction of nitrite by the purified dissimilatory nitrite reductase of *Pseudomonas aeruginosa*. J. Biol. Chem. 258: 4861–4863

Knowles CJ and Bunch AW (1986) Microbial cyanide metabolism. Adv. in Mic. Physiol. 27: 73–111

Knowles R (1982) Denitrification. Microbiol. Rev. 46: 43–70

Krebs HA and Henseleit K (1932) Untersuchungen uber die Harnstoffbildung im Tierkorper. Klin. Whehnschr. 11: 757–759

Lancaster JR (1992) Nitric oxide in cells. American Scientist 80: 248–259

Liu MY, Liu MC, Payne WJ and LeGall J (1986) Properties and electron transfer specificity of copper proteins from the denitrifier *"Achromobacter cycloclastes"*. J. Bacteriol. 166: 604–608

Macadam AM and Knowles CJ (1984) Purification and properties of β-cyano-L-alanine synthase from the cyanide producing bacterium, *Chromatium violaceum*. Biochim. Biophys. Acta 786: 123–132

Macdonald RMcL (1986) Nitrification in soil: An introductory history. In: JI Prosser (ed) Nitrification. Special Publications of the Society for General Microbiology, Vol 20 (pp 1–16). IRL Press, Oxford, U.K.

Masuko M, Iwasaki H, Sakurai S, Susuki S and Nakahara J (1984) Characterization of nitrite reductase from a denitrifier, *Alcaligenes* sp. NCIB 11015. A novel copper protein. J. Biochem. 96: 447–454

Meng S-Y and Bennett GN (1992) Nucleotide sequence of the *Escherichia coli cad* operon: a system for neutralization of low extracellular pH. J. Bacteriol. 174: 2659–2669

Meyer TE and Cusanovich MA (1989) Structure, function and distribution of soluble bacterial redox proteins. Biochim. Biophys. Acta 975: 1–28

Moncada SR, Palmer MJ and Higgs EA (1991) Nitric oxide physiology, pathophysiology, and pharmacology. Pharmacological Rev. 43: 109–142

Munro GF, Bell CA and Ledermen M (1974) Multiple transport components for putrescine in *Escherichia coli*. J. Bacteriol. 118: 952–963

Page L, Griffiths L and Cole JA (1990) Different physiological roles for two independent pathways for nitrite reduction to ammonia by enteric bacteria. Arch. Microbiol. 154: 349–354

Parekh NR, Welch SJ, Roberts SJ and Walker A (1992) Rapid degradation of the herbicide metamitron by an isolated soil bacterium. Abstracts of the 122nd Meeting of the Society for General Microbiology, 73

Parr SR, Barber D, Greenwood C and Brunori M (1977) The electron transfer reaction between azurin and the cytochrome c oxidase from *Pseudomonas aeruginosa*. Biochem. J. 167: 447–455

Payne WJ (1973) Reduction of nitrogenous oxides by microorganisms. Bacteriol. Rev. 37: 409–452

Pistorius EK, Gewitz HS, Voss H and Vennesland B (1977a) Cyanide formation from histidine in *Chlorella*. Biochim. Biophys. Acta 481: 384–394

Pistorius EK, Gewitz HS, Voss H and Vennesland B (1977b) Reversible inactivation of nitrate reductase in *Chlorella vulgaris in vivo*. Planta 128: 73–80

Poth M (1986) Dinitrogen production from nitrite by a *Nitrosomonas* isolate. Appl. Environ. Microbiol. 52: 957–959

Prosser JI (1986) Experimental and theoretical models of nitrification. In: JI Prosser (ed) Nitrification. Special Publications of the Society for General Microbiology, Vol 20 (pp 63–78). IRL Press, Oxford, U.K.

Prosser JI (1989) Autotrophic nitrification in bacteria. Adv. in Microbial Physiol. 30: 125–181

Rho J (1986) Microbial interactions in heterotrophic nitrification. Canad. J Microbiol. 32: 243–247

Ritchie GAF and Nicholas DJD (1972) Identification of the sources of nitrous oxide produced by oxidative and reductive processes in *Nitrosomonas europaea*. Biochem. J. 126: 1181–1191

Roberts SJ, Walker A, Waddington MJ and Welch SJ (1991) Isolation of a bacterial culture capable of degrading linuron. BCPC Monograph 47 Pesticides in Soils and Water, 51–58

Robertson LA and Kuenen JG (1984) Aerobic denitrification: a controversy revisited. Arch. Microbiol. 139: 351–354.

Rodgers PB and Knowles CJ (1978) Cyanide production and degradation during growth of *Chromobacterium violaceum*. J. Gen. Microbiol. 108: 261–267

Romermann D and Friedrich B (1985) Denitrification by *Alcaligenes eutrophus* is plasmid dependent. J. Bacteriol. 160: 852–854

Silvestrini MC, Galotti CL, Gervais M, Schinina E, Barra D, Bossa F and Brunori M (1989) Nitrite reductase from *Pseudomonas aeruginosa*: sequence of the gene and the protein. FEBS Lett. 254: 33–38

Silvestrini MC, Tardi MG, Musci G and Brunori M (1990) The reaction of *Pseudomonas* nitrite reductase and nitrite. J. Biol. Chem. 265: 11783–11787

Smith MS (1982) Dissimilatory reduction of NO_2^- to NH_4^+ and N_2O by a soil *Citrobacter* sp. Appl. Environm. Microbiol. 43: 854–860

Smith MS (1983) Nitrous oxide production by *Escherichia coli* is correlated with nitrate reductase activity. Appl. Environm. Microbiol. 45: 1545–1547

Solomonson LP and Spehar AM (1979) Stimulation of cyanide formatioin by ADP and its possible role in the regulation of nitrate reductase. J. Biol. Chem. 254: 2176–2180

Steenkamp DJ and Peck HD (1981) Proton translocation associated with nitrite respiration in *Desulfovibrio desulfuricans*. J. Biol. Chem. 256: 5450–5458

Stryer L (1990) Biochemistry. Freeman and Company. San Francisco

Tabor CW and Tabor H (1984) Polyamines. Ann. Rev. Biochem. 53: 749–790

Timkovich R (1986) ^{1}H NMR investigation of cytochrome cd_1: complexes with electron-donor proteins. Biochemistry 25: 1089–1093

Unden G, Hackenberg H and Kroger A (1980) Isolation and functional aspects of the fumarate reductase involved in the phosphorylative electron transport of *Vibrio succinogenes*. Biochim. Biophys. Acta 591: 275–288

van der Graaf A, Mulder A, Slijkhuis H, Robertson LA and Kuenen JG (1990) Anoxic ammonium oxidation. In: C Christiansen, L Munck and J Villadsen (eds) Proceedings of the 5th European Congress of Biotechnology (pp 388–391). Munksgaard International Publisher, Copenhagen

van't Riet J, Wientjes FB, van Doorn J and Planta RJ (1979) Purification and characterization of the respiratory nitrate reductase of *Bacillus licheniformis*. Biochim. Biophys. Acta 576: 347–360

Walker N and Wickramasinghe KN (1979) Nitrification and autotrophic nitrifying bacteria in acid teasoils. Soil Biology and Biochemistry 11: 231–236

Ward EWB and Thorn GD (1961) Evidence for the formation of HCN from glycine by a snow mould fungus. Can. J Botany 44: 95–104

Wood PM (1986) Nitrification as a bacterial energy source. In: JI Prosser (ed) Nitrification (pp 39–62). IRL Press, Oxford, U.K.

Wood PM (1988) Monooxygenase and free radical mechanisms for biological ammonia oxidation. In: JA Cole and SJ Ferguson (eds) The Nitrogen and Sulphur Cycles. Symposium 40 of the Society for General Microbiology (pp 219–243). Cambridge University Press, Cambridge

Xie Q, Cho HJ, Calaycay J, Mumford RA, Swiderek KM, Lee TD, Ding A, Troso T and Nathan C (1992) Cloning and characterization of inducible nitric oxide synthase from mouse macrophages. Science 256: 225–228

Zhang C, Hollocher TC, Kolodziej AF and Orme-Johnson WH (1991) Electron paramagnetic resonance observations on the cytochrome c-containing nitrous oxide reductase from *Wolinella succinogenes*. J. Biol. Chem. 266: 2199–2202

Zumft WG, Gotzman DJ and Kroneck PM (1987). Type 1, blue copper proteins constitute a respiratory nitrite reducing system in *Pseudomonas aureofaciens*. Eur. J. Biochem. 168: 301–307

16. Biochemistry of anaerobic biodegradation of aromatic compounds

GEORG FUCHS, MAGDY EL SAID MOHAMED,
UWE ALTENSCHMIDT, JÜRGEN KOCH, ACHIM LACK,
RUTH BRACKMANN, CHRISTA LOCHMEYER and
BRIGITTE OSWALD
Abteilung Angewandte Mikrobiologie, Universität Ulm, D-89069 Ulm, Germany

I. Introduction

Aromatic compounds comprise the second largest group of natural products (Fig. 1); in addition a variety of xenobiotics are man-made aromatic pollutants. They are metabolized by microorganisms following two fundamentally different strategies: Under aerobic conditions, aromatic compounds are transformed by monooxygenases and dioxygenases into a few central intermediates such as catechol, protocatechuate and gentisate. These dihydroxylated compounds are suitable for an oxidative cleavage of the aromatic ring (the pertinent aerobic literature will not be covered here but see Chapter 11 by Smith). Under anoxic conditions, aromatic compounds have to be transformed by other means than by oxygenases. It is now proven that numerous low molecular weight aromatic compounds are degraded by different groups of anaerobic bacteria (early reports by Tarvin and Buswell 1934; Dutton and Evans 1967) and that the aromatic ring structures are reductively attacked, as proposed by the late W.C. Evans (Dutton and Evans 1969; Balba and Evans 1977; Evans 1977; Williams and Evans 1975; Evans and Fuchs 1988). It has to be pointed out, however, that there is no evidence for a significant anaerobic degradation of polymeric high molecular weight aromatics such as lignins, which represent probably more than half of the aromatic compounds (Hackett et al. 1977; Zeikus 1980; Young and Frazer 1987).

In this contribution an outline of the biochemistry of anaerobic degradation of soluble, low molecular weight, aromatic compounds will be given. Emphasis is put on the novel reactions and enzymes involved in the oxygen-independent degradation. Unfortunately, it is premature to discuss mechanisms of the novel enzyme reactions since the experimental evidence is either poor or completely lacking. The limited number of studies performed with pure cultures and *in vitro* will preferentially be covered. Several comprehensive reviews, each with a different focus, are available (Evans 1977; Dutton and Evans 1978; Kaiser and Hanselmann 1982; Sleat and Robinson 1984; Young 1984; Gibson and Subramanian 1984; Berry et al 1987a; Evans and Fuchs 1988; Schink and

C. Ratledge (ed.), Biochemistry of Microbial Degradation, 513–553.
© 1994 *Kluwer Academic Publishers. Printed in the Netherlands.*

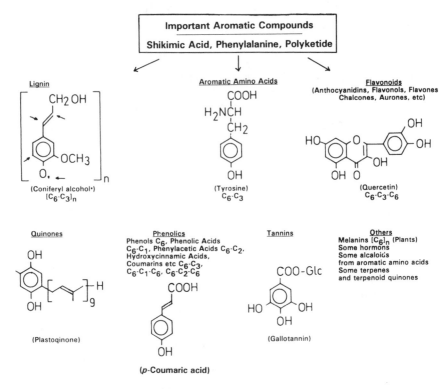

Fig. 1. Important natural aromatic compounds and specific examples of each type.

Tschech 1988; Hegeman 1988; Tschech 1989; Schink et al 1992). Other relevant contributions in this volume concern dehalogenation of haloaromatics and degradation of alicyclic and of heterocyclic compounds, of lignin, and of aromatic hydrocarbons.

This overview will not treat the following aspects: most degradations, dehalogenations and other transformations in non-defined cultures or enrichments; degradation of the aliphatic side chains except for a few mechanistically interesting examples; degradation of biaryl, polycyclic, and heterocyclic compounds; aspects of energy conservation coupled to the anaerobic degradation of aromatic compounds; transport of substrates and chemotaxis for aromatic substrates; evolutionary aspects of oxygen-independent degradation of aromatic and other inert compounds which normally require molecular oxygen for breakdown; ecological aspects and adaptation to changing environments; medical, toxicological and nutritional aspects of transformation and degradation of aromatics in the intestinal tract; microbial lignocellulose degradation e.g. in termites; biotransformations resulting in incomplete metabolism; applied aspects, e.g. bioremediation of contaminated ground or ground water by anaerobic or facultative anaerobes or by combined aerobic and anaerobic processes; metabolic regulation and molecular biology of anaerobic aromatic metabolism; taxonomic affiliation of organisms enabled to attack the aromatic nucleus anaerobically; distribution of the capability to degrade aromatics anaerobically in procaryotes and eucaryotes.

II. Principles

Anaerobic biodegradation of soluble aromatic compounds proceeds in three steps (Fig. 2):

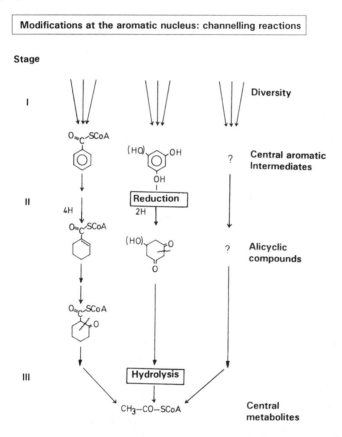

Fig. 2. Simplified scheme of anaerobic metabolism of aromatic compounds in three steps.

First, chemically inert compounds need to be activated; characteristic examples for anaerobic activation reactions are carboxylations, anaerobic hydroxylations, and CoA thioster formation of aromatic acids. Furthermore, the enormous diversity of natural and synthetic aromatic compounds has to be directed into a few central intermediates which are suitable for a reductive attack on the aromatic nucleus. Examples are reductive dehydroxylations or transhydroxylations.

Second, in keeping with this, the central aromatic intermediates are attacked enzymatically by reductases; these intermediates are benzoyl-CoA rather than benzoate, resorcinol (1,3-benzenediol), phloroglucinol (1,3,5-benzenetriol), and probably a few others. The resulting alicyclic compounds are easily converted by β-oxidation to 3-oxo-compounds, or the reduction directly provides the 3-oxo-compound. The 3-oxo-compounds have been shown or postulated to become hydrolytically cleaved. The reducing equivalents initially put into the aromatic system will subsequently be gained back.

Third, the resulting non-cyclic compounds have to be converted into central metabolites using rather conventional pathways. The products of C_6, C_6-C_1 and C_6-C_2 compounds are probably three acetyl-CoA and CO_2.

Since the reactive oxygen molecule is lacking, the critical first two steps are restricted to the use of the following cosubstrates and energy forms: CO_2, water, reduced or oxidized coenzymes, ATP, CoA thioesters of aromatic acids, and possibly others. So far, there is no example for the use of the electrochemical proton potential for driving endergonic redox reactions (reversed electron transport) in aromatic metabolism; however, some reduction reactions, in which aromatic compounds function as electron acceptors, may be coupled to ATP synthesis via a chemiosmotic mechanism (see sections IV.A and IV.K).

The metabolism of aromatic compounds and of other inert compounds such as alkanes (Aeckersberg et al. 1991) in the absence of molecular oxygen poses some intriguing mechanistic problems. The most evident question is how the aromatic nucleus can become reduced. This task requires drastic conditions in chemistry. The reduction of benzene with molecular hydrogen (two reducing equivalents) to cyclohexadiene is endergonic by approximately 60 kJ per mol at pH 7. The overall reaction becomes energetically favourable only if the diene is further reduced to cyclohexene (Harman and Taube 1988). The biochemical reduction of the aromatic nucleus therefore has to use one of the following mechanisms: (a) four-electron reduction of the aromatic ring; (b) two-electron reduction and hydration or a different energetically favourable reaction of the diene; (c) two sequential two-electron steps in which the diene intermediate is not a free intermediate, but is stabilized with the enzyme; (d) the aromatic nucleus is partially dearomatized before the reduction step such that the diene is resonance-stabilized by functional group(s).

III. Anaerobic microorganisms

Anaerobic microorganisms are confronted with energetic constraints unless an exogenous electron acceptor with a relatively positive midpoint potential is

```
Oxidation                                                                      free
                                                                               energy
                                                                               change

          C7H6O2 + 12 H2O          ————> 7 CO2 + 15 H2              + 332 kJ

          C7H6O2 + 6 H2O           ————> 3 CH3COOH + 3 H2 + CO2     +  47 kJ


Coupled to the reduction of:                                                   total
                                                                               free
                                                                               energy
Nitrate
          15 H2 + 15 NO3⁻          ————> 15 NO2⁻ + 15 H2O           -2116 kJ

Sulphate
          15 H2 + 3.75 SO4= + 7.5 H⁺ ————> 3.75 H2S + 15 H2O        - 327 kJ

CO2
          15 H2 + 3.75 CO2         ————> 3.75 CH4 + 7.5 H2O         - 159 kJ
```

Fig. 3. Energetics of anaerobic benzoate metabolism.

available (Fig. 3).

In most anaerobic marine environments, sufficient sulphate (27 mM) is available which becomes reduced to hydrogen sulphide; aromatic metabolism by pure cultures of sulphate reducing bacteria is well documented (review by Widdel 1988). In soil and fresh water, electron acceptors such as oxidized nitrogen or sulphur compounds or oxidized metal ions such as Fe(III) or Mn(IV) (Lovley 1991) at low concentrations can serve as oxidants in anaerobic respirations. Facultative microorganisms may be able to degrade many aromatic compounds both aerobically and anaerobically which requires the expression of two quite different sets of enzymes. The spectrum of aromatic substrates for anaerobic growth of a denitrifying *Pseudomonas* strain K172 is illustrated in Fig. 4. In all these cases aromatic compounds can be completely oxidized to CO_2 and serve as source for cell carbon.

Most anoxic terrestrial environments do not provide sufficient external electron acceptors; aromatics have to be fermented (e.g. Mountfort et al. 1984; Barik et al. 1985; Krumholz and Bryant 1986; for review see Schink and Tschech 1988) which sometimes requires the presence of strongly reducing co-substrates such as hydrogen or formate for the initial reduction of the ring. Many aromatic substrates cannot simply be fermented for energetic reasons (for thermodynamic data see Stull et al. 1969; Karapet'yants and Karapet'yants 1970; Thauer et al. 1977; Stouthamer 1988). Often, fermentation products have to be removed and disproportionated to methane and CO_2; this obligate syntrophy requires the cooperation of fermenting and methanogenic bacteria (review by Schink 1992). Hence the process cannot be studied in pure culture, and growth is slow, which makes this system rather difficult to study.

Finally, phototrophs obtain energy anaerobically in the light by photosynthetic electron transport phosphorylation; these bacteria may use aromatic compounds only as source of cell carbon provided that surplus

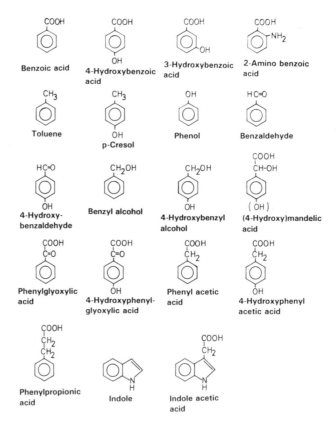

Fig. 4. Substrate spectrum for anaerobic growth under denitrifying conditions of *Pseudomonas* strain K172.

reducing equivalents can be disposed by either being used for the assimilation of CO_2 or by reduction of oxidized sulphur compounds.

In principle, aromatic compounds can be oxidized to CO_2 by pure cultures of bacteria with some sort of anaerobic respiration [nitrate, sulphate, ferric iron, Mn(IV) etc. respirations] or by phototrophic bacteria, whereas the methanogenic fermentation requires mixed cultures. Only a limited number of phenolic compounds may be fermented in pure culture. For thermodynamic reasons aromatic hydrocarbons cannot completely be metabolized under methanogenic conditions. However, one cannot predict the growth rate on different aromatic substrates on theoretical grounds. For instance, the generation time of the denitrifying *Pseudomonas* strain K172 with phenol is 14 h, with toluene 6 h, with benzoate 4 h (Lack and Fuchs 1992; Altenschmidt and Fuchs 1992b).

Although there are common biochemical principles of anaerobic aromatic metabolism, there must exist also many variations. First, mixed aerobic and anaerobic strategies as well as common aerobic and anaerobic intermediates may occur in facultative anaerobes. Second, one has to expect biochemical variations solely due to the fact that the organisms belong to distinct and sometimes distant phylogenetic groups. Third, the different electron acceptors, energy sources and amounts of free energy available may necessitate different biochemical solutions for one problem. Fourth, the strategies used may be quite different in a specialist which can metabolize only one aromatic compound, compared to an organism which can handle as much as, say, 50 different compounds, both aerobically or anaerobically. One has to keep this warning in mind when biochemical findings in one group of organisms are generalized.

Up to now, very limited insights into the biochemistry were obtained mostly from the study of nitrate reducing and phototrophic bacteria; they are less energy limited and therefore are easier to grow and to study. Since this review will focus on the new principles, enzymes and biochemical reactions, we will not discuss whether the findings can be generalized, except for those instances where differences are already evident.

IV. Channelling reactions

The conversion of different aromatic structures into a few central reactive aromatic intermediates (channelling reactions) has to serve two functions: to maximize the chemical reactivity of the aromatic nucleus whilst minimizing the number of enzymes required to deal with hundreds of similar compounds and to oxidize all of them to acetyl-CoA or CO_2. Central aromatic intermediates recognized so far are benzoyl-CoA, resorcinol, phloroglucinol and possibly others (Fig. 5 and Fig. 6). These two sides of the anaerobic strategy are equally important when one considers the individual reactions described below.

Aromatic ring structures are biosynthetically derived mostly from phenylalanine via the shikimate pathway, some via the polyketide or via the isoprene pathway. Rarely, complex aromatic compounds in the biomass are hydrolyzed into simpler aromatic monomers which can be transported ($M_r \leq$ 600). Smaller ones like the flavonoids may be still rather complex, containing several aromatic rings or heterocycles. At present, very little is known about the biochemistry of anaerobic degradation of complex aromatic compounds. The degradation of aliphatic side chains attached to aromatic rings and the cleavage of ester bonds offers no principal difficulties for anaerobes and therefore will be discussed only briefly, if at all. Many of the simpler aromatic substrates still contain several substituents, e.g. hydroxyl, O-methyl and carboxyl or carboxymethyl substituents, which are modified or removed in a certain order; we will not discuss this order of events, but concentrate on the individual modifying reactions at C_6, C_6-C_1, and C_6-C_2 compounds.

Under natural conditions, many aromatic compounds are transformed or co-metabolized to products which are not intermediates in the degradative pathway; however, they represent intermediates of complete degradation in

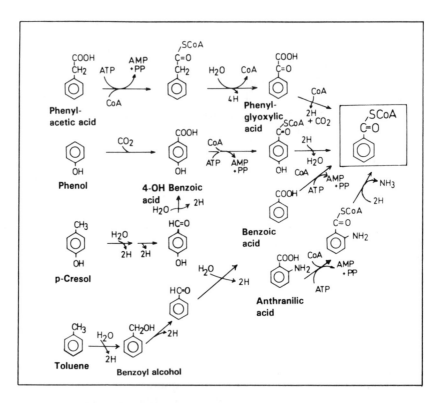

Fig. 5. Some channelling reactions leading to benzoyl-CoA.

the anaerobic food chain of the biotop. As an example, aromatic amino acids are converted into a bewildering variety of aromatic and alicyclic products (Barker 1981, further examples reviewed by Evans and Fuchs 1988). Compounds such as phenol, indole, toluene, aniline, or *p*-cresol are rather common natural intermediates in anaerobic habitats. This may explain why microorganisms can be isolated from non-contaminated soil which completely degrade, for example, phenol but are unable to degrade the aromatic amino acid tyrosine (Seyfried et al. 1991).

Quite a few of the channelling and aromatic ring reducing reactions have no precedent in the biochemical literature and are new, not only in terms of the individual reaction catalysed, but they also represent new general types of enzymes that are likely to use uncommon mechanisms. These reactions are summarized in Fig. 7.

A. Lyase reactions and β-oxidation of C_6-C_3 compounds

The cleavage of C-C bonds, by which aliphatic side chains are removed from the aromatic ring, are well known examples of textbook biochemistry:

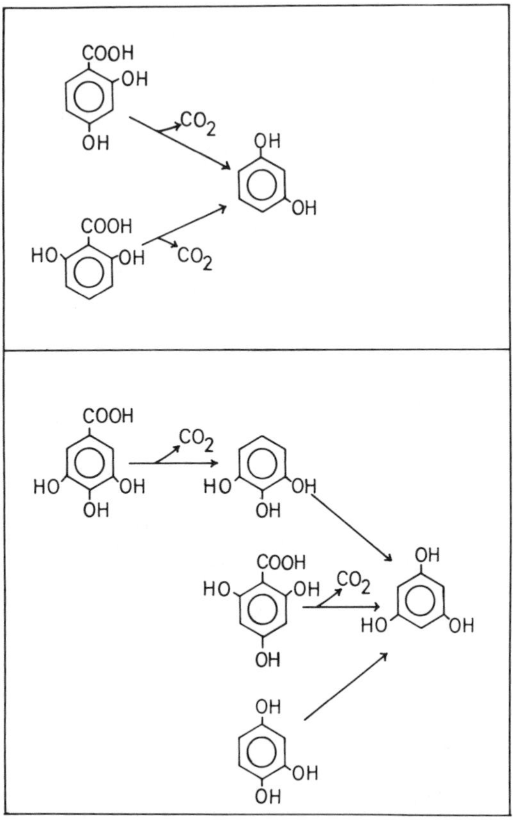

Fig. 6. Some channelling reactions leading to resorcinol and phloroglucinol.

biochemistry: for example tyrosine phenol lyase or tryptophane indole lyase (Elsden et al. 1976). Interestingly, some bacteria that can degrade the aromatic nucleus are unable to use tyrosine or tryptophan, but depend on phenol and indole, respectively, which are provided probably by fermenting bacteria (Barker 1981). The shortening of aliphatic side chains with three or more carbons probably proceeds via reactions of β-oxidation (Zenk et al. 1980; Elder et al. 1992). This requires oxidation of alcohol or aldehyde functions either directly to activated acids, or activation of the free acids via CoA ligases or possibly via CoA transferases. Under natural conditions, a variety of different transformation products are excreted such as phenylpropionate, phenylacetate, cresols, phenol, toluene, hydroquinone, catechol, and others. Also, transformation products, in which the aromatic ring is reduced to a cyclohexane ring, have been detected (Grbic-Galic 1986; for review see Evans and Fuchs 1988). However, if the aromatic ring is to be degraded, either benzoic acids, phenylacetic acids or the corresponding CoA thioesters are

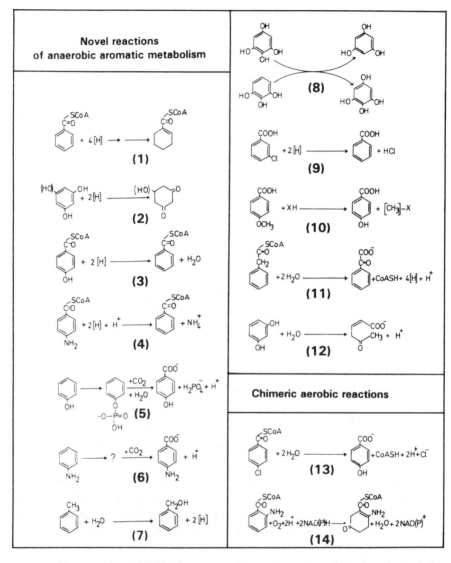

Fig. 7. Some new types of enzyme reactions involved in the anaerobic metabolism of aromatic compounds. (1) Benzoyl-CoA reduction, (2) resorcinol (phloroglucinol) reduction, (3) reductive dehydroxylation, (4) reductive deamination, (5) phenol carboxylation, (6) anilin carboxylation, (7) toluene methylhydroxylation, (8) transhydroxylation, (9) reductive dehalogenation, (10) *O*-demethylation, (11) anaerobic alpha-oxidation, (12) resorcinol hydration. New aerobic reactions: (13) hydrolytic dehalogenation, (14) monooxygenase/reductase.

being formed as intermediates. The reduction of unsaturated aliphatic side chains can be coupled to ATP synthesis (Tschech and Pfennig 1984; Hansen et al. 1988) via a chemiosmotic mechanism, similar to fumarate reduction.

B. Decarboxylation

The decarboxylation of aromatic acids is common in transformations that are not part of a degradation pathway. There are numerous reports in the literature which, however, do not provide any biochemical evidence that these decarboxylations play a role in the complete degradation of the respective aromatic acids. They play an important role in the food chain since they remove organic acids or provide CO_2 for acetogenesis (Hsu et al. 1990a,b). For instance, decarboxylation of phenylacetic acids would lead to cresols, xylene or toluene. These reactions appear not to be useful if the aromatic ring is to be degraded (D'Ari and Barker 1985; Grbic-Galic 1986; Ward et al. 1987; Knoll and Winter 1987). However, aromatic acids with two or more hydroxyl functions (or methylated hydroxyl functions after being O-demethylated), may become decarboxylated if the resulting product contains a 1,3-diol grouping. These compounds have little aromaticity, as one can judge from their UV spectra, and can easily be reduced both chemically and enzymatically. Examples are the decarboxylation of resorcylic acids (2,4-dihydroxybenzoic acid and 2,6-dihydroxybenzoic acid) to resorcinol (1,3-benzenediol) (Kluge et al. 1990); the fate of 3,5-dihydroxybenzoic acid is unknown. Similarly, gallic acid (3,4,5-trihydroxybenzoic acid) is decarboxylated to pyrogallol (1,2,3-benzenetriol) and phloroglucinic acid (2,4,6-trihydroxybenzoic acid) to phloroglucinol (1,3,5-benzenetriol) (see section V.B). These reactions are catalysed by specific, inducible and soluble enzymes which have not yet been purified. So far, no example of coupling the decarboxylation of aromatic acids, especially hydroxylatic aromatic acids such as 4-hydroxybenzoate, to the generation of a Na^+ or proton motive force has been reported; this possibility would require a membrane-associated enzyme, analogous to some sodium ion transport decarboxylases of dicarboxylic acids (Dimroth 1987; Beatrix et al. 1990). Membrane association of some of these decarboxylases has been reported but has not been proven nor has energy generation been shown. This question requires further investigations.

4-Hydroxybenzoate decarboxylation to phenol, gentisate decarboxylation to hydroquinone, or protocatechuate decarboxylation to catechol are further examples of decarboxylation reactions. Whereas 4-hydroxybenzoate appears not to be degraded via phenol by pure cultures (for enrichment cultures see Knoll and Winter 1987), the other two decarboxylation reactions may represent the first steps in the respective pathways. In general, the decarboxylation of aromatic acids with a hydroxyl function *para* to the carboxyl group is a chemically favoured reaction and has been observed many times in biological systems. Phthalic acids (benzenedicarboxylic acids) appear to be first converted to the CoA mono thioester by CoA ligases, followed by decarboxylation of *o*-, *m*-, or *p*-phthaloyl-CoA to benzoyl-CoA (Aftring and Taylor, 1981, Nozawa and Maruyama 1988a,b); direct decarboxylation may be mechanistically difficult. An initial reduction and subsequent oxidative decarboxylation of free phthalic acids has been proposed; this possibility though remains hypothetical (Taylor and Ribbons 1983).

C. O-Demethylation, aryl ether cleavage

Many naturally occurring aromatic compounds contain phenolic hydroxyl functions which are protected against undesired oxidation side reactions by methyl ether formation. These O-methyl ether linkages can be cleaved by an O-demethylation reaction (Fig. 7) which does not require molecular oxygen (Bache and Pfennig 1981; Taylor 1983; Frazer and Young 1985, 1986; Kreikenbohm and Pfennig 1985; Krumholz and Bryant 1985; Mountfort and Asher 1986; Daniel et al. 1988; Genthner and Bryant 1987; DeWeerd et al. 1988; Wu et al. 1988; Cocaign et al. 1991). In the known aerobic pathways, the methyl group is transformed in a monooxygenase reaction to a memiacetal [formal] from which the methyl carbon is released as formaldehyde. In the anaerobic degradation, no free methanol seems to be formed (Bache and Pfennig 1981), and the ether oxygen remains with the aromatic residue (DeWeerd et al. 1988).

The demethylation reaction was studied *in vitro* with acetogenic bacteria which form the methyl group of acetate from it (Bache and Pfennig 1981; R. Konle and E. Stupperich, personal communication). Cells grown on 3,4-dimethoxybenzoic acid (veratric acid) form 3,4-dihydroxybenzoic acid (protocatechuic acid) via 4-hydroxy-3-methoxybenzoic acid (vanillic acid) and contain a characteristic 23 kDa protein. It is thought that the methyl group is transferred to a corrinoid protein forming an enzyme bound methyl-corrinoid which acts as methyl donor in the synthesis of acetyl-CoA. It is unknown whether the 23 kDa protein is a corrinoid protein, whether a separate methyl transferase unit is required, and how specific the methyl transferase is with respect to the methoxylated aromatic substrate. Cobalt (I) in the corrin ring is a strong nucleophile which seems to be predestined to perform this chemically intriguing O-demethylation reaction. *In vitro* hydroxy-vitamin B_{12} functions as methyl-accepting cofactor in veratric acid transformation forming methyl-vitamin B_{12}.

Berman and Frazer (1992) studied *in vitro* the O-demethylation of phenylmethyl ethers with extracts of acetogenic bacteria. The reaction required tetrahydrofolate which is plausible to act as the methyl accepting coenzyme forming methyltetrahydrofolate. The reaction also depends on ATP. ATP may be required in non-stoichiometric amounts possibly for the activation of a corrinoid containing methyl transferase; ATP may be used in the endergonic reduction of Co(II) to the active Co(I) form in coenzyme B_{12}. The enzyme system appears to be soluble and is oxygen sensitive.

The cleavage of other aromatic ether linkages under anaerobic conditions has not been studied *in vitro*. However, an interesting reductive cleavage of a β-aryl ether has been described in the aerobic *Pseudomonas paucimobilis* (Masai et al. 1989, 1991).

D. α-Oxidation of phenylacetic acids and oxidation of mandelic acids to benzoylformate (phenylglyoxylate)

The oxidation of D- and L-mandelic acid and analogues to phenylglyoxylic acid or benzaldehyde and analogues has been studied in great detail in aerobes (for review see Fewson 1988). Anaerobically, mandelic acids can be oxidized via phenylglyoxylate to benzoyl-CoA (Dörner and Schink 1991), as shown for phenylglyoxylate (Dangel et al. 1991); the stereospecificity of the dehydrogenase(s) is unknown. The metabolism of phenylacetic acids represents an interesting case. On the one hand, even if the carboxyl group would be converted to a coenzyme A thioester, the methylene bridge would still prevent the activation of the aromatic ring for subsequent reduction. On the other hand, β-oxidation of the carboxymethyl side chain is not possible because of the aromatic ring. This dilemma led to the suggestion that phenylacetic acids are reduced to cyclohexylacetic acids or cyclohexanone (Balba and Evans 1979). Although such compounds have been reported in enrichment cultures, recent studies with pure cultures and *in vitro* studies point to a different solution. Phenylacetic acid is first converted to phenylacetyl-CoA by an inducible, soluble and specific CoA ligase which has been purified (Dangel et al. 1991; M. Mohamed and G. Fuchs unpublished). It is a monomer of 60 kDa which is rather labile. 4-Hydroxyphenylacetate seems to be activated by a separate enzyme. The thioester apparently activates the α-methylene carbon strongly enough to allow its dehydrogenation and hydroxylation with water as the oxygen source, i.e. the anaerobic α-oxidation (Fig. 7). The first product detected is phenylglyoxylate or 4-hydroxyphenylglyoxylate, respectively (Seyfried 1989; Seyfried et al. 1991; M. Mohamed and G. Fuchs unpublished); free mandelic acid may not be an intermediate in this process but was observed in a sulphate reducing bacterium (Sembiring and Winter 1989a). This intriguing α-oxidation, in which four electrons are withdrawn and oxygen is introduced from water, remains to be studied.

E. Oxidative decarboxylation

(4-Hydroxy)Phenylglyoxylate is oxidative decarboxylated to give (4-hydroxy)benzoyl-CoA (Dangel et al. 1991). The oxygen-sensitive enzyme (4-hydroxy)phenylglyoxylate dehydrogenase (CoA acylating) which requires CoA as co-substrate has not been purified. The natural electron acceptor may be ferredoxin, *in vitro* viologen dyes are active. Although the enzyme reaction resembles pyruvate:ferredoxin oxidoreductase, e.g. in catalysing a $^{14}CO_2$:pyruvate isotope exchange (Kerscher and Oesterhelt 1982), this 2-oxoacid oxidoreductase is specific for (4-hydroxy)phenylglyoxylate and is induced only when cells are grown on the corresponding C_6-C_2 acids.

F. Aromatic alcohol and aldehyde dehydrogenases

Aromatic alcohols and aldehydes are oxidized to the corresponding benzoic acids by pyridine nucleotide-dependent dehydrogenases, either NAD^+- or $NADP^+$-specific. These enzymes have been purified from aerobes (Chalmers and Fewson 1989; Chalmers et al 1990, 1991; Shaw and Harayama 1990), but not from anaerobes. They may be more or less specific for aromatic alcohols and aldehydes (Mountfort 1990; Lux et al. 1990) and are found also in many bacteria that cannot metabolize the aromatic acid further (e.g. Zellner et al., 1990). In all cases studied *in vitro*, benzaldehyde and analogues appear to be oxidized in a CoA-independent reaction to the free acids; in fermenting bacteria it may well be that the corresponding acyl-CoA compound is formed in a CoA-acylating aldehyde dehydrogenase reaction (Gross, 1972). Since none of the anaerobic enzymes has been purified, the substrate and co-substrate specificity remains unclear. In addition, pyridine nucleotide-independent alcohol dehydrogenases occur which *in vitro* couple with artificial one-electron carriers (Dangel et al. 1991).

G. Carboxylation

Carboxylation of the aromatic ring is the first step in the degradation of phenol (Fig. 7) (Tschech and Fuchs 1987, 1989; Dangel et al. 1991), of *o*-cresol and possibly of other phenolic compounds substituted in *ortho*-position (Bisaillon et al. 1991b; Rudolphi et al. 1991). Also *m*-cresol seems to be carboxylated in some organisms (Roberts et al. 1990; Ramanand and Suflita 1991) whereas in others it is not (Bisaillon et al. 1991b; Rudolphi et al. 1991). Phenol carboxylation is well known in chemistry as Kolbe-Schmitt reaction.

Phenol carboxylation has been studied *in vitro* in phenol-degrading, denitrifying bacteria (Tschech and Fuchs 1987, 1989; Lack et al. 1991; Lack and Fuchs 1992). It may occur also in sulphate reducing bacteria (Bak and Widdel 1986b). The soluble enzyme system referred to as 'phenol carboxylase' catalyses an isotope exchange between $^{14}CO_2$ and the carboxyl of 4-hydroxybenzoate; unexpectedly, ^{14}C-phenol is not exchanged. The enzyme also catalyses the carboxylation of phenylphosphate (phosphoric acid monophenyl ester) to 4-hydroxybenzoate and phosphate whereas phenol is not carboxylated. The enzyme has been enriched with little success because of its oxygen sensitivity, with a half life time in air of 30 s. It is a 280 kDa protein which consists of probably 3 different subunits and is induced by growth of the organism on phenol. The enzyme requires Mn^{2+} for activity and the isotope exchange reaction in addition requires K^+. The actual substrate is CO_2 rather than bicarbonate, and the carboxylase is probably not biotin-dependent. The carboxylase is absolutely *para*-specific; 2-hydroxybenzoate was not detected. It remains to be shown whether phenylphosphate or another phenol derivative is the physiological intermediate and, if so, how it is formed from phenol. The use of phenylphosphate instead of phenol renders the carboxylation reaction exergonic under natural CO_2 and phenol concentrations; even more, phosphorylation would facilitate the cellular accumulation of this toxic substrate in a non-toxic reactive form.

The carboxylation of phenol is supported by other evidence. The expression of the phenol carboxylase, determined by the isotope exchange activity, was strictly regulated (Dangel et al. 1991). Growth on phenol was CO_2-dependent, in contrast to growth on 4-hydroxybenzoate, and phenol-grown cells incorporated four times more $^{14}CO_2$ into cell material as compared to cells grown on 4-hydroxybenzoate (Tschech and Fuchs 1987, 1989). Similar conclusions were derived from feeding fluorinated phenols to cell suspensions; benzoic acids were formed which were fluorinated at the expected carbon if the carboxylation were in *para*-position (Genthner et al. 1990). [1-^{13}C]Phenol was converted to [4-^{13}C]benzoate (Zhang et al. 1990). Enrichment cultures catalysed an H/D exchange in D_2O at C-4 of phenol, the site of carboxylation (Gallert et al. 1991), or excreted 4-hydroxybenzoate (Genthner et al. 1991). It is to be expected that 4-hydroxybenzoate is an intermediate in all fermenting bacteria and that benzoate is not formed directly; direct benzoate formation from phenol would require carboxylation and immediate dehydroxylation (Knoll and Winter 1989; Kobayashi et al. 1989; Zhang et al. 1990; Bechard et al. 1990; Gallert et al. 1991; Bisaillon et al. 1991a). In denitrifying bacteria, two additional enzymes are required to convert 4-hydroxybenzoate to benzoyl-CoA without free benzoate being an intermediate (see sections IV.H and IV.I). The evidence for the role of *para*-carboxylation in *o*-cresol metabolism is less direct (Bisaillon et al. 1991b; Rudolphi et al. 1991), the same holds true for *para*-carboxylation of *m*-cresol (Roberts et al. 1990) and catechol (Schnell et al. 1989) and *ortho*-carboxylation of hydroquinone (Szewzyk and Schink 1989).

Another interesting case for carboxylation is *para*-carboxylation of aniline to 4-hydroxybenzoate (Schnell et al. 1989; Schnell and Schink 1991). This soluble enzyme system does not catalyse a $^{14}CO_2$:4-aminobenzoate isotope exchange with the carboxyl group. It will be interesting to know whether aniline is modified as phenol appears to be activated. For the aerobic *ortho*-carboxylation of aniline, see Aoki et al. (1985).

It has been proposed that phenol is reduced to cyclohexanol or cyclohexanone (Bakker 1977; Balba et al. 1979). However, in the cases studied more intensively this is definitely not the case. However, this possibility cannot be excluded and has recently been suggested to be realized in another phenol-degrading, denitrifying bacterium (O'Connor and Young 1990).

H. Coenzyme A thioester formation

Benzoic acid and analogues, if they are not decarboxylated, are converted into their coenzyme A thioesters by soluble, relatively specific, inducible coenzyme ligases, before they are metabolized further (Hutber and Ribbons 1983). No exception to this rule is known so far. Enzymes for benzoic acid (Geissler et al.

1988; Altenschmidt et al. 1991), 2-aminobenzoic acid (Altenschmidt et al. 1991), phenylacetic acid (Martinez-Blanco et al. 1990; M. Mohamed and G. Fuchs unpublished), and 4-hydroxybenzoic acid (Fogg and Gibson 1990; T. Biegert, U. Altenschmidt and G. Fuchs unpublished) have been purified from *Rhodopseudomonas palustris* and a denitrifying *Pseudomonas* species, respectively. ATP is cleaved into AMP and pyrophosphate which suggests the occurence of an intermediate acyl-AMP. Pyrophosphate is thought to become hydrolysed. This renders the overall reaction strongly exergonic and allows the bacteria to trap aromatic acids after being taken up by passive diffusion (Harwood and Gibson 1986, 1988). The regulation of benzoate-CoA ligase has been studied in *Rhodopseudomonas*. The enzyme was induced not only by benzoate, but also by hydroxyl- and methyl-substituted benzoates and partly reduced alicyclic compounds which are thought to be intermediates of the benzoate pathway (Kim and Harwood 1991). Other ligases may act on 2-hydroxybenzoic acid, 3-hydroxybenzoic acid (Heising et al. 1991), *o-*, *m-*, and *p*-phthalic acids (Nozawa and Maruyama 1988b), 3-methylbenzoic acid (Rudolphi et al. 1991), 3-aminobenzoic acid (Schnell and Schink 1992), 4-aminobenzoic acid (Schnell and Schink 1991), and others (Ziegler et al. 1989). It would not be surprising if energy-limited fermenting bacteria would activate aromatic acids by CoA transfer from the metabolic end product acetyl-CoA; this would save one energy-rich phosphate anhydride bond.

I. Reductive dehydroxylation

The reductive dehydroxylation of aromatic hydroxyl functions, notably with hydroxyl functions *para* to a carboxyl group, have long been known (Scheline et al. 1960; Booth and Williams 1963a,b; Perez-Silva et al. 1966; Taylor et al. 1970; Scheline 1973; Szewzyk et al. 1985; Grbic-Galic 1986; Smolenski and Suflita 1978). The only case studied in some detail is the reductive dehydroxylation of 4-hydroxybenzoyl-CoA to benzoyl-CoA by 4-hydroxybenzoyl-CoA reductase (dehydroxylating) (Fig. 7) (Glöckler et al. 1989; R. Brackmann and G. Fuchs unpublished). This reaction plays an important role in the metabolism of phenol, 4-hydroxybenzoate, *p*-cresol and 4-hydroxyphenylacetate (Dangel et al. 1991). The enzyme has been purified. It is an iron-sulfur protein which contains 12Fe and $12S^=$ per 260 kDa and consists of three subunits of 75, 35, and 17 kDa. The subunit composition is likely to be $a_2 b_2 c_2$. The reaction requires a reduced electron donor such as reduced viologen dyes; no other cocatalysts are required, the product is benzoyl-CoA and oxidized dye. It is inactivated by oxygen (half life of 15 min in air). The inactivation by low concentrations of cyanide or azide suggests that it may contain a transition metal in an oxidation state which reacts with these ligands.

Similar enzymes must be widespread and require coenzyme A thioester formation of the aromatic acids to be dehydroxylated, e.g. in 3-hydroxybenzoate metabolism (Heising et al. 1991) or in *ortho*-cresol metabolism (Bisaillon et al. 1991b; Rudolphi et al. 1991; Tschech and Schink 1986). The role of reductive dehydroxylation in catechol, hydroquinone, and gentisic acid degradation remains to be proven (Balba and Evans 1980; Szewzyk et al. 1985; Young and Rivera 1985; Szewzyk and Pfennig 1987; Schnell et al. 1989; for alternative pathways see section IV.G).

J. Reductive deamination

Not only can aromatic hydroxyl groups be reductively removed but also aromatic amino functions. The first case reported was the reductive deamination of 4-aminobenzoyl-CoA to benzoyl-CoA, catalysed by 4-aminobenzoyl-CoA reductase (deaminating) (Fig. 7) (Schnell and Schink 1991). The reaction plays a role in the degradation of aniline and 4-aminobenzoate, the enzyme is active with different reductants. A second case was the reductive deamination of 2-aminobenzoyl (anthranoyl)-CoA to benzoyl-CoA by a different enzyme, 2-aminobenzoyl-CoA reductase (deaminating) (Tschech and Schink 1988; Lochmeyer et al. 1991). The reaction probably plays a role in the anaerobic degradation of indole and indolic compounds (Bak and Widdel 1986a; Berry et al. 1987b; Madsen and Bollag 1989), that are thought to be metabolized via anthranilate, as well as being involved in anthranilate degradation. The 2-aminobenzoyl-CoA reductase (deaminating) probably consists of several subunits as deduced from comparing the protein pattern of 2-aminobenzoate- and benzoate-grown cells. The protein can be resolved in two protein components of 180 and 260 kDa (C. Lochmeyer, G. Fuchs, unpublished).

No example of a hydrolytic removal of an amino substituent has been reported. It has been reported that 3-aminobenzoyl-CoA is reduced by NAD(P)H to a not yet identified, reduced product, although cyanide inhibited cell suspensions produced benzoate from 3-aminobenzoate (Schnell and Schink 1992). These evidently conflicting results remain unsolved.

K. Reductive dehalogenation

The reductive dehalogenation at the aromatic ring, which is referred to as hydrogenolysis, has been demonstrated many times and is covered in greater detail in Chapter 13 by Commandeur and Parsons. It has been studied only in a few cases with pure cultures or *in vitro*. A sulphate reducing bacterium, *Desulfomonile tiedjei*, has been characterised as being able to reductively dechlorinate 3-chlorobenzoic acid to benzoic acid with hydrogen or formate as reductant (Fig. 7) (den Tweel et al. 1987; Dolfing and Tiedje 1987, 1991a,b; Stevens et al. 1988; Stevens and Tiedje 1988; DeWeerd et al. 1990; DeWeerd and Suflita 1990; Mohn and Tiedje 1990a) and can obtain energy from this redox couple via a chemiosmotic mechanism (Apajalahti et al. 1989; Dolfing 1990; Mohn and Tiedje 1990b, 1991; DeWeerd et al. 1991; for review see Tiedje et al. 1987 and also chapter 13). Dechlorination by cell extracts depends on reduced methyl viologen, is membrane associated and inducible (DeWeerd and Suflita 1990). A *Staphylococcus epidermidis* strain slowly dechlorinates 1,2,4-trichlorobenzene to dichlorobenzenes and chlorobenzene when hydrogen is present (Tsuchiya and Yamaha 1984). A *Pseudomonas*-like strain is able to grow rapidly on 2-fluorobenzoate (Schennen et al. 1985) and we have indirect evidence suggesting that 2-fluorobenzoate is metabolized like benzoate and that the loss of fluoride is a late event which may occur spontaneously only when the aromatic ring has been reduced (for recent aerobic examples see Reineke and Knackmuss 1988; Schlömann et al. 1990). Several enrichment cultures have been studied with respect to reductive dehalogenation, e.g. of 1,2,3-trichlorobenzene to 1,3-dichlorobenzene. So far no hydrolytic removal of a halogen substituent has been reported in an anaerobic organism (for an aerobic *Pseudomonas* strain see section VII).

L. Transhydroxylation

It has already been mentioned that gallic acid (3,4,5-trihydroxybenzoic acid) is first decarboxylated to pyrogallol (1,2,3-benzenetriol). This compound is not directly reduced but is isomerized to phloroglucinol (1,3,5-benzenetriol) (Krumholz and Bryant 1988). In this interesting reaction, a tetrahydroxybenzene molecule, preferentially 1,2,3,5-tetrahydroxybenzene, acts as co-substrate (Fig. 7) (Brune and Schink 1989). Since this co-substrate is cyclically regenerated in the course of the reaction, it has to be considered as a co-catalyst. Its function is to donate one hydroxyl group to C5 and concomitantly accept the hydroxyl group from C2 of pyrogallol. Three similar, yet different hydroxyl transfers seem to be involved in hydroxyhydroquinone (1,2,4-benzenetriol) degradation which also leads to 1,3,5-benzenetriol (Schnell et al. 1991). First, the substrate is disproportionated to 1,3-dihydroxy- and 1,2,4,5-tetrahydroxybenzene. Then, the tetrahydroxybenzene is isomerized to the 1,2,3,5-tetrahydroxy isomer by an enzyme activity which is induced in hydroxyhydroquinone cells only. Finally, 1,2,3,5-tetrahydroxybenzene forms the product phloroglucinol by transferring its 2-hydroxyl group to either hydroxyhydroquinone or resorcinol, thus also regenerating the cosubstrates involved in earlier reactions of the sequence (Brune et al. 1992).

M. Methyl hydroxylation of p-cresol

The methyl group in *p*-cresol and other phenolic compounds, that have in common a methyl group *para* to a hydroxyl group, can be oxidized via the alcohol and aldehyde (Hopper 1978). The corresponding acid is formed without participation of molecular oxygen. The methyl hydroxylation catalysed by *p*-cresol dehydrogenase (methyl hydroxylating) appears to be virtually the same in aerobic and anaerobic organisms (Bossert and Young 1986; Smolenski and Suflita 1987; Bossert et al. 1989; Hopper et al. 1991;

Rudolphi et al. 1991) and has been studied in detail in aerobic bacteria (for review see Hopper 1988). The substrate is first oxidized to a quinone methide intermediate; then water is added to the carbon of the methylene substituent probably by an ionic mechanism. The product appears to be benzyl alcohol which is oxidized further by an aromatic alcohol dehydrogenase and aldehyde dehydrogenase. The same, or a similar enzyme, which is not absolutely specific for *p*-cresol (Hopper 1988) but requires a *para*-hydroxyl function, oxidizes dimethylphenols provided they contain a methyl *para* to a hydroxyl function (Rudolphi et al. 1991).

N. *Methyl hydroxylation of toluene*

Several bacteria degrading toluene in the absence of molecular oxygen have been reported (Dolfing et al. 1990; Lovley and Lonergan 1990; Evans et al. 1991a; Gorby and Lovley 1991; Altenschmidt and Fuchs 1991; Schocher et al. 1991). The first reaction in the anaerobic metabolism of toluene is a denitrifying *Pseudomonas* is the oxidation of the methyl group to the hydroxymethyl with water as hydroxyl source (Altenschmidt and Fuchs 1992b). This reaction must be different from *p*-cresol methylhydroxylation in which the dehydrogenation product is stabilized as a quinone methide due to the *para* hydroxyl group. Benzyl alcohol is the first product, whereas other researchers maintain that benzyl alcohol cannot be metabolized by their cultures (Schocher et al. 1991). Suspensions of cells, that were metabolically blocked by iodoacetamide, converted ^{14}C toluene first to ^{14}C-benzyl alcohol and then to ^{14}C-benzaldehyde (Altenschmidt and Fuchs 1992b).

Although no *in vitro* investigations have yet been reported on toluene methylhydroxylase, it promises to be a fascinating enzyme. It has been suggested that the aromatic ring in toluene becomes hydroxylated and that *p*-cresol is an intermediate; in enrichment cultures the incorporation of oxygen from water into *p*-cresol has been demonstrated (Vogel and Grbic-Galic 1986; Grbic-Galic and Vogel 1987). In all cases studied with pure cultures, *p*-cresol seems not to be intermediate and *p*-cresol methylhydroxylase is almost totally lacking in toluene-grown cells (Altenschmidt and Fuchs 1991); however, this possibility and others cannot be excluded.

The metabolism of the three xylene isomers has not been studied but enrichment cultures or single organisms oxidizing these aromatic hydrocarbons to CO_2 have been obtained (Dolfing et al. 1990; Alvarez and Vogel 1991; Evans et al. 1991b, 1992). Evans et al. (1992) recently found two dead-end products of toluene metabolism, benzylsuccinic acid and benzylfumaric acid. *o*-Xylene is transformed to analogous dead-end metabolites, (2-methylbenzyl)-succinic acid and (2-methylbenzyl)-fumaric acid. This reaction requires the oxidative addition, for instance, of succinyl-CoA, a process that requires the oxidation of the methyl group of toluene or of *o*-xylene to a methylene substituent. These authors reason that the main pathway for toluene oxidation involves a mechanism similar to that for the formation of the dead-end metabolites and involves an attack of the methyl group of toluene by acetyl-CoA yielding phenylpropionate. Phenylpropionate would then be oxidized to benzoyl-CoA by β-oxidation. This intriguing proposal remains to be tested.

Occasionally, the demethylation of p-cresol to phenol has been reported in enrichment cultures (Young and Rivera 1985). This reaction almost certainly is the result of conventional p-cresol oxidation to 4-hydroxybenzoate followed by decarboxylation to this dead-end product. The metabolic fate of the methyl group in methylbenzoic acids (Roberts et al. 1990; Rudolphi et al. 1991) and of m-cresol in some organisms (Roberts et al. 1987; Rudolphi et al. 1991) is unclear.

O. Aromatic ring hydroxylation

The hydroxylation of the aromatic ring may play a role in the anaerobic metabolism of toluene in some organisms; in enrichment cultures traces of p-cresol were found (Vogel and Grbic-Galic 1986). Such a reaction, however, must occur if benzene is anaerobically oxidized (Leahy and Colwell 1990). In enrichment cultures, phenol was formed in which the oxygen was derived from water (Vogel and Grbic-Galic 1986). The transformation of benzene, however, was extremely slow and has not been further studied. These enrichment cultures were conducted under methanogenic or denitrifying conditions (Grbic-Galic and Vogel 1987; Hutchins 1991). For thermodynamic reasons, the conversion of benzene to CO_2 and methane is unlikely. While the significance of these observations under methanogenic conditions is not doubted it is difficult to explain the findings. Future approaches using denitrifying conditions appear promising and may lead to pure cultures of nitrate reducing bacteria capable of degrading benzene.

P. Nitro group reduction

In nitro-aromatic compounds, the nitro substituents can be released by several mechanisms either as nitrite or NH_4^+, or it is reduced to amino substituents. The oxidative elimination of nitrite, whereby an hydroxyl function is introduced, requires molecular oxygen; this process therefore is confined to aerobes. The nitro substituent can be eliminated reductively as nitrite via the intermediary formation of a Meisenheimer complex (Lenke 1990). The most common anaerobic process appears to be the reduction of the nitro substituent via the nitroso and hydroxylamine compounds to the amino compound. The subsequent anaerobic, reductive elimination of amino substituents has already been discussed; however, the corresponding deaminating enzymes involved in anaerobic nitro-aromatic degradation have not been described. Aerobically, the oxidative elimination of NH_4^+ from amino-substituted aromatic compounds requires molecular oxygen.

The reduction of nitro compounds to amino compounds has been described in several aerobic organisms (e.g. Parrish 1977; Amerkhanova and Naumova

1979; Kinouchi and Onishi 1983; Bryant and DeLuca 1991; Schackmann and Müller 1991), but seems to be more common in anaerobic organisms (O'Brien and Morris 1971; Lindmark and Müller 1976; McCormick et al. 1978; Angermaier and Simon 1983a,b; Naumova et al. 1989; Braun et al. 1991; Raffi et al. 1991; Oren et al. 1991). This transformation may lead to the accumulation of potentially carcinogenic nitroso compounds. In many cases, notably in higher organisms, the reduction appears to be unspecific due to the action of various oxido-reductases which have a different physiological function; even reduced ferredoxin or $FADH_2$ may bring about this reaction. A nitro(aryl) reductase was purified from *Neurospora crassa* (Zucker and Nason 1955) and from *Nocardia* (Villanueva 1964). Recently, a NADPH-dependent nitroreductase was found in *Pseudomonas CBS 3* (Schackmann and Müller 1991) and a NAD(P)H-dependent enzyme in *Enterobacter cloacae* (Bryant and DeLuca 1991). Pure cultures of anaerobic bacteria seem to contain different nitroreductase activities, e.g. reduced ferredoxin itself or NAD(P)H-dependent enzymes (Angermaier and Simon 1983a,b). The specificity of the enzymes with respect to the rest of the molecule appears to be low, and consistently the nitro group is reduced to the amino function probably via three separate two electron transfer reactions. Unstable nitroso and hydroxylamino substituents are the most likely intermediates. Whereas the first two reduction steps to the hydroxylamino group appear to be fast and often unspecific, the reduction of the hydroxylamino substituent appears to be slow and generally enzyme catalysed.

Q. Removal of sulpho or sulphonic acid substituents

These compounds could serve either as sulphur and/or carbon source. Nothing is known about their anaerobic metabolism.

R. Polycyclic and oligonuclear compounds

The complete anaerobic degradation of polycyclic and oligonuclear compounds has not been studied (for heteroaromatic compounds see Chapter 14). Transformations of *o*-phenylphenol (Sembiring and Winter 1989b) and condensed aromatics (Mihelcic and Luthy 1988a,b) have been reported but biochemical studies are lacking.

V. Ring reduction and hydration

The different channelling reactions are combined such that one of several reactive central aromatic intermediates is obtained. The aromatic ring in these compounds is activated in such a manner that it can be reduced by specific enzymes. The most general intermediate is benzoyl-CoA, in which the aromatic ring is activated by the adjacent thioesterified carboxyl group (Fig. 8);

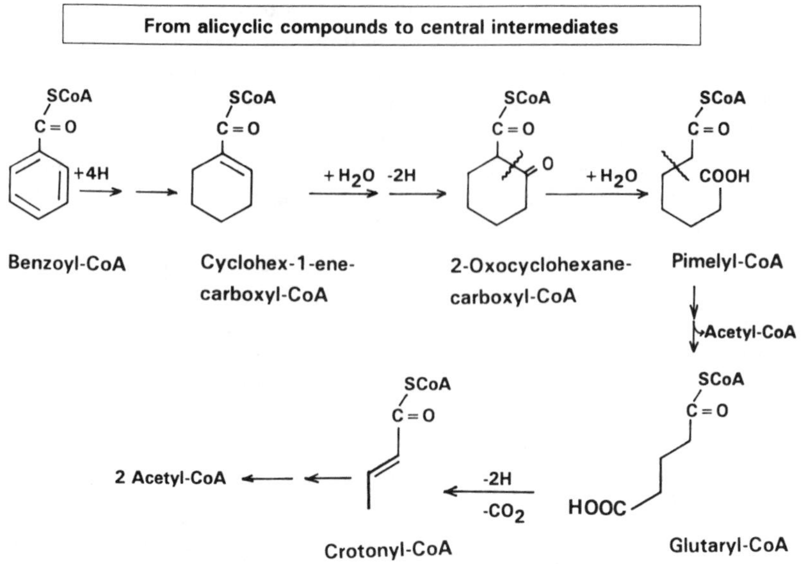

Fig. 8. Reactions leading from benzoyl-CoA to central metabolites.

less common are resorcinol and phloroglucinol, in which the aromatic ring is activated by two of three *m*-hydroxyl functions (keto-enol tautomery) (Fig. 9).

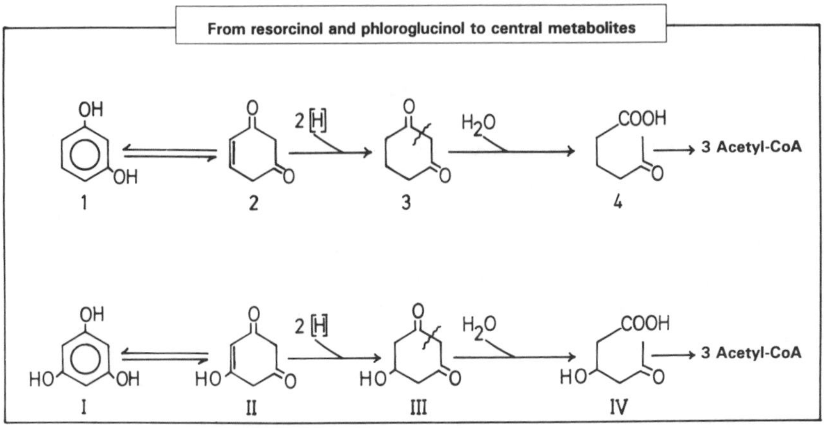

Fig. 9. Reactions leading from resorcinol and phloroglucinol to central metabolites (I, II) Phloroglucinol (enol-keto forms), (iii) Dihydrophloroglucinol, (IV) 5-oxo-3-hydroxycyclohexane carboxylic acid. 1, 2 Resorcinol (enol-keto forms), 3 cyclohexan-1,3-dion, 45-oxocyclohexane carboxylic acid.

A. Benzoyl-CoA reduction

This intermediate is formed from a large variety of different compounds, such as phenol, 4-hydroxybenzoate, *p*-cresol, phenylacetate, 4-hydroxyphenylacetate, mandelate, hydroxymandelate, toluene, 2-aminobenzoate, aniline, 4-aminobenzoate and many others. It is now clear that it is not benzoate itself but benzoyl-CoA which is the substrate of the ring reducing enzyme. Benzoyl-CoA is formed from benzoate by benzoate-CoA ligase (Geissler et al. 1988; Altenschmidt et al. 1991) or is indirectly formed from an already CoA-activated aromatic acid or from phenylglyoxylate. The reduction of the aromatic ring was first documented with whole cells of phototrophic bacteria by Evans and coworkers (Dutton and Evans 1968); this work was confirmed with denitrifying bacters (Williams and Evans 1975; Nozawa and Maruyama 1988a,b; Häggblom et al. 1988) and by studies of mutants (Guyer and Hegeman 1969). Cultures produced cyclohex-1-enecarboxylic acid, *trans*-2-hydroxycyyclohexanecarboxylic acid and pimelic acid, among others, when anaerobically grown on benzoate.

The reduction of benzoyl-CoA has only recently been demonstrated *in vitro* using a denitrifying *Pseudomonas* species (Koch and Fuchs 1992). The reaction requires benzoyl-CoA and a strong reductant such as Ti(III), $Na_2S_2O_4$, or $NaBH_4$; electron mediating redox dyes do not stimulate. Ti(III) is oxidized to TI(IV). The fact that this one electron donating compound functions as electron donor demonstrates that the protons are derived from water. Under these conditions, benzoyl-CoA is reduced at a specific activity of 15 $nmol \cdot min^{-1} \cdot mg^{-1}$ protein, and seven products are formed (Fig. 10). All products seem to be CoA-thioesters which were separated after alkaline hydrolysis. Three products are less polar than benzoic acid. The least polar was preliminarily identified as cyclohex-1-enecarboxylic acid; this compound appeared relatively late in the reaction sequence (Fig. 7, Fig. 8). Cyclohexanecarboxylic acid was not detected. The early formation of two other non-polar compounds suggests that either cyclohexenecarboxylic acid isomers or cyclohexadiencarboxylic acid(s) are formed first.

Four compounds that are more polar than benzoic acid were also found in the reaction. One compound, which appeared last, was *trans*-2-hydroxycyclohexanecarboxylic acid; the *cis*-stereoisomer was not formed. Very early in the reaction, rather polar compounds were observed which indicates that they are hydroxylated, non-aromatic intermediates; it is unclear whether they are natural products or artifacts of alkaline hydrolysis of the thioesters. Since the reaction proceeds only under strictly anaerobic conditions, and up to 0.5 mM products are formed within 10 min, the oxygen must come from water. The two most polar products were not seen in the presence of borate buffer. The products will be identified and the sites and stereospecificity of the reduction and the hydration of the ring be determined by ^1H- and ^{13}C-NMR spectroscopy. It is evident that several enzymes must be involved in the production of *trans*-2-hydroxycyclohexanecarboxylic acid from

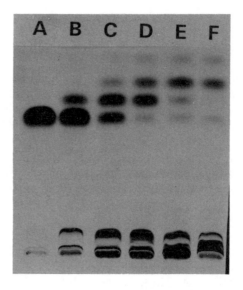

Fig. 10. Separation by TLC of some products of ^{14}C-benzoyl-CoA (0.2 mM) reduction observed with cell extracts of *Pseudomonas* K172 (3 mg protein·ml^{-1}). The panels represent: (A) Benzoate standard; in the following samples (lanes B to F) the coenzyme A thioesters were treated with alkali and the free acids formed after 5, 15, 30, 60, and 120 min were separated. For details see Koch and Fuchs (1992).

benzoic acid. It is premature to decide whether two sequential 2-electron or one 4-electron reduction takes place. A *trans*-2-hydroxycyclohexanecarboxyl-CoA dehydrogenase would be expected to form 2-oxocyclohexanecarboxyl-CoA as the ultimate alicyclic compound. Virtually the same product pattern was observed when extracts of *Rhodopseudomonas palustris* were tested (J. Koch and G. Fuchs unpublished). This shows that the enzymes involved in benzoyl-CoA de-aromatization and further metabolism are similar in these bacteria. In some organisms the enzymes responsible for anaerobic benzoate metabolism appear to be located on a plasmid (Blake and Hegeman 1987).

Recently, the early intermediates in anaerobic benzoate metabolism were analysed in *Rhodopseudomonas palustris* (Gibson and Gibson 1992). Alkali-treated extracts of whole cells growing photosynthetically on benzoate were examined by gas chromatography/mass spectrometry for partially reduced benzoate derivatives. Two cyclic dienes, cyclohexa-2,5-diene-1-carboxylate and the 1,4-diene isomer, were detected. Each compound supported growth as effectively as benzoate. These interesting results suggest that these cyclohexadienecarboxylates, probably as their coenzyme A thioesters, are the initial reduction products. This important finding suggests that the unpolar labelled products found in the *in vitro* ^{14}C-benzoyl-CoA reduction by both *Pseudomonas* K172 and *Rhodopseudomonas palustris* may be identical with these intermediates (see also "Note added in proof").

B. *Phloroglucinol reduction*

Phloroglucinol, an intermediate in the degradation of a limited number of trihydroxybenzene molecules, is directly reduced (e.g. Whittle et al. 1976; Patel et al. 1981; Schink and Pfennig 1982; Samain et al. 1986; Krumholz et al. 1987). It has almost no aromatic character due to three *m*-hydroxyl groups and can easily become reduced chemically by mild reducing agents. A soluble, oxygen-insensitive NADPH-dependent phloroglucinol reductase has been purified from *Eubacterium oxidoreducens* (Haddock and Ferry 1989). It is a homodimer of 78 kDa and forms in a reversible reaction dihydrophloroglucinol (1,3-dioxo-5-hydroxycyclohexane) as the ultimate alicyclic product plus $NADP^+$ (Fig. 7, Fig. 9). This ring reduction mechanism has even been found in aerobic organisms (Blackwood et al. 1970; Patel et al. 1990).

C. *Resorcinol reduction*

Resorcinol is formed from some dihydroxybenzoic acids (Tschech and Schink 1985; Kluge et al. 1990) or from 1,2,4-trihydroxybenzene (Schnell et al. 1991) and is reduced to 1,3-dioxocyclohexane as the ultimate alicyclic compound in a fermenting *Clostridium* (Fig. 7, Fig. 9). The reaction is catalysed by a soluble, oxygen-sensitive reductase that is active with reduced viologen dyes but not with pyridine nucleotides (Kluge et al. 1990). The enzyme has not been purified. Interestingly, the anaerobic degradation of cyclohexanol and cyclohexanone in a denitrifying *Pseudomonas* strain also proceeds via 1,3-dioxocyclohexane (Dangel et al. 1988, 1989).

D. *Reduction of other aromatic intermediates*

It is likely that besides benzoyl-CoA, phloroglucinol and resorcinol a few other central aromatic intermediates are formed that are directly attacked. One proposed case is 4-hydroxybenzoyl-CoA which is thought to be directly reduced in *Rhodospeudomonas palustris* (Harwood and Gibson 1988). The experimental results suggesting this possibility, however, may be interpreted differently. Furthermore, the organism contains sufficient amounts of 4-hydroxybenzoyl-CoA reductase (R. Brackmann and G. Fuchs unpublished); this enzyme forms benzoyl-CoA as shown in the denitrifying *Pseudomonas* strains (Glöckler et al. 1989). 3-Methylbenzoyl-CoA, a postulated intermediate in anaerobic o-cresol metabolism in a denitrifying bacterium, may be another example (Rudolphi et al. 1991); however, the 3-methyl group could be further oxidized to a carboxyl group which then becomes decarboxylated. Another candidate is 2-fluorobenzoyl-CoA which may become dehalogenated only after ring reduction by the normal benzoyl-CoA reducing system (Schennen et al. 1985); the fluorine atom has a similar size as the hydrogen atom and may not inhibit the ring reduction at this position. 2-Hydroxybenzoyl-CoA has been proposed to be directly reduced to a non-aromatic product (Tschech and Schink 1986) but experimental evidence is lacking. Phenol has been proposed to be reduced directly to cyclohexanol (Bakker 1977; Balba and Evans 1977). Although in most cases studied this does not happen, it may occur in methanogenic enrichment cultures as well as in other denitrifying bacteria (O'Connor and Young 1990). Whole cells formed cyclohexanol, cyclohexanone, 2-cyclohexene-1-ol, and 2-cyclohexene-1-one from phenol. Other ring reductions may occur as well (Grbic-Galic 1986) but have not been studied in pure cultures nor *in vitro*, or the observed *in vitro* rates are extremely low. In summary, no evidence for possible other central aromatic intermediates exists so far.

Many reductive catalytic mechanisms require the absence of molecular oxygen, and the enzymes are easily denatured by oxygen. This requires rapid adaptation processes in facultative anaerobic bacteria. The oxygen sensitivity of the most interesting enzymes has also prevented their purification and further study.

E. Resorcinol hydration

Evidence has been presented that, in a denitrifying bacterium, resorcinol is degraded in a pathway which is different from that in fermenting bacteria (Kluge et al. 1990). Recently, it was found that resorcinol is hydrolytically transformed by an obligate, denitrifying bacterium in one step reaction to a non-cyclic product, 5-oxo-hex-2-enecarboxylic acid (Fig. 7) (Gorny et al. 1991). This finding represents the first case of a hydrolytic rather than reductive attack of the aromatic nucleus which is facilitated by the 1,3-dioxo structure (Fig. 9). It remains to be tested whether the enzyme combines the reduction of the substrate with its hydrolysis and subsequent oxidation; this would require activation of the enzyme by reduction.

VI. From alicyclic compounds to central metabolites

In all cases studied, the cleavage of the aromatic ring via 3-oxo-compounds is by water rather than by coenzyme A and finally yields 3 molecules of acetyl-CoA. (The term '3-oxo' refers to the position of the oxo function relative to another electron withdrawing group, either a keto group or a carboxyl-CoA group.)

The ultimate alicyclic intermediate in benzoyl-CoA degradation appears to be 2-oxocyclohexane-1-carboxyl-CoA. This compound is likely to be hydrolytically cleaved to pimelyl-CoA which could be shortened by conventional β-oxidation to glutaryl-CoA (Fig. 8). This hypothesis was corroborated from studies with denitrifying bacteria. When grown on a variety of different aromatic acids, which are metabolized via benzoyl-CoA, cells contain high enzyme activities of a soluble, NAD^+-specific biotin-independent glutaryl-CoA   dehydrogenase   (decarboxylating),   in   contrast   to cells grown on acetate (J. Koch, U. Härtel, W. Buckel and G. Fuchs unpublished). NAD^+-dependent glutaryl-CoA dehydrogenase has been purified from two denitrifying *Pseudomonas* strains KB740 and K172 grown anaerobically on benzoate as sole carbon source (U. Härtel and W. Buckel, personal communication). The enzymes from these strains have almost identical N-terminal amino acid sequences and are homo-tetramers with a molecular mass of 170,000. The labile dehydrogenation product, glutaconyl-CoA, is subsequently decarboxylated to crotonyl-CoA. This conventional metabolite is further oxidized to two mols of acetyl-CoA. The final products of benzoyl-CoA degradation, therefore, are three acetyl-CoA, CO_2 from C2/C6 of the ring carbon, and six reducing equivalents (Fig. 8). Early experiments in which ^{14}C-benzoate conversion to CO_2 and methane were studied are consistent with this notion (Fig. 11) (Clark and Fina 1952; Fina and Fiskin 1960; Fina et al. 1978; Keith et al. 1978; Shlomi et al. 1978) although other fermentation products may occur. In some aerobic bacteria, cyclohexanecarboxylic acid seems to be similarly oxidized (Blakley 1978).

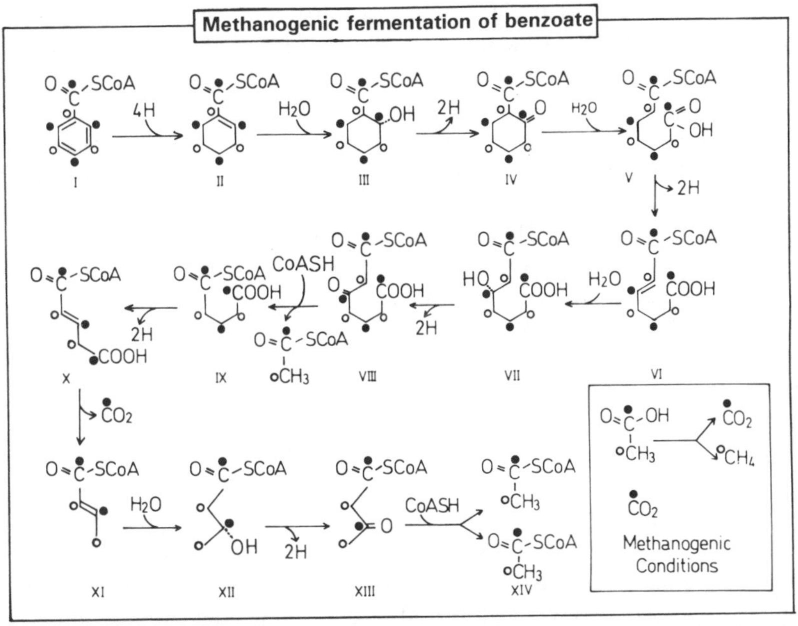

Fig. 11. Proposed scheme illustrating the observed labelling pattern of CO_2 and CH_4 when specifically ^{14}C-labelled benzoate was fermented to CO_2 and CH_4 in methanogenic enrichment cultures. The scheme is identical with that proposed in Fig. 5, which was derived from studies of nitrate reducing pure cultures.
(I) Benzoyl-CoA, (II) cyclohex-1-ene carboxyl-CoA, (III) trans-2-hydroxycyclo-hexane carboxyl-CoA, (IV) 2-oxocyclohexane carboxyl CoA, (V) pimelyl-CoA, (VI) 2,3 dehydropimelyl-CoA, (VII) 3-hydroxypimelyl-CoA, (VIII) 3-oxopimelyl-CoA, (IX) glutaryl-CoA, (X) glutaconyl-CoA, (XI) crotonyl-CoA, (XII) 3-hydroxybutyryl-CoA, (XIII) 3-oxobutyryl-CoA, (XIV) acetyl-CoA.

1,3-Dioxo-5-hydroxyyclohexane   and   1,3-dioxocyclohexane   are
hydrolytically cleaved by two different, soluble enzymes to 3-hydroxy-5-oxo-
hexanoic acid and 5-oxocyclohexanoic acid, respectively (Krumholz et al.
1987; Kluge et al. 1990). They are further oxidized to three acetyl-CoA
(Krumholz et al. 1987; Brune and Schink 1992). 3-Hydroxy-5-oxo-hexanoic
acid is oxidized via triacetic acid (3,5-dioxohexanoic acid), activation to the
coenzyme A thioester by CoA transfer, and then converted to 3 acetyl-CoA by
two subsequent β-ketothiolase reactions. Phloroglucinol conversion to acetyl-
CoA yields no net reducing equivalents, resorcinol yields two reducing
equivalents.

Acetyl-CoA therefore serves as the central biosynthetic intermediate
in all organisms which can use aromatic compounds anaerobically as principal
or sole carbon source. Acetyl-CoA can be assimilated via the well-known
glyoxylate bypass. Alternatively, it may directly be converted into pyruvate
through pyruvate synthase; this enzyme catalyses the reductive carboxylation
of acetyl-CoA to pyruvate, normally with reduced ferredoxin as electron
donor. Acetyl-CoA may be oxidized to 2 CO_2 via the common oxidative
citric acid cycle (Krebs cycle). An alternative pathway for the complete
oxidation of acetyl-CoA would be the oxidative acetyl-CoA pathway
(Schauder et al. 1986, reviewed by Thauer et al. 1989); in this pathway, acetyl-
CoA is first cleaved into a carbonyl and a methyl group which are then oxidized
to 2 CO_2 via separate reactions. Interestingly, in denitrifying bacteria capable
of degrading aromatic compounds anaerobically, an $NADP^+$-specific
2-oxoglutarate dehydrogenase was present, in contrast to an NAD^+-specific
pyruvate dehydrogenase (Lochmeyer and Fuchs 1990); the reason for
this is unknown. Fermenting bacteria using aromatic substrates form acetate,
CO_2 and H_2 as main fermentation products which are converted into
methane and CO_2 by synthrophic methanogenic bacteria (e.g. Ferry and Wolfe
1976).

VII. Chimeric pathways

Chimeric pathways, which combine typical anaerobic and aerobic characters in
one pathway, may result under certain conditions (for reviews on the specific
problems of anaerobes see Thauer et al. 1977; Morris 1984). Examples are
changing environments with different types of aromatic compounds and
varying O_2 levels; the capability of facultative bacteria to degrade aromatic
compounds with and without O_2; or O_2 limitation in microaerophilic
organisms. Two recent examples may be representative for others which
remain to be discovered.

2-Aminobenzoate is degraded aerobically in a denitrifying *Pseudomonas*
species (Braun and Gibson 1984) via activation to 2-aminobenzoyl-CoA
catalysed by a specific, aerobically induced 2-aminobenzoate-CoA ligase
isoenzyme (Altenschmidt et al. 1991; Altenschmidt and Fuchs 1992a). The

aromatic ring in 2-aminobenzoyl-CoA is hydroxylated and reduced by a flavoenzyme requiring 2 NAD(P)H, 2-aminobenzoyl-CoA monooxygenase/ reductase (Buder and Fuchs 1989; Buder et al. 1989; Altenschmidt et al. 1990). The product is 5-oxo-2-aminocyclohex-1-enecarboxyl-CoA (Fig. 7) (Langkau et al. 1990). These two enzymes have been purified. The whole pathway is coded on a very small 8.2 kb plasmid that has been sequenced, including the genes for these two enzymes (Altenschmidt and Fuchs 1990, 1992; Altenschmidt et al. 1992). After transforming this plasmid into *Escherichia coli, E. coli* expressed these enzymes and gained the ability to grow aerobically with 2-aminobenzoate as the sole carbon source (Altenschmidt and Fuchs 1991).

4-Chlorobenzoate is aerobically metabolized in a *Pseudomonas* strain CBS 3 via 4-chlorobenzoyl-CoA which is formed by action of 4-chlorobenzoate-CoA ligase (Oltmanns et al. 1989; Elsner et al. 1991a,b; Löffler and Müller 1991). The coenzymer A-ligase has been purified and sequenced. 4-Chlorobenzoyl-CoA is hydrolytically dehalogenated to HCl and 4-hydroxybenzoate; in this reaction the thioester bond becomes cleaved and CoA is released (Fig. 7). The 4-chlorobenzoyl-CoA hydrolase (dehalogenating) has been purified and sequenced (Scholten et al. 1991). The further metabolism of 4-hydroxybenzoate is conventional.

VIII. Comparison of aerobic and anaerobic pathways

This chapter can be summarized as follows (Fig. 12). The biochemistry of anaerobic biodegradation of aromatic compounds is based on a totally different strategy compared to the aerobic biodegradation (see Chapter 13). This strategy is simply based on the necessity to substitute molecular oxygen, and derived reactive oxygen species, by less reactive co-substrates. As a consequence the channelling part of metabolism employs a variety of biochemical reactions that are not encountered in aerobic metabolism because there O_2 activates the ring. The anaerobic, central aromatic intermediates are designed to become reduced. This requires benzoic acid to become activated as a thioester, and its reduction requires a strong physiological reductant. Oxygenolytic aromatic ring cleavage results in non-aromatic products which differ from the products of hydrolysis of alicyclic compounds formed upon reduction of the aromatic ring. Therefore, the final steps and end products also differ between aerobic and anaerobic organisms. O_2 sensitivity of enzymes is probably a consequence of the enzyme mechanisms; many of the enzyme reactions have no direct precedent in the literature and promise to reveal interesting mechanisms upon closer inspection and study (Fig. 7). The next decade will shed light into the biochemistry of anaerobic aromatic metabolism of which only the outlines are visible now.

Comparison of aerobic and anaerobic aromatic metabolism

	anaerobic	aerobic
Channelling reactions	$+H_2O$, $+2[H]$, $-2[H]+H_2O$ $+CO_2$, $+$ CoA $+$ ATP	$+O_2$
Central intermediates	Benzoyl-CoA Resorcinol Phloroglucinol	Catechol Protocatechuate Gentisate
Properties of central intermediates	easy to reduce (hydrate)	easy to oxidize (cleave)
Attack at the ring	2 or 4[H] $(+H_2O)$	O_2
Ring cleavage	Hydrolysis of 3-oxo compound	Oxygenolysis of aromatic
Pathway to central metabolites	ß-Oxidation, e.g. ——> Glutaryl-CoA ——> Acetyl-CoA	3-Oxoadipate pathway, e.g. ——> Succinat + Acetyl-CoA

Fig. 12. Comparison of anaerobic and aerobic pathways of degradation of aromatic compounds.

Acknowledgements

Work from our laboratory was financially supported by the Fonds der Chemischen Industrie and Deutsche Forschungsgemeinschaft through Schwerpunktprogramm "Neuartige Reaktionen und Katalysemechanismen in anaeroben Bakterien".

Note added in proof
This review covers the literature up to May 1992. Recent experiments from our laboratory (Eur. J. Biochem. 211: 649–661 (1993)) indicate that benzoyl-CoA is reduced in a two-electron step to cyclohex-1.5-diene-1-carboxyl-CoA. Fig. 2, Fig. 7, Fig. 8, and Fig. 11 therefore need to be modified correspondingly. 3-Hydroxypimelyl-CoA is the major product formed in vitro from benzoyl-CoA under reducing conditions.

References

Aeckersberg F, Bak F and Widdel F (1991) Anaerobic oxidation of saturated hydrocarbons to CO_2 by a new type of sulphate reducing bacteria. Arch. Microbiol. 156: 5–14

Aftring RP and Taylor BF (1981) Aerobic and anaerobic catabolism of phthalic acid by a nitrate-respiring bacterium. Arch. Microbiol. 130: 101–104

Altenschmidt U and Fuchs G (1991) Anaerobic degradation of toluene in denitrifying *Pseudomonas* sp.: indication for toluene methylhydroxylation and benzoyl-CoA as central aromatic intermediate. Arch. Microbiol. 156: 152–158

Altenschmidt U and Fuchs G (1992a) Novel aerobic 2-aminobenzoate metabolism. Purification and characterization of 2-aminobenzoate-CoA ligase, localisation of the gene on a 8 kbp plasmid, and cloning and sequencing of the gene. Eur. J. Biochem. 205: 721–727

Altenschmidt U and Fuchs G (1992b) Demonstration of anaerobic toluene oxidation to benzyl alcohol and benzaldehyde in a denitrifying *Pseudomonas* species. J. Bacteriol. 174: 4860-4862

Altenschmidt U, Eckerskorn C and Fuchs G (1990) Evidence that enzymes of a novel aerobic 2-aminobenzoate metabolism in denitrifying *Pseudomonas* are coded on a small plasmid. Eur. J. Biochem. 194: 647–653

Altenschmidt U, Oswald B and Fuchs G (1991) Purification and characterization of benzoate-coenzyme A ligase and 2-aminobenzoate-coenzyme A ligases from a denitrifying *Pseudomonas* sp. J. Bacteriol. 173: 5494–5501

Altenschmidt U, Bokranz M and Fuchs G (1992) Novel aerobic 2-aminobenzoate metabolism. Nucleotide sequence of the plasmid carrying the gene for the flavoprotein 2-aminobenzoyl-CoA monooxygenase/reductase in a denitrifying *Pseudomonas* sp. Eur. J. Biochem. 207: 715–722

Alvarez PJJ, Vogel TM (1991) Substrate interactions of benzene, toluene, and *para*-xylene during microbial degradation by pure cultures and mixed culture aquifer slurries. Appl. Environ. Microbiol. 57: 2981–2985

Amerkhanova NN, Naumova RP (1979) 2,4,6-Trinitrotoluene as a source of nutrition for bacteria. Microbiologiya 47: 393–395

Angermaier L, Simon H (1983a) On nitroaryl reductase activities in several *Clostridia*. Hoppe-Seyler's Z. Physiol. Chem. 364: 1653–1663

Angermaier L, Simon H (1983b) On the reduction of aliphatic and aromatic nitro compounds by *Clostridia*, the role of ferredoxin and its stabilization. Hoppe Seyler's Z. Physiol. Chem. 364: 961–975

Aoki K, Vemori T, Shinke R, Nishira H (1985) Further characterization of bacterial production of anthranilic acid from aniline. Agric. Biol. Chem. 49: 1151–1158

Apajalahti J, Cole J, Tiedje J (1989) Characterization of a dechlorination cofactor: an essential activator for 3-chlorobenzoate dechlorination by the bacterium DCB-1. ASM Annu. Meet. Abstr.: 336

Bache R, Pfennig N (1981) Selective isolation of *Acetobacterium woodii* on methoxylated aromatic compounds. Arch. Microbiol. 130: 255–261

Bak F, Widdel F (1986a) Anaerobic degradation of indolic compounds by sulphate-reducing enrichment cultures, and description of *Desulfobacterium indolicum* gen. nov., sp. nov. Arch. Microbiol. 146: 170–176

Bak F, Widdel F (1986b) Anaerobic degradation of phenol and phenol derivatives by *Desulfobacterium phenolicum* sp. nov. Arch. Microbiol 146: 177–180

Bakker G (1977) Anaerobic degradation of aromatic compounds in the presence of nitrate. FEMS Microbiol. Lett. 1: 103–108.

Balba MT, Evans WC (1977) The methanogenic fermentation of aromatic substrates. Biochem. Soc. Trans. 5: 302–305

Balba MT, Evans WC (1979) The methanogenic fermentation of omega-phenylalkane carboxylic acids. Biochem. Soc. Trans. 7: 403–405

Balba MT, Evans WC (1980) The methanogenic biodegradation of catechol by a microbial consortium: evidence for the production of phenol through *cis*-benzenediol. Biochem. Soc. Trans. 8: 452–454

Balba MT, Clarke NA, Evans WC (1979) The methanogenic fermentation of plant phenolics. Biochem. Soc. Trans. 7: 1115–1116

Barik S, Brulla WJ, Bryant MP (1985) PA-1, a versatile anaerobe obtained in pure culture, catabolizes benzenoids and other compounds in syntropy with hydrogenotrophs, and P-2 plus *Wolinella* sp. degrades benzenoids. Appl. Environ. Microbiol. 50: 304–310

Barker HA (1981) Amino acid degradation by anaerobic bacteria. Ann. Rev. Biochem. 50: 23–40

Beatrix B, Bendrat K, Rospert S, Buckel W (1990) The biotin-dependent sodium ion pump glutaconyl-CoA decarboxylase from *Fusobacterium nucleatum* (subsp. nucleatum). Comparison with the glutaconyl-CoA decarboxylases from Gram-positive bacteria. Arch. Microbiol. 156: 362–369

Béchard G. Bisaillon JG, Beaudet R (1990) Degradation of phenol by a bacterial consortium under methanogenic conditions. Can. J. Microbiol. 36: 573–578

Berman MH, Frazer AC (1992) Importance of tetrahydrofolate and ATP in the anaerobic demethylation reaction for phenylmethyl ethers. Appl. Environ. Microbiol. 58: 925-931

Berry DF, Francis AJ, Bollag J-M (1987a) Microbial metabolism of homocyclic and heterocyclic aromatic compounds under anaerobic conditions. Microbiol. Rev. 51: 43–59

Berry DF, Madsen EL, Bollag J-M (1987b) Conversion of indole to oxindole under methanogenic conditions. App. Environ. Microbiol. 53: 180–182

Bisaillon J-G, Lépine F, Beaudet R (1991a) Study of the methanogenic degradation of phenol via carboxylation to benzoate. Can. J. Microbiol. 37: 573–576

Bisaillon J-G, Lépine F, Beaudet R, Sylvestre M (1991b) Carboxylation of o-cresol by an anaerobic consortium under methanogenic conditions. Appl. Environ. Microbiol. 57: 2131–2134

Blackwood AC, Hang YD, Robern H, Mathur DK (1970) Reductive pathway for the degradation of phloroglucinol by a pseudomonad. Bacteriol. Proc. 70: 124

Blake CK, Hegeman GD (1987) Plasmid pCBI carries genes for anaerobic benzoate catabolism in *Alcaligenes xylosoxidans* subsp. denitrificans PN 1. J. Bacteriol. 169; 4878–4883

Blakley ER (1978) The microbial degradation of cyclohexanecarboxylic acid by a beta-oxidation pathway with simultaneous induction to the utilization of benzoate. Can. J. Microbiol. 24: 847–855

Booth AN, Williams RT (1963a) Dehydroxylation of caffeic acid by rat and rabbit caecal contents and sheep rumen liquor. Nature 198: 684–685

Booth AN, Williams RT (1963b) Dehydroxylation of catechol acids by intestinal contents. Biochem. J. 88: 66P–67P

Bossert ID, Young LY (1976) Anaerobic oxidation of p-cresol by a denitrifying bacterium. Appl. Environ. Microbiol. 52: 1117–1122

Bossert ID, Whited G, Gibson DT, Young LY (1989) Anaerobic oxidation of p-cresol mediated by a partially purified methylhydroxylase from a denitrifying bacterium. J. Bacteriol. 171: 2956–2962

Braun H, Schmidtchen FP, Schneider A, Simon H (1991) Microbial reduction of N-allylhydroxylamines to N-allylamines using *Clostridia*. Tetrahedron 47: 3329–3334

Braun K, Gibson DT (1984) Anaerobic degradation of 2-aminobenzoate (anthranilic acid) by denitrifying bacteria. Appl. Environ. Microbiol. 48: 102–107

Brune A, Schink B (1989) Pyrogallol-to-phloroglucinol conversion and other hydroxyl-transfer reactions catalysed by cell extracts of *Pelobacter acidigallici*. J. Bacteriol. 172: 1070–1076

Brune A, Schink B (1992) Phloroglucinol pathway in the strictly anaerobic *Pelobacter acidigallici*: fermentation of trihydroxybenzenes to acetate via triacetic acid. Arch. Microbiol. 157: 417–424

Brune A, Schnell S, Schink B (1992) Sequential transhydroxylation coverting hydroxyhydroquinone to phloroglucinol in the strictly anaerobic, fermenting bacterium *Pelobacter massiliensis*. Appl. Environm. Microbiol., 58: 1861–1868

Bryant C, DeLuca (1991) Purification and characterization of an oxygen-insensitive NAD(P)H nitroreductase from *Enterobacter cloacae*. J. Biol. Chem. 266: 4119–4125

Buder R, Fuchs G (1989) 2-Aminobenzoyl-CoA monooxygenase/reductase, a novel type of flavoenzyme. Purification and some properties of the enzyme. Eur. J. Biochem. 185: 629–635

Buder R, Ziegler K, Fuchs G, Langkau B, Ghisla S (1989) 2-Aminobenzoyl-CoA monooxygenase/ reductase, a novel type of flavoenzyme. Studies on the stoichiometry and the course of the reaction. Eur. J. Biochem. 185: 637–643

Chalmers RM, Fewson CA (1989) Purification and characterization of benzaldehyde dehydrogenase I from *Acinetobacter calcoaceticus*. Biochem. J. 263: 913–919

Chalmers RM, Scott AJ, Fewson CA (1990) Purification of the benzyl alcohol dehydrogenase and benzaldehyde dehydrogenase encoded by the TOL plasmid pWW53 of *Pseudomonas putida* MT53 and their preliminary comparison with benzyl alcohol dehydrogenase and benzaldehyde dehydrogenases I and II from *Acinetobacter calcoaceticus*. J. Gen. Microbiol. 136: 637–643

Chalmers RM, Keen JN, Fewson CA (1991) Comparison of benzyl alcohol dehydrogenases and benzaldehyde dehydrogenases from the benzyl alcohol and mandelate pathways in *Acinetobacter calcoaceticus* and from the TOL-plasmid-encoded toluene pathway in *Pseudomonas putida*. Biochem. J. 273: 99–107

Clark FM, Fina LR (1952) The anaerobic decomposition of benzoic acid during methane fermentation. Arch. Biochem. 36: 26–32

Cocaign M, Wilberg E, Lindley ND (1991) Sequential demethoxylation reactions during methylotrophic growth of methoxylated aromatic substrates with *Eubacterium limosum*. Arch. Microbiol. 155: 496–499

D'Ari L, Barker HA (1985) *p*-Cresol formation by cell-free extracts of *Clostridum difficile*. Arch. Microbiol. 143: 311–312

Dangel W, Tschech A, Fuchs G (1988) Anaerobic metabolism of cyclohexanol by denitrifying bacteria. Arch. Microbiol. 150: 358–362

Dangel W, Tschech A, Fuchs G (1989) Enzyme reactions involved in anaerobic cyclohexanol metabolism by a denitrifying *Pseudomonas* species. Arch. Microbiol. 152: 273–279

Dangel W, Brackmann R, Lack A, Magdy M, Koch J, Oswald B, Seyfried B, Tschech A, Fuchs G (1991) Differential expression of enzymes initiating anoxic metabolism of various aromatic compounds via benzoyl-CoA. Arch. Microbiol. 155: 256–262

Daniel SL, Wu Z, Drake HL (1988) Growth of thermophilic acetogenic bacteria on methoxylated aromatic acids. FEMS Microbiol. Lett. 52: 25–28

den Tweel WJJv, Kok JB, de Bont JAM (1987) Reductive dechlorination of 2,4-dichlorobenzoate to 4-chlorobenzoate and hydrolytic dehalogenation of 4-chloro-, 4-bromo-, and 4-iodobenzoate by *Alcaligenes denitrificans*. Appl. Environ. Microbiol. 53: 810–815

DeWeerd KA, Suflita JM (1990) Anaerobic aryl reductive dehalogenation of halobenzoates by cell extracts of "*Desulfomonile tiedjei*". Appl. Environ. Microbiol. 56: 2999–3005

DeWeerd KA, Concannon F, Suflita JM (1991) Relationship between hydrogen consumption, dehalogenation, and the reduction of sulphur oxyanions by *Desulfomonile tiedjei*. Appl. Environ. Microbiol. 57: 1929–1934

DeWeerd KA, Mandelco L, Tanner RS, Woese CR, J., Suflita M (1990) *Desulfomonile tiedjei* gen. nov. and sp. nov., a novel anaerobic, dehalogenating, sulfate-reducing bacterium. Arch. Microbiol. 154: 23–30

DeWeerd KA, Saxena A, Nagle DP, Suflita JM (1988) Metabolism of the ^{18}O-methoxy substituent of 3-methoxybenzoic acid and other unlabeled methoxybenzoic acids by anaerobic bacteria. Appl. Environ. Microbiol. 54: 1237–1242

Dimroth P (1987) Sodium ion transport decarboxylase and other aspects of sodium ion cycling in bacteria. Microbiol. Rev. 51: 320–340

Dolfing J (1990) Reductive dechlorination of 3-chlorobenzoate is coupled to ATP production and growth in an anaerobic bacterium strain DCB-1. Arch. Microbiol. 153: 264–266

Dolfing J, Tiedje TM (1987) Growth yield increase linked to reductive dechlorination in a defined 3-chlorobenzoate degrading methanogenic coculture. Arch. Microbiol. 149: 102–105

Dolfing J, Tiedje JM (1991a) Acetate as a source of reducing equivalents in the reductive dechlorination of 2,5-dichlorobenzoate. Arch. Microbiol. 156: 356–361

Dolfing J, Tiedje JM (1991b) Influence of substituents on reductive dehalogenation of 3-chlorobenzoate analogs. Appl. Environ. Microbiol. 57: 820–824

Dolfing J, Zeyer J, Binder-Eicher P, Schwarzenbach RP (1990) Isolation and characterization of a bacterium that mineralizes toluene in the absence of molecular oxygen. Arch. Microbiol. 154: 336–341

Dörner C, Schink B (1991) Fermentation of mandelate to benzoate and acetate by a homoacetogenic bacterium. Arch. Microbiol. 156: 302–306

Dutton PL, Evans WC (1967) Dissimilation of aromatic substrates by *Rhodopseudomonas palustris*. Biochem. J. 104: 30–31

Dutton PL, Evans WC (1968) The photometabolism of benzoic acid by *Rhodopseudomonas palustris*: a new pathway of aromatic ring metabolism. Biochem. J. 109: 5P.

Dutton PL, Evans WC (1969) The metabolism of aromatic compounds by *Rhodopseudomonas palustris*: a new reductive method of aromatic ring metabolism. Biochem. J. 113: 525–536

Dutton PL, Evans WC (1978) Metabolism of aromatic compounds by Rhodospirillaceae. In: RK Clayton and WR Sistrom (eds) The Photosynthetic bacteria (pp 719–726). Plenum Publishing Corp., New York

Elder DJE, Morgan P, Kelly DJ (1992) Anaerobic degradation of *trans*-cinnamate and omega-phenylalkane carboxylic acids by the photosynthetic bacterium *Rhodopseudomonas palustris*: evidence for a β-oxidation mechanism. Arch. Microbiol. 157: 148–154

Elsden SR, Hilton MG, Waller JM (1976) The end products of the metabolism of aromatic amino acids by *Clostridia*. Arch. Microbiol. 107: 283–288

Elsner A, Löffler F, Miyashita K, Müller R, Lingens F (1991a) Resolution of 4-chlorobenzoate dehalogenase from *Pseudomonas* sp. strain CBS3 into three components. Appl. Environ. Microbiol. 57: 324–326

Elsner A, Müller R, Lingens F (1991b) Separate cloning and expression analysis of two protein components of 4-chlorobenzoate dehalogenase from *Pseudomonas* sp. C8S3. J. Gen. Microbiol. 137: 477–481

Evans PJ, Ling W, Goldschmidt B, Ritter ED, Young LY (1992) Metabolites formed during anaerobic transformation of toluene and *o*-xylene and their proposed relationship to the initial steps of toluene mineralization. Appl. Environ. Microbiol. 58: 496–501

Evans PJ, Mang DT, Kim KS, Young LY (1991a) Anaerobic degradation of toluene by a denitrifying bacterium. Appl. Environ. Microbiol. 57: 1139–1145

Evans PJ, Mang DT, Young LY (1991b) Degradation of toluene and *m*-xylene and transformation of *o*-xylene by denitrifying enrichment cultures. Appl. Environ. Microbiol. 57: 450–454

Evans WC (1977) Biochemistry of the bacterial catabolism of aromatic compounds in anaerobic environments. Nature (London) 270: 17–22

Evans WC, Fuchs G (1988) Anaerobic degradation of aromatic compounds. Ann. Rev. Microbiol. 42: 289–317

Ferry JG, Wolfe RS (1976) Anaerobic degradation of benzoate to methane by a microbial consortium. Arch. Microbiol. 107: 33–40

Fewson CA (1988) Microbial metabolism of mandelate: a microcosm of diversity. FEMS Microbiol. Rev. 54: 85–110

Fina LR, Fiskin AM (1960) The anaerobic decomposition of benzoic acid during methane fermentation. II. Fate of carbons one and seven. Arch. Biochem. Biophys. 91: 163–165

Fina LR, Bridges RL, Coblentz TH, Roberts FF (1978) The anaerobic decomposition of benzoic acid during methane fermentation. III. The fate of carbon four and the identification of propanoic acid. Arch. Microbiol. 118: 169–172

Fogg GC, Gibson J (1990) 4-Hydroxybenzoate-coenzyme A ligase from *Rhodopseudomonas palustris*. ASM Annu. Meet. Abstr: 242

Frazer AC, Young LY (1985) A Gram-negative anaerobic bacterium that utilizes O-methyl substituents of aromatic acids. Appl. Environ. Microbiol. 49: 1345–1347

Frazer AC, Young LY (1986) Anaerobic C_1 metabolism of the O-methyl-^{14}C-labeled substituent of vanillate. Appl. Environ. Microbiol. 51: 84–87

Gallert C, Knoll G, Winter J (1991) Anaerobic carboxylation of phenol to benzoate: use of deuterated phenols revealed carboxylation exclusively in the C4-position. Appl. Microbiol. Biotechnol. 36: 124–129

Geissler JF, Harwood CS, Gibson J (1988) Purification and properties of benzoate-coenzyme A ligase, a *Rhodopseudomonas palustris* enzyme involved in the anaerobic degradation of benzoate. J. Bacteriol. 170: 1709–1714

Genthner BRS, Bryant MP (1987) Additional characteristics of one-carbon compound utilization by *Eubacterium limosum* and *Acetobacterium woodii*. Appl. Environ. Microbiol. 53: 471–476

Genthner BRS, Townsend GT, Chapman PJ (1990) Effect of fluorinated analogues of phenol and hydroxybenzoates on the anaerobic transformation of phenol to benzoate. Biodegradation 1: 65–74

Genthner BRS, Townsend GT, Chapman PJ (1991) *para*-Hydroxybenzoate as an intermediate in the anaerobic transformation of phenol to benzoate. FEMS Microbiol. Lett. 78: 265–270

Gibson DT, Subramanian V (1984) Microbial degradation of aromatic hydrocarbons. In: DT Gibson (ed) Microbial Degradation of Organic Compounds (pp 181–252). Marcel Dekker Inc., New York

Gibson KJ, Gibson J (1992) Potential early intermediates in anaerobic benzoate degradation by *Rhodopseudomonas palustris*. Appl. Environm. Microbiol. 58: 696–698

Glöckler R, Tschech A, Fuchs G (1989) Reductive dehydroxylation of 4-hydroxybenzoyl-CoA to benzoyl-CoA in a denitrifying, phenol degrading *Pseudomonas* species. FEBS Lett. 251: 237–240

Gorby YA, Lovley DR (1991) Electron transport in the dissimilatory iron reducer, GS-15. Appl. Environ. Microbiol. 57: 867–870

Gorny N, Wahl G, Brune A, Schink, B (1992) A strictly anaerobic nitrate-reducing bacterium growing with resorcinol and other aromatic compounds. Arch. Microbiol. 158: 48–53

Grbić-Galić (1986) O-Demethylation, dehydroxylation, ring-reduction and cleavage of aromatic substrates by Enterobacteriaceae under anaerobic conditions. J. Appl. Bacteriol. 61: 491–497

Grbić-Galić D, Vogel TM (1987) Transformation of toluene and benzene by mixed methanogenic cultures. Appl. Environ. Microbiol. 53: 254–260

Gross GG (1972) Formation and reduction of intermediate acyladenylate by aryl-aldehyde: NADP oxidoreductase from *Neurospora crassa*. Eur. J. Biochem. 31: 585–592

Guyer M, Hegeman G (1969) Evidence for a reductive pathway for the anaerobic metabolism of benzoate. J. Bacteriol. 99: 906–907

Hackett WF, Connors WJ, Kirk TK, Zeikus JG (1977) Microbial decomposition of synthetic ^{14}C-labeled lignins in nature: lignin biodegradation in a variety of natural materials. Appl. Environ. Microbiol. 33: 43–51

Haddock JD, Ferry JG (1989) Purification and properties of phloroglucinol reductase from *Eubacterium oxidoreducens* G-41. J. Biol. Chem. 264: 4423–4427

Häggblom MM, Nohynek LJ, Salkinoja-Salonen MS (1988) Degradation and O-methylation of chlorinated phenolic compounds by *Rhodococcus* and *Mycobacterium* strains. Appl. Environ. Microbiol. 54: 3043–3052

Hansen B, Bokranz M, Schönheit P, Kröger A (1988) ATP formation coupled to caffeate reduction by H_2 in *Acetobacterium woodii* NZva16. Arch. Microbiol. 150: 447–451

Harman WD, Taube H (1988) The selective hydrogenation of benzene to cyclohexene on pentaamineosmium(II). J. Am. Chem. Soc. 110: 7906–7907

Harwood CS, Gibson J (1986) Uptake of benzoate by *Rhodopseudomonas palustris* grown anaerobically in light. J. Bacteriol. 165: 504–509

Harwood CS, Gibson J (1988) Anaerobic and aerobic metabolism of diverse aromatic compounds by the photosynthetic bacterium *Rhodopseudomonas palustris*. Appl. Environ. Microbiol. 54: 712–717

Hegeman G (1988) Anaerobic growth of bacteria on benzoate under denitrifying conditions. In: SR Hagedorn RS Hanson and DA Kunz (eds) Microbial Metabolism and the Carbon Cycle (pp 181–190). Harwood Academic Publishers, Chur, Switzerland

Heising S, Brune A, Schink, B (1991) Anaerobic degradation of 3-hydroxybenzoate by a newly isolated nitrate-reducing bacterium. FEMS Microbiol. Letters 84: 267–272

Hopper DJ (1978) Incorporation of [^{18}O]water in the formation of *p*-hydroxybenzyl alcohol by the *p*-cresol methylhydroxylase from *Pseudomonas putida*. Biochem. J. 175: 345–347

Hopper DJ (1988) Properties of *p*-cresol methylhydroxylases. In: SR Hagedorn RS Hanson and DA Kunz (eds) Microbial Metabolism and the Carbon Cycle (pp 247–258). Harwood Academic Publishers, Chur, Switzerland

Hopper DJ, Bossert ID, Rhodes-Roberts ME (1991) *p*-Cresol methylhydroxylase from a denitrifying bacterium involved in anaerobic degradation of *p*-cresol. J. Bacteriol. 173: 1298–1301

Hsu T, Daniel SL, Lux MF, Drake HL (1990a) Biotransformations of carboxylated aromatic compounds by the acetogen *Clostridium thermoaceticum*: generation of growth-supportive CO_2 equivalents under CO_2-limited conditions. J. bacteriol. 172: 212–217

Hsu T, Lux MF, Drake HL (1990b) Expression of an aromatic-dependent decarboxylase which provides growth-essential CO_2 equivalents for the acetogenic (Wood) pathway of *Clostridium thermoaceticum*. J. Bacteriol. 172: 5901–5907

Hutber GN, Ribbons DW (1983) Involvement of coenzyme A esters in metabolism of benzoate and cyclohexanecarboxylate by *Rhodopseudomonas palustris*. J. Gen. Microbiol. 129: 2413–2420

Hutchins SR (1991) Biodegradation of monoaromatic hydrocarbons by aquifer micro-organisms using oxygen, nitrate, or nitrous oxide as the terminal electron acceptor. Appl. Environ. Microbiol. 57: 2403–2407

Kaiser JP, Hanselmann KW (1982) Aromatic chemicals through anaerobic microbial conversion of lignin monomers. Experientia 38: 167–175

Karapet'yants MK, Karapet'yants ML (1970) Thermodynamic Constants of Inorganic and Organic Compounds. Ann Arbor Humphrey Science Publishers, Ann Arbor

Keith CL, Bridges RL, Fina LR, Iverson KL, Cloran JA (1978) The anaerobic decomposition of benzoic acid during methane fermentation. IV. Dearomatization of the ring and volatile fatty acids formed on ring rupture. Arch. Microbiol. 118: 173–176

Kerscher L, Oesterhelt D (1982) Pyruvate: ferredoxin oxidoreductase – new findings on an ancient enzyme. TIBS 7: 371–374

Kim M-K, Harwood CS (1991) Regulation of benzoate-CoA ligase in *Rhodopseudomonas palustris*. FEMS Microbiol. Lett. 83: 199–204

Kinouchi T, Ohnishi Y (1983) Purification and characterization of 1-nitropyrene nitroreductase from *Bacteroides fragilis*. Appl. Environ. Microbiol. 46: 596–604

Kluge C, Tschech A, Fuchs G (1990) Anaerobic metabolism of resorcylic acids (m-dihydroxybenzoic acids) and resorcinol (1,3-benzenediol) in a fermenting and in a denitrifying bacterium. Arch. Microbiol. 155: 68–74

Knoll G, Winter J (1987) Anaerobic degradation of phenol in sewage sludge. Benzoate formation of phenol and CO_2 in the presence of hydrogen. Appl. Microbiol. Biotechnol. 25: 384–391

Knoll G, Winter J (1989) Degradation of phenol via carboxylation to benzoate by a defined, obligate syntrophic consortium of anaerobic bacteria. Appl. Microbiol. Biotechnol. 30: 318–324

Kobayashi T, Hashinuga T, Mikami E, Suschi T (1989) Methanogenic degradation of phenol and benzoate in acclimated sludges. Water Sci. Technol. 21: 55–65

Koch J, Fuchs G (1992) Enzymatic reduction of benzoyl-CoA to alicyclic compounds, a key reaction in anaerobic aromatic metabolism. Eur. J. Biochem. 205: 195–202

Kreikenbohm R, Pfennig N (1985) Anaerobic degradation of 3,4,5-trimethoxy-benzoate by a defined mixed culture of *Acetobacterium woodii*, *Pelobacter acidigallici*, and *Desulfobacter postgatei*. FEMS Microbiol. Ecology 31: 29–38

Krumholz LR, Bryant MP (1985) *Clostridium pfennigii* sp. nov. uses methoxyl groups of monobenzenoids and produces butyrate. Int. J. Syst. bacteriol. 35: 454–456

Krumholz LR, Bryant MP (1986) *Eubacterium oxidoreducens* sp. nov. requiring H_2 or formate to degrade gallate, pyrogallol, phloroglucinol and quercetin. Arch. Microbiol. 144: 8–14

Krumholz LR, Bryant MP (1988) Characterization of the pyrogallol-phloroglucinol isomerase of *Eubacterium oxidoreducens*. J. Bacteriol. 170: 2472–2479

Krumholz LR, Crawford RL, Hemling ME, Bryant MP (1987) Metabolism of gallate and phloroglucinol in *Eubacterium oxidoreducens* via 3-hydroxy-5-oxohexanoate. J. Bacteriol. 169: 1886–1890

Lack A, Fuchs G (1992) Carboxylation of phenylphosphate by "phenol carboxylase", an enzyme system of anaerobic phenol metabolism. J. Bacteriol. 174: 3629–3636

Lack A, Tommasi I, Aresta M, Fuchs G (1991) Catalytic properties of phenol carboxylase. *In vitro* study of CO_2: 4-hydroxybenzoate isotope exchange reaction. Eur. J. Biochem. 197: 473–479

Langkau B, Ghisla S, Buder R, Fuchs G (1990) 2-Aminobenzoyl-CoA monooxygenase/reductase, a novel type of flavoenzyme. Identification of the reaction products. Eur. J. Biochem. 191: 365–371

Leahy JG, Colwell RR (1990) Microbial degradation of hydrocarbons in the environment. Microbiol. Rev. 54: 305–315

Lenke H (1990) Mikrobieller Abbau von Nitrophenolen: 2,4-Dinitrophenole und 2,4,6-Trinitrophenole. Ph.D. thesis, Universität Stuttgart, Germany.

Lindmark DG, Müller M (1976) Antitrichomonad action, mutagenecity, and reduction of metronidazole and other nitroimidazoles. Antimicr. Agents Chemother. 10: 476–482.

Lochmeyer C, Fuchs G (1990) $NADP^+$-specific 2-oxoglutarate dehydrogenase in denitrifying *Pseudomonas* species. Arch. Microbiol. 153: 226–229.

Lochmeyer C, Koch J, Fuchs G (1992) Anaerobic degradation of 2-aminobenzoic acid (anthranilic acid) via benzoyl-CoA and cyclohex-1-enecarboxyl-CoA in a denitrifying bacterium. J.

Bacteriol. 174: 3621–3628

Löffler F, Müller R (1991) Identification of 4-chlorobenzoyl-coenzyme A as intermediate in the dehalogenation catalysed by 4-chlorobenzoate dehalogenase from *Pseudomonas* sp. CBS3. FEBS Lett. 290: 224–226.

Lovley DR (1991) Dissimilatory Fe(III) and Mn(IV) reduction. Microbiol. Rev. 55: 259–287.

Lovley DR, Lonergan DJ (1990) Anaerobic oxidation of toluene, phenol, and *p*-cresol by the dissimilatory iron-reducing organism, GS-15. Appl. Environ. Microbiol. 56: 1858–1864.

Lux MF, Keith E, Hsu T, Drake HL (1990) Biotransformations of aromatic aldehydes by acetogenic bacteria. FEMS Microbiol. Lett. 67: 73–78.

Madsen EL, Bollag JM (1989) Pathway of indole metabolism by a denitrifying microbial community. Arch. Microbiol. 151: 71–76.

Martinez-Blanco H, Reglero A, Rodriguez-Aparicio LB, Luengo JM (1990) Purification and biochemical characterization of phenylacetyl-CoA ligase from *Pseudomonas putida*. A specific enzyme for the catabolism of phenylacetic acid. J. Biol. Chem. 265: 7084–7090.

Masai E, Katayama Y, Nishikawa S, Yamasaki M, Morohoshi N, Haraguchi T (1989) Detection and localization of a new enzyme catalyzing β-aryl ether cleavage in the soil bacterium (*Pseudomonas paucimobilis* SYK-6). FEBS Lett. 249: 348–352.

Masai E, Katayama Y, Kawai S, Nishikawa S, Yamasaki M, Morohoshi N (1991) Cloning and sequencing of the gene for a *Pseudomonas paucimobilis* enzyme that cleaves β-aryl ether. J. Bacteriol. 173: 7950–7955.

McCormick NG, Cornell JH, Kaplan AM (1978) Identification of biotransformation products from 2,4-dinitrotoluene. Appl. Environ. Microbiol. 35: 945–948.

Mihelcic JR, Luthy RG (1988a) Degradation of polycyclic aromatic hydrocarbon compounds under various redox conditions in soil-water systems. Appl. Environ. Microbiol. 54: 1182–1187.

Mihelcic JR, Luthy RG (1988b) Microbial degradation of acenaphthene and naphthalene under denitrification conditions in soil-water systems. Appl. Environ. Microbiol. 54: 1188–1198.

Mohn WW, Tiedje JM (1990a) Catabolic thiosulfate disproportionation and carbon dioxide reduction in strain DCB-1, a reductively dechlorinating anaerobe. J. Bacteriol. 172: 2065–2070.

Mohn WW, Tiedje JM (1990b) Strain DCB-1 conserves energy for growth from reductive dechlorination coupled to formate oxidation. Arch. Microbiol. 153: 267–271.

Mohn WW, Tiedje JM (1991) Evidence for chemiosmotic coupling of reductive dechlorination and ATP synthesis in *Desulfomonile tiedjei*. Arch. Microbiol. 1991: 1–8.

Morris JG (1984) Changes in oxygen tension and the microbial metabolism of organic carbon. In: Aspects of Microbial Metabolism and Ecology (pp 59–96). Spec. Publ. Soc. Gen. Microbiol., London.

Mountfort DO (1990) Oxidation of aromatic alcohols by purified methanol dehydrogenase from *Methylosinus trichosporium*. J. Bacteriol. 172: 3690–3694.

Mountfort DO, Asher RA (1986) Isolation from a methanogenic ferulate degrading consortium of an anaerobe that converts methoxyl groups of aromatic acids to volatile fatty acids. Arch. Microbiol. 144: 55–61.

Mountfort DO, Brulla WJ, Krumholz LR, Bryant MP (1984) *Syntrophus buswelli* gen. nov., sp. nov.: a benzoate catabolizer from methanogenic ecosystems. Int. J. Syst. Bacteriol. 34: 216–217.

Naumova RP, Selivanovskaya SY, Cherepneva IE (1989) Conversion of 2,4,6-trinitrotoluene under conditions of oxygen and nitrate respiration of *Pseudomonas fluorescens*. Appl. Biochem. Microbiol. 24: 409–413.

Nozawa T, Maruyama Y (1988a) Anaerobic metabolism of phthalate and other aromatic compounds by a denitrifying bacterium. J. Bacteriol. 170: 5778–5784.

Nozawa T, Maruyama Y (1988b) Denitrification by a soil bacterium with phthalate and other aromatic compounds as substrates. J. Bacteriol. 170: 2501–2505.

O'Brien RW, Morris JG (1971) The ferredoxin-dependent reduction of chloramphenicol by *Clostridium acetobutylicum*. J. Gen. Microbiol. 67: 265–271.

O'Connor OA, Young LY (1990) Metabolism of phenol by a denitrifying pure culture. ASM Annu. Meet. Abstr: 295.

Oltmanns RH, Müller R, Otto MK, Lingens F (1989) Evidence for a new pathway in the bacterial degradation of 4-fluorobenzoate. Appl. Environ. Microbiol. 55: 2499–2504.

Oren A, Gurevich P, Henis Y (1991) Reduction of nitrosubstituted aromatic compounds by the halophilic anaerobic eubacteria *Haloanaerobium praevalens* and *Sporohalobacter marismortui*. Appl. Environ. Microbiol. 57: 3367–3370.

Parrish FW (1977) Fungal transformation of 2,4-dinitrotoluene and 2,4,6-trinitrotoluene. Appl. Environ. Microbiol. 34: 232–233.

Patel TR, Hameed N, Martin AM (1990) Initial steps of phloroglucinol metabolism in *Penicillium simplicissimum*. Arch. Microbiol. 153: 438–443.

Patel TR, Jure KG, Jones GA (1981) Catabolism of phloroglucinol by the rumen anaerobe *Coprococcus*. Appl. Environ. Microbiol. 42: 1010–1017.

Perez-Silva G, Rodriguez D, Perez-Silva J (1966) Dehydroxylation of caffeic acid by a bacterium isolated from rat faeces. Nature 212: 303–304.

Raffi F, Franklin W, Heflich RH, Cerniglia CE (1991) Reduction of nitroaromatic compounds by anaerobic bacteria isolated from the human gastrointestinal tract. Appl. Environ. Microbiol. 57: 962–968.

Ramanand K, Suflita JM (1991) Anaerobic degradation of *m*-cresol in anoxic aquifer slurries: carboxylation reactions in a sulfate-reducing bacterial enrichment. Appl. Environ. Microbiol. 57: 1689–1695.

Reineke W, Knackmuss H-J (1988) Microbial degradation of haloaromatics. Ann. Rev. Microbiol. 42: 263–287.

Roberts DJ, Fedorak PM, Hrudey SE (1987) Comparison of the fates of the methyl carbons of *m*-cresol and *p*-cresol in methanogenic consortia. Can. J. Microbiol. 33: 335–338.

Roberts J, Fedorak PM, Hrudey SE (1990) CO$_2$ incorporation and 4-hydroxy-2-methylbenzoic acid formation during anaeorobic metabolism of *m*-cresol by a methanogenic consortium. Appl. Environ. Microbiol. 56: 472–478.

Rudolphi A, Tschech A, Fuchs G (1991) Anaerobic degradation of cresols by denitrifying bacteria. Arch. Microbiol. 155: 238–248.

Samain E, Albagnac G, Dubourgier H-C (1986) Initial steps of catabolism of trihydroxybenzenes in *Pelobacter acidigallici*. Arch. Microbiol. 144: 242–244.

Schackmann A, Müller R (1991) Reduction of nitroaromatic compounds by different *Pseudomonas* species under aerobic conditions. Appl. Microbiol. Biotechnol. 34: 809–813.

Schauder R, Eikmanns B, Thauer RK, Widdel F, Fuchs G (1986) Acetate oxidation to CO$_2$ in anaerobic bacteria via a novel pathway not involving reactions of the citric acid cycle. Arch. Microbiol. 145: 162–172.

Scheline RR (1973) Metabolism of foreign compounds by gastro-intestinal microorganisms. Microbiol. Rev. 25: 451–523.

Scheline RR, Williams RT, Wit JG (1960) Biological dehydroxylation. Nature 188: 849–850.

Schennen UK, Braun K, Knackmuss H-J (1985) Anaerobic degradation of 2-fluoro-benzoate by benzoate-degrading denitrifying bacteria. J. Bacteriol. 161: 321–325.

Schink B (1992) Syntrophism among prokaryotes. In: A Balows et al. (eds) The Prokaryotes, 2nd edition (pp 276–299). Springer-Verlag, Heidelberg.

Schink B, Pfennig N (1982) Fermentation of trihydroxybenzenes by *Pelobacter acidigallici* gen. nov. sp. nov., a new strictly anaerobic, non-sporeforming bacterium. Arch. Microbiol. 133: 195–201.

Schink B, Tschech A (1988) Fermentative degradation of aromatic compounds. In: SR Hagedorn RS Hanson and DA Kunz (eds) Microbial Metabolism and the Carbon Cycle (pp 213–226). Harwood Academic Publishers, Chur, Switzerland.

Schink B, Brune A, Schnell, S (1992) Anaerobic degradation of aromatic compounds. In: G Winkelmann (ed) Microbial Degradation of Natural Products (pp 219–242). Verlag Chemie, Weinheim.

Schlömann M, Fischer P, Schmidt E, Knackmuss HJ (1990) Enzymatic formation, stability, and spontaneous reactions of 4-fluoromuconolactone, a metabolite of the bacterial degradation of 4-fluorobenzoate. J. Bacteriol. 172: 5119–5129.

Schnell S, Schink B (1991) Anaerobic aniline degradation via reductive deamination of 4-aminobenzoyl-CoA in *Desulfobacterium anilini*. Arch. Microbiol. 155: 183–190.

Schnell S, Schink B (1992) Anaerobic degradation of 3-aminobenzoate by a newly isolated sulfate reducer and a methanogenic enrichment culture. Arch. Microbiol. submitted.

Schnell S, Bak F, Pfennig N (1989) Anaerobic degradation of aniline and dihydroxybenzenes by newly isolated sulfate-reducing bacteria and description of *Desulfobacterium anilini*. Arch. Microbiol. 152: 556–563.

Schnell S, Brune A, Schink B (1991) Degradation of hydroxyhydroquinone by the strictly anaerobic fermenting bacterium *Pelobacter massiliensis* sp. nov., Arch. Microbiol. 155: 511–516.

Schocher RJ, Seyfried B, Vazquez F, Zeyer J (1991) Anaerobic degradation of toluene by pure cultures of denitrifying bacteria. Arch. Microbiol. 157: 7–12

Scholten J, Chang K, Babitt P, Charest H, Sylvestre M, Dunaway-Mariano D (1991) Novel enzymic hydrolytic dehalogenation of a chlorinated aromatic. Science 253: 182–185

Sembiring T, Winter J (1989a) Anaerobic degradation of phenylacetic acid by mixed and pure cultures. Appl. Microbiol. Biotechnol. 31: 84–88

Sembiring T, Winter J (1989b) Anaerobic degradation of *ortho*-phenylphenol by mixed and pure cultures. Appl. Microbiol. Biotechnol. 31: 89–92

Seyfried B (1989) Anaerober Abbau von Phenylacetat über alpha-Oxidation durch denitrifizierende Bakterien. Ph.D. thesis, Universität Ulm, Germany

Seyfried B, Tschech A, Fuchs G (1991) Anaerobic degradation of phenylacetate and 4-hydroxyphenylacetate by denitrifying bacteria. Arch. Microbiol. 155: 249–255.

Shaw JP, Harayama S (1990) Purification and characterization of TOL plasmid-encoded benzyl alcohol dehydrogenase and benzaldehyde dehydrogenase of *Pseudomonas putida*. Eur. J. Biochem. 191: 705–714.

Shlomi EK, Lankhorst A, Prins RA (1978) Methanogenic fermentation of benzoate in an enrichment culture. Microbial Ecol. 4: 249–261.

Sleat R, Robinson JP (1984) The bacteriology of anaerobic degradation of aromatic compounds. J. Appl. Bacteriol. 57: 381–394.

Smolenski WJ, Suflita JM (1987) Biodegradation of cresol isomers in anoxic aquifers. Appl. Environ. Microbiol. 53: 710–716.

Stevens TO, Tiedje JM (1988) Carbon dioxide fixation and mixotrophic metabolism by strain DCB-1, a dehalogenating anaerobic bacterium. Appl. Environ. Microbiol. 54: 2944–2948.

Stevens TO, Linkfield TG, Tiedje JM (1988) Physiological characterization of strain DCB-1, a unique dehalogenating sulfidogenic bacterium. Appl. Environ. Microbiol. 54: 2938–2943.

Stouthamer AH (1988) Bioenergetics and yields with electron acceptors other than oxygen. In: LE Erickson and D Yee-Chak Fung (eds) Handbook on Anaerobic Fermentations (pp 345–437). Marcel Dekker Inc., New York.

Stull DR, Westrum EF, Sinke GC (1969) The Chemical Thermodynamics of Organic Compounds. John Wiley and Sons Inc., New York.

Szewzyk R, Pfennig N (1987) Complete oxidation of catechol by the strictly anaerobic sulfate-reducing *Desulfobacterium catecholicum* sp. nov., Arch. Microbiol. 147: 163–168.

Szewzyk U, Schink B (1989) Degradation of hydroquinone, gentisate, and benzoate by a fermenting bacterium in pure or defined mixed culture. Arch. Microbiol. 151: 541–545.

Szewzyk U, Szewzyk R, Schink B (1985) Methanogenic degradation of hydroquinone and catechol via reductive dehydroxylation to phenol. FEMS Microbiol. Ecol. 31: 79–87.

Tarvin D, Buswell AM (1934) The methane fermentation of organic acids and carbohydrates, J. Amer. Chem. Soc. 56: 1751–1755.

Taylor BF (1983) Aerobic and anaerobic catabolism of vanillic acid and some other methoxy-aromatic compounds by *Pseudomonas* sp. strain PN-1. Appl. Environ. Microbiol. 46: 1286–1292

Taylor BF, Ribbons DW (1983) Bacterial decarboxylation of *o*-phthalic acids. Appl. Environ. Microbiol. 46: 1276–1281

Taylor BF, Campbell WL, Chinoy I (1970) Anaerobic degradation of the benzene nucleus by a facultatively anaerobic microorganism. J. Bacteriol. 102: 430–437

Thauer RK, Jungermann K, Decker K (1977) Energy conservation in chemotrophic anaerobic bacteria. Bacteriol. Rev. 41: 100–180

Thauer RK, Möller-Zinkhan D, Spormann AM (1989) Biochemistry of acetate catabolism in anaerobic chemotrophic bacteria. Annu. Rev. Microbiol. 43: 43–67

Tiedje JM, Boyd SA, Fathepure BZ (1987) Anaerobic degradation of chlorinated aromatic hydrocarbons. Dev. Indust. Microbiol. 27: 117–127

Tschech A (1989) Anaerober Abbau von aromatischen Verbindungen. Forum Mikrobiologie 5: 251–264

Tschech A, Fuchs G (1987) Anaerobic degradation of phenol by pure cultures of newly isolated denitrifying pseudomonads. Arch. Microbiol. 148: 213–217

Tschech A, Fuchs G (1989) Anaerobic degradation of phenol via carboxylation to 4-hydroxybenzoate: in vitro study of isotope exchange between $^{14}CO_2$ and 4-hydroxybenzoate. Arch. Microbiol. 152: 594–599

Tschech A, Pfennig N (1984) Growth yield increase linked to caffeate reduction in *Acetobacterium woodii*. Arch. Microbiol. 137: 163–167

Tschech A, Schink B (1985) Fermentative degradation of resorcinol and resorcylic acids. Arch. Microbiol. 143: 52–59

Tschech A, Schink B (1986) Fermentative degradation of monohydroxybenzoates by defined syntrophic cocultures. Arch. Microbiol. 145: 396–402

Tschech A, Schink B (1988) Methanogenic degradation of anthranilate (2-aminobenzoate). System. Appl. Microbiol. 11: 9–12

Tsuchiya T, Yamaha T (1984) Reductive dechlorination of 1,2,4-trichlorobenzene by *Staphylococcus epidermidis* isolated from intestinal contents of rats. Agric. Biol. Chem. 48: 1545–1550

Villanueva JR (1964) The purification of a nitroreductase from *Nocardia V.* J. Biol. Chem. 239: 773–776.

Vogel TM, Grbic-Galic D (1986) Incorporation of oxygen from water into toluene and benzene during anaerobic fermentative transformation. Appl. Environ. Microbiol. 52: 200–202

Ward LA, Johnson KA, Robinson JM, Yokoyama MT (1987) Isolation from swine feces of a bacterium which decarboxylates-*p*-hydroxyphenylacetic acid to 4-methylphenol (*p*-cresol). Appl. Environ. Microbiol. 53: 189–192

Whittle PJ, Lunt DO, Evans WC (1976) Anaerobic photometabolism of aromatic compounds by *Rhodopseudomonas* sp. Biochem. Soc. Trans. 4: 490–491

Widdel F (1988) Microbiology and ecology of sulfate- and sulfur-reducing bacteria. In: AJB Zehnder (eds) Biology of Anaerobic Microorganisms (pp 469–586). John Wiley and Sons Inc., London.

Williams RJ, Evans WC (1975) The metabolism of benzoate by *Moraxella* species through anaerobic nitrate respiration. Evidence for a reductive pathway. Biochem. J. 14—: 1–10

Wu Z, Daniel SL, Drake HL (1988) Characterization of a CO-dependent O-demethylating enzyme system from the acetogen *Clostridium thermoaceticum*. J. Bacteriol. 170: 5747–5750

Young LY (1984) Anaerobic degradation of aromatic compounds. In: DT Gibson (eds) Microbial Degradation of Organic Compounds (pp 487–523). Marcel Dekker Inc., New York

Young LY, Frazer AC (1987) The fate of lignin and lignin-derived compounds in anaerobic ecosystems. Geomicrobiol. J. 5: 261–293

Young LY, Rivera MD (1985) Methanogenic degradation of four phenolic compounds. Water Res. 19: 1325–1332

Zeikus JG (1980) Fate of lignin and related aromatics in anaerobic environments. In: T Kirk, T Higushi and HM Chung (eds) Lignin biodegradation: Microbiology, Chemistry and Potential Applications (pp 101–110). CRC Press, Boca Raton

Zellner G, Kneifel H, Winter J (1990) Oxidation of benzaldehydes to benzoic acid derivatives by three *Desulfovibrio* strains. Appl. Environ. Microbiol. 56: 2228–2233.

Zenk MH, Ulbrich B, Brusse J, Stöckigt J (1980) Procedure for the enzymatic synthesis and isolation of cinnamoyl-CoA thioesters using a bacterial system. Anal. Biochem. 101: 182–187

Zhang X, Morgan TV, Wiegel J (1990) Conversion of ^{13}C-1 phenol to ^{13}C-4 benzoate, an intermediate step in the anaerobic degradation of chlorophenols. FEMS Microbiol. Lett. 67: 63–66

Ziegler K, Braun K, Böckler A, Fuchs G (1987) Studies on the anaerobic degradation of benzoic acid and 2-aminobenzoic acid by a denitrifying *Pseudomonas* strain. Arch. Microbiol. 149: 62–69

Ziegler K, Buder R, Winter J, Fuchs G (1989) Activation of aromatic acids and aerobic 2-aminobenzoate metabolism in a denitrifying *Pseudomonas* strain. Arch. Microbiol. 151: 171–176

Zucker M, Nason A (1955) Nitroaryl reductase from *Neurospora crassa*. Methods in Enzymology 2: 406–411

17. Biocorrosion: the action of sulphate-reducing bacteria

W.A. HAMILTON
Department of Molecular and Cell Biology, Marischal College, University, Aberdeen AB9 1AS, Scotland, U.K.

I. Introduction

One of the most characteristic features of microbial degradation is that it involves not individual organisms growing in axenic culture but rather mixed communities of species with differing but complementary metabolic capabilities. Often such consortia demonstrate structural as well as functional organisation and exist in the form of flocs or biofilms within which there are likely to be localized microenvironments, each with a particular combination of organisms and physico-chemical conditions. Such biofilms are ubiquitous in nature being found, for example, on soil and sediment particles, and on the surfaces of teeth, ships and off-shore oil production platforms. In controlled processes of biodegradation and biotransformation they are the active component of trickling filters and the various forms of immobilized cell and fixed bed reactors.

In the case of biocorrosion, biofilms are again of central importance to the processes involved, with the following individual features of their structure and function assuming particular importance.

a) The substratum on which the biofilm is built may become also a substrate in respect of acting as a source of metabolic energy (H_2 oxidation).

b) Heterogeneities in the horizontal dimension (patchiness) can establish localized electrochemical corrosion cells through the creation of oxygen concentration or differential aeration cells.

c) The most significant vertical heterogeneity arises from the development of anaerobic regions at the base of the biofilm which can support the growth of sulphate-reducing bacteria.

d) The extracellular polymeric substances (EPS), which constitute the main mass of the biofilm and underpin the maintenance of these heterogeneities and microenvironments, can also influence corrosion more directly by metal binding and/or retention of corrosion products.

e) In a manner directly complementary to the stimulation of primary biodegradative activity within consortia by the terminal oxidative steps involving either methanogenic or sulphate-reducing bacteria, the aerobic

C. Ratledge (ed.), Biochemistry of Microbial Degradation, 555–570.

and facultative heterotrophic species in the surface regions of the biofilm can create the nutrient and physico-chemical conditions necessary for the growth and activity of sulphate-reducing bacteria at the base of the biofilm where they then initiate corrosion of the metal substratum.

Biocorrosion can therefore be thought of as a particular form of biodegradation, sharing many of the characteristics common to other biodegradative processes. In biocorrosion the normal catabolic breakdown of the primary substrate is no longer the key biodegradative event. It is often required, however, to create the necessary conditions and to fuel those reactions that do constitute corrosion *per se*, in which it is the sulphate-reducing bacteria that play the central role.

Setting biocorrosion by sulphate-reducing bacteria in the general context of biodegradation allows us to identify three central issues that require to be more fully considered. These are: biofilms, the physiology and ecology of the sulphate-reducing bacteria, and the mechanisms of microbially influenced corrosion. Each of these areas has been the focus of considerable activity in recent years and there are available a number of high quality reviews and monographs (Widdel 1988; Characklis and Wilderer 1989; Characklis and Marshall 1990; Flemming and Geesey 1991; Dowling et al. 1991; Videla and Gaylarde 1992). This chapter will take as its starting point an earlier review by the present author in which the three main themes identified above were for the first time drawn together in an attempt to gain a more complete understanding of the processes involved (Hamilton 1985).

II. Biofilms

A biofilm can be defined as a collection of microorganisms and their extracellular products bound to a solid substratum.

Biofilms are ubiquitous in nature and are found associated with such diverse substrata as rumen and intestinal epithelia; replacement joints, heart valves and pacemakers; urinary catheters; teeth; sediment particles, stones, and man-made structures in aquatic envionments; fluid transportation lines, heat exchangers, and cooling towers (Costerton et al. 1987). In these environments, they are major sites of biological activity and are responsible for cellulose digestion, bacterial infection and mechanical failure, dental caries, general biodegradation, corrosion, increased fluid frictional resistance, and impaired heat transfer.

Although much of the laboratory analyses to date has been carried out with biofilms involving single species under closely defined conditions, in most circumstances natural biofilms contain a mixed community of microorganisms. Two consequences follow from this which are of major significance with respect to the mechanisms of biocorrosion: a) the organisms show close interdependencies so that the biofilm as a whole behaves as a unified consortium; b) the organisms, and the microenvironments created by their activities, generate structural and temporal heterogeneities within the biofilm.

A. Microbial consortia

There are now many examples in the literature of microbial consortia in which groups of otherwise unrelated organisms interact by a mechanism of nutrient succession that ensures both cooperation between species and maximum utilization of the primary nutrients. Such consortia inevitably have a structural component and they exist as aggregates in the form of flocs or biofilms. Diffusional resistance to oxygen penetration, coupled with its uptake by aerobic species, can result in the creation of anoxic and reduced microenvironments within the central regions of the consortium. A general model for a microbial consortium would therefore be a nutritionally-linked group of aerobic, facultative, and anaerobic bacterial species displaying heterotrophic hydrolytic and fermentative activity, acetogenesis, and a terminal oxidative stage involving either methanogenesis or sulphate reduction (Fig. 1) (Parkes 1987). The initial heterotrophic species will be determined by the nature and availability of the primary substrates and by the prevailing physico-chemical conditions in the liquid bulk phase. The nutritionally-dependent facultative and anaerobic organisms will be more constant in character, with the key difference between terminal oxidation by methanogens or sulphate-reducing bacteria being largely determined by the availability of sulphate. Whereas the methanogens are restricted to the oxidation of only methylated amines and methanol in addition to hydrogen and acetate which are their normal substates, the sulphate-reducing bacteria are now known to be capable of growth on an extensive range of short and long chain fatty acids and alcohols, aromatic and heterocyclic compounds, and even hydrocarbons (Widdel 1988; Aeckersberg et al. 1991). Nevertheless, the balance of evidence suggests that in most natural environments, sulphate reducers also oxidize hydrogen and acetate as their key energy and carbon substrates (Jørgensen 1982). This implies, therefore, that in biofilms and other microbial consortia the processes of acetogenesis and interspecies hydrogen transfer are of central importance (Zeikus 1983; Hamilton 1988).

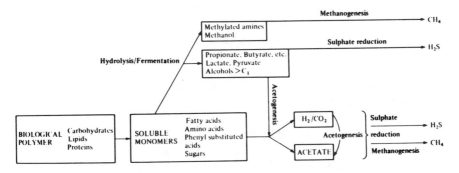

Fig. 1. A scheme of carbon flow through an anaerobic ecosystem showing some of the functional metabolic groups and their relationships (from Parkes 1987, with permission).

B. Heterogeneities

Biofilms are dynamic structures. The pattern of their development can be divided into phases of transport of molecules and cells to the substratum, attachment, increase in biomass (ie. cells and EPS), and detachment (Characklis and Marshall 1990). This holds true for each individual species, and a mature biofilm is considered to be in a steady state when the rate of growth is equal to the rate of detachment. In a multispecies biofilm this situation may be complicated by the requirement for organism A to establish the necessary conditions, for example of anaerobiosis, before organism B can become established within the biofilm. Other less well characterized features of changes with time which may be of particular significance with respect to corrosion mechanisms are the retention of corrosion products within the polymer matrix of the biofilm, and the random sloughing of whole areas of biofilm with the loss of cells, polymers and any inclusions.

Reference has already been made to the development of anaerobic conditions within biofilms. This is a common occurrence and it has been demonstrated even within thin biofilms of only 12 μm depth. Oxygen limitation results from the combination of mass transfer and diffusional resistances to the entry and passage of O_2 through the biofilm, and its utilization by aerobic organisms in the biofilm surface layers. Where an illuminated biofilm contains also diatoms and/or cyanobacteria, the effect may be partially alleviated by oxygenic photosynthesis. While the development of an aerobic/anaerobic (O_2/ AnO_2) interface is the most striking vertical heterogeneity noted in biofilms, the phenomenon is a general one and can equally affect carbon substrate or sulphate limitation in the deeper layers of particular biofilms. It is clear, therefore, that there is a close interdependence between these discontinuities throughout the depth of the biofilm and the essential character of the biofilm as a microbial consortium. Particular organisms exist within appropriate microenvironments created by their own and other organisms' activities and are maintained by virtue of the polymeric matrix within which they are imbedded.

It may appear surprising to stress the importance of anaerobic regions within a biofilm existing in an aerobic bulk phase, when one might expect the so-called anaerobic corrosion caused by the sulphate-reducing bacteria to be more directly associated with anoxic and reducing environments. In fact, it is general experience that maximum levels of corrosion are most usually linked to environments where the necessary reducing conditions are in close proximity to an aerobic phase. In addition to any stimulation of the sulphate-reducing bacteria themselves (Jørgensen 1988; Rosser and Hamilton 1983), there appears to be a direct effect of O_2 on the actual corrosion reactions.

Biofilms also demonstrate a degree of heterogeneity in the plane horizontal to the underlying substratum. This so-called patchiness arises from localized or colonial growth of individual species with their associated exopolymers but can also occur when significant areas of the biofilm slough off, in what is a well documented but poorly understood phenomenon. Since the essential characteristic of corrosion processes is the establishment and operation of electrochemical cells with separately identifiable anodic and cathodic regions on the metal substratum, discontinuities across the surface of the biofilm are likely to be of direct and major importance.

III. Sulphate-reducing bacteria

The sulphate-reducing bacteria are a relatively diverse taxonomic group of bacteria which share a common property that dominates both their physiology and their ecology. They are strict anaerobes with the capacity to reduce sulphate as the terminal electron acceptor in a respiratory mode of metabolism. Other oxidized forms of sulphur, such as thiosulphate, can also fulfil this role, and a few species reduce sulphur itself rather than sulphate. The group as a whole is sometimes referred to as the sulphidogens which has the merit of including the sulphur-reducing organisms, paralleling the nomenclature of the functionally-related acetogens and methanogens, and identifying that feature which very largely determines the ecological and economic importance of the group. The excellent monograph by Postgate (1984) reviews the status of our knowledge of this intriquing group of bacteria up to that time. In the last twenty years, however, there has been a dramatic increase in the degree to which the sulphate-reducing bacteria have become the subject of major studies in a number of internationally important laboratories. To a certain extent at least, this activity has been stimulated by the recognition of the importance of this group of organisms to problems experienced by the oil industry, notably corrosion.

In comparison with O_2 as terminal electron acceptor (E_o' +820 mV), the energy yield from substrate oxidation is considerably lower with sulphate (E_o' −60 mV). This effect is even more marked as sulphate itself cannot act directly as electron acceptor in biological reaction sequences (E_o' −516 mV) but requires to be activated, at the net cost of 2 ATP, to adenosine phosphosulphate (APS). This energetic burden carried by the sulphate reducers underlies both their relatively restricted nutrition and certain unusual features of their metabolism.

Traditionally, the sulphate-reducing bacteria have been grown on a narrow range of organic acids and alcohols such as lactate, pyruvate or ethanol. More recently, it has become evident that the group as a whole has a greater degree of nutritional versatility and two sub-groupings have been identified (Table 1). The first sub-group includes those organisms capable of only the partial oxidation of a limited range of nutrients such as lactate, with acetate being produced as a metabolic end-product; organisms in the second sub-group carry out the complete oxidation to CO_2 of a much more extensive range of sources of carbon and energy, including short and long chain fatty acids from acetate to stearate, and more complex organic compounds such as benzoate and nicotinate

(Widdel 1988; Hamilton 1988). Most strikingly, an organism capable of the complete oxidation of saturated alkanes has recently been isolated (Aeckersberg et al. 1991).

Particularly amongst the genera and species of the first nutritional group, the capacity to oxidize H_2 as a source of energy is widespread (Nethe-Jaenchen and Thauer 1984). Although in contrast the capability to oxidize acetate as an energy source is relatively restricted, it is none the less widely assimilated, in association with CO_2, as a source of carbon. The anaplerotic route is through a ferredoxin-dependent reductive carboxylation of acetyl-CoA to pyruvate, and thence to oxaloacetate. Where acetate is used as both carbon and energy source by certain organisms in the second sub-group, it is notable that two quite different routes of oxidative catabolism have been identified. Earlier studies with *Desulfobacter postgatei* and the sulphur reducing *Desulfuromonas acetoxidans* found that a modified tricarboxylic acid cycle operates and is able to circumvent the thermodynamic problem associated with the use of sulphate as electron acceptor (E_o' $-60\,mV$) for the succinic dehydrogenase reaction (E_o' $+33\,mV$) (Brandis-Heep et al. 1983; Gebhardt et al. 1985; Kroger et al. 1988). A more detailed analysis of the operation of the tricarboxylic acid cycle in *D. postgatei* has shown that citrate synthesis occurs by the action of an ATP citrate-lyase and is accompanied by a net gain of 1 ATP by substrate phosphorylation (Thauer 1988). Additionally, acetyl-CoA is formed by succinyl-CoA: acetate CoA transferase without the intermediate involvement of either substrate phosphorylation or ATP-driven activation. The reaction sequence is shown in Fig. 2. In all other species, however, acetate oxidation is by the acetyl-CoA pathway which is the mechanism found also in the acetogens and methanogens (Fuchs 1986; Wood et al. 1986; Hamilton 1988).

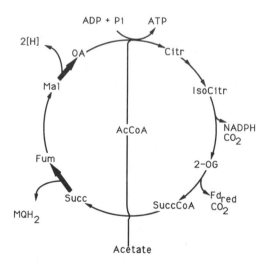

Fig. 2. Tricarboxylic acid cycle in the modification operative in *Desulfobacter postgatei*. Bold arrows indicate reactions catalyzed by membrane fractions in cell extracts and *thin arrows* reactions catalyzed by soluble fractions. (From Thauer 1988, with permission.)

Table 1. Characteristics of representative sulphate-reducing bacteria. (from Hamilton, 1988, with permission).

	Cell form	Approximate optimum temperature for growth (°C)	Compounds oxidized			Lactate	Others
			H_2	Acetate	Fatty acids		
Incomplete oxidation							
Desulfovibrio							
desulfuricans	Curved	30	+	–	–	+	Ethanol
vulgaris	Curved	30	+	–	–	+	Ethanol
gigas	Curved	30	+	–	–	+	Ethanol
salexigens	Curved	30	+	–	–	+	Ethanol
sapovorans	Curved	30	–	–	C_4 through C_{16}	+	
thermophilus	Rod-shaped	70	+	–	–	+	
Desulfotomaculum orientis	Rod-shaped	30 to 35	+	–	–	+	Methanol
ruminis	Rod-shaped	37	+	–	–	+	
nigrificans	Rod-shaped	55	+	–	–	+	
Desulfobulbus propionicus	Oval	30 to 38	+	–	C_3	+	Ethanol
Complete oxidation							
Desulfobacter postgatei	Oval	30	–	+	–	–	
Desulfovibrio baarsii	Curved	30 to 38	–	(+)	C_3 through C_{18}	–	
Desulfotomaculum acetoxidans	Rod-shaped	35	–	+	C_4, C_5	–	Ethanol
Desulfococcus multivorans	Spherical	35	–	(+)	C_3 through C_{14}	+	Ethanol, benzoate
niacini	Spherical	30	+	(+)	C_3 through C_{14}	–	Ethanol, nicotinate, glutarate
Desulfosarcina variabilis	Cell packets	30	+	(+)	C_3 through C_{14}	+	Ethanol, benzoate
Desulfobacterium phenolicum	Oval	30	–	(+)	C_4	–	Phenol, *p*-cresol, benzoate, glutarate
Desulfonema limicola	Filamentous	30	+	(+)	C_3 through C_{12}	+	Succinate

Symbols: + = utilized; (+) = slowly utilized; – = not utilized.

Our knowledge of the sulphate reducers, however, remains heavily biased by the continuing widespread practice of enriching from natural environments by growth in lactate-based media, although there is accumulating evidence to suggest that those genera and species capable of growth on acetate, propionate, butyrate or benzoate are likely to be of major importance in particular environments (Cord-Ruwisch et al. 1987; Parkes 1987). This general situation is almost certainly greatly influenced by the tendency, as discussed above, for sulphate reducers to exist within microbial consortia rather than to grow in monospecies culture. Sulphate-reducing bacteria are, in any case, often difficult to grow, and their isolation and purification from natural environments can be a tedious process. This fact also arises, at least in part, from the close associations that characterize the growth of microorganisms in consortia. As a consequence of these factors, therefore, our analyses of environments by the traditional cultural techniques has given us only a partial view of those sulphate-reducing bacteria that are present and has provided little or no quantitative data to relate microbial cause to environmental effect.

A number of attempts have been made to overcome this problem. These have variously been stimulated either by the need to have a more rapid quantitative measure of sulphate reducing activity in a particular environment, or by the desire to have a more meaningful understanding of the nature of individual species and of their phylogenetic relationships.

Two detection kits are presently available commercially. Conoco Speciality Products of Houston market RapidChek II which works on the principle of chromogenic detection of antibody against the enzyme APS reductase (Tatnall and Horacek 1991). Although this enzyme is present in other organisms, its key role in dissimilatory sulphate reduction means that as a quantitative measure it is highly specific for sulphate-reducing bacteria. It is important to appreciate, however, that the assay determines the amount of enzyme protein present in a given sample and so can only give an indirect measure of bacterial numbers, viability or metabolic activity *in situ*.

The Hydrogenase Test manufactured by Caproco of Edmonton operates on the principle of measuring hydrogenase enzyme activity in a given sample. Rather than an assay for sulphate-reducing bacteria as such, however, the test is designed to detect areas of active microbial corrosion, which coincidentally will most often be caused by sulphate reducers. The Hydrogenase Test therefore pre-supposes the central importance of hydrogenase activity in the mechanism of microbially influenced corrosion (Bryant et al. 1991).

The powerful technologies of 16S rRNA sequencing and nucleic acid probes have also now been applied extensively to the study of sulphate-reducing bacteria and have given rise to a much more informed picture of phylogenetic relationships within the group (Devereux et al. 1989; Devereux et al. 1990). Specific efforts are currently being made to extend such approaches with the objective of developing probes that will allow more accurate and detailed characterization of natural populations, without the bias introduced by cultural techniques based on a particular medium composition. Voordouw and his colleagues have found that a probe for the [NiFe] hydrogenase gene can identify species from the genus *Desulfovibrio* but not other sulphate reducers (Voordouw et al. 1990). [Three hydrogenase enzymes have been found variously distributed across the group and individually characterized by their metal cofactors: Fe, NiFe and NiFeSe (Peck and Lissolo 1988)]. Recently, these workers have developed a technique using whole genomic DNA to differentiate among at least 20 genotypically individual sulphate-reducing bacteria from a range of oil field samples (Voordouw et al. 1991). Interestingly, there appears to be little or no relationship between this series of natural isolates and the supposedly representative species held in culture collections. Somewhat complementary to this observation is the fact that at no stage has any suggestion been made that corrosion might be the direct consequence of any one or more particular species of sulphate-reducing bacteria, rather than a general property of the group as a whole. At a higher degree of resolution, it is at least theoretically possible that only the occurrence of different species at different locations within a sulphate reducing biofilm could give rise to the establishment of electrochemical corrosion cells. A first attempt to approach such a question has been made using polymerase chain reaction amplification and fluorescence microscopy to study the distribution within a multispecies biofilm of *Deslfovibrio vulgaris*-like and *Desulfuromonas*-like organisms (Amann et al. 1992).

A further significant development in our knowledge of sulphidogenic bacteria has been the increasing incidence of the isolation of both eubacterial and archaebacterial thermophilic species from a wide range of natural and man-made environments (Stetter et al. 1987; Cochrane et al. 1988; Rosnes et al. 1991). Although such organisms might have a significant effect on corrosion in oil field production systems, it is likely that their major economic impact will arise from their potential to cause souring of petroleum reservoirs.

IV. Microbially influenced corrosion

As our appreciation of the true nature of biological corrosion mechanisms has increased, it is interesting to note that the firm terminology of "biocorrosion" or "microbial corrosion" has been replaced by the rather more circumspect "microbially influenced corrosion" (MIC). This is a direct reflection of there being a range of quite different microbial processes at least potentially involved, and that they operate primarily by stimulating pre-existing standard electrochemical mechanisms of corrosion rather than by introducing any novel reactions schemes of their own.

Further, it must be pointed out that although the major thrust of this presentation is focussed on the activities of sulphate-reducing bacteria within multispecies biofilms, there are signifcant examples of MIC that involve neither sulphate-reducing bacteria nor biofilms (Tatnall 1991). Stainless steels, for example, are generally resistant to corrosion by virtue of a protective iron and chromium oxide-rich inorganic film. The integrity of this film can be disrupted by high levels of Cl^- or H_2S, or through colonial growth of *Gallionella* or *Siderocapsa* giving, respectively, iron- and iron and manganese-rich deposits. Separate instances of pitting corrosion in stainless steels have also been recorded in association with the growth of unidentified aerobic, slime-forming bacteria. In one case the pits had a tunnelling character and were associated with a grey-brown manganese-rich deposit, while in the other the biodeposits were chloride-rich, with no involvement of either manganese or iron. Sulphate-reducing bacteria were reported to be the cause of one case of crevice corrosion which developed under joint gaskets. This appeared, however, to be unique to type 304 stainless steel which is low in molybdenum and the problem was resolved by changing the specification to molybdenum-containing type 316 stainless.

The main instances of MIC resulting from the action of the sulphate-reducing bacteria are noted with cast iron and mild steel. In the case of buried cast iron pipes, the effect can be dramatic with the complete removal of the iron leaving behind an apparently unaltered structure which is, in fact, composed solely of graphite and retains none of the strength of the original material. It is, however, with the mild steels, universally used in all construction work, that the sulphate-reducing bacteria have their major impact. Here the corrosion is of a pitting character and is invariably linked to the production of black iron sulphide corrosion products. Although the pitting implies a localized corrosion mechanism, it is generally associated with a degree of confluent growth in the form of a biofilm. It is these aspects we can now explore further with reference to the various points already raised in our general considerations of biofilms and of the sulphate-reducing bacteria.

A. Generalised mechanisms of corrosion

Metal loss during corrosion occurs by passage into solution of positively charged metal ions at the anode of an electrochemical cell. This must be balanced by transfer of the excess electrons to that part of the metal substratum which is cathodic and where a second reaction can absorb the electrons. The simple, non-stoichiometric, equations describing aerobic corrosion or the rusting of iron would be:

$$Fe \rightleftharpoons Fe^{2+} + 2e^- \qquad \text{anode}$$
$$1/2 O_2 + H_2O + 2e \rightleftharpoons 2OH^- \qquad \text{cathode}$$

with the subsequent formation of a complex of iron oxides and hydroxides as corrosion products.

In the absence of O_2, it has been hypothesized that protons (von Wolzogen Kuhr and van der Vlught 1934) or H_2S (Costello 1974) might serve as electron acceptor at the cathode:

a) $2H^+ + 2e \rightleftharpoons 2H \rightleftharpoons H_2$
b) $2H_2S + 2e \rightleftharpoons 2HS^- + H_2$

with sulphate-reducing bacteria subsequently oxidizing the H_2, thus both preventing polarization of the cathode by adsorbed H_2, and giving rise to S^{2-} and the potential for metal sulphide corrosion products:

$$4H_2 + SO_4^{2-} \rightleftharpoons 4H_2O + S^{2-}$$
$$Fe^{2+} + S^{2-} \rightleftharpoons FeS$$

That sulphate-reducing bacteria can oxidize cathodic H_2 and use it as a source of metabolic energy has now been demonstrated in several laboratories (Hardy 1983; Pankhania et al. 1986; Cord-Ruwisch et al. 1987). Additionally, it has been noted that the application of an impressed current to confer protection against corrosion weight loss, also has the effect of stimulating sulphate reducer growth and activity in direct response to the increased tendency to produce H_2 at the cathode (Guezennec et al. 1991).

Most discussions of MIC involving the sulphate-reducing bacteria have concluded, however, that the major factor determining the rate and extent of corrosion is the presence of the iron sulphide corrosion products themselves (King and Miller 1981). Even Bryant et al. (1991) noted that the highest rates of corrosion occurred after the period of microbial activity, and suggested that iron sulphide corrosion products may have a role to play in corrosion mechanisms in addition to that of hydrogenase. In purely electrochemical terms, iron sulphide is cathodic to unreacted iron. The deposition of corrosion product might also be expected to offer an increased surface area for the formation and subsequent oxidation of cathodic H_2. More detailed studies, however, have identified that thin adherent layers of the primary corrosion product, mackinawite, are generally protective but that time-dependent rupture, or high iron concentration-dependent bulky precipitation, can lead to high non-transient corrosion rates that are independent of microbial growth or hydrogenase activity (Mara and Williams 1972; King et al. 1973).

B. Aerobic/anaerobic interface

As was noted above, not only are sulphate-reducing bacteria dependent upon the development of anaerobic regions within sediments or biofilms, but they also generally show their highest activities close to the interface between aerobic and anaerobic microenvironments. Almost certainly this reflects nutrient input from the metabolic activities of other aerobic and facultative organisms also present but with greater biodegradative capability than the sulphate reducers, as well as the availability of biotic and abiotic oxidation of the produced sulphide, to which the sulphate reducers are themselves sensitive. It is clear, however, that O_2 can also play a much more direct role in MIC, by at least two quite separate mechanisms.

Microbial growth at a metal surface may facilitate the establishment of separate anodic and cathodic regions by a number of specific reactions within the general mechanism of establishing concentration gradients. These may arise from the local production of protons or other metabolic products, or from differential binding of dissolved metal ions by the EPS produced by particular cellular species in colonial growth (Costerton et al. 1987). A particularly important example of this mechanism is the O_2 concentration or differential aeration cell arising from localized growth of aerobic organisms (Fig. 3). Here the region immediately below the area of active growth becomes anaerobic and hence anodic relative to the surrounding metallic surface exposed to the air, and is the site of metal dissolution. It is likely that this constitutes the most common and widespread mechanism of MIC associated with generalised microbial growth, either as a discontinuous biofilm or in discrete colonies. The further microbial oxidation of iron from the ferrous to the ferric state can occasionally increase the extent of corrosion and lead to the formation of tubercles. The anaerobic regions created by such microbial activity serve, however, not only as anodes in electrochemical corrosion cells but also as ideal microenvironments for the growth of anaerobic species, including sulphate-reducing bacteria. By this means therefore, sulphate reducer MIC generally occurs alongside more generalized mechanisms and may even be, to some extent, dependent upon them.

OXYGEN CONCENTRATION CELL:

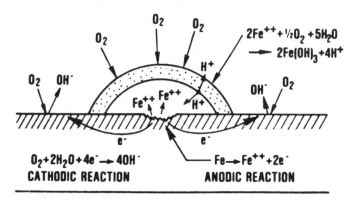

Fig. 3. Differential aeration cell (from Schaschl 1980, with permission).

Most of the earlier studies of the so-called anaerobic corrosion caused by sulphate-reducing bacteria were indeed carried out under strictly anaerobic conditions but, in field situations, it is widely recognized that the most extensive corrosion occurs where there is ready access of air. Under such conditions, penetration rates of up to 5 mm per year have been recorded (Tatnall 1991). At least two studies suggest that this stimulation by O_2 is neither on the metabolic activity of the sulphate-reducing bacteria nor even on the formation of the iron sulphide corrosion products themselves. Hardy and Bown (1984) showed that if a metal surface that had previously been subjected to growth of sulphate reducers was subsequently exposed to pulses of air, there was a marked increase in the instantaneous corrosion rate. A similar effect of oxygenation on preformed iron sulphides was noted by Braithwaite and Lichti (1980) in their examination of corrosion in geothermal power stations where the principal corrodent is H_2S in the high temperature steam, with no involvement of any microbial activity.

These findings have been confirmed and extended by some of our own work with corroding steel coupons in sea bed sediments associated with oil production platforms, and in a laboratory simulation of a stratified marine system (Moosavi et al. 1991; McKenzie and Hamilton 1992). Again, maximal rates of corrosion were noted in those bulk environments where conditions were aerobic. Two factors were considered to be of possible significance under these circumstances. Firstly, visual observation revealed three discrete layers of corrosion product; a thin black adherent layer, surmounted by first a bulky layer of looser material, again black, and finally an overlying layer of brown, oxidized products. Preliminary attempts at chemical analysis served only to establish an extremely complex mixture of sulphides, oxides and carbonates. It should be noted at this point that iron sulphide exists in a wide range of forms, each with its characteristic iron/sulphur stoichiometry, and each with a unique physical form, some being crystalline and others amorphous. Much of the early seminal work on corrosion mechanisms suggested that the transition of one form of iron sulphide to another was the direct cause of a thin adherent layer of mackinawite (FeS_{1-x}) losing both its chemical and physical form, and consequently its protective character (King et al. 1976). Secondly, it was noted that in the more oxidizing regions associated with the higher corrosion rates, up to 92% of the sulphides were in a form that did not give rise to H_2S on treatment with cold acid. The principal forms of non-acid volatile sulphur compounds have been identified as pyrite (FeS_2) and sulphur itself, the formation of each of which is favoured by oxidizing conditions. Pyrite has been implicated previously in corrosion mechanisms and sulphur is well known to be highly corrosive (Schaschl 1980). These data confirm therefore, that it is the effect of environmental variables, particulary O_2, on the nature and form of the sulphide corrosion products that determine the true nature and extent of the corrosion which occurs as a result of the growth of sulphate-reducing bacteria.

At the present time, work is continuing at the Center for Interfacial Microbial Process Engineering in Bozeman, Montana, U.S.A. with the objective of defining more closely these effects (W. Lee, personal communication). It has been noted that the characteristic pitting corrosion of sulphate reducers is increased with the presence of greater O_2 concentrations (up to 7 mg.l^{-1}) in the bulk phase, with Auger electron spectroscopy analysis showing a loose but definite spatial correlation between the oxygen and sulphur signals within the pits.

V. Concluding remarks

A full appreciation of the mechanisms of microbially influenced corrosion requires consideration of both the biological and the physical or engineering aspects of the phenomenon. With respect to the sulphate reducers this is especially so as the balance of evidence suggests that the true corrosion, as a mechanism that continues with time and can lead to extensive structural damage, is a process firmly in the realm of materials science. None the less, the reaction triggering these events is the microbial production of sulphide. In order to understand that reaction and the factors, biological and physical, that control it, one must examine the physiology, and particularly the energetics of the sulphate-reducing bacteria, and consider the nature of their growth as component members of microbial consortia and biofilms.

References

Aeckersberg F, Bak F, Widdel F (1991) Anaerobic oxidation of saturated hydrocarbons to CO_2 by a new type of sulfate-reducing bacterium. Arch. Microbiol. 156: 5–14

Amann RI, Stromely J, Devereux R, Key R and Stahl DA (1992) Molecular and microscopic identification of sulfate-reducing bacteria in multispecies biofilms. Appl. Environ. Microbiol. 58: 614–623

Braithwaite WR and Lichti KA (1980) Surface corrosion of metals in geothermal fluids at Broadlands, New Zealand. In: LA Casper and TR Pinchback (eds) Geothermal Scaling and Corrosion (pp 81–121). American Society for Testing of Materials, Washington

Brandis-Heep A, Gebhardt NA, Thauer RK, Widdel F and Pfennig N (1983) Anaerobic acetate oxidation to CO_2 by *Desulfobacter postgatei* 1. Demonstration of all enzymes required for the operation of the citric acid cycle. Arch. Microbiol. 136: 222–229

Bryant RD, Jansen W, Boivin J, Laishley EJ and Costerton JW (1991) Effect of hydrogenase and mixed sulfate-reducing bacterial populations on the corrosion of steel. Appl. Environ. Microbiol. 57: 2804–2809

Characklis WG and Marshall KC (eds) (1990) Biofilms. John Wiley and Sons, New York

Characklis WG and Wilderer PA (eds) (1989) Structure and Function of Biofilms. John Wiley and Sons, Chichester

Cochrane WJ, Jones PS, Sanders PF, Holt DM and Moseley MJ (1988) Studies on the thermophilic sulfate-reducing bacteria from a souring North Sea oil field. Society of Petroleum Engineering, Paper 18368: 301–316

Cord-Ruwisch R, Kleinitz W and Widdel F (1987) Sulfate-reducing bacteria and their activities in oil production. J. Pet. Technol. Jan: 97–106

Costello JA (1974) Cathodic depolarisation by sulphate-reducing bacteria. S. Afr. J. Sci. 70: 202–204

Costerton JW, Cheng K-J, Geesey GG, Ladd TI, Nickel JC, Dasgupta M and Marrie TJ (1987) Bacterial biofilms in nature and disease. Ann. Rev. Microbiol. 41: 435–464

Devereux R, Delany M, Widdel F and Stahl DA (1989) Natural relationships among sulfate-reducing eubacteria. J. Bacteriol. 171: 6689–6695

Devereux R, He S-H, Doyle CL, Orkland S, Stahl DA, LeGall J and Whitman WB (1990) Diversity and origin of *Desulfovibrio* species: phylogenetic definition of a family. J. Bacteriol. 172: 3609–3619

Dowling NJ, Mittleman MW and Danko JC (eds) (1991) Microbially Influenced Corrosion and Biodeterioration. National Association of Corrosion Engineers, Washington

Flemming H-C and Geesey GG (eds) (1991) Biofouling and Biocorrosion in Industrial Water Systems. Springer-Verlag, Berlin

Fuchs G (1986) CO_2 fixation in acetogenic bacteria: variations on a theme. FEMS Microbiol. Revs. 39: 181–213

Gebhardt NA, Thauer RK, Linder D, Kaulfers PM and Pfennig N (1985) Mechanism of acetate oxidation to CO_2 with elemental sulfur in *Desulfuromonas acetoxidans*. Arch. Microbiol. 141: 392–398

Guezennec J, Dowling NJ, Conte M, Antoine E and Fiksdal L (1991) Cathodic protection in marine sediments and the aerated seawater column. In: NJ Dowling, MW Mittleman and JC Danko (eds) Microbially Influencd Corrosion and Biodeterioration (pp 6.43–6.50) National Association of Corrosion Engineers, Washington

Hamilton WA (1985) Sulphate-reducing bacteria and anaerobic corrosion. Ann. Rev. Microbiol. 39: 195–217

Hamilton WA (1988) Energy transduction in anaerobic bacteria. In: C Anthony (ed) Bacterial Energy Transduction (pp 83–149). Academic Press, London

Hardy JA (1983) Utilisation of cathodic hydrogen by sulphate-reducing bacteria. Br. Corros. J. 18: 190–193

Hardy JA and Bown J (1984) The corrosion of mild steel by biogenic sulfide films exposed to air. Corrosion 40: 650–654

Jørgensen BB (1982) Ecology of the bacteria of the sulphur cycle with special reference to anoxic-oxic interface environments. Phil. Trans. R. Soc. Lond. Ser B 298: 543–561

Jørgensen BB (1988) Ecology of the sulphur cycle: oxidative pathways in sediments. Symp. Soc. Gen. Microbiol. 42: 31–63

King RA and Miller JDA (1981) Corrosion by sulphate-reducing bacteria. Nature 233: 491–492

King RA, Miller JDA and Wakerley DS (1973) Corrosion of mild steel in cultures of sulphate-reducing bacteria: effect of changing the soluble iron concentration during growth. Br. Corros. J. 8: 89–93

King RA, Dittmer CK and Miller JDA (1976) Effect of ferrous iron concentration on the corrosion of iron in semicontinuous cultures of sulphate-reducing bacteria. Br. Corros. J. 11: 105–107

Kroger A, Schroder J, Paulsen J and Beilmann (1988) Acetate oxidation with sulphur and sulphate as terminal electron acceptors. Symp. Soc. Gen. Microbiol. 42: 133–145

McKenzie J and Hamilton WA (1992) The assay of in-situ activities of sulphate-reducing bacteria in a laboratory marine corrosion model. Int. Biodeterior. Biodegrad. 29: 285–297

Mara DD and Williams DJA (1972) The mechanism of sulphide corrosion by sulphate-reducing bacteria. In: AM Walters and EH Hueck van der Plas (eds) Biodeterioration of Materials, Vol 2 (pp 103–113). Applied Science Publishers, London

Moosavi AN, Pirrie RS and Hamilton WA (1991) Effect of sulphate-reducing bacteria activity on performance of sacrificial anodes. In: NJ Dowling, MW Mittleman and JC Danko (eds) Microbially Influenced Corrosion and Biodeterioration (pp 3.13–3.27). National Association of Corrosion Engineers, Washington

Nethe-Jaenchen R and Thauer RK (1984) Growth yields and saturation constant of *Desulfovibrio vulgaris* in chemostat culture. Arch. Microbiol. 137: 236–240

Pankhania IP, Moosavi AN and Hamilton WA (1986) Utilization of cathodic hydrogen by *Desulfovibrio vulgaris* (Hildenborough). J. Gen. Microbiol. 132: 3357–3365

Parkes RJ (1987) Analysis of microbial communities within sediments using biomarkers. Symp. Soc. Gen. Microbiol. 41: 147–177

Peck HD and Lissolo T (1988) Assimilatory and dissimilatory sulphate reduction: enzymology and bioenergetics. Symp. Soc. Gen. Microbiol. 42: 99–132

Postgate JR (1984) The Sulphate-reducing Bacteria, 2nd Edition. Cambridge University Press, Cambridge

Rosnes JT, Torsvik T and Lien T (1991) Spore-forming thermophilic sulfate-reducing bacteria isolated fron North Sea oil field waters. Appl. Environ. Microbiol. 57: 2302–2307

Rosser HR and Hamilton WA (1983) Simple assay for accurate determination of [^{35}S] sulfate reduction activity. Appl. Environ. Microbiol. 45: 1956–1959

Schaschl E (1980) Elemental sulfur as a corrodent in deaerated, neutral aqueous solutions. Materials Performance 19: 9–12

Stetter KO, Lauerer G, Thomm M and Neuner A (1987) Isolation of extremely thermophilic sulfate reducers: evidence for a novel branch of Archaebacteria. Science 236: 822–824

Tatnall RE (1991) Case histories: biocorrosion. In: H-C Flemming and GG Geesey (eds) Biofouling and Biocorrosion in Industrial Water Systems (pp 165–185). Springer-Verlag, Berlin

Tatnall RE and Horacek GL (1991) New perspectives on testing for sulfate-reducing bacteria. In: NJ Dowling, MW Mittleman and JC Danko (eds) Microbially Influenced Corrosion and Biodeterioration (pp 5.17–5.32). National Association of Corrosion Engineers, Washington

Thauer RK (1988) Citric-acid cycle, 50 years on. Modifications and an alternative pathway in anaerobic bacteria. Eur. J. Biochem. 176: 497–508

Videla HA and Gaylarde CC (1992) (eds) Microbially Influenced Corrosion. Int. Biodeterior. Biodegrad. 29: 193–375

von Wolzogen Kuhr CAM and van der Vlught IS (1934) The graphitization of cast iron as an electrobiochemical process in anaerobic soils. Water 18: 147–165

Voordouw G, Niviere V, Ferris FG, Fedorak PM and Westlake DWS (1990) Distribution of hydrogenase genes in *Desulfovibrio* spp. and their use in identification of species from the oil field environment. Appl. Environ. Microbiol. 56: 3748–3754

Voordouw G, Voordouw JK, Karkhoff-Schweizer RR, Fedorak PM and Westlake DWS (1991) Reverse sample genome probing, a new technique for identification of bacteria in environmental samples by DNA hybridization, and its application to the identification of sulfate-reducing bacteria in oil field samples. Appl. Environ. Microbiol. 57: 3070–3078

Widdel F (1988) Microbiology and ecology of sulfate- and sulfur-reducing bacteria. In: AJB Zehnder (ed) Biology of Anaerobic Microorganisms (pp 469–585). John Wiley and Sons, London

Wood HG, Ragsdale SW and Pezacka E (1986) The acetyl-CoA pathway of autotrophic growth. FEMS Microbiol. Rev. 39: 345–362

Zeikus JG (1983) Metabolic communication between biodegradative populations in nature. Symp. Soc. Gen. Microbiol. 34: 423–462

Index of compounds

Index of organisms